时代教育·国外高校优秀教材精选
格里菲斯系列

量子力学概论

（英文注释版　原书第3版）

［美］大卫·J. 格里菲斯（David J. Griffiths）
达雷尔·F. 施勒特（Darrell F. Schroeter）著

贾　瑜　注释

机械工业出版社

本书是一本适合本科生学习的量子力学教材，内容分为两部分。第 I 部分为理论，包含波函数、定态薛定谔方程、形式理论、三维空间中的量子力学、全同粒子、对称性和守恒律；第 II 部分为应用，包含定态微扰理论、变分原理、WKB 近似、散射、量子动力学。本书构思新颖，取材前沿，作者抛开量子力学发展历史的负担，直接从薛定谔方程开始讲授，通过穿插对一些经典问题的讨论，让学生从大量具体问题中体会到量子力学的精髓。本书注重把现代物理前沿引入教学，把量子力学问题扩展到多个前沿的研究领域，如统计物理、固体物理、天体物理、粒子物理、计算物理等。此外，作者着重于交互式的写作，用第一人称“I”以对话式的语言进行叙述，简明扼要，文笔流畅，使人耳目一新。

本书为高校物理专业学生学习量子力学的教材和参考书，也可作为材料、化学等专业学生的教材和相关教师的参考书。

图书在版编目（CIP）数据

量子力学概论：英文注释版：原书第 3 版：英文 /（美）大卫 • J. 格里菲斯 (David J. Griffiths),（美）达雷尔 • F. 施勒特 (Darrell F. Schroeter) 著；贾瑜注释．—北京：机械工业出版社，2020.8（2024.1 重印）
（时代教育）
书名原文：Introduction to Quantum Mechanics, 3rd Edition
国外高校优秀教材精选
ISBN 978-7-111-65751-4

Ⅰ．①量… Ⅱ．①大… ②达… ③贾… Ⅲ．①量子力学 – 高等学校 – 教材 – 英文
Ⅳ．① O413.1

中国版本图书馆 CIP 数据核字（2020）第 096159 号

机械工业出版社（北京市百万庄大街 22 号 邮政编码 100037）
策划编辑：李永联 责任编辑：李永联
责任校对：张 薇 封面设计：马精明
责任印制：郜 敏
三河市宏达印刷有限公司印刷
2024 年 1 月第 1 版第 4 次印刷
187mm × 260mm · 33.5 印张 · 2 插页 · 558 千字
标准书号：ISBN 978-7-111-65751-4
定价：138.00 元

电话服务	网络服务
客服电话：010-88361066	机 工 官 网：www.cmpbook.com
010-88379833	机 工 官 博：weibo.com/cmp1952
010-68326294	金 书 网：www.golden-book.com
封底无防伪标均为盗版	机工教育服务网：www.cmpedu.com

注释者序 / The Annotator's Preface

格里菲斯教授所著《量子力学概论》是一本适合本科生学习的量子力学教材。第 1 版、第 2 版分别于 1994 年、2005 年由美国培生出版集团出版。英文版的《量子力学概论》出版以来，一直是美国和欧洲许多一流理工科大学，包括麻省理工学院、斯坦福大学、加州大学洛杉矶分校等高校物理系学生的教材和指定教学用书，在欧美则被认为是较合适、较现代的教材之一。现在的第 3 版为格里菲斯教授和施勒特教授两人合著，改由英国剑桥大学出版社于 2018 年出版。在第 2 版出版后，我国机械工业出版社积极引进英文版，并组织翻译出版中文版，出版以后深受国内广大读者的欢迎，多次重印。值第 3 版出版之际，机械工业出版社和剑桥大学出版社协商，继续引进英文版、翻译版在国内出版。

量子力学是现代科技的基础，做好量子力学的教学也是基础物理教学现代化的具体体现。目前国际、国内已出版的量子力学教材估计有数百种，其中不乏一些著名科学家、教育家的名著，可谓繁杂多样，也各具特色。但在如何处理好量子力学内容“教”与“学”的关系上，仍存在这样那样的问题，使得量子力学一直是物理系学生最难学的课程。教材内容如果过繁，往往让学生摸不着头脑，陷入繁杂的公式推导，无法理解其含义；如果过简，也容易让学生感到空无一物，很难把所学的知识应用到具体的问题中去。格里菲斯教授在这方面做出了很好的尝试，不仅如此，他做到了不仅使量子力学的内容简明扼要，条理分明，而且把量子力学的知识拓展到物理学分支的各个方面，甚至可以和现在的学术前沿研究接轨。总结起来，本书有以下的特点：

（1）立足于“量子力学入门水平”。本书包含了大学量子力学最主要的内容，有两部分：第 I 部分讲述基本理论，其中也含有相关的应用举例；第 II 部分讲述基本应用，介绍研究工作中需要的一些近似方法、散射理论和量子物理实验基础。这样的处理不仅便于具有各种不同培养目标的各类学校灵活选择讲授内容（例如只讲授基本理论，或另外讲授部分或全部基本应用部分），也不破坏量子力学作为一门学科课程应当具有的基本完整性。

（2）打破常规，敢于抛开量子力学发展历史的负担。作者避免了很多量子力学发展史和实验现象的介绍（事实上，学生在开始学习时很难把量子力学的一些实验现象较好地联系起来，过多地介绍实验结果也就可能引起学生更多的迷惑），讲解直接从薛定谔方程开始，力图改变量子力学难于理解、难于接受的教学状况。本书注重量子力学的基本思想和基本方法的讲授，而不是把量子力学引入繁琐的理论推导，而内容又涉及各个研究领域，非一般的量子力学教材所能及，这在量子力学教学现代化方面的确是很成功的尝试。

（3）不局限于知识的讲授，而是让读者真正从大量具体问题中体会量子力学的精髓。针对量子力学不易理解的特点，本书先从简单的概率论和微分方程入手，通过穿插对一些经典问题的讨论，让学生能迅速对一些简单的量子力学问题“上手”，而不仅仅是望着深奥的知识兴叹。

（4）充分体现现代物理内容。本书在讲述量子力学的同时，把问题扩展到多个前沿研究领域，如统计物理、固体物理、天体物理、粒子物理等。在物理学各个分支中常用的部分既有精辟的叙述，又有实际举例。

（5）作者通过把一些知识移到课外习题的方式来缩减正文内容，使学生可以通过自学来掌握量子力学相当大的一部分内容，使得本书主线清晰、内容简练。为此，作者在练习题选择上特别下功夫。例题与习题对数学的要求并不高。习题分为容易、中等和较难三个层次，可供不同基础的学生选择。对一些较难的题目还附有提示。同时，作者在使用计算机进行量子力学数值模拟教学上面也下了很大功夫，设计了不少新颖的题目。特别值得指出的是，书中许多例题和习题都属于原创，大部分素材来源于科研文献，这拉近了所学知识与科研的距离。

（6）以对话式的语言进行叙述。作者从务实的角度出发，着重于交互式的写作，用第一人称“I”以对话式的语言进行叙述，简明扼要，文笔流畅，使人耳目一新。他山之石，可以攻玉，我认为这本书非常适合作为我国大学生在学习量子力学时参考，尤其适合作为量子力学英语教学的教材。毫无疑问，引进出版《量子力学概论》将会对我国量子力学教学水平的提高起到积极的作用。

注释者

2020 年 2 月 10 日于郑州

前　言

与牛顿力学、麦克斯韦电动力学或者爱因斯坦相对论不同，量子力学不是由个别人建立的（或者是明确地归结于某个人），量子力学在令人振奋但又悲壮的发展初期留下的那些瘢痕直到现还保留着。对于它的基本原理是什么，如何去教，它到底“意味”着什么，至今没有形成普遍、一致的共识。任何一个有能力的物理学家可以“谈论”量子力学，但是我们告诉自己关于我们正在做什么的故事就像舍赫拉查德（Scheherazade）传说一样千变万化，几乎是难以置信的。玻尔曾说过，“如果你没有被量子力学搞迷惑，则你根本就没有真正地理解量子力学”。费曼评述道，“我想我可以有把握地说没有人明白量子力学”。

本书的目的是教你如何学习量子力学。除了在第 1 章中一些必备的基础知识外，更深的准 - 哲学问题将留在书尾。我们不相信一个人在对量子力学是干什么的有一个透彻的理解之前，他可以明智地讨论量子力学的意义。但是，如果你急不可待，在学习过第 1 章后可立即阅读跋。

量子理论不仅概念丰富，在技术上处理起来也十分困难。量子力学的实际问题能够严格求解是十分罕见的，更多的是课本上人为编的一些题目，因此发展处理实际问题的特殊技术十分必要。相应地，本书分为两部分[㊀]：第 I 部分涵盖了基本理论，第 II 部分汇集了近似方法，同时配以直观的有启发性的应用示例。尽管在逻辑上保持两部分的独立是重要的，但在学习时也不一定按照目前的次序。例如，有些教师可能希望在学习第 2 章之后能立即开始接触定态微扰理论的学习。

本书供大学三年级或四年级学生一学期或一学年的课程之用。一学期的课程应主要集中在第 I 部分的学习；一学年的课程在第 II 部分之外还可以学习一些补充材料。读者必须具备线性代数（总结在附录之中）、复数、微积分的基础知识，若能熟悉一些傅里叶变换和狄拉克 δ 函数的知识则有帮助。当然，基本经典力学基础是必要的，一些电动力学的知识也会很有帮助。在通常情况下，你的物理和数学知识越多，学习起来就越容易，获得的知识就越多。但量子力学不是以前理论自然平滑过渡的产物。相反，它代表着对经典思想的一种急剧的革命性变革，

㊀ 这种结构受 David Park 的经典教材的启发，《量子理论导论》，第 3 版，麦格劳 - 希尔（McGraw-Hill）出版公司，纽约（1992）。

唤起一种全新的、和直觉完全相反的思考自然世界的方法，这也正是使它成为一个如此有魅力的学科的原因所在。

初看起来，本书留给你的印象是可怕的数学。我们会用到勒让德多项式、厄密多项式、拉盖尔多项式、球谐函数、贝塞尔函数、诺伊曼函数、汉克尔函数、艾里函数，甚至是黎曼 ζ 函数——更不用说傅里叶变换、希尔伯特空间、厄密算符、CG（Clebsch-Gordan）系数。所有这些东西都是必要的吗？也许是不必要的，但是物理学像木匠活一样：使用正确的工具使工作简易，减少困难，学习量子力学而没有适当的数学工具就像让学生用螺丝刀去挖地基一样——它纵然是可能的，但着实痛苦。（另一方面，教师在讲授课程时感到每一个复杂课程都必须使用完善的数学工具，这样学习就会变得枯燥乏味，而且会使教学重点偏移。我本人的经验是把铁铲交给学生，告诉他们自己去开始挖掘。开始时也许他们手上会磨起水泡（遇到困难），但是我一直认为这种方式是最有效、最激励的。）不管怎样，我可以向你保证本书没有涉及很深的数学，如果你遇到不熟悉的数学知识，并且对我们给出的解释觉得不充分，务必请教他人，或钻研它。关于数学方法有很多优秀的书籍——我特别推荐玛丽·博厄斯（Mary Boas）的《物理科学中的数学方法》[㊀] 第 3 版，Wiley 出版社，纽约（2006）；或者乔治·阿夫肯（George Arfken）和汉斯 - 玖根·韦伯（Hans-Jurgen Weber）的《物理学家所用的数学方法》第 7 版，美国学术出版社，奥兰多（2013）。但是无论如何，不要让数学——对我们来说它仅是工具——把物理变得模糊。

一些读者已经注意到，和通常的教科书相比，本书中例题较少，并且一些重要的内容放在了习题中，这绝非偶然。我不认为可以通过不做大量的习题而学懂量子力学。如果时间允许，教师理应在课堂上给出更多的例题。但是学生们应当意识到这不是一个任何人都有直观感觉的课题——这里你们正在开发的是一个全新的肌体，根本就没有运动的替代物。马克·西蒙（Mark Semon）建议我对习题给出一个“米歇尔（Michelin）导引”，用不同数目的星号标出其重要性和难度。这的确是一个好主意（尽管这样，像一个饭店的质量一样，一个习题的重要性部分是取决于口味），我将采取下面的分级方案：

- *　每个读者都应该研究的基本问题。
- **　有点难度或次要的问题。
- ***　极有挑战性的问题，可能花费 1 小时以上的时间来解决。

（没有星号的意味着快餐：好的，如果你饿了，这可以解决你的问题，但是营养不太丰富。）大多数标“*”的习题出现在相关一节的后面，而大多数标

㊀ 机械工业出版社已引进该书中文版版权，即将出版。——编辑注

“***” 的习题出现在相关章的后面。如果解题过程中需要使用计算机，我们在题前页面空白处标注一个鼠标。

在准备第 3 版时，我们试图尽可能地保持第 1 版和第 2 版的特色和意图。尽管新版现在是两个作者，我们仍延续使用单人称我（“I”）向读者介绍，这样会感到更加亲密。毕竟是每次仅我们当中只有一个人来讲授（文中 “我们” 指读者你自己和作者我本人，我们一起工作）。施勒特的参与带来了一个固体物理学家的新视角，他主要负责新增的一章内容——对称性的编写。我们增加了一些习题，澄清了许多解释，重新修订了跋。但我们决定不让这本书变厚，正是出于这样一个原因，我们删除了绝热近似一章（这一章中重要的观点已经整合在第 11 章中），删除了第 5 章中统计物理的相关内容（这部分属于热物理）。毫无疑问，如果教师感觉一些内容合适，很欢迎他将认为合适的部分包含在课程内，但我们仅是让教材包含本课程最为核心的内容。

许多同事的建议和评论使我受益匪浅，他们阅读了初稿，指出前两版中的不足之处（或错误），提出在一些叙述上的改进，提供有趣的习题。我们对如下同事表示特别感谢：P.K. Aravind（伍斯特理工学院）、Greg Benesh（贝勒学院）、James Bernhard（普吉桑德大学）、Burt Brody（巴德学院）、Ash Carter（德鲁大学）、Edward Chang（麻省大学）、Peter Collings（斯沃斯莫尔学院）、Richard Crandall（里德学院）、Jeff Dunham（明德学院）、Greg Elliott（普吉桑德大学）、John Essick（里德学院）、Gregg Franklin（卡耐基梅隆大学）、Joel Franklin（里德学院）、Henry Greenside（杜克大学）、Paul Haines（达特茅斯学院）、J. R. Huddle（商船学院）、Larry Hunter（阿默斯特学院）、David Kaplan（华盛顿学院）、Don Koks（阿德莱德大学）、Peter Leung（波特兰州立大学）、Tony Liss（伊利诺伊学院）、Jeffry Mallow（芝加哥洛约拉大学）、James McTavish（利物浦大学）、James Nearing（迈阿密大学）、Dick Palas、Johnny Powell（里德学院）、Krishna Rajagopal（麻省理工学院）、Brian Raue（佛罗里达国际大学）、Robert Reynolds（里德学院）、Keith Riles（密歇根大学）、Klaus Schmidt-Rohr（布兰迪斯大学）、Kenny Scott（伦敦大学）、Dan Schroeder（韦伯州立大学）、Mark Semon（贝茨学院）、Herschel Snodgrass（路易克拉克大学）、John Taylor（科罗拉多大学）、Stavros Theodorakis（塞浦路斯学院）、A. S. Tremsin（伯克利学院）、Dan Velleman（阿默斯特学院）、Nicholas Wheeler（里德学院）、Scott Willenbrock（伊利诺伊学院）、William Wootters（威廉姆斯学院）和 Jens Zorn（密歇根大学）。

Preface

Unlike Newton's mechanics, or Maxwell's electrodynamics, or Einstein's relativity, quantum theory was not created—or even definitively packaged—by one individual, and it retains to this day some of the scars of its exhilarating but traumatic youth. There is no general consensus as to what its fundamental principles are, how it should be taught, or what it really "means." Every competent physicist can "do" quantum mechanics, but the stories we tell ourselves about what we are doing are as various as the tales of Scheherazade, and almost as implausible. Niels Bohr said, "If you are not confused by quantum physics then you haven't really understood it"; Richard Feynman remarked, "I think I can safely say that nobody understands quantum mechanics."

The purpose of this book is to teach you how to *do* quantum mechanics. Apart from some essential background in Chapter 1, the deeper quasi-philosophical questions are saved for the end. We do not believe one can intelligently discuss what quantum mechanics *means* until one has a firm sense of what quantum mechanics *does*. But if you absolutely cannot wait, by all means read the Afterword immediately after finishing Chapter 1.

Not only is quantum theory conceptually rich, it is also technically difficult, and exact solutions to all but the most artificial textbook examples are few and far between. It is therefore essential to develop special techniques for attacking more realistic problems. Accordingly, this book is divided into two parts;[1] Part I covers the basic theory, and Part II assembles an arsenal of approximation schemes, with illustrative applications. Although it is important to keep the two parts *logically* separate, it is not necessary to study the material in the order presented here. Some instructors, for example, may wish to treat time-independent perturbation theory right after Chapter 2.

This book is intended for a one-semester or one-year course at the junior or senior level. A one-semester course will have to concentrate mainly on Part I; a full-year course should have room for supplementary material beyond Part II. The reader must be familiar with the rudiments of linear algebra (as summarized in the Appendix), complex numbers, and calculus up through partial derivatives; some acquaintance with Fourier analysis and the Dirac delta function would help. Elementary classical mechanics is essential, of course, and a little electrodynamics would be useful in places. As always, the more physics and math you know the easier it will be, and the more you will get out of your study. But quantum mechanics is not something that flows smoothly and naturally from earlier theories. On the contrary, it represents an abrupt and revolutionary departure from classical ideas, calling forth a wholly new and radically counterintuitive way of thinking about the world. That, indeed, is what makes it such a fascinating subject.

---

[1] This structure was inspired by David Park's classic text *Introduction to the Quantum Theory*, 3rd edn, McGraw-Hill, New York (1992).

At first glance, this book may strike you as forbiddingly mathematical. We encounter Legendre, Hermite, and Laguerre polynomials, spherical harmonics, Bessel, Neumann, and Hankel functions, Airy functions, and even the Riemann zeta function—not to mention Fourier transforms, Hilbert spaces, hermitian operators, and Clebsch–Gordan coefficients. Is all this baggage really necessary? Perhaps not, but physics is like carpentry: Using the right tool makes the job *easier*, not more difficult, and teaching quantum mechanics without the appropriate mathematical equipment is like having a tooth extracted with a pair of pliers—it's possible, but painful. (On the other hand, it can be tedious and diverting if the instructor feels obliged to give elaborate lessons on the proper use of each tool. Our instinct is to hand the students shovels and tell them to start digging. They may develop blisters at first, but we still think this is the most efficient and exciting way to learn.) At any rate, we can assure you that there is no *deep* mathematics in this book, and if you run into something unfamiliar, and you don't find our explanation adequate, by all means *ask* someone about it, or look it up. There are many good books on mathematical methods—we particularly recommend Mary Boas, *Mathematical Methods in the Physical Sciences*, 3rd edn, Wiley, New York (2006), or George Arfken and Hans-Jurgen Weber, *Mathematical Methods for Physicists*, 7th edn, Academic Press, Orlando (2013). But whatever you do, don't let the mathematics—which, for us, is only a *tool*—obscure the physics.

Several readers have noted that there are fewer worked examples in this book than is customary, and that some important material is relegated to the problems. This is no accident. We don't believe you can learn quantum mechanics without doing many exercises for yourself. Instructors should of course go over as many problems in class as time allows, but students should be warned that this is not a subject about which *any*one has natural intuitions—you're developing a whole new set of muscles here, and there is simply no substitute for calisthenics. Mark Semon suggested that we offer a "Michelin Guide" to the problems, with varying numbers of stars to indicate the level of difficulty and importance. This seemed like a good idea (though, like the quality of a restaurant, the significance of a problem is partly a matter of taste); we have adopted the following rating scheme:

- * an *essential* problem that every reader should study;
- ** a somewhat more difficult or peripheral problem;
- *** an unusually challenging problem, that may take over an hour.

(No stars at all means fast food: OK if you're hungry, but not very nourishing.) Most of the one-star problems appear at the end of the relevant section; most of the three-star problems are at the end of the chapter. If a computer is required, we put a mouse in the margin. A solution manual is available (to instructors only) from the publisher.

In preparing this third edition we have tried to retain as much as possible the spirit of the first and second. Although there are now two authors, we still use the singular ("I") in addressing the reader—it feels more intimate, and after all only one of us can speak at a time ("we" in the text means you, the reader, and I, the author, working together). Schroeter brings the fresh perspective of a solid state theorist, and he is largely responsible for the new chapter on symmetries. We have added a number of problems, clarified many explanations, and revised the Afterword. But we were determined not to allow the book to grow fat, and for that reason we have eliminated the chapter on the adiabatic approximation (significant insights from that chapter have been incorporated into Chapter 11), and removed material from Chapter 5 on statistical mechanics (which properly belongs in a book on thermal physics). It goes without saying that instructors are welcome to cover such other topics as they see fit, but we want the textbook itself to represent the essential core of the subject.

We have benefitted from the comments and advice of many colleagues, who read the original manuscript, pointed out weaknesses (or errors) in the first two editions, suggested improvements in the presentation, and supplied interesting problems. We especially thank P. K. Aravind (Worcester Polytech), Greg Benesh (Baylor), James Bernhard (Puget Sound), Burt Brody (Bard), Ash Carter (Drew), Edward Chang (Massachusetts), Peter Collings (Swarthmore), Richard Crandall (Reed), Jeff Dunham (Middlebury), Greg Elliott (Puget Sound), John Essick (Reed), Gregg Franklin (Carnegie Mellon), Joel Franklin (Reed), Henry Greenside (Duke), Paul Haines (Dartmouth), J. R. Huddle (Navy), Larry Hunter (Amherst), David Kaplan (Washington), Don Koks (Adelaide), Peter Leung (Portland State), Tony Liss (Illinois), Jeffry Mallow (Chicago Loyola), James McTavish (Liverpool), James Nearing (Miami), Dick Palas, Johnny Powell (Reed), Krishna Rajagopal (MIT), Brian Raue (Florida International), Robert Reynolds (Reed), Keith Riles (Michigan), Klaus Schmidt-Rohr (Brandeis), Kenny Scott (London), Dan Schroeder (Weber State), Mark Semon (Bates), Herschel Snodgrass (Lewis and Clark), John Taylor (Colorado), Stavros Theodorakis (Cyprus), A. S. Tremsin (Berkeley), Dan Velleman (Amherst), Nicholas Wheeler (Reed), Scott Willenbrock (Illinois), William Wootters (Williams), and Jens Zorn (Michigan).

目 录

Contents

导 读

导 读

第 I 部分
理　论

I
THEORY

导读 / Guidance

第 1 章　波函数

本章介绍了量子力学的基本思想和描述微观粒子状态所需要的基本工具。与其他一些量子力学教材不同，本书直接回避对量子力学发展过程中一些实验现象的介绍和发展历史，开门见山地指出微观粒子具有波动性和不确定性，微观粒子的状态由波函数描写，波函数可以通过解薛定谔方程得到。接下来讨论几率的概念，给出离散变量和连续变量的中值、平均值、最概然值、方差的计算公式，在此基础上给出玻恩关于波函数的统计诠释和波函数的归一化要求，并以力学量——动量为例，给出量子力学计算力学量期待值的计算方法，最后，以一维平面波为例，讨论微观粒子力学量的不确定原理。

习题特色

（1）用经典力学的概念来讨论几率的问题，把经典力学的一些问题重新用量子力学处理；试图建立经典力学和量子力学的过渡和衔接。见例题 1.2、习题 1.11、习题 1.13。

（2）引入一些物理概念和物理定律，如几率流的定义和计算（见习题 1.14）、埃伦费斯特定理（见习题 1.7）。

（3）通过估算，说明在一些物理体系中哪些体系需要用量子力学处理，哪些体系需要用经典力学处理，给出使用量子力学和经典力学处理问题的简单判据，见习题 1.18。

1 THE WAVE FUNCTION

1.1 THE SCHRÖDINGER EQUATION

Imagine a particle of mass m, constrained to move along the x axis, subject to some specified force $F(x, t)$ (Figure 1.1). The program of *classical* mechanics is to determine the position of the particle at any given time: $x(t)$. Once we know that, we can figure out the velocity ($v = dx/dt$), the momentum ($p = mv$), the kinetic energy $\left(T = (1/2)mv^2\right)$, or any other dynamical variable of interest. And how do we go about determining $x(t)$? We apply Newton's second law: $F = ma$. (For *conservative* systems—the only kind we shall consider, and, fortunately, the only kind that *occur* at the microscopic level—the force can be expressed as the derivative of a potential energy function,[1] $F = -\partial V/\partial x$, and Newton's law reads $m\, d^2x/dt^2 = -\partial V/\partial x$.) This, together with appropriate initial conditions (typically the position and velocity at $t = 0$), determines $x(t)$.

Quantum mechanics approaches this same problem quite differently. In this case what we're looking for is the particle's **wave function**, $\Psi(x, t)$, and we get it by solving the **Schrödinger equation**:

$$\boxed{i\hbar\frac{\partial\Psi}{\partial t} = -\frac{\hbar^2}{2m}\frac{\partial^2\Psi}{\partial x^2} + V\Psi.} \tag{1.1}$$

Here i is the square root of -1, and $\hbar$ is Planck's constant—or rather, his *original* constant (h) divided by 2π:

$$\hbar = \frac{h}{2\pi} = 1.054573 \times 10^{-34}\ \text{J s}. \tag{1.2}$$

The Schrödinger equation plays a role logically analogous to Newton's second law: Given suitable initial conditions (typically, $\Psi(x, 0)$), the Schrödinger equation determines $\Psi(x, t)$ for all future time, just as, in classical mechanics, Newton's law determines $x(t)$ for all future time.[2]

1.2 THE STATISTICAL INTERPRETATION

But what exactly *is* this "wave function," and what does it do for you once you've *got* it? After all, a particle, by its nature, is localized at a point, whereas the wave function (as its name

[1] Magnetic forces are an exception, but let's not worry about them just yet. By the way, we shall assume throughout this book that the motion is nonrelativistic ($v \ll c$).

[2] For a delightful first-hand account of the origins of the Schrödinger equation see the article by Felix Bloch in *Physics Today*, December 1976.

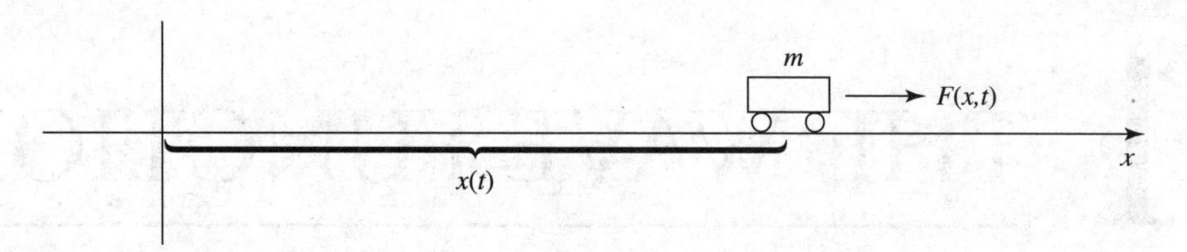

Figure 1.1: A "particle" constrained to move in one dimension under the influence of a specified force.

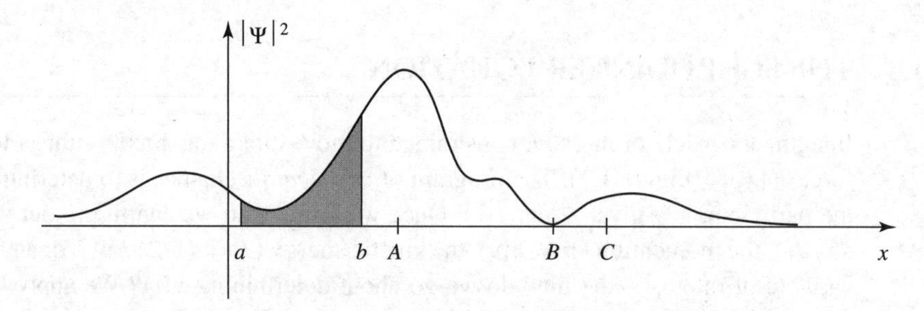

Figure 1.2: A typical wave function. The shaded area represents the probability of finding the particle between a and b. The particle would be relatively likely to be found near A, and unlikely to be found near B.

suggests) is spread out in space (it's a function of x, for any given t). How can such an object represent the state of a *particle*? The answer is provided by Born's **statistical interpretation**, which says that $|\Psi(x, t)|^2$ gives the *probability* of finding the particle at point x, at time t—or, more precisely,[3]

$$\int_a^b |\Psi(x,t)|^2\, dx = \left\{ \begin{array}{c} \text{probability of finding the particle} \\ \text{between } a \text{ and } b, \text{ at time } t. \end{array} \right\} \tag{1.3}$$

Probability is the *area* under the graph of $|\Psi|^2$. For the wave function in Figure 1.2, you would be quite likely to find the particle in the vicinity of point A, where $|\Psi|^2$ is large, and relatively *un*likely to find it near point B.

The statistical interpretation introduces a kind of **indeterminacy** into quantum mechanics, for even if you know everything the theory has to tell you about the particle (to wit: its wave function), still you cannot predict with certainty the outcome of a simple experiment to measure its position—all quantum mechanics has to offer is *statistical* information about the *possible* results. This indeterminacy has been profoundly disturbing to physicists and philosophers alike, and it is natural to wonder whether it is a fact of nature, or a defect in the theory.

Suppose I *do* measure the position of the particle, and I find it to be at point C.[4] *Question:* Where was the particle just *before* I made the measurement? There are three plausible answers

[3] The wave function itself is complex, but $|\Psi|^2 = \Psi^*\Psi$ (where Ψ^* is the complex conjugate of Ψ) is real and non-negative—as a probability, of course, *must* be.

[4] Of course, no measuring instrument is perfectly precise; what I *mean* is that the particle was found *in the vicinity* of C, as defined by the precision of the equipment.

to this question, and they serve to characterize the main schools of thought regarding quantum indeterminacy:

1. The **realist** position: *The particle was at C.* This certainly seems reasonable, and it is the response Einstein advocated. Note, however, that if this is true then quantum mechanics is an *incomplete* theory, since the particle *really was* at C, and yet quantum mechanics was unable to tell us so. To the realist, indeterminacy is not a fact of nature, but a reflection of our ignorance. As d'Espagnat put it, "the position of the particle was never indeterminate, but was merely unknown to the experimenter."[5] Evidently Ψ is not the whole story—some additional information (known as a **hidden variable**) is needed to provide a complete description of the particle.

2. The **orthodox** position: *The particle wasn't really anywhere.* It was the act of measurement that forced it to "take a stand" (though how and why it decided on the point C we dare not ask). Jordan said it most starkly: "Observations not only *disturb* what is to be measured, they *produce* it . . . We *compel* [the particle] to assume a definite position."[6] This view (the so-called **Copenhagen interpretation**), is associated with Bohr and his followers. Among physicists it has always been the most widely accepted position. Note, however, that if it is correct there is something very peculiar about the act of measurement—something that almost a century of debate has done precious little to illuminate.

3. The **agnostic** position: *Refuse to answer.* This is not quite as silly as it sounds—after all, what sense can there be in making assertions about the status of a particle *before* a measurement, when the only way of knowing whether you were right is precisely to *make* a measurement, in which case what you get is no longer "before the measurement"? It is metaphysics (in the pejorative sense of the word) to worry about something that cannot, by its nature, be tested. Pauli said: "One should no more rack one's brain about the problem of whether something one cannot know anything about exists all the same, than about the ancient question of how many angels are able to sit on the point of a needle."[7] For decades this was the "fall-back" position of most physicists: they'd try to sell you the orthodox answer, but if you were persistent they'd retreat to the agnostic response, and terminate the conversation.

Until fairly recently, all three positions (realist, orthodox, and agnostic) had their partisans. But in 1964 John Bell astonished the physics community by showing that it makes an *observable* difference whether the particle had a precise (though unknown) position prior to

[5] Bernard d'Espagnat, "The Quantum Theory and Reality" (*Scientific American*, November 1979, p. 165).

[6] Quoted in a lovely article by N. David Mermin, "Is the moon there when nobody looks?" (*Physics Today*, April 1985, p. 38).

[7] Ibid., p. 40.

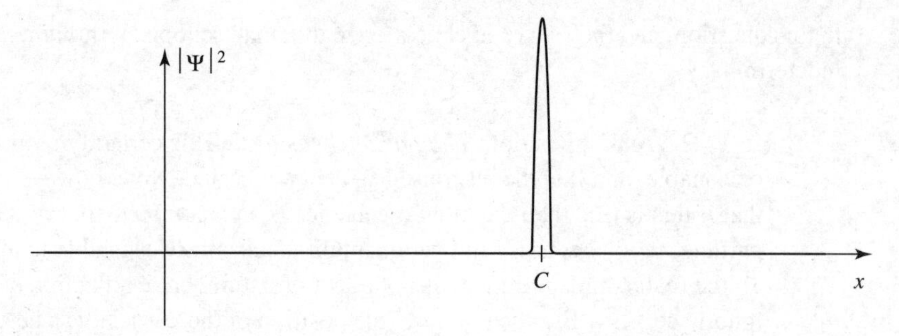

Figure 1.3: Collapse of the wave function: graph of $|\Psi|^2$ immediately *after* a measurement has found the particle at point C.

the measurement, or not. Bell's discovery effectively eliminated agnosticism as a viable option, and made it an *experimental* question whether 1 or 2 is the correct choice. I'll return to this story at the end of the book, when you will be in a better position to appreciate Bell's argument; for now, suffice it to say that the experiments have decisively confirmed the orthodox interpretation:[8] a particle simply does not have a precise position prior to measurement, any more than the ripples on a pond do; it is the measurement process that insists on one particular number, and thereby in a sense *creates* the specific result, limited only by the statistical weighting imposed by the wave function.

What if I made a *second* measurement, immediately after the first? Would I get C again, or does the act of measurement cough up some completely new number each time? On this question everyone is in agreement: A repeated measurement (on the same particle) must return the same value. Indeed, it would be tough to prove that the particle was really found at C in the first instance, if this could not be confirmed by immediate repetition of the measurement. How does the orthodox interpretation account for the fact that the second measurement is bound to yield the value C? It must be that the first measurement radically alters the wave function, so that it is now sharply peaked about C (Figure 1.3). We say that the wave function **collapses**, upon measurement, to a spike at the point C (it soon spreads out again, in accordance with the Schrödinger equation, so the second measurement must be made quickly). There are, then, two entirely distinct kinds of physical processes: "ordinary" ones, in which the wave function evolves in a leisurely fashion under the Schrödinger equation, and "measurements," in which Ψ suddenly and discontinuously collapses.[9]

[8] This statement is a little too strong: there exist viable nonlocal hidden variable theories (notably David Bohm's), and other formulations (such as the **many worlds** interpretation) that do not fit cleanly into any of my three categories. But I think it is wise, at least from a pedagogical point of view, to adopt a clear and coherent platform at this stage, and worry about the alternatives later.

[9] The role of measurement in quantum mechanics is so critical and so bizarre that you may well be wondering what precisely *constitutes* a measurement. I'll return to this thorny issue in the Afterword; for the moment let's take the naive view: a measurement is the kind of thing that a scientist in a white coat does in the laboratory, with rulers, stopwatches, Geiger counters, and so on.

Example 1.1

Electron Interference. I have asserted that particles (electrons, for example) have a wave nature, encoded in Ψ. How might we check this, in the laboratory?

The classic signature of a wave phenomenon is *interference*: two waves *in phase* interfere constructively, and out of phase they interfere destructively. The wave nature of light was confirmed in 1801 by Young's famous double-slit experiment, showing interference "fringes" on a distant screen when a monochromatic beam passes through two slits. If essentially the same experiment is done with *electrons*, the same pattern develops,[10] confirming the wave nature of electrons.

Now suppose we decrease the intensity of the electron beam, until only one electron is present in the apparatus at any particular time. According to the statistical interpretation each electron will produce a spot on the screen. Quantum mechanics cannot predict the precise *location* of that spot—all it can tell us is the *probability* of a given electron landing at a particular place. But if we are patient, and wait for a hundred thousand electrons—one at a time—to make the trip, the accumulating spots reveal the classic two-slit interference pattern (Figure 1.4).[11]

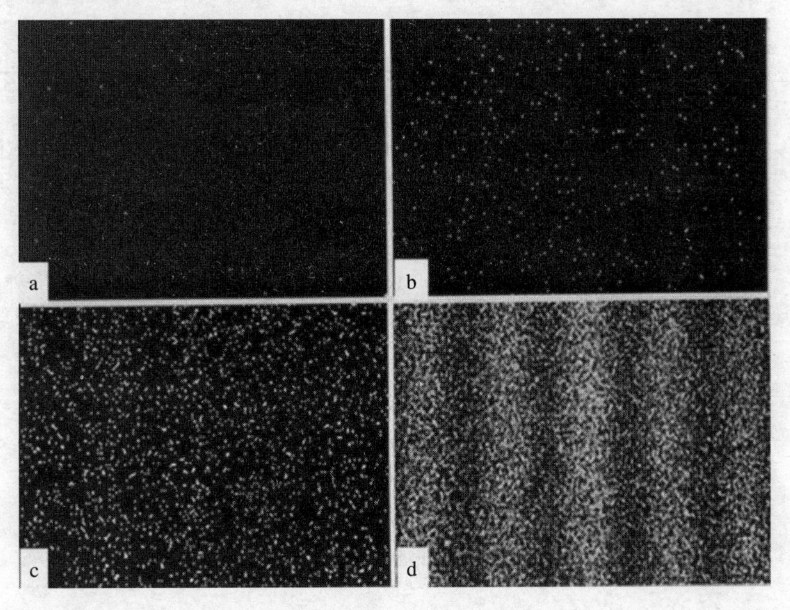

Figure 1.4: Build-up of the electron interference pattern. (a) Eight electrons, (b) 270 electrons, (c) 2000 electrons, (d) 160,000 electrons. Reprinted courtesy of the Central Research Laboratory, Hitachi, Ltd., Japan.

[10] Because the wavelength of electrons is typically very small, the slits have to be extremely close together. Historically, this was first achieved by Davisson and Germer, in 1925, using the atomic layers in a crystal as "slits." For an interesting account, see R. K. Gehrenbeck, *Physics Today*, January 1978, page 34.

[11] See Tonomura et al., *American Journal of Physics*, Volume 57, Issue 2, pp. 117–120 (1989), and the amazing associated video at www.hitachi.com/rd/portal/highlight/quantum/doubleslit/. This experiment can now be done with much more massive particles, including "Bucky-balls"; see M. Arndt, et al., *Nature* **40**, 680 (1999). Incidentally, the same thing can be done with light: turn the intensity so low that only one "photon" is present at a time and you get an identical point-by-point assembly of the interference pattern. See R. S. Aspden, M. J. Padgett, and G. C. Spalding, *Am. J. Phys.* **84**, 671 (2016).

Of course, if you close off one slit, or somehow contrive to detect which slit each electron passes through, the interference pattern disappears; the wave function of the emerging particle is now entirely different (in the first case because the boundary conditions for the Schrödinger equation have been changed, and in the second because of the collapse of the wave function upon measurement). But with both slits open, and no interruption of the electron in flight, each electron interferes with itself; it didn't pass through one slit or the other, but through both at once, just as a water wave, impinging on a jetty with two openings, interferes with itself. There is nothing mysterious about this, once you have accepted the notion that particles obey a wave equation. The truly *astonishing* thing is the blip-by-blip assembly of the pattern. In any classical wave theory the pattern would develop smoothly and continuously, simply getting more intense as time goes on. The quantum process is more like the pointillist painting of Seurat: The picture emerges from the cumulative contributions of all the individual dots.[12]

1.3 PROBABILITY

1.3.1 Discrete Variables

Because of the statistical interpretation, probability plays a central role in quantum mechanics, so I digress now for a brief discussion of probability theory. It is mainly a question of introducing some notation and terminology, and I shall do it in the context of a simple example.

Imagine a room containing fourteen people, whose ages are as follows:

one person aged 14,
one person aged 15,
three people aged 16,
two people aged 22,
two people aged 24,
five people aged 25.

If we let $N(j)$ represent the number of people of age j, then

$$
\begin{aligned}
N(14) &= 1,\\
N(15) &= 1,\\
N(16) &= 3,\\
N(22) &= 2,\\
N(24) &= 2,\\
N(25) &= 5,
\end{aligned}
$$

[12] I think it is important to distinguish things like interference and diffraction that would hold for any wave theory from the uniquely quantum mechanical features of the measurement process, which derive from the statistical interpretation.

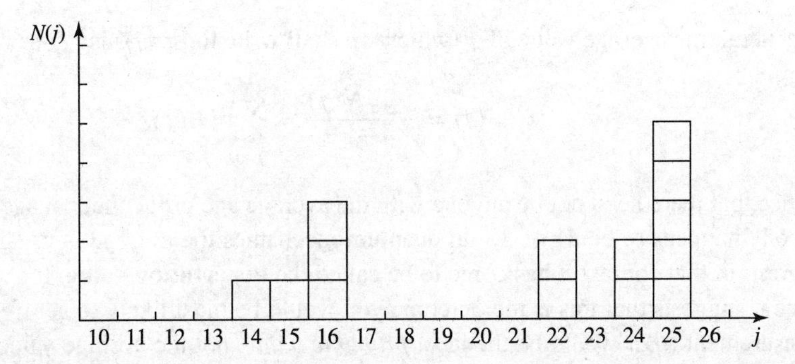

Figure 1.5: Histogram showing the number of people, $N(j)$, with age j, for the example in Section 1.3.1.

while $N(17)$, for instance, is zero. The *total* number of people in the room is

$$N = \sum_{j=0}^{\infty} N(j). \tag{1.4}$$

(In the example, of course, $N = 14$.) Figure 1.5 is a histogram of the data. The following are some questions one might ask about this distribution.

Question 1 If you selected one individual at random from this group, what is the **probability** that this person's age would be 15?

Answer One chance in 14, since there are 14 possible choices, all equally likely, of whom only one has that particular age. If $P(j)$ is the probability of getting age j, then $P(14) = 1/14$, $P(15) = 1/14$, $P(16) = 3/14$, and so on. In general,

$$P(j) = \frac{N(j)}{N}. \tag{1.5}$$

Notice that the probability of getting *either* 14 *or* 15 is the *sum* of the individual probabilities (in this case, 1/7). In particular, the sum of *all* the probabilities is 1—the person you select must have *some* age:

$$\sum_{j=0}^{\infty} P(j) = 1. \tag{1.6}$$

Question 2 What is the **most probable** age?

Answer 25, obviously; five people share this age, whereas at most three have any other age. The most probable j is the j for which $P(j)$ is a maximum.

Question 3 What is the **median** age?

Answer 23, for 7 people are younger than 23, and 7 are older. (The median is that value of j such that the probability of getting a larger result is the same as the probability of getting a smaller result.)

Question 4 What is the **average** (or **mean**) age?

Answer

$$\frac{(14) + (15) + 3(16) + 2(22) + 2(24) + 5(25)}{14} = \frac{294}{14} = 21.$$

In general, the average value of j (which we shall write thus: $\langle j\rangle$) is

$$\langle j\rangle = \frac{\sum jN(j)}{N} = \sum_{j=0}^{\infty} jP(j). \tag{1.7}$$

Notice that there need not be anyone with the average age or the median age—in this example nobody happens to be 21 or 23. In quantum mechanics the average is usually the quantity of interest; in that context it has come to be called the **expectation value**. It's a misleading term, since it suggests that this is the outcome you would be most likely to get if you made a single measurement (*that* would be the *most probable value*, not the average value)—but I'm afraid we're stuck with it.

Question 5 What is the average of the *squares* of the ages?
Answer You could get $14^2 = 196$, with probability 1/14, or $15^2 = 225$, with probability 1/14, or $16^2 = 256$, with probability 3/14, and so on. The average, then, is

$$\left\langle j^2\right\rangle = \sum_{j=0}^{\infty} j^2 P(j). \tag{1.8}$$

In general, the average value of some *function* of j is given by

$$\boxed{\langle f(j)\rangle = \sum_{j=0}^{\infty} f(j)P(j).} \tag{1.9}$$

(Equations 1.6, 1.7, and 1.8 are, if you like, special cases of this formula.) *Beware:* The average of the squares, $\left\langle j^2\right\rangle$, is *not* equal, in general, to the square of the average, $\langle j\rangle^2$. For instance, if the room contains just two babies, aged 1 and 3, then $\left\langle j^2\right\rangle = 5$, but $\langle j\rangle^2 = 4$.

Now, there is a conspicuous difference between the two histograms in Figure 1.6, even though they have the same median, the same average, the same most probable value, and the same number of elements: The first is sharply peaked about the average value, whereas the second is broad and flat. (The first might represent the age profile for students in a big-city classroom, the second, perhaps, a rural one-room schoolhouse.) We need a numerical measure

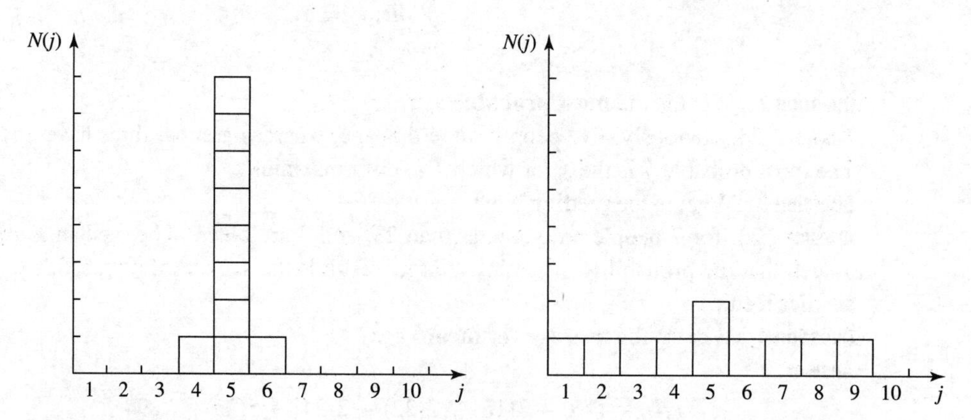

Figure 1.6: Two histograms with the same median, same average, and same most probable value, but different standard deviations.

of the amount of "spread" in a distribution, with respect to the average. The most obvious way to do this would be to find out how far each individual is from the average,

$$\Delta j = j - \langle j \rangle, \tag{1.10}$$

and compute the average of Δj. Trouble is, of course, that you get *zero*:

$$\begin{aligned}\langle \Delta j \rangle &= \sum (j - \langle j \rangle) P(j) = \sum j P(j) - \langle j \rangle \sum P(j) \\ &= \langle j \rangle - \langle j \rangle = 0.\end{aligned}$$

(Note that $\langle j \rangle$ is constant—it does not change as you go from one member of the sample to another—so it can be taken outside the summation.) To avoid this irritating problem you might decide to average the *absolute value* of Δj. But absolute values are nasty to work with; instead, we get around the sign problem by *squaring* before averaging:

$$\sigma^2 \equiv \left\langle (\Delta j)^2 \right\rangle. \tag{1.11}$$

This quantity is known as the **variance** of the distribution; σ itself (the square root of the average of the square of the deviation from the average—gulp!) is called the **standard deviation**. The latter is the customary measure of the spread about $\langle j \rangle$.

There is a useful little theorem on variances:

$$\begin{aligned}\sigma^2 &= \left\langle (\Delta j)^2 \right\rangle = \sum (\Delta j)^2 P(j) = \sum (j - \langle j \rangle)^2 P(j) \\ &= \sum \left(j^2 - 2j \langle j \rangle + \langle j \rangle^2 \right) P(j) \\ &= \sum j^2 P(j) - 2 \langle j \rangle \sum j P(j) + \langle j \rangle^2 \sum P(j) \\ &= \left\langle j^2 \right\rangle - 2 \langle j \rangle \langle j \rangle + \langle j \rangle^2 = \left\langle j^2 \right\rangle - \langle j \rangle^2 .\end{aligned}$$

Taking the square root, the standard deviation itself can be written as

$$\sigma = \sqrt{\left\langle j^2 \right\rangle - \langle j \rangle^2}. \tag{1.12}$$

In practice, this is a much faster way to get σ than by direct application of Equation 1.11: simply calculate $\left\langle j^2 \right\rangle$ and $\langle j \rangle^2$, subtract, and take the square root. Incidentally, I warned you a moment ago that $\left\langle j^2 \right\rangle$ is not, in general, equal to $\langle j \rangle^2$. Since σ^2 is plainly non-negative (from its definition 1.11), Equation 1.12 implies that

$$\left\langle j^2 \right\rangle \geq \langle j \rangle^2 , \tag{1.13}$$

and the two are equal only when $\sigma = 0$, which is to say, for distributions with no spread at all (every member having the same value).

1.3.2 Continuous Variables

So far, I have assumed that we are dealing with a *discrete* variable—that is, one that can take on only certain isolated values (in the example, j had to be an integer, since I gave ages only in years). But it is simple enough to generalize to *continuous* distributions. If I select a random person off the street, the probability that her age is *precisely* 16 years, 4 hours, 27 minutes, and 3.333... seconds is *zero*. The only sensible thing to speak about is the probability that her age lies in some *interval*—say, between 16 and 17. If the interval is sufficiently short, this probability is *proportional to the length of the interval*. For example, the chance that her age is between 16 and 16 plus *two* days is presumably twice the probability that it is between 16 and

16 plus *one* day. (Unless, I suppose, there was some extraordinary baby boom 16 years ago, on exactly that day—in which case we have simply chosen an interval too long for the rule to apply. If the baby boom lasted six hours, we'll take intervals of a second or less, to be on the safe side. Technically, we're talking about *infinitesimal* intervals.) Thus

$$\left\{\begin{array}{l}\text{probability that an individual (chosen} \\ \text{at random) lies between } x \text{ and } (x+dx)\end{array}\right\} = \rho(x)\,dx. \tag{1.14}$$

The proportionality factor, $\rho(x)$, is often loosely called "the probability of getting x," but this is sloppy language; a better term is **probability density**. The probability that x lies between a and b (a *finite* interval) is given by the integral of $\rho(x)$:

$$P_{ab} = \int_a^b \rho(x)\,dx, \tag{1.15}$$

and the rules we deduced for discrete distributions translate in the obvious way:

$$\int_{-\infty}^{+\infty} \rho(x)\,dx = 1, \tag{1.16}$$

$$\langle x\rangle = \int_{-\infty}^{+\infty} x\rho(x)\,dx, \tag{1.17}$$

$$\langle f(x)\rangle = \int_{-\infty}^{+\infty} f(x)\,\rho(x)\,dx, \tag{1.18}$$

$$\sigma^2 \equiv \left\langle(\Delta x)^2\right\rangle = \left\langle x^2\right\rangle - \langle x\rangle^2. \tag{1.19}$$

Example 1.2

Suppose someone drops a rock off a cliff of height h. As it falls, I snap a million photographs, at random intervals. On each picture I measure the distance the rock has fallen. *Question:* What is the *average* of all these distances? That is to say, what is the *time average* of the distance traveled?[13]

Solution: The rock starts out at rest, and picks up speed as it falls; it spends more time near the top, so the average distance will surely be less than $h/2$. Ignoring air resistance, the distance x at time t is

$$x(t) = \frac{1}{2}gt^2.$$

The velocity is $dx/dt = gt$, and the total flight time is $T = \sqrt{2h/g}$. The probability that a particular photograph was taken between t and $t+dt$ is dt/T, so the probability that it shows a distance in the corresponding range x to $x+dx$ is

$$\frac{dt}{T} = \frac{dx}{gt}\sqrt{\frac{g}{2h}} = \frac{1}{2\sqrt{hx}}\,dx.$$

[13] A statistician will complain that I am confusing the average of a *finite sample* (a million, in this case) with the "true" average (over the whole continuum). This can be an awkward problem for the experimentalist, especially when the sample size is small, but here I am only concerned with the *true* average, to which the sample average is presumably a good approximation.

Thus the probability *density* (Equation 1.14) is

$$\rho(x) = \frac{1}{2\sqrt{hx}}, \quad (0 \leq x \leq h)$$

(outside this range, of course, the probability density is zero).

We can check this result, using Equation 1.16:

$$\int_0^h \frac{1}{2\sqrt{hx}}\,dx = \frac{1}{2\sqrt{h}}\left(2x^{1/2}\right)\Big|_0^h = 1.$$

The *average* distance (Equation 1.17) is

$$\langle x \rangle = \int_0^h x\frac{1}{2\sqrt{hx}}\,dx = \frac{1}{2\sqrt{h}}\left(\frac{2}{3}x^{3/2}\right)\Big|_0^h = \frac{h}{3},$$

which is somewhat *less* than $h/2$, as anticipated.

Figure 1.7 shows the graph of $\rho(x)$. Notice that a probability *density* can be infinite, though probability itself (the *integral* of ρ) must of course be finite (indeed, less than or equal to 1).

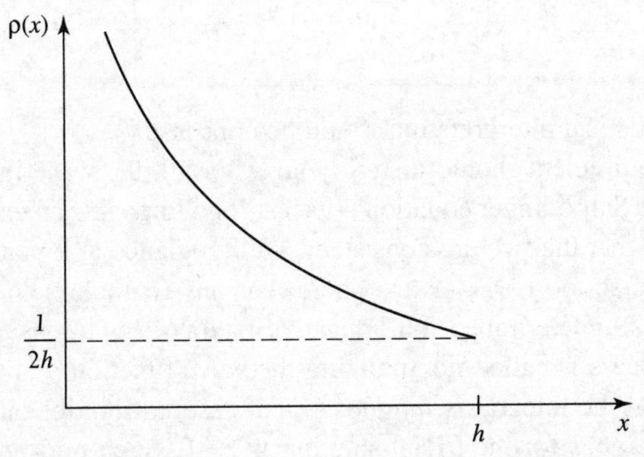

Figure 1.7: The probability density in Example 1.2: $\rho(x) = 1\Big/\left(2\sqrt{hx}\right)$.

* **Problem 1.1** For the distribution of ages in the example in Section 1.3.1:

(a) Compute $\langle j^2 \rangle$ and $\langle j \rangle^2$.

(b) Determine Δj for each j, and use Equation 1.11 to compute the standard deviation.

(c) Use your results in (a) and (b) to check Equation 1.12.

Problem 1.2

(a) Find the standard deviation of the distribution in Example 1.2.

(b) What is the probability that a photograph, selected at random, would show a distance x more than one standard deviation away from the average?

* **Problem 1.3** Consider the **gaussian** distribution

$$\rho(x) = Ae^{-\lambda(x-a)^2},$$

where A, a, and λ are positive real constants. (The necessary integrals are inside the back cover.)

(a) Use Equation 1.16 to determine A.

(b) Find $\langle x \rangle$, $\langle x^2 \rangle$, and σ.

(c) Sketch the graph of $\rho(x)$.

1.4 NORMALIZATION

We return now to the statistical interpretation of the wave function (Equation 1.3), which says that $|\Psi(x,t)|^2$ is the probability density for finding the particle at point x, at time t. It follows (Equation 1.16) that the integral of $|\Psi|^2$ over *all* x must be 1 (the particle's got to be *some*where):

$$\boxed{\int_{-\infty}^{+\infty} |\Psi(x,t)|^2 \, dx = 1.} \tag{1.20}$$

Without this, the statistical interpretation would be nonsense.

However, this requirement should disturb you: After all, the wave function is supposed to be determined by the Schrödinger equation—we can't go imposing an extraneous condition on Ψ without checking that the two are consistent. Well, a glance at Equation 1.1 reveals that if $\Psi(x,t)$ is a solution, so too is $A\Psi(x,t)$, where A is any (complex) constant. What we must do, then, is pick this undetermined multiplicative factor so as to ensure that Equation 1.20 is satisfied. This process is called **normalizing** the wave function. For some solutions to the Schrödinger equation the integral is *infinite*; in that case *no* multiplicative factor is going to make it 1. The same goes for the trivial solution $\Psi = 0$. Such **non-normalizable** solutions cannot represent particles, and must be rejected. Physically realizable states correspond to the **square-integrable** solutions to Schrödinger's equation.[14]

But wait a minute! Suppose I have normalized the wave function at time $t = 0$. How do I know that it will *stay* normalized, as time goes on, and Ψ evolves? (You can't keep *re*normalizing the wave function, for then A becomes a function of t, and you no longer have a solution to the Schrödinger equation.) Fortunately, the Schrödinger equation has the remarkable property that it automatically preserves the normalization of the wave function—without this crucial feature the Schrödinger equation would be incompatible with the statistical interpretation, and the whole theory would crumble.

This is important, so we'd better pause for a careful proof. To begin with,

$$\frac{d}{dt}\int_{-\infty}^{+\infty} |\Psi(x,t)|^2 \, dx = \int_{-\infty}^{+\infty} \frac{\partial}{\partial t} |\Psi(x,t)|^2 \, dx. \tag{1.21}$$

[14] Evidently $\Psi(x,t)$ must go to zero faster than $1/\sqrt{|x|}$, as $|x| \to \infty$. Incidentally, normalization only fixes the *modulus* of A; the *phase* remains undetermined. However, as we shall see, the latter carries no physical significance anyway.

(Note that the *integral* is a function only of t, so I use a *total* derivative (d/dt) on the left, but the *integrand* is a function of x as well as t, so it's a *partial* derivative $(\partial/\partial t)$ on the right.) By the product rule,

$$\frac{\partial}{\partial t}|\Psi|^2 = \frac{\partial}{\partial t}\left(\Psi^*\Psi\right) = \Psi^*\frac{\partial \Psi}{\partial t} + \frac{\partial \Psi^*}{\partial t}\Psi. \tag{1.22}$$

Now the Schrödinger equation says that

$$\frac{\partial \Psi}{\partial t} = \frac{i\hbar}{2m}\frac{\partial^2 \Psi}{\partial x^2} - \frac{i}{\hbar}V\Psi, \tag{1.23}$$

and hence also (taking the complex conjugate of Equation 1.23)

$$\frac{\partial \Psi^*}{\partial t} = -\frac{i\hbar}{2m}\frac{\partial^2 \Psi^*}{\partial x^2} + \frac{i}{\hbar}V\Psi^*, \tag{1.24}$$

so

$$\frac{\partial}{\partial t}|\Psi|^2 = \frac{i\hbar}{2m}\left(\Psi^*\frac{\partial^2 \Psi}{\partial x^2} - \frac{\partial^2 \Psi^*}{\partial x^2}\Psi\right) = \frac{\partial}{\partial x}\left[\frac{i\hbar}{2m}\left(\Psi^*\frac{\partial \Psi}{\partial x} - \frac{\partial \Psi^*}{\partial x}\Psi\right)\right]. \tag{1.25}$$

The integral in Equation 1.21 can now be evaluated explicitly:

$$\frac{d}{dt}\int_{-\infty}^{+\infty}|\Psi(x,t)|^2\,dx = \frac{i\hbar}{2m}\left(\Psi^*\frac{\partial \Psi}{\partial x} - \frac{\partial \Psi^*}{\partial x}\Psi\right)\Bigg|_{-\infty}^{+\infty}. \tag{1.26}$$

But $\Psi(x,t)$ must go to zero as x goes to $(\pm)$ infinity—otherwise the wave function would not be normalizable.[15] It follows that

$$\frac{d}{dt}\int_{-\infty}^{+\infty}|\Psi(x,t)|^2\,dx = 0, \tag{1.27}$$

and hence that the integral is *constant* (independent of time); if Ψ is normalized at $t = 0$, it *stays* normalized for all future time. QED

Problem 1.4 At time $t = 0$ a particle is represented by the wave function

$$\Psi(x,0) = \begin{cases} A(x/a), & 0 \le x \le a, \\ A(b-x)/(b-a), & a \le x \le b, \\ 0, & \text{otherwise,} \end{cases}$$

where A, a, and b are (positive) constants.

(a) Normalize Ψ (that is, find A, in terms of a and b).

(b) Sketch $\Psi(x,0)$, as a function of x.

(c) Where is the particle most likely to be found, at $t = 0$?

(d) What is the probability of finding the particle to the left of a? Check your result in the limiting cases $b = a$ and $b = 2a$.

(e) What is the expectation value of x?

---

15 A competent mathematician can supply you with pathological counterexamples, but they do not arise in physics; for us the wave function and all its derivatives go to zero at infinity.

* **Problem 1.5** Consider the wave function

$$\Psi(x,t) = Ae^{-\lambda|x|}e^{-i\omega t},$$

where A, λ, and ω are positive real constants. (We'll see in Chapter 2 for what potential (V) this wave function satisfies the Schrödinger equation.)

(a) Normalize Ψ.

(b) Determine the expectation values of x and x^2.

(c) Find the standard deviation of x. Sketch the graph of $|\Psi|^2$, as a function of x, and mark the points $(\langle x\rangle+\sigma)$ and $(\langle x\rangle-\sigma)$, to illustrate the sense in which σ represents the "spread" in x. What is the probability that the particle would be found outside this range?

1.5 MOMENTUM

For a particle in state Ψ, the expectation value of x is

$$\boxed{\langle x\rangle = \int_{-\infty}^{+\infty} x\,|\Psi(x,t)|^2\,dx.} \tag{1.28}$$

What exactly does this mean? It emphatically does *not* mean that if you measure the position of one particle over and over again, $\int x\,|\Psi|^2\,dx$ is the average of the results you'll get. On the contrary: The first measurement (whose outcome is indeterminate) will collapse the wave function to a spike at the value actually obtained, and the subsequent measurements (if they're performed quickly) will simply repeat that same result. Rather, $\langle x\rangle$ is the average of measurements performed on particles *all in the state* Ψ, which means that either you must find some way of returning the particle to its original state after each measurement, or else you have to prepare a whole **ensemble** of particles, each in the same state Ψ, and measure the positions of all of them: $\langle x\rangle$ is the average of *these* results. I like to picture a row of bottles on a shelf, each containing a particle in the state Ψ (relative to the center of the bottle). A graduate student with a ruler is assigned to each bottle, and at a signal they all measure the positions of their respective particles. We then construct a histogram of the results, which should match $|\Psi|^2$, and compute the average, which should agree with $\langle x\rangle$. (Of course, since we're only using a finite sample, we can't expect perfect agreement, but the more bottles we use, the closer we ought to come.) In short, *the expectation value is the average of measurements on an ensemble of identically-prepared systems,* not the average of repeated measurements on one and the same system.

Now, as time goes on, $\langle x\rangle$ will change (because of the time dependence of Ψ), and we might be interested in knowing how fast it moves. Referring to Equations 1.25 and 1.28, we see that[16]

$$\frac{d\langle x\rangle}{dt} = \int x\frac{\partial}{\partial t}|\Psi|^2\,dx = \frac{i\hbar}{2m}\int x\frac{\partial}{\partial x}\left(\Psi^*\frac{\partial\Psi}{\partial x} - \frac{\partial\Psi^*}{\partial x}\Psi\right)dx. \tag{1.29}$$

[16] To keep things from getting too cluttered, I'll suppress the limits of integration ($\pm\infty$).

This expression can be simplified using integration-by-parts:[17]

$$\frac{d\langle x\rangle}{dt} = -\frac{i\hbar}{2m}\int\left(\Psi^*\frac{\partial\Psi}{\partial x} - \frac{\partial\Psi^*}{\partial x}\Psi\right)dx. \tag{1.30}$$

(I used the fact that $\partial x/\partial x = 1$, and threw away the boundary term, on the ground that Ψ goes to zero at ($\pm$) infinity.) Performing another integration-by-parts, on the second term, we conclude:

$$\frac{d\langle x\rangle}{dt} = -\frac{i\hbar}{m}\int\Psi^*\frac{\partial\Psi}{\partial x}\,dx. \tag{1.31}$$

What are we to make of this result? Note that we're talking about the "velocity" of the *expectation* value of x, which is not the same thing as the velocity of the *particle*. Nothing we have seen so far would enable us to calculate the velocity of a particle. It's not even clear what velocity *means* in quantum mechanics: If the particle doesn't have a determinate position (prior to measurement), neither does it have a well-defined velocity. All we could reasonably ask for is the *probability* of getting a particular value. We'll see in Chapter 3 how to construct the probability density for velocity, given Ψ; for the moment it will suffice to postulate that the *expectation value of the velocity is equal to the time derivative of the expectation value of position*:

$$\langle v\rangle = \frac{d\langle x\rangle}{dt}. \tag{1.32}$$

Equation 1.31 tells us, then, how to calculate $\langle v\rangle$ directly from Ψ.

Actually, it is customary to work with **momentum** ($p = mv$), rather than velocity:

$$\boxed{\langle p\rangle = m\frac{d\langle x\rangle}{dt} = -i\hbar\int\left(\Psi^*\frac{\partial\Psi}{\partial x}\right)dx.} \tag{1.33}$$

Let me write the expressions for $\langle x\rangle$ and $\langle p\rangle$ in a more suggestive way:

$$\langle x\rangle = \int\Psi^*\,[x]\,\Psi\,dx, \tag{1.34}$$

$$\langle p\rangle = \int\Psi^*\left[-i\hbar\,(\partial/\partial x)\right]\Psi\,dx. \tag{1.35}$$

We say that the **operator**[18] x "represents" position, and the operator $-i\hbar\,(\partial/\partial x)$ "represents" momentum; to calculate expectation values we "sandwich" the appropriate operator between Ψ^* and Ψ, and integrate.

---

[17] The product rule says that

$$\frac{d}{dx}(fg) = f\frac{dg}{dx} + \frac{df}{dx}g,$$

from which it follows that

$$\int_a^b f\frac{dg}{dx}\,dx = -\int_a^b \frac{df}{dx}g\,dx + fg\Big|_a^b.$$

Under the integral sign, then, you can peel a derivative off one factor in a product, and slap it onto the other one—it'll cost you a minus sign, and you'll pick up a boundary term.

[18] An "operator" is an instruction to *do something* to the function that follows; it takes in one function, and spits out some other function. The position operator tells you to *multiply* by x; the momentum operator tells you to *differentiate* with respect to x (and multiply the result by $-i\hbar$).

That's cute, but what about other quantities? The fact is, *all* classical dynamical variables can be expressed in terms of position and momentum. Kinetic energy, for example, is

$$T = \frac{1}{2}mv^2 = \frac{p^2}{2m},$$

and angular momentum is

$$\mathbf{L} = \mathbf{r} \times m\mathbf{v} = \mathbf{r} \times \mathbf{p}$$

(the latter, of course, does not occur for motion in one dimension). To calculate the expectation value of *any* such quantity, $Q(x, p)$, we simply replace every p by $-i\hbar(\partial/\partial x)$, insert the resulting operator between Ψ^* and Ψ, and integrate:

$$\boxed{\langle Q(x, p)\rangle = \int \Psi^* \left[Q(x, -i\hbar\,\partial/\partial x)\right] \Psi \, dx.} \tag{1.36}$$

For example, the expectation value of the kinetic energy is

$$\langle T\rangle = -\frac{\hbar^2}{2m}\int \Psi^* \frac{\partial^2\Psi}{\partial x^2}\, dx. \tag{1.37}$$

Equation 1.36 is a recipe for computing the expectation value of any dynamical quantity, for a particle in state Ψ; it subsumes Equations 1.34 and 1.35 as special cases. I have tried to make Equation 1.36 seem plausible, given Born's statistical interpretation, but in truth this represents such a radically new way of doing business (as compared with classical mechanics) that it's a good idea to get some practice *using* it before we come back (in Chapter 3) and put it on a firmer theoretical foundation. In the mean time, if you prefer to think of it as an *axiom*, that's fine with me.

Problem 1.6 Why can't you do integration-by-parts directly on the middle expression in Equation 1.29—pull the time derivative over onto x, note that $\partial x/\partial t = 0$, and conclude that $d\langle x\rangle/dt = 0$?

* **Problem 1.7** Calculate $d\langle p\rangle/dt$. *Answer*:

$$\frac{d\langle p\rangle}{dt} = \left\langle -\frac{\partial V}{\partial x}\right\rangle. \tag{1.38}$$

This is an instance of **Ehrenfest's theorem**, which asserts that *expectation values obey the classical laws.*[19]

Problem 1.8 Suppose you add a constant V_0 to the potential energy (by "constant" I mean independent of x as well as t). In *classical* mechanics this doesn't change anything, but what about *quantum* mechanics? Show that the wave function picks up a time-dependent phase factor: $\exp(-iV_0t/\hbar)$. What effect does this have on the expectation value of a dynamical variable?

[19] Some authors limit the term to the pair of equations $\langle p\rangle = m\,d\langle x\rangle/dt$ and $\langle -\partial V/\partial x\rangle = d\langle p\rangle/dt$.

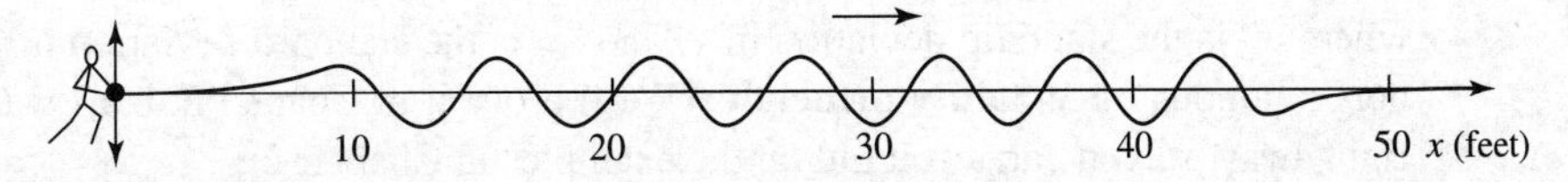

Figure 1.8: A wave with a (fairly) well-defined *wavelength*, but an ill-defined *position*.

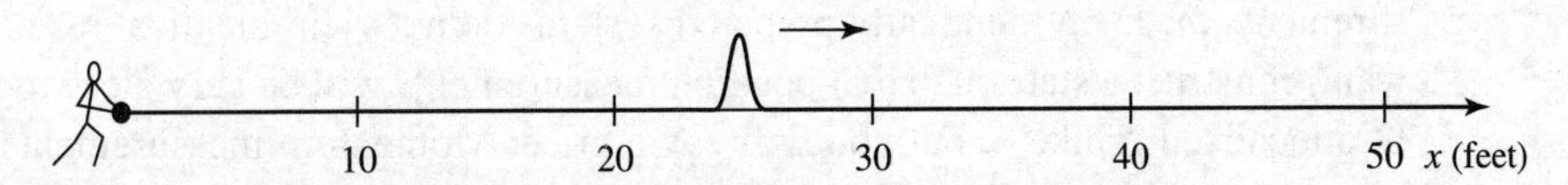

Figure 1.9: A wave with a (fairly) well-defined *position*, but an ill-defined *wavelength*.

1.6 THE UNCERTAINTY PRINCIPLE

Imagine that you're holding one end of a very long rope, and you generate a wave by shaking it up and down rhythmically (Figure 1.8). If someone asked you "Precisely where *is* that wave?" you'd probably think he was a little bit nutty: The wave isn't precisely *any*where—it's spread out over 50 feet or so. On the other hand, if he asked you what its *wavelength* is, you could give him a reasonable answer: it looks like about 6 feet. By contrast, if you gave the rope a sudden jerk (Figure 1.9), you'd get a relatively narrow bump traveling down the line. This time the first question (Where precisely is the wave?) is a sensible one, and the second (What is its wavelength?) seems nutty—it isn't even vaguely periodic, so how can you assign a wavelength to it? Of course, you can draw intermediate cases, in which the wave is *fairly* well localized and the wavelength is *fairly* well defined, but there is an inescapable trade-off here: the more precise a wave's position is, the less precise is its wavelength, and vice versa.[20] A theorem in Fourier analysis makes all this rigorous, but for the moment I am only concerned with the qualitative argument.

This applies, of course, to *any* wave phenomenon, and hence in particular to the quantum mechanical wave function. But the wavelength of Ψ is related to the *momentum* of the particle by the **de Broglie formula**:[21]

$$p = \frac{h}{\lambda} = \frac{2\pi\hbar}{\lambda}. \tag{1.39}$$

Thus a spread in *wavelength* corresponds to a spread in *momentum*, and our general observation now says that the more precisely determined a particle's position is, the less precisely is its momentum. Quantitatively,

$$\boxed{\sigma_x \sigma_p \geq \frac{\hbar}{2},} \tag{1.40}$$

[20] That's why a piccolo player must be right on pitch, whereas a double-bass player can afford to wear garden gloves. For the piccolo, a sixty-fourth note contains many full cycles, and the frequency (we're working in the time domain now, instead of space) is well defined, whereas for the bass, at a much lower register, the sixty-fourth note contains only a few cycles, and all you hear is a general sort of "oomph," with no very clear pitch.

[21] I'll explain this in due course. Many authors take the de Broglie formula as an *axiom*, from which they then deduce the association of momentum with the operator $-i\hbar(\partial/\partial x)$. Although this is a conceptually cleaner approach, it involves diverting mathematical complications that I would rather save for later.

where σ_x is the standard deviation in x, and σ_p is the standard deviation in p. This is Heisenberg's famous **uncertainty principle**. (We'll prove it in Chapter 3, but I wanted to mention it right away, so you can test it out on the examples in Chapter 2.)

Please understand what the uncertainty principle *means*: Like position measurements, momentum measurements yield precise answers—the "spread" here refers to the fact that measurements made on identically prepared systems do not yield identical results. You can, if you want, construct a state such that position measurements will be very close together (by making Ψ a localized "spike"), but you will pay a price: Momentum measurements on this state will be widely scattered. Or you can prepare a state with a definite momentum (by making Ψ a long sinusoidal wave), but in that case position measurements will be widely scattered. And, of course, if you're in a really bad mood you can create a state for which neither position nor momentum is well defined: Equation 1.40 is an *inequality*, and there's no limit on how *big* σ_x and σ_p can be—just make Ψ some long wiggly line with lots of bumps and potholes and no periodic structure.

* **Problem 1.9** A particle of mass m has the wave function

$$\Psi(x,t) = Ae^{-a\left[(mx^2/\hbar)+it\right]},$$

where A and a are positive real constants.

(a) Find A.

(b) For what potential energy function, $V(x)$, is this a solution to the Schrödinger equation?

(c) Calculate the expectation values of x, x^2, p, and p^2.

(d) Find σ_x and σ_p. Is their product consistent with the uncertainty principle?

FURTHER PROBLEMS ON CHAPTER 1

Problem 1.10 Consider the first 25 digits in the decimal expansion of π (3, 1, 4, 1, 5, 9, ...).

(a) If you selected one number at random, from this set, what are the probabilities of getting each of the 10 digits?

(b) What is the most probable digit? What is the median digit? What is the average value?

(c) Find the standard deviation for this distribution.

Problem 1.11 [This problem generalizes Example 1.2.] Imagine a particle of mass m and energy E in a potential well $V(x)$, sliding frictionlessly back and forth between the classical turning points (a and b in Figure 1.10). Classically, the probability of finding the particle in the range dx (if, for example, you took a snapshot at a random time t) is equal to the fraction of the time T it takes to get from a to b that it spends in the interval dx:

$$\rho(x)\,dx = \frac{dt}{T} = \frac{(dt/dx)\,dx}{T} = \frac{1}{v(x)\,T}\,dx, \tag{1.41}$$

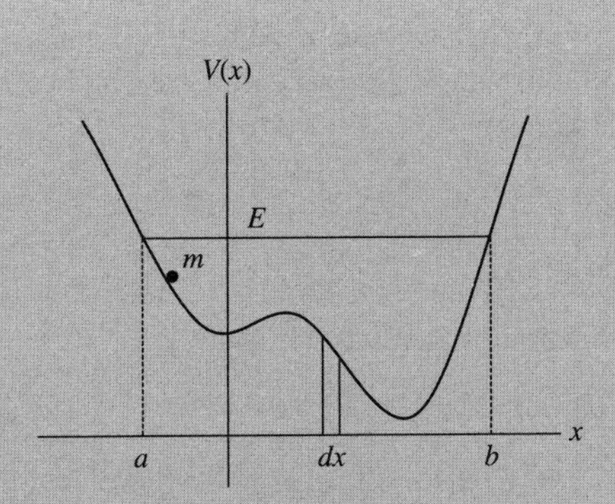

Figure 1.10: Classical particle in a potential well.

where $v(x)$ is the speed, and

$$T=\int_0^T dt=\int_a^b \frac{1}{v(x)}\,dx. \tag{1.42}$$

Thus

$$\rho(x)=\frac{1}{v(x)T}. \tag{1.43}$$

This is perhaps the closest classical analog[22] to $|\Psi|^2$.

(a) Use conservation of energy to express $v(x)$ in terms of E and $V(x)$.

(b) As an example, find $\rho(x)$ for the simple harmonic oscillator, $V(x)=kx^2/2$. Plot $\rho(x)$, and check that it is correctly normalized.

(c) For the classical harmonic oscillator in part (b), find $\langle x\rangle$, $\langle x^2\rangle$, and σ_x.

** **Problem 1.12** What if we were interested in the distribution of *momenta* ($p=mv$), for the classical harmonic oscillator (Problem 1.11(b)).

(a) Find the classical probability distribution $\rho(p)$ (note that p ranges from $-\sqrt{2mE}$ to $+\sqrt{2mE}$).

(b) Calculate $\langle p\rangle$, $\langle p^2\rangle$, and σ_p.

(c) What's the *classical* uncertainty product, $\sigma_x\sigma_p$, for this system? Notice that this product can be as small as you like, classically, simply by sending $E\to 0$. But in quantum mechanics, as we shall see in Chapter 2, the energy of a simple harmonic oscillator cannot be less than $\hbar\omega/2$, where $\omega=\sqrt{k/m}$ is the classical frequency. In that case what can you say about the product $\sigma_x\sigma_p$?

Problem 1.13 Check your results in Problem 1.11(b) with the following "numerical experiment." The position of the oscillator at time t is

$$x(t)=A\cos(\omega t). \tag{1.44}$$

[22] If you like, instead of photos of *one* system at random times, picture an ensemble of such systems, all with the same energy but with random starting positions, and photograph them all at the *same* time. The analysis is identical, but this interpretation is closer to the quantum notion of indeterminacy.

You might as well take $\omega = 1$ (that sets the scale for time) and $A = 1$ (that sets the scale for length). Make a plot of x at 10,000 random times, and compare it with $\rho(x)$. *Hint:* In Mathematica, first define

```
x[t_] := Cos[t]
```

then construct a table of positions:

```
snapshots = Table[x[π RandomReal[j]], {j, 10000}]
```

and finally, make a histogram of the data:

```
Histogram[snapshots, 100, "PDF", PlotRange → {0,2}]
```

Meanwhile, make a plot of the density function, $\rho(x)$, and, using **Show**, superimpose the two.

Problem 1.14 Let $P_{ab}(t)$ be the probability of finding the particle in the range ($a < x < b$), at time t.

(a) Show that

$$\frac{dP_{ab}}{dt} = J(a,t) - J(b,t),$$

where

$$J(x,t) \equiv \frac{i\hbar}{2m}\left(\Psi\frac{\partial\Psi^*}{\partial x} - \Psi^*\frac{\partial\Psi}{\partial x}\right).$$

What are the units of $J(x,t)$? *Comment:* J is called the **probability current**, because it tells you the rate at which probability is "flowing" past the point x. If $P_{ab}(t)$ is increasing, then more probability is flowing into the region at one end than flows out at the other.

(b) Find the probability current for the wave function in Problem 1.9. (This is not a very pithy example, I'm afraid; we'll encounter more substantial ones in due course.)

Problem 1.15 Show that

$$\frac{d}{dt}\int_{-\infty}^{\infty}\Psi_1^*\Psi_2\,dx = 0$$

for any two (normalizable) solutions to the Schrödinger equation (with the same $V(x)$), Ψ_1 and Ψ_2.

Problem 1.16 A particle is represented (at time $t = 0$) by the wave function

$$\Psi(x,0) = \begin{cases} A\left(a^2 - x^2\right), & -a \le x \le +a, \\ 0, & \text{otherwise.} \end{cases}$$

(a) Determine the normalization constant A.

(b) What is the expectation value of x?

(c) What is the expectation value of p? (Note that you *cannot* get it from $\langle p\rangle = md\langle x\rangle/dt$. Why not?)

(d) Find the expectation value of x^2.
(e) Find the expectation value of p^2.
(f) Find the uncertainty in x (σ_x).
(g) Find the uncertainty in p (σ_p).
(h) Check that your results are consistent with the uncertainty principle.

** **Problem 1.17** Suppose you wanted to describe an **unstable particle**, that spontaneously disintegrates with a "lifetime" τ. In that case the total probability of finding the particle somewhere should *not* be constant, but should decrease at (say) an exponential rate:

$$P(t) \equiv \int_{-\infty}^{+\infty} |\Psi(x,t)|^2\, dx = e^{-t/\tau}.$$

A crude way of achieving this result is as follows. In Equation 1.24 we tacitly assumed that V (the potential energy) is *real*. That is certainly reasonable, but it leads to the "conservation of probability" enshrined in Equation 1.27. What if we assign to V an imaginary part:

$$V = V_0 - i\Gamma,$$

where V_0 is the true potential energy and Γ is a positive real constant?
(a) Show that (in place of Equation 1.27) we now get

$$\frac{dP}{dt} = -\frac{2\Gamma}{\hbar} P.$$

(b) Solve for $P(t)$, and find the lifetime of the particle in terms of Γ.

Problem 1.18 Very roughly speaking, quantum mechanics is relevant when the de Broglie wavelength of the particle in question (h/p) is greater than the characteristic size of the system (d). In thermal equilibrium at (Kelvin) temperature T, the average kinetic energy of a particle is

$$\frac{p^2}{2m} = \frac{3}{2} k_B T$$

(where k_B is Boltzmann's constant), so the typical de Broglie wavelength is

$$\lambda = \frac{h}{\sqrt{3mk_BT}}. \tag{1.45}$$

The purpose of this problem is to determine which systems will have to be treated quantum mechanically, and which can safely be described classically.
(a) Solids. The lattice spacing in a typical solid is around $d = 0.3$ nm. Find the temperature below which the unbound[23] *electrons* in a solid are quantum mechanical. Below what temperature are the *nuclei* in a solid quantum mechanical? (Use silicon as an example.)

[23] In a solid the inner electrons are attached to a particular nucleus, and for them the relevant size would be the radius of the atom. But the outer-most electrons are not attached, and for them the relevant distance is the lattice spacing. This problem pertains to the *outer* electrons.

Moral: The free electrons in a solid are *always* quantum mechanical; the nuclei are generally *not* quantum mechanical. The same goes for liquids (for which the interatomic spacing is roughly the same), with the exception of helium below 4 K.

(b) Gases. For what temperatures are the atoms in an ideal gas at pressure P quantum mechanical? *Hint:* Use the ideal gas law ($PV = Nk_BT$) to deduce the interatomic spacing.
Answer: $T < (1/k_B)\left(h^2/3m\right)^{3/5} P^{2/5}$. Obviously (for the gas to show quantum behavior) we want m to be as *small* as possible, and P as *large* as possible. Put in the numbers for helium at atmospheric pressure. Is hydrogen in outer space (where the interatomic spacing is about 1 cm and the temperature is at least 3 K) quantum mechanical? (Assume it's monatomic hydrogen, not H_2.)

导读 / Guidance

第 2 章　定态薛定谔方程

本章讨论定态的概念，求解定态薛定谔方程，给出体系的能级和波函数；以几种具有简单势的一维体系为例，通过分离变量法求解其薛定谔方程的解，给出其粒子运动的能级和波函数的特征；引入分立谱、连续谱、束缚态、散射态等概念。讨论的体系包括：一维无限深方势阱（分立谱、束缚态）、一维谐振子（分立谱、束缚态）、一维自由粒子（连续谱、散射态）、一维 δ 函数势阱（束缚态、散射态）、有限深方势阱（束缚态、散射态）等。对一维谐振子的处理采用代数法和解析法两种方法，在代数法求解中引入升降算符的概念，对自由粒子的讨论引入群速、相速、波包等概念。本章讨论的内容是一般的量子力学教科书上都有的，作者没有回避复杂的数学物理方法，而是把必要的结果都给出来；在处理谐振子问题时，把代数方法和分析方法两种解法放在一起对比讨论。

习题特色

（1）把同一个问题在经典情况下和量子情况下进行对比求解，加深对经典物理和量子物理区别的理解。如习题 2.35、习题 2.38、习题 2.51、习题 2.59。

（2）引入一些新的概念和物理定律。如普朗克尔定理（见习题 2.19、习题 2.26）、高斯波包（见习题 2.21、习题 2.42）、能量简并（见习题 2.44）、散射矩阵（见习题 2.53）、传递矩阵（习题 2.54）。

（3）增加扩展性的定性讨论题目和数值计算题目。定性讨论题，如习题 2.47；数值计算题，如利用谐振子的数值解方法，通过 mathematic 方法计算基态、激发态的能量，如习题 2.55、习题 2.56、习题 2.57、习题 2.60、习题 2.61、习题 2.62。

2 TIME-INDEPENDENT SCHRÖDINGER EQUATION

2.1 STATIONARY STATES

In Chapter 1 we talked a lot about the wave function, and how you use it to calculate various quantities of interest. The time has come to stop procrastinating, and confront what is, logically, the prior question: How do you *get* $\Psi(x, t)$ in the *first* place? We need to solve the Schrödinger equation,

$$i\hbar\frac{\partial\Psi}{\partial t} = -\frac{\hbar^2}{2m}\frac{\partial^2\Psi}{\partial x^2} + V\Psi, \tag{2.1}$$

for a specified potential[1] $V(x, t)$. In this chapter (and most of this book) I shall assume that V is *independent of* t. In that case the Schrödinger equation can be solved by the method of **separation of variables** (the physicist's first line of attack on any partial differential equation): We look for solutions that are *products*,

$$\Psi(x, t) = \psi(x)\,\varphi(t), \tag{2.2}$$

where ψ (*lower*-case) is a function of x alone, and φ is a function of t alone. On its face, this is an absurd restriction, and we cannot hope to obtain more than a tiny subset of all solutions in this way. But hang on, because the solutions we *do* get turn out to be of great interest. Moreover (as is typically the case with separation of variables) we will be able at the end to patch together the separable solutions in such a way as to construct the most general solution.

For separable solutions we have

$$\frac{\partial\Psi}{\partial t} = \psi\frac{d\varphi}{dt}, \quad \frac{\partial^2\Psi}{\partial x^2} = \frac{d^2\psi}{dx^2}\varphi$$

(*ordinary* derivatives, now), and the Schrödinger equation reads

$$i\hbar\psi\frac{d\varphi}{dt} = -\frac{\hbar^2}{2m}\frac{d^2\psi}{dx^2}\varphi + V\psi\varphi.$$

[1] It is tiresome to keep saying "potential energy function," so most people just call V the "potential," even though this invites occasional confusion with *electric* potential, which is actually potential energy *per unit charge*.

Or, dividing through by $\psi\varphi$:

$$i\hbar\frac{1}{\varphi}\frac{d\varphi}{dt} = -\frac{\hbar^2}{2m}\frac{1}{\psi}\frac{d^2\psi}{dx^2} + V. \tag{2.3}$$

Now, the left side is a function of t alone, and the right side is a function of x alone.[2] The only way this can possibly be true is if both sides are in fact *constant*—otherwise, by varying t, I could change the left side without touching the right side, and the two would no longer be equal. (That's a subtle but crucial argument, so if it's new to you, be sure to pause and think it through.) For reasons that will appear in a moment, we shall call the separation constant E. Then

$$i\hbar\frac{1}{\varphi}\frac{d\varphi}{dt} = E,$$

or

$$\frac{d\varphi}{dt} = -\frac{iE}{\hbar}\varphi, \tag{2.4}$$

and

$$-\frac{\hbar^2}{2m}\frac{1}{\psi}\frac{d^2\psi}{dx^2} + V = E,$$

or

$$\boxed{-\frac{\hbar^2}{2m}\frac{d^2\psi}{dx^2} + V\psi = E\psi.} \tag{2.5}$$

Separation of variables has turned a *partial* differential equation into two *ordinary* differential equations (Equations 2.4 and 2.5). The first of these is easy to solve (just multiply through by dt and integrate); the general solution is $C\exp(-iEt/\hbar)$, but we might as well absorb the constant C into ψ (since the quantity of interest is the product $\psi\varphi$). Then[3]

$$\varphi(t) = e^{-iEt/\hbar}. \tag{2.6}$$

The second (Equation 2.5) is called the **time-independent Schrödinger equation**; we can go no further with it until the potential $V(x)$ is specified.

The rest of this chapter will be devoted to solving the time-independent Schrödinger equation, for a variety of simple potentials. But before I get to that you have every right to ask: *What's so great about separable solutions?* After all, *most* solutions to the (time *de*pendent) Schrödinger equation do *not* take the form $\psi(x)\varphi(t)$. I offer three answers—two of them physical, and one mathematical:

---

[2] Note that this would *not* be true if V were a function of t as well as x.

[3] Using **Euler's formula**,

$$e^{i\theta} = \cos\theta + i\sin\theta,$$

you could equivalently write

$$\varphi(t) = \cos(Et/\hbar) + i\sin(Et/\hbar);$$

the real and imaginary parts oscillate sinusoidally. Mike Casper (of Carleton College) dubbed φ the "wiggle factor"—it's the characteristic time dependence in quantum mechanics.

1. They are **stationary states**. Although the wave function itself,

$$\Psi(x, t) = \psi(x)e^{-iEt/\hbar}, \tag{2.7}$$

does (obviously) depend on t, the *probability density*,

$$|\Psi(x, t)|^2 = \Psi^*\Psi = \psi^* e^{+iEt/\hbar}\psi e^{-iEt/\hbar} = |\psi(x)|^2, \tag{2.8}$$

does *not*—the time-dependence cancels out.[4] The same thing happens in calculating the expectation value of any dynamical variable; Equation 1.36 reduces to

$$\langle Q(x, p)\rangle = \int \psi^* \left[Q\left(x, -i\hbar\frac{d}{dx}\right)\right] \psi \, dx. \tag{2.9}$$

Every expectation value is constant in time; we might as well drop the factor $\varphi(t)$ altogether, and simply use ψ in place of Ψ. (Indeed, it is common to refer to ψ as "the wave function," but this is sloppy language that can be dangerous, and it is important to remember that the *true* wave function always carries that time-dependent wiggle factor.) In particular, $\langle x\rangle$ is constant, and hence (Equation 1.33) $\langle p\rangle = 0$. Nothing ever *happens* in a stationary state.

2. They are states of *definite total energy*. In classical mechanics, the total energy (kinetic plus potential) is called the **Hamiltonian**:

$$H(x, p) = \frac{p^2}{2m} + V(x). \tag{2.10}$$

The corresponding Hamiltonian *operator*, obtained by the canonical substitution $p \rightarrow -i\hbar(\partial/\partial x)$, is therefore[5]

$$\hat{H} = -\frac{\hbar^2}{2m}\frac{\partial^2}{\partial x^2} + V(x). \tag{2.11}$$

Thus the time-independent Schrödinger equation (Equation 2.5) can be written

$$\hat{H}\psi = E\psi, \tag{2.12}$$

and the expectation value of the total energy is

$$\langle H\rangle = \int \psi^*\hat{H}\psi \, dx = E\int |\psi|^2 \, dx = E\int |\Psi|^2 \, dx = E. \tag{2.13}$$

(Notice that the normalization of Ψ entails the normalization of ψ.) Moreover,

$$\hat{H}^2\psi = \hat{H}\left(\hat{H}\psi\right) = \hat{H}\left(E\psi\right) = E\left(\hat{H}\psi\right) = E^2\psi,$$

and hence

$$\left\langle H^2\right\rangle = \int \psi^*\hat{H}^2\psi \, dx = E^2\int |\psi|^2 \, dx = E^2.$$

So the variance of H is

$$\sigma_H^2 = \left\langle H^2\right\rangle - \langle H\rangle^2 = E^2 - E^2 = 0. \tag{2.14}$$

---

[4] For normalizable solutions, E must be *real* (see Problem 2.1(a)).

[5] Whenever confusion might arise, I'll put a "hat" (^) on the operator, to distinguish it from the dynamical variable it represents.

But remember, if $\sigma = 0$, then every member of the sample must share the same value (the distribution has zero spread). *Conclusion:* A separable solution has the property that *every measurement of the total energy is certain to return the value E*. (That's why I chose that letter for the separation constant.)

3. The general solution is a **linear combination** of separable solutions. As we're about to discover, the time-independent Schrödinger equation (Equation 2.5) yields an infinite collection of solutions ($\psi_1(x), \psi_2(x), \psi_3(x), \ldots$, which we write as $\{\psi_n(x)\}$), each with its associated separation constant ($E_1, E_2, E_3, \ldots = \{E_n\}$); thus there is a different wave function for each **allowed energy**:

$$\Psi_1(x,t) = \psi_1(x)e^{-iE_1 t/\hbar}, \quad \Psi_2(x,t) = \psi_2(x)e^{-iE_2 t/\hbar}, \ \ldots.$$

Now (as you can easily check for yourself) the (time-*de*pendent) Schrödinger equation (Equation 2.1) has the property that any linear combination[6] of solutions is itself a solution. Once we have found the separable solutions, then, we can immediately construct a much more general solution, of the form

$$\Psi(x,t) = \sum_{n=1}^{\infty} c_n \psi_n(x) e^{-iE_n t/\hbar}. \tag{2.15}$$

It so happens that *every* solution to the (time-dependent) Schrödinger equation can be written in this form—it is simply a matter of finding the right constants ($c_1, c_2, \ldots$) so as to fit the initial conditions for the problem at hand. You'll see in the following sections how all this works out in practice, and in Chapter 3 we'll put it into more elegant language, but the main point is this: Once you've solved the time-*in*dependent Schrödinger equation, you're essentially *done;* getting from there to the general solution of the time-*de*pendent Schrödinger equation is, in principle, simple and straightforward.

A lot has happened in the past four pages, so let me recapitulate, from a somewhat different perspective. Here's the generic problem: You're given a (time-independent) potential $V(x)$, and the starting wave function $\Psi(x,0)$; your job is to find the wave function, $\Psi(x,t)$, for any subsequent time t. To do this you must solve the (time-dependent) Schrödinger equation (Equation 2.1). The strategy is first to solve the time-*in*dependent Schrödinger equation (Equation 2.5); this yields, in general, an infinite set of solutions, $\{\psi_n(x)\}$, each with its own associated energy, $\{E_n\}$. To fit $\Psi(x,0)$ you write down the general linear combination of these solutions:

$$\Psi(x,0) = \sum_{n=1}^{\infty} c_n \psi_n(x); \tag{2.16}$$

[6] A **linear combination** of the functions $f_1(z), f_2(z), \ldots$ is an expression of the form

$$f(z) = c_1 f_1(z) + c_2 f_2(z) + \cdots,$$

where $c_1, c_2, \ldots$ are (possibly complex) constants.

the miracle is that you can *always* match the specified initial state[7] by appropriate choice of the constants $\{c_n\}$. To construct $\Psi(x,t)$ you simply tack onto each term its characteristic time dependence (its "wiggle factor"), $\exp(-iE_nt/\hbar)$:[8]

$$\boxed{\Psi(x,t) = \sum_{n=1}^{\infty} c_n \psi_n(x) e^{-iE_nt/\hbar} = \sum_{n=1}^{\infty} c_n \Psi_n(x,t).} \tag{2.17}$$

The separable solutions themselves,

$$\Psi_n(x,t) = \psi_n(x) e^{-iE_nt/\hbar}, \tag{2.18}$$

are *stationary* states, in the sense that all probabilities and expectation values are independent of time, but this property is emphatically *not* shared by the general solution (Equation 2.17): the energies are different, for different stationary states, and the exponentials do not cancel, when you construct $|\Psi|^2$.

Example 2.1

Suppose a particle starts out in a linear combination of just *two* stationary states:

$$\Psi(x,0) = c_1\psi_1(x) + c_2\psi_2(x).$$

(To keep things simple I'll assume that the constants c_n and the states $\psi_n(x)$ are *real*.) What is the wave function $\Psi(x,t)$ at subsequent times? Find the probability density, and describe its motion.

Solution: The first part is easy:

$$\Psi(x,t) = c_1\psi_1(x)e^{-iE_1t/\hbar} + c_2\psi_2(x)\, e^{-iE_2t/\hbar},$$

where E_1 and E_2 are the energies associated with ψ_1 and ψ_2. It follows that

$$\begin{aligned} |\Psi(x,t)|^2 &= \left(c_1\psi_1 e^{iE_1t/\hbar} + c_2\psi_2 e^{iE_2t/\hbar}\right)\left(c_1\psi_1 e^{-iE_1t/\hbar} + c_2\psi_2 e^{-iE_2t/\hbar}\right) \\ &= c_1^2\psi_1^2 + c_2^2\psi_2^2 + 2c_1c_2\psi_1\psi_2 \cos\left[(E_2 - E_1)\,t/\hbar\right]. \end{aligned}$$

The probability density *oscillates* sinusoidally, at an angular frequency $\omega = (E_2 - E_1)/\hbar$; this is certainly *not* a stationary state. But notice that it took a *linear combination* of stationary states (with different energies) to produce motion.[9]

---

[7] In principle, *any* normalized function $\Psi(x,0)$ is fair game—it need not even be continuous. How you might actually *get* a particle into that state is a different question, and one (curiously) we seldom have occasion to ask.

[8] If this is your first encounter with the method of separation of variables, you may be disappointed that the solution takes the form of an infinite series. Occasionally it is possible to sum the series, or to solve the time-dependent Schrödinger equation without recourse to separation of variables—see, for instance, Problems 2.49, 2.50, and 2.51. But such cases are extremely rare.

[9] This is nicely illustrated in an applet by Paul Falstad, at www.falstad.com/qm1d/.

You may be wondering what the coefficients $\{c_n\}$ represent *physically*. I'll tell you the answer, though the explanation will have to await Chapter 3:

$$|c_n|^2 \quad \text{is the } \textit{probability} \text{ that a measurement of the energy would return the value } E_n. \tag{2.19}$$

A competent measurement will always yield *one* of the "allowed" values (hence the name), and $|c_n|^2$ is the probability of getting the *particular* value E_n.[10] Of course, the *sum* of these probabilities should be 1:

$$\sum_{n=1}^{\infty} |c_n|^2 = 1, \tag{2.20}$$

and the expectation value of the energy must be

$$\langle H \rangle = \sum_{n=1}^{\infty} |c_n|^2 E_n. \tag{2.21}$$

We'll soon see how this works out in some concrete examples. Notice, finally, that because the constants $\{c_n\}$ are independent of time, so too is the probability of getting a particular energy, and, *a fortiori*, the expectation value of H. These are manifestations of **energy conservation** in quantum mechanics.

* **Problem 2.1** Prove the following three theorems:

(a) For normalizable solutions, the separation constant E must be *real*. *Hint:* Write E (in Equation 2.7) as $E_0 + i\Gamma$ (with E_0 and Γ real), and show that if Equation 1.20 is to hold for all t, Γ must be zero.

(b) The time-independent wave function $\psi(x)$ can always be taken to be *real* (unlike $\Psi(x,t)$, which is necessarily complex). This doesn't mean that every solution to the time-independent Schrödinger equation *is* real; what it says is that if you've got one that is *not*, it can always be expressed as a linear combination of solutions (with the same energy) that *are*. So you *might as well* stick to ψs that are real. *Hint:* If $\psi(x)$ satisfies Equation 2.5, for a given E, so too does its complex conjugate, and hence also the real linear combinations $(\psi + \psi^*)$ and $i\,(\psi - \psi^*)$.

(c) If $V(x)$ is an **even function** (that is, $V(-x) = V(x)$) then $\psi(x)$ can always be taken to be either even or odd. *Hint:* If $\psi(x)$ satisfies Equation 2.5, for a given E, so too does $\psi(-x)$, and hence also the even and odd linear combinations $\psi(x) \pm \psi(-x)$.

---

[10] Some people will tell you that $|c_n|^2$ is "the probability that the particle is in the nth stationary state," but this is bad language: the particle is in the state Ψ, *not* Ψ_n, and anyhow, in the laboratory you don't "find the particle to be in a particular state," you measure some observable, and what you get is a *number*, not a wave function.

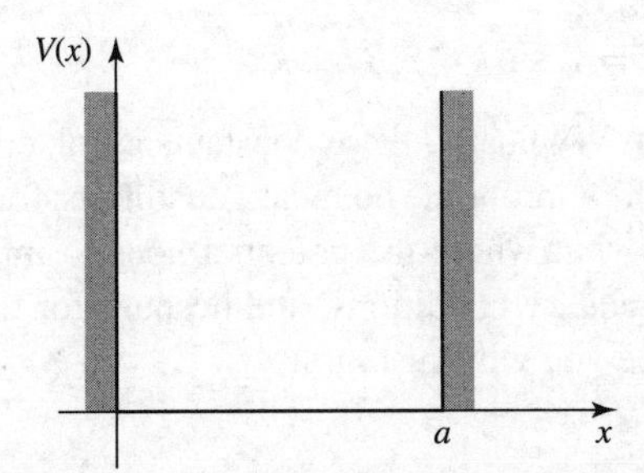

Figure 2.1: The infinite square well potential (Equation 2.22).

* **Problem 2.2** Show that E must exceed the minimum value of $V(x)$, for every normalizable solution to the time-independent Schrödinger equation. What is the classical analog to this statement? *Hint:* Rewrite Equation 2.5 in the form

$$\frac{d^2\psi}{dx^2} = \frac{2m}{\hbar^2}\left[V(x) - E\right]\psi;$$

if $E < V_{\min}$, then ψ and its second derivative always have the *same sign*—argue that such a function cannot be normalized.

2.2 THE INFINITE SQUARE WELL

Suppose

$$V(x) = \begin{cases} 0, & 0 \leq x \leq a, \\ \infty, & \text{otherwise} \end{cases} \tag{2.22}$$

(Figure 2.1). A particle in this potential is completely free, except at the two ends ($x = 0$ and $x = a$), where an infinite force prevents it from escaping. A classical model would be a cart on a frictionless horizontal air track, with perfectly elastic bumpers—it just keeps bouncing back and forth forever. (This potential is artificial, of course, but I urge you to treat it with respect. Despite its simplicity—or rather, precisely *because* of its simplicity—it serves as a wonderfully accessible test case for all the fancy machinery that comes later. We'll refer back to it frequently.)

Outside the well, $\psi(x) = 0$ (the probability of finding the particle there is zero). *Inside* the well, where $V = 0$, the time-independent Schrödinger equation (Equation 2.5) reads

$$-\frac{\hbar^2}{2m}\frac{d^2\psi}{dx^2} = E\psi, \tag{2.23}$$

or

$$\frac{d^2\psi}{dx^2} = -k^2\psi, \quad \text{where } k \equiv \frac{\sqrt{2mE}}{\hbar}. \tag{2.24}$$

(By writing it in this way, I have tacitly assumed that $E \geq 0$; we know from Problem 2.2 that $E < 0$ won't work.) Equation 2.24 is the classical **simple harmonic oscillator** equation; the general solution is

$$\psi(x) = A \sin kx + B \cos kx, \tag{2.25}$$

where A and B are arbitrary constants. Typically, these constants are fixed by the **boundary conditions** of the problem. What *are* the appropriate boundary conditions for $\psi(x)$? Ordinarily, *both* ψ *and* $d\psi/dx$ *are continuous,*[11] but where the potential goes to infinity only the first of these applies. (I'll *justify* these boundary conditions, and account for the exception when $V = \infty$, in Section 2.5; for now I hope you will trust me.)

Continuity of $\psi(x)$ requires that

$$\psi(0) = \psi(a) = 0, \tag{2.26}$$

so as to join onto the solution outside the well. What does this tell us about A and B? Well,

$$\psi(0) = A \sin 0 + B \cos 0 = B,$$

so $B = 0$, and hence

$$\psi(x) = A \sin kx. \tag{2.27}$$

Then $\psi(a) = A \sin ka$, so either $A = 0$ (in which case we're left with the trivial—non-normalizable—solution $\psi(x) = 0$), or else $\sin ka = 0$, which means that

$$ka = 0,\ \pm\pi,\ \pm 2\pi,\ \pm 3\pi,\ \ldots. \tag{2.28}$$

But $k = 0$ is no good (again, that would imply $\psi(x) = 0$), and the negative solutions give nothing new, since $\sin(-\theta) = -\sin(\theta)$ and we can absorb the minus sign into A. So the *distinct* solutions are

$$k_n = \frac{n\pi}{a}, \quad \text{with } n = 1,\ 2,\ 3,\ \ldots. \tag{2.29}$$

Curiously, the boundary condition at $x = a$ does not determine the constant A, but rather the constant k, and hence the possible values of E:

$$\boxed{E_n = \frac{\hbar^2 k_n^2}{2m} = \frac{n^2\pi^2\hbar^2}{2ma^2}.} \tag{2.30}$$

In radical contrast to the classical case, a quantum particle in the infinite square well cannot have just *any* old energy—it has to be one of these special ("allowed") values.[12] To find A, we *normalize* ψ:[13]

$$\int_0^a |A|^2 \sin^2(kx)\, dx = |A|^2 \frac{a}{2} = 1, \quad \text{so} \quad |A|^2 = \frac{2}{a}.$$

This only determines the *magnitude* of A, but it is simplest to pick the positive real root: $A = \sqrt{2/a}$ (the phase of A carries no physical significance anyway). Inside the well, then, the solutions are

$$\boxed{\psi_n(x) = \sqrt{\frac{2}{a}} \sin\left(\frac{n\pi}{a}x\right).} \tag{2.31}$$

[11] That's right: $\psi(x)$ is a continuous function of x, even though $\Psi(x, t)$ need not be.

[12] Notice that the quantization of energy emerges as a rather technical consequence of the boundary conditions on solutions to the time-independent Schrödinger equation.

[13] Actually, it's $\Psi(x, t)$ that must be normalized, but in view of Equation 2.7 this entails the normalization of $\psi(x)$.

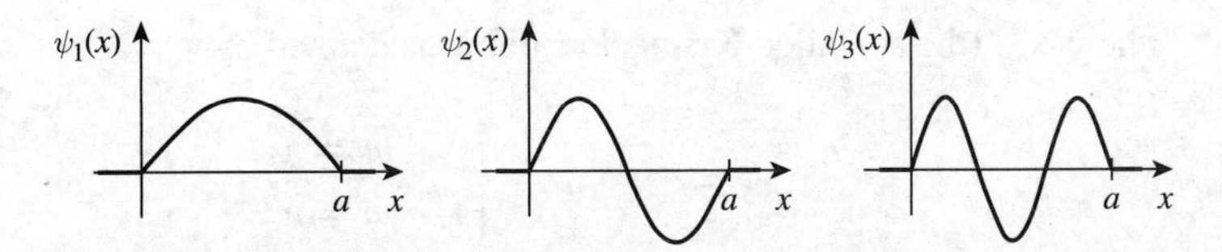

Figure 2.2: The first three stationary states of the infinite square well (Equation 2.31).

As promised, the time-independent Schrödinger equation has delivered an infinite set of solutions (one for each positive integer n). The first few of these are plotted in Figure 2.2. They look just like the standing waves on a string of length a; ψ_1, which carries the lowest energy, is called the **ground state**, the others, whose energies increase in proportion to n^2, are called **excited states**. As a collection, the functions $\psi_n(x)$ have some interesting and important properties:

1. They are alternately **even** and **odd**, with respect to the center of the well: ψ_1 is even, ψ_2 is odd, ψ_3 is even, and so on.[14]
2. As you go up in energy, each successive state has one more **node** (zero-crossing): ψ_1 has none (the end points don't count), ψ_2 has one, ψ_3 has two, and so on.
3. They are mutually **orthogonal**, in the sense that[15]

$$\int \psi_m(x)^* \, \psi_n(x) \, dx = 0, \quad (m \neq n). \tag{2.32}$$

Proof:

$$\begin{aligned}\int \psi_m(x)^* \psi_n(x) \, dx &= \frac{2}{a} \int_0^a \sin\left(\frac{m\pi}{a} x\right) \sin\left(\frac{n\pi}{a} x\right) dx \\ &= \frac{1}{a} \int_0^a \left[\cos\left(\frac{m-n}{a} \pi x\right) - \cos\left(\frac{m+n}{a} \pi x\right)\right] dx \\ &= \left\{\frac{1}{(m-n)\pi} \sin\left(\frac{m-n}{a} \pi x\right) - \frac{1}{(m+n)\pi} \sin\left(\frac{m+n}{a} \pi x\right)\right\}\Bigg|_0^a \\ &= \frac{1}{\pi}\left\{\frac{\sin[(m-n)\pi]}{(m-n)} - \frac{\sin[(m+n)\pi]}{(m+n)}\right\} = 0.\end{aligned}$$

Note that this argument does *not* work if $m = n$. (Can you spot the point at which it fails?) In that case normalization tells us that the integral is 1. In fact, we can combine orthogonality and normalization into a single statement:

$$\boxed{\int \psi_m(x)^* \, \psi_n(x) \, dx = \delta_{mn},} \tag{2.33}$$

[14] To make this symmetry more apparent, some authors center the well at the origin (running it now from $-a$ to $+a$). The even functions are then cosines, and the odd ones are sines. See Problem 2.36.

[15] In this case the ψs are *real*, so the complex conjugation (*) of ψ_m is unnecessary, but for future purposes it's a good idea to get in the habit of putting it there.

where δ_{mn} (the so-called **Kronecker delta**) is defined by

$$\delta_{mn} = \begin{cases} 0, & m \neq n, \\ 1, & m = n. \end{cases} \tag{2.34}$$

We say that the ψs are **orthonormal**.

4. They are **complete**, in the sense that any *other* function, $f(x)$, can be expressed as a linear combination of them:

$$f(x) = \sum_{n=1}^{\infty} c_n \psi_n(x) = \sqrt{\frac{2}{a}} \sum_{n=1}^{\infty} c_n \sin\left(\frac{n\pi}{a}x\right). \tag{2.35}$$

I'm not about to *prove* the completeness of the functions $\sqrt{2/a}\sin(n\pi x/a)$, but if you've studied advanced calculus you will recognize that Equation 2.35 is nothing but the **Fourier series** for $f(x)$, and the fact that "any" function can be expanded in this way is sometimes called **Dirichlet's theorem.**[16]

The coefficients c_n can be evaluated—for a given $f(x)$—by a method I call **Fourier's trick**, which beautifully exploits the orthonormality of $\{\psi_n\}$: Multiply both sides of Equation 2.35 by $\psi_m(x)^*$, and integrate.

$$\int \psi_m(x)^* f(x)\,dx = \sum_{n=1}^{\infty} c_n \int \psi_m(x)^* \psi_n(x)\,dx = \sum_{n=1}^{\infty} c_n \delta_{mn} = c_m. \tag{2.36}$$

(Notice how the Kronecker delta kills every term in the sum except the one for which $n = m$.) Thus the nth coefficient in the expansion of $f(x)$ is[17]

$$\boxed{c_n = \int \psi_n(x)^* f(x)\,dx.} \tag{2.37}$$

These four properties are extremely powerful, and they are not peculiar to the infinite square well. The first is true whenever the potential itself is a symmetric function; the second is universal, regardless of the shape of the potential.[18] Orthogonality is also quite general—I'll show you the proof in Chapter 3. Completeness holds for all the potentials you are likely to encounter, but the proofs tend to be nasty and laborious; I'm afraid most physicists simply *assume* completeness, and hope for the best.

The stationary states (Equation 2.18) of the infinite square well are

$$\Psi_n(x,t) = \sqrt{\frac{2}{a}} \sin\left(\frac{n\pi}{a}x\right) e^{-i(n^2\pi^2\hbar/2ma^2)t}. \tag{2.38}$$

I claimed (Equation 2.17) that the most general solution to the (time-dependent) Schrödinger equation is a linear combination of stationary states:

---

[16] See, for example, Mary Boas, *Mathematical Methods in the Physical Sciences*, 3rd edn (New York: John Wiley, 2006), p. 356; $f(x)$ can even have a finite number of finite discontinuities.

[17] It doesn't matter whether you use m or n as the "dummy index" here (as long as you are consistent on the two sides of the equation, of course); *whatever* letter you use, it just stands for "any positive integer."

[18] Problem 2.45 explores this property. For further discussion, see John L. Powell and Bernd Crasemann, *Quantum Mechanics* (Addison-Wesley, Reading, MA, 1961), Section 5-7.

$$\Psi(x,t)=\sum_{n=1}^{\infty}c_n\sqrt{\frac{2}{a}}\sin\left(\frac{n\pi}{a}x\right)e^{-i(n^2\pi^2\hbar/2ma^2)t}. \tag{2.39}$$

(If you doubt that this *is* a solution, by all means *check* it!) It remains only for me to demonstrate that I can fit any prescribed initial wave function, $\Psi(x,0)$ by appropriate choice of the coefficients c_n:

$$\Psi(x,0)=\sum_{n=1}^{\infty}c_n\psi_n(x).$$

The completeness of the ψs (confirmed in this case by Dirichlet's theorem) guarantees that I can always express $\Psi(x,0)$ in this way, and their orthonormality licenses the use of Fourier's trick to determine the actual coefficients:

$$c_n=\sqrt{\frac{2}{a}}\int_0^a\sin\left(\frac{n\pi}{a}x\right)\Psi(x,0)\,dx. \tag{2.40}$$

That *does* it: Given the initial wave function, $\Psi(x,0)$, we first compute the expansion coefficients c_n, using Equation 2.40, and then plug these into Equation 2.39 to obtain $\Psi(x,t)$. Armed with the wave function, we are in a position to compute any dynamical quantities of interest, using the procedures in Chapter 1. And this same ritual applies to *any* potential—the only things that change are the functional form of the ψs and the equation for the allowed energies.

Example 2.2

A particle in the infinite square well has the initial wave function

$$\Psi(x,0)=Ax\,(a-x),\quad (0\le x\le a),$$

for some constant A (see Figure 2.3). *Outside* the well, of course, $\Psi=0$. Find $\Psi(x,t)$.

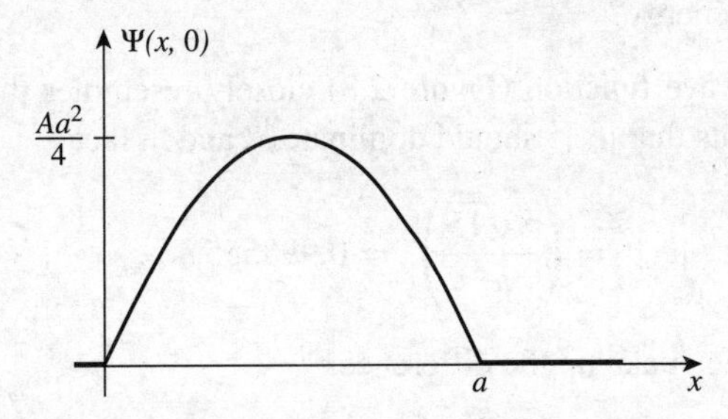

Figure 2.3: The starting wave function in Example 2.2.

Solution: First we need to determine A, by normalizing $\Psi(x,0)$:

$$1=\int_0^a|\Psi(x,0)|^2\,dx=|A|^2\int_0^a x^2\,(a-x)^2\,dx=|A|^2\frac{a^5}{30},$$

so

$$A=\sqrt{\frac{30}{a^5}}.$$

The nth coefficient is (Equation 2.40)

$$\begin{aligned}
c_n &= \sqrt{\frac{2}{a}} \int_0^a \sin\left(\frac{n\pi}{a}x\right) \sqrt{\frac{30}{a^5}}\, x\,(a-x)\, dx \\
&= \frac{2\sqrt{15}}{a^3}\left[a \int_0^a x \sin\left(\frac{n\pi}{a}x\right) dx - \int_0^a x^2 \sin\left(\frac{n\pi}{a}x\right) dx\right] \\
&= \frac{2\sqrt{15}}{a^3}\left\{a\left[\left(\frac{a}{n\pi}\right)^2 \sin\left(\frac{n\pi}{a}x\right) - \frac{ax}{n\pi}\cos\left(\frac{n\pi}{a}x\right)\right]\Bigg|_0^a \right. \\
&\quad \left. - \left[2\left(\frac{a}{n\pi}\right)^2 x \sin\left(\frac{n\pi}{a}x\right) - \frac{(n\pi x/a)^2 - 2}{(n\pi/a)^3}\cos\left(\frac{n\pi}{a}x\right)\right]\Bigg|_0^a\right\} \\
&= \frac{2\sqrt{15}}{a^3}\left[-\frac{a^3}{n\pi}\cos(n\pi) + a^3\frac{(n\pi)^2 - 2}{(n\pi)^3}\cos(n\pi) + a^3\frac{2}{(n\pi)^3}\cos(0)\right] \\
&= \frac{4\sqrt{15}}{(n\pi)^3}\left[\cos(0) - \cos(n\pi)\right] \\
&= \begin{cases} 0, & n \text{ even}, \\ 8\sqrt{15}/(n\pi)^3, & n \text{ odd}. \end{cases}
\end{aligned}$$

Thus (Equation 2.39):

$$\Psi(x,t) = \sqrt{\frac{30}{a}}\left(\frac{2}{\pi}\right)^3 \sum_{n=1,3,5\ldots} \frac{1}{n^3}\sin\left(\frac{n\pi}{a}x\right) e^{-in^2\pi^2\hbar t/2ma^2}.$$

Example 2.3

Check that Equation 2.20 is satisfied, for the wave function in Example 2.2. If you measured the energy of a particle in this state, what is the most probable result? What is the expectation value of the energy?

Solution: The starting wave function (Figure 2.3) closely resembles the ground state ψ_1 (Figure 2.2). This suggests that $|c_1|^2$ should dominate,[19] and in fact

$$|c_1|^2 = \left(\frac{8\sqrt{15}}{\pi^3}\right)^2 = 0.998555\ldots.$$

The rest of the coefficients make up the difference:[20]

---

[19] Loosely speaking, c_n tells you the "amount of ψ_n that is contained in Ψ."

[20] You can look up the series

$$\frac{1}{1^6} + \frac{1}{3^6} + \frac{1}{5^6} + \cdots = \frac{\pi^6}{960}$$

and

$$\frac{1}{1^4} + \frac{1}{3^4} + \frac{1}{5^4} + \cdots = \frac{\pi^4}{96}$$

in math tables, under "Sums of Reciprocal Powers" or "Riemann Zeta Function."

$$\sum_{n=1}^{\infty} |c_n|^2 = \left(\frac{8\sqrt{15}}{\pi^3}\right)^2 \sum_{n=1,3,5,\ldots}^{\infty} \frac{1}{n^6} = 1.$$

The most likely outcome of an energy measurement is $E_1 = \pi^2\hbar^2/2ma^2$—more than 99.8% of all measurements will yield this value. The expectation value of the energy (Equation 2.21) is

$$\langle H \rangle = \sum_{n=1,3,5,\ldots}^{\infty} \left(\frac{8\sqrt{15}}{n^3\pi^3}\right)^2 \frac{n^2\pi^2\hbar^2}{2ma^2} = \frac{480\hbar^2}{\pi^4 ma^2} \sum_{n=1,3,5,\ldots}^{\infty} \frac{1}{n^4} = \frac{5\hbar^2}{ma^2}.$$

As one would expect, it is very close to E_1 (5 in place of $\pi^2/2 \approx 4.935$)—slightly *larger*, because of the admixture of excited states.

Of course, it's no accident that Equation 2.20 came out right in Example 2.3. Indeed, this follows from the normalization of Ψ (the c_ns are independent of time, so I'm going to do the proof for $t = 0$; if this bothers you, you can easily generalize the argument to arbitrary t).

$$\begin{aligned} 1 &= \int |\Psi(x,0)|^2\, dx = \int \left(\sum_{m=1}^{\infty} c_m \psi_m(x)\right)^* \left(\sum_{n=1}^{\infty} c_n \psi_n(x)\right) dx \\ &= \sum_{m=1}^{\infty}\sum_{n=1}^{\infty} c_m^* c_n \int \psi_m(x)^* \psi_n(x)\, dx \\ &= \sum_{n=1}^{\infty}\sum_{m=1}^{\infty} c_m^* c_n \delta_{mn} = \sum_{n=1}^{\infty} |c_n|^2 . \end{aligned}$$

(Again, the Kronecker delta picks out the term $m = n$ in the summation over m.) Similarly, the expectation value of the energy (Equation 2.21) can be checked explicitly: The time-independent Schrödinger equation (Equation 2.12) says

$$\hat{H}\psi_n = E_n \psi_n, \tag{2.41}$$

so

$$\begin{aligned} \langle H \rangle &= \int \Psi^* \hat{H} \Psi\, dx = \int \left(\sum c_m \psi_m\right)^* \hat{H} \left(\sum c_n \psi_n\right) dx \\ &= \sum\sum c_m^* c_n E_n \int \psi_m^* \psi_n\, dx = \sum |c_n|^2 E_n. \end{aligned}$$

Problem 2.3 Show that there is no acceptable solution to the (time-independent) Schrödinger equation for the infinite square well with $E = 0$ or $E < 0$. (This is a special case of the general theorem in Problem 2.2, but this time do it by explicitly solving the Schrödinger equation, and showing that you cannot satisfy the boundary conditions.)

* **Problem 2.4** Calculate $\langle x \rangle$, $\langle x^2 \rangle$, $\langle p \rangle$, $\langle p^2 \rangle$, σ_x, and σ_p, for the nth stationary state of the infinite square well. Check that the uncertainty principle is satisfied. Which state comes closest to the uncertainty limit?

* **Problem 2.5** A particle in the infinite square well has as its initial wave function an even mixture of the first two stationary states:

$$\Psi(x,0) = A\left[\psi_1(x) + \psi_2(x)\right].$$

(a) Normalize $\Psi(x,0)$. (That is, find A. This is very easy, if you exploit the orthonormality of ψ_1 and ψ_2. Recall that, having normalized Ψ at $t = 0$, you can rest assured that it *stays* normalized—if you doubt this, check it explicitly after doing part (b).)

(b) Find $\Psi(x,t)$ and $|\Psi(x,t)|^2$. Express the latter as a sinusoidal function of time, as in Example 2.1. To simplify the result, let $\omega \equiv \pi^2\hbar/2ma^2$.

(c) Compute $\langle x\rangle$. Notice that it oscillates in time. What is the angular frequency of the oscillation? What is the amplitude of the oscillation? (If your amplitude is greater than $a/2$, go directly to jail.)

(d) Compute $\langle p\rangle$. (As Peter Lorre would say, "Do it ze *kveek* vay, Johnny!")

(e) If you measured the energy of this particle, what values might you get, and what is the probability of getting each of them? Find the expectation valuc of H. How docs it compare with E_1 and E_2?

Problem 2.6 Although the *overall* phase constant of the wave function is of no physical significance (it cancels out whenever you calculate a measurable quantity), the *relative* phase of the coefficients in Equation 2.17 *does* matter. For example, suppose we change the relative phase of ψ_1 and ψ_2 in Problem 2.5:

$$\Psi(x,0) = A\left[\psi_1(x) + e^{i\phi}\psi_2(x)\right],$$

where ϕ is some constant. Find $\Psi(x,t)$, $|\Psi(x,t)|^2$, and $\langle x\rangle$, and compare your results with what you got before. Study the special cases $\phi = \pi/2$ and $\phi = \pi$. (For a graphical exploration of this problem see the applet in footnote 9 of this chapter.)

* **Problem 2.7** A particle in the infinite square well has the initial wave function

$$\Psi(x,0) = \begin{cases} Ax, & 0 \le x \le a/2, \\ A(a-x), & a/2 \le x \le a. \end{cases}$$

(a) Sketch $\Psi(x,0)$, and determine the constant A.

(b) Find $\Psi(x,t)$.

(c) What is the probability that a measurement of the energy would yield the value E_1?

(d) Find the expectation value of the energy, using Equation 2.21.[21]

[21] Remember, there is no restriction in principle on the *shape* of the starting wave function, as long as it is normalizable. In particular, $\Psi(x,0)$ need not have a continuous derivative. However, if you try to calculate $\langle H\rangle$ using $\int \Psi(x,0)^* \hat{H}\Psi(x,0)\,dx$ in such a case, you may encounter technical difficulties, because the second derivative of $\Psi(x,0)$ is ill defined. It works in Problem 2.9 because the discontinuities occur at the end points, where the wave function is zero anyway. In Problem 2.39 you'll see how to manage cases like Problem 2.7.

Problem 2.8 A particle of mass m in the infinite square well (of width a) starts out in the state

$$\Psi(x,0) = \begin{cases} A, & 0 \leq x \leq a/2, \\ 0, & a/2 \leq x \leq a, \end{cases}$$

for some constant A, so it is (at $t = 0$) equally likely to be found at any point in the left half of the well. What is the probability that a measurement of the energy (at some later time t) would yield the value $\pi^2\hbar^2/2ma^2$?

Problem 2.9 For the wave function in Example 2.2, find the expectation value of H, at time $t = 0$, the "old fashioned" way:

$$\langle H \rangle = \int \Psi(x,0)^* \, \hat{H} \Psi(x,0) \, dx.$$

Compare the result we got in Example 2.3. *Note:* Because $\langle H \rangle$ is independent of time, there is no loss of generality in using $t = 0$.

2.3 THE HARMONIC OSCILLATOR

The paradigm for a classical harmonic oscillator is a mass m attached to a spring of force constant k. The motion is governed by **Hooke's law**,

$$F = -kx = m\frac{d^2x}{dt^2}$$

(ignoring friction), and the solution is

$$x(t) = A\sin(\omega t) + B\cos(\omega t),$$

where

$$\omega \equiv \sqrt{\frac{k}{m}} \tag{2.42}$$

is the (angular) frequency of oscillation. The potential energy is

$$V(x) = \frac{1}{2}kx^2; \tag{2.43}$$

its graph is a parabola.

Of course, there's no such thing as a *perfect* harmonic oscillator—if you stretch it too far the spring is going to break, and typically Hooke's law fails long before that point is reached. But practically any potential is *approximately* parabolic, in the neighborhood of a local minimum (Figure 2.4). Formally, if we expand $V(x)$ in a **Taylor series** about the minimum:

$$V(x) = V(x_0) + V'(x_0)(x - x_0) + \frac{1}{2}V''(x_0)(x - x_0)^2 + \cdots,$$

subtract $V(x_0)$ (you can add a constant to $V(x)$ with impunity, since that doesn't change the force), recognize that $V'(x_0) = 0$ (since x_0 is a minimum), and drop the higher-order terms (which are negligible as long as $(x - x_0)$ stays small), we get

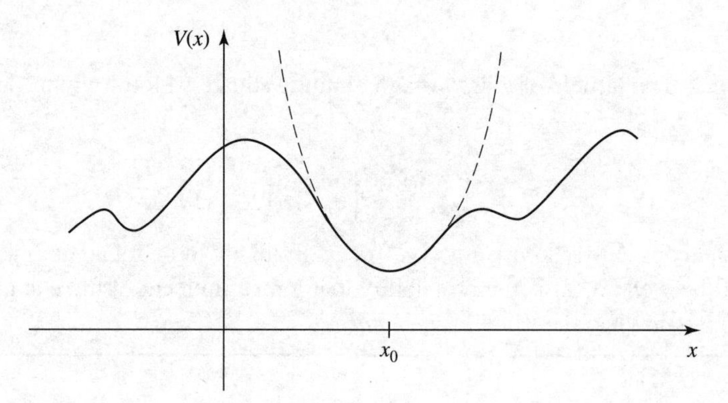

Figure 2.4: Parabolic approximation (dashed curve) to an arbitrary potential, in the neighborhood of a local minimum.

$$V(x) \approx \frac{1}{2} V''(x_0)\,(x - x_0)^2,$$

which describes simple harmonic oscillation (about the point x_0), with an effective spring constant $k = V''(x_0)$. That's why the simple harmonic oscillator is so important: Virtually *any* oscillatory motion is approximately simple harmonic, as long as the amplitude is small.[22]

The *quantum* problem is to solve the Schrödinger equation for the potential

$$V(x) = \frac{1}{2} m\omega^2 x^2 \tag{2.44}$$

(it is customary to eliminate the spring constant in favor of the classical frequency, using Equation 2.42). As we have seen, it suffices to solve the time-independent Schrödinger equation:

$$-\frac{\hbar^2}{2m}\frac{d^2\psi}{dx^2} + \frac{1}{2} m\omega^2 x^2 \psi = E\psi. \tag{2.45}$$

In the literature you will find two entirely different approaches to this problem. The first is a straightforward "brute force" solution to the differential equation, using the **power series method**; it has the virtue that the same strategy can be applied to many other potentials (in fact, we'll use it in Chapter 4 to treat the hydrogen atom). The second is a diabolically clever algebraic technique, using so-called **ladder operators**. I'll show you the algebraic method first, because it is quicker and simpler (and a lot more fun);[23] if you want to skip the power series method for now, that's fine, but you should certainly plan to study it at some stage.

2.3.1 Algebraic Method

To begin with, let's rewrite Equation 2.45 in a more suggestive form:

$$\frac{1}{2m}\left[\hat{p}^2 + (m\omega x)^2\right]\psi = E\psi, \tag{2.46}$$

[22] Note that $V''(x_0) \geq 0$, since by assumption x_0 is a *minimum*. Only in the rare case $V''(x_0) = 0$ is the oscillation not even approximately simple harmonic.

[23] We'll encounter some of the same strategies in the theory of angular momentum (Chapter 4), and the technique generalizes to a broad class of potentials in **supersymmetric quantum mechanics** (Problem 3.47; see also Richard W. Robinett, *Quantum Mechanics* (Oxford University Press, New York, 1997), Section 14.4).

where $\hat{p} \equiv -i\hbar\, d/dx$ is the momentum operator.[24] The basic idea is to *factor* the Hamiltonian,

$$\hat{H} = \frac{1}{2m}\left[\hat{p}^2 + (m\omega x)^2\right]. \tag{2.47}$$

If these were *numbers*, it would be easy:

$$u^2 + v^2 = (iu + v)(-iu + v).$$

Here, however, it's not quite so simple, because $\hat{p}$ and x are *operators*, and operators do not, in general, **commute** ($x\hat{p}$ is not the same as $\hat{p}x$, as we'll see in a moment—though you might want to stop right now and think it through for yourself). Still, this does motivate us to examine the quantities

$$\hat{a}_\pm \equiv \frac{1}{\sqrt{2\hbar m\omega}}\left(\mp i\hat{p} + m\omega x\right) \tag{2.48}$$

(the factor in front is just there to make the final results look nicer).

Well, what *is* the product $\hat{a}_-\hat{a}_+$?

$$\begin{aligned}\hat{a}_-\hat{a}_+ &= \frac{1}{2\hbar m\omega}\left(i\hat{p} + m\omega x\right)\left(-i\hat{p} + m\omega x\right)\\ &= \frac{1}{2\hbar m\omega}\left[\hat{p}^2 + (m\omega x)^2 - im\omega\left(x\hat{p} - \hat{p}x\right)\right].\end{aligned}$$

As anticipated, there's an extra term, involving $\left(x\hat{p} - \hat{p}x\right)$. We call this the **commutator** of x and $\hat{p}$; it is a measure of how badly they *fail* to commute. In general, the commutator of operators $\hat{A}$ and $\hat{B}$ (written with square brackets) is

$$\left[\hat{A}, \hat{B}\right] \equiv \hat{A}\hat{B} - \hat{B}\hat{A}. \tag{2.49}$$

In this notation,

$$\hat{a}_-\hat{a}_+ = \frac{1}{2\hbar m\omega}\left[\hat{p}^2 + (m\omega x)^2\right] - \frac{i}{2\hbar}\left[x, \hat{p}\right]. \tag{2.50}$$

We need to figure out the commutator of x and $\hat{p}$. *Warning:* Operators are notoriously slippery to work with in the abstract, and you are bound to make mistakes unless you give them a "test function," $f(x)$, to act on. At the end you can throw away the test function, and you'll be left with an equation involving the operators alone. In the present case we have:

$$\begin{aligned}\left[x, \hat{p}\right] f(x) &= \left[x(-i\hbar)\frac{d}{dx}(f) - (-i\hbar)\frac{d}{dx}(xf)\right] = -i\hbar\left(x\frac{df}{dx} - x\frac{df}{dx} - f\right)\\ &= i\hbar f(x).\end{aligned} \tag{2.51}$$

Dropping the test function, which has served its purpose,

$$\boxed{\left[x, \hat{p}\right] = i\hbar.} \tag{2.52}$$

This lovely and ubiquitous formula is known as the **canonical commutation relation**.[25]

---

[24] Put a hat on x, too, if you like, but since $\hat{x} = x$ we usually leave it off.

[25] In a deep sense all of the mysteries of quantum mechanics can be traced to the fact that position and momentum do not commute. Indeed, some authors take the canonical commutation relation as an *axiom* of the theory, and use it to *derive* $\hat{p} = -i\hbar\, d/dx$.

With this, Equation 2.50 becomes

$$\hat{a}_-\hat{a}_+ = \frac{1}{\hbar\omega}\hat{H} + \frac{1}{2}, \tag{2.53}$$

or

$$\hat{H} = \hbar\omega\left(\hat{a}_-\hat{a}_+ - \frac{1}{2}\right). \tag{2.54}$$

Evidently the Hamiltonian does *not* factor perfectly—there's that extra $-1/2$ on the right. Notice that the ordering of $\hat{a}_+$ and $\hat{a}_-$ is important here; the same argument, with $\hat{a}_+$ on the left, yields

$$\hat{a}_+\hat{a}_- = \frac{1}{\hbar\omega}\hat{H} - \frac{1}{2}. \tag{2.55}$$

In particular,

$$\left[\hat{a}_-, \hat{a}_+\right] = 1. \tag{2.56}$$

Meanwhile, the Hamiltonian can equally well be written

$$\hat{H} = \hbar\omega\left(\hat{a}_+\hat{a}_- + \frac{1}{2}\right). \tag{2.57}$$

In terms of $\hat{a}_\pm$, then, the Schrödinger equation[26] for the harmonic oscillator takes the form

$$\hbar\omega\left(\hat{a}_\pm\hat{a}_\mp \pm \frac{1}{2}\right)\psi = E\psi \tag{2.58}$$

(in equations like this you read the upper signs all the way across, or else the lower signs).

Now, here comes the crucial step: I claim that:

> If ψ satisfies the Schrödinger equation with energy E (that is: $\hat{H}\psi = E\psi$), then $\hat{a}_+\psi$ satisfies the Schrödinger equation with energy $(E + \hbar\omega)$: $\hat{H}\left(\hat{a}_+\psi\right) = (E + \hbar\omega)\left(\hat{a}_+\psi\right)$.

Proof:

$$\begin{aligned}\hat{H}\left(\hat{a}_+\psi\right) &= \hbar\omega\left(\hat{a}_+\hat{a}_- + \frac{1}{2}\right)\left(\hat{a}_+\psi\right) = \hbar\omega\left(\hat{a}_+\hat{a}_-\hat{a}_+ + \frac{1}{2}\hat{a}_+\right)\psi \\ &= \hbar\omega\hat{a}_+\left(\hat{a}_-\hat{a}_+ + \frac{1}{2}\right)\psi = \hat{a}_+\left[\hbar\omega\left(\hat{a}_+\hat{a}_- + 1 + \frac{1}{2}\right)\psi\right] \\ &= \hat{a}_+\left(\hat{H} + \hbar\omega\right)\psi = \hat{a}_+(E + \hbar\omega)\psi = (E + \hbar\omega)\left(\hat{a}_+\psi\right). \quad \text{QED}\end{aligned}$$

(I used Equation 2.56 to replace $\hat{a}_-\hat{a}_+$ by $\left(\hat{a}_+\hat{a}_- + 1\right)$ in the second line. Notice that whereas the ordering of $\hat{a}_+$ and $\hat{a}_-$ *does* matter, the ordering of $\hat{a}_\pm$ and any *constants*—such as $\hbar$, ω, and E—does *not*; an operator commutes with any constant.)

By the same token, $\hat{a}_-\psi$ is a solution with energy $(E - \hbar\omega)$:

$$\begin{aligned}\hat{H}\left(\hat{a}_-\psi\right) &= \hbar\omega\left(\hat{a}_-\hat{a}_+ - \frac{1}{2}\right)\left(\hat{a}_-\psi\right) = \hbar\omega\hat{a}_-\left(\hat{a}_+\hat{a}_- - \frac{1}{2}\right)\psi \\ &= \hat{a}_-\left[\hbar\omega\left(\hat{a}_-\hat{a}_+ - 1 - \frac{1}{2}\right)\psi\right] = \hat{a}_-\left(\hat{H} - \hbar\omega\right)\psi = \hat{a}_-(E - \hbar\omega)\psi \\ &= (E - \hbar\omega)\left(\hat{a}_-\psi\right).\end{aligned}$$

[26] I'm getting tired of writing "time-independent Schrödinger equation," so when it's clear from the context which one I mean, I'll just call it the "Schrödinger equation."

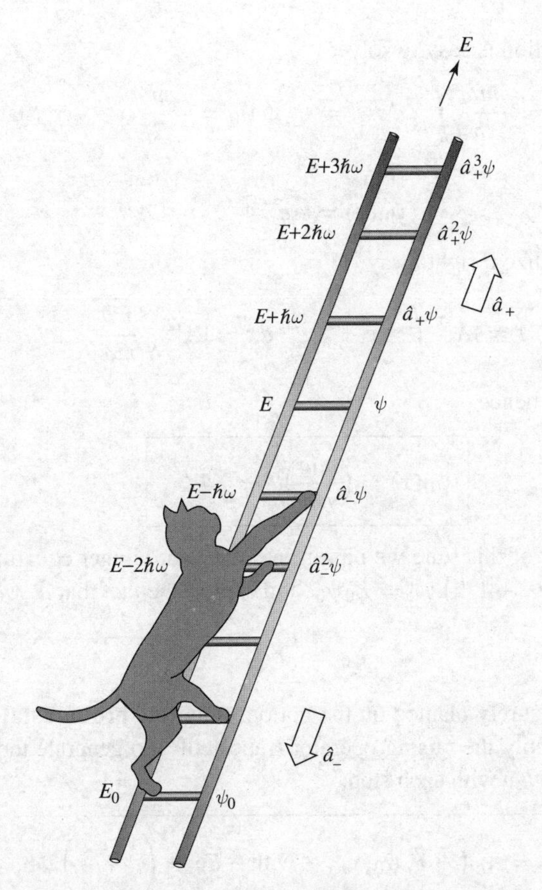

Figure 2.5: The "ladder" of states for the harmonic oscillator.

Here, then, is a wonderful machine for generating new solutions, with higher and lower energies—if we could just find *one* solution, to get started! We call $\hat{a}_\pm$ **ladder operators**, because they allow us to climb up and down in energy; $\hat{a}_+$ is the **raising operator**, and $\hat{a}_-$ the **lowering operator**. The "ladder" of states is illustrated in Figure 2.5.

But wait! What if I apply the lowering operator repeatedly? Eventually I'm going to reach a state with energy less than zero, which (according to the general theorem in Problem 2.2) does not exist! At some point the machine must fail. How can that happen? We know that $\hat{a}_-\psi$ is a new solution to the Schrödinger equation, but *there is no guarantee that it will be normalizable*—it might be zero, or its square-integral might be infinite. In practice it is the former: There occurs a "lowest rung" (call it ψ_0) such that

$$\hat{a}_-\psi_0 = 0. \tag{2.59}$$

We can use this to determine $\psi_0(x)$:

$$\frac{1}{\sqrt{2\hbar m\omega}}\left(\hbar\frac{d}{dx} + m\omega x\right)\psi_0 = 0,$$

or

$$\frac{d\psi_0}{dx} = -\frac{m\omega}{\hbar}x\psi_0.$$

This differential equation is easy to solve:

$$\int \frac{d\psi_0}{\psi_0} = -\frac{m\omega}{\hbar}\int x\,dx \quad\Rightarrow\quad \ln\psi_0 = -\frac{m\omega}{2\hbar}x^2 + \text{constant},$$

so

$$\psi_0(x) = Ae^{-\frac{m\omega}{2\hbar}x^2}.$$

We might as well normalize it right away:

$$1 = |A|^2\int_{-\infty}^{\infty} e^{-m\omega x^2/\hbar}\,dx = |A|^2\sqrt{\frac{\pi\hbar}{m\omega}},$$

so $A^2 = \sqrt{m\omega/\pi\hbar}$, and hence

$$\boxed{\psi_0(x) = \left(\frac{m\omega}{\pi\hbar}\right)^{1/4} e^{-\frac{m\omega}{2\hbar}x^2}.} \tag{2.60}$$

To determine the energy of this state we plug it into the Schrödinger equation (in the form of Equation 2.58), $\hbar\omega\left(\hat{a}_+\hat{a}_- + 1/2\right)\psi_0 = E_0\psi_0$, and exploit the fact that $\hat{a}_-\psi_0 = 0$:

$$E_0 = \frac{1}{2}\hbar\omega. \tag{2.61}$$

With our foot now securely planted on the bottom rung (the **ground state** of the quantum oscillator), we simply apply the raising operator (repeatedly) to generate the excited states,[27] increasing the energy by $\hbar\omega$ with each step:

$$\boxed{\psi_n(x) = A_n\left(\hat{a}_+\right)^n\psi_0(x), \quad \text{with} \quad E_n = \left(n + \frac{1}{2}\right)\hbar\omega,} \tag{2.62}$$

where A_n is the normalization constant. By applying the raising operator (repeatedly) to ψ_0, then, we can (in principle) construct all[28] the stationary states of the harmonic oscillator. Meanwhile, without ever doing that explicitly, we have determined the allowed energies!

Example 2.4

Find the first excited state of the harmonic oscillator.

Solution: Using Equation 2.62,

$$\psi_1(x) = A_1\hat{a}_+\psi_0 = \frac{A_1}{\sqrt{2\hbar m\omega}}\left(-\hbar\frac{d}{dx} + m\omega x\right)\left(\frac{m\omega}{\pi\hbar}\right)^{1/4} e^{-\frac{m\omega}{2\hbar}x^2}$$

$$= A_1\left(\frac{m\omega}{\pi\hbar}\right)^{1/4}\sqrt{\frac{2m\omega}{\hbar}}\,x e^{-\frac{m\omega}{2\hbar}x^2}. \tag{2.63}$$

[27] In the case of the harmonic oscillator it is customary, for some reason, to depart from the usual practice, and number the states starting with $n = 0$, instead of $n = 1$. Of course, the lower limit on the sum in a formula such as Equation 2.17 should be altered accordingly.

[28] Note that we obtain *all* the (normalizable) solutions by this procedure. For if there were some *other* solution, we could generate from it a second ladder, by repeated application of the raising and lowering operators. But the bottom rung of this new ladder would have to satisfy Equation 2.59, and since that leads inexorably to Equation 2.60, the bottom rungs would be the same, and hence the two ladders would in fact be identical.

We can normalize it "by hand":

$$\int |\psi_1|^2\, dx = |A_1|^2 \sqrt{\frac{m\omega}{\pi\hbar}} \left(\frac{2m\omega}{\hbar}\right) \int_{-\infty}^{\infty} x^2 e^{-\frac{m\omega}{\hbar}x^2} dx = |A_1|^2,$$

so, as it happens, $A_1 = 1$.

I wouldn't want to calculate ψ_{50} this way (applying the raising operator fifty times!), but never mind: In *principle* Equation 2.62 does the job—except for the normalization.

You can even get the normalization algebraically, but it takes some fancy footwork, so watch closely. We know that $\hat{a}_\pm \psi_n$ is *proportional* to $\psi_{n\pm 1}$,

$$\hat{a}_+ \psi_n = c_n \psi_{n+1}, \quad \hat{a}_- \psi_n = d_n \psi_{n-1} \tag{2.64}$$

but what are the proportionality factors, c_n and d_n? First note that for "any"[29] functions $f(x)$ and $g(x)$,

$$\int_{-\infty}^{\infty} f^* \left(\hat{a}_\pm g\right) dx = \int_{-\infty}^{\infty} \left(\hat{a}_\mp f\right)^* g\, dx. \tag{2.65}$$

In the language of linear algebra, $\hat{a}_\mp$ is the **hermitian conjugate** (or **adjoint**) of $\hat{a}_\pm$.

Proof:

$$\int_{-\infty}^{\infty} f^* \left(\hat{a}_\pm g\right) dx = \frac{1}{\sqrt{2\hbar m\omega}} \int_{-\infty}^{\infty} f^* \left(\mp\hbar\frac{d}{dx} + m\omega x\right) g\, dx,$$

and integration by parts takes $\int f^* (dg/dx)\, dx$ to $-\int (df/dx)^* g\, dx$ (the boundary terms vanish, for the reason indicated in footnote 29), so

$$\int_{-\infty}^{\infty} f^* \left(\hat{a}_\pm g\right) dx = \frac{1}{\sqrt{2\hbar m\omega}} \int_{-\infty}^{\infty} \left[\left(\pm\hbar\frac{d}{dx} + m\omega x\right) f\right]^* g\, dx$$
$$= \int_{-\infty}^{\infty} \left(\hat{a}_\mp f\right)^* g\, dx. \quad \text{QED}$$

In particular,

$$\int_{-\infty}^{\infty} \left(\hat{a}_\pm \psi_n\right)^* \left(\hat{a}_\pm \psi_n\right) dx = \int_{-\infty}^{\infty} \left(\hat{a}_\mp \hat{a}_\pm \psi_n\right)^* \psi_n\, dx.$$

But (invoking Equations 2.58 and 2.62)

$$\hat{a}_+ \hat{a}_- \psi_n = n\psi_n, \quad \hat{a}_- \hat{a}_+ \psi_n = (n+1)\, \psi_n, \tag{2.66}$$

so

$$\int_{-\infty}^{\infty} \left(\hat{a}_+ \psi_n\right)^* \left(\hat{a}_+ \psi_n\right) dx = |c_n|^2 \int |\psi_{n+1}|^2\, dx = (n+1) \int_{-\infty}^{\infty} |\psi_n|^2\, dx,$$
$$\int_{-\infty}^{\infty} \left(\hat{a}_- \psi_n\right)^* \left(\hat{a}_- \psi_n\right) dx = |d_n|^2 \int |\psi_{n-1}|^2\, dx = n \int_{-\infty}^{\infty} |\psi_n|^2\, dx.$$

[29] Of course, the integrals must *exist*, and this means that $f(x)$ and $g(x)$ must go to zero at $\pm\infty$.

But since ψ_n and $\psi_{n\pm1}$ are normalized, it follows that $|c_n|^2 = n+1$ and $|d_n|^2 = n$, and hence[30]

$$\boxed{\hat{a}_+\psi_n = \sqrt{n+1}\,\psi_{n+1}, \quad \hat{a}_-\psi_n = \sqrt{n}\,\psi_{n-1}.} \tag{2.67}$$

Thus

$$\psi_1 = \hat{a}_+\psi_0, \quad \psi_2 = \frac{1}{\sqrt{2}}\hat{a}_+\psi_1 = \frac{1}{\sqrt{2}}\left(\hat{a}_+\right)^2\psi_0,$$

$$\psi_3 = \frac{1}{\sqrt{3}}\hat{a}_+\psi_2 = \frac{1}{\sqrt{3\cdot 2}}\left(\hat{a}_+\right)^3\psi_0, \quad \psi_4 = \frac{1}{\sqrt{4}}\hat{a}_+\psi_3 = \frac{1}{\sqrt{4\cdot 3\cdot 2}}\left(\hat{a}_+\right)^4\psi_0,$$

and so on. Clearly

$$\boxed{\psi_n = \frac{1}{\sqrt{n!}}\left(\hat{a}_+\right)^n\psi_0,} \tag{2.68}$$

which is to say that the normalization factor in Equation 2.62 is $A_n = 1/\sqrt{n!}$ (in particular, $A_1 = 1$, confirming our result in Example 2.4).

As in the case of the infinite square well, the stationary states of the harmonic oscillator are orthogonal:

$$\int_{-\infty}^{\infty}\psi_m^*\psi_n\,dx = \delta_{mn}. \tag{2.69}$$

This can be proved using Equation 2.66, and Equation 2.65 twice—first moving $\hat{a}_+$ and then moving $\hat{a}_-$:

$$\begin{aligned}\int_{-\infty}^{\infty}\psi_m^*\left(\hat{a}_+\hat{a}_-\right)\psi_n\,dx &= n\int_{-\infty}^{\infty}\psi_m^*\psi_n\,dx \\ &= \int_{-\infty}^{\infty}\left(\hat{a}_-\psi_m\right)^*\left(\hat{a}_-\psi_n\right)dx = \int_{-\infty}^{\infty}\left(\hat{a}_+\hat{a}_-\psi_m\right)^*\psi_n\,dx \\ &= m\int_{-\infty}^{\infty}\psi_m^*\psi_n\,dx.\end{aligned}$$

Unless $m = n$, then, $\int\psi_m^*\psi_n\,dx$ must be zero. Orthonormality means that we can again use Fourier's trick (Equation 2.37) to evaluate the coefficients c_n, when we expand $\Psi(x,0)$ as a linear combination of stationary states (Equation 2.16). As always, $|c_n|^2$ is the probability that a measurement of the energy would yield the value E_n.

---

[30] Of course, we could multiply c_n and d_n by phase factors, amounting to a different *definition* of the ψ_n; but this choice keeps the wave functions *real*.

Example 2.5

Find the expectation value of the potential energy in the nth stationary state of the harmonic oscillator.

Solution:

$$\langle V\rangle = \left\langle \frac{1}{2}m\omega^2 x^2\right\rangle = \frac{1}{2}m\omega^2 \int_{-\infty}^{\infty} \psi_n^* x^2 \psi_n \, dx.$$

There's a beautiful device for evaluating integrals of this kind (involving powers of x or $\hat{p}$): Use the definition (Equation 2.48) to express x and $\hat{p}$ in terms of the raising and lowering operators:

$$\boxed{x = \sqrt{\frac{\hbar}{2m\omega}}\left(\hat{a}_+ + \hat{a}_-\right); \quad \hat{p} = i\sqrt{\frac{\hbar m\omega}{2}}\left(\hat{a}_+ - \hat{a}_-\right).} \tag{2.70}$$

In this example we are interested in x^2:

$$x^2 = \frac{\hbar}{2m\omega}\left[\left(\hat{a}_+\right)^2 + \left(\hat{a}_+\hat{a}_-\right) + \left(\hat{a}_-\hat{a}_+\right) + \left(\hat{a}_-\right)^2\right].$$

So

$$\langle V\rangle = \frac{\hbar\omega}{4}\int_{-\infty}^{\infty} \psi_n^* \left[\left(\hat{a}_+\right)^2 + \left(\hat{a}_+\hat{a}_-\right) + \left(\hat{a}_-\hat{a}_+\right) + \left(\hat{a}_-\right)^2\right] \psi_n \, dx.$$

But $\left(\hat{a}_+\right)^2 \psi_n$ is (apart from normalization) ψ_{n+2}, which is orthogonal to ψ_n, and the same goes for $\left(\hat{a}_-\right)^2 \psi_n$, which is proportional to ψ_{n-2}. So those terms drop out, and we can use Equation 2.66 to evaluate the remaining two:

$$\langle V\rangle = \frac{\hbar\omega}{4}(n + n + 1) = \frac{1}{2}\hbar\omega\left(n + \frac{1}{2}\right).$$

As it happens, the expectation value of the potential energy is exactly *half* the total (the other half, of course, is kinetic). This is a peculiarity of the harmonic oscillator, as we'll see later on (Problem 3.37).

* **Problem 2.10**

(a) Construct $\psi_2(x)$.

(b) Sketch ψ_0, ψ_1, and ψ_2.

(c) Check the orthogonality of ψ_0, ψ_1, and ψ_2, by explicit integration. *Hint:* If you exploit the even-ness and odd-ness of the functions, there is really only one integral left to do.

* **Problem 2.11**

(a) Compute $\langle x\rangle$, $\langle p\rangle$, $\langle x^2\rangle$, and $\langle p^2\rangle$, for the states ψ_0 (Equation 2.60) and ψ_1 (Equation 2.63), by explicit integration. *Comment:* In this and other problems involving the harmonic oscillator it simplifies matters if you introduce the variable $\xi \equiv \sqrt{m\omega/\hbar}\,x$ and the constant $\alpha \equiv (m\omega/\pi\hbar)^{1/4}$.

(b) Check the uncertainty principle for these states.

(c) Compute $\langle T\rangle$ and $\langle V\rangle$ for these states. (No new integration allowed!) Is their sum what you would expect?

* **Problem 2.12** Find $\langle x \rangle$, $\langle p \rangle$, $\langle x^2 \rangle$, $\langle p^2 \rangle$, and $\langle T \rangle$, for the nth stationary state of the harmonic oscillator, using the method of Example 2.5. Check that the uncertainty principle is satisfied.

Problem 2.13 A particle in the harmonic oscillator potential starts out in the state

$$\Psi(x,0) = A\left[3\psi_0(x) + 4\psi_1(x)\right].$$

(a) Find A.

(b) Construct $\Psi(x,t)$ and $|\Psi(x,t)|^2$. Don't get too excited if $|\Psi(x,t)|^2$ oscillates at exactly the classical frequency; what would it have been had I specified $\psi_2(x)$, instead of $\psi_1(x)$?[31]

(c) Find $\langle x \rangle$ and $\langle p \rangle$. Check that Ehrenfest's theorem (Equation 1.38) holds, for this wave function.

(d) If you measured the energy of this particle, what values might you get, and with what probabilities?

2.3.2 Analytic Method

We return now to the Schrödinger equation for the harmonic oscillator,

$$-\frac{\hbar^2}{2m}\frac{d^2\psi}{dx^2} + \frac{1}{2}m\omega^2 x^2 \psi = E\psi, \tag{2.71}$$

and solve it directly, by the power series method. Things look a little cleaner if we introduce the dimensionless variable

$$\xi \equiv \sqrt{\frac{m\omega}{\hbar}}\,x; \tag{2.72}$$

in terms of ξ the Schrödinger equation reads

$$\frac{d^2\psi}{d\xi^2} = \left(\xi^2 - K\right)\psi, \tag{2.73}$$

where K is the energy, in units of $(1/2)\,\hbar\omega$:

$$K \equiv \frac{2E}{\hbar\omega}. \tag{2.74}$$

Our problem is to solve Equation 2.73, and in the process obtain the "allowed" values of K (and hence of E).

To begin with, note that at very large ξ (which is to say, at very large x), ξ^2 completely dominates over the constant K, so in this regime

$$\frac{d^2\psi}{d\xi^2} \approx \xi^2\psi, \tag{2.75}$$

which has the approximate solution (check it!)

$$\psi(\xi) \approx Ae^{-\xi^2/2} + Be^{+\xi^2/2}. \tag{2.76}$$

The B term is clearly not normalizable (it blows up as $|x| \to \infty$); the physically acceptable solutions, then, have the asymptotic form

[31] However, $\langle x \rangle$ *does* oscillate at the classical frequency—see Problem 3.40.

$$\psi(\xi) \rightarrow (\quad)e^{-\xi^2/2}, \quad \text{at large } \xi. \tag{2.77}$$

This suggests that we "peel off" the exponential part,

$$\psi(\xi) = h(\xi)\, e^{-\xi^2/2}, \tag{2.78}$$

in hopes that what remains, $h(\xi)$, has a simpler functional form than $\psi(\xi)$ itself.[32] Differentiating Equation 2.78,

$$\frac{d\psi}{d\xi} = \left(\frac{dh}{d\xi} - \xi h\right) e^{-\xi^2/2},$$

and

$$\frac{d^2\psi}{d\xi^2} = \left(\frac{d^2h}{d\xi^2} - 2\xi\frac{dh}{d\xi} + (\xi^2 - 1)h\right) e^{-\xi^2/2},$$

so the Schrödinger equation (Equation 2.73) becomes

$$\frac{d^2h}{d\xi^2} - 2\xi\frac{dh}{d\xi} + (K-1)\, h = 0. \tag{2.79}$$

I propose to look for solutions to Equation 2.79 in the form of *power series* in ξ:[33]

$$h(\xi) = a_0 + a_1\xi + a_2\xi^2 + \cdots = \sum_{j=0}^{\infty} a_j \xi^j. \tag{2.80}$$

Differentiating the series term by term,

$$\frac{dh}{d\xi} = a_1 + 2a_2\xi + 3a_3\xi^2 + \cdots = \sum_{j=0}^{\infty} j a_j \xi^{j-1},$$

and

$$\frac{d^2h}{d\xi^2} = 2a_2 + 2\cdot 3a_3\xi + 3\cdot 4a_4\xi^2 + \cdots = \sum_{j=0}^{\infty} (j+1)\,(j+2)\, a_{j+2}\xi^j.$$

Putting these into Equation 2.79, we find

$$\sum_{j=0}^{\infty} \left[(j+1)\,(j+2)\, a_{j+2} - 2j a_j + (K-1)\, a_j\right] \xi^j = 0. \tag{2.81}$$

It follows (from the uniqueness of power series expansions[34]) that the coefficient of *each power* of ξ must vanish,

$$(j+1)\,(j+2)\, a_{j+2} - 2j a_j + (K-1)\, a_j = 0,$$

---

[32] Note that although we invoked some approximations to *motivate* Equation 2.78, what follows is *exact*. The device of stripping off the asymptotic behavior is the standard first step in the power series method for solving differential equations—see, for example, Boas (footnote 16), Chapter 12.

[33] According to Taylor's theorem, *any* reasonably well-behaved function can be expressed as a power series, so Equation 2.80 ordinarily involves no loss of generality. For conditions on the applicability of the method, see Boas (footnote 16) or George B. Arfken and Hans-Jurgen Weber, *Mathematical Methods for Physicists*, 7th edn, Academic Press, Orlando (2013), Section 7.5.

[34] See, for example, Arfken and Weber (footnote 33), Section 1.2.

and hence that

$$a_{j+2} = \frac{(2j+1-K)}{(j+1)(j+2)} a_j. \tag{2.82}$$

This **recursion formula** is entirely equivalent to the Schrödinger equation. Starting with a_0, it generates all the even-numbered coefficients:

$$a_2 = \frac{(1-K)}{2} a_0, \quad a_4 = \frac{(5-K)}{12} a_2 = \frac{(5-K)(1-K)}{24} a_0, \quad \ldots,$$

and starting with a_1, it generates the odd coefficients:

$$a_3 = \frac{(3-K)}{6} a_1, \quad a_5 = \frac{(7-K)}{20} a_3 = \frac{(7-K)(3-K)}{120} a_1, \quad \ldots.$$

We write the complete solution as

$$h(\xi) = h_{\text{even}}(\xi) + h_{\text{odd}}(\xi), \tag{2.83}$$

where

$$h_{\text{even}}(\xi) \equiv a_0 + a_2\xi^2 + a_4\xi^4 + \cdots$$

is an even function of ξ, built on a_0, and

$$h_{\text{odd}}(\xi) \equiv a_1\xi + a_3\xi^3 + a_5\xi^5 + \cdots$$

is an odd function, built on a_1. Thus Equation 2.82 determines $h(\xi)$ in terms of two arbitrary constants (a_0 and a_1)—which is just what we would expect, for a second-order differential equation.

However, not all the solutions so obtained are *normalizable*. For at very large j, the recursion formula becomes (approximately)

$$a_{j+2} \approx \frac{2}{j} a_j,$$

with the (approximate) solution

$$a_j \approx \frac{C}{(j/2)!},$$

for some constant C, and this yields (at large ξ, where the higher powers dominate)

$$h(\xi) \approx C \sum \frac{1}{(j/2)!} \xi^j \approx C \sum \frac{1}{j!} \xi^{2j} \approx C e^{\xi^2}.$$

Now, if h goes like $\exp(\xi^2)$, then ψ (remember ψ?—that's what we're trying to calculate) goes like $\exp(\xi^2/2)$ (Equation 2.78), which is precisely the asymptotic behavior we *didn't* want.[35] There is only one way to wiggle out of this: For normalizable solutions *the power series must terminate*. There must occur some "highest" j (call it n), such that the recursion formula spits out $a_{n+2} = 0$ (this will truncate *either* the series h_{even} *or* the series h_{odd}; the *other* one must be zero from the start: $a_1 = 0$ if n is even, and $a_0 = 0$ if n is odd). For physically acceptable solutions, then, Equation 2.82 requires that

$$K = 2n + 1,$$

[35] It's no surprise that the ill-behaved solutions are still contained in Equation 2.82; this recursion relation is equivalent to the Schrödinger equation, so it's *got* to include both the asymptotic forms we found in Equation 2.76.

for some positive integer n, which is to say (referring to Equation 2.74) that the *energy* must be

$$E_n = \left(n + \frac{1}{2}\right)\hbar\omega, \quad \text{for } n = 0, 1, 2, \ldots. \tag{2.84}$$

Thus we recover, by a completely different method, the fundamental quantization condition we found algebraically in Equation 2.62.

It seems at first rather surprising that the quantization of energy should emerge from a technical detail in the power series solution to the Schrödinger equation, but let's look at it from a different perspective. Equation 2.71 has solutions, of course, for *any* value of E (in fact, it has *two* linearly independent solutions for every E). But almost all of these solutions blow up exponentially at large x, and hence are not normalizable. Imagine, for example, using an E that is slightly *less* than one of the allowed values (say, $0.49\hbar\omega$), and plotting the solution: Figure 2.6(a). Now try an E slightly *larger* (say, $0.51\hbar\omega$); the "tail" now blows up in the *other* direction (Figure 2.6(b)). As you tweak the parameter in tiny increments from 0.49 to 0.51, the graph "flips over" at precisely the value 0.5—only here does the solution escape the exponential asymptotic growth that renders it physically unacceptable.[36]

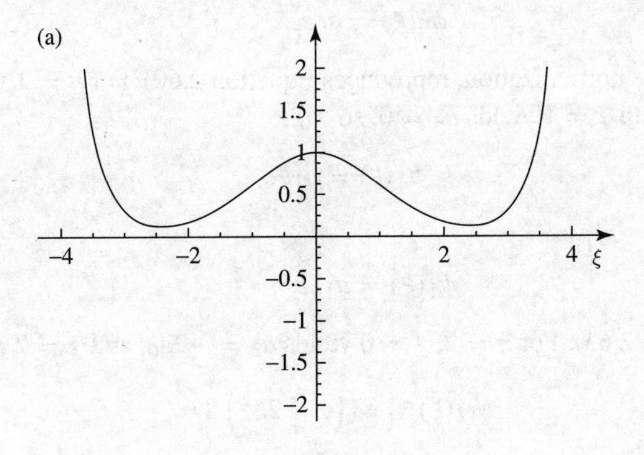

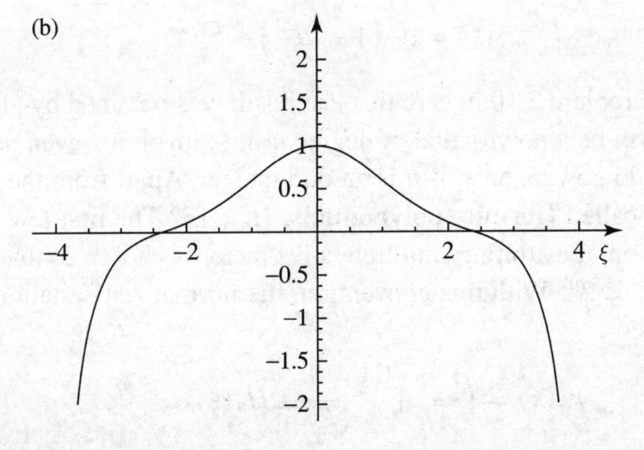

Figure 2.6: Solutions to the Schrödinger equation for (a) $E = 0.49\hbar\omega$, and (b) $E = 0.51\hbar\omega$.

[36] It is possible to set this up on a computer, and discover the allowed energies "experimentally." You might call it the **wag the dog** method: When the tail wags, you know you've just passed over an allowed value. Computer scientists call it the **shooting** method (Nicholas Giordano, *Computational Physics*, Prentice Hall, Upper Saddle River, NJ (1997), Section 10.2). See Problems 2.55–2.57.

Table 2.1: *The first few Hermite polynomials, $H_n(\xi)$.*

$$
\begin{aligned}
H_0 &= 1, \\
H_1 &= 2\xi, \\
H_2 &= 4\xi^2 - 2, \\
H_3 &= 8\xi^3 - 12\xi, \\
H_4 &= 16\xi^4 - 48\xi^2 + 12, \\
H_5 &= 32\xi^5 - 160\xi^3 + 120\xi.
\end{aligned}
$$

For the allowed values of K, the recursion formula reads

$$a_{j+2} = \frac{-2(n-j)}{(j+1)(j+2)} a_j. \tag{2.85}$$

If $n = 0$, there is only one term in the series (we must pick $a_1 = 0$ to kill h_{odd}, and $j = 0$ in Equation 2.85 yields $a_2 = 0$):

$$h_0(\xi) = a_0,$$

and hence

$$\psi_0(\xi) = a_0 e^{-\xi^2/2}$$

(which, apart from the normalization, reproduces Equation 2.60). For $n = 1$ we take $a_0 = 0$,[37] and Equation 2.85 with $j = 1$ yields $a_3 = 0$, so

$$h_1(\xi) = a_1 \xi,$$

and hence

$$\psi_1(\xi) = a_1 \xi e^{-\xi^2/2}$$

(confirming Equation 2.63). For $n = 2$, $j = 0$ yields $a_2 = -2a_0$, and $j = 2$ gives $a_4 = 0$, so

$$h_2(\xi) = a_0 \left(1 - 2\xi^2\right),$$

and

$$\psi_2(\xi) = a_0 \left(1 - 2\xi^2\right) e^{-\xi^2/2},$$

and so on. (Compare Problem 2.10, where this last result was obtained by algebraic means).

In general, $h_n(\xi)$ will be a polynomial of degree n in ξ, involving even powers only, if n is an even integer, and odd powers only, if n is an odd integer. Apart from the overall factor (a_0 or a_1) they are the so-called **Hermite polynomials**, $H_n(\xi)$.[38] The first few of them are listed in Table 2.1. By tradition, the arbitrary multiplicative factor is chosen so that the coefficient of the highest power of ξ is 2^n. With this convention, the normalized[39] stationary states for the harmonic oscillator are

$$\psi_n(x) = \left(\frac{m\omega}{\pi\hbar}\right)^{1/4} \frac{1}{\sqrt{2^n n!}} H_n(\xi)\, e^{-\xi^2/2}. \tag{2.86}$$

They are identical (of course) to the ones we obtained algebraically in Equation 2.68.

---

37 Note that there is a completely different set of coefficients a_j for each value of n.

38 The Hermite polynomials have been studied extensively in the mathematical literature, and there are many tools and tricks for working with them. A few of these are explored in Problem 2.16.

39 I shall not work out the normalization constant here; if you are interested in knowing how it is done, see for example Leonard Schiff, *Quantum Mechanics*, 3rd edn, McGraw-Hill, New York (1968), Section 13.

In Figure 2.7(a) I have plotted $\psi_n(x)$ for the first few ns. The quantum oscillator is strikingly different from its classical counterpart—not only are the energies quantized, but the position distributions have some bizarre features. For instance, the probability of finding the particle outside the classically allowed range (that is, with x greater than the classical amplitude for

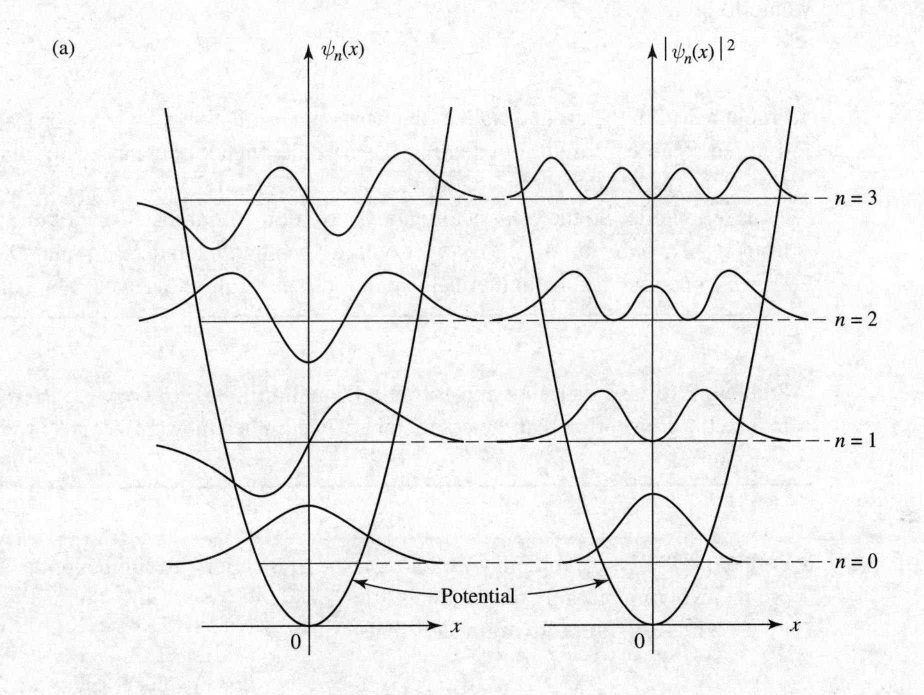

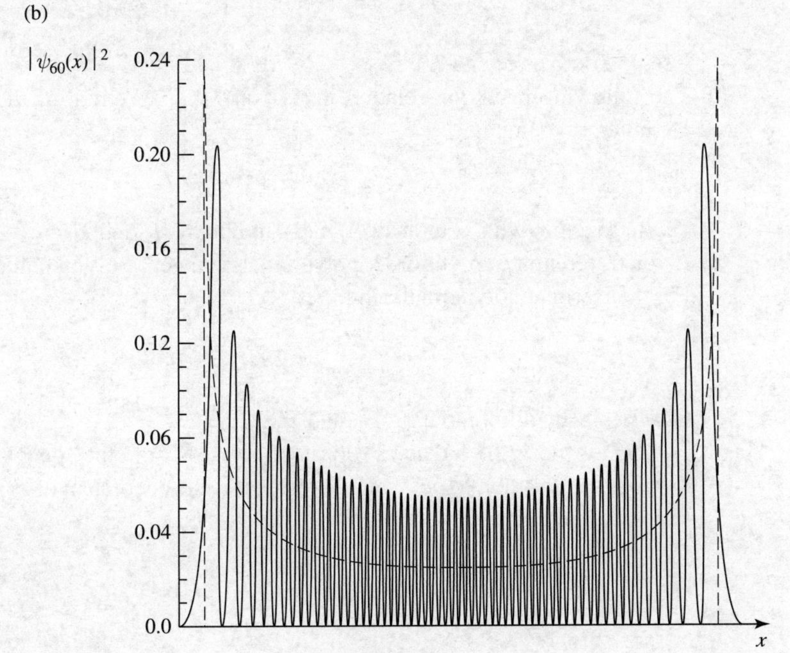

Figure 2.7: (a) The first four stationary states of the harmonic oscillator. (b) Graph of $|\psi_{60}|^2$, with the classical distribution (dashed curve) superimposed.

the energy in question) is *not* zero (see Problem 2.14), and in all odd states the probability of finding the particle at the center is zero. Only at large n do we begin to see some resemblance to the classical case. In Figure 2.7(b) I have superimposed the classical position distribution (Problem 1.11) on the quantum one (for $n = 60$); if you smoothed out the bumps, the two would fit pretty well.

Problem 2.14 In the ground state of the harmonic oscillator, what is the probability (correct to three significant digits) of finding the particle outside the classically allowed region? *Hint:* Classically, the energy of an oscillator is $E = (1/2)\,ka^2 = (1/2)\,m\omega^2a^2$, where a is the amplitude. So the "classically allowed region" for an oscillator of energy E extends from $-\sqrt{2E/m\omega^2}$ to $+\sqrt{2E/m\omega^2}$. Look in a math table under "Normal Distribution" or "Error Function" for the numerical value of the integral, or evaluate it by computer.

Problem 2.15 Use the recursion formula (Equation 2.85) to work out $H_5(\xi)$ and $H_6(\xi)$. Invoke the convention that the coefficient of the highest power of ξ is 2^n to fix the overall constant.

** **Problem 2.16** In this problem we explore some of the more useful theorems (stated without proof) involving Hermite polynomials.

(a) The **Rodrigues formula** says that

$$H_n(\xi) = (-1)^n\, e^{\xi^2} \left(\frac{d}{d\xi}\right)^n e^{-\xi^2}. \tag{2.87}$$

Use it to derive H_3 and H_4.

(b) The following recursion relation gives you H_{n+1} in terms of the two preceding Hermite polynomials:

$$H_{n+1}(\xi) = 2\xi H_n(\xi) - 2nH_{n-1}(\xi). \tag{2.88}$$

Use it, together with your answer in (a), to obtain H_5 and H_6.

(c) If you differentiate an nth-order polynomial, you get a polynomial of order $(n-1)$. For the Hermite polynomials, in fact,

$$\frac{dH_n}{d\xi} = 2nH_{n-1}(\xi). \tag{2.89}$$

Check this, by differentiating H_5 and H_6.

(d) $H_n(\xi)$ is the nth z-derivative, at $z = 0$, of the **generating function** $\exp\left(-z^2 + 2z\xi\right)$; or, to put it another way, it is the coefficient of $z^n/n!$ in the Taylor series expansion for this function:

$$e^{-z^2+2z\xi} = \sum_{n=0}^{\infty} \frac{z^n}{n!} H_n(\xi). \tag{2.90}$$

Use this to obtain H_1, H_2, and H_3.

2.4 THE FREE PARTICLE

We turn next to what *should* have been the simplest case of all: the free particle ($V(x) = 0$ everywhere). Classically this would just be motion at constant velocity, but in quantum mechanics the problem is surprisingly subtle. The time-independent Schrödinger equation reads

$$-\frac{\hbar^2}{2m}\frac{d^2\psi}{dx^2} = E\psi, \tag{2.91}$$

or

$$\frac{d^2\psi}{dx^2} = -k^2\psi, \quad \text{where } k \equiv \frac{\sqrt{2mE}}{\hbar}. \tag{2.92}$$

So far, it's the same as inside the infinite square well (Equation 2.24), where the potential is also zero; this time, however, I prefer to write the general solution in exponential form (instead of sines and cosines), for reasons that will appear in due course:

$$\psi(x) = Ae^{ikx} + Be^{-ikx}. \tag{2.93}$$

Unlike the infinite square well, there are no boundary conditions to restrict the possible values of k (and hence of E); the free particle can carry *any* (positive) energy. Tacking on the standard time dependence, $\exp(-iEt/\hbar)$,

$$\Psi(x,t) = Ae^{ik\left(x - \frac{\hbar k}{2m}t\right)} + Be^{-ik\left(x + \frac{\hbar k}{2m}t\right)}. \tag{2.94}$$

Now, *any* function of x and t that depends on these variables in the special combination $(x \pm vt)$ (for some constant v) represents a wave of unchanging shape, traveling in the $\mp x$-direction at speed v: A fixed point on the waveform (for example, a maximum or a minimum) corresponds to a fixed value of the argument, and hence to x and t such that

$$x \pm vt = \text{constant}, \quad \text{or} \quad x = \mp vt + \text{constant}.$$

Since every point on the waveform moves with the same velocity, its *shape* doesn't change as it propagates. Thus the first term in Equation 2.94 represents a wave traveling to the *right*, and the second represents a wave (of the same energy) going to the *left*. By the way, since they only differ by the *sign* in front of k, we might as well write

$$\Psi_k(x,t) = Ae^{i\left(kx - \frac{\hbar k^2}{2m}t\right)}, \tag{2.95}$$

and let k run negative to cover the case of waves traveling to the left:

$$k \equiv \pm\frac{\sqrt{2mE}}{\hbar}, \quad \text{with} \begin{cases} k > 0 \Rightarrow & \text{traveling to the right,} \\ k < 0 \Rightarrow & \text{traveling to the left.} \end{cases} \tag{2.96}$$

Evidently the "stationary states" of the free particle are propagating waves; their wavelength is $\lambda = 2\pi/|k|$, and, according to the de Broglie formula (Equation 1.39), they carry momentum

$$p = \hbar k. \tag{2.97}$$

The speed of these waves (the coefficient of t over the coefficient of x) is

$$v_{\text{quantum}} = \frac{\hbar|k|}{2m} = \sqrt{\frac{E}{2m}}. \tag{2.98}$$

On the other hand, the *classical* speed of a free particle with energy E is given by $E = (1/2)\,mv^2$ (pure kinetic, since $V = 0$), so

$$v_{\text{classical}} = \sqrt{\frac{2E}{m}} = 2v_{\text{quantum}}. \tag{2.99}$$

Apparently the quantum mechanical wave function travels at *half* the speed of the particle it is supposed to represent! We'll return to this paradox in a moment—there is an even more serious problem we need to confront first: *This wave function is not normalizable*:

$$\int_{-\infty}^{+\infty} \Psi_k^* \Psi_k \, dx = |A|^2 \int_{-\infty}^{+\infty} dx = |A|^2 \, (\infty). \tag{2.100}$$

In the case of the free particle, then, the separable solutions do not represent physically realizable states. A free particle cannot exist in a stationary state; or, to put it another way, *there is no such thing as a free particle with a definite energy.*

But that doesn't mean the separable solutions are of no use to us. For they play a *mathematical* role that is entirely independent of their *physical* interpretation: The general solution to the time-dependent Schrödinger equation is still a linear combination of separable solutions (only this time it's an *integral* over the *continuous* variable k, instead of a *sum* over the *discrete* index n):

$$\boxed{\Psi(x,t) = \frac{1}{\sqrt{2\pi}} \int_{-\infty}^{+\infty} \phi(k) e^{i\left(kx - \frac{\hbar k^2}{2m} t\right)} \, dk.} \tag{2.101}$$

(The quantity $1/\sqrt{2\pi}$ is factored out for convenience; what plays the role of the coefficient c_n in Equation 2.17 is the combination $\left(1/\sqrt{2\pi}\right)\phi(k)\,dk$.) Now *this* wave function *can* be normalized (for appropriate $\phi(k)$). But it necessarily carries a *range* of ks, and hence a range of energies and speeds. We call it a **wave packet**.[40]

In the generic quantum problem, we are *given* $\Psi(x,0)$, and we are asked to *find* $\Psi(x,t)$. For a free particle the solution takes the form of Equation 2.101; the only question is how to determine $\phi(k)$ so as to match the initial wave function:

$$\Psi(x,0) = \frac{1}{\sqrt{2\pi}} \int_{-\infty}^{+\infty} \phi(k) e^{ikx} \, dk. \tag{2.102}$$

This is a classic problem in Fourier analysis; the answer is provided by **Plancherel's theorem** (see Problem 2.19):

$$\boxed{f(x) = \frac{1}{\sqrt{2\pi}} \int_{-\infty}^{+\infty} F(k)\, e^{ikx} \, dk \iff F(k) = \frac{1}{\sqrt{2\pi}} \int_{-\infty}^{+\infty} f(x)\, e^{-ikx} \, dx.} \tag{2.103}$$

$F(k)$ is called the **Fourier transform** of $f(x)$; $f(x)$ is the **inverse Fourier transform** of $F(k)$ (the only difference is the sign in the exponent).[41] There is, of course, some restriction on the

---

[40] Sinusoidal waves extend out to infinity, and they are not normalizable. But *superpositions* of such waves lead to interference, which allows for localization and normalizability.

[41] Some people define the Fourier transform without the factor of $1/\sqrt{2\pi}$. Then the inverse transform becomes $f(x) = (1/2\pi) \int_{-\infty}^{\infty} F(k) e^{ikx} dk$, spoiling the symmetry of the two formulas.

allowable functions: The integrals have to *exist*.[42] For our purposes this is guaranteed by the physical requirement that $\Psi(x,0)$ itself be normalized. So the solution to the generic quantum problem, for the free particle, is Equation 2.101, with

$$\boxed{\phi(k)=\frac{1}{\sqrt{2\pi}}\int_{-\infty}^{+\infty}\Psi(x,0)\,e^{-ikx}\,dx.} \tag{2.104}$$

Example 2.6

A free particle, which is initially localized in the range $-a<x<a$, is released at time $t=0$:

$$\Psi(x,0)=\begin{cases}A, & -a<x<a,\\ 0, & \text{otherwise,}\end{cases}$$

where A and a are positive real constants. Find $\Psi(x,t)$.

Solution: First we need to normalize $\Psi(x,0)$:

$$1=\int_{-\infty}^{\infty}|\Psi(x,0)|^2\,dx=|A|^2\int_{-a}^{a}dx=2a\,|A|^2\;\Rightarrow\;A=\frac{1}{\sqrt{2a}}.$$

Next we calculate $\phi(k)$, using Equation 2.104:

$$\phi(k)=\frac{1}{\sqrt{2\pi}}\frac{1}{\sqrt{2a}}\int_{-a}^{a}e^{-ikx}\,dx=\frac{1}{2\sqrt{\pi a}}\left.\frac{e^{-ikx}}{-ik}\right|_{-a}^{a}$$

$$=\frac{1}{k\sqrt{\pi a}}\left(\frac{e^{ika}-e^{-ika}}{2i}\right)=\frac{1}{\sqrt{\pi a}}\frac{\sin(ka)}{k}.$$

Finally, we plug this back into Equation 2.101:

$$\Psi(x,t)=\frac{1}{\pi\sqrt{2a}}\int_{-\infty}^{\infty}\frac{\sin(ka)}{k}e^{i\left(kx-\frac{\hbar k^2}{2m}t\right)}\,dk. \tag{2.105}$$

Unfortunately, this integral cannot be solved in terms of elementary functions, though it can of course be evaluated numerically (Figure 2.8). (There are, in fact, precious few cases in which the integral for $\Psi(x,t)$ (Equation 2.101) *can* be carried out explicitly; see Problem 2.21 for a particularly beautiful example.)

In Figure 2.9 I have plotted $\Psi(x,0)$ and $\phi(k)$. Note that for small a, $\Psi(x,0)$ is narrow (in x), while $\phi(k)$ is broad (in k), and vice versa for large a. But k is related to momentum, by Equation 2.97, so this is a manifestation of the uncertainty principle: the position can be well defined (small a), or the momentum (large a), but not both.

[42] The necessary and sufficient condition on $f(x)$ is that $\int_{-\infty}^{\infty}|f(x)|^2\,dx$ be *finite*. (In that case $\int_{-\infty}^{\infty}|F(k)|^2\,dk$ is also finite, and in fact the two integrals are equal. Some people call *this* Plancherel's theorem, leaving Equation 2.102 without a name.) See Arfken and Weber (footnote 33), Section 20.4.

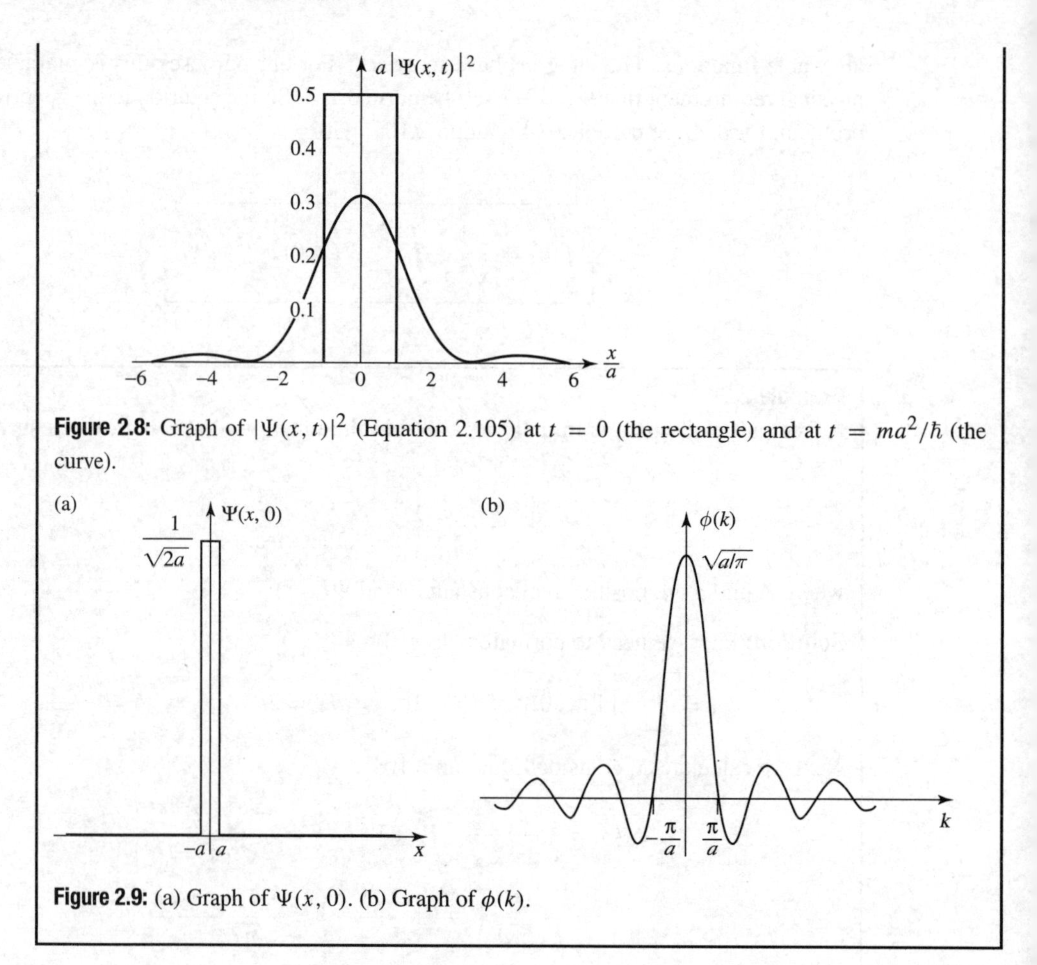

Figure 2.8: Graph of $|\Psi(x,t)|^2$ (Equation 2.105) at $t = 0$ (the rectangle) and at $t = ma^2/\hbar$ (the curve).

Figure 2.9: (a) Graph of $\Psi(x,0)$. (b) Graph of $\phi(k)$.

I return now to the paradox noted earlier: the fact that the separable solution $\Psi_k(x,t)$ travels at the "wrong" speed for the particle it ostensibly represents. Strictly speaking, the problem evaporated when we discovered that Ψ_k is not a physically realizable state. Nevertheless, it is of interest to figure out how information about the particle velocity *is* contained in the wave function (Equation 2.101). The essential idea is this: A wave packet is a superposition of sinusoidal functions whose amplitude is modulated by ϕ (Figure 2.10); it consists of "ripples" contained within an "envelope." What corresponds to the particle velocity is not the speed of the individual ripples (the so-called **phase velocity**), but rather the speed of the envelope (the **group velocity**)—which, depending on the nature of the waves, can be greater than, less than, or equal to, the velocity of the ripples that go to make it up. For waves on a string, the group velocity is the same as the phase velocity. For water waves it is one-half the phase velocity, as you may have noticed when you toss a rock into a pond (if you concentrate on a particular ripple, you will see it build up from the rear, move forward through the group, and fade away at the front, while the group as a whole propagates out at half that speed). What I need to show is that for the wave function of a free particle in quantum mechanics the group velocity is *twice* the phase velocity—just right to match the classical particle speed.

The problem, then, is to determine the group velocity of a wave packet with the generic form

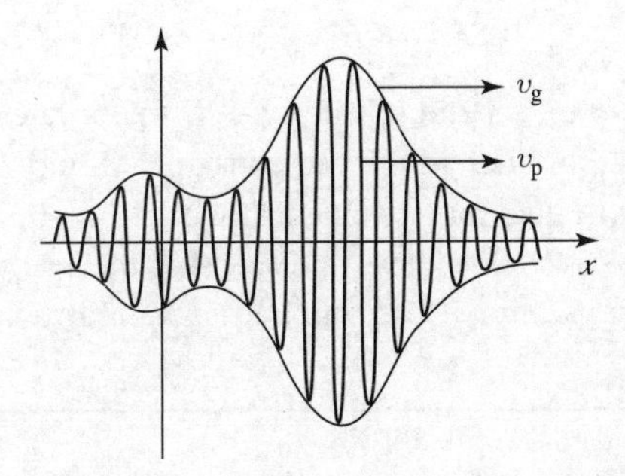

Figure 2.10: A wave packet. The "envelope" travels at the group velocity; the "ripples" travel at the phase velocity.

$$\Psi(x,t) = \frac{1}{\sqrt{2\pi}} \int_{-\infty}^{+\infty} \phi(k)\, e^{i(kx-\omega t)}\, dk. \tag{2.106}$$

In our case $\omega = \left(\hbar k^2/2m\right)$, but what I have to say now applies to *any* kind of wave packet, regardless of its **dispersion relation** (the formula for ω as a function of k). Let us assume that $\phi(k)$ is narrowly peaked about some particular value k_0. (There is nothing *illegal* about a broad spread in k, but such wave packets change shape rapidly—different components travel at different speeds, so the whole notion of a "group," with a well-defined velocity, loses its meaning.) Since the integrand is negligible except in the vicinity of k_0, we may as well Taylor-expand the function $\omega(k)$ about that point, and keep only the leading terms:

$$\omega(k) \approx \omega_0 + \omega_0' \left(k - k_0\right),$$

where ω_0' is the derivative of ω with respect to k, at the point k_0.

Changing variables from k to $s \equiv k - k_0$ (to center the integral at k_0), we have

$$\begin{aligned}\Psi(x,t) &\approx \frac{1}{\sqrt{2\pi}} \int_{-\infty}^{+\infty} \phi(k_0+s)\, e^{i[(k_0+s)x-(\omega_0+\omega_0's)t]}\, ds \\ &= \frac{1}{\sqrt{2\pi}} e^{i(k_0x-\omega_0 t)} \int_{-\infty}^{+\infty} \phi(k_0+s)\, e^{is(x-\omega_0't)}\, ds.\end{aligned} \tag{2.107}$$

The term in front is a sinusoidal wave (the "ripples"), traveling at speed ω_0/k_0. It is modulated by the integral (the "envelope"), which is a function of $x - \omega_0't$, and therefore propagates at the speed ω_0'. Thus the *phase* velocity is

$$v_{\text{phase}} = \frac{\omega}{k}, \tag{2.108}$$

while the *group* velocity is

$$v_{\text{group}} = \frac{d\omega}{dk} \tag{2.109}$$

(both of them evaluated at $k = k_0$).

In our case, $\omega = \left(\hbar k^2/2m\right)$, so $\omega/k = (\hbar k/2m)$, whereas $d\omega/dk = (\hbar k/m)$, which is twice as great. This confirms that the group velocity of the wave packet matches the classical particle velocity:

$$v_{\text{classical}} = v_{\text{group}} = 2v_{\text{phase}}. \tag{2.110}$$

* **Problem 2.17** Show that $\left[Ae^{ikx} + Be^{-ikx}\right]$ and $[C \cos kx + D \sin kx]$ are equivalent ways of writing the same function of x, and determine the constants C and D in terms of A and B, and vice versa. *Comment:* In quantum mechanics, when $V = 0$, the exponentials represent *traveling* waves, and are most convenient in discussing the free particle, whereas sines and cosines correspond to *standing* waves, which arise naturally in the case of the infinite square well.

Problem 2.18 Find the probability current, J (Problem 1.14) for the free particle wave function Equation 2.95. Which direction does the probability flow?

** **Problem 2.19** This problem is designed to guide you through a "proof" of Plancherel's theorem, by starting with the theory of ordinary Fourier series on a *finite* interval, and allowing that interval to expand to infinity.

(a) Dirichlet's theorem says that "any" function $f(x)$ on the interval $[-a, +a]$ can be expanded as a Fourier series:

$$f(x) = \sum_{n=0}^{\infty}\left[a_n \sin\left(\frac{n\pi x}{a}\right) + b_n \cos\left(\frac{n\pi x}{a}\right)\right].$$

Show that this can be written equivalently as

$$f(x) = \sum_{n=-\infty}^{\infty} c_n e^{in\pi x/a}.$$

What is c_n, in terms of a_n and b_n?

(b) Show (by appropriate modification of Fourier's trick) that

$$c_n = \frac{1}{2a}\int_{-a}^{+a} f(x)\, e^{-in\pi x/a}\, dx.$$

(c) Eliminate n and c_n in favor of the new variables $k = (n\pi/a)$ and $F(k) = \sqrt{2/\pi}\, a c_n$. Show that (a) and (b) now become

$$f(x) = \frac{1}{\sqrt{2\pi}} \sum_{n=-\infty}^{\infty} F(k)\, e^{ikx}\, \Delta k; \quad F(k) = \frac{1}{\sqrt{2\pi}}\int_{-a}^{+a} f(x)\, e^{-ikx}\, dx,$$

where Δk is the increment in k from one n to the next.

(d) Take the limit $a \to \infty$ to obtain Plancherel's theorem. *Comment:* In view of their quite different origins, it is surprising (and delightful) that the two formulas—one for $F(k)$ in terms of $f(x)$, the other for $f(x)$ in terms of $F(k)$—have such a similar structure in the limit $a \to \infty$.

Problem 2.20 A free particle has the initial wave function

$$\Psi(x, 0) = Ae^{-a|x|},$$

where A and a are positive real constants.

(a) Normalize $\Psi(x, 0)$.

(b) Find $\phi(k)$.

(c) Construct $\Psi(x, t)$, in the form of an integral.

(d) Discuss the limiting cases (a very large, and a very small).

* **Problem 2.21 The gaussian wave packet.** A free particle has the initial wave function

$$\Psi(x, 0) = Ae^{-ax^2},$$

where A and a are (real and positive) constants.

(a) Normalize $\Psi(x, 0)$.

(b) Find $\Psi(x, t)$. *Hint:* Integrals of the form

$$\int_{-\infty}^{+\infty} e^{-(ax^2+bx)}\, dx$$

can be handled by "completing the square": Let $y \equiv \sqrt{a}\left[x + (b/2a)\right]$, and note that $\left(ax^2 + bx\right) = y^2 - \left(b^2/4a\right)$. *Answer:*

$$\Psi(x, t) = \left(\frac{2a}{\pi}\right)^{1/4} \frac{1}{\gamma} e^{-ax^2/\gamma^2}, \quad \text{where} \quad \gamma \equiv \sqrt{1 + (2i\hbar at/m)}. \tag{2.111}$$

(c) Find $|\Psi(x, t)|^2$. Express your answer in terms of the quantity

$$w \equiv \sqrt{a/\left[1 + (2\hbar at/m)^2\right]}.$$

Sketch $|\Psi|^2$ (as a function of x) at $t = 0$, and again for some very large t. Qualitatively, what happens to $|\Psi|^2$, as time goes on?

(d) Find $\langle x\rangle$, $\langle p\rangle$, $\langle x^2\rangle$, $\langle p^2\rangle$, σ_x, and σ_p. *Partial answer:* $\langle p^2\rangle = a\hbar^2$, but it may take some algebra to reduce it to this simple form.

(e) Does the uncertainty principle hold? At what time t does the system come closest to the uncertainty limit?

2.5 THE DELTA-FUNCTION POTENTIAL

2.5.1 Bound States and Scattering States

We have encountered two very different kinds of solutions to the time-independent Schrödinger equation: For the infinite square well and the harmonic oscillator they are *normalizable*, and labeled by a *discrete index n*; for the free particle they are *non-normalizable*, and labeled by a *continuous variable k*. The former represent physically realizable states in their own right, the latter do not; but in both cases the general solution to the time-dependent Schrödinger equation is a linear combination of stationary states—for the first type this combination takes the form of a *sum* (over n), whereas for the second it is an *integral* (over k). What is the physical significance of this distinction?

In *classical* mechanics a one-dimensional time-independent potential can give rise to two rather different kinds of motion. If $V(x)$ rises higher than the particle's total energy (E) on either side (Figure 2.11(a)), then the particle is "stuck" in the potential well—it rocks back

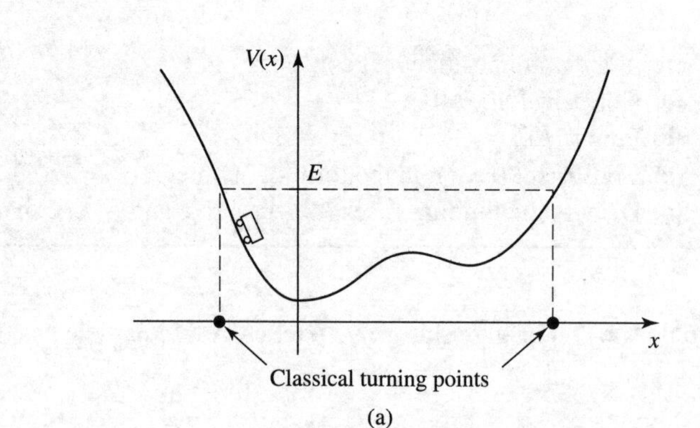

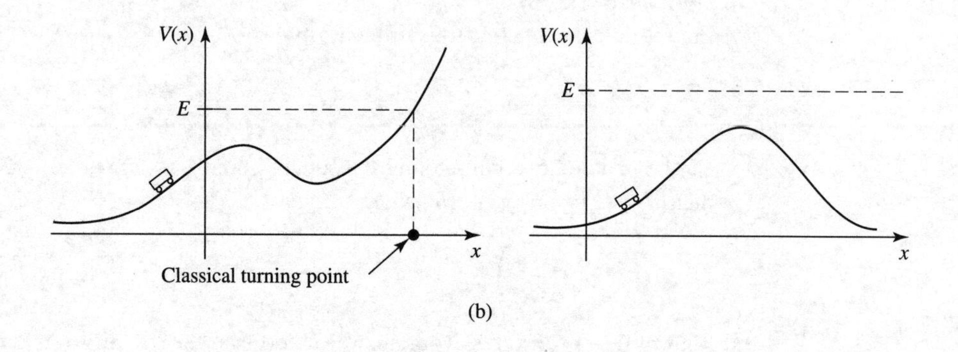

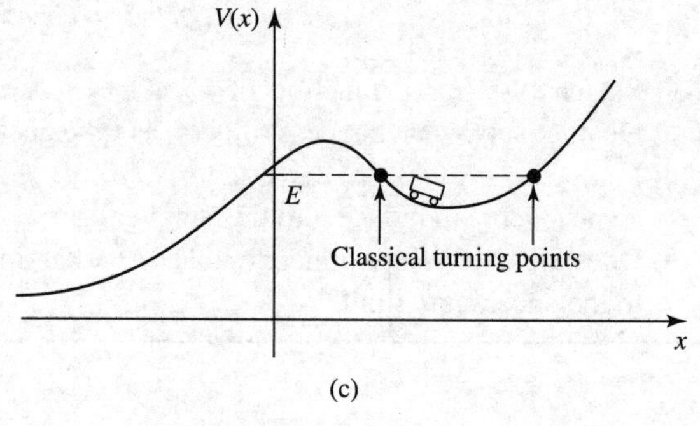

Figure 2.11: (a) A bound state. (b) Scattering states. (c) A *classical* bound state, but a quantum scattering state.

and forth between the **turning points**, but it cannot escape (unless, of course, you provide it with a source of extra energy, such as a motor, but we're not talking about that). We call this a **bound state**. If, on the other hand, E exceeds $V(x)$ on one side (or both), then the particle comes in from "infinity," slows down or speeds up under the influence of the potential, and returns to infinity (Figure 2.11(b)). (It can't get trapped in the potential unless there is some mechanism, such as friction, to *dissipate* energy, but again, we're not talking about that.) We call this a **scattering state**. Some potentials admit only bound states (for instance, the harmonic oscillator); some allow only scattering states (a potential hill with no dips in it, for example); some permit both kinds, depending on the energy of the particle.

The two kinds of solutions to the Schrödinger equation correspond precisely to bound and scattering states. The distinction is even cleaner in the quantum domain, because the phenomenon of **tunneling** (which we'll come to shortly) allows the particle to "leak" through any finite potential barrier, so the only thing that matters is the potential at infinity (Figure 2.11(c)):

$$\begin{cases} E < V(-\infty) \text{ and } V(+\infty) \Rightarrow & \text{bound state,} \\ E > V(-\infty) \text{ or } V(+\infty) \Rightarrow & \text{scattering state.} \end{cases} \tag{2.112}$$

In real life most potentials go to *zero* at infinity, in which case the criterion simplifies even further:

$$\begin{cases} E < 0 \Rightarrow & \text{bound state,} \\ E > 0 \Rightarrow & \text{scattering state.} \end{cases} \tag{2.113}$$

Because the infinite square well and harmonic oscillator potentials go to infinity as $x \to \pm\infty$, they admit bound states only; because the free particle potential is zero everywhere, it only allows scattering states.[43] In this section (and the following one) we shall explore potentials that support both kinds of states.

2.5.2 The Delta-Function Well

The **Dirac delta function** is an infinitely high, infinitesimally narrow spike at the origin, whose *area* is 1 (Figure 2.12):

$$\delta(x) \equiv \left\{ \begin{array}{ll} 0, & \text{if } x \neq 0 \\ \infty, & \text{if } x = 0 \end{array} \right\}, \text{ with } \int_{-\infty}^{+\infty} \delta(x)\, dx = 1. \tag{2.114}$$

Technically, it isn't a function at all, since it is not finite at $x = 0$ (mathematicians call it a **generalized function**, or **distribution**).[44] Nevertheless, it is an extremely useful construct in theoretical physics. (For example, in electrodynamics the charge *density* of a point charge is a

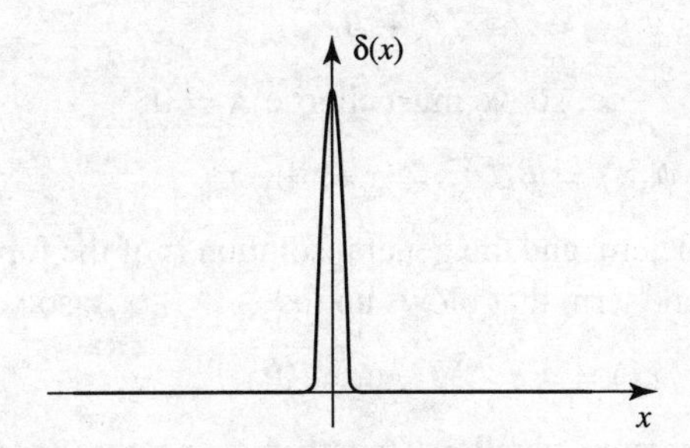

Figure 2.12: The Dirac delta function (Equation 2.114).

[43] If you are irritatingly observant, you may have noticed that the general theorem requiring $E > V_{\text{min}}$ (Problem 2.2) does not really apply to scattering states, since they are not normalizable. If this bothers you, try solving the Schrödinger equation with $E \leq 0$, for the free particle, and note that *even linear combinations* of these solutions cannot be normalized. The positive energy solutions by themselves constitute a complete set.

[44] The delta function can be thought of as the *limit* of a *sequence* of functions, such as rectangles (or triangles) of ever-increasing height and ever-decreasing width.

delta function.) Notice that $\delta(x-a)$ would be a spike of area 1 at the point a. If you multiply $\delta(x-a)$ by an *ordinary* function $f(x)$, it's the same as multiplying by $f(a)$,

$$f(x)\,\delta(x-a) = f(a)\,\delta(x-a), \tag{2.115}$$

because the product is zero anyway except at the point a. In particular,

$$\int_{-\infty}^{+\infty} f(x)\,\delta(x-a)\,dx = f(a)\int_{-\infty}^{+\infty} \delta(x-a)\,dx = f(a)\,. \tag{2.116}$$

That's the most important property of the delta function: Under the integral sign it serves to "pick out" the value of $f(x)$ at the point a. (Of course, the integral need not go from $-\infty$ to $+\infty$; all that matters is that the domain of integration include the point a, so $a-\epsilon$ to $a+\epsilon$ would do, for any $\epsilon > 0$.)

Let's consider a potential of the form

$$V(x) = -\alpha\delta(x)\,, \tag{2.117}$$

where α is some positive constant.[45] This is an artificial potential, to be sure (so was the infinite square well), but it's delightfully simple to work with, and illuminates the basic theory with a minimum of analytical clutter. The Schrödinger equation for the delta-function well reads

$$-\frac{\hbar^2}{2m}\frac{d^2\psi}{dx^2} - \alpha\delta(x)\,\psi = E\psi; \tag{2.118}$$

it yields both bound states ($E < 0$) and scattering states ($E > 0$).

We'll look first at the bound states. In the region $x < 0$, $V(x) = 0$, so

$$\frac{d^2\psi}{dx^2} = -\frac{2mE}{\hbar^2}\psi = \kappa^2\psi, \tag{2.119}$$

where

$$\kappa \equiv \frac{\sqrt{-2mE}}{\hbar}. \tag{2.120}$$

(E is negative, by assumption, so κ is real and positive.) The general solution to Equation 2.119 is

$$\psi(x) = Ae^{-\kappa x} + Be^{\kappa x}, \tag{2.121}$$

but the first term blows up as $x \to -\infty$, so we must choose $A = 0$:

$$\psi(x) = Be^{\kappa x}, \quad (x < 0)\,. \tag{2.122}$$

In the region $x > 0$, $V(x)$ is again zero, and the general solution is of the form $F\exp(-\kappa x) + G\exp(\kappa x)$; this time it's the second term that blows up (as $x \to +\infty$), so

$$\psi(x) = Fe^{-\kappa x}, \quad (x > 0)\,. \tag{2.123}$$

It remains only to stitch these two functions together, using the appropriate boundary conditions at $x = 0$. I quoted earlier the standard boundary conditions for ψ:

$$\begin{cases} 1.\ \psi & \text{is always continuous;} \\ 2.\ d\psi/dx & \text{is continuous except at points where the potential is infinite.} \end{cases} \tag{2.124}$$

[45] The delta function itself carries units of $1/\textit{length}$ (see Equation 2.114), so α has the dimensions *energy*×*length*.

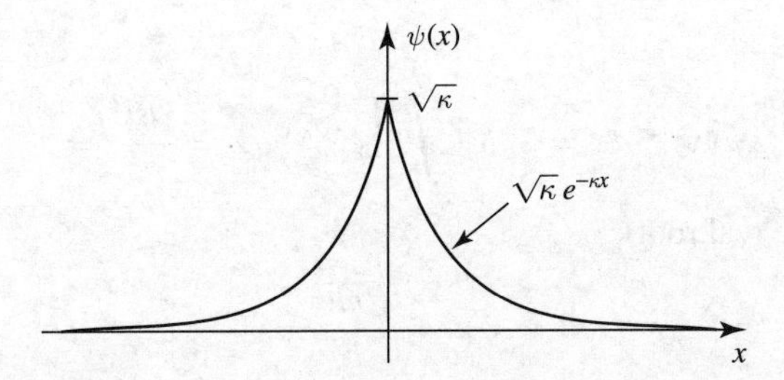

Figure 2.13: Bound state wave function for the delta-function potential (Equation 2.125).

In this case the first boundary condition tells us that $F = B$, so

$$\psi(x) = \begin{cases} Be^{\kappa x}, & (x \leq 0), \\ Be^{-\kappa x}, & (x \geq 0); \end{cases} \tag{2.125}$$

$\psi(x)$ is plotted in Figure 2.13. The second boundary condition tells us nothing; this is (like the walls of the infinite square well) the exceptional case where V is infinite at the join, and it's clear from the graph that this function has a kink at $x = 0$. Moreover, up to this point the delta function has not come into the story at all. It turns out that the delta function determines the *discontinuity in the derivative* of ψ, at $x = 0$. I'll show you now how this works, and as a byproduct we'll see why $d\psi/dx$ is ordinarily continuous.

The idea is to *integrate* the Schrödinger equation, from $-\epsilon$ to $+\epsilon$, and then take the limit as $\epsilon \to 0$:

$$-\frac{\hbar^2}{2m}\int_{-\epsilon}^{+\epsilon}\frac{d^2\psi}{dx^2}\,dx + \int_{-\epsilon}^{+\epsilon} V(x)\,\psi(x)\,dx = E\int_{-\epsilon}^{+\epsilon}\psi(x)\,dx \tag{2.126}$$

The first integral is nothing but $d\psi/dx$, evaluated at the two end points; the last integral is *zero*, in the limit $\epsilon \to 0$, since it's the area of a sliver with vanishing width and finite height. Thus

$$\Delta\left(\frac{d\psi}{dx}\right) \equiv \lim_{\epsilon\to 0}\left(\left.\frac{\partial\psi}{\partial x}\right|_{+\epsilon} - \left.\frac{\partial\psi}{\partial x}\right|_{-\epsilon}\right) = \frac{2m}{\hbar^2}\lim_{\epsilon\to 0}\int_{-\epsilon}^{+\epsilon} V(x)\,\psi(x)\,dx. \tag{2.127}$$

Ordinarily, the limit on the right is again zero, and that's why $d\psi/dx$ is ordinarily continuous. But when $V(x)$ is *infinite* at the boundary, this argument fails. In particular, if $V(x) = -\alpha\delta(x)$, Equation 2.116 yields

$$\Delta\left(\frac{d\psi}{dx}\right) = -\frac{2m\alpha}{\hbar^2}\psi(0). \tag{2.128}$$

For the case at hand (Equation 2.125),

$$\begin{cases} d\psi/dx = -B\kappa e^{-\kappa x}, \text{ for } (x > 0), & \text{so } d\psi/dx\big|_{+} = -B\kappa, \\ d\psi/dx = +B\kappa e^{+\kappa x}, \text{ for } (x < 0), & \text{so } d\psi/dx\big|_{-} = +B\kappa, \end{cases}$$

and hence $\Delta\,(d\psi/dx) = -2B\kappa$. And $\psi(0) = B$. So Equation 2.128 says

$$\kappa = \frac{m\alpha}{\hbar^2}, \tag{2.129}$$

and the allowed energy (Equation 2.120) is

$$E = -\frac{\hbar^2\kappa^2}{2m} = -\frac{m\alpha^2}{2\hbar^2}. \tag{2.130}$$

Finally, we normalize ψ:

$$\int_{-\infty}^{+\infty} |\psi(x)|^2 \, dx = 2\,|B|^2 \int_0^{\infty} e^{-2\kappa x} \, dx = \frac{|B|^2}{\kappa} = 1,$$

so (choosing the positive real root):

$$B = \sqrt{\kappa} = \frac{\sqrt{m\alpha}}{\hbar}. \tag{2.131}$$

Evidently the delta function well, regardless of its "strength" α, has *exactly one* bound state:

$$\boxed{\psi(x) = \frac{\sqrt{m\alpha}}{\hbar} e^{-m\alpha|x|/\hbar^2}; \quad E = -\frac{m\alpha^2}{2\hbar^2}.} \tag{2.132}$$

What about *scattering* states, with $E > 0$? For $x < 0$ the Schrödinger equation reads

$$\frac{d^2\psi}{dx^2} = -\frac{2mE}{\hbar^2}\psi = -k^2\psi,$$

where

$$k \equiv \frac{\sqrt{2mE}}{\hbar} \tag{2.133}$$

is real and positive. The general solution is

$$\psi(x) = Ae^{ikx} + Be^{-ikx}, \tag{2.134}$$

and this time we cannot rule out either term, since neither of them blows up. Similarly, for $x > 0$,

$$\psi(x) = Fe^{ikx} + Ge^{-ikx}. \tag{2.135}$$

The continuity of $\psi(x)$ at $x = 0$ requires that

$$F + G = A + B. \tag{2.136}$$

The derivatives are

$$\begin{cases} d\psi/dx = ik\left(Fe^{ikx} - Ge^{-ikx}\right), \text{ for } (x > 0), & \text{so } d\psi/dx\big|_{+} = ik\,(F - G), \\ d\psi/dx = ik\left(Ae^{ikx} - Be^{-ikx}\right), \text{ for } (x < 0), & \text{so } d\psi/dx\big|_{-} = ik\,(A - B), \end{cases}$$

and hence $\Delta\,(d\psi/dx) = ik\,(F - G - A + B)$. Meanwhile, $\psi(0) = (A + B)$, so the second boundary condition (Equation 2.128) says

$$ik\,(F - G - A + B) = -\frac{2m\alpha}{\hbar^2}\,(A + B), \tag{2.137}$$

or, more compactly,

$$F - G = A\,(1 + 2i\beta) - B\,(1 - 2i\beta), \quad \text{where } \beta \equiv \frac{m\alpha}{\hbar^2 k}. \tag{2.138}$$

Having imposed both boundary conditions, we are left with two equations (Equations 2.136 and 2.138) in four unknowns (A, B, F, and G)—*five*, if you count k. Normalization won't help—this isn't a normalizable state. Perhaps we'd better pause, then, and examine the physical significance of these various constants. Recall that $\exp(ikx)$ gives rise (when coupled with the wiggle factor $\exp(-iEt/\hbar)$) to a wave function propagating to the *right*, and $\exp(-ikx)$ leads to a wave propagating to the *left*. It follows that A (in Equation 2.134) is the amplitude of a

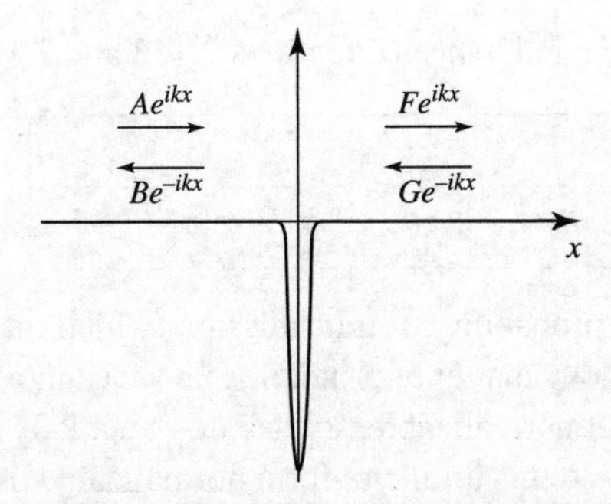

Figure 2.14: Scattering from a delta function well.

wave coming in from the left, B is the amplitude of a wave returning to the left; F (Equation 2.135) is the amplitude of a wave traveling off to the right, and G is the amplitude of a wave coming in from the right (see Figure 2.14). In a typical scattering experiment particles are fired in from one direction—let's say, from the left. In that case the amplitude of the wave coming in from the *right* will be *zero*:

$$G = 0 \quad \text{(for scattering from the left)}\,; \tag{2.139}$$

A is the amplitude of the **incident wave**, B is the amplitude of the **reflected wave**, and F is the amplitude of the **transmitted wave**. Solving Equations 2.136 and 2.138 for B and F, we find

$$B = \frac{i\beta}{1 - i\beta}A, \quad F = \frac{1}{1 - i\beta}A. \tag{2.140}$$

(If you want to study scattering from the *right*, set $A = 0$; then G is the incident amplitude, F is the reflected amplitude, and B is the transmitted amplitude.)

Now, the probability of finding the particle at a specified location is given by $|\Psi|^2$, so the *relative*[46] probability that an incident particle will be reflected back is

$$R \equiv \frac{|B|^2}{|A|^2} = \frac{\beta^2}{1 + \beta^2}. \tag{2.141}$$

R is called the **reflection coefficient**. (If you have a *beam* of particles, it tells you the *fraction* of the incoming number that will bounce back.) Meanwhile, the probability that a particle will continue right on through is the **transmission coefficient**[47]

$$T \equiv \frac{|F|^2}{|A|^2} = \frac{1}{1 + \beta^2}. \tag{2.142}$$

Of course, the *sum* of these probabilities should be 1—and it *is*:

$$R + T = 1. \tag{2.143}$$

[46] This is not a normalizable wave function, so the *absolute* probability of finding the particle at a particular location is not well defined; nevertheless, the *ratio* of probabilities for the incident and reflected waves *is* meaningful. More on this in the next paragraph.

[47] Note that the particle's velocity is the same on both sides of the well. Problem 2.34 treats the general case.

Notice that R and T are functions of β, and hence (Equations 2.133 and 2.138) of E:

$$R = \frac{1}{1 + \left(2\hbar^2 E/m\alpha^2\right)}, \quad T = \frac{1}{1 + \left(m\alpha^2/2\hbar^2 E\right)}. \tag{2.144}$$

The higher the energy, the greater the probability of transmission (which makes sense).

This is all very tidy, but there is a sticky matter of principle that we cannot altogether ignore: These scattering wave functions are not normalizable, so they don't actually represent possible particle states. We know the resolution to this problem: form normalizable linear combinations of the stationary states, just as we did for the free particle—true physical particles are represented by the resulting wave packets. Though straightforward in principle, this is a messy business in practice, and at this point it is best to turn the problem over to a computer.[48] Meanwhile, since it is impossible to create a normalizable free-particle wave function without involving a *range* of energies, R and T should be interpreted as the *approximate* reflection and transmission probabilities for particles with energies in the *vicinity* of E.

Incidentally, it might strike you as peculiar that we were able to analyze a quintessentially time-dependent problem (particle comes in, scatters off a potential, and flies off to infinity) using *stationary* states. After all, ψ (in Equations 2.134 and 2.135) is simply a complex, time-independent, sinusoidal function, extending (with constant amplitude) to infinity in both directions. And yet, by imposing appropriate boundary conditions on this function we were able to determine the probability that a particle (represented by a *localized* wave packet) would bounce off, or pass through, the potential. The mathematical miracle behind this is, I suppose, the fact that by taking linear combinations of states spread over all space, and with essentially trivial time dependence, we can *construct* wave functions that are concentrated about a (moving) point, with quite elaborate behavior in time (see Problem 2.42).

As long as we've got the relevant equations on the table, let's look briefly at the case of a delta-function *barrier* (Figure 2.15). Formally, all we have to do is change the sign of α. This kills the bound state, of course (Problem 2.2). On the other hand, the reflection and transmission coefficients, which depend only on α^2, are unchanged. Strange to say, the particle is just as likely to pass through the barrier as to cross over the well! *Classically*, of course, a particle cannot make it over an infinitely high barrier, regardless of its energy. In fact, classical scattering problems are pretty dull: If $E > V_{max}$, then $T = 1$ and $R = 0$—the particle

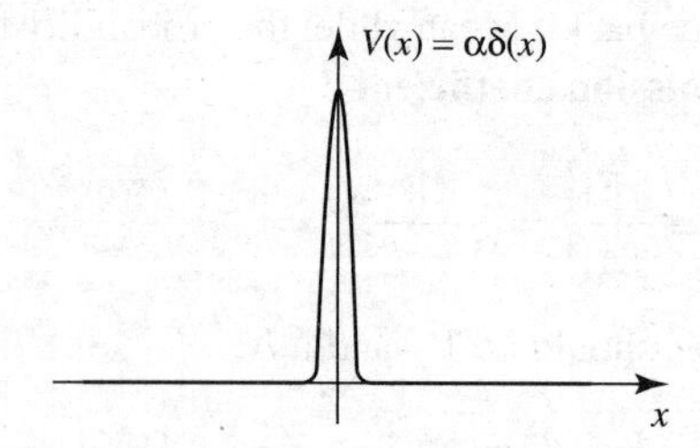

Figure 2.15: The delta-function barrier.

[48] There exist some powerful programs for analyzing the scattering of a wave packet from a one-dimensional potential; see, for instance, "Quantum Tunneling and Wave Packets," at PhET Interactive Simulations, University of Colorado Boulder, https://phet.colorado.edu.

certainly makes it over; if $E < V_{max}$ then $T = 0$ and $R = 1$—it rides up the hill until it runs out of steam, and then returns the same way it came. *Quantum* scattering problems are much richer: The particle has some nonzero probability of passing through the potential even if $E < V_{max}$. We call this phenomenon **tunneling**; it is the mechanism that makes possible much of modern electronics—not to mention spectacular advances in microscopy. Conversely, even if $E > V_{max}$ there is a possibility that the particle will bounce back—though I wouldn't advise driving off a cliff in the hope that quantum mechanics will save you (see Problem 2.35).

* **Problem 2.22** Evaluate the following integrals:

(a) $\int_{-3}^{+1} \left(x^3 - 3x^2 + 2x - 1\right) \delta(x+2)\, dx$.

(b) $\int_0^{\infty} [\cos(3x) + 2]\, \delta(x - \pi)\, dx$.

(c) $\int_{-1}^{+1} \exp(|x| + 3) \delta(x - 2)\, dx$.

* **Problem 2.23** Delta functions live under integral signs, and two expressions ($D_1(x)$ and $D_2(x)$) involving delta functions are said to be equal if

$$\int_{-\infty}^{+\infty} f(x)\, D_1(x)\, dx = \int_{-\infty}^{+\infty} f(x)\, D_2(x)\, dx,$$

for every (ordinary) function $f(x)$.

(a) Show that

$$\delta(cx) = \frac{1}{|c|}\delta(x), \tag{2.145}$$

where c is a real constant. (Be sure to check the case where c is negative.)

(b) Let $\theta(x)$ be the **step function**:

$$\theta(x) \equiv \begin{cases} 1, & x > 0, \\ 0, & x < 0. \end{cases} \tag{2.146}$$

(In the rare case where it actually matters, we define $\theta(0)$ to be 1/2.) Show that $d\theta/dx = \delta(x)$.

** **Problem 2.24** Check the uncertainty principle for the wave function in Equation 2.132. *Hint:* Calculating $\langle p^2 \rangle$ can be tricky, because the derivative of ψ has a step discontinuity at $x = 0$. You may want to use the result in Problem 2.23(b). *Partial answer:* $\langle p^2 \rangle = (m\alpha/\hbar)^2$.

Problem 2.25 Check that the bound state of the delta-function well (Equation 2.132) is orthogonal to the scattering states (Equations 2.134 and 2.135).

* **Problem 2.26** What is the Fourier transform of $\delta(x)$? Using Plancherel's theorem, show that

$$\delta(x) = \frac{1}{2\pi}\int_{-\infty}^{+\infty} e^{ikx}\,dk. \tag{2.147}$$

Comment: This formula gives any respectable mathematician apoplexy. Although the integral is clearly infinite when $x = 0$, it doesn't converge (to zero or anything else) when $x \neq 0$, since the integrand oscillates forever. There are ways to patch it up (for instance, you can integrate from $-L$ to $+L$, and interpret Equation 2.147 to mean the *average* value of the finite integral, as $L \to \infty$). The source of the problem is that the delta function doesn't meet the requirement (square-integrability) for Plancherel's theorem (see footnote 42). In spite of this, Equation 2.147 can be extremely useful, if handled with care.

** **Problem 2.27** Consider the *double* delta-function potential

$$V(x) = -\alpha\left[\delta(x+a) + \delta(x-a)\right],$$

where α and a are positive constants.

(a) Sketch this potential.

(b) How many bound states does it possess? Find the allowed energies, for $\alpha = \hbar^2/ma$ and for $\alpha = \hbar^2/4ma$, and sketch the wave functions.

(c) What are the bound state energies in the limiting cases (i) $a \to 0$ and (ii) $a \to \infty$ (holding α fixed)? Explain why your answers are reasonable, by comparison with the single delta-function well.

** **Problem 2.28** Find the transmission coefficient, for the potential in Problem 2.27.

2.6 THE FINITE SQUARE WELL

As a last example, consider the *finite* square well

$$V(x) = \begin{cases} -V_0, & -a \le x \le a, \\ 0, & |x| > a, \end{cases} \tag{2.148}$$

where V_0 is a (positive) constant (Figure 2.16). Like the delta-function well, this potential admits both bound states (with $E < 0$) and scattering states (with $E > 0$). We'll look first at the bound states.

In the region $x < -a$ the potential is zero, so the Schrödinger equation reads

$$-\frac{\hbar^2}{2m}\frac{d^2\psi}{dx^2} = E\psi, \quad \text{or} \quad \frac{d^2\psi}{dx^2} = \kappa^2\psi,$$

where

$$\kappa \equiv \frac{\sqrt{-2mE}}{\hbar} \tag{2.149}$$

is real and positive. The general solution is $\psi(x) = A\exp(-\kappa x) + B\exp(\kappa x)$, but the first term blows up (as $x \to -\infty$), so the physically admissible solution is

$$\psi(x) = Be^{\kappa x}, \quad (x < -a). \tag{2.150}$$

In the region $-a < x < a$, $V(x) = -V_0$, and the Schrödinger equation reads

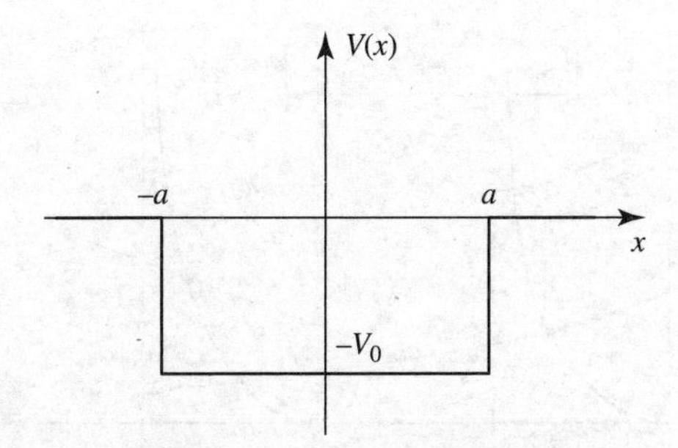

Figure 2.16: The finite square well (Equation 2.148).

$$-\frac{\hbar^2}{2m}\frac{d^2\psi}{dx^2} - V_0\psi = E\psi, \quad \text{or} \quad \frac{d^2\psi}{dx^2} = -l^2\psi,$$

where

$$l \equiv \frac{\sqrt{2m\,(E+V_0)}}{\hbar}. \tag{2.151}$$

Although E is negative, for bound states, it must be greater than $-V_0$, by the old theorem $E > V_{\text{min}}$ (Problem 2.2); so l is also real and positive. The general solution is[49]

$$\psi(x) = C\sin(lx) + D\cos(lx)\,, \quad (-a < x < a)\,, \tag{2.152}$$

where C and D are arbitrary constants. Finally, in the region $x > a$ the potential is again zero; the general solution is $\psi(x) = F\exp(-\kappa x) + G\exp(\kappa x)$, but the second term blows up (as $x \to \infty$), so we are left with

$$\psi(x) = Fe^{-\kappa x}, \quad (x > a)\,. \tag{2.153}$$

The next step is to impose boundary conditions: ψ and $d\psi/dx$ continuous at $-a$ and $+a$. But we can save a little time by noting that this potential is an even function, so we can assume with no loss of generality that the solutions are either even or odd (Problem 2.1(c)). The advantage of this is that we need only impose the boundary conditions on one side (say, at $+a$); the other side is then automatic, since $\psi(-x) = \pm\psi(x)$. I'll work out the even solutions; you get to do the odd ones in Problem 2.29. The cosine is even (and the sine is odd), so I'm looking for solutions of the form

$$\psi(x) = \begin{cases} Fe^{-\kappa x}, & (x > a)\,, \\ D\cos(lx)\,, & (0 < x < a)\,, \\ \psi(-x)\,, & (x < 0)\,. \end{cases} \tag{2.154}$$

The continuity of $\psi(x)$, at $x = a$, says

$$Fe^{-\kappa a} = D\cos(la)\,, \tag{2.155}$$

and the continuity of $d\psi/dx$ says

$$-\kappa Fe^{-\kappa a} = -lD\sin(la)\,. \tag{2.156}$$

Dividing Equation 2.156 by Equation 2.155, we find that

[49] You can, if you like, write the general solution in exponential form $(C'e^{ilx} + D'e^{-ilx})$. This leads to the same final result, but since the potential is symmetric, we know the solutions will be either even or odd, and the sine/cosine notation allows us to exploit this right from the start.

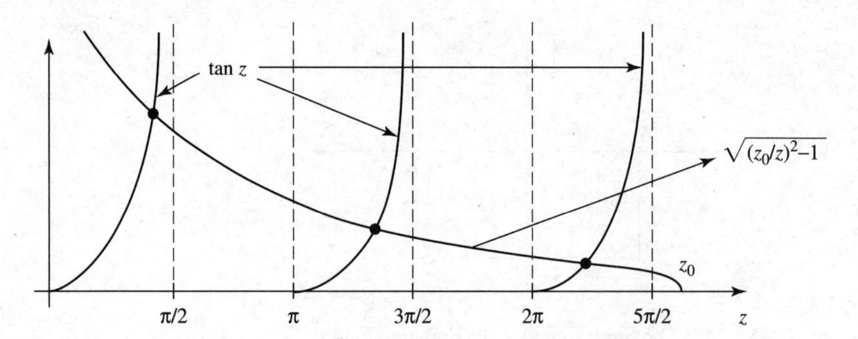

Figure 2.17: Graphical solution to Equation 2.159, for $z_0 = 8$ (*even* states).

$$\kappa = l\tan(la). \tag{2.157}$$

This is a formula for the allowed energies, since κ and l are both functions of E. To solve for E, we first adopt some nicer notation: Let

$$z \equiv la, \quad \text{and} \quad z_0 \equiv \frac{a}{\hbar}\sqrt{2mV_0}. \tag{2.158}$$

According to Equations 2.149 and 2.151, $\left(\kappa^2 + l^2\right) = 2mV_0/\hbar^2$, so $\kappa a = \sqrt{z_0^2 - z^2}$, and Equation 2.157 reads

$$\tan z = \sqrt{(z_0/z)^2 - 1}. \tag{2.159}$$

This is a transcendental equation for z (and hence for E) as a function of z_0 (which is a measure of the "size" of the well). It can be solved numerically, using a computer, or graphically, by plotting $\tan z$ and $\sqrt{(z_0/z)^2 - 1}$ on the same grid, and looking for points of intersection (see Figure 2.17). Two limiting cases are of special interest:

1. **Wide, deep well.** If z_0 is very large (pushing the curve $\sqrt{(z_0/z)^2 - 1}$ upward on the graph, and sliding the zero crossing, z_0, to the right) the intersections occur just slightly below $z_n = n\pi/2$, with n odd; it follows (Equations 2.158 and 2.151) that

$$E_n + V_0 \approx \frac{n^2\pi^2\hbar^2}{2m(2a)^2} \quad (n = 1, 3, 5, \ldots). \tag{2.160}$$

   But $E + V_0$ is the energy *above the bottom of the well*, and on the right side we have precisely the infinite square well energies, for a well of width $2a$ (see Equation 2.30)—or rather, *half* of them, since this n is odd. (The other ones, of course, come from the *odd* wave functions, as you'll discover in Problem 2.29.) So the finite square well goes over to the infinite square well, as $V_0 \to \infty$; however, for any *finite* V_0 there are only a finite number of bound states.

2. **Shallow, narrow well.** As z_0 decreases, there are fewer and fewer bound states, until finally, for $z_0 < \pi/2$, only one remains. It is interesting to note, however, that there is always *one* bound state, no matter *how* "weak" the well becomes.

You're welcome to normalize ψ (Equation 2.154), if you're interested (Problem 2.30), but I'm going to move on now to the scattering states ($E > 0$). To the left, where $V(x) = 0$, we have

$$\psi(x) = Ae^{ikx} + Be^{-ikx}, \quad \text{for } (x < -a), \tag{2.161}$$

where (as usual)

$$k \equiv \frac{\sqrt{2mE}}{\hbar}. \tag{2.162}$$

Inside the well, where $V(x) = -V_0$,

$$\psi(x) = C\sin(lx) + D\cos(lx), \quad \text{for } (-a < x < a), \tag{2.163}$$

where, as before,

$$l \equiv \frac{\sqrt{2m(E+V_0)}}{\hbar}. \tag{2.164}$$

To the right, assuming there is no incoming wave in this region, we have

$$\psi(x) = Fe^{ikx}. \tag{2.165}$$

Here A is the incident amplitude, B is the reflected amplitude, and F is the transmitted amplitude.[50]

There are four boundary conditions: Continuity of $\psi(x)$ at $-a$ says

$$Ae^{-ika} + Be^{ika} = -C\sin(la) + D\cos(la), \tag{2.166}$$

continuity of $d\psi/dx$ at $-a$ gives

$$ik\left[Ae^{-ika} - Be^{ika}\right] = l\left[C\cos(la) + D\sin(la)\right] \tag{2.167}$$

continuity of $\psi(x)$ at $+a$ yields

$$C\sin(la) + D\cos(la) = Fe^{ika}, \tag{2.168}$$

and continuity of $d\psi/dx$ at $+a$ requires

$$l\left[C\cos(la) - D\sin(la)\right] = ikFe^{ika}. \tag{2.169}$$

We can use two of these to eliminate C and D, and solve the remaining two for B and F (see Problem 2.32):

$$B = i\frac{\sin(2la)}{2kl}\left(l^2 - k^2\right)F, \tag{2.170}$$

$$F = \frac{e^{-2ika}A}{\cos(2la) - i\frac{(k^2+l^2)}{2kl}\sin(2la)}. \tag{2.171}$$

The transmission coefficient $\left(T = |F|^2/|A|^2\right)$, expressed in terms of the original variables, is given by

$$T^{-1} = 1 + \frac{V_0^2}{4E(E+V_0)}\sin^2\left(\frac{2a}{\hbar}\sqrt{2m(E+V_0)}\right). \tag{2.172}$$

Notice that $T = 1$ (the well becomes "transparent") whenever the sine is zero, which is to say, when

[50] We *could* look for even and odd functions, as we did in the case of bound states, but the scattering problem is inherently asymmetric, since the waves come in from one side only, and the exponential notation (representing traveling waves) is more natural in this context.

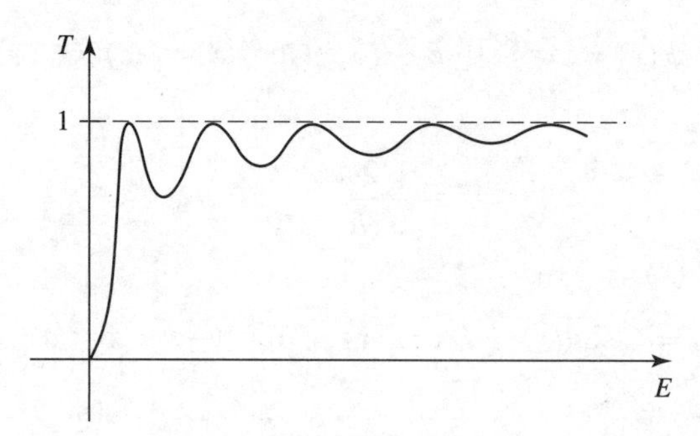

Figure 2.18: Transmission coefficient as a function of energy (Equation 2.172).

$$\frac{2a}{\hbar}\sqrt{2m\left(E_n+V_0\right)}=n\pi, \tag{2.173}$$

where n is any integer. The energies for perfect transmission, then, are given by

$$E_n+V_0=\frac{n^2\pi^2\hbar^2}{2m\left(2a\right)^2}, \tag{2.174}$$

which happen to be precisely the allowed energies for the *infinite* square well. T is plotted in Figure 2.18, as a function of energy.[51]

* **Problem 2.29** Analyze the *odd* bound state wave functions for the finite square well. Derive the transcendental equation for the allowed energies, and solve it graphically. Examine the two limiting cases. Is there always an odd bound state?

Problem 2.30 Normalize $\psi(x)$ in Equation 2.154, to determine the constants D and F.

Problem 2.31 The Dirac delta function can be thought of as the limiting case of a rectangle of area 1, as the height goes to infinity and the width goes to zero. Show that the delta-function well (Equation 2.117) is a "weak" potential (even though it is infinitely deep), in the sense that $z_0 \to 0$. Determine the bound state energy for the delta-function potential, by treating it as the limit of a finite square well. Check that your answer is consistent with Equation 2.132. Also show that Equation 2.172 reduces to Equation 2.144 in the appropriate limit.

Problem 2.32 Derive Equations 2.170 and 2.171. *Hint:* Use Equations 2.168 and 2.169 to solve for C and D in terms of F:

$$C=\left[\sin(la)+i\frac{k}{l}\cos(la)\right]e^{ika}F;\quad D=\left[\cos(la)-i\frac{k}{l}\sin(la)\right]e^{ika}F.$$

Plug these back into Equations 2.166 and 2.167. Obtain the transmission coefficient, and confirm Equation 2.172.

[51] This remarkable phenomenon was observed in the laboratory before the advent of quantum mechanics, in the form of the **Ramsauer–Townsend effect**. For an illuminating discussion see Richard W. Robinett, *Quantum Mechanics*, Oxford University Press, 1997, Section 12.4.1.

** **Problem 2.33** Determine the transmission coefficient for a rectangular *barrier* (same as Equation 2.148, only with $V(x) = +V_0 > 0$ in the region $-a < x < a$). Treat separately the three cases $E < V_0$, $E = V_0$, and $E > V_0$ (note that the wave function inside the barrier is different in the three cases). *Partial answer:* for $E < V_0$,[52]

$$T^{-1} = 1 + \frac{V_0^2}{4E(V_0 - E)} \sinh^2\left(\frac{2a}{\hbar}\sqrt{2m(V_0 - E)}\right).$$

* **Problem 2.34** Consider the "step" potential:[53]

$$V(x) = \begin{cases} 0, & x \leq 0, \\ V_0, & x > 0. \end{cases}$$

(a) Calculate the reflection coefficient, for the case $E < V_0$, and comment on the answer.

(b) Calculate the reflection coefficient for the case $E > V_0$.

(c) For a potential (such as this one) that does not go back to zero to the right of the barrier, the transmission coefficient is *not* simply $|F|^2 / |A|^2$ (with A the incident amplitude and F the transmitted amplitude), because the transmitted wave travels at a different *speed*. Show that

$$T = \sqrt{\frac{E - V_0}{E}} \frac{|F|^2}{|A|^2}, \tag{2.175}$$

for $E > V_0$. *Hint:* You can figure it out using Equation 2.99, or—more elegantly, but less informatively—from the probability current (Problem 2.18). What is T, for $E < V_0$?

(d) For $E > V_0$, calculate the transmission coefficient for the step potential, and check that $T + R = 1$.

Problem 2.35 A particle of mass m and kinetic energy $E > 0$ approaches an abrupt potential drop V_0 (Figure 2.19).[54]

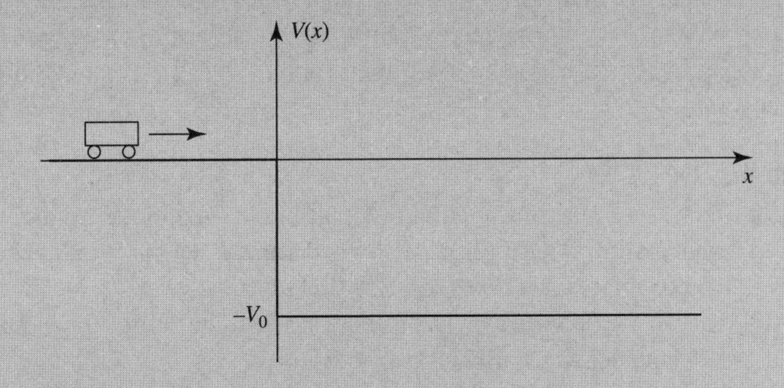

Figure 2.19: Scattering from a "cliff" (Problem 2.35).

[52] This is a good example of tunneling—*classically* the particle would bounce back.

[53] For interesting commentary see C. O. Dib and O. Orellana, *Eur. J. Phys.* **38**, 045403 (2017).

[54] For further discussion see P. L. Garrido, et al., *Am. J. Phys.* **79**, 1218 (2011).

(a) What is the probability that it will "reflect" back, if $E = V_0/3$? *Hint:* This is just like Problem 2.34, except that the step now goes *down*, instead of up.

(b) I drew the figure so as to make you think of a car approaching a cliff, but obviously the probability of "bouncing back" from the edge of a cliff is *far* smaller than what you got in (a)—unless you're Bugs Bunny. Explain why this potential does *not* correctly represent a cliff. *Hint:* In Figure 2.20 the potential energy of the car drops *discontinuously* to $-V_0$, as it passes $x = 0$; would this be true for a falling car?

(c) When a free neutron enters a nucleus, it experiences a sudden drop in potential energy, from $V = 0$ outside to around -12 MeV (million electron volts) inside. Suppose a neutron, emitted with kinetic energy 4 MeV by a fission event, strikes such a nucleus. What is the probability it will be absorbed, thereby initiating another fission? *Hint:* You calculated the probability of *reflection* in part (a); use $T = 1 - R$ to get the probability of transmission through the surface.

FURTHER PROBLEMS ON CHAPTER 2

Problem 2.36 Solve the time-independent Schrödinger equation with appropriate boundary conditions for the "centered" infinite square well: $V(x) = 0$ (for $-a < x < +a$), $V(x) = \infty$ (otherwise). Check that your allowed energies are consistent with mine (Equation 2.30), and confirm that your ψs can be obtained from mine (Equation 2.31) by the substitution $x \to (x + a)/2$ (and appropriate renormalization). Sketch your first three solutions, and compare Figure 2.2. Note that the width of the well is now $2a$.

Problem 2.37 A particle in the infinite square well (Equation 2.22) has the initial wave function

$$\Psi(x, 0) = A \sin^3(\pi x/a) \quad (0 \le x \le a).$$

Determine A, find $\Psi(x, t)$, and calculate $\langle x \rangle$, as a function of time. What is the expectation value of the energy? *Hint:* $\sin^n \theta$ and $\cos^n \theta$ can be reduced, by repeated application of the trigonometric sum formulas, to linear combinations of $\sin(m\theta)$ and $\cos(m\theta)$, with $m = 0, 1, 2, \ldots, n$.

Problem 2.38

(a) Show that the wave function of a particle in the infinite square well returns to its original form after a quantum **revival time** $T = 4ma^2/\pi\hbar$. That is: $\Psi(x, T) = \Psi(x, 0)$ for any state (*not* just a stationary state).

(b) What is the *classical* revival time, for a particle of energy E bouncing back and forth between the walls?

(c) For what energy are the two revival times equal?[55]

[55] The fact that the classical and quantum revival times bear no obvious relation to one another (and the quantum one doesn't even depend on the energy) is a curious paradox; see D. F. Styer, *Am. J. Phys.* **69**, 56 (2001).

Problem 2.39 In Problem 2.7(d) you got the expectation value of the energy by summing the series in Equation 2.21, but I warned you (in footnote 21) not to try it the "old fashioned way," $\langle H \rangle = \int \Psi(x,0)^* \hat{H} \Psi(x,0)\, dx$, because the discontinuous first derivative of $\Psi(x,0)$ renders the second derivative problematic. Actually, you *could* have done it using integration by parts, but the Dirac delta function affords a much cleaner way to handle such anomalies.

(a) Calculate the first derivative of $\Psi(x,0)$ (in Problem 2.7), and express the answer in terms of the step function, $\theta(x-a/2)$, defined in Equation 2.146.

(b) Exploit the result of Problem 2.23(b) to write the second derivative of $\Psi(x,0)$ in terms of the delta function.

(c) Evaluate the integral $\int \Psi(x,0)^* \hat{H} \Psi(x,0)\, dx$, and check that you get the same answer as before.

** **Problem 2.40** A particle of mass m in the harmonic oscillator potential (Equation 2.44) starts out in the state

$$\Psi(x,0) = A\left(1 - 2\sqrt{\frac{m\omega}{\hbar}}x\right)^2 e^{-\frac{m\omega}{2\hbar}x^2},$$

for some constant A.

(a) Determine A and the coefficients c_n in the expansion of this state in terms of the stationary states of the harmonic oscillator.

(b) In a measurement of the particle's energy, what results could you get, and what are their probabilities? What is the expectation value of the energy?

(c) At a later time T the wave function is

$$\Psi(x,T) = B\left(1 + 2\sqrt{\frac{m\omega}{\hbar}}x\right)^2 e^{-\frac{m\omega}{2\hbar}x^2},$$

for some constant B. What is the smallest possible value of T?

Problem 2.41 Find the allowed energies of the *half* harmonic oscillator

$$V(x) = \begin{cases} (1/2)\, m\omega^2 x^2, & x > 0, \\ \infty, & x < 0. \end{cases}$$

(This represents, for example, a spring that can be stretched, but not compressed.) *Hint:* This requires some careful thought, but very little actual calculation.

*** **Problem 2.42** In Problem 2.21 you analyzed the *stationary* gaussian free particle wave packet. Now solve the same problem for the *traveling* gaussian wave packet, starting with the initial wave function

$$\Psi(x,0) = Ae^{-ax^2}e^{ilx},$$

where l is a (real) constant. [*Suggestion:* In going from $\phi(k)$ to $\Psi(x,t)$, change variables to $u \equiv k - l$ before doing the integral.] *Partial answer:*

$$\Psi(x,t) = \left(\frac{2a}{\pi}\right)^{1/4} \frac{1}{\gamma} e^{-a(x-\hbar lt/m)^2/\gamma^2} e^{il(x-\hbar lt/2m)}$$

where $\gamma \equiv \sqrt{1+2ia\hbar t/m}$, as before. Notice that $\Psi(x,t)$ has the structure of a gaussian "envelope" modulating a traveling sinusoidal wave. What is the speed of the envelope? What is the speed of the traveling wave?

** **Problem 2.43** Solve the time-independent Schrödinger equation for a centered infinite square well with a delta-function barrier in the middle:

$$V(x) = \begin{cases} \alpha\delta(x), & -a < x < +a, \\ \infty, & |x| \geq a. \end{cases}$$

Treat the even and odd wave functions separately. Don't bother to normalize them. Find the allowed energies (graphically, if necessary). How do they compare with the corresponding energies in the absence of the delta function? Explain why the odd solutions are not affected by the delta function. Comment on the limiting cases $\alpha \to 0$ and $\alpha \to \infty$.

Problem 2.44 If two (or more) distinct[56] solutions to the (time-independent) Schrödinger equation have the same energy E, these states are said to be **degenerate**. For example, the free particle states are doubly degenerate—one solution representing motion to the right, and the other motion to the left. But we have never encountered *normalizable* degenerate solutions, and this is no accident. Prove the following theorem: *In one dimension*[57] ($-\infty < x < \infty$) *there are no degenerate bound states.* [*Hint*: Suppose there are *two* solutions, ψ_1 and ψ_2, with the same energy E. Multiply the Schrödinger equation for ψ_1 by ψ_2, and the Schrödinger equation for ψ_2 by ψ_1, and subtract, to show that $(\psi_2 d\psi_1/dx - \psi_1 d\psi_2/dx)$ is a constant. Use the fact that for normalizable solutions $\psi \to 0$ at $\pm\infty$ to demonstrate that this constant is in fact zero. Conclude that ψ_2 is a multiple of ψ_1, and hence that the two solutions are not distinct.]

Problem 2.45 In this problem you will show that the number of nodes of the stationary states of a one-dimensional potential always increases with energy.[58] Consider two (real, normalized) solutions (ψ_n and ψ_m) to the time-independent Schrödinger equation (for a given potential $V(x)$), with energies $E_n > E_m$.

(a) Show that

$$\frac{d}{dx}\left(\frac{d\psi_m}{dx}\psi_n - \psi_m\frac{d\psi_n}{dx}\right) = \frac{2m}{\hbar^2}(E_n - E_m)\,\psi_m\psi_n.$$

---

[56] If two solutions differ only by a multiplicative constant (so that, once normalized, they differ only by a phase factor $e^{i\phi}$), they represent the same physical state, and in this sense they are *not* distinct solutions. Technically, by "distinct" I mean "linearly independent."

[57] In higher dimensions such degeneracy is very common, as we shall see in Chapters 4 and 6. Assume that the potential does not consist of isolated pieces separated by regions where $V = \infty$—two isolated infinite square wells, for instance, would give rise to degenerate bound states, for which the particle is either in one well or in the other.

[58] M. Moriconi, *Am. J. Phys.* **75**, 284 (2007).

(b) Let x_1 and x_2 be two adjacent nodes of the function $\psi_m(x)$. Show that

$$\psi_m'(x_2)\,\psi_n(x_2) - \psi_m'(x_1)\,\psi_n(x_1) = \frac{2m}{\hbar^2}\,(E_n - E_m)\int_{x_1}^{x_2} \psi_m \psi_n \, dx.$$

(c) If $\psi_n(x)$ has no nodes between x_1 and x_2, then it must have the same sign everywhere in the interval. Show that (b) then leads to a contradiction. Therefore, between every pair of nodes of $\psi_m(x)$, $\psi_n(x)$ must have *at least* one node, and in particular the number of nodes increases with energy.

Problem 2.46 Imagine a bead of mass m that slides frictionlessly around a circular wire ring of circumference L. (This is just like a free particle, except that $\psi(x + L) = \psi(x)$.) Find the stationary states (with appropriate normalization) and the corresponding allowed energies. Note that there are (with one exception) *two* independent solutions for each energy E_n—corresponding to clockwise and counter-clockwise circulation; call them $\psi_n^+(x)$ and $\psi_n^-(x)$. How do you account for this degeneracy, in view of the theorem in Problem 2.44 (why does the theorem fail, in this case)?

** **Problem 2.47** *Attention*: This is a *strictly qualitative* problem—no calculations allowed! Consider the "double square well" potential (Figure 2.20). Suppose the depth V_0 and the width a are fixed, and large enough so that several bound states occur.

(a) Sketch the ground state wave function ψ_1 and the first excited state ψ_2, (i) for the case $b = 0$, (ii) for $b \approx a$, and (iii) for $b \gg a$.

(b) Qualitatively, how do the corresponding energies (E_1 and E_2) vary, as b goes from 0 to ∞? Sketch $E_1(b)$ and $E_2(b)$ on the same graph.

(c) The double well is a very primitive one-dimensional model for the potential experienced by an electron in a diatomic molecule (the two wells represent the attractive force of the nuclei). If the nuclei are free to move, they will adopt the configuration of minimum energy. In view of your conclusions in (b), does the electron tend to draw the nuclei together, or push them apart? (Of course, there is also the internuclear repulsion to consider, but that's a separate problem.)

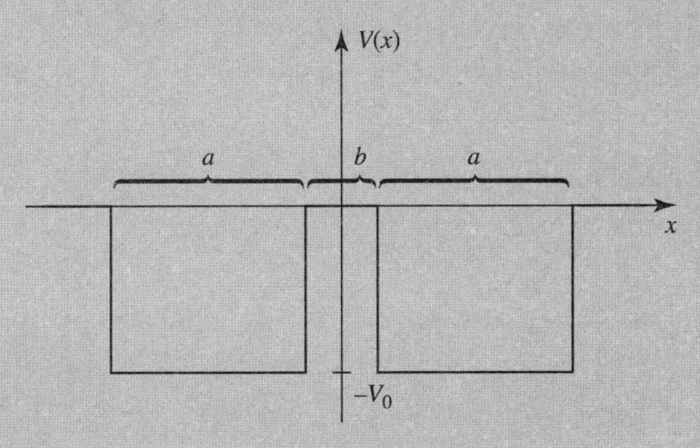

Figure 2.20: The double square well (Problem 2.47).

Problem 2.48 Consider a particle of mass m in the potential

$$V(x) = \begin{cases} \infty & x < 0, \\ -32\hbar^2/ma^2 & 0 \le x \le a, \\ 0 & x > a. \end{cases}$$

(a) How many bound states are there?

(b) In the highest-energy bound state, what is the probability that the particle would be found *outside* the well ($x > a$)? *Answer:* 0.542, so even though it is "bound" by the well, it is more likely to be found outside than inside!

∗∗∗ **Problem 2.49**

(a) Show that

$$\Psi(x,t) = \left(\frac{m\omega}{\pi\hbar}\right)^{1/4} \exp\left[-\frac{m\omega}{2\hbar}\left(x^2 + \frac{x_0^2}{2}\left(1 + e^{-2i\omega t}\right) + \frac{i\hbar t}{m} - 2x_0 x e^{-i\omega t}\right)\right]$$

satisfies the time-*dependent* Schrödinger equation for the harmonic oscillator potential (Equation 2.44). Here x_0 is any real constant with the dimensions of length.[59]

(b) Find $|\Psi(x,t)|^2$, and describe the motion of the wave packet.

(c) Compute $\langle x \rangle$ and $\langle p \rangle$, and check that Ehrenfest's theorem (Equation 1.38) is satisfied.

∗∗ **Problem 2.50** Consider the *moving* delta-function well:

$$V(x,t) = -\alpha\delta(x - vt),$$

where v is the (constant) velocity of the well.

(a) Show that the time-dependent Schrödinger equation admits the exact solution[60]

$$\Psi(x,t) = \frac{\sqrt{m\alpha}}{\hbar} e^{-m\alpha|x-vt|/\hbar^2} e^{-i\left[(E+(1/2)mv^2)t - mvx\right]/\hbar},$$

where $E = -m\alpha^2/2\hbar^2$ is the bound-state energy of the *stationary* delta function. *Hint:* Plug it in and *check* it! Use the result of Problem 2.23(b).

(b) Find the expectation value of the Hamiltonian in this state, and comment on the result.

∗∗ **Problem 2.51 Free fall.** Show that

$$\Psi(x,t) = \Psi_0\left(x + \frac{1}{2}gt^2, t\right) \exp\left[-i\frac{mgt}{\hbar}\left(x + \frac{1}{6}gt^2\right)\right] \tag{2.176}$$

satisfies the time-dependent Schrödinger equation for a particle in a uniform gravitational field,

$$V(x) = mgx, \tag{2.177}$$

---

59 This rare example of an exact closed-form solution to the time-dependent Schrödinger equation was discovered by Schrödinger himself, in 1926. One way to obtain it is explored in Problem 6.30. For a discussion of this and related problems see W. van Dijk, et al., *Am. J. Phys.* **82**, 955 (2014).

60 See Problem 6.35 for a derivation.

where $\Psi_0(x, t)$ is the free gaussian wave packet (Equation 2.111). Find $\langle x \rangle$ as a function of time, and comment on the result.[61]

*** **Problem 2.52** Consider the potential

$$V(x) = -\frac{\hbar^2 a^2}{m} \operatorname{sech}^2(ax),$$

where a is a positive constant, and "sech" stands for the hyperbolic secant.

(a) Graph this potential.

(b) Check that this potential has the ground state

$$\psi_0(x) = A \operatorname{sech}(ax),$$

and find its energy. Normalize ψ_0, and sketch its graph.

(c) Show that the function

$$\psi_k(x) = A\left(\frac{ik - a\tanh(ax)}{ik + a}\right)e^{ikx},$$

(where $k \equiv \sqrt{2mE}/\hbar$, as usual) solves the Schrödinger equation for any (positive) energy E. Since $\tanh z \to -1$ as $z \to -\infty$,

$$\psi_k(x) \approx Ae^{ikx}, \quad \text{for large negative } x.$$

This represents, then, a wave coming in from the left with *no accompanying reflected wave* (i.e. no term $\exp(-ikx)$). What is the asymptotic form of $\psi_k(x)$ at large *positive* x? What are R and T, for this potential? *Comment:* This is a famous example of a **reflectionless potential**—every incident particle, regardless its energy, passes right through.[62]

Problem 2.53 The Scattering Matrix. The theory of scattering generalizes in a pretty obvious way to arbitrary localized potentials (Figure 2.21). To the left (Region I), $V(x) = 0$, so

$$\psi(x) = Ae^{ikx} + Be^{-ikx}, \quad \text{where } k \equiv \frac{\sqrt{2mE}}{\hbar}. \tag{2.178}$$

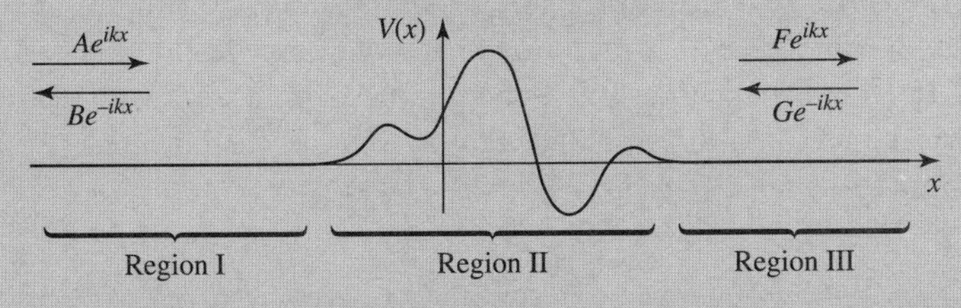

Figure 2.21: Scattering from an arbitrary localized potential ($V(x) = 0$ except in Region II); Problem 2.53.

---

[61] For illuminating discussion see M. Nauenberg, *Am. J. Phys.* **84**, 879 (2016).

[62] R. E. Crandall and B. R. Litt, *Annals of Physics*, **146**, 458 (1983).

To the right (Region III), $V(x)$ is again zero, so

$$\psi(x) = Fe^{ikx} + Ge^{-ikx}. \tag{2.179}$$

In between (Region II), of course, I can't tell you what ψ is until you specify the potential, but because the Schrödinger equation is a linear, second-order differential equation, the general solution has got to be of the form

$$\psi(x) = Cf(x) + Dg(x),$$

where $f(x)$ and $g(x)$ are two linearly independent particular solutions.[63] There will be four boundary conditions (two joining Regions I and II, and two joining Regions II and III). Two of these can be used to eliminate C and D, and the other two can be "solved" for B and F in terms of A and G:

$$B = S_{11}A + S_{12}G, \quad F = S_{21}A + S_{22}G.$$

The four coefficients S_{ij}, which depend on k (and hence on E), constitute a 2×2 matrix **S**, called the **scattering matrix** (or **S-matrix**, for short). The S-matrix tells you the outgoing amplitudes (B and F) in terms of the incoming amplitudes (A and G):

$$\begin{pmatrix} B \\ F \end{pmatrix} = \begin{pmatrix} S_{11} & S_{12} \\ S_{21} & S_{22} \end{pmatrix} \begin{pmatrix} A \\ G \end{pmatrix}. \tag{2.180}$$

In the typical case of scattering from the left, $G = 0$, so the reflection and transmission coefficients are

$$R_l = \left.\frac{|B|^2}{|A|^2}\right|_{G=0} = |S_{11}|^2, \quad T_l = \left.\frac{|F|^2}{|A|^2}\right|_{G=0} = |S_{21}|^2. \tag{2.181}$$

For scattering from the right, $A = 0$, and

$$R_r = \left.\frac{|F|^2}{|G|^2}\right|_{A=0} = |S_{22}|^2, \quad T_r = \left.\frac{|B|^2}{|G|^2}\right|_{A=0} = |S_{12}|^2. \tag{2.182}$$

(a) Construct the S-matrix for scattering from a delta-function well (Equation 2.117).
(b) Construct the S-matrix for the finite square well (Equation 2.148). *Hint:* This requires no new work, if you carefully exploit the symmetry of the problem.

*** **Problem 2.54 The transfer matrix.**[64] The S-matrix (Problem 2.53) tells you the *outgoing* amplitudes (B and F) in terms of the *incoming* amplitudes (A and G)—Equation 2.180. For some purposes it is more convenient to work with the **transfer matrix**, **M**, which gives you the amplitudes to the *right* of the potential (F and G) in terms of those to the *left* (A and B):

$$\begin{pmatrix} F \\ G \end{pmatrix} = \begin{pmatrix} M_{11} & M_{12} \\ M_{21} & M_{22} \end{pmatrix} \begin{pmatrix} A \\ B \end{pmatrix}. \tag{2.183}$$

---

[63] See any book on differential equations—for example, John L. Van Iwaarden, *Ordinary Differential Equations with Numerical Techniques*, Harcourt Brace Jovanovich, San Diego, 1985, Chapter 3.

[64] For applications of this method see, for instance, D. J. Griffiths and C. A. Steinke, *Am. J. Phys.* **69**, 137 (2001) or S. Das, *Am. J. Phys.* **83**, 590 (2015).

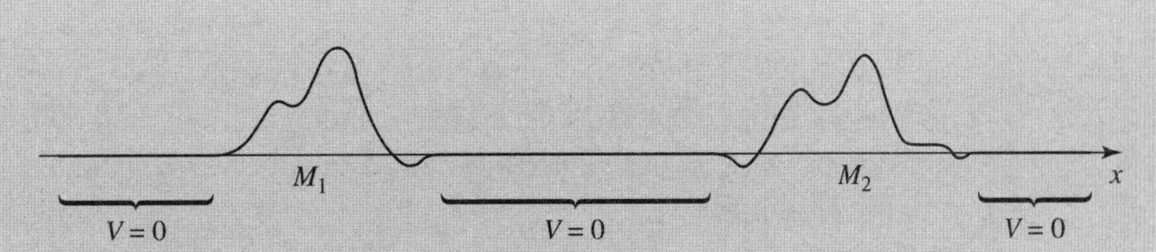

Figure 2.22: A potential consisting of two isolated pieces (Problem 2.54).

(a) Find the four elements of the M-matrix, in terms of the elements of the S-matrix, and vice versa. Express R_l, T_l, R_r, and T_r (Equations 2.181 and 2.182) in terms of elements of the M-matrix.

(b) Suppose you have a potential consisting of two isolated pieces (Figure 2.22). Show that the M-matrix for the combination is the *product* of the two M-matrices for each section separately:

$$\mathsf{M} = \mathsf{M}_2\mathsf{M}_1. \tag{2.184}$$

(This obviously generalizes to any number of pieces, and accounts for the usefulness of the M-matrix.)

(c) Construct the M-matrix for scattering from a single delta-function potential at point a:

$$V(x) = -\alpha\delta(x-a).$$

(d) By the method of part (b), find the M-matrix for scattering from the double delta-function

$$V(x) = -\alpha\left[\delta(x+a) + \delta(x-a)\right].$$

What is the transmission coefficient for this potential?

Problem 2.55 Find the ground state energy of the harmonic oscillator, to five significant digits, by the "wag-the-dog" method. That is, solve Equation 2.73 numerically, varying K until you get a wave function that goes to zero at large ξ. In Mathematica, appropriate input code would be

```
Plot[
  Evaluate[
    u[x] /.
      NDSolve[
        {u''[x] -(x^2 - K)*u[x] == 0,  u[0] == 1, u'[0] == 0},
        u[x], {x, 0, b}
      ]
    ],
  {x, a, b}, PlotRange -> {c, d}
]
```

(Here (a, b) is the horizontal range of the graph, and (c, d) is the vertical range—start with $a = 0$, $b = 10$, $c = -10$, $d = 10$.) We know that the correct solution is $K = 2n + 1$, so you might start with a "guess" of $K = 0.9$. Notice what the "tail" of the wave function does. Now try $K = 1.1$, and note that the tail flips over. Somewhere in between those values lies the correct solution. Zero in on it by bracketing K tighter and tighter. As you do so, you may want to adjust a, b, c, and d, to zero in on the cross-over point.

Problem 2.56 Find the first three excited state energies (to five significant digits) for the harmonic oscillator, by wagging the dog (Problem 2.55). For the first (and third) excited state you will need to set **u[0] == 0, u′[0] == 1**.)

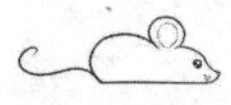

Problem 2.57 Find the first four allowed energies (to five significant digits) for the infinite square well, by wagging the dog. *Hint:* Refer to Problem 2.55, making appropriate changes to the differential equation. This time the condition you are looking for is $u(1) = 0$.

Problem 2.58 In a monovalent metal, one electron per atom is free to roam throughout the object. What holds such a material together—why doesn't it simply fall apart into a pile of individual atoms? Evidently the energy of the composite structure must be *less* than the energy of the isolated atoms. This problem offers a crude but illuminating explanation for the cohesiveness of metals.

(a) Estimate the energy of N isolated atoms, by treating each one as an electron in the ground state of an infinite square well of width a (Figure 2.23(a)).

(b) When these atoms come together to form a metal, we get N electrons in a much larger infinite square well of width Na (Figure 2.23(b)). Because of the Pauli exclusion principle (which we will discuss in Chapter 5) there can only be one electron (two, if you include spin, but let's ignore that) in each allowed state. What is the lowest energy for this system (Figure 2.23(b))?

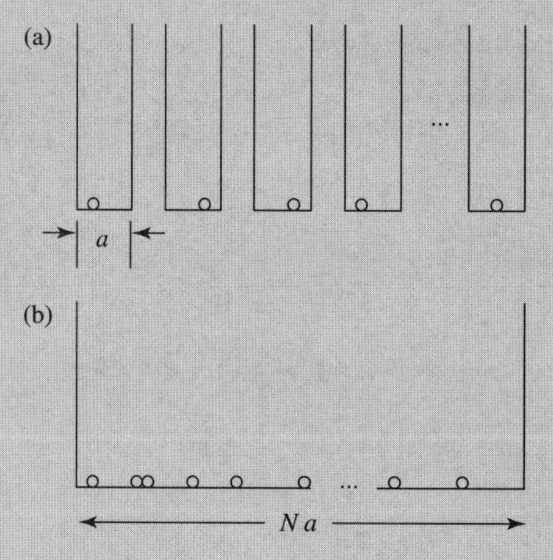

Figure 2.23: (a) N electrons in individual wells of width a. (b) N electrons in a single well of width Na.

(c) The difference of these two energies is the **cohesive energy** of the metal—the energy it would take to tear it apart into isolated atoms. Find the cohesive energy per atom, in the limit of large N.

(d) A typical atomic separation in a metal is a few Ångström (say, $a \approx 4$ Å). What is the numerical value of the cohesive energy per atom, in this model? (Measured values are in the range of 2–4 eV.)

* 

Problem 2.59 The "bouncing ball."[65] Suppose

$$V(x) = \begin{cases} mgx, & x > 0, \\ \infty, & x \le 0. \end{cases} \tag{2.185}$$

(a) Solve the (time-independent) Schrödinger equation for this potential. *Hint:* First convert it to dimensionless form:

$$-\,y''(z) + z\,y(z) = \epsilon y(z) \tag{2.186}$$

by letting $z \equiv ax$ and $y(z) \equiv \left(1/\sqrt{a}\right)\psi(x)$ (the $\sqrt{a}$ is just so $y(z)$ is normalized with respect to z when $\psi(x)$ is normalized with respect to x). What are the constants a and ϵ? Actually, we might as well set $a \to 1$—this amounts to a convenient choice for the unit of length. Find the general solution to this equation (in Mathematica **DSolve** will do the job). The result is (of course) a linear combination of two (probably unfamiliar) functions. Plot each of them, for $(-15 < z < 5)$. One of them clearly does not go to zero at large z (more precisely, it's not normalizable), so discard it. The allowed values of ϵ (and hence of E) are determined by the condition $\psi(0) = 0$. Find the ground state ϵ_1 numerically (in Mathematica **FindRoot** will do it), and also the 10th, ϵ_{10}. Obtain the corresponding normalization factors. Plot $\psi_1(x)$ and $\psi_{10}(x)$, for $0 \le z < 16$. Just as a check, confirm that $\psi_1(x)$ and $\psi_{10}(x)$ are orthogonal.

(b) Find (numerically) the uncertainties σ_x and σ_p for these two states, and check that the uncertainty principle is obeyed.

(c) The probability of finding the ball in the neighborhood dx of height x is (of course) $\rho_Q(x)\,dx = |\psi(x)|^2\,dx$. The nearest classical analog would be the fraction of *time* an elastically bouncing ball (with the same energy, E) spends in the neighborhood dx of height x (see Problem 1.11). Show that this is

$$\rho_C(x)\,dx = \frac{mg}{2\sqrt{E\,(E - mgx)}}\,dx, \tag{2.187}$$

or, in our units (with $a = 1$),

$$\rho_C(x) = \frac{1}{2\sqrt{\epsilon\,(\epsilon - x)}}. \tag{2.188}$$

Plot $\rho_Q(x)$ and $\rho_C(x)$ for the state $\psi_{10}(x)$, on the range $0 \le x \le 12.5$; superimpose the graphs (**Show**, in Mathematica), and comment on the result.

---

[65] This problem was suggested by Nicholas Wheeler.

***  **Problem 2.60 The $1/x^2$ potential.** Suppose

$$V(x) = \begin{cases} -\alpha/x^2, & x > 0, \\ \infty, & x \leq 0. \end{cases} \tag{2.189}$$

where α is some positive constant with the appropriate dimensions. We'd like to find the bound states—solutions to the time-independent Schrödinger equation

$$-\frac{\hbar^2}{2m}\frac{d^2\psi}{dx^2} - \frac{\alpha}{x^2}\psi = E\psi \tag{2.190}$$

with negative energy ($E < 0$).

(a) Let's first go for the ground state energy, E_0. Prove, on dimensional grounds, that there is *no possible formula* for E_0—no way to construct (from the available constants m, $\hbar$, and α) a quantity with the units of energy. That's weird, but it gets worse ....

(b) For convenience, rewrite Equation 2.190 as

$$\frac{d^2\psi}{dx^2} + \frac{\beta}{x^2}\psi = \kappa^2\psi, \text{ where } \beta \equiv \frac{2m\alpha}{\hbar^2} \text{ and } \kappa \equiv \frac{\sqrt{-2mE}}{\hbar}. \tag{2.191}$$

Show that if $\psi(x)$ satisfies this equation with energy E, then so too does $\psi(\lambda x)$, with energy $E' = \lambda^2 E$, for any positive number λ. [This is a catastrophe: if there exists any solution at all, then there's a solution for *every* (negative) energy! Unlike the square well, the harmonic oscillator, and every other potential well we have encountered, there are no discrete allowed states—and no ground state. A system with no ground state—no lowest allowed energy—would be wildly unstable, cascading down to lower and lower levels, giving off an unlimited amount of energy as it falls. It might solve our energy problem, but we'd all be fried in the process.] Well, perhaps there simply are *no solutions at all* ....

(c) (Use a computer for the remainder of this problem.) Show that

$$\psi_\kappa(x) = A\sqrt{x}\,K_{ig}(\kappa x), \tag{2.192}$$

satisfies Equation 2.191 (here K_{ig} is the modified Bessel function of order ig, and $g \equiv \sqrt{\beta - 1/4}$). Plot this function, for $g = 4$ (you might as well let $\kappa = 1$ for the graph; this just sets the scale of length). Notice that it goes to 0 as $x \to 0$ and as $x \to \infty$. And it's normalizable: determine A.[66] How about the old rule that the number of nodes counts the number of lower-energy states? This function has an *infinite* number of nodes, regardless of the energy (i.e. of κ). I guess that's consistent, since for any E there are always an infinite number of states with even lower energy.

(d) This potential confounds practically everything we have come to expect. The problem is that it blows up too violently as $x \to 0$. If you move the "brick wall" over a hair,

[66] $\psi_k(x)$ is normalizable as long as g is real—which is to say, provided $\beta > 1/4$. For more on this strange problem see A. M. Essin and D. J. Griffiths, *Am. J. Phys.* **74**, 109 (2006), and references therein.

$$V(x)=\begin{cases}-\alpha/x^2, & x>\epsilon>0,\\ \infty, & x\le\epsilon,\end{cases} \tag{2.193}$$

it's suddenly perfectly normal. Plot the ground state wave function, for $g=4$ and $\epsilon=1$ (you'll first need to determine the appropriate value of κ), from $x=0$ to $x=6$. Notice that we have introduced a new parameter (ϵ), with the dimensions of length, so the argument in (a) is out the window. Show that the ground state energy takes the form

$$E_0=-\frac{\alpha}{\epsilon^2}f(\beta), \tag{2.194}$$

for some function f of the dimensionless quantity β.

* **Problem 2.61** One way to obtain the allowed energies of a potential well numerically is to turn the Schrödinger equation into a *matrix* equation, by discretizing the variable x. Slice the relevant interval at evenly spaced points $\{x_j\}$, with $x_{j+1}-x_j\equiv\Delta x$, and let $\psi_j\equiv\psi(x_j)$ (likewise $V_j\equiv V(x_j)$). Then

$$\frac{d\psi}{dx}\approx\frac{\psi_{j+1}-\psi_j}{\Delta x},\quad \frac{d^2\psi}{dx^2}\approx\frac{(\psi_{j+1}-\psi_j)-(\psi_j-\psi_{j-1})}{(\Delta x)^2}=\frac{\psi_{j+1}-2\psi_j+\psi_{j-1}}{(\Delta x)^2}. \tag{2.195}$$

(The approximation presumably improves as Δx decreases.) The discretized Schrödinger equation reads

$$-\frac{\hbar^2}{2m}\left(\frac{\psi_{j+1}-2\psi_j+\psi_{j-1}}{(\Delta x)^2}\right)+V_j\psi_j=E\psi_j, \tag{2.196}$$

or

$$-\lambda\psi_{j+1}+(2\lambda+V_j)\,\psi_j-\lambda\psi_{j-1}=E\psi_j,\quad\text{where}\quad\lambda\equiv\frac{\hbar^2}{2m\,(\Delta x)^2}. \tag{2.197}$$

In matrix form,

$$\mathsf{H}\Psi=E\Psi \tag{2.198}$$

where (letting $v_j\equiv V_j/\lambda$)

$$\mathsf{H}\equiv\lambda\begin{pmatrix}\ddots & & & & & \\ & -1 & (2+v_{j-1}) & -1 & 0 & 0 \\ & 0 & -1 & (2+v_j) & -1 & 0 \\ & 0 & 0 & -1 & (2+v_{j+1}) & -1 \\ & & & & & \ddots\end{pmatrix} \tag{2.199}$$

and

$$\Psi\equiv\begin{pmatrix}\vdots\\ \psi_{j-1}\\ \psi_j\\ \psi_{j+1}\\ \vdots\end{pmatrix} \tag{2.200}$$

(what goes in the upper left and lower right corners of H depends on the boundary conditions, as we shall see). Evidently the allowed energies are the *eigenvalues* of the matrix H (or *would* be, in the limit $\Delta x \to 0$).[67]

Apply this method to the infinite square well. Chop the interval $(0 \leq x \leq a)$ into $N+1$ equal segments (so that $\Delta x = a/(N+1)$), letting $x_0 \equiv 0$ and $x_{N+1} \equiv a$. The boundary conditions fix $\psi_0 = \psi_{N+1} = 0$, leaving

$$\Psi \equiv \begin{pmatrix} \psi_1 \\ \vdots \\ \psi_N \end{pmatrix}. \tag{2.201}$$

(a) Construct the $N \times N$ matrix H, for $N = 1$, $N = 2$, and $N = 3$. (Make sure you are correctly representing Equation 2.197 for the special cases $j = 1$ and $j = N$.)
(b) Find the eigenvalues of H for these three cases "by hand," and compare them with the exact allowed energies (Equation 2.30).
(c) Using a computer (Mathematica's **Eigenvalues** package will do it) find the five lowest eigenvalues numerically for $N = 10$ and $N = 100$, and compare the exact energies.
(d) Plot (by hand) the eigen*vectors* for $N = 1$, 2, and 3, and (by computer, **Eigenvectors**) the first three eigenvectors for $N = 10$ and $N = 100$.

** 

Problem 2.62 Suppose the bottom of the infinite square well is not flat ($V(x) = 0$), but rather

$$V(x) = 500\, V_0 \sin\left(\frac{\pi x}{a}\right), \quad \text{where} \quad V_0 \equiv \frac{\hbar^2}{2ma^2}.$$

Use the method of Problem 2.61 to find the three lowest allowed energies numerically, and plot the associated wave functions (use $N = 100$).

Problem 2.63 The Boltzmann equation[68]

$$P(n) = \frac{1}{Z} e^{-\beta E_n}, \quad Z \equiv \sum_n e^{-\beta E_n}, \quad \beta \equiv \frac{1}{k_B T}, \tag{2.202}$$

gives the probability of finding a system in the state n (with energy E_n), at temperature T (k_B is Boltzmann's constant). *Note:* The probability here refers to the random thermal distribution, and has nothing to do with quantum indeterminacy. Quantum mechanics will only enter this problem through quantization of the energies E_n.
(a) Show that the thermal average of the system's energy can be written as

$$\bar{E} = \sum_n E_n P(n) = -\frac{\partial}{\partial \beta} \ln(Z). \tag{2.203}$$

[67] For further discussion see Joel Franklin, *Computational Methods for Physics* (Cambridge University Press, Cambridge, UK, 2013), Section 10.4.2.

[68] See, for instance, Daniel V. Schroeder, *An Introduction to Thermal Physics*, Pearson, Boston (2000), Section 6.1.

(b) For a quantum simple harmonic oscillator the index n is the familiar quantum number, and $E_n = (n + 1/2)\,\hbar\omega$. Show that in this case the **partition function** Z is

$$Z = \frac{e^{-\beta\hbar\omega/2}}{1 - e^{-\beta\hbar\omega}}. \tag{2.204}$$

You will need to sum a geometric series. Incidentally, for a *classical* simple harmonic oscillator it can be shown that $Z_{\text{classical}} = 2\pi/(\omega\beta)$.

(c) Use your results from parts (a) and (b) to show that for the quantum oscillator

$$\bar{E} = \left(\frac{\hbar\omega}{2}\right)\frac{1 + e^{-\beta\hbar\omega}}{1 - e^{-\beta\hbar\omega}}. \tag{2.205}$$

For a *classical* oscillator the same reasoning would give $\bar{E}_{\text{classical}} = 1/\beta = k_B T$.

(d) A crystal consisting of N atoms can be thought of as a collection of $3N$ oscillators (each atom is attached by springs to its 6 nearest neighbors, along the x, y, and z directions, but those springs are shared by the atoms at the two ends). The **heat capacity** of the crystal (per atom) will therefore be

$$C = 3\frac{\partial \bar{E}}{\partial T}. \tag{2.206}$$

Show that (in this model)

$$C = 3k_B\left(\frac{\theta_E}{T}\right)^2 \frac{e^{\theta_E/T}}{\left(e^{\theta_E/T} - 1\right)^2}, \tag{2.207}$$

where $\theta_E \equiv \hbar\omega/k_B$ is the so-called **Einstein temperature**. The same reasoning using the *classical* expression for $\bar{E}$ yields $C_{\text{classical}} = 3k_B$, independent of temperature.

(e) Sketch the graph of C/k_B versus T/θ_E. Your result should look something like the data for diamond in Figure 2.24, and nothing like the classical prediction.

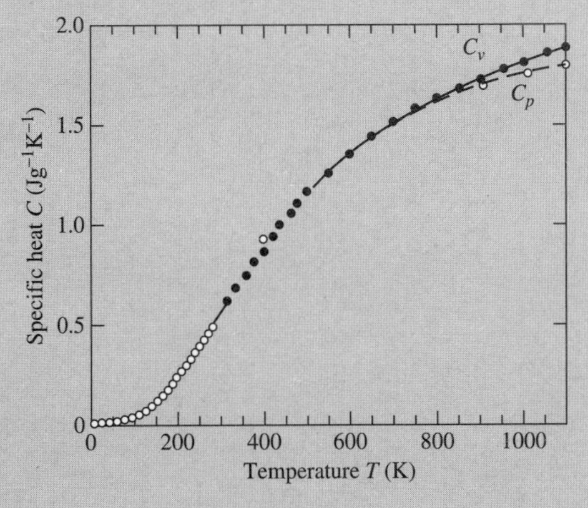

Figure 2.24: Specific heat of diamond (for Problem 2.63). From *Semiconductors on NSM* (http://www.ioffe.rssi.ru/SVA/NSM/Semicond/).

Problem 2.64 Legendre's differential equation reads

$$\left(1-x^2\right)\frac{d^2 f}{dx^2}-2x\frac{df}{dx}+\ell(\ell+1)\,f=0, \tag{2.208}$$

where ℓ is some (non-negative) real number.

(a) Assume a power series solution,

$$f(x)=\sum_{n=0}^{\infty} a_n\,x^n,$$

and obtain a recursion relation for the constants a_n.

(b) Argue that unless the series truncates (which can only happen if ℓ is an integer), the solution will diverge at $x=1$.

(c) When ℓ *is* an integer, the series for one of the two linearly independent solutions (either f_{even} or f_{odd} depending on whether ℓ is even or odd) will truncate, and those solutions are called **Legendre polynomials** $P_\ell(x)$. Find $P_0(x)$, $P_1(x)$, $P_2(x)$, and $P_3(x)$ from the recursion relation. Leave your answer in terms of either a_0 or a_1.[69]

[69] By convention Legendre polynomials are normalized such that $P_\ell(1)=1$. Note that the nonvanishing coefficients will take different values for different ℓ.

导读 / Guidance

第 3 章　形式理论

本章讨论量子力学的形式理论，开始先介绍希尔伯特空间的概念，波函数作为态矢存在于希尔伯特空间中；接下来讨论厄密算符和厄密算符的本征函数问题，讨论分立谱和连续谱这两种情况，给出广义的统计诠释。在此基础上，给出不确定性原理的一般性证明，并推广到能量 - 时间的不确定性原理的情况，强调不确定性原理是统计诠释的结果。最后，引入狄拉克符号、左矢、右矢、投影算符、对偶空间等概念。

习题特色

（1）利用不确定性原理讨论粒子物理中高能粒子的质量问题，如例题 3.7；介绍中微子振荡的简化模型，如例题 3.8。

（2）在习题中引入 Baker-Campbell-Hausdorff 公式的讨论，见习题 3.29；引入维里定理的证明，如习题 3.37；以谐振子为例讨论相干态的概念，见习题 3.42；对扩展不确定性原理进行深入讨论，见习题 3.43。

（3）拓展至高能物理对超对称的讨论，见习题 3.47；拓展对自伴算子的讨论，见习题 3.48。

3 FORMALISM

3.1 HILBERT SPACE

In the previous two chapters we have stumbled on a number of interesting properties of simple quantum systems. Some of these are "accidental" features of specific potentials (the even spacing of energy levels for the harmonic oscillator, for example), but others seem to be more general, and it would be nice to prove them once and for all (the uncertainty principle, for instance, and the orthogonality of stationary states). The purpose of this chapter is to recast the theory in more powerful form, with that in mind. There is not much here that is genuinely *new*; the idea, rather, is to make coherent sense of what we have already discovered in particular cases.

Quantum theory is based on two constructs: wave functions and operators. The **state** of a system is represented by its wave function, **observables** are represented by operators. Mathematically, wave functions satisfy the defining conditions for abstract **vectors**, and operators act on them as **linear transformations**. So the natural language of quantum mechanics is **linear algebra**.[1]

But it is not, I suspect, a form of linear algebra with which you may be familiar. In an N-dimensional space it is simplest to represent a vector, $|\alpha\rangle$, by the N-tuple of its components, $\{a_n\}$, with respect to a specified orthonormal basis:

$$|\alpha\rangle \rightarrow \mathsf{a} = \begin{pmatrix} a_1 \\ a_2 \\ \vdots \\ a_N \end{pmatrix}; \tag{3.1}$$

the **inner product**, $\langle\alpha|\beta\rangle$, of two vectors (generalizing the dot product in three dimensions) is a complex number,

$$\langle\alpha|\beta\rangle = a_1^* b_1 + a_2^* b_2 + \cdots + a_N^* b_N; \tag{3.2}$$

linear transformations, T, are represented by **matrices** (with respect to the specified basis), which act on vectors (to produce new vectors) by the ordinary rules of matrix multiplication:

$$|\beta\rangle = \hat{T}|\alpha\rangle \rightarrow \mathsf{b} = \mathsf{Ta} = \begin{pmatrix} t_{11} & t_{12} & \cdots & t_{1N} \\ t_{21} & t_{22} & \cdots & t_{2N} \\ \vdots & \vdots & & \vdots \\ t_{N1} & t_{N2} & \cdots & t_{NN} \end{pmatrix} \begin{pmatrix} a_1 \\ a_2 \\ \vdots \\ a_N \end{pmatrix}. \tag{3.3}$$

But the "vectors" we encounter in quantum mechanics are (for the most part) *functions*, and they live in *infinite*-dimensional spaces. For them the N-tuple/matrix notation is awkward, at best, and manipulations that are well behaved in the finite-dimensional case can be

[1] If you have never studied linear algebra, you should read the Appendix before continuing.

problematic. (The underlying reason is that whereas the *finite* sum in Equation 3.2 always exists, an *infinite* sum—or an integral—may not converge, in which case the inner product does not exist, and any argument involving inner products is immediately suspect.) So even though most of the terminology and notation should be familiar, it pays to approach this subject with caution.

The collection of *all* functions of x constitutes a vector space, but for our purposes it is much too large. To represent a possible physical state, the wave function Ψ must be *normalized*:

$$\int |\Psi|^2 \, dx = 1.$$

The set of all **square-integrable** functions, on a specified interval,[2]

$$f(x) \quad \text{such that} \quad \int_a^b |f(x)|^2 \, dx < \infty, \tag{3.4}$$

constitutes a (much smaller) vector space (see Problem 3.1(a)). Mathematicians call it $L^2(a, b)$; physicists call it **Hilbert space**.[3] In quantum mechanics, then:

$$\boxed{\textbf{Wave functions live in Hilbert space.}} \tag{3.5}$$

We define the **inner product of two functions**, $f(x)$ and $g(x)$, as follows:

$$\boxed{\langle f|g\rangle \equiv \int_a^b f(x)^* g(x)\, dx.} \tag{3.6}$$

If f and g are both square-integrable (that is, if they are both in Hilbert space), their inner product is guaranteed to exist (the integral in Equation 3.6 converges to a finite number).[4] This follows from the integral **Schwarz inequality**:[5]

---

[2] For us, the limits (a and b) will almost always be $\pm\infty$, but we might as well keep things more general for the moment.

[3] Technically, a Hilbert space is a **complete inner product space**, and the collection of square-integrable functions is only *one example* of a Hilbert space—indeed, every finite-dimensional vector space is trivially a Hilbert space. But since L^2 is the arena of quantum mechanics, it's what physicists generally *mean* when they say "Hilbert space." By the way, the word **complete** here means that any Cauchy sequence of functions in Hilbert space converges to a function that is also in the space: it has no "holes" in it, just as the set of all real numbers has no holes (by contrast, the space of all *polynomials*, for example, like the set of all *rational* numbers, certainly *does* have holes in it). The completeness of a *space* has nothing to do with the completeness (same word, unfortunately) of a *set of functions*, which is the property that any other function can be expressed as a linear combination of them. For an accessible introduction to Hilbert space see Daniel T. Gillespie, *A Quantum Mechanics Primer* (International Textbook Company, London, 1970), Sections 2.3 and 2.4.

[4] In Chapter 2 we were obliged on occasion to work with functions that were *not* normalizable. Such functions lie *outside* Hilbert space, and we are going to have to handle them with special care. For the moment, I shall assume that all the functions we encounter *are* in Hilbert space.

[5] For a proof, see Frigyes Riesz and Bela Sz.-Nagy, *Functional Analysis* (Dover, Mineola, NY, 1990), Section 21. In a *finite*-dimensional vector space the Schwarz inequality, $|\langle\alpha|\beta\rangle|^2 \le \langle\alpha|\alpha\rangle\,\langle\beta|\beta\rangle$, is easy to prove (see Problem A.5). But that proof *assumes* the existence of the inner products, which is precisely what we are trying to *establish* here.

$$\left| \int_a^b f(x)^* g(x) \, dx \right| \le \sqrt{\int_a^b |f(x)|^2 \, dx \int_a^b |g(x)|^2 \, dx}. \tag{3.7}$$

You can check for yourself that definition (Equation 3.6) satisfies all the conditions for an inner product (Problem 3.1(b)). Notice in particular that

$$\langle g \mid f \rangle = \langle f \mid g \rangle^* . \tag{3.8}$$

Moreover, the inner product of $f(x)$ with *itself*,

$$\langle f \mid f \rangle = \int_a^b |f(x)|^2 \, dx, \tag{3.9}$$

is *real* and non-negative; it's *zero* only when $f(x) = 0$.[6]

A function is said to be **normalized** if its inner product with itself is 1; two functions are **orthogonal** if their inner product is 0; and a *set* of functions, $\{f_n\}$, is **orthonormal** if they are normalized and mutually orthogonal:

$$\langle f_m \mid f_n \rangle = \delta_{mn}. \tag{3.10}$$

Finally, a set of functions is **complete** if any *other* function (in Hilbert space) can be expressed as a linear combination of them:

$$f(x) = \sum_{n=1}^{\infty} c_n f_n(x). \tag{3.11}$$

If the functions $\{f_n(x)\}$ are orthonormal, the coefficients are given by Fourier's trick:

$$c_n = \langle f_n \mid f \rangle , \tag{3.12}$$

as you can check for yourself. I anticipated this terminology, of course, back in Chapter 2. (The stationary states of the infinite square well (Equation 2.31) constitute a complete orthonormal set on the interval $(0, a)$; the stationary states for the harmonic oscillator (Equation 2.68 or 2.86) are a complete orthonormal set on the interval $(-\infty, \infty)$.)

Problem 3.1

(a) Show that the set of all square-integrable functions is a vector space (refer to Section A.1 for the definition). *Hint:* The main point is to show that the sum of two square-integrable functions is itself square-integrable. Use Equation 3.7. Is the set of all *normalized* functions a vector space?

(b) Show that the integral in Equation 3.6 satisfies the conditions for an inner product (Section A.2).

[6] What about a function that is zero everywhere except at a few isolated points? The integral (Equation 3.9) would still vanish, even though the function itself does not. If this bothers you, you should have been a math major. In physics such pathological functions do not occur, but in any case, in Hilbert space two functions are considered equivalent if the integral of the absolute square of their difference vanishes. Technically, vectors in Hilbert space represent **equivalence classes** of functions.

∗ **Problem 3.2**

(a) For what range of ν is the function $f(x) = x^{\nu}$ in Hilbert space, on the interval $(0, 1)$? Assume ν is real, but not necessarily positive.

(b) For the specific case $\nu = 1/2$, is $f(x)$ in this Hilbert space? What about $xf(x)$? How about $(d/dx)\, f(x)$?

3.2 OBSERVABLES

3.2.1 Hermitian Operators

The expectation value of an observable $Q(x, p)$ can be expressed very neatly in inner-product notation:[7]

$$\langle Q \rangle = \int \Psi^* \hat{Q} \Psi \, dx = \langle \Psi | \hat{Q} \Psi \rangle. \tag{3.13}$$

Now, the outcome of a measurement has got to be *real*, and so, *a fortiori*, is the *average* of many measurements:

$$\langle Q \rangle = \langle Q \rangle^* . \tag{3.14}$$

But the complex conjugate of an inner product reverses the order (Equation 3.8), so

$$\langle \Psi | \hat{Q} \Psi \rangle = \langle \hat{Q} \Psi | \Psi \rangle, \tag{3.15}$$

and this must hold true for any wave function Ψ. Thus operators representing *observables* have the very special property that

$$\langle f | \hat{Q} f \rangle = \langle \hat{Q} f | f \rangle \quad \text{for all } f(x). \tag{3.16}$$

We call such operators **hermitian.**[8]

Actually, most books require an ostensibly stronger condition:

$$\langle f | \hat{Q} g \rangle = \langle \hat{Q} f | g \rangle \quad \text{for all } f(x) \text{ and all } g(x). \tag{3.17}$$

But it turns out, in spite of appearances, that this is perfectly equivalent to my definition (Equation 3.16), as you will prove in Problem 3.3. So use whichever you like. The essential point is that a hermitian operator can be applied either to the first member of an inner product or to

[7] Remember that $\hat{Q}$ is the operator constructed from Q by the replacement $p \to -i\hbar d/dx$. These operators are **linear**, in the sense that

$$\hat{Q}\left[af(x) + bg(x)\right] = a\,\hat{Q}f(x) + b\,\hat{Q}g(x),$$

for any functions f and g and any complex numbers a and b. They constitute *linear transformations* (Section A.3) on the space of all functions. However, they sometimes carry a function *inside* Hilbert space into a function *outside* it (see Problem 3.2(b)), and in that case the **domain** of the operator (the set of functions on which it acts) may have to be restricted (see Problem 3.48).

[8] In a *finite*-dimensional vector space hermitian operators are represented by hermitian *matrices*; a hermitian matrix is equal to its transpose conjugate: $\mathsf{T} = \mathsf{T}^{\dagger} = \tilde{\mathsf{T}}^{*}$. If this is unfamiliar to you please see the Appendix.

the second, with the same result, and hermitian operators naturally arise in quantum mechanics because their expectation values are real:

Observables are represented by hermitian operators. (3.18)

Well, let's *check* this. Is the momentum operator, for example, hermitian?

$$\langle f|\hat{p}g\rangle = \int_{-\infty}^{\infty} f^* (-i\hbar) \frac{dg}{dx}\, dx = -i\hbar f^* g\Big|_{-\infty}^{\infty} + \int_{-\infty}^{\infty} \left(-i\hbar\frac{df}{dx}\right)^* g\, dx = \langle \hat{p}f \mid g\rangle. \quad (3.19)$$

I used integration by parts, of course, and threw away the boundary term for the usual reason: If $f(x)$ and $g(x)$ are square integrable, they must go to zero at $\pm\infty$.[9] Notice how the complex conjugation of i compensates for the minus sign picked up from integration by parts—the operator d/dx (without the i) is *not* hermitian, and it does not represent a possible observable.

The **hermitian conjugate** (or **adjoint**) of an operator $\hat{Q}$ is the operator $\hat{Q}^\dagger$ such that

$$\langle f|\hat{Q}g\rangle = \langle \hat{Q}^\dagger f|g\rangle \quad \text{(for all } f \text{ and } g). \quad (3.20)$$

A hermitian operator, then, is equal to its hermitian conjugate: $\hat{Q} = \hat{Q}^\dagger$.

* **Problem 3.3** Show that if $\langle h|\hat{Q}h\rangle = \langle \hat{Q}h|h\rangle$ for all h (in Hilbert space), then $\langle f|\hat{Q}g\rangle = \langle \hat{Q}f|g\rangle$ for all f and g (i.e. the two definitions of "hermitian" —Equations 3.16 and 3.17— are equivalent). *Hint:* First let $h = f + g$, and then let $h = f + ig$.

Problem 3.4

(a) Show that the *sum* of two hermitian operators is hermitian.

(b) Suppose $\hat{Q}$ is hermitian, and α is a complex number. Under what condition (on α) is $\alpha\hat{Q}$ hermitian?

(c) When is the *product* of two hermitian operators hermitian?

(d) Show that the position operator $(\hat{x})$ and the Hamiltonian operator $\left(\hat{H} = -(\hbar^2/2m)d^2/dx^2 + V(x)\right)$ are hermitian.

* **Problem 3.5**

(a) Find the hermitian conjugates of x, i, and d/dx.

(b) Show that $\left(\hat{Q}\hat{R}\right)^\dagger = \hat{R}^\dagger\hat{Q}^\dagger$ (note the reversed order), $\left(\hat{Q}+\hat{R}\right)^\dagger = \hat{Q}^\dagger + \hat{R}^\dagger$, and $(c\hat{Q})^\dagger = c^*\hat{Q}^\dagger$ for a complex number c.

(c) Construct the hermitian conjugate of a_+ (Equation 2.48).

[9] As I mentioned in Chapter 1, there exist pathological functions that are square-integrable but do *not* go to zero at infinity. However, such functions do not arise in physics, and if you are worried about it we will simply restrict the domain of our operators to exclude them. On *finite* intervals, though, you really *do* have to be more careful with the boundary terms, and an operator that is hermitian on $(-\infty, \infty)$ may *not* be hermitian on $(0, \infty)$ or $(-\pi, \pi)$. (If you're wondering about the infinite square well, it's safest to think of those wave functions as residing on the infinite line—they just happen to be *zero* outside $(0, a)$.) See Problem 3.48.

3.2.2 Determinate States

Ordinarily, when you measure an observable Q on an ensemble of identically prepared systems, all in the same state Ψ, you do *not* get the same result each time—this is the *indeterminacy* of quantum mechanics. *Question:* Would it be possible to prepare a state such that *every* measurement of Q is certain to return the *same* value (call it q)? This would be, if you like, a **determinate state**, for the observable Q. (Actually, we already know one example: Stationary states are determinate states of the Hamiltonian; a measurement of the energy, on a particle in the stationary state Ψ_n, is certain to yield the corresponding "allowed" energy E_n.)

Well, the standard deviation of Q, in a determinate state, would be *zero*, which is to say,

$$\sigma^2 = \left\langle (Q - \langle Q \rangle)^2 \right\rangle = \left\langle \Psi \middle| (\hat{Q} - q)^2 \Psi \right\rangle = \langle (\hat{Q} - q)\Psi | (\hat{Q} - q)\Psi \rangle = 0. \tag{3.21}$$

(Of course, if every measurement gives q, their average is also q: $\langle Q \rangle = q$. I used the fact that $\hat{Q}$ (and hence also $\hat{Q} - q$) is a *hermitian* operator, to move one factor over to the first term in the inner product.) But the only vector whose inner product with itself vanishes is 0, so

$$\hat{Q}\Psi = q\Psi. \tag{3.22}$$

This is the **eigenvalue equation** for the operator $\hat{Q}$; Ψ is an **eigenfunction** of $\hat{Q}$, and q is the corresponding **eigenvalue**:

$$\boxed{\textbf{Determinate states of } Q \textbf{ are eigenfunctions of } \hat{Q}.} \tag{3.23}$$

Measurement of Q on such a state is certain to yield the eigenvalue, q.[10]

Note that the eigen*value* is a *number* (not an operator or a function). You can multiply any eigenfunction by a constant, and it is still an eigenfunction, with the same eigenvalue. Zero does not count as an eigenfunction (we exclude it by definition—otherwise *every* number would be an eigenvalue, since $\hat{Q}\,0 = q\,0 = 0$ for any linear operator $\hat{Q}$ and all q). But there's nothing wrong with zero as an eigen*value*. The collection of all the eigenvalues of an operator is called its **spectrum**. Sometimes two (or more) linearly independent eigenfunctions share the same eigenvalue; in that case the spectrum is said to be **degenerate**. (You encountered this term already, for the case of energy eigenstates, if you worked Problems 2.44 or 2.46.)

For example, determinate states of the total energy are eigenfunctions of the Hamiltonian:

$$\hat{H}\psi = E\psi, \tag{3.24}$$

which is precisely the time-independent Schrödinger equation. In this context we use the letter E for the eigenvalue, and the lower case ψ for the eigenfunction (tack on the wiggle factor $\exp\left(-iEt/\hbar\right)$ to make it Ψ, if you like; it's still an eigenfunction of $\hat{H}$).

---

[10] I'm talking about a *competent* measurement, of course—it's always possible to make a *mistake*, and simply get the wrong answer, but that's not the fault of quantum mechanics.

Example 3.1
Consider the operator

$$\hat{Q} \equiv i\frac{d}{d\phi}, \tag{3.25}$$

where ϕ is the usual polar coordinate in two dimensions. (This operator might arise in a physical context if we were studying the bead-on-a-ring; see Problem 2.46.) Is $\hat{Q}$ hermitian? Find its eigenfunctions and eigenvalues.

Solution: Here we are working with functions $f(\phi)$ on the *finite* interval $0 \le \phi \le 2\pi$, with the property that

$$f(\phi + 2\pi) = f(\phi), \tag{3.26}$$

since ϕ and $\phi + 2\pi$ describe the same physical point. Using integration by parts,

$$\langle f|\hat{Q}g\rangle = \int_0^{2\pi} f^* \left(i\frac{dg}{d\phi}\right) d\phi = if^*g\Big|_0^{2\pi} - \int_0^{2\pi} i\left(\frac{df^*}{d\phi}\right) g\, d\phi = \langle \hat{Q} f|g\rangle,$$

so $\hat{Q}$ *is* hermitian (this time the boundary term disappears by virtue of Equation 3.26).

The eigenvalue equation,

$$i\frac{d}{d\phi} f(\phi) = q f(\phi), \tag{3.27}$$

has the general solution

$$f(\phi) = Ae^{-iq\phi}. \tag{3.28}$$

Equation 3.26 restricts the possible values of the q:

$$e^{-iq2\pi} = 1 \quad \Rightarrow \quad q = 0, \pm 1, \pm 2, \ldots. \tag{3.29}$$

The spectrum of this operator is the set of all integers, and it is nondegenerate.

Problem 3.6 Consider the operator $\hat{Q} = d^2/d\phi^2$, where (as in Example 3.1) ϕ is the azimuthal angle in polar coordinates, and the functions are subject to Equation 3.26. Is $\hat{Q}$ hermitian? Find its eigenfunctions and eigenvalues. What is the spectrum of $\hat{Q}$? Is the spectrum degenerate?

3.3 EIGENFUNCTIONS OF A HERMITIAN OPERATOR

Our attention is thus directed to the *eigenfunctions of hermitian operators* (physically: determinate states of observables). These fall into two categories: If the spectrum is **discrete** (i.e. the eigenvalues are separated from one another) then the eigenfunctions lie in Hilbert space and they constitute physically realizable states. If the spectrum is **continuous** (i.e. the eigenvalues fill out an entire range) then the eigenfunctions are not normalizable, and they do not represent possible wave functions (though *linear combinations* of them—involving necessarily a spread in eigenvalues—may be normalizable). Some operators have a discrete spectrum only (for example, the Hamiltonian for the harmonic oscillator), some have only a continuous spectrum (for example, the free particle Hamiltonian), and some have both a discrete part and a continuous part (for example, the Hamiltonian for a finite square well). The discrete case is easier to handle, because the relevant inner products are guaranteed to exist—in fact, it is very similar to the finite-dimensional theory (the eigenvectors of a hermitian *matrix*). I'll treat the discrete case first, and then the continuous one.

3.3.1 Discrete Spectra

Mathematically, the normalizable eigenfunctions of a hermitian operator have two important properties:

Theorem 1 Their eigen*values* are *real*.

Proof: Suppose

$$\hat{Q}f = qf,$$

(i.e. $f(x)$ is an eigenfunction of $\hat{Q}$, with eigenvalue q), and[11]

$$\left\langle f \middle| \hat{Q}f \right\rangle = \left\langle \hat{Q}f \middle| f \right\rangle$$

$\left(\hat{Q}\text{ is hermitian}\right)$. Then

$$q\langle f \mid f\rangle = q^*\langle f \mid f\rangle$$

(q is a *number*, so it comes outside the integral, and because the first function in the inner product is complex conjugated (Equation 3.6), so too is the q on the right). But $\langle f|f\rangle$ cannot be zero ($f(x) = 0$ is not a legal eigenfunction), so $q = q^*$, and hence q is real. QED

This is comforting: If you measure an observable on a particle in a determinate state, you will at least get a real number.

Theorem 2 Eigenfunctions belonging to distinct eigenvalues are *orthogonal*.

Proof: Suppose

$$\hat{Q}f = qf, \quad \text{and} \quad \hat{Q}g = q'g,$$

and $\hat{Q}$ is hermitian. Then $\langle f|\hat{Q}g\rangle = \langle \hat{Q}f|g\rangle$, so

$$q'\langle f \mid g\rangle = q^*\langle f \mid g\rangle$$

(again, the inner products exist because the eigenfunctions are in Hilbert space). But q is real (from Theorem 1), so if $q' \neq q$ it must be that $\langle f|g\rangle = 0$. QED

That's why the stationary states of the infinite square well, for example, or the harmonic oscillator, are orthogonal—they are eigenfunctions of the Hamiltonian with distinct eigenvalues. But this property is not peculiar to them, or even to the Hamiltonian—the same holds for determinate states of *any* observable.

Unfortunately, Theorem 2 tells us nothing about degenerate states $\left(q' = q\right)$. However, if two (or more) eigenfunctions share the same eigenvalue, any linear combination of them is itself an eigenfunction, with the same eigenvalue (Problem 3.7), and we can use the **Gram–Schmidt orthogonalization procedure** (Problem A.4) to *construct* orthogonal eigenfunctions within each degenerate subspace. It is almost never necessary to do this explicitly (thank God!), but it can always be done in principle. So *even in the presence of degeneracy* the eigenfunctions

[11] It is here that we assume the eigenfunctions are in Hilbert space—otherwise the inner product might not exist at all.

can be *chosen* to be orthonormal, and we shall always assume that this has been done. That licenses the use of Fourier's trick, which depends on the orthonormality of the basis functions.

In a *finite*-dimensional vector space the eigenvectors of a hermitian matrix have a third fundamental property: They span the space (every vector can be expressed as a linear combination of them). Unfortunately, the proof does not generalize to infinite-dimensional spaces. But the property itself is essential to the internal consistency of quantum mechanics, so (following Dirac[12]) we will take it as an *axiom* (or, more precisely, as a restriction on the class of hermitian operators that can represent observables):

> **Axiom:** The eigenfunctions of an observable operator are *complete*: Any function (in Hilbert space) can be expressed as a linear combination of them.[13]

Problem 3.7

(a) Suppose that $f(x)$ and $g(x)$ are two eigenfunctions of an operator $\hat{Q}$, with the same eigenvalue q. Show that any linear combination of f and g is itself an eigenfunction of $\hat{Q}$, with eigenvalue q.

(b) Check that $f(x) = \exp(x)$ and $g(x) = \exp(-x)$ are eigenfunctions of the operator d^2/dx^2, with the same eigenvalue. Construct two linear combinations of f and g that are *orthogonal* eigenfunctions on the interval $(-1, 1)$.

Problem 3.8

(a) Check that the eigenvalues of the hermitian operator in Example 3.1 are real. Show that the eigenfunctions (for distinct eigenvalues) are orthogonal.

(b) Do the same for the operator in Problem 3.6.

3.3.2 Continuous Spectra

If the spectrum of a hermitian operator is *continuous*, the eigenfunctions are not normalizable, and the proofs of Theorems 1 and 2 fail, because the inner products may not exist. Nevertheless, there is a sense in which the three essential properties (reality, orthogonality, and completeness) still hold. I think it's best to approach this case through specific examples.

Example 3.2

Find the eigenfunctions and eigenvalues of the momentum operator (on the interval $-\infty < x < \infty$).

Solution: Let $f_p(x)$ be the eigenfunction and p the eigenvalue:

$$-i\hbar\frac{d}{dx}f_p(x) = pf_p(x). \tag{3.30}$$

The general solution is

$$f_p(x) = Ae^{ipx/\hbar}.$$

---

[12] P. A. M. Dirac, *The Principles of Quantum Mechanics*, Oxford University Press, New York (1958).

[13] In some specific cases completeness is provable (we know that the stationary states of the infinite square well, for example, are complete, because of Dirichlet's theorem). It is a little awkward to call something an "axiom" that is *provable* in some cases, but I don't know a better way to do it.

This is not square-integrable for *any* (complex) value of p—the momentum operator has *no* eigenfunctions in Hilbert space.

And yet, if we restrict ourselves to *real* eigenvalues, we do recover a kind of *ersatz* "orthonormality." Referring to Problems 2.23(a) and 2.26,

$$\int_{-\infty}^{\infty} f_{p'}^*(x)\, f_p(x)\, dx = |A|^2 \int_{-\infty}^{\infty} e^{i(p-p')x/\hbar}\, dx = |A|^2\, 2\pi\hbar\, \delta(p-p'). \tag{3.31}$$

If we pick $A = 1/\sqrt{2\pi\hbar}$, so that

$$f_p(x) = \frac{1}{\sqrt{2\pi\hbar}} e^{ipx/\hbar}, \tag{3.32}$$

then

$$\langle f_{p'} | f_p \rangle = \delta(p-p'), \tag{3.33}$$

which is reminiscent of *true* orthonormality (Equation 3.10)—the indices are now continuous variables, and the Kronecker delta has become a Dirac delta, but otherwise it looks just the same. I'll call Equation 3.33 **Dirac orthonormality**.

Most important, the eigenfunctions (with real eigenvalues) are *complete*, with the sum (in Equation 3.11) replaced by an integral: Any (square-integrable) function $f(x)$ can be written in the form

$$f(x) = \int_{-\infty}^{\infty} c(p)\, f_p(x)\, dp = \frac{1}{\sqrt{2\pi\hbar}} \int_{-\infty}^{\infty} c(p)\, e^{ipx/\hbar}\, dp. \tag{3.34}$$

The "coefficients" (now a *function*, $c(p)$) are obtained, as always, by Fourier's trick:

$$\langle f_{p'} \,|\, f \rangle = \int_{-\infty}^{\infty} c(p) \langle f_{p'} \,|\, f_p \rangle\, dp = \int_{-\infty}^{\infty} c(p)\, \delta(p-p')\, dp = c(p'). \tag{3.35}$$

Alternatively, you can get them from Plancherel's theorem (Equation 2.103); indeed, the expansion (Equation 3.34) is nothing but a Fourier transform.

The eigenfunctions of momentum (Equation 3.32) are sinusoidal, with wavelength

$$\lambda = \frac{2\pi\hbar}{p}. \tag{3.36}$$

This is the old de Broglie formula (Equation 1.39), which I promised to justify at the appropriate time. It turns out to be a little more subtle than de Broglie imagined, because we now know that there is actually *no such thing* as a particle with determinate momentum. But we could make a normalizable wave *packet* with a narrow range of momenta, and it is to such an object that the de Broglie relation applies.

What are we to make of Example 3.2? Although none of the eigenfunctions of $\hat{p}$ lives in Hilbert space, a certain family of them (those with real eigenvalues) resides in the nearby "suburbs," with a kind of quasi-normalizability. They do not represent possible physical states, but they are still very useful (as we have already seen, in our study of one-dimensional scattering).[14]

---

[14] What about the eigenfunctions with *non*real eigenvalues? These are not merely non-normalizable—they actually blow up at $\pm\infty$. Functions in what I called the "suburbs" of Hilbert space (the entire metropolitan area is sometimes called a "rigged Hilbert space"; see, for example, Leslie Ballentine's *Quantum Mechanics: A Modern Development*,

Example 3.3

Find the eigenfunctions and eigenvalues of the position operator.

Solution: Let $g_y(x)$ be the eigenfunction and y the eigenvalue:

$$\hat{x}\, g_y(x) = x\, g_y(x) = y\, g_y(x). \tag{3.37}$$

Here y is a fixed number (for any given eigenfunction), but x is a continuous variable. What function of x has the property that multiplying it by x is the same as multiplying it by the constant y? Obviously it's got to be *zero*, except at the one point $x = y$; in fact, it is nothing but the Dirac delta function:

$$g_y(x) = A\delta(x - y).$$

This time the eigenvalue *has* to be real; the eigenfunctions are not square integrable, but again they admit *Dirac* orthonormality:

$$\int_{-\infty}^{\infty} g_{y'}^*(x)\, g_y(x)\, dx = |A|^2 \int_{-\infty}^{\infty} \delta(x - y')\delta(x - y)\, dx = |A|^2\, \delta(y - y'). \tag{3.38}$$

If we pick $A = 1$, so

$$g_y(x) = \delta(x - y), \tag{3.39}$$

then

$$\langle g_{y'} | g_y \rangle = \delta(y - y'). \tag{3.40}$$

These eigenfunctions are also *complete*:

$$f(x) = \int_{-\infty}^{\infty} c(y)\, g_y(x)\, dy = \int_{-\infty}^{\infty} c(y)\delta(x - y)\, dy, \tag{3.41}$$

with

$$c(y) = f(y) \tag{3.42}$$

(trivial, in this case, but you can get it from Fourier's trick if you insist).

If the spectrum of a hermitian operator is *continuous* (so the eigenvalues are labeled by a continuous variable—p or y, in the examples; z, generically, in what follows), the eigenfunctions are not normalizable, they are not in Hilbert space and they do not represent possible physical states; nevertheless, the eigenfunctions with real eigenvalues are *Dirac* orthonormalizable and complete (with the sum now an integral). Luckily, this is all we really require.

World Scientific, 1998) have the property that although they have no (finite) inner product with *themselves*, they *do* admit inner products with all members of Hilbert space. This is *not* true for eigenfunctions of $\hat{p}$ with nonreal eigenvalues. In particular, I showed that the momentum operator is hermitian *for functions in Hilbert space*, but the argument depended on dropping the boundary term (in Equation 3.19). That term is still zero if g is an eigenfunction of $\hat{p}$ with a real eigenvalue (as long as f is in Hilbert space), but not if the eigenvalue has an imaginary part. In this sense *any* complex number is an eigenvalue of the operator $\hat{p}$, but only *real* numbers are eigenvalues of the *hermitian* operator $\hat{p}$—the others lie outside the space over which $\hat{p}$ is hermitian.

Problem 3.9

(a) Cite a Hamiltonian from Chapter 2 (*other* than the harmonic oscillator) that has only a *discrete* spectrum.

(b) Cite a Hamiltonian from Chapter 2 (*other* than the free particle) that has only a *continuous* spectrum.

(c) Cite a Hamiltonian from Chapter 2 (*other* than the finite square well) that has both a discrete and a continuous part to its spectrum.

Problem 3.10 Is the ground state of the infinite square well an eigenfunction of momentum? If so, what is its momentum? If not, *why* not? [For further discussion, see Problem 3.34.]

3.4 GENERALIZED STATISTICAL INTERPRETATION

In Chapter 1 I showed you how to calculate the probability that a particle would be found in a particular location, and how to determine the expectation value of any observable quantity. In Chapter 2 you learned how to find the possible outcomes of an energy measurement, and their probabilities. I am now in a position to state the **generalized statistical interpretation**, which subsumes all of this, and enables you to figure out the possible results of *any* measurement, and their probabilities. Together with the Schrödinger equation (which tells you how the wave function evolves in time) it is the foundation of quantum mechanics.

Generalized statistical interpretation: If you measure an observable $Q(x, p)$ on a particle in the state $\Psi(x, t)$, you are certain to get *one of the eigenvalues* of the hermitian operator $\hat{Q}(x, -i\hbar d/dx)$.[15] If the spectrum of $\hat{Q}$ is discrete, the probability of getting the particular eigenvalue q_n associated with the (orthonormalized) eigenfunction $f_n(x)$ is

$$|c_n|^2, \quad \text{where} \quad c_n = \langle f_n | \Psi \rangle . \tag{3.43}$$

If the spectrum is continuous, with real eigenvalues $q(z)$ and associated (Dirac-orthonormalized) eigenfunctions $f_z(x)$, the probability of getting a result in the range dz is

$$|c(z)|^2 \, dz \quad \text{where} \quad c(z) = \langle f_z | \Psi \rangle . \tag{3.44}$$

Upon measurement, the wave function "collapses" to the corresponding eigenstate.[16]

[15] You may have noticed that there is an ambiguity in this prescription, if $Q(x, p)$ involves the product xp. Because $\hat{x}$ and $\hat{p}$ do not commute (Equation 2.52)—whereas the classical variables x and p, of course, *do*—it is not clear whether we should write $\hat{x}\hat{p}$ or $\hat{p}\hat{x}$ (or perhaps some linear combination of the two). Luckily, such observables are very rare, but when they do occur some other consideration must be invoked to resolve the ambiguity.

[16] In the case of continuous spectra the collapse is to a narrow *range* about the measured value, depending on the precision of the measuring device.

The statistical interpretation is radically different from anything we encounter in classical physics. A somewhat different perspective helps to make it plausible: The eigenfunctions of an observable operator are *complete*, so the wave function can be written as a linear combination of them:

$$\Psi(x,t) = \sum_n c_n(t)\, f_n(x). \tag{3.45}$$

(For simplicity, I'll assume that the spectrum is discrete; it's easy to generalize this discussion to the continuous case.) Because the eigenfunctions are *orthonormal*, the coefficients are given by Fourier's trick:[17]

$$c_n(t) = \langle f_n|\Psi\rangle = \int f_n(x)^*\, \Psi(x,t)\, dx. \tag{3.46}$$

Qualitatively, c_n tells you "how much f_n is contained in Ψ," and given that a measurement has to return one of the eigenvalues of $\hat{Q}$, it seems reasonable that the probability of getting the particular eigenvalue q_n would be determined by the "amount of f_n" in Ψ. But because probabilities are determined by the absolute *square* of the wave function, the precise measure is actually $|c_n|^2$. That's the essential message of the generalized statistical interpretation.[18]

Of course, the *total* probability (summed over all possible outcomes) has got to be *one*:

$$\sum_n |c_n|^2 = 1, \tag{3.47}$$

and sure enough, this follows from the normalization of the wave function:

$$\begin{aligned} 1 = \langle\Psi\mid\Psi\rangle &= \left\langle\left(\sum_{n'} c_{n'} f_{n'}\right)\middle|\left(\sum_n c_n f_n\right)\right\rangle = \sum_{n'}\sum_n c_{n'}^* c_n \langle f_{n'}|f_n\rangle \\ &= \sum_{n'}\sum_n c_{n'}^* c_n \delta_{n'n} = \sum_n c_n^* c_n = \sum_n |c_n|^2. \end{aligned} \tag{3.48}$$

Similarly, the expectation value of Q should be the sum over all possible outcomes of the eigenvalue times the probability of getting that eigenvalue:

$$\langle Q\rangle = \sum_n q_n\, |c_n|^2. \tag{3.49}$$

Indeed,

$$\langle Q\rangle = \langle\Psi|\hat{Q}\Psi\rangle = \left\langle\left(\sum_{n'} c_{n'} f_{n'}\right)\middle|\left(\hat{Q}\sum_n c_n f_n\right)\right\rangle, \tag{3.50}$$

---

[17] Notice that the time dependence—which is not at issue here—is carried by the coefficients; to make this explicit I write $c_n(t)$. In the special case of the Hamiltonian ($\hat{Q} = \hat{H}$), when the potential energy is time independent, the absolute value of each coefficient is in fact constant, as we saw in Section 2.1.

[18] Again, I am scrupulously avoiding the all-too-common claim "$|c_n|^2$ is the probability that the particle is in the state f_n." This is nonsense: The particle is in the state Ψ, *period*. Rather, $|c_n|^2$ is the probability that a *measurement* of Q would yield the value q_n. It is true that such a measurement will collapse the state to the eigenfunction f_n, so one might correctly say "$|c_n|^2$ is the probability that a particle which is *now* in the state Ψ *will be* in the state f_n subsequent to a measurement of Q" ... but that's a quite different assertion.

but $\hat{Q} f_n = q_n f_n$, so

$$\langle Q\rangle = \sum_{n'}\sum_{n} c_{n'}^* c_n q_n \langle f_{n'}|f_n\rangle = \sum_{n'}\sum_{n} c_{n'}^* c_n q_n \delta_{n'n} = \sum_{n} q_n |c_n|^2. \tag{3.51}$$

So far, at least, everything looks consistent.

Can we reproduce, in this language, the original statistical interpretation for position measurements? Sure—it's overkill, but worth checking. A measurement of x on a particle in state Ψ must return one of the eigenvalues of the position operator. Well, in Example 3.3 we found that every (real) number y is an eigenvalue of x, and the corresponding (Dirac-orthonormalized) eigenfunction is $g_y(x) = \delta(x-y)$. Evidently

$$c(y) = \langle g_y|\Psi\rangle = \int_{-\infty}^{\infty} \delta(x-y)\Psi(x,t)\,dx = \Psi(y,t), \tag{3.52}$$

so the probability of getting a result in the range dy is $|\Psi(y,t)|^2\,dy$, which is precisely the original statistical interpretation.

What about momentum? In Example 3.2 we found the (Dirac-orthonormalized) eigenfunctions of the momentum operator, $f_p(x) = (1/\sqrt{2\pi\hbar})\exp(ipx/\hbar)$, so

$$c(p) = \langle f_p|\Psi\rangle = \frac{1}{\sqrt{2\pi\hbar}}\int_{-\infty}^{\infty} e^{-ipx/\hbar}\Psi(x,t)\,dx. \tag{3.53}$$

This is such an important quantity that we give it a special name and symbol: the **momentum space wave function**, $\Phi(p,t)$. It is essentially the *Fourier transform* of the (**position space**) wave function $\Psi(x,t)$—which, by Plancherel's theorem, is its *inverse* Fourier transform:

$$\Phi(p,t) = \frac{1}{\sqrt{2\pi\hbar}}\int_{-\infty}^{\infty} e^{-ipx/\hbar}\Psi(x,t)\,dx; \tag{3.54}$$

$$\Psi(x,t) = \frac{1}{\sqrt{2\pi\hbar}}\int_{-\infty}^{\infty} e^{ipx/\hbar}\Phi(p,t)\,dp. \tag{3.55}$$

According to the generalized statistical interpretation, the probability that a measurement of momentum would yield a result in the range dp is

$$|\Phi(p,t)|^2\,dp. \tag{3.56}$$

Example 3.4

A particle of mass m is bound in the delta function well $V(x) = -\alpha\delta(x)$. What is the probability that a measurement of its momentum would yield a value greater than $p_0 = m\alpha/\hbar$?

Solution: The (position space) wave function is (Equation 2.132)

$$\Psi(x,t) = \frac{\sqrt{m\alpha}}{\hbar}e^{-m\alpha|x|/\hbar^2}e^{-iEt/\hbar}$$

(where $E = -m\alpha^2/2\hbar^2$). The momentum space wave function is therefore

$$\Phi(p,t) = \frac{1}{\sqrt{2\pi\hbar}}\frac{\sqrt{m\alpha}}{\hbar}e^{-iEt/\hbar}\int_{-\infty}^{\infty} e^{-ipx/\hbar}e^{-m\alpha|x|/\hbar^2}\,dx = \sqrt{\frac{2}{\pi}}\frac{p_0^{3/2}e^{-iEt/\hbar}}{p^2+p_0^2}$$

(I looked up the integral). The probability, then, is

$$\frac{2}{\pi}p_0^3\int_{p_0}^{\infty}\frac{1}{\left(p^2+p_0^2\right)^2}\,dp=\frac{1}{\pi}\left[\frac{pp_0}{p^2+p_0^2}+\tan^{-1}\left(\frac{p}{p_0}\right)\right]\Bigg|_{p_0}^{\infty}=\frac{1}{4}-\frac{1}{2\pi}=0.0908$$

(again, I looked up the integral).

Problem 3.11 Find the momentum-space wave function, $\Phi(p,t)$, for a particle in the ground state of the harmonic oscillator. What is the probability (to two significant digits) that a measurement of p on a particle in this state would yield a value outside the classical range (for the same energy)? *Hint:* Look in a math table under "Normal Distribution" or "Error Function" for the numerical part—or use Mathematica.

Problem 3.12 Find $\Phi(p,t)$ for the free particle in terms of the function $\phi(k)$ introduced in Equation 2.101. Show that for the free particle $|\Phi(p,t)|^2$ is independent of time. *Comment*: the time independence of $|\Phi(p,t)|^2$ for the free particle is a manifestation of momentum conservation in this system.

* **Problem 3.13** Show that

$$\langle x\rangle=\int\Phi^*\left(i\hbar\frac{\partial}{\partial p}\right)\Phi\,dp. \tag{3.57}$$

Hint: Notice that $x\exp(ipx/\hbar)=-i\hbar(\partial/\partial p)\exp(ipx/\hbar)$, and use Equation 2.147. In momentum space, then, the position operator is $i\hbar\,\partial/\partial p$. More generally,

$$\langle Q(x,p,t)\rangle=\begin{cases}\int\Psi^*\hat{Q}\left(x,-i\hbar\frac{\partial}{\partial x},t\right)\Psi\,dx, & \text{in position space;}\\ \int\Phi^*\hat{Q}\left(i\hbar\frac{\partial}{\partial p},p,t\right)\Phi\,dp, & \text{in momentum space.}\end{cases} \tag{3.58}$$

In principle you can do all calculations in momentum space just as well (though not always as *easily*) as in position space.

3.5 THE UNCERTAINTY PRINCIPLE

I stated the uncertainty principle (in the form $\sigma_x\sigma_p\ge\hbar/2$), back in Section 1.6, and you have checked it several times, in the problems. But we have never actually *proved* it. In this section I will prove a more general version of the uncertainty principle, and explore some of its ramifications. The argument is beautiful, but rather abstract, so watch closely.

3.5.1 Proof of the Generalized Uncertainty Principle

For any observable A, we have (Equation 3.21):

$$\sigma_A^2=\left\langle\left(\hat{A}-\langle A\rangle\right)\Psi\,\middle|\,\left(\hat{A}-\langle A\rangle\right)\Psi\right\rangle=\langle f|f\rangle$$

where $f \equiv \left(\hat{A} - \langle A \rangle\right)\Psi$. Likewise, for any *other* observable, B,

$$\sigma_B^2 = \langle g \mid g \rangle, \quad \text{where } g \equiv \left(\hat{B} - \langle B \rangle\right)\Psi.$$

Therefore (invoking the Schwarz inequality, Equation 3.7),

$$\sigma_A^2 \sigma_B^2 = \langle f \mid f \rangle \langle g \mid g \rangle \geq |\langle f \mid g \rangle|^2. \tag{3.59}$$

Now, for any complex number z,

$$|z|^2 = [\text{Re}(z)]^2 + [\text{Im}(z)]^2 \geq [\text{Im}(z)]^2 = \left[\frac{1}{2i}(z - z^*)\right]^2. \tag{3.60}$$

Therefore, letting $z = \langle f | g \rangle$,

$$\sigma_A^2 \sigma_B^2 \geq \left(\frac{1}{2i}\left[\langle f | g \rangle - \langle g | f \rangle\right]\right)^2. \tag{3.61}$$

But (exploiting the hermiticity of $\left(\hat{A} - \langle A \rangle\right)$ in the first line)

$$\begin{aligned}
\langle f \mid g \rangle &= \left\langle \left(\hat{A} - \langle A \rangle\right)\Psi \middle| \left(\hat{B} - \langle B \rangle\right)\Psi \right\rangle = \left\langle \Psi \middle| \left(\hat{A} - \langle A \rangle\right)\left(\hat{B} - \langle B \rangle\right)\Psi \right\rangle \\
&= \left\langle \Psi \middle| \left(\hat{A}\hat{B} - \hat{A}\langle B \rangle - \hat{B}\langle A \rangle + \langle A \rangle \langle B \rangle\right)\Psi \right\rangle \\
&= \left\langle \Psi \middle| \hat{A}\hat{B}\Psi \right\rangle - \langle B \rangle \left\langle \Psi \middle| \hat{A}\Psi \right\rangle - \langle A \rangle \left\langle \Psi \middle| \hat{B}\Psi \right\rangle + \langle A \rangle \langle B \rangle \langle \Psi \mid \Psi \rangle \\
&= \langle \hat{A}\hat{B} \rangle - \langle B \rangle \langle A \rangle - \langle A \rangle \langle B \rangle + \langle A \rangle \langle B \rangle \\
&= \langle \hat{A}\hat{B} \rangle - \langle A \rangle \langle B \rangle.
\end{aligned}$$

(Remember, $\langle A \rangle$ and $\langle B \rangle$ are *numbers*, not operators, so you can write them in either order.) Similarly,

$$\langle g \mid f \rangle = \langle \hat{B}\hat{A} \rangle - \langle A \rangle \langle B \rangle,$$

so

$$\langle f | g \rangle - \langle g | f \rangle = \langle \hat{A}\hat{B} \rangle - \langle \hat{B}\hat{A} \rangle = \langle [\hat{A}, \hat{B}] \rangle,$$

where

$$\left[\hat{A}, \hat{B}\right] \equiv \hat{A}\hat{B} - \hat{B}\hat{A}$$

is the commutator of the two operators (Equation 2.49). *Conclusion:*

$$\boxed{\sigma_A^2 \sigma_B^2 \geq \left(\frac{1}{2i}\left\langle\left[\hat{A}, \hat{B}\right]\right\rangle\right)^2.} \tag{3.62}$$

This is the (generalized) **uncertainty principle**. (You might think the i makes it trivial—isn't the right side *negative*? No, for the commutator of two hermitian operators carries its own factor of i, and the two cancel out;[19] the quantity in parentheses is *real*, and its square is *positive*.)

[19] More precisely, the commutator of two hermitian operators is itself *anti*-hermitian $\left(\hat{Q}^\dagger = -\hat{Q}\right)$, and its expectation value is imaginary (Problem 3.32).

As an example, suppose the first observable is position $\left(\hat{A} = x\right)$, and the second is momentum $\left(\hat{B} = -i\hbar d/dx\right)$. We worked out their commutator back in Chapter 2 (Equation 2.52):

$$\left[\hat{x}, \hat{p}\right] = i\hbar.$$

So

$$\sigma_x^2 \sigma_p^2 \geq \left(\frac{1}{2i} i\hbar\right)^2 = \left(\frac{\hbar}{2}\right)^2,$$

or, since standard deviations are by their nature positive,

$$\sigma_x \sigma_p \geq \frac{\hbar}{2}. \tag{3.63}$$

That's the original Heisenberg uncertainty principle, but we now see that it is just one application of a much more general theorem.

There is, in fact, an "uncertainty principle" for *every pair of observables whose operators do not commute*—we call them **incompatible observables**. Incompatible observables do not have shared eigenfunctions—at least, they cannot have a *complete set* of common eigenfunctions (see Problem 3.16). By contrast, *compatible* (commuting) observables *do* admit complete sets of simultaneous eigenfunctions (that is: states that are determinate for both observables).[20] For example, in the hydrogen atom (as we shall see in Chapter 4) the Hamiltonian, the magnitude of the angular momentum, and the z component of angular momentum are mutually compatible observables, and we will construct simultaneous eigenfunctions of all three, labeled by their respective eigenvalues. But there is *no* eigenfunction of position that is also an eigenfunction of momentum; these operators are *in*compatible.

Note that the uncertainty principle is not an *extra* assumption in quantum theory, but rather a *consequence* of the statistical interpretation. You might wonder how it is enforced in the laboratory—*why* can't you determine (say) both the position and the momentum of a particle? You can certainly measure the position of the particle, but the act of measurement collapses the wave function to a narrow spike, which necessarily carries a broad range of wavelengths (hence momenta) in its Fourier decomposition. If you now measure the momentum, the state will collapse to a long sinusoidal wave, with (now) a well-defined wavelength—but the particle no longer has the position you got in the first measurement.[21] The problem, then, is that the second measurement renders the outcome of the first measurement obsolete. Only if the wave function were simultaneously an eigenstate of both observables would it be possible to

[20] This corresponds to the fact that noncommuting matrices cannot be simultaneously diagonalized (that is, they cannot both be brought to diagonal form by the same similarity transformation), whereas commuting hermitian matrices *can* be simultaneously diagonalized. See Section A.5.

[21] Bohr and Heisenberg were at pains to track down the *mechanism* by which the measurement of x (for instance) destroys the previously existing value of p. The crux of the matter is that in order to determine the position of a particle you have to poke it with something—shine light on it, say. But these photons impart to the particle a momentum you cannot control. You now know the position, but you no longer know the momentum. Bohr's famous debates with Einstein include many delightful examples, showing in detail how experimental constraints enforce the uncertainty principle. For an inspired account see Bohr's article in *Albert Einstein: Philosopher-Scientist*, edited by Paul A. Schilpp, Open Court Publishing Co., Peru, IL (1970). In recent years the Bohr/Heisenberg explanation has been called into question; for a nice discussion see G. Brumfiel, *Nature News* https://doi.org/10.1038/nature.2012.11394.

make the second measurement without disturbing the state of the particle (the second collapse wouldn't change anything, in that case). But this is only possible, in general, if the two observables are compatible.

* **Problem 3.14**

(a) Prove the following commutator identities:

$$\left[\hat{A}+\hat{B},\hat{C}\right]=\left[\hat{A},\hat{C}\right]+\left[\hat{B},\hat{C}\right], \tag{3.64}$$

$$\left[\hat{A}\hat{B},\hat{C}\right]=\hat{A}\left[\hat{B},\hat{C}\right]+\left[\hat{A},\hat{C}\right]\hat{B}. \tag{3.65}$$

(b) Show that

$$\left[x^n,\hat{p}\right]=i\hbar n x^{n-1}.$$

(c) Show more generally that

$$\left[f(x),\hat{p}\right]=i\hbar\frac{df}{dx}, \tag{3.66}$$

for any function $f(x)$ that admits a Taylor series expansion.

(d) Show that for the simple harmonic oscillator

$$\left[\hat{H},\hat{a}_\pm\right]=\pm\hbar\omega\hat{a}_\pm. \tag{3.67}$$

Hint: Use Equation 2.54.

* **Problem 3.15** Prove the famous "(your name) uncertainty principle," relating the uncertainty in position $(A=x)$ to the uncertainty in energy $\left(B=p^2/2m+V\right)$:

$$\sigma_x\sigma_H\geq\frac{\hbar}{2m}\left|\langle p\rangle\right|.$$

For stationary states this doesn't tell you much—why not?

Problem 3.16 Show that two noncommuting operators cannot have a complete set of common eigenfunctions. *Hint:* Show that *if* $\hat{P}$ and $\hat{Q}$ have a complete set of common eigenfunctions, then $\left[\hat{P},\hat{Q}\right]f=0$ for any function in Hilbert space.

3.5.2 The Minimum-Uncertainty Wave Packet

We have twice encountered wave functions that *hit* the position-momentum uncertainty limit $\left(\sigma_x\sigma_p=\hbar/2\right)$: the ground state of the harmonic oscillator (Problem 2.11) and the Gaussian wave packet for the free particle (Problem 2.21). This raises an interesting question: What is the *most general* minimum-uncertainty wave packet? Looking back at the proof of the uncertainty principle, we note that there were two points at which *in*equalities came into the argument: Equation 3.59 and Equation 3.60. Suppose we require that each of these be an *equality*, and see what this tells us about Ψ.

The Schwarz inequality becomes an equality when one function is a multiple of the other: $g(x)=cf(x)$, for some complex number c (see Problem A.5). Meanwhile, in Equation

3.60 I threw away the real part of z; equality results if $\text{Re}(z) = 0$, which is to say, if $\text{Re}\langle f|g\rangle = \text{Re}(c\langle f|f\rangle) = 0$. Now, $\langle f|f\rangle$ is certainly real, so this means the constant c must be pure imaginary—let's call it ia. The necessary and sufficient condition for minimum uncertainty, then, is

$$g(x) = iaf(x), \quad \text{where } a \text{ is real.} \tag{3.68}$$

For the position-momentum uncertainty principle this criterion becomes:

$$\left(-i\hbar\frac{d}{dx} - \langle p\rangle\right)\Psi = ia(x - \langle x\rangle)\Psi, \tag{3.69}$$

which is a differential equation for Ψ as a function of x. Its general solution (see Problem 3.17) is

$$\Psi(x) = Ae^{-a(x-\langle x\rangle)^2/2\hbar}e^{i\langle p\rangle x/\hbar}. \tag{3.70}$$

Evidently the minimum-uncertainty wave packet is a *gaussian*—and, sure enough, the two examples we encountered earlier *were* gaussians.[22]

Problem 3.17 Solve Equation 3.69 for $\Psi(x)$. Note that $\langle x\rangle$ and $\langle p\rangle$ are *constants* (independent of x).

3.5.3 The Energy-Time Uncertainty Principle

The position-momentum uncertainty principle is often written in the form

$$\Delta x\,\Delta p \geq \frac{\hbar}{2}; \tag{3.71}$$

Δx (the "uncertainty" in x) is loose notation (and sloppy language) for the standard deviation of the results of repeated measurements on identically prepared systems.[23] Equation 3.71 is often paired with the **energy-time uncertainty principle**,

$$\Delta t\,\Delta E \geq \frac{\hbar}{2}. \tag{3.72}$$

Indeed, in the context of special relativity the energy-time form might be thought of as a *consequence* of the position-momentum version, because x and t (or rather, ct) go together in the position-time four-vector, while p and E (or rather, E/c) go together in the energy-momentum four-vector. So in a relativistic theory Equation 3.72 would be a necessary concomitant to Equation 3.71. But we're not doing relativistic quantum mechanics. The Schrödinger equation is explicitly nonrelativistic: It treats t and x on a very unequal footing (as a differential equation it is *first*-order in t, but *second*-order in x), and Equation 3.72 is emphatically *not* implied by Equation 3.71. My purpose now is to *derive* the energy-time uncertainty principle, and in

[22] Note that it is only the dependence of Ψ on x that is at issue here—the "constants" A, a, $\langle x\rangle$, and $\langle p\rangle$ may all be functions of time, and for that matter Ψ may evolve away from the minimal form. All I'm asserting is that if, at some instant, the wave function is gaussian in x, then (at that instant) the uncertainty product is minimal.

[23] Many casual applications of the uncertainty principle are actually based (often inadvertently) on a completely different—and sometimes quite unjustified—measure of "uncertainty." See J. Hilgevoord, *Am. J. Phys.* **70**, 983 (2002).

the course of that derivation to persuade you that it is really an altogether different beast, whose superficial resemblance to the position-momentum uncertainty principle is actually quite misleading.

After all, position, momentum, and energy are all dynamical variables—measurable characteristics of the system, at any given time. But time itself is not a dynamical variable (not, at any rate, in a nonrelativistic theory): You don't go out and measure the "time" of a particle, as you might its position or its energy. Time is the *independent* variable, of which the dynamical quantities are *functions*. In particular, the Δt in the energy-time uncertainty principle is not the standard deviation of a collection of time measurements; roughly speaking (I'll make this more precise in a moment) it is the *time it takes the system to change substantially*.

As a measure of how fast the system is changing, let us compute the time derivative of the expectation value of some observable, $Q(x, p, t)$:

$$\frac{d}{dt}\langle Q\rangle = \frac{d}{dt}\left\langle \Psi \middle| \hat{Q}\Psi\right\rangle = \left\langle \frac{\partial\Psi}{\partial t} \middle| \hat{Q}\Psi\right\rangle + \left\langle \Psi \middle| \frac{\partial\hat{Q}}{\partial t}\Psi\right\rangle + \left\langle \Psi \middle| \hat{Q}\frac{\partial\Psi}{\partial t}\right\rangle.$$

Now, the Schrödinger equation says

$$i\hbar\frac{\partial\Psi}{\partial t} = \hat{H}\Psi$$

(where $H = p^2/2m + V$ is the Hamiltonian). So

$$\frac{d}{dt}\langle Q\rangle = -\frac{1}{i\hbar}\left\langle \hat{H}\Psi \middle| \hat{Q}\Psi\right\rangle + \frac{1}{i\hbar}\left\langle \Psi \middle| \hat{Q}\hat{H}\Psi\right\rangle + \left\langle \frac{\partial\hat{Q}}{\partial t}\right\rangle.$$

But $\hat{H}$ is hermitian, so $\langle\hat{H}\Psi|\hat{Q}\Psi\rangle = \langle\Psi|\hat{H}\hat{Q}\Psi\rangle$, and hence

$$\boxed{\frac{d}{dt}\langle Q\rangle = \frac{i}{\hbar}\left\langle\left[\hat{H}, \hat{Q}\right]\right\rangle + \left\langle \frac{\partial\hat{Q}}{\partial t}\right\rangle.} \tag{3.73}$$

This is an interesting and useful result in its own right (see Problems 3.18 and 3.37). It has no name, though it surely deserves one; I'll call it the **generalized Ehrenfest theorem**. In the typical case where the operator does not depend explicitly on time,[24] it tells us that the rate of change of the expectation value is determined by the commutator of the operator with the Hamiltonian. In particular, if $\hat{Q}$ *commutes* with $\hat{H}$, then $\langle Q\rangle$ is constant, and in this sense Q is a *conserved* quantity.

Now, suppose we pick $A = H$ and $B = Q$, in the generalized uncertainty principle (Equation 3.62), and assume that Q does not depend explicitly on t:

$$\sigma_H^2\sigma_Q^2 \geq \left(\frac{1}{2i}\left\langle\left[\hat{H}, \hat{Q}\right]\right\rangle\right)^2 = \left(\frac{1}{2i}\frac{\hbar}{i}\frac{d\langle Q\rangle}{dt}\right)^2 = \left(\frac{\hbar}{2}\right)^2\left(\frac{d\langle Q\rangle}{dt}\right)^2.$$

Or, more simply,

$$\sigma_H\sigma_Q \geq \frac{\hbar}{2}\left|\frac{d\langle Q\rangle}{dt}\right|. \tag{3.74}$$

---

[24] Operators that depend explicitly on t are quite rare, so *almost always* $\partial\hat{Q}/\partial t = 0$. As an example of *explicit* time dependence, consider the potential energy of a harmonic oscillator whose spring constant is changing (perhaps the temperature is rising, so the spring becomes more flexible): $Q = (1/2)\,m\,[\omega(t)]^2\,x^2$.

Let's define $\Delta E \equiv \sigma_H$, and

$$\Delta t \equiv \frac{\sigma_Q}{|d\langle Q\rangle/dt|}. \tag{3.75}$$

Then

$$\Delta E\,\Delta t \geq \frac{\hbar}{2}, \tag{3.76}$$

and that's the energy-time uncertainty principle. But notice what is meant by Δt, here: Since

$$\sigma_Q = \left|\frac{d\langle Q\rangle}{dt}\right|\Delta t,$$

Δt represents the *amount of time it takes the expectation value of Q to change by one standard deviation*.[25] In particular, Δt depends entirely on what observable (Q) you care to look at—the change might be rapid for one observable and slow for another. But if ΔE is small, then the rate of change of *all* observables must be very gradual; or, to put it the other way around, if *any* observable changes rapidly, the "uncertainty" in the energy must be large.

Example 3.5

In the extreme case of a stationary state, for which the energy is uniquely determined, all expectation values are constant in time ($\Delta E = 0 \Rightarrow \Delta t = \infty$)—as in fact we noticed some time ago (see Equation 2.9). To make something *happen* you must take a linear combination of at least two stationary states—say:

$$\Psi(x,t) = a\psi_1(x)e^{-iE_1t/\hbar} + b\psi_2(x)e^{-iE_2t/\hbar}.$$

If a, b, ψ_1, and ψ_2 are real,

$$|\Psi(x,t)|^2 = a^2\left(\psi_1(x)\right)^2 + b^2\left(\psi_2(x)\right)^2 + 2ab\psi_1(x)\psi_2(x)\cos\left(\frac{E_2-E_1}{\hbar}t\right).$$

The period of oscillation is $\tau = 2\pi\hbar/\left(E_2 - E_1\right)$. Roughly speaking, $\Delta E = E_2 - E_1$ and $\Delta t = \tau$ (for the *exact* calculation see Problem 3.20), so

$$\Delta E\,\Delta t = 2\pi\hbar,$$

which is indeed $\geq \hbar/2$.

Example 3.6

Let Δt be the time it takes a free-particle wave packet to pass a particular point (Figure 3.1). Qualitatively (an exact version is explored in Problem 3.21), $\Delta t = \Delta x/v = m\Delta x/p$. But $E = p^2/2m$, so $\Delta E = p\Delta p/m$, and therefore,

$$\Delta E\,\Delta t = \frac{p\Delta p}{m}\frac{m\Delta x}{p} = \Delta x\Delta p,$$

which is $\geq \hbar/2$ by the position-momentum uncertainty principle.

[25] This is sometimes called the "Mandelstam–Tamm" formulation of the energy-time uncertainty principle. For a review of alternative approaches see P. Busch, *Found. Phys.* **20**, 1 (1990).

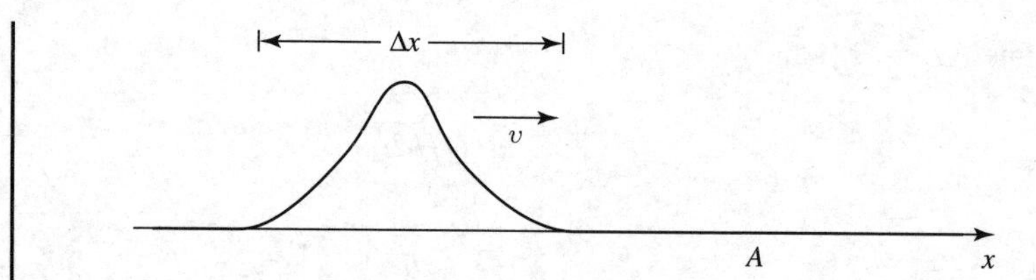

Figure 3.1: A free particle wave packet approaches the point A (Example 3.6).

Example 3.7

The Δ particle lasts about 10^{-23} s, before spontaneously disintegrating. If you make a histogram of all measurements of its mass, you get a kind of bell-shaped curve centered at 1232 MeV/c^2, with a width of about 120 MeV/c^2 (Figure 3.2). Why does the rest energy (mc^2) sometimes come out higher than 1232, and sometimes lower? Is this experimental error? No, for if we take Δt to be the lifetime of the particle (certainly *one* measure of "how long it takes the system to change appreciably"),

$$\Delta E\,\Delta t = \left(\frac{120}{2}\text{ MeV}\right)\left(10^{-23}\text{ s}\right) = 6 \times 10^{-22}\text{ MeV s},$$

whereas $\hbar/2 = 3 \times 10^{-22}$ MeV s. So the spread in m is about as small as the uncertainty principle allows—a particle with so short a lifetime just doesn't *have* a very well-defined mass.[26]

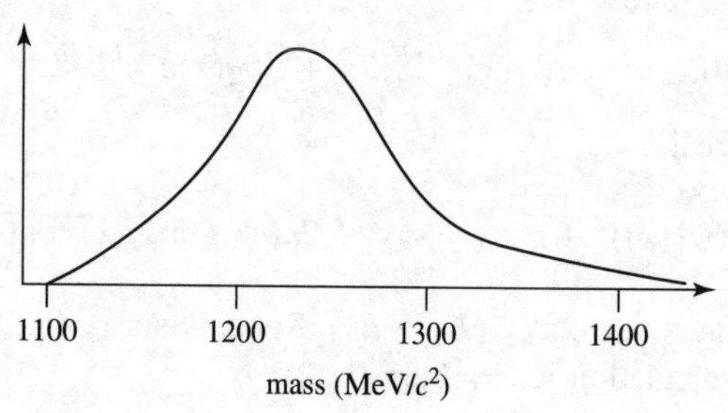

Figure 3.2: Measurements of the Δ mass (Example 3.7).

Notice the variety of specific meanings attaching to the term Δt in these examples: In Example 3.5 it's a period of oscillation; in Example 3.6 it's the time it takes a particle to pass a point; in Example 3.7 it's the lifetime of an unstable particle. In every case, however, Δt is the time it takes for the system to undergo "substantial" change.

It is often said that the uncertainty principle means energy is not strictly conserved in quantum mechanics—that you're allowed to "borrow" energy ΔE, as long as you "pay it back"

[26] In truth, Example 3.7 is a bit of a fraud. You can't measure 10^{-23} s on a stop-watch, and in practice the lifetime of such a short-lived particle is *inferred* from the width of the mass plot, using the uncertainty principle as *input*. However, the point is valid, even if the logic is backwards. Moreover, if you assume the Δ is about the same size as a proton $\left(\sim 10^{-15}\text{ m}\right)$, then 10^{-23} sec is roughly the time it takes light to cross the particle, and it's hard to imagine that the lifetime could be much *less* than that.

in a time $\Delta t \approx \hbar/(2\Delta E)$; the greater the violation, the briefer the period over which it can occur. Now, there are many legitimate readings of the energy-time uncertainty principle, but this is not one of them. Nowhere does quantum mechanics license violation of energy conservation, and certainly no such authorization entered into the derivation of Equation 3.76. But the uncertainty principle is extraordinarily robust: It can be misused without leading to seriously incorrect results, and as a consequence physicists are in the habit of applying it rather carelessly.

* **Problem 3.18** Apply Equation 3.73 to the following special cases: (a) $Q = 1$; (b) $Q = H$; (c) $Q = x$; (d) $Q = p$. In each case, comment on the result, with particular reference to Equations 1.27, 1.33, 1.38, and conservation of energy (see remarks following Equation 2.21).

Problem 3.19 Use Equation 3.73 (or Problem 3.18 (c) and (d)) to show that:

(a) For any (normalized) wave packet representing a free particle ($V(x) = 0$), $\langle x \rangle$ moves at constant velocity (this is the quantum analog to Newton's first law). *Note:* You showed this for a *gaussian* wave packet in Problem 2.42, but it is completely general.

(b) For any (normalized) wave packet representing a particle in the harmonic oscillator potential $\left(V(x) = \frac{1}{2}m\omega^2 x^2\right)$, $\langle x \rangle$ oscillates at the classical frequency. *Note:* You showed this for a particular *gaussian* wave packet in Problem 2.49, but it is completely general.

Problem 3.20 Test the energy-time uncertainty principle for the wave function in Problem 2.5 and the observable x, by calculating σ_H, σ_x, and $d\langle x \rangle/dt$ exactly.

Problem 3.21 Test the energy-time uncertainty principle for the free particle wave packet in Problem 2.42 and the observable x, by calculating σ_H, σ_x, and $d\langle x \rangle/dt$ exactly.

Problem 3.22 Show that the energy-time uncertainty principle reduces to the "your name" uncertainty principle (Problem 3.15), when the observable in question is x.

3.6 VECTORS AND OPERATORS

3.6.1 Bases in Hilbert Space

Imagine an ordinary vector **A** in two dimensions (Fig. 3.3(a)). How would you describe this vector to someone? You might tell them "It's about an inch long, and it points 20° clockwise from straight up, with respect to the page." But that's pretty awkward. A better way would be

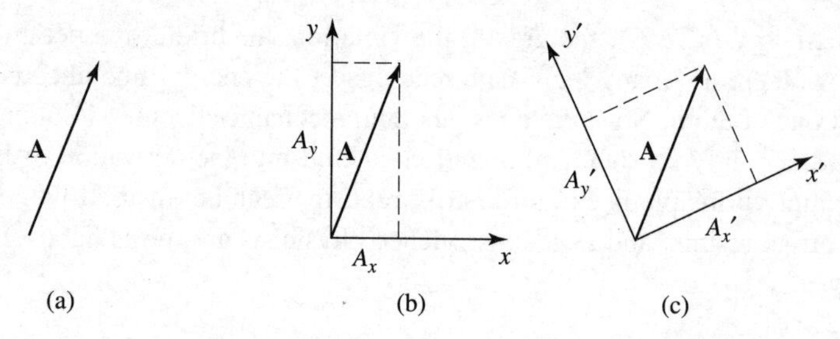

Figure 3.3: (a) Vector **A**. (b) Components of **A** with respect to xy axes. (c) Components of **A** with respect to $x'y'$ axes.

to introduce cartesian axes, x and y, and specify the *components* of **A**: $A_x = \hat{\imath} \cdot \mathbf{A},\ A_y = \hat{\jmath} \cdot \mathbf{A}$ (Fig. 3.3(b)). Of course, your sister might draw a different set of axes, x' and y', and she would report different components: $A'_x = \hat{\imath}' \cdot \mathbf{A},\ A'_y = \hat{\jmath}' \cdot \mathbf{A}$ (Fig. 3.3(c)) ... but it's all the same *vector*—we're simply expressing it with respect to two different **bases** $\left(\{\hat{\imath}, \hat{\jmath}\}\right.$ and $\left.\{\hat{\imath}', \hat{\jmath}'\}\right)$. The vector itself lives "out there in space," independent of anybody's (arbitrary) choice of coordinates.

The same is true for the state of a system in quantum mechanics. It is represented by a *vector*, $|\mathcal{S}(t)\rangle$, that lives "out there in Hilbert space," but we can *express* it with respect to any number of different *bases*. The wave function $\Psi(x,t)$ is actually the x "component" in the expansion of $|\mathcal{S}(t)\rangle$ in the basis of position eigenfunctions:

$$\boxed{\Psi(x,t) = \langle x|\mathcal{S}(t)\rangle\,,} \tag{3.77}$$

(the analog to $\hat{\imath} \cdot \mathbf{A}$) with $|x\rangle$ standing for the eigenfunction of $\hat{x}$ with eigenvalue x.[27] The momentum space wave function $\Phi(p,t)$ is the p component in the expansion of $|\mathcal{S}(t)\rangle$ in the basis of momentum eigenfunctions:

$$\Phi(p,t) = \langle p|\mathcal{S}(t)\rangle \tag{3.78}$$

(with $|p\rangle$ standing for the eigenfunction of $\hat{p}$ with eigenvalue p).[28] Or we could expand $|\mathcal{S}(t)\rangle$ in the basis of energy eigenfunctions (supposing for simplicity that the spectrum is discrete):

$$c_n(t) = \langle n|\mathcal{S}(t)\rangle \tag{3.79}$$

(with $|n\rangle$ standing for the nth eigenfunction of $\hat{H}$—Equation 3.46). But it's all the same state; the functions Ψ and Φ, and the collection of coefficients $\{c_n\}$, contain exactly the same information—they are simply three different ways of identifying the same vector:

---

[27] I hesitate to call it g_x (Equation 3.39), because that is its form in the position basis, and the whole point here is to free ourselves from any particular basis. Indeed, when I first defined Hilbert space as the set of square-integrable functions—over x—that was already too restrictive, committing us to a specific representation (the position basis). I want now to think of it as an abstract vector space, whose members can be expressed with respect to any basis you like.

[28] In position space it would be f_p (Equation 3.32).

$$|\mathcal{S}(t)\rangle \to \int \Psi(y,t)\delta(x-y)\,dy = \int \Phi(p,t)\frac{1}{\sqrt{2\pi\hbar}}e^{ipx/\hbar}\,dp$$
$$= \sum c_n e^{-iE_n t/\hbar}\psi_n(x). \tag{3.80}$$

Operators (representing observables) are linear transformations on Hilbert space—they "transform" one vector into another:

$$|\beta\rangle = \hat{Q}|\alpha\rangle. \tag{3.81}$$

Just as vectors are represented, with respect to an orthonormal basis $\{|e_n\rangle\}$,[29] by their components,

$$|\alpha\rangle = \sum_n a_n|e_n\rangle, \quad |\beta\rangle = \sum_n b_n|e_n\rangle, \quad a_n = \langle e_n|\alpha\rangle, \quad b_n = \langle e_n \mid \beta\rangle, \tag{3.82}$$

operators are represented (with respect to a particular basis) by their **matrix elements**[30]

$$\left\langle e_m \middle| \hat{Q} \middle| e_n \right\rangle \equiv Q_{mn}. \tag{3.83}$$

In this notation Equation 3.81 says

$$\sum_n b_n\,|e_n\rangle = \sum_n a_n\,\hat{Q}|e_n\rangle, \tag{3.84}$$

or, taking the inner product with $|e_m\rangle$,

$$\sum_n b_n\,\langle e_m \mid e_n\rangle = \sum_n a_n \left\langle e_m \middle| \hat{Q} \middle| e_n \right\rangle, \tag{3.85}$$

and hence (since $\langle e_m|e_n\rangle = \delta_{mn}$)

$$b_m = \sum_n Q_{mn}a_n. \tag{3.86}$$

Thus the matrix elements of $\hat{Q}$ tell you how the components transform.[31]

Later on we will encounter systems that admit only a finite number N of linearly independent states. In that case $|\mathcal{S}(t)\rangle$ lives in an N-dimensional vector space; it can be represented as a column of (N) components (with respect to a given basis), and operators take the form of ordinary $(N \times N)$ matrices. These are the simplest quantum systems—none of the subtleties associated with infinite-dimensional vector spaces arise. Easiest of all is the two-state system, which we explore in the following example.

---

[29] I'll assume the basis is discrete; otherwise n becomes a continuous index and the sums are replaced by integrals.

[30] This terminology is inspired, obviously, by the finite-dimensional case, but the "matrix" will now typically have an infinite (maybe even uncountable) number of elements.

[31] In matrix notation Equation 3.86 becomes $\mathsf{b} = \mathsf{Qa}$ (with the vectors expressed as columns), by the ordinary rules of matrix multiplication—see Equation A.42.

Example 3.8

Imagine a system in which there are just *two* linearly independent states:[32]

$$|1\rangle = \begin{pmatrix}1\\0\end{pmatrix} \quad \text{and} \quad |2\rangle = \begin{pmatrix}0\\1\end{pmatrix}.$$

The most general state is a normalized linear combination:

$$|S\rangle = a|1\rangle + b|2\rangle = \begin{pmatrix}a\\b\end{pmatrix}, \quad \text{with } |a|^2 + |b|^2 = 1.$$

The Hamiltonian can be expressed as a (hermitian) matrix (Equation 3.83); suppose it has the specific form

$$\mathsf{H} = \begin{pmatrix}h & g\\ g & h\end{pmatrix},$$

where g and h are real constants. If the system starts out (at $t = 0$) in state $|1\rangle$, what is its state at time t?

Solution: The (time-dependent) Schrödinger equation[33] says

$$i\hbar \frac{d}{dt}|\mathcal{S}(t)\rangle = \hat{H}|\mathcal{S}(t)\rangle. \tag{3.87}$$

As always, we begin by solving the time-*in*dependent Schrödinger equation:

$$\hat{H}|s\rangle = E|s\rangle; \tag{3.88}$$

that is, we look for the eigenvectors and eigenvalues of $\hat{H}$. The characteristic equation determines the eigenvalues:

$$\det\begin{pmatrix}h-E & g\\ g & h-E\end{pmatrix} = (h-E)^2 - g^2 = 0 \Rightarrow h - E = \mp g \Rightarrow E_{\pm} = h \pm g.$$

Evidently the allowed energies are $(h+g)$ and $(h-g)$. To determine the eigenvectors, we write

$$\begin{pmatrix}h & g\\ g & h\end{pmatrix}\begin{pmatrix}\alpha\\ \beta\end{pmatrix} = (h \pm g)\begin{pmatrix}\alpha\\ \beta\end{pmatrix} \Rightarrow h\alpha + g\beta = (h \pm g)\alpha \Rightarrow \beta = \pm\alpha,$$

so the normalized eigenvectors are

$$|s_{\pm}\rangle = \frac{1}{\sqrt{2}}\begin{pmatrix}1\\ \pm 1\end{pmatrix}.$$

Next we expand the initial state as a linear combination of eigenvectors of the Hamiltonian:

$$|\mathcal{S}(0)\rangle = \begin{pmatrix}1\\0\end{pmatrix} = \frac{1}{\sqrt{2}}(|s_+\rangle + |s_-\rangle).$$

---

[32] Technically, the "equals" signs here mean "is represented by," but I don't think any confusion will arise if we adopt the customary informal notation.

[33] We began, back in Chapter 1, with the Schrödinger equation for the wave function in position space; here we generalize it to the state vector in Hilbert space.

Finally, we tack on the standard time-dependence (the wiggle factor) $\exp(-iE_nt/\hbar)$:

$$|\mathcal{S}(t)\rangle = \frac{1}{\sqrt{2}}\left[e^{-i(h+g)t/\hbar}|s_+\rangle + e^{-i(h-g)t/\hbar}|s_-\rangle\right]$$
$$= \frac{1}{2}e^{-iht/\hbar}\left[e^{-igt/\hbar}\begin{pmatrix}1\\1\end{pmatrix} + e^{igt/\hbar}\begin{pmatrix}1\\-1\end{pmatrix}\right]$$
$$= \frac{1}{2}e^{-iht/\hbar}\begin{pmatrix}e^{-igt/\hbar}+e^{igt/\hbar}\\ e^{-igt/\hbar}-e^{igt/\hbar}\end{pmatrix} = e^{-iht/\hbar}\begin{pmatrix}\cos(gt/\hbar)\\ -i\sin(gt/\hbar)\end{pmatrix}.$$

If you doubt this result, by all means *check* it: Does it satisfy the time-dependent Schrödinger equation (Equation 3.87)? Does it match the initial state when $t = 0$?[34]

Just as vectors look different when expressed in different bases, so too do operators (or, in the discrete case, the matrices that represent them). We have already encountered a particularly nice example:

$$\hat{x}\text{ (the position operator)} \rightarrow \begin{cases} x & \text{(in position space)},\\ i\hbar\partial/\partial p & \text{(in momentum space)}; \end{cases}$$

$$\hat{p}\text{ (the momentum operator)} \rightarrow \begin{cases} -i\hbar\partial/\partial x & \text{(in position space)},\\ p & \text{(in momentum space)}. \end{cases}$$

("Position space" is nothing but the position basis; "momentum space" is the momentum basis.) If someone asked you, "What is the operator, $\hat{x}$, representing position, in quantum mechanics?" you would probably answer "Just x itself." But an equally correct reply would be "$i\hbar\,\partial/\partial p$," and the best response would be "With respect to what basis?"

I have often said "the state of a system is represented by its wave function, $\Psi(x,t)$," and this is true, in the same sense that an ordinary vector in three dimensions is "represented by" the triplet of its components; but really, I should always add "in the position basis." After all, the state of the system is a vector in Hilbert space, $|\mathcal{S}(t)\rangle$; it makes no reference to any particular basis. Its connection to $\Psi(x,t)$ is given by Equation 3.77: $\Psi(x,t) = \langle x|\mathcal{S}(t)\rangle$. Having said that, for the most part we *do* in fact work in position space, and no serious harm comes from referring to the wave function as "the state of the system."

3.6.2 Dirac Notation

Dirac proposed to chop the bracket notation for the inner product, $\langle\alpha|\beta\rangle$, into two pieces, which he called **bra**, $\langle\alpha|$, and **ket**, $|\beta\rangle$ (I don't know what happened to the c). The latter is a vector, but what exactly is the former? It's a *linear function* of vectors, in the sense that when it hits a vector (to its right) it yields a (complex) number—the inner product. (When an *operator* hits a vector, it delivers another vector; when a *bra* hits a vector, it delivers a number.) In a function space, the bra can be thought of as an instruction to integrate:

$$\langle f| = \int f^*[\cdots]\,dx,$$

[34] This is a crude model for (among other things) **neutrino oscillations**. In that context $|1\rangle$ represents (say) the electron neutrino, and $|2\rangle$ the muon neutrino; if the Hamiltonian has a nonvanishing off-diagonal term (g) then in the course of time the electron neutrino will turn into a muon neutrino (and back again).

with the ellipsis $[\cdots]$ waiting to be filled by whatever function the bra encounters in the ket to its right. In a finite-dimensional vector space, with the kets expressed as columns (of components with respect to some basis),

$$|\alpha\rangle \rightarrow \begin{pmatrix} a_1 \\ a_2 \\ \vdots \\ a_n \end{pmatrix}, \tag{3.89}$$

the bras are rows:

$$\langle\beta| \rightarrow \left(b_1^* \; b_2^* \; \ldots \; b_n^*\right), \tag{3.90}$$

and $\langle\beta|\alpha\rangle = b_1^* a_1 + b_2^* a_2 + \cdots + b_n^* a_n$ is the matrix product. The collection of all bras constitutes another vector space—the so-called **dual space**.

The license to treat bras as separate entities in their own right allows for some powerful and pretty notation. For example, if $|\alpha\rangle$ is a normalized vector, the operator

$$\hat{P} \equiv |\alpha\rangle\langle\alpha| \tag{3.91}$$

picks out the portion of any other vector that "lies along" $|\alpha\rangle$:

$$\hat{P}\,|\beta\rangle = (\langle\alpha|\,\beta\rangle)\,|\alpha\rangle;$$

we call it the **projection operator** onto the one-dimensional subspace spanned by $|\alpha\rangle$. If $\{|e_n\rangle\}$ is a discrete orthonormal basis,

$$\langle e_m|e_n\rangle = \delta_{mn}, \tag{3.92}$$

then

$$\boxed{\sum_n |e_n\rangle\langle e_n| = 1} \tag{3.93}$$

(the identity operator). For if we let this operator act on any vector $|\alpha\rangle$, we recover the expansion of $|\alpha\rangle$ in the $\{|e_n\rangle\}$ basis:

$$\sum_n (\langle e_n\,|\alpha\rangle)|\,e_n\rangle = |\alpha\rangle. \tag{3.94}$$

Similarly, if $\{|e_z\rangle\}$ is a *Dirac* orthonormalized continuous basis,

$$\langle e_z|e_{z'}\rangle = \delta(z - z'), \tag{3.95}$$

then

$$\boxed{\int |e_z\rangle\langle e_z|\,dz = 1.} \tag{3.96}$$

Equations 3.93 and 3.96 are the tidiest ways to express *completeness*.

Technically, the guts of a ket or a bra (the ellipsis in $|\cdots\rangle$ or $\langle\cdots|$) is a *name*—the name of the vector in question: "α," or "n," or for that matter "Alice," or "Bob." It is endowed with no intrinsic mathematical attributes. Of course, it may be helpful to choose an evocative name—for instance, if you're working in the space L^2 of square-integrable functions, it is natural to name each vector after the function it represents: $|f\rangle$. Then, for example, we can write the definition of a hermitian operator as we did in Equation 3.17:

$$\left\langle f \,\middle|\, \hat{Q} f\right\rangle = \left\langle \hat{Q} f \,\middle|\, f\right\rangle.$$

Strictly speaking, in Dirac notation this is a nonsense expression: f here is a *name*, and operators act on *vectors*, not on *names*. The left side should properly be written as

$$\left\langle f \middle| \hat{Q} \middle| f \right\rangle,$$

but what are we to make of the right side? $\left\langle \hat{Q} f \right|$ really means "the bra dual to $\hat{Q} \,| f \rangle$," but what is its name? I suppose we could say

$$\left\langle \left(\text{the name of the vector } \hat{Q}\,| f \rangle \right) \right|,$$

but that's a mouthful. However, since we have chosen to name each vector after the function it represents, and since we *do* know how $\hat{Q}$ acts on the *function* (as opposed to the *name*) f, this in fact becomes[35]

$$\langle \hat{Q} f |,$$

and we are OK after all.[36]

An operator takes one vector in Hilbert space and delivers another:

$$\hat{Q} |\alpha\rangle = |\beta\rangle . \tag{3.97}$$

The sum of two operators is defined in the obvious way,

$$\left(\hat{Q} + \hat{R} \right) |\alpha\rangle = \hat{Q}\,|\alpha\rangle + \hat{R}\,|\alpha\rangle , \tag{3.98}$$

and the product of two operators is

$$\hat{Q}\hat{R}\,|\alpha\rangle = \hat{Q} \left(\hat{R}\,|\alpha\rangle \right) \tag{3.99}$$

(first apply $\hat{R}$ to $|\alpha\rangle$, and then apply $\hat{Q}$ to what you got—being careful, of course, to respect their *ordering*). Occasionally we shall encounter *functions* of operators. They are typically defined by the power series expansion:

$$e^{\hat{Q}} \equiv 1 + \hat{Q} + \frac{1}{2}\hat{Q}^2 + \frac{1}{3!}\hat{Q}^3 + \cdots \tag{3.100}$$

$$\frac{1}{1-\hat{Q}} \equiv 1 + \hat{Q} + \hat{Q}^2 + \hat{Q}^3 + \hat{Q}^4 + \cdots \tag{3.101}$$

$$\ln\left(1 + \hat{Q}\right) \equiv \hat{Q} - \frac{1}{2}\hat{Q}^2 + \frac{1}{3}\hat{Q}^3 - \frac{1}{4}\hat{Q}^4 + \cdots \tag{3.102}$$

and so on. On the right-hand side we have only sums and products, and we *know* how to handle *them*.

Problem 3.23 Show that projection operators are **idempotent**: $\hat{P}^2 = \hat{P}$. Determine the eigenvalues of $\hat{P}$, and characterize its eigenvectors.

---

[35] Note that $\langle \hat{Q} f | = \langle f |\, \hat{Q}^\dagger$, by virtue of Equation 3.20.

[36] Like his delta function, Dirac's notation is beautiful, powerful, and obedient. You can abuse it (everyone does), and it won't bite. But once in a while you should pause to ask yourself what the symbols really mean.

Problem 3.24 Show that if an operator $\hat{Q}$ is hermitian, then its matrix elements in any orthonormal basis satisfy $Q_{mn} = Q^*_{nm}$. That is, the corresponding matrix is equal to its transpose conjugate.

Problem 3.25 The Hamiltonian for a certain two-level system is

$$\hat{H} = \epsilon \left(|1\rangle\langle 1| - |2\rangle\langle 2| + |1\rangle\langle 2| + |2\rangle\langle 1|\right),$$

where $|1\rangle, |2\rangle$ is an orthonormal basis and ϵ is a number with the dimensions of energy. Find its eigenvalues and eigenvectors (as linear combinations of $|1\rangle$ and $|2\rangle$). What is the matrix H representing $\hat{H}$ with respect to this basis?

* **Problem 3.26** Consider a three-dimensional vector space spanned by an orthonormal basis $|1\rangle, |2\rangle, |3\rangle$. Kets $|\alpha\rangle$ and $|\beta\rangle$ are given by

$$|\alpha\rangle = i|1\rangle - 2|2\rangle - i|3\rangle, \quad |\beta\rangle = i|1\rangle + 2|3\rangle.$$

(a) Construct $\langle\alpha|$ and $\langle\beta|$ (in terms of the dual basis $\langle 1|, \langle 2|, \langle 3|$).
(b) Find $\langle\alpha|\beta\rangle$ and $\langle\beta|\alpha\rangle$, and confirm that $\langle\beta|\alpha\rangle = \langle\alpha|\beta\rangle^*$.
(c) Find all nine matrix elements of the operator $\hat{A} \equiv |\alpha\rangle\langle\beta|$, in this basis, and construct the matrix A. Is it hermitian?

Problem 3.27 Let $\hat{Q}$ be an operator with a complete set of orthonormal eigenvectors:

$$\hat{Q}|e_n\rangle = q_n|e_n\rangle \quad (n = 1, 2, 3, \ldots).$$

(a) Show that $\hat{Q}$ can be written in terms of its **spectral decomposition**:

$$\hat{Q} = \sum_n q_n |e_n\rangle\langle e_n|. \tag{3.103}$$

Hint: An operator is characterized by its action on all possible vectors, so what you must show is that

$$\hat{Q}|\alpha\rangle = \left\{\sum_n q_n|e_n\rangle\langle e_n|\right\}|\alpha\rangle,$$

for any vector $|\alpha\rangle$.

(b) Another way to define a function of $\hat{Q}$ is via the spectral decomposition:

$$f(\hat{Q}) = \sum_n f(q_n)|e_n\rangle\langle e_n|. \tag{3.104}$$

Show that this is equivalent to Equation 3.100 in the case of $e^{\hat{Q}}$.

Problem 3.28 Let $\hat{D} = d/dx$ (the derivative operator). Find

(a) $\left(\sin \hat{D}\right) x^5$.

(b) $\left(\dfrac{1}{1-\hat{D}/2}\right) \cos(x)$.

** **Problem 3.29** Consider operators $\hat{A}$ and $\hat{B}$ that do not commute with each other $\left(\hat{C} = \left[\hat{A}, \hat{B}\right]\right)$ but do commute with their commutator: $\left[\hat{A}, \hat{C}\right] = \left[\hat{B}, \hat{C}\right] = 0$ (for instance, $\hat{x}$ and $\hat{p}$).

(a) Show that

$$\left[\hat{A}^n, \hat{B}\right] = n\,\hat{A}^{n-1}\,\hat{C}\,.$$

Hint: You can prove this by induction on n, using Equation 3.65.

(b) Show that

$$\left[e^{\lambda \hat{A}}, \hat{B}\right] = \lambda\, e^{\lambda \hat{A}}\,\hat{C}\,,$$

where λ is any complex number. *Hint:* Express $e^{\lambda \hat{A}}$ as a power series.

(c) Derive the **Baker–Campbell–Hausdorff formula**:[37]

$$e^{\hat{A}+\hat{B}} = e^{\hat{A}}\,e^{\hat{B}}\,e^{-\hat{C}/2}\,.$$

Hint: Define the functions

$$\hat{f}(\lambda) = e^{\lambda\left(\hat{A}+\hat{B}\right)}, \qquad \hat{g}(\lambda) = e^{\lambda \hat{A}}\,e^{\lambda \hat{B}}\,e^{-\lambda^2 \hat{C}/2}\,.$$

Note that these functions are equal at $\lambda = 0$, and show that they satisfy the same differential equation: $d\hat{f}/d\lambda = \left(\hat{A}+\hat{B}\right)\hat{f}$ and $d\hat{g}/d\lambda = \left(\hat{A}+\hat{B}\right)\hat{g}$. Therefore, the functions are themselves equal for all λ.[38]

3.6.3 Changing Bases in Dirac Notation

The advantage of Dirac notation is that it frees us from working in any particular basis, and makes transforming between bases seamless. Recall that the identity operator can be written as a projection onto a complete set of states (Equations 3.93 and 3.96); of particular interest are the position eigenstates $|x\rangle$, the momentum eigenstates $|p\rangle$, and the energy eigenstates (we will assume those are discrete) $|n\rangle$:

---

[37] This is a special case of a more general formula that applies when $\hat{A}$ and $\hat{B}$ do *not* commute with $\hat{C}$. See, for example, Eugen Merzbacher, *Quantum Mechanics,* 3rd edn, Wiley, New York (1998), page 40.

[38] The product rule holds for differentiating operators as long as you respect their order:

$$\frac{d}{d\lambda}\left[\hat{A}(\lambda)\;\hat{B}(\lambda)\right] = \hat{A}'(\lambda)\;\hat{B}(\lambda) + \hat{A}(\lambda)\;\hat{B}'(\lambda)\,. \tag{3.105}$$

$$1 = \int dx\,|x\rangle\,\langle x|,$$

$$1 = \int dp\,|p\rangle\,\langle p|,$$

$$1 = \sum |n\rangle\,\langle n|\,. \tag{3.106}$$

Acting on the state vector $|\mathcal{S}(t)\rangle$ with each of these resolutions of the identity gives

$$|\mathcal{S}(t)\rangle = \int dx\,|x\rangle\,\langle x\,|\,\mathcal{S}(t)\rangle \equiv \int \Psi(x,t)\;|x\rangle\;dx,$$

$$|\mathcal{S}(t)\rangle = \int dp\,|p\rangle\,\langle p\,|\,\mathcal{S}(t)\rangle \equiv \int \Phi(p,t)\;|p\rangle\;dp,$$

$$|\mathcal{S}(t)\rangle = \sum_n |n\rangle\,\langle n\,|\,\mathcal{S}(t)\rangle \equiv \sum c_n(t)\;|n\rangle\,. \tag{3.107}$$

Here we recognize the position-space, momentum-space, and "energy-space" wave functions (Equations 3.77–3.79) as the *components* of the vector $|\mathcal{S}(t)\rangle$ in the respective bases.

Example 3.9

Derive the transformation from the position-space wave function to the momentum-space wave function. (We already know the answer, of course, but I want to show you how this works out in Dirac notation.)

Solution: We want to find $\Phi(p,t) = \langle p|\mathcal{S}(t)\rangle$ given $\Psi(x,t) = \langle x|\mathcal{S}(t)\rangle$. We can relate the two by inserting a resolution of the identity:

$$\begin{aligned}\Phi(p,t) &= \langle p\,|\,\mathcal{S}(t)\rangle \\ &= \left\langle p\,\middle|\left(\int dx\,|x\rangle\,\langle x|\right)\middle|\,\mathcal{S}(t)\right\rangle \\ &= \int \langle p\,|\,x\rangle\;\langle x\,|\,\mathcal{S}(t)\rangle\;dx \\ &= \int \langle p\,|\,x\rangle\;\Psi(x,t)\;dx.\end{aligned} \tag{3.108}$$

Now, $\langle x\,|\,p\rangle$ is the momentum eigenstate (with eigenvalue p) in the position basis—what we called $f_p(x)$, in Equation 3.32. So

$$\langle p|x\rangle = \langle x|p\rangle^* = \left[f_p(x)\right]^* = \frac{1}{\sqrt{2\pi\hbar}}\,e^{-i\,p\,x/\hbar}.$$

Plugging this into Equation 3.108 gives

$$\Phi(p,t) = \int \frac{1}{\sqrt{2\pi\hbar}}\,e^{-i\,p\,x/\hbar}\;\Psi(x,t)\;dx,$$

which is precisely Equation 3.54.

Just as the wave function takes different forms in different bases, so do operators. The position operator is given by

$$\hat{x} \to x$$

in the position basis, or

$$\hat{x} \to i\hbar \frac{\partial}{\partial p}$$

in the momentum basis. However, Dirac notation allows us to do away with the arrows and stick to equalities. Operators act on kets (for instance, $\hat{x}\ |\mathcal{S}(t)\rangle$); the outcome of this operation can be expressed in any basis by taking the inner product with an appropriate basis vector. That is,

$$\langle x\,|\hat{x}|\,\mathcal{S}(t)\rangle = \text{action of position operator in } x \text{ basis} = x\,\Psi(x,t)\,, \tag{3.109}$$

or

$$\langle p\,|\hat{x}|\,\mathcal{S}(t)\rangle = \text{action of position operator in } p \text{ basis} = i\hbar \frac{\partial \Phi}{\partial p}\,. \tag{3.110}$$

In this notation it is straightforward to transform operators between bases, as the following example illustrates.

Example 3.10

Obtain the position operator in the momentum basis (Equation 3.110) by inserting a resolution of the identity on the left-hand side.

Solution:

$$\begin{aligned}\langle p\,|\hat{x}|\,\mathcal{S}(t)\rangle &= \left\langle p \left| \hat{x} \int dx \left| x \right\rangle \left\langle x \right| \right| \mathcal{S}(t) \right\rangle \\ &= \int \langle p\,|x|\,x\rangle\ \langle x\,|\,\mathcal{S}(t)\rangle\ dx\,,\end{aligned}$$

where I've used the fact that $|x\rangle$ is an eigenstate of $\hat{x}\big(\hat{x}\ |x\rangle = x\ |x\rangle\big)$; x can then be pulled out of the inner product (it's just a number) and

$$\begin{aligned}\langle p\,|\hat{x}|\,\mathcal{S}(t)\rangle &= \int x\ \langle p\,|\,x\rangle\ \Psi(x,t)\ dx \\ &= \int x\,\frac{e^{-ipx/\hbar}}{\sqrt{2\pi\hbar}}\Psi(x,t)\ dx \\ &= i\hbar\frac{\partial}{\partial p}\int \frac{e^{-ipx/\hbar}}{\sqrt{2\pi\hbar}}\Psi(x,t)\ dx\,.\end{aligned}$$

Finally we recognize the integral as $\Phi(p,t)$ (Equation 3.54).

Problem 3.30 Derive the transformation from the position-space wave function to the "energy-space" wave function ($c_n(t)$) using the technique of Example 3.9. Assume that the energy spectrum is discrete, and the potential is time-independent.

FURTHER PROBLEMS ON CHAPTER 3

* **Problem 3.31 Legendre polynomials.** Use the Gram–Schmidt procedure (Problem A.4) to orthonormalize the functions $1,\ x,\ x^2$, and x^3, on the interval $-1 \leq x \leq 1$. You may recognize the results—they are (apart from normalization)[39] **Legendre polynomials** (Problem 2.64 and Table 4.1).

Problem 3.32 An **anti-hermitian** (or **skew-hermitian**) operator is equal to *minus* its hermitian conjugate:

$$\hat{Q}^\dagger = -\hat{Q}. \tag{3.111}$$

(a) Show that the expectation value of an anti-hermitian operator is imaginary.
(b) Show that the eigenvalues of an anti-hermitian operator are imaginary.
(c) Show that the eigenvectors of an anti-hermitian operator belonging to distinct eigenvalues are orthogonal.
(d) Show that the commutator of two hermitian operators is anti-hermitian. How about the commutator of two *anti*-hermitian operators?
(e) Show that any operator $\hat{Q}$ can be written as a sum of a hermitian operator $\hat{A}$ and an anti-hermitian operator $\hat{B}$, and give expressions for $\hat{A}$ and $\hat{B}$ in terms of $\hat{Q}$ and its adjoint $\hat{Q}^\dagger$.

Problem 3.33 Sequential measurements. An operator $\hat{A}$, representing observable A, has two (normalized) eigenstates ψ_1 and ψ_2, with eigenvalues a_1 and a_2, respectively. Operator $\hat{B}$, representing observable B, has two (normalized) eigenstates ϕ_1 and ϕ_2, with eigenvalues b_1 and b_2. The eigenstates are related by

$$\psi_1 = (3\phi_1 + 4\phi_2)/5, \quad \psi_2 = (4\phi_1 - 3\phi_2)/5.$$

(a) Observable A is measured, and the value a_1 is obtained. What is the state of the system (immediately) after this measurement?
(b) If B is now measured, what are the possible results, and what are their probabilities?
(c) Right after the measurement of B, A is measured again. What is the probability of getting a_1? (Note that the answer would be quite different if I had told you the outcome of the B measurement.)

***  **Problem 3.34**

(a) Find the momentum-space wave function $\Phi_n(p,t)$ for the nth stationary state of the infinite square well.
(b) Find the probability density $|\Phi_n(p,t)|^2$. Graph this function, for $n = 1$, $n = 2$, $n = 5$, and $n = 10$. What are the most probable values of p, for large n? Is this what you would have expected?[40] Compare your answer to Problem 3.10.
(c) Use $\Phi_n(p,t)$ to calculate the expectation value of p^2, in the nth state. Compare your answer to Problem 2.4.

---

[39] Legendre didn't know what the best convention would be; he picked the overall factor so that all his functions would go to 1 at $x = 1$, and we're stuck with his unfortunate choice.

[40] See F. L. Markley, *Am. J. Phys.* **40**, 1545 (1972).

Problem 3.35 Consider the wave function

$$\Psi(x,0) = \begin{cases} \frac{1}{\sqrt{2n\lambda}} e^{i2\pi x/\lambda}, & -n\lambda < x < n\lambda, \\ 0, & \text{otherwise,} \end{cases}$$

where n is some positive integer. This function is purely sinusoidal (with wavelength λ) on the interval $-n\lambda < x < n\lambda$, but it still carries a *range* of momenta, because the oscillations do not continue out to infinity. Find the momentum space wave function $\Phi(p,0)$. Sketch the graphs of $|\Psi(x,0)|^2$ and $|\Phi(p,0)|^2$, and determine their widths, w_x and w_p (the distance between zeros on either side of the main peak). Note what happens to each width as $n \to \infty$. Using w_x and w_p as estimates of Δx and Δp, check that the uncertainty principle is satisfied. *Warning:* If you try calculating σ_p, you're in for a rude surprise. Can you diagnose the problem?

Problem 3.36 Suppose

$$\Psi(x,0) = \frac{A}{x^2 + a^2}, \quad (-\infty < x < \infty)$$

for constants A and a.

(a) Determine A, by normalizing $\Psi(x,0)$.
(b) Find $\langle x \rangle$, $\langle x^2 \rangle$, and σ_x (at time $t = 0$).
(c) Find the momentum space wave function $\Phi(p,0)$, and check that it is normalized.
(d) Use $\Phi(p,0)$ to calculate $\langle p \rangle$, $\langle p^2 \rangle$, and σ_p (at time $t = 0$).
(e) Check the Heisenberg uncertainty principle for this state.

* **Problem 3.37 Virial theorem.** Use Equation 3.73 to show that

$$\frac{d}{dt}\langle xp \rangle = 2\langle T \rangle - \left\langle x \frac{\partial V}{\partial x} \right\rangle, \tag{3.112}$$

where T is the kinetic energy ($H = T + V$). In a *stationary* state the left side is zero (why?) so

$$2\langle T \rangle = \left\langle x \frac{dV}{dx} \right\rangle. \tag{3.113}$$

This is called the **virial theorem**. Use it to prove that $\langle T \rangle = \langle V \rangle$ for stationary states of the harmonic oscillator, and check that this is consistent with the results you got in Problems 2.11 and 2.12.

Problem 3.38 In an interesting version of the energy-time uncertainty principle[41] $\Delta t = \tau/\pi$, where τ is the time it takes $\Psi(x,t)$ to evolve into a state orthogonal to $\Psi(x,0)$. Test this out, using a wave function that is a linear combination of two (orthonormal) stationary states of some (arbitrary) potential: $\Psi(x,0) = \left(1/\sqrt{2}\right)[\psi_1(x) + \psi_2(x)]$.

[41] See L. Vaidman, *Am. J. Phys.* **60**, 182 (1992) for a proof.

∗∗ **Problem 3.39** Find the matrix elements $\langle n|x|n'\rangle$ and $\langle n|p|n'\rangle$ in the (orthonormal) basis of stationary states for the harmonic oscillator (Equation 2.68). You already calculated the "diagonal" elements $(n = n')$ in Problem 2.12; use the same technique for the general case. Construct the corresponding (infinite) matrices, X and P. Show that $(1/2m)\,\mathsf{P}^2 + (m\omega^2/2)\mathsf{X}^2 = \mathsf{H}$ is *diagonal*, in this basis. Are its diagonal elements what you would expect? *Partial answer:*

$$\langle n \,|\, x \,|\, n'\rangle = \sqrt{\frac{\hbar}{2m\omega}}\left(\sqrt{n'}\delta_{n,n'-1} + \sqrt{n}\delta_{n',n-1}\right). \tag{3.114}$$

∗∗ **Problem 3.40** The most general wave function of a particle in the simple harmonic oscillator potential is

$$\Psi(x,t) = \sum_n c_n\, \psi_n(x)\, e^{-iE_nt/\hbar}.$$

Show that the expectation value of position is

$$\langle x\rangle = C\cos(\omega t - \phi),$$

where the real constants C and ϕ are given by

$$Ce^{-i\phi} = \left(\sqrt{\frac{2\hbar}{m\omega}}\right)\sum_{n=0}^{\infty}\sqrt{n+1}\; c_{n+1}^*\, c_n.$$

Thus the expectation value of position for a particle in the harmonic oscillator oscillates at the classical frequency ω (as you would expect from Ehrenfest's theorem; see problem 3.19(b)). *Hint:* Use Equation 3.114. As an example, find C and ϕ for the wave function in Problem 2.40.

Problem 3.41 A harmonic oscillator is in a state such that a measurement of the energy would yield either $(1/2)\,\hbar\omega$ or $(3/2)\,\hbar\omega$, with equal probability. What is the largest possible value of $\langle p\rangle$ in such a state? If it assumes this maximal value at time $t = 0$, what is $\Psi(x,t)$?

∗∗∗ **Problem 3.42 Coherent states of the harmonic oscillator.** Among the stationary states of the harmonic oscillator (Equation 2.68) only $n = 0$ hits the uncertainty limit $(\sigma_x\sigma_p = \hbar/2)$; in general, $\sigma_x\sigma_p = (2n+1)\,\hbar/2$, as you found in Problem 2.12. But certain *linear combinations* (known as **coherent states**) also minimize the uncertainty product. They are (as it turns out) *eigenfunctions of the lowering operator:*[42]

$$a_-\,|\alpha\rangle = \alpha\,|\,\alpha\rangle$$

(the eigenvalue α can be any complex number).

[42] There are no normalizable eigenfunctions of the *raising* operator.

(a) Calculate $\langle x\rangle$, $\langle x^2\rangle$, $\langle p\rangle$, $\langle p^2\rangle$ in the state $|\alpha\rangle$. *Hint:* Use the technique in Example 2.5, and remember that a_+ is the hermitian conjugate of a_-. Do *not* assume α is real.

(b) Find σ_x and σ_p; show that $\sigma_x\sigma_p = \hbar/2$.

(c) Like any other wave function, a coherent state can be expanded in terms of energy eigenstates:

$$|\alpha\rangle = \sum_{n=0}^{\infty} c_n\, |n\rangle\,.$$

Show that the expansion coefficients are

$$c_n = \frac{\alpha^n}{\sqrt{n!}} c_0.$$

(d) Determine c_0 by normalizing $|\alpha\rangle$. *Answer:* $\exp\left(-|\alpha|^2/2\right)$.

(e) Now put in the time dependence:

$$|n\rangle \to e^{-iE_n t/\hbar}\, |n\rangle\,,$$

and show that $|\alpha(t)\rangle$ remains an eigenstate of a_-, but the eigen*value* evolves in time:

$$\alpha(t) = e^{-i\omega t}\alpha.$$

So a coherent state *stays* coherent, and continues to minimize the uncertainty product.

(f) Based on your answers to (a), (b), and (e), find $\langle x\rangle$ and σ_x as functions of time. It helps if you write the complex number α as

$$\alpha = C\sqrt{\frac{m\,\omega}{2\,\hbar}}\, e^{i\,\phi}$$

for real numbers C and ϕ. *Comment:* In a sense, coherent states behave quasi-classically.

(g) Is the ground state ($|n=0\rangle$) itself a coherent state? If so, what is the eigenvalue?

Problem 3.43 Extended uncertainty principle.[43] The generalized uncertainty principle (Equation 3.62) states that

$$\sigma_A^2\sigma_B^2 \geq \frac{1}{4}\langle C\rangle^2\,,$$

where $\hat{C} \equiv -i\left[\hat{A}, \hat{B}\right]$.

(a) Show that it can be strengthened to read

$$\sigma_A^2\sigma_B^2 \geq \frac{1}{4}\left(\langle C\rangle^2 + \langle D\rangle^2\right), \tag{3.115}$$

where $\hat{D} \equiv \hat{A}\hat{B} + \hat{B}\hat{A} - 2\langle A\rangle\langle B\rangle$. *Hint:* Keep the Re$(z)$ term in Equation 3.60.

[43] For interesting commentary and references, see R. R. Puri, *Phys. Rev. A* **49**, 2178 (1994).

(b) Check Equation 3.115 for the case $B = A$ (the standard uncertainty principle is trivial, in this case, since $\hat{C} = 0$; unfortunately, the extended uncertainty principle doesn't help much either).

Problem 3.44 The Hamiltonian for a certain three-level system is represented by the matrix

$$\mathsf{H} = \begin{pmatrix} a & 0 & b \\ 0 & c & 0 \\ b & 0 & a \end{pmatrix},$$

where a, b, and c are real numbers.

(a) If the system starts out in the state

$$|\mathcal{S}(0)\rangle = \begin{pmatrix} 0 \\ 1 \\ 0 \end{pmatrix},$$

what is $|\mathcal{S}(t)\rangle$?

(b) If the system starts out in the state

$$|\mathcal{S}(0)\rangle = \begin{pmatrix} 1 \\ 0 \\ 0 \end{pmatrix},$$

what is $|\mathcal{S}(t)\rangle$?

Problem 3.45 Find the position operator in the basis of simple harmonic oscillator energy states. That is, express

$$\langle n \,|\hat{x}|\, \mathcal{S}(t)\rangle$$

in terms of $c_n(t) = \langle n|\mathcal{S}(t)\rangle$. *Hint*: Use Equation 3.114.

Problem 3.46 The Hamiltonian for a certain three-level system is represented by the matrix

$$\mathsf{H} = \hbar\omega \begin{pmatrix} 1 & 0 & 0 \\ 0 & 2 & 0 \\ 0 & 0 & 2 \end{pmatrix}.$$

Two other observables, A and B, are represented by the matrices

$$\mathsf{A} = \lambda \begin{pmatrix} 0 & 1 & 0 \\ 1 & 0 & 0 \\ 0 & 0 & 2 \end{pmatrix}, \quad \mathsf{B} = \mu \begin{pmatrix} 2 & 0 & 0 \\ 0 & 0 & 1 \\ 0 & 1 & 0 \end{pmatrix},$$

where ω, λ, and μ are positive real numbers.

(a) Find the eigenvalues and (normalized) eigenvectors of H, A, and B.

(b) Suppose the system starts out in the generic state

$$|\mathcal{S}(0)\rangle = \begin{pmatrix} c_1 \\ c_2 \\ c_3 \end{pmatrix},$$

with $|c_1|^2 + |c_2|^2 + |c_3|^2 = 1$. Find the expectation values (at $t = 0$) of H, A, and B.

(c) What is $|\mathcal{S}(t)\rangle$? If you measured the energy of this state (at time t), what values might you get, and what is the probability of each? Answer the same questions for observables A and for B.

** **Problem 3.47 Supersymmetry.** Consider the two operators

$$\hat{A} = i\frac{\hat{p}}{\sqrt{2m}} + W(x) \quad \text{and} \quad \hat{A}^\dagger = -i\frac{\hat{p}}{\sqrt{2m}} + W(x), \tag{3.116}$$

for some function $W(x)$. These may be multiplied in either order to construct two Hamiltonians:

$$\hat{H}_1 \equiv \hat{A}^\dagger \hat{A} = \frac{\hat{p}^2}{2m} + V_1(x) \quad \text{and} \quad \hat{H}_2 \equiv \hat{A}\hat{A}^\dagger = \frac{\hat{p}^2}{2m} + V_2(x); \tag{3.117}$$

V_1 and V_2 are called **supersymmetric partner potentials**. The energies and eigenstates of $\hat{H}_1$ and $\hat{H}_2$ are related in interesting ways.[44]

(a) Find the potentials $V_1(x)$ and $V_2(x)$, in terms of the **superpotential**, $W(x)$.

(b) Show that if $\psi_n^{(1)}$ is an eigenstate of $\hat{H}_1$ with eigenvalue $E_n^{(1)}$, then $\hat{A}\psi_n^{(1)}$ is an eigenstate of $\hat{H}_2$ with the same eigenvalue. Similarly, show that if $\psi_n^{(2)}(x)$ is an eigenstate of $\hat{H}_2$ with eigenvalue $E_n^{(2)}$, then $\hat{A}^\dagger\psi_n^{(2)}$ is an eigenstate of $\hat{H}_1$ with the same eigenvalue. The two Hamiltonians therefore have essentially identical spectra.

(c) One ordinarily chooses $W(x)$ such that the ground state of $\hat{H}_1$ satisfies

$$\hat{A}\psi_0^{(1)}(x) = 0, \tag{3.118}$$

and hence $E_0^{(1)} = 0$. Use this to find the superpotential $W(x)$, in terms of the ground state wave function, $\psi_0^{(1)}(x)$. (The fact that $\hat{A}$ annihilates $\psi_0^{(1)}$ means that $\hat{H}_2$ actually has one less eigenstate than $\hat{H}_1$, and is missing the eigenvalue $E_0^{(1)}$.)

(d) Consider the Dirac delta function well,

$$V_1(x) = \frac{m\alpha^2}{2\hbar^2} - \alpha\delta(x), \tag{3.119}$$

(the constant term, $m\alpha^2/2\hbar^2$, is included so that $E_0^{(1)} = 0$). It has a single bound state (Equation 2.132)

$$\psi_0^{(1)}(x) = \frac{\sqrt{m\alpha}}{\hbar}\exp\left[-\frac{m\alpha}{\hbar^2}|x|\right]. \tag{3.120}$$

Use the results of parts (a) and (c), and Problem 2.23(b), to determine the superpotential $W(x)$ and the partner potential $V_2(x)$. This partner potential is one that you will likely recognize, and while it has no bound states, the supersymmetry between these two systems explains the fact that their reflection and transmission coefficients are identical (see the last paragraph of Section 2.5.2).

---

[44] Fred Cooper, Avinash Khare, and Uday Sukhatme, *Supersymmetry in Quantum Mechanics*, World Scientific, Singapore, 2001.

** **Problem 3.48** An operator is defined not just by its *action* (what it *does* to the vector it is applied to) but its *domain* (the set of vectors on which it acts). In a *finite*-dimensional vector space the domain is the entire space, and we don't need to worry about it. But for most operators in Hilbert space the domain is restricted. In particular, only functions such that $\hat{Q} f(x)$ remains in Hilbert space are allowed in the domain of $\hat{Q}$. (As you found in Problem 3.2, the derivative operator can knock a function out of L^2.)

A hermitian operator is one whose *action* is the same as that of its adjoint[45] (Problem 3.5). But what is required to represent observables is actually something more: the *domains* of $\hat{Q}$ and $\hat{Q}^\dagger$ must *also* be identical. Such operators are called **self-adjoint.**[46]

(a) Consider the momentum operator, $\hat{p} = -i\hbar d/dx$, on the finite interval $0 \leq x \leq a$. With the infinite square well in mind, we might define its domain as the set of functions $f(x)$ such that $f(0) = f(a) = 0$ (it goes without saying that $f(x)$ and $\hat{p} f(x)$ are in $L^2(0, a)$). Show that $\hat{p}$ is hermitian: $\langle g | \hat{p} f \rangle = \langle \hat{p}^\dagger g | f \rangle$, with $\hat{p}^\dagger = \hat{p}$. But is it self-adjoint? *Hint:* as long as $f(0) = f(a) = 0$, there is no restriction on $g(0)$ or $g(a)$—the domain of $\hat{p}^\dagger$ is much larger than the domain of $\hat{p}$.[47]

(b) Suppose we *extend* the domain of $\hat{p}$ to include all functions of the form $f(a) = \lambda f(0)$, for some fixed complex number λ. What condition must we then impose on the domain of $\hat{p}^\dagger$ in order that $\hat{p}$ be hermitian? What value(s) of λ will render $\hat{p}$ self-adjoint? *Comment:* Technically, then, there *is* no momentum operator on the finite interval—or rather, there are infinitely many, and no way to decide which of them is "correct." (In Problem 3.34 we avoided the issue by working on the infinite interval.)

(c) What about the semi-infinite interval, $0 \leq x < \infty$? Is there a self-adjoint momentum operator in this case?[48]

** **Problem 3.49**

(a) Write down the time-dependent "Schrödinger equation" in momentum space, for a free particle, and solve it. *Answer:* $\exp\left(-ip^2 t/2m\hbar\right) \Phi(p, 0)$.

(b) Find $\Phi(p, 0)$ for the traveling gaussian wave packet (Problem 2.42), and construct $\Phi(p, t)$ for this case. Also construct $|\Phi(p, t)|^2$, and note that it is independent of time.

(c) Calculate $\langle p \rangle$ and $\langle p^2 \rangle$ by evaluating the appropriate integrals involving Φ, and compare your answers to Problem 2.42.

(d) Show that $\langle H \rangle = \langle p \rangle^2 / 2m + \langle H \rangle_0$ (where the subscript 0 denotes the *stationary* gaussian), and comment on this result.

---

[45] Mathematicians call them "symmetric" operators.

[46] Because the distinction rarely intrudes, physicists tend to use the word "hermitian" indiscriminately; technically, we should always say "self-adjoint," meaning $\hat{Q} = \hat{Q}^\dagger$ both in action and in domain.

[47] The domain of $\hat{Q}$ is something we *stipulate*; that *determines* the domain of $\hat{Q}^\dagger$.

[48] J. von Neumann introduced machinery for generating **self-adjoint extensions** of hermitian operators—or in some cases proving that they cannot exist. For an accessible introduction see G. Bonneau, J. Faraut, and B. Valent, *Am. J. Phys.* **69**, 322 (2001); for an interesting application see M. T. Ahari, G. Ortiz, and B. Seradjeh, *Am. J. Phys.* **84**, 858 (2016).

导读 / Guidance

第 4 章　三维空间中的量子力学

本章讨论三维空间的量子力学解析求解问题。首先通过建立球坐标系下的薛定谔方程和分离变量，给出径向、角动量方程的形式和其解析解；接下来以氢原子为例，给出其薛定谔方程解的波函数和对应的能级大小，并解释实验上氢原子的光谱问题；其次讨论角动量的本征值和本征函数，定义角动量的产生和湮灭算符；接下来引入电子的自旋概念，给出自旋算符（泡利自旋矩阵）相应分量的本征矢和本征值，讨论电子在磁场中受到作用引起的能级分裂，解释施特恩 - 格拉赫实验现象；最后讨论自旋角动量的叠加和阿哈罗诺夫 - 玻姆效应。本章较多地涉及数学物理方法求解偏微分方程的内容，如分离变量法、渐近形式求解、用幂级数求和表示方程的解；给出一些特殊函数形式，如关联勒让德函数、球贝塞尔函数、球诺依曼函数、拉盖尔函数等。

习题特色

（1）把经典物理问题和量子力学问题做对比研究，如把地球 - 太阳引力体系类比氢原子体系，见习题 4.20；把电子看成一个经典固体球，见习题 4.28。

（2）拓展一氧化碳刚性分子转动光谱的求解问题，见习题 4.27；讨论拉莫尔进动问题，见例题 4.3；解释施特恩 - 格拉赫实验的结果，见例题 4.4；讨论三维位力定理形式，见习题 4.48。

（3）讨论基本粒子（如夸克、重子、介子）的自旋问题，见习题 4.35；推广自旋大于 1/2 值的粒子的自旋矩阵形式，见习题 4.62。

（4）讨论磁涨落问题和一些其他前沿问题，见习题 4.67、习题 4.72、习题 4.74、习题 4.75 等。

（5）给出一些利用计算机数值计算的题目，如习题 4.52、习题 4.68、习题 4.69、习题 4.71。

4 QUANTUM MECHANICS IN THREE DIMENSIONS

4.1 THE SCHRÖDINGER EQUATION

The generalization to three dimensions is straightforward. Schrödinger's equation says

$$i\hbar\frac{\partial\Psi}{\partial t} = \hat{H}\Psi; \tag{4.1}$$

the Hamiltonian operator $\hat{H}$ is obtained from the classical energy

$$\frac{1}{2}mv^2 + V = \frac{1}{2m}\left(p_x^2 + p_y^2 + p_z^2\right) + V$$

by the standard prescription (applied now to y and z, as well as x):

$$p_x \to -i\hbar\frac{\partial}{\partial x}, \quad p_y \to -i\hbar\frac{\partial}{\partial y}, \quad p_z \to -i\hbar\frac{\partial}{\partial z}, \tag{4.2}$$

or

$$\boxed{\mathbf{p} \to -i\hbar\nabla,} \tag{4.3}$$

for short. Thus

$$\boxed{i\hbar\frac{\partial\Psi}{\partial t} = -\frac{\hbar^2}{2m}\nabla^2\Psi + V\Psi,} \tag{4.4}$$

where

$$\nabla^2 \equiv \frac{\partial^2}{\partial x^2} + \frac{\partial^2}{\partial y^2} + \frac{\partial^2}{\partial z^2}, \tag{4.5}$$

is the **Laplacian**, in cartesian coordinates.

The potential energy V and the wave function Ψ are now functions of $\mathbf{r} = (x, y, z)$ and t. The probability of finding the particle in the infinitesimal volume $d^3\mathbf{r} = dx\,dy\,dz$ is $|\Psi(\mathbf{r}, t)|^2\,d^3\mathbf{r}$, and the normalization condition reads

$$\int |\Psi|^2\,d^3\mathbf{r} = 1, \tag{4.6}$$

with the integral taken over all space. If V is independent of time, there will be a complete set of stationary states,

$$\Psi_n(\mathbf{r}, t) = \psi_n(\mathbf{r})\,e^{-iE_nt/\hbar}, \tag{4.7}$$

where the spatial wave function ψ_n satisfies the time-*independent* Schrödinger equation:

$$\boxed{-\frac{\hbar^2}{2m}\nabla^2\psi + V\psi = E\psi.} \tag{4.8}$$

The general solution to the (time-*dependent*) Schrödinger equation is

$$\Psi(\mathbf{r}, t) = \sum c_n \psi_n(\mathbf{r})\, e^{-iE_n t/\hbar}, \tag{4.9}$$

with the constants c_n determined by the initial wave function, $\Psi(\mathbf{r}, 0)$, in the usual way. (If the potential admits continuum states, then the sum in Equation 4.9 becomes an integral.)

* **Problem 4.1**

(a) Work out all of the **canonical commutation relations** for components of the operators **r** and **p**: $[x, y]$, $[x, p_y]$, $[x, p_x]$, $[p_y, p_z]$, and so on. *Answer:*

$$[r_i, p_j] = -[p_i, r_j] = i\hbar\delta_{ij}, \quad [r_i, r_j] = [p_i, p_j] = 0, \tag{4.10}$$

where the indices stand for x, y, or z, and $r_x = x$, $r_y = y$, and $r_z = z$.

(b) Confirm the three-dimensional version of **Ehrenfest's theorem,**

$$\frac{d}{dt}\langle\mathbf{r}\rangle = \frac{1}{m}\langle\mathbf{p}\rangle, \quad \text{and} \quad \frac{d}{dt}\langle\mathbf{p}\rangle = \langle-\nabla V\rangle. \tag{4.11}$$

(Each of these, of course, stands for *three* equations—one for each component.) *Hint:* First check that the "generalized" Ehrenfest theorem, Equation 3.73, is valid in three dimensions.

(c) Formulate **Heisenberg's uncertainty principle** in three dimensions. *Answer:*

$$\sigma_x\sigma_{p_x} \geq \hbar/2, \quad \sigma_y\sigma_{p_y} \geq \hbar/2, \quad \sigma_z\sigma_{p_z} \geq \hbar/2, \tag{4.12}$$

but there is no restriction on, say, $\sigma_x\sigma_{p_y}$.

* **Problem 4.2** Use separation of variables in cartesian coordinates to solve the infinite *cubical* well (or "particle in a box"):

$$V(x, y, z) = \begin{cases} 0, & x, y, z \text{ all between 0 and } a; \\ \infty, & \text{otherwise.} \end{cases}$$

(a) Find the stationary states, and the corresponding energies.

(b) Call the distinct energies $E_1, E_2, E_3, \ldots$, in order of increasing energy. Find E_1, E_2, E_3, E_4, E_5, and E_6. Determine their degeneracies (that is, the number of different states that share the same energy). *Comment*: In *one* dimension degenerate bound states do not occur (see Problem 2.44), but in three dimensions they are very common.

(c) What is the degeneracy of E_{14}, and why is this case interesting?

4.1.1 Spherical Coordinates

Most of the applications we will encounter involve **central potentials**, for which V is a function only of the distance from the origin, $V(\mathbf{r}) \to V(r)$. In that case it is natural to adopt

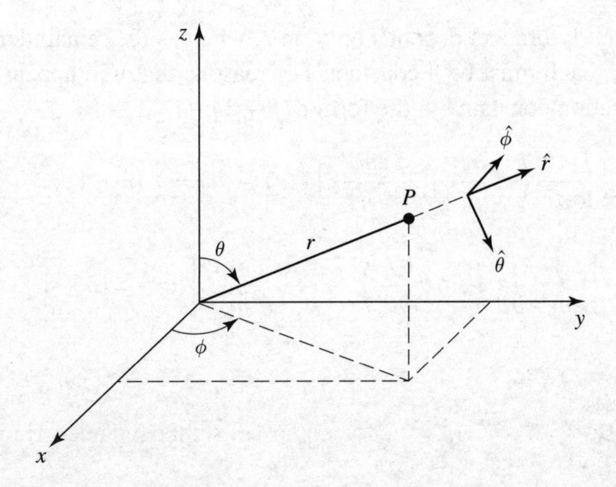

Figure 4.1: Spherical coordinates: radius r, polar angle θ, and azimuthal angle ϕ.

spherical coordinates, (r, θ, ϕ) (Figure 4.1). In spherical coordinates the Laplacian takes the form[1]

$$\nabla^2 = \frac{1}{r^2}\frac{\partial}{\partial r}\left(r^2\frac{\partial}{\partial r}\right) + \frac{1}{r^2\sin\theta}\frac{\partial}{\partial\theta}\left(\sin\theta\frac{\partial}{\partial\theta}\right) + \frac{1}{r^2\sin^2\theta}\left(\frac{\partial^2}{\partial\phi^2}\right). \tag{4.13}$$

In spherical coordinates, then, the time-independent Schrödinger equation reads

$$-\frac{\hbar^2}{2m}\left[\frac{1}{r^2}\frac{\partial}{\partial r}\left(r^2\frac{\partial\psi}{\partial r}\right) + \frac{1}{r^2\sin\theta}\frac{\partial}{\partial\theta}\left(\sin\theta\frac{\partial\psi}{\partial\theta}\right) + \frac{1}{r^2\sin^2\theta}\left(\frac{\partial^2\psi}{\partial\phi^2}\right)\right] + V\psi = E\psi. \tag{4.14}$$

We begin by looking for solutions that are separable into products (a function of r times a function of θ and ϕ):

$$\psi(r, \theta, \phi) = R(r)\,Y(\theta, \phi). \tag{4.15}$$

Putting this into Equation 4.14, we have

$$-\frac{\hbar^2}{2m}\left[\frac{Y}{r^2}\frac{d}{dr}\left(r^2\frac{dR}{dr}\right) + \frac{R}{r^2\sin\theta}\frac{\partial}{\partial\theta}\left(\sin\theta\frac{\partial Y}{\partial\theta}\right) + \frac{R}{r^2\sin^2\theta}\frac{\partial^2 Y}{\partial\phi^2}\right] + VRY = ERY.$$

Dividing by YR and multiplying by $-2mr^2/\hbar^2$:

$$\left\{\frac{1}{R}\frac{d}{dr}\left(r^2\frac{dR}{dr}\right) - \frac{2mr^2}{\hbar^2}[V(r) - E]\right\} + \frac{1}{Y}\left\{\frac{1}{\sin\theta}\frac{\partial}{\partial\theta}\left(\sin\theta\frac{\partial Y}{\partial\theta}\right) + \frac{1}{\sin^2\theta}\frac{\partial^2 Y}{\partial\phi^2}\right\} = 0.$$

[1] In principle, this can be obtained by change of variables from the cartesian expression 4.5. However, there are much more efficient ways of getting it; see, for instance, M. Boas, *Mathematical Methods in the Physical Sciences* 3rd edn, Wiley, New York (2006), Chapter 10, Section 9.

The term in the first curly bracket depends only on r, whereas the remainder depends only on θ and ϕ; accordingly, each must be a constant. For reasons that will appear in due course,[2] I will write this "separation constant" in the form $\ell(\ell+1)$:

$$\frac{1}{R}\frac{d}{dr}\left(r^2\frac{dR}{dr}\right)-\frac{2mr^2}{\hbar^2}\left[V(r)-E\right]=\ell(\ell+1); \tag{4.16}$$

$$\frac{1}{Y}\left\{\frac{1}{\sin\theta}\frac{\partial}{\partial\theta}\left(\sin\theta\frac{\partial Y}{\partial\theta}\right)+\frac{1}{\sin^2\theta}\frac{\partial^2 Y}{\partial\phi^2}\right\}=-\ell(\ell+1). \tag{4.17}$$

Problem 4.3

(a) Suppose $\psi(r,\theta,\phi)=Ae^{-r/a}$, for some constants A and a. Find E and $V(r)$, assuming $V(r)\to 0$ as $r\to\infty$.

(b) Do the same for $\psi(r,\theta,\phi)=Ae^{-r^2/a^2}$, assuming $V(0)=0$.

4.1.2 The Angular Equation

Equation 4.17 determines the dependence of ψ on θ and ϕ; multiplying by $Y\sin^2\theta$, it becomes:

$$\sin\theta\frac{\partial}{\partial\theta}\left(\sin\theta\frac{\partial Y}{\partial\theta}\right)+\frac{\partial^2 Y}{\partial\phi^2}=-\ell(\ell+1)\sin^2\theta\, Y. \tag{4.18}$$

You might recognize this equation—it occurs in the solution to Laplace's equation in classical electrodynamics. As always, we solve it by separation of variables:

$$Y(\theta,\phi)=\Theta(\theta)\,\Phi(\phi). \tag{4.19}$$

Plugging this in, and dividing by $\Theta\Phi$,

$$\left\{\frac{1}{\Theta}\left[\sin\theta\frac{d}{d\theta}\left(\sin\theta\frac{d\Theta}{d\theta}\right)\right]+\ell(\ell+1)\sin^2\theta\right\}+\frac{1}{\Phi}\frac{d^2\Phi}{d\phi^2}=0.$$

The first term is a function only of θ, and the second is a function only of ϕ, so each must be a constant. This time[3] I'll call the separation constant m^2:

$$\frac{1}{\Theta}\left[\sin\theta\frac{d}{d\theta}\left(\sin\theta\frac{d\Theta}{d\theta}\right)\right]+\ell(\ell+1)\sin^2\theta=m^2; \tag{4.20}$$

$$\frac{1}{\Phi}\frac{d^2\Phi}{d\phi^2}=-m^2. \tag{4.21}$$

The ϕ equation is easy:

$$\frac{d^2\Phi}{d\phi^2}=-m^2\Phi\ \Rightarrow\ \Phi(\phi)=e^{im\phi}. \tag{4.22}$$

---

[2] Note that there is no loss of generality here—at this stage ℓ could be any complex number. Later on we'll discover that ℓ must in fact be an *integer*, and it is in anticipation of that result that I express the separation constant in a way that looks peculiar now.

[3] Again, there is no loss of generality here, since at this stage m could be any complex number; in a moment, though, we will discover that m must in fact be an *integer*. *Beware:* The letter m is now doing double duty, as *mass* and as a separation constant. There is no graceful way to avoid this confusion, since both uses are standard. Some authors now switch to M or μ for mass, but I hate to change notation in mid-stream, and I don't think confusion will arise, a long as you are aware of the problem.

Actually, there are *two* solutions: $\exp(im\phi)$ and $\exp(-im\phi)$, but we'll cover the latter by allowing m to run negative. There could also be a constant factor in front, but we might as well absorb that into Θ. Incidentally, in electrodynamics we would write the azimuthal function (Φ) in terms of sines and cosines, instead of exponentials, because electric fields are *real*. But there is no such constraint on the wave function, and exponentials are a lot easier to work with. Now, when ϕ advances by 2π, we return to the same point in space (see Figure 4.1), so it is natural to require that[4]

$$\Phi(\phi + 2\pi) = \Phi(\phi). \tag{4.23}$$

In other words, $\exp[im(\phi + 2\pi)] = \exp(im\phi)$, or $\exp(2\pi im) = 1$. From this it follows that m must be an *integer*:

$$m = 0, \pm 1, \pm 2, \ldots . \tag{4.24}$$

The θ equation,

$$\sin\theta \frac{d}{d\theta}\left(\sin\theta \frac{d\Theta}{d\theta}\right) + \left[\ell(\ell+1)\sin^2\theta - m^2\right]\Theta = 0, \tag{4.25}$$

may not be so familiar. The solution is

$$\Theta(\theta) = A P_\ell^m(\cos\theta), \tag{4.26}$$

where P_ℓ^m is the **associated Legendre function**, defined by[5]

$$P_\ell^m(x) \equiv (-1)^m \left(1 - x^2\right)^{m/2} \left(\frac{d}{dx}\right)^m P_\ell(x), \tag{4.27}$$

and $P_\ell(x)$ is the ℓth **Legendre polynomial**, defined by the **Rodrigues formula**:

$$P_\ell(x) \equiv \frac{1}{2^\ell \ell!}\left(\frac{d}{dx}\right)^\ell \left(x^2 - 1\right)^\ell. \tag{4.28}$$

For example,

$$P_0(x) = 1, \quad P_1(x) = \frac{1}{2}\frac{d}{dx}\left(x^2 - 1\right) = x,$$

$$P_2(x) = \frac{1}{4 \cdot 2}\left(\frac{d}{dx}\right)^2 \left(x^2 - 1\right)^2 = \frac{1}{2}\left(3x^2 - 1\right),$$

and so on. The first few Legendre polynomials are listed in Table 4.1. As the name suggests, $P_\ell(x)$ is a polynomial (of degree ℓ) in x, and is even or odd according to the parity of ℓ. But $P_\ell^m(x)$ is not, in general, a polynomial[6]—if m is odd it carries a factor of $\sqrt{1 - x^2}$:

[4] This is more slippery than it looks. After all, the *probability* density ($|\Phi|^2$) is single valued *regardless* of m. In Section 4.3 we'll obtain the condition on m by an entirely different—and more compelling—argument.

[5] Some books (including earlier editions of this one) do not include the factor $(-1)^m$ in the definition of P_ℓ^m. Equation 4.27 assumes that $m \geq 0$; for negative values we define

$$P_\ell^{-m}(x) = (-1)^m \frac{(\ell - m)!}{(\ell + m)!} P_\ell^m(x).$$

A few books (including earlier versions of this one) define $P_\ell^{-m} = P_\ell^m$. I am adopting now the more standard convention used by Mathematica.

[6] Nevertheless, some authors call them (confusingly) "associated Legendre polynomials."

Table 4.1: *The first few Legendre polynomials, $P_\ell(x)$: (a) functional form, (b) graph.*

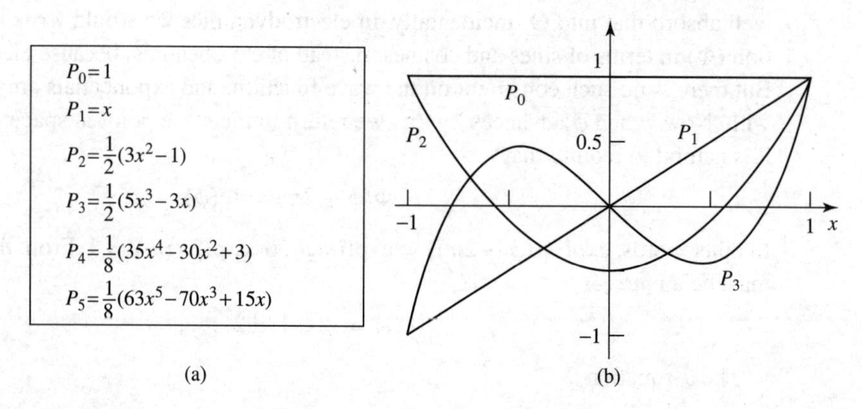

$$P_2^0(x) = \frac{1}{2}\left(3x^2 - 1\right), \quad P_2^1(x) = -\left(1 - x^2\right)^{1/2} \frac{d}{dx}\left[\frac{1}{2}\left(3x^2 - 1\right)\right] = -3x\sqrt{1 - x^2},$$

$$P_2^2(x) = \left(1 - x^2\right)\left(\frac{d}{dx}\right)^2\left[\frac{1}{2}\left(3x^2 - 1\right)\right] = 3\left(1 - x^2\right),$$

etc. (On the other hand, what *we* need is $P_\ell^m(\cos\theta)$, and $\sqrt{1-\cos^2\theta} = \sin\theta$, so $P_\ell^m(\cos\theta)$ is always a polynomial in $\cos\theta$, multiplied—if m is odd—by $\sin\theta$. Some associated Legendre functions of $\cos\theta$ are listed in Table 4.2.)

Notice that ℓ must be a non-negative *integer*, for the Rodrigues formula to make any sense; moreover, if $m > \ell$, then Equation 4.27 says $P_\ell^m = 0$. For any given ℓ, then, there are $(2\ell + 1)$ possible values of m:

$$\ell = 0,\ 1,\ 2, \ldots; \quad m = -\ell,\ -\ell + 1,\ \ldots,\ -1,\ 0,\ 1, \ldots,\ \ell - 1,\ \ell. \tag{4.29}$$

But wait! Equation 4.25 is a second-order differential equation: It should have *two* linearly independent solutions, for *any old* values of ℓ and m. Where are all the other solutions? *Answer*: They *exist*, of course, as mathematical solutions to the equation, but they are *physically* unacceptable, because they blow up at $\theta = 0$ and/or $\theta = \pi$ (see Problem 4.5).

Now, the volume element in spherical coordinates[7] is

$$d^3\mathbf{r} = r^2 \sin\theta\, dr\, d\theta\, d\phi = r^2\, dr\, d\Omega, \quad \text{where} \quad d\Omega \equiv \sin\theta\, d\theta\, d\phi, \tag{4.30}$$

so the normalization condition (Equation 4.6) becomes

$$\int |\psi|^2 r^2 \sin\theta\, dr\, d\theta\, d\phi = \int |R|^2 r^2\, dr \int |Y|^2\, d\Omega = 1.$$

It is convenient to normalize R and Y separately:

$$\int_0^\infty |R|^2 r^2\, dr = 1 \quad \text{and} \quad \int_0^{2\pi}\int_0^\pi |Y|^2 \sin\theta\, d\theta\, d\phi = 1. \tag{4.31}$$

---

[7] See, for instance, Boas (footnote 1), Chapter 5, Section 4.

Table 4.2: *Some associated Legendre functions, $P_\ell^m(\cos\theta)$: (a) functional form, (b) graphs of $r = \left|P_\ell^m(\cos\theta)\right|$ (in these plots r tells you the magnitude of the function in the direction θ; each figure should be rotated about the z axis).*

$P_0^0 = 1$	$P_2^0 = \frac{1}{2}(3\cos^2\theta - 1)$
$P_1^1 = -\sin\theta$	$P_3^3 = -15\sin\theta(1-\cos^2\theta)$
$P_1^0 = \cos\theta$	$P_3^2 = 15\sin^2\theta\cos\theta$
$P_2^2 = 3\sin^2\theta$	$P_3^1 = -\frac{3}{2}\sin\theta(5\cos^2\theta - 1)$
$P_2^1 = -3\sin\theta\cos\theta$	$P_3^0 = \frac{1}{2}(5\cos^3\theta - 3\cos\theta)$

(a)

(b)

Table 4.3: *The first few spherical harmonics, $Y_\ell^m(\theta,\phi)$.*

$Y_0^0 = \left(\frac{1}{4\pi}\right)^{1/2}$	$Y_2^{\pm2} = \left(\frac{15}{32\pi}\right)^{1/2}\sin^2\theta e^{\pm 2i\phi}$
$Y_1^0 = \left(\frac{3}{4\pi}\right)^{1/2}\cos\theta$	$Y_3^0 = \left(\frac{7}{16\pi}\right)^{1/2}(5\cos^3\theta - 3\cos\theta)$
$Y_1^{\pm1} = \mp\left(\frac{3}{8\pi}\right)^{1/2}\sin\theta e^{\pm i\phi}$	$Y_3^{\pm1} = \mp\left(\frac{21}{64\pi}\right)^{1/2}\sin\theta(5\cos^2\theta - 1)e^{\pm i\phi}$
$Y_2^0 = \left(\frac{5}{16\pi}\right)^{1/2}(3\cos^2\theta - 1)$	$Y_3^{\pm2} = \left(\frac{105}{32\pi}\right)^{1/2}\sin^2\theta\cos\theta e^{\pm 2i\phi}$
$Y_2^{\pm1} = \mp\left(\frac{15}{8\pi}\right)^{1/2}\sin\theta\cos\theta e^{\pm i\phi}$	$Y_3^{\pm3} = \mp\left(\frac{35}{64\pi}\right)^{1/2}\sin^3\theta e^{\pm 3i\phi}$

The normalized angular wave functions[8] are called **spherical harmonics**:

$$\boxed{Y_\ell^m(\theta,\phi) = \sqrt{\frac{(2\ell+1)}{4\pi}\frac{(\ell-m)!}{(\ell+m)!}}\, e^{im\phi}\, P_\ell^m(\cos\theta),} \tag{4.32}$$

As we shall prove later on, they are automatically orthogonal:

$$\int_0^{2\pi}\int_0^{\pi}\left[Y_\ell^m(\theta,\phi)\right]^*\left[Y_{\ell'}^{m'}(\theta,\phi)\right]\sin\theta\,d\theta\,d\phi = \delta_{\ell\ell'}\delta_{mm'}, \tag{4.33}$$

In Table 4.3 I have listed the first few spherical harmonics.

* **Problem 4.4** Use Equations 4.27, 4.28, and 4.32, to construct Y_0^0 and Y_2^1. Check that they are normalized and orthogonal.

[8] The normalization factor is derived in Problem 4.63.

Problem 4.5 Show that

$$\Theta(\theta) = A \ln\left[\tan(\theta/2)\right]$$

satisfies the θ equation (Equation 4.25), for $\ell = m = 0$. This is the unacceptable "second solution"—what's *wrong* with it?

Problem 4.6 Using Equation 4.32 and footnote 5, show that

$$Y_\ell^{-m} = (-1)^m \left(Y_\ell^m\right)^*.$$

* **Problem 4.7** Using Equation 4.32, find $Y_\ell^\ell(\theta, \phi)$ and $Y_3^2(\theta, \phi)$. (You can take P_3^2 from Table 4.2, but you'll have to work out P_ℓ^ℓ from Equations 4.27 and 4.28.) Check that they satisfy the angular equation (Equation 4.18), for the appropriate values of ℓ and m.

** **Problem 4.8** Starting from the Rodrigues formula, derive the orthonormality condition for Legendre polynomials:

$$\int_{-1}^{1} P_\ell(x) P_{\ell'}(x)\, dx = \left(\frac{2}{2\ell + 1}\right) \delta_{\ell\ell'}. \tag{4.34}$$

Hint: Use integration by parts.

4.1.3 The Radial Equation

Notice that the angular part of the wave function, $Y(\theta, \phi)$, is the same for *all* spherically symmetric potentials; the actual *shape* of the potential, $V(r)$, affects only the *radial* part of the wave function, $R(r)$, which is determined by Equation 4.16:

$$\frac{d}{dr}\left(r^2 \frac{dR}{dr}\right) - \frac{2mr^2}{\hbar^2}\left[V(r) - E\right] R = \ell(\ell + 1) R. \tag{4.35}$$

This simplifies if we change variables: Let

$$u(r) \equiv r R(r), \tag{4.36}$$

so that $R = u/r$, $dR/dr = \left[r\,(du/dr) - u\right]/r^2$, $(d/dr)\left[r^2\,(dR/dr)\right] = r d^2u/dr^2$, and hence

$$\boxed{-\frac{\hbar^2}{2m}\frac{d^2u}{dr^2} + \left[V + \frac{\hbar^2}{2m}\frac{\ell(\ell + 1)}{r^2}\right] u = Eu.} \tag{4.37}$$

This is called the **radial equation**;[9] it is *identical in form* to the one-dimensional Schrödinger equation (Equation 2.5), except that the **effective potential**,

$$V_{\text{eff}} = V + \frac{\hbar^2}{2m}\frac{\ell(\ell + 1)}{r^2}, \tag{4.38}$$

[9] Those ms are *masses*, of course—the separation constant m does not appear in the radial equation.

contains an extra piece, the so-called **centrifugal term**, $\left(\hbar^2/2m\right)\left[\ell\left(\ell+1\right)/r^2\right]$. It tends to throw the particle outward (away from the origin), just like the centrifugal (pseudo-)force in classical mechanics. Meanwhile, the normalization condition (Equation 4.31) becomes

$$\int_0^\infty |u|^2 \, dr = 1. \tag{4.39}$$

That's as far as we can go until a specific potential $V(r)$ is provided.

Example 4.1

Consider the **infinite spherical well**,

$$V(r) = \begin{cases} 0, & r \leq a; \\ \infty, & r > a. \end{cases} \tag{4.40}$$

Find the wave functions and the allowed energies.

Solution: Outside the well the wave function is zero; inside the well, the radial equation says

$$\frac{d^2u}{dr^2} = \left[\frac{\ell\left(\ell+1\right)}{r^2} - k^2\right]u, \tag{4.41}$$

where

$$k \equiv \frac{\sqrt{2mE}}{\hbar}. \tag{4.42}$$

Our problem is to solve Equation 4.41, subject to the boundary condition $u(a) = 0$. The case $\ell = 0$ is easy:

$$\frac{d^2u}{dr^2} = -k^2u \quad \Rightarrow u(r) = A\sin(kr) + B\cos(kr).$$

But remember, the actual radial wave function is $R(r) = u(r)/r$, and $\left[\cos(kr)\right]/r$ blows up as $r \to 0$. So[10] $B = 0$. The boundary condition then requires $\sin(ka) = 0$, and hence $ka = N\pi$, for some integer N. The allowed energies are

$$E_{N0} = \frac{N^2\pi^2\hbar^2}{2ma^2}, \quad (N = 1, 2, 3, \ldots) \tag{4.43}$$

(same as for the one-dimensional infinite square well, Equation 2.30). Normalizing $u(r)$ yields $A = \sqrt{2/a}$:

$$u_{N0} = \sqrt{\frac{2}{a}}\sin\left(\frac{N\pi r}{a}\right). \tag{4.44}$$

Notice that the radial wave function has $N-1$ nodes (or, if you prefer, N "lobes").

The general solution to Equation 4.41 (for an *arbitrary* integer ℓ) is not so familiar:

$$u(r) = Arj_\ell(kr) + Brn_\ell(kr), \tag{4.45}$$

[10] Actually, all we require is that the wave function be *normalizable*, not that it be *finite*: $R(r) \sim 1/r$ at the origin *is* normalizable (because of the r^2 in Equation 4.31). For a compelling general argument that $u(0) = 0$, see Ramamurti Shankar, *Principles of Quantum Mechanics*, 2nd edn (Plenum, New York, 1994), p. 342. For further discussion see F. A. B. Coutinho and M. Amaku, *Eur. J. Phys.* **30**, 1015 (2009).

where $j_\ell(x)$ is the **spherical Bessel function** of order ℓ, and $n_\ell(x)$ is the **spherical Neumann function** of order ℓ. They are defined as follows:

$$j_\ell(x) \equiv (-x)^\ell \left(\frac{1}{x}\frac{d}{dx}\right)^\ell \frac{\sin x}{x}; \quad n_\ell(x) \equiv -(-x)^\ell \left(\frac{1}{x}\frac{d}{dx}\right)^\ell \frac{\cos x}{x}. \tag{4.46}$$

For example,

$$j_0(x) = \frac{\sin x}{x}; \quad n_0(x) = -\frac{\cos x}{x};$$

$$j_1(x) = (-x)\frac{1}{x}\frac{d}{dx}\left(\frac{\sin x}{x}\right) = \frac{\sin x}{x^2} - \frac{\cos x}{x};$$

$$j_2(x) = (-x)^2\left(\frac{1}{x}\frac{d}{dx}\right)^2 \frac{\sin x}{x} = x^2\left(\frac{1}{x}\frac{d}{dx}\right)\frac{x\cos x - \sin x}{x^3}$$
$$= \frac{3\sin x - 3x\cos x - x^2\sin x}{x^3};$$

and so on. The first few spherical Bessel and Neumann functions are listed in Table 4.4. For small x (where $\sin x = x - x^3/3! + x^5/5! - \cdots$ and $\cos x = 1 - x^2/2 + x^4/4! - \cdots$),

$$j_0(x) \approx 1; \quad n_0(x) \approx \frac{1}{x}; \quad j_1(x) \approx \frac{x}{3}; \quad j_2(x) \approx \frac{x^2}{15};$$

etc. Notice that Bessel functions are finite at the origin, but *Neumann* functions blow up at the origin. Accordingly, $B = 0$, and hence

$$R(r) = A\, j_\ell(kr). \tag{4.47}$$

There remains the boundary condition, $R(a) = 0$. Evidently k must be chosen such that

$$j_\ell(ka) = 0; \tag{4.48}$$

that is, (ka) is a zero of the ℓth-order spherical Bessel function. Now, the Bessel functions are oscillatory (see Figure 4.2); each one has an infinite number of zeros. But (unfortunately for us) they are not located at nice sensible points (such as multiples of π); they have to be computed numerically.[11] At any rate, the boundary condition requires that

$$k = \frac{1}{a}\beta_{N\ell}, \tag{4.49}$$

where $\beta_{N\ell}$ is the Nth zero of the ℓth spherical Bessel function. The allowed energies, then, are given by

$$E_{N\ell} = \frac{\hbar^2}{2ma^2}\beta_{N\ell}^2. \tag{4.50}$$

It is customary to introduce the **principal quantum number**, n, which simply orders the allowed energies, starting with 1 for the ground state (see Figure 4.3). The wave functions are

[11] Milton Abramowitz and Irene A. Stegun, eds., *Handbook of Mathematical Functions*, Dover, New York (1965), Chapter 10, provides an extensive listing.

Table 4.4 *The first few spherical Bessel and Neumann functions, $j_\ell(x)$ and $n_\ell(x)$; asymptotic forms for small x.*

$$j_0 = \frac{\sin x}{x} \qquad\qquad n_0 = -\frac{\cos x}{x}$$

$$j_1 = \frac{\sin x}{x^2} - \frac{\cos x}{x} \qquad\qquad n_1 = -\frac{\cos x}{x^2} - \frac{\sin x}{x}$$

$$j_2 = \left(\frac{3}{x^3} - \frac{1}{x}\right)\sin x - \frac{3}{x^2}\cos x \qquad\qquad n_2 = -\left(\frac{3}{x^3} - \frac{1}{x}\right)\cos x - \frac{3}{x^2}\sin x$$

$$j_\ell \to \frac{2^\ell \ell!}{(2\ell+1)!}x^\ell, \qquad\qquad n_\ell \to -\frac{(2\ell)!}{2^\ell \ell!}\frac{1}{x^{\ell+1}}, \quad \text{for } x \ll 1.$$

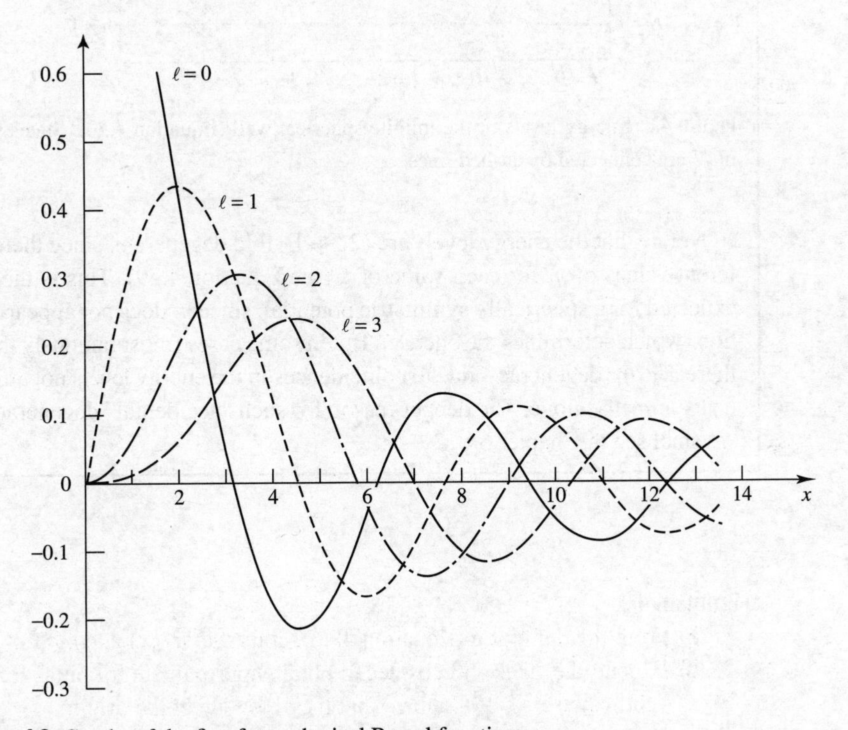

Figure 4.2: Graphs of the first four spherical Bessel functions.

$$\psi_{n\ell m}(r, \theta, \phi) = A_{n\ell}\, j_\ell\left(\beta_{N\ell}\frac{r}{a}\right) Y_\ell^m(\theta, \phi), \tag{4.51}$$

with the constant $A_{n\ell}$ to be determined by normalization. As before, the wave function has $N - 1$ radial nodes.[12]

[12] We shall use this notation ($N - 1$ as a count of the number of radial nodes, n for the order of the energy) with all central potentials. Both n and N are by their nature integers $(1, 2, 3, \ldots)$; n is determined by N and ℓ (conversely, N is determined by n and ℓ), but the actual relation can (as here) be complicated. In the special case of the Coulomb potential, as we shall see, there is a delightfully simple formula relating the two.

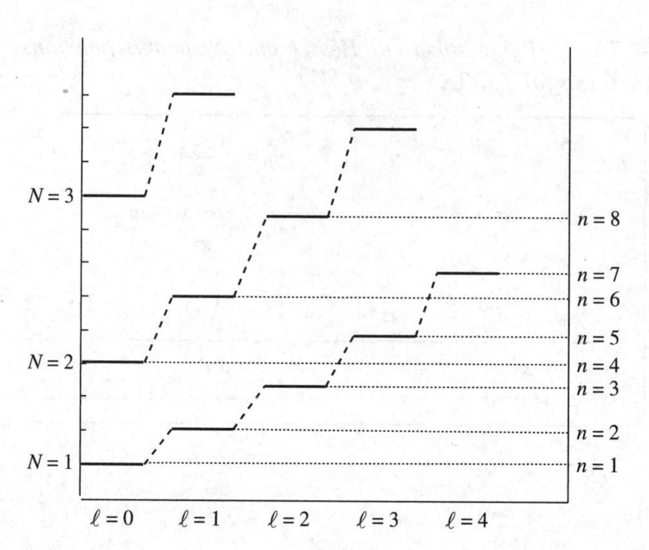

Figure 4.3: Energy levels of the infinite spherical well (Equation 4.50). States with the same value of N are connected by dashed lines.

Notice that the energy levels are $(2\ell + 1)$-fold degenerate, since there are $(2\ell + 1)$ different values of m for each value of ℓ (see Equation 4.29). This is the degeneracy to be expected for a spherically symmetric potential, since m does not appear in the radial equation (which determines the energy). But in some cases (most famously the hydrogen atom) there is *extra* degeneracy, due to coincidences in the energy levels not attributable to spherical symmetry alone. The deeper reason for such "accidental" degeneracy is intriguing, as we shall see in Chapter 6.

Problem 4.9

(a) From the definition (Equation 4.46), construct $n_1(x)$ and $n_2(x)$.

(b) Expand the sines and cosines to obtain approximate formulas for $n_1(x)$ and $n_2(x)$, valid when $x \ll 1$. Confirm that they blow up at the origin.

Problem 4.10

(a) Check that $Arj_1(kr)$ satisfies the radial equation with $V(r) = 0$ and $\ell = 1$.

(b) Determine graphically the allowed energies for the infinite spherical well, when $\ell = 1$. Show that for large N, $E_{N1} \approx \left(\hbar^2\pi^2/2ma^2\right)(N + 1/2)^2$. *Hint:* First show that $j_1(x) = 0 \Rightarrow x = \tan x$. Plot x and $\tan x$ on the same graph, and locate the points of intersection.

** **Problem 4.11** A particle of mass m is placed in a *finite* spherical well:

$$V(r) = \begin{cases} -V_0, & r \leq a; \\ 0, & r > a. \end{cases}$$

Find the ground state, by solving the radial equation with $\ell = 0$. Show that there is no bound state if $V_0 a^2 < \pi^2 \hbar^2 / 8m$.

4.2 THE HYDROGEN ATOM

The hydrogen atom consists of a heavy, essentially motionless proton (we may as well put it at the origin), of charge e, together with a much lighter electron (mass m_e, charge $-e$) that orbits around it, bound by the mutual attraction of opposite charges (see Figure 4.4). From Coulomb's law, the potential energy of the electron[13] (in SI units) is

$$V(r) = -\frac{e^2}{4\pi\epsilon_0}\frac{1}{r}, \tag{4.52}$$

and the radial equation (Equation 4.37) says

$$-\frac{\hbar^2}{2m_e}\frac{d^2u}{dr^2} + \left[-\frac{e^2}{4\pi\epsilon_0}\frac{1}{r} + \frac{\hbar^2}{2m_e}\frac{\ell(\ell+1)}{r^2}\right]u = Eu. \tag{4.53}$$

(The effective potential—the term in square brackets—is shown in Figure 4.5.) Our problem is to solve this equation for $u(r)$, and determine the allowed energies. The hydrogen atom is such an important case that I'm not going to hand you the solutions this time—we'll work them out in detail, by the method we used in the analytical solution to the harmonic oscillator. (If any step in this process is unclear, you may want to refer back to Section 2.3.2 for a more complete explanation.) Incidentally, the Coulomb potential (Equation 4.52) admits *continuum* states (with $E > 0$), describing electron-proton scattering, as well as discrete *bound* states, representing the hydrogen atom, but we shall confine our attention to the latter.[14]

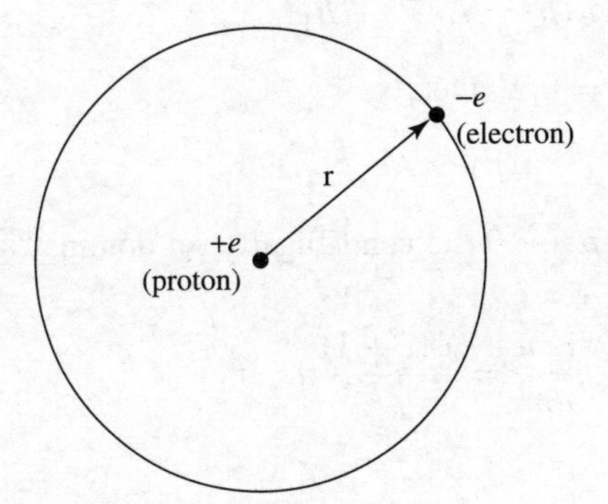

Figure 4.4: The hydrogen atom.

[13] This is what goes into the Schrödinger equation—not the electric *potential* $(e/4\pi\epsilon_0 r)$.

[14] Note, however, that the bound states by themselves are not complete.

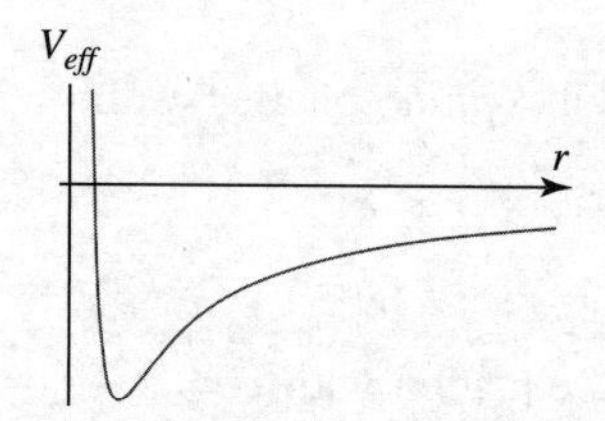

Figure 4.5: The effective potential for hydrogen (Equation 4.53), if $\ell > 0$.

4.2.1 The Radial Wave Function

Our first task is to tidy up the notation. Let

$$\kappa \equiv \frac{\sqrt{-2m_e E}}{\hbar}. \tag{4.54}$$

(For bound states, E is negative, so κ is *real*.) Dividing Equation 4.53 by E, we have

$$\frac{1}{\kappa^2}\frac{d^2u}{dr^2} = \left[1 - \frac{m_e e^2}{2\pi\epsilon_0\hbar^2\kappa}\frac{1}{(\kappa r)} + \frac{\ell(\ell+1)}{(\kappa r)^2}\right]u.$$

This suggests that we introduce

$$\rho \equiv \kappa r, \quad \text{and} \quad \rho_0 \equiv \frac{m_e e^2}{2\pi\epsilon_0\hbar^2\kappa}, \tag{4.55}$$

so that

$$\frac{d^2u}{d\rho^2} = \left[1 - \frac{\rho_0}{\rho} + \frac{\ell(\ell+1)}{\rho^2}\right]u. \tag{4.56}$$

Next we examine the asymptotic form of the solutions. As $\rho \to \infty$, the constant term in the brackets dominates, so (approximately)

$$\frac{d^2u}{d\rho^2} = u.$$

The general solution is

$$u(\rho) = Ae^{-\rho} + Be^{\rho}, \tag{4.57}$$

but e^{ρ} blows up (as $\rho \to \infty$), so $B = 0$. Evidently,

$$u(\rho) \sim Ae^{-\rho}, \tag{4.58}$$

for large ρ. On the other hand, as $\rho \to 0$ the centrifugal term dominates;[15] approximately, then:

$$\frac{d^2u}{d\rho^2} = \frac{\ell(\ell+1)}{\rho^2}u.$$

The general solution (check it!) is

$$u(\rho) = C\rho^{\ell+1} + D\rho^{-\ell},$$

[15] This argument does not apply when $\ell = 0$ (although the conclusion, Equation 4.59, is in fact valid for that case too). But never mind: All I am trying to do is provide some *motivation* for a change of variables (Equation 4.60).

but $\rho^{-\ell}$ blows up (as $\rho \to 0$), so $D = 0$. Thus

$$u(\rho) \sim C\rho^{\ell+1}, \tag{4.59}$$

for small ρ.

The next step is to peel off the asymptotic behavior, introducing the new function $v(\rho)$:

$$u(\rho) = \rho^{\ell+1} e^{-\rho} v(\rho), \tag{4.60}$$

in the hope that $v(\rho)$ will turn out to be simpler than $u(\rho)$. The first indications are not auspicious:

$$\frac{du}{d\rho} = \rho^{\ell} e^{-\rho} \left[(\ell + 1 - \rho) v + \rho \frac{dv}{d\rho} \right],$$

and

$$\frac{d^2u}{d\rho^2} = \rho^{\ell} e^{-\rho} \left\{ \left[-2\ell - 2 + \rho + \frac{\ell(\ell+1)}{\rho} \right] v + 2(\ell + 1 - \rho) \frac{dv}{d\rho} + \rho \frac{d^2v}{d\rho^2} \right\}.$$

In terms of $v(\rho)$, then, the radial equation (Equation 4.56) reads

$$\rho \frac{d^2v}{d\rho^2} + 2(\ell + 1 - \rho) \frac{dv}{d\rho} + [\rho_0 - 2(\ell+1)] v = 0. \tag{4.61}$$

Finally, we assume the solution, $v(\rho)$, can be expressed as a power series in ρ:

$$v(\rho) = \sum_{j=0}^{\infty} c_j \rho^j. \tag{4.62}$$

Our problem is to determine the coefficients $(c_0, c_1, c_2, \ldots)$. Differentiating term by term:

$$\frac{dv}{d\rho} = \sum_{j=0}^{\infty} j c_j \rho^{j-1} = \sum_{j=0}^{\infty} (j+1) c_{j+1} \rho^j.$$

(In the second summation I have renamed the "dummy index": $j \to j + 1$. If this troubles you, write out the first few terms explicitly, and *check* it. You may object that the sum should now begin at $j = -1$, but the factor $(j + 1)$ kills that term anyway, so we might as well start at zero.) Differentiating again,

$$\frac{d^2v}{d\rho^2} = \sum_{j=0}^{\infty} j(j+1) c_{j+1} \rho^{j-1}.$$

Inserting these into Equation 4.61,

$$\sum_{j=0}^{\infty} j(j+1) c_{j+1} \rho^j + 2(\ell+1) \sum_{j=0}^{\infty} (j+1) c_{j+1} \rho^j$$

$$-2 \sum_{j=0}^{\infty} j c_j \rho^j + [\rho_0 - 2(\ell+1)] \sum_{j=0}^{\infty} c_j \rho^j = 0.$$

Equating the coefficients of like powers yields

$$j(j+1) c_{j+1} + 2(\ell+1)(j+1) c_{j+1} - 2j c_j + [\rho_0 - 2(\ell+1)] c_j = 0,$$

or:

$$c_{j+1} = \left\{ \frac{2(j+\ell+1) - \rho_0}{(j+1)(j+2\ell+2)} \right\} c_j. \tag{4.63}$$

This recursion formula determines the coefficients, and hence the function $v(\rho)$: We start with c_0 (this becomes an overall constant, to be fixed eventually by normalization), and Equation 4.63 gives us c_1; putting this back in, we obtain c_2, and so on.[16]

Now let's see what the coefficients look like for large j (this corresponds to large ρ, where the higher powers dominate). In this regime the recursion formula says[17]

$$c_{j+1} \approx \frac{2j}{j(j+1)} c_j = \frac{2}{j+1} c_j,$$

so

$$c_j \approx \frac{2^j}{j!} c_0. \tag{4.64}$$

Suppose for a moment that this were the *exact* result. Then

$$v(\rho) = c_0 \sum_{j=0}^{\infty} \frac{2^j}{j!} \rho^j = c_0 e^{2\rho},$$

and hence

$$u(\rho) = c_0 \rho^{l+1} e^{\rho}, \tag{4.65}$$

which blows up at large ρ. The positive exponential is precisely the asymptotic behavior we *didn't* want, in Equation 4.57. (It's no accident that it reappears here; after all, it *does* represent the asymptotic form of *some* solutions to the radial equation—they just don't happen to be the ones we're interested in, because they aren't normalizable.)

There is only one escape from this dilemma: *The series must terminate*. There must occur some integer N such that

$$c_{N-1} \neq 0 \quad \text{but} \quad c_N = 0 \tag{4.66}$$

(beyond this all coefficients vanish automatically).[18] In that case Equation 4.63 says

$$2(N+\ell) - \rho_0 = 0.$$

Defining

$$n \equiv N + \ell, \tag{4.67}$$

---

[16] You might wonder why I didn't use the series method directly on $u(\rho)$—why factor out the asymptotic behavior before applying this procedure? Well, the reason for peeling off $\rho^{\ell+1}$ is largely aesthetic: Without this, the sequence would begin with a long string of zeros (the first nonzero coefficient being $c_{\ell+1}$); by factoring out $\rho^{\ell+1}$ we obtain a series that starts out with ρ^0. The $e^{-\rho}$ factor is more critical—if you *don't* pull that out, you get a three-term recursion formula, involving c_{j+2}, c_{j+1} and c_j (*try* it!), and that is enormously more difficult to work with.

[17] Why not drop the 1 in $j+1$? After all, I'm ignoring $2(\ell+1) - \rho_0$ in the numerator, and $2\ell+2$ in the denominator. In this approximation it would be fine to drop the 1 as well, but keeping it makes the argument a little cleaner. Try doing it without the 1, and you'll see what I mean.

[18] This makes $v(\rho)$ a polynomial of order $(N-1)$, with (therefore) $N-1$ roots, and hence the radial wave function has $N-1$ nodes.

we have

$$\rho_0 = 2n. \tag{4.68}$$

But ρ_0 determines E (Equations 4.54 and 4.55):

$$E = -\frac{\hbar^2\kappa^2}{2m_e} = -\frac{m_e e^4}{8\pi^2\epsilon_0^2\hbar^2\rho_0^2}, \tag{4.69}$$

so the allowed energies are

$$\boxed{E_n = -\left[\frac{m_e}{2\hbar^2}\left(\frac{e^2}{4\pi\epsilon_0}\right)^2\right]\frac{1}{n^2} = \frac{E_1}{n^2}, \quad n = 1, 2, 3, \ldots.} \tag{4.70}$$

This is the famous **Bohr formula**—by any measure the most important result in all of quantum mechanics. Bohr obtained it in 1913 by a serendipitous mixture of inapplicable classical physics and premature quantum theory (the Schrödinger equation did not come until 1926).

Combining Equations 4.55 and 4.68, we find that

$$\kappa = \left(\frac{m_e e^2}{4\pi\epsilon_0\hbar^2}\right)\frac{1}{n} = \frac{1}{an}, \tag{4.71}$$

where

$$\boxed{a \equiv \frac{4\pi\epsilon_0\hbar^2}{m_e e^2} = 0.529 \times 10^{-10}\ \text{m}} \tag{4.72}$$

is the so-called **Bohr radius**.[19] It follows (again, from Equation 4.55) that

$$\rho = \frac{r}{an}. \tag{4.73}$$

The spatial wave functions are labeled by three quantum numbers (n, ℓ, and m):[20]

$$\psi_{n\ell m}(r,\theta,\phi) = R_{n\ell}(r)\,Y_\ell^m(\theta,\phi), \tag{4.74}$$

where (referring back to Equations 4.36 and 4.60)

$$R_{n\ell}(r) = \frac{1}{r}\rho^{\ell+1}e^{-\rho}v(\rho), \tag{4.75}$$

and $v(\rho)$ is a polynomial of degree $n-\ell-1$ in ρ, whose coefficients are determined (up to an overall normalization factor) by the recursion formula

$$c_{j+1} = \frac{2(j+\ell+1-n)}{(j+1)(j+2\ell+2)}c_j. \tag{4.76}$$

[19] It is customary to write the Bohr radius with a subscript: a_0. But this is cumbersome and unnecessary, so I prefer to leave the subscript off.

[20] Again, n is the **principal quantum number**; it tells you the energy of the electron (Equation 4.70). For unfortunate historical reasons ℓ is called the **azimuthal quantum number** and m the **magnetic quantum number**; as we'll see in Section 4.3, they are related to the angular momentum of the electron.

The **ground state** (that is, the state of lowest energy) is the case $n = 1$; putting in the accepted values for the physical constants, we get:[21]

$$E_1 = -\left[\frac{m_e}{2\hbar^2}\left(\frac{e^2}{4\pi\epsilon_0}\right)^2\right] = -13.6 \text{ eV}. \tag{4.77}$$

In other words, the **binding energy** of hydrogen (the amount of energy you would have to impart to the electron in its ground state in order to ionize the atom) is 13.6 eV. Equation 4.67 forces $\ell = 0$, whence also $m = 0$ (see Equation 4.29), so

$$\psi_{100}(r, \theta, \phi) = R_{10}(r) Y_0^0(\theta, \phi). \tag{4.78}$$

The recursion formula truncates after the first term (Equation 4.76 with $j = 0$ yields $c_1 = 0$), so $v(\rho)$ is a constant (c_0), and

$$R_{10}(r) = \frac{c_0}{a} e^{-r/a}. \tag{4.79}$$

Normalizing it, in accordance with Equation 4.31:

$$\int_0^\infty |R_{10}|^2 r^2 \, dr = \frac{|c_0|^2}{a^2} \int_0^\infty e^{-2r/a} r^2 \, dr = |c_0|^2 \frac{a}{4} = 1,$$

so $c_0 = 2/\sqrt{a}$. Meanwhile, $Y_0^0 = 1/\sqrt{4\pi}$, and hence the ground state of hydrogen is

$$\psi_{100}(r, \theta, \phi) = \frac{1}{\sqrt{\pi a^3}} e^{-r/a}. \tag{4.80}$$

If $n = 2$ the energy is

$$E_2 = \frac{-13.6 \text{ eV}}{4} = -3.40 \text{ eV}; \tag{4.81}$$

this is the first excited state—or rather, *states*, since we can have either $\ell = 0$ (in which case $m = 0$) or $\ell = 1$ (with $m = -1$, 0, or +1); evidently four different states share this same energy. If $\ell = 0$, the recursion relation (Equation 4.76) gives

$$c_1 = -c_0 \text{ (using } j = 0), \quad \text{and } c_2 = 0 \text{ (using } j = 1),$$

so $v(\rho) = c_0(1 - \rho)$, and therefore

$$R_{20}(r) = \frac{c_0}{2a}\left(1 - \frac{r}{2a}\right) e^{-r/2a}. \tag{4.82}$$

(Notice that the expansion coefficients (c_j) are completely different for different quantum numbers n and ℓ.) If $\ell = 1$ the recursion formula terminates the series after a single term; $v(\rho)$ is a constant, and we find

$$R_{21}(r) = \frac{c_0}{4a^2} r e^{-r/2a}. \tag{4.83}$$

(In each case the constant c_0 is to be determined by normalization—see Problem 4.13.)

[21] An **electron volt** is the energy acquired by an electron when accelerated through an electric potential of 1 volt: 1 eV = 1.60×10^{-19} J.

For arbitrary n, the possible values of ℓ (consistent with Equation 4.67) are

$$\ell = 0, 1, 2, \dots, n-1, \tag{4.84}$$

and for each ℓ there are $(2\ell + 1)$ possible values of m (Equation 4.29), so the total degeneracy of the energy level E_n is

$$d(n) = \sum_{\ell=0}^{n-1} (2\ell + 1) = n^2. \tag{4.85}$$

In Figure 4.6 I plot the energy levels for hydrogen. Notice that different values of ℓ carry the same energy (for a given n)—contrast the infinite spherical well, Figure 4.3. (With Equation 4.67, ℓ dropped out of sight, in the derivation of the allowed energies, though it does still affect the wave functions.) This is what gives rise to the "extra" degeneracy of the Coulomb potential, as compared to what you would expect from spherical symmetry alone ($n^2 = 1, 4, 9, 16, \dots$, as opposed to $(2\ell + 1) = 1, 3, 5, 7, \dots$).

The polynomial $v(\rho)$ (defined by the recursion formula, Equation 4.76) is a function well known to applied mathematicians; apart from normalization, it can be written as

$$v(\rho) = L_{n-\ell-1}^{2\ell+1}(2\rho), \tag{4.86}$$

where

$$L_q^p(x) \equiv (-1)^p \left(\frac{d}{dx}\right)^p L_{p+q}(x) \tag{4.87}$$

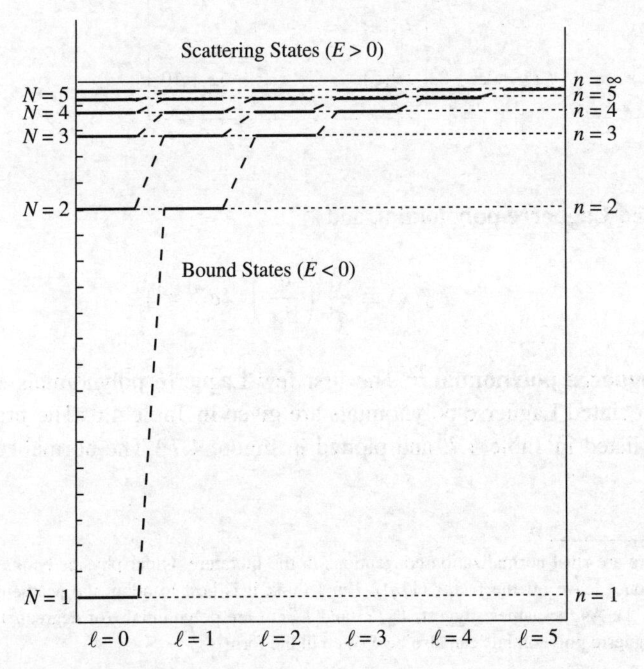

Figure 4.6: Energy levels for hydrogen (Equation 4.70); $n = 1$ is the ground state, with $E_1 = -13.6$ eV; an infinite number of states are squeezed in between $n = 5$ and $n = \infty$; $E_\infty = 0$ separates the bound states from the scattering states. Compare Figure 4.3, and note the extra ("accidental") degeneracy of the hydrogen energies.

Table 4.5: *The first few Laguerre polynomials.*

$L_0(x)$	$=$	1
$L_1(x)$	$=$	$-x+1$
$L_2(x)$	$=$	$\frac{1}{2}x^2-2x+1$
$L_3(x)$	$=$	$-\frac{1}{6}x^3+\frac{3}{2}x^2-3x+1$
$L_4(x)$	$=$	$\frac{1}{24}x^4-\frac{2}{3}x^3+3x^2-4x+1$
$L_5(x)$	$=$	$-\frac{1}{120}x^5+\frac{5}{24}x^4-\frac{5}{3}x^3+5x^2-5x+1$
$L_6(x)$	$=$	$\frac{1}{720}x^6-\frac{1}{20}x^5+\frac{5}{8}x^4-\frac{10}{3}x^3+\frac{15}{2}x^2-6x+1$

Table 4.6: *Some associated Laguerre polynomials.*

$L_0^0(x)$	$=$	1	$L_0^2(x)$	$=$	1
$L_1^0(x)$	$=$	$-x+1$	$L_1^2(x)$	$=$	$-x+3$
$L_2^0(x)$	$=$	$\frac{1}{2}x^2-2x+1$	$L_2^2(x)$	$=$	$\frac{1}{2}x^2-4x+6$
$L_0^1(x)$	$=$	1	$L_0^3(x)$	$=$	1
$L_1^1(x)$	$=$	$-x+2$	$L_1^3(x)$	$=$	$-x+4$
$L_2^1(x)$	$=$	$\frac{1}{2}x^2-3x+3$	$L_2^3(x)$	$=$	$\frac{1}{2}x^2-5x+10$

is an **associated Laguerre polynomial**, and

$$L_q(x) \equiv \frac{e^x}{q!}\left(\frac{d}{dx}\right)^q \left(e^{-x}x^q\right) \tag{4.88}$$

is the qth **Laguerre polynomial.**[22] The first few Laguerre polynomials are listed in Table 4.5; some associated Laguerre polynomials are given in Table 4.6. The first few radial wave functions are listed in Table 4.7, and plotted in Figure 4.7.) The normalized hydrogen wave functions are[23]

---

[22] As usual, there are rival normalization conventions in the literature. Older physics books (including earlier editions of this one) leave off the factor $(1/q!)$. But I think it is best to adopt the Mathematica standard (which sets $L_q(0)=1$). As the names suggest, $L_q(x)$ and $L_q^p(x)$ are polynomials (of degree q) in x. Incidentally, the associated Laguerre polynomials can also be written in the form

$$L_q^p(x) = \frac{x^{-p}e^x}{q!}\left(\frac{d}{dx}\right)^q \left(e^{-x}x^{p+q}\right).$$

[23] If you want to see how the normalization factor is calculated, study (for example), Leonard I. Schiff, *Quantum Mechanics*, 2nd edn, McGraw-Hill, New York, 1968, page 93. In books using the older normalization convention for the Laguerre polynomials (see footnote 22) the factor $(n+\ell)!$ under the square root will be cubed.

Table 4.7: *The first few radial wave functions for hydrogen,* $R_{n\ell}(r)$.

$R_{10} = 2a^{-3/2}\exp(-r/a)$
$R_{20} = \frac{1}{\sqrt{2}}a^{-3/2}\left(1 - \frac{1}{2}\frac{r}{a}\right)\exp(-r/2a)$ $R_{21} = \frac{1}{2\sqrt{6}}a^{-3/2}\left(\frac{r}{a}\right)\exp(-r/2a)$
$R_{30} = \frac{2}{3\sqrt{3}}a^{-3/2}\left(1 - \frac{2}{3}\frac{r}{a} + \frac{2}{27}\left(\frac{r}{a}\right)^2\right)\exp(-r/3a)$ $R_{31} = \frac{8}{27\sqrt{6}}a^{-3/2}\left(1 - \frac{1}{6}\frac{r}{a}\right)\left(\frac{r}{a}\right)\exp(-r/3a)$ $R_{32} = \frac{4}{81\sqrt{30}}a^{-3/2}\left(\frac{r}{a}\right)^2\exp(-r/3a)$
$R_{40} = \frac{1}{4}a^{-3/2}\left(1 - \frac{3}{4}\frac{r}{a} + \frac{1}{8}\left(\frac{r}{a}\right)^2 - \frac{1}{192}\left(\frac{r}{a}\right)^3\right)\exp(-r/4a)$ $R_{41} = \frac{5}{16\sqrt{15}}a^{-3/2}\left(1 - \frac{1}{4}\frac{r}{a} + \frac{1}{80}\left(\frac{r}{a}\right)^2\right)\left(\frac{r}{a}\right)\exp(-r/4a)$ $R_{42} = \frac{1}{64\sqrt{5}}a^{-3/2}\left(1 - \frac{1}{12}\frac{r}{a}\right)\left(\frac{r}{a}\right)^2\exp(-r/4a)$ $R_{43} = \frac{1}{768\sqrt{35}}a^{-3/2}\left(\frac{r}{a}\right)^3\exp(-r/4a)$

$$\boxed{\psi_{n\ell m} = \sqrt{\left(\frac{2}{na}\right)^3 \frac{(n-\ell-1)!}{2n\,(n+\ell)!}}\; e^{-r/na}\left(\frac{2r}{na}\right)^{\ell}\left[L_{n-\ell-1}^{2\ell+1}(2r/na)\right]Y_\ell^m(\theta,\phi).} \tag{4.89}$$

They are not pretty, but don't complain—this is one of the very few realistic systems that can be solved at all, in exact closed form. The wave functions are mutually orthogonal:

$$\int \psi^*_{n\ell m}\,\psi_{n'\ell'm'}\,r^2\,dr\,d\Omega = \delta_{nn'}\delta_{\ell\ell'}\delta_{mm'}. \tag{4.90}$$

This follows from the orthogonality of the spherical harmonics (Equation 4.33) and (for $n \neq n'$) from the fact that they are eigenfunctions of $\hat{H}$ with distinct eigenvalues.

Visualizing the hydrogen wave functions is not easy. Chemists like to draw **density plots**, in which the brightness of the cloud is proportional to $|\psi|^2$ (Figure 4.8). More quantitative (but perhaps harder to decipher) are surfaces of constant probability density (Figure 4.9). The quantum numbers n, ℓ, and m can be identified from the nodes of the wave function. The number of radial nodes is, as always, given by $N-1$ (for hydrogen this is $n-\ell-1$). For each radial node the wave function vanishes on a sphere, as can be seen in Figure 4.8. The quantum

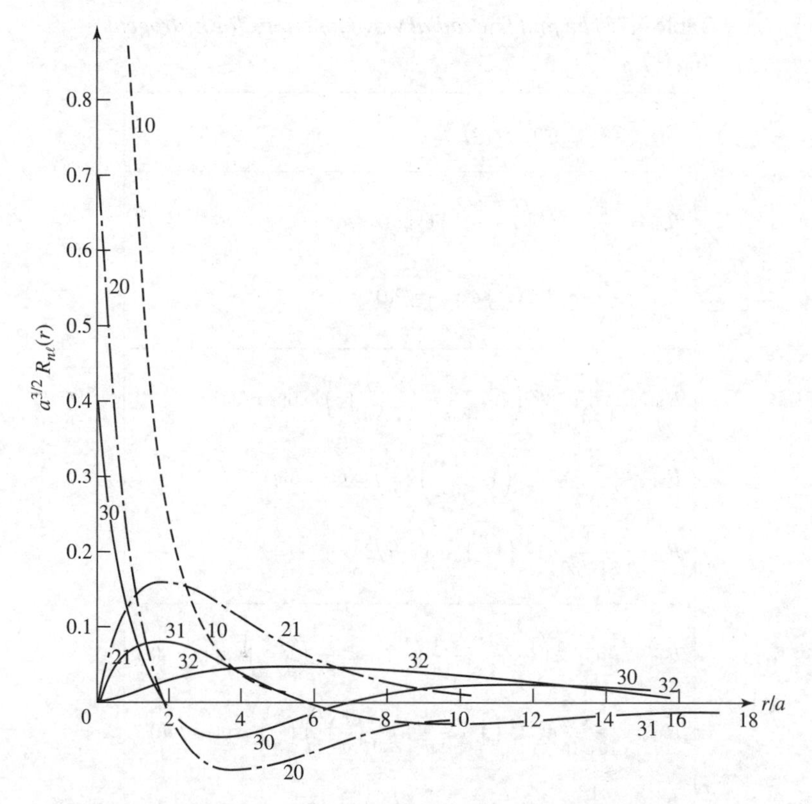

Figure 4.7: Graphs of the first few hydrogen radial wave functions, $R_{n\ell}(r)$.

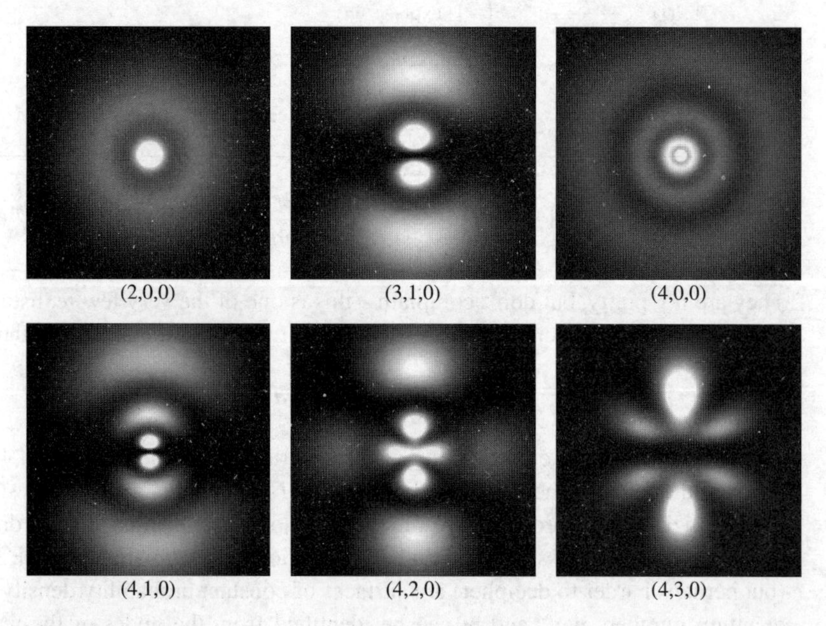

Figure 4.8: Density plots for the first few hydrogen wave functions, labeled by (n, ℓ, m). Printed by permission using "Atom in a Box" by Dauger Research. You can make your own plots by going to: http://dauger.com.

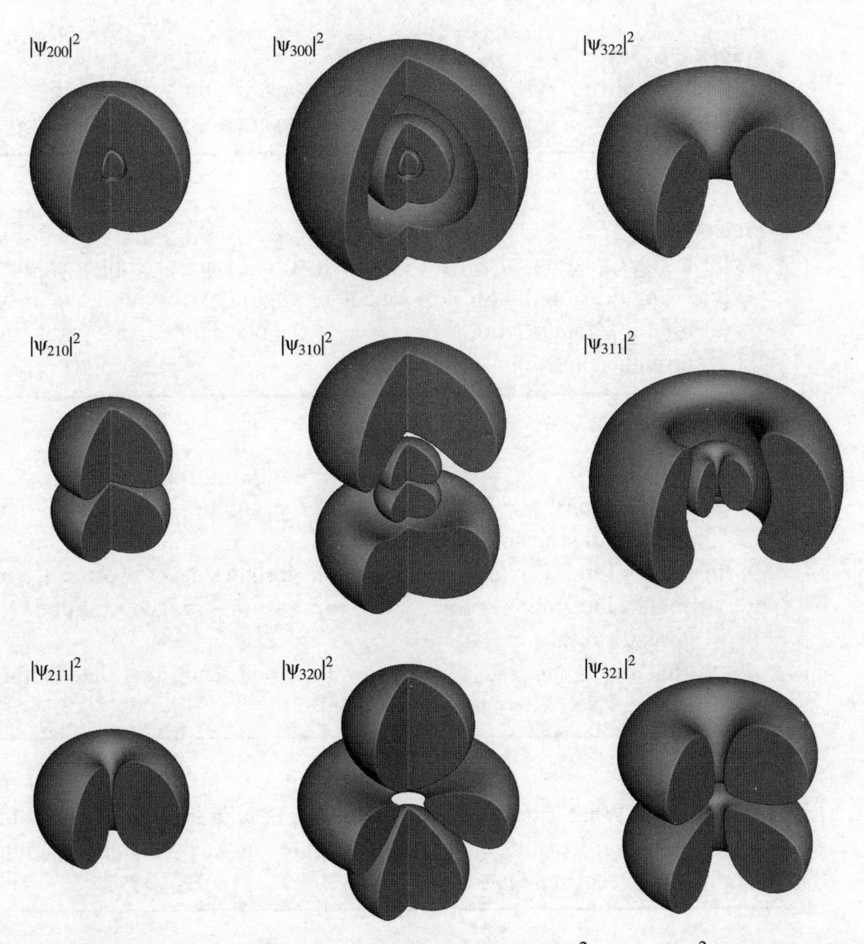

Figure 4.9: Shaded regions indicate significant electron density ($|\psi|^2 > 0.25/\text{nm}^3$) for the first few hydrogen wave functions. The region $0 < \phi < \pi/2$ has been cut away; $|\psi|^2$ has azimuthal symmetry in all cases.

number m counts the number of nodes of the real (or imaginary) part of the wave function in the ϕ direction. These nodes are planes containing the z axis on which the real or imaginary part of ψ vanishes.[24] Finally, $\ell - m$ gives the number of nodes in the θ direction. These are cones about the z axis on which ψ vanishes (note that a cone with opening angle $\pi/2$ is the $x - y$ plane itself).

* **Problem 4.12** Work out the radial wave functions R_{30}, R_{31}, and R_{32}, using the recursion formula (Equation 4.76). Don't bother to normalize them.

[24] These planes aren't visible in Figure 4.8 or 4.9, since these figures show the absolute value of ψ, and the real and imaginary parts of the wave function vanish on different sets of planes. However, since both sets contain the z axis, the wave function itself must vanish on the z axis for $m \neq 0$ (see Figure 4.9).

∗ **Problem 4.13**

(a) Normalize R_{20} (Equation 4.82), and construct the function ψ_{200}.

(b) Normalize R_{21} (Equation 4.83), and construct ψ_{211}, ψ_{210}, and ψ_{21-1}.

∗ **Problem 4.14**

(a) Using Equation 4.88, work out the first four Laguerre polynomials.

(b) Using Equations 4.86, 4.87, and 4.88, find $v(\rho)$, for the case $n = 5$, $\ell = 2$.

(c) Find $v(\rho)$ again (for the case $n = 5$, $\ell = 2$), but this time get it from the recursion formula (Equation 4.76).

∗ **Problem 4.15**

(a) Find $\langle r\rangle$ and $\langle r^2\rangle$ for an electron in the ground state of hydrogen. Express your answers in terms of the Bohr radius.

(b) Find $\langle x\rangle$ and $\langle x^2\rangle$ for an electron in the ground state of hydrogen. *Hint*: This requires no new integration—note that $r^2 = x^2 + y^2 + z^2$, and exploit the symmetry of the ground state.

(c) Find $\langle x^2\rangle$ in the state $n = 2$, $\ell = 1$, $m = 1$. *Hint*: this state is *not* symmetrical in x, y, z. Use $x = r\sin\theta\cos\phi$.

Problem 4.16 What is the *most probable* value of r, in the ground state of hydrogen? (The answer is *not* zero!) *Hint:* First you must figure out the probability that the electron would be found between r and $r + dr$.

Problem 4.17 Calculate $\left\langle z\hat{H}z\right\rangle$, in the ground state of hydrogen. *Hint:* This takes two pages and six integrals, or four lines and no integrals, depending on how you set it up. To do it the quick way, start by noting that $[z, [H, z]] = 2zHz - Hz^2 - z^2H$.[25]

Problem 4.18 A hydrogen atom starts out in the following linear combination of the stationary states $n = 2$, $\ell = 1$, $m = 1$ and $n = 2$, $\ell = 1$, $m = -1$:

$$\Psi(\mathbf{r}, 0) = \frac{1}{\sqrt{2}}\left(\psi_{211} + \psi_{21-1}\right).$$

(a) Construct $\Psi(\mathbf{r}, t)$. Simplify it as much as you can.

(b) Find the expectation value of the potential energy, $\langle V\rangle$. (Does it depend on t?) Give both the formula and the actual number, in electron volts.

[25] The idea is to reorder the operators in such a way that $\hat{H}$ appears either to the left or to the right, because we know (of course) what $\hat{H}\psi_{100}$ is.

4.2.2 The Spectrum of Hydrogen

In principle, if you put a hydrogen atom into some stationary state $\Psi_{n\ell m}$, it should stay there forever. However, if you *tickle* it slightly (by collision with another atom, say, or by shining light on it), the atom may undergo a **transition** to some other stationary state—either by *absorbing* energy, and moving up to a higher-energy state, or by *giving off* energy (typically in the form of electromagnetic radiation), and moving down.[26] In practice such perturbations are *always* present; transitions (or, as they are sometimes called, **quantum jumps**) are constantly occurring, and the result is that a container of hydrogen gives off light (**photons**), whose energy corresponds to the *difference* in energy between the initial and final states:

$$E_\gamma = E_i - E_f = -13.6\,\mathrm{eV}\left(\frac{1}{n_i^2} - \frac{1}{n_f^2}\right). \tag{4.91}$$

Now, according to the **Planck formula,**[27] the energy of a photon is proportional to its frequency:

$$E_\gamma = h\nu. \tag{4.92}$$

Meanwhile, the *wavelength* is given by $\lambda = c/\nu$, so

$$\frac{1}{\lambda} = \mathcal{R}\left(\frac{1}{n_f^2} - \frac{1}{n_i^2}\right), \tag{4.93}$$

where

$$\mathcal{R} \equiv \frac{m_e}{4\pi c\hbar^3}\left(\frac{e^2}{4\pi\epsilon_0}\right)^2 = 1.097 \times 10^7\,\mathrm{m}^{-1} \tag{4.94}$$

is known as the **Rydberg constant**. Equation 4.93 is the **Rydberg formula** for the spectrum of hydrogen; it was discovered empirically in the nineteenth century, and the greatest triumph of Bohr's theory was its ability to account for this result—and to calculate $\mathcal{R}$ in terms of the fundamental constants of nature. Transitions to the ground state $(n_f = 1)$ lie in the ultraviolet; they are known to spectroscopists as the **Lyman series**. Transitions to the first excited state $(n_f = 2)$ fall in the visible region; they constitute the **Balmer series**. Transitions to $n_f = 3$ (the **Paschen series**) are in the infrared; and so on (see Figure 4.10). (At room temperature, most hydrogen atoms are in the ground state; to obtain the emission spectrum you must first populate the various excited states; typically this is done by passing an electric spark through the gas.)

* **Problem 4.19** A **hydrogenic atom** consists of a single electron orbiting a nucleus with Z protons. ($Z = 1$ would be hydrogen itself, $Z = 2$ is ionized helium, $Z = 3$ is doubly ionized lithium, and so on.) Determine the Bohr energies $E_n(Z)$, the binding energy $E_1(Z)$, the Bohr radius $a(Z)$, and the Rydberg constant $\mathcal{R}(Z)$ for a hydrogenic atom. (Express

---

[26] By its nature, this involves a time-*dependent* potential, and the details will have to await Chapter 11; for our present purposes the actual mechanism involved is immaterial.

[27] The photon is a quantum of electromagnetic radiation; it's a relativistic object if there ever was one, and therefore outside the scope of nonrelativistic quantum mechanics. It will be useful in a few places to speak of photons, and to invoke the Planck formula for their energy, but please bear in mind that this is external to the theory we are developing.

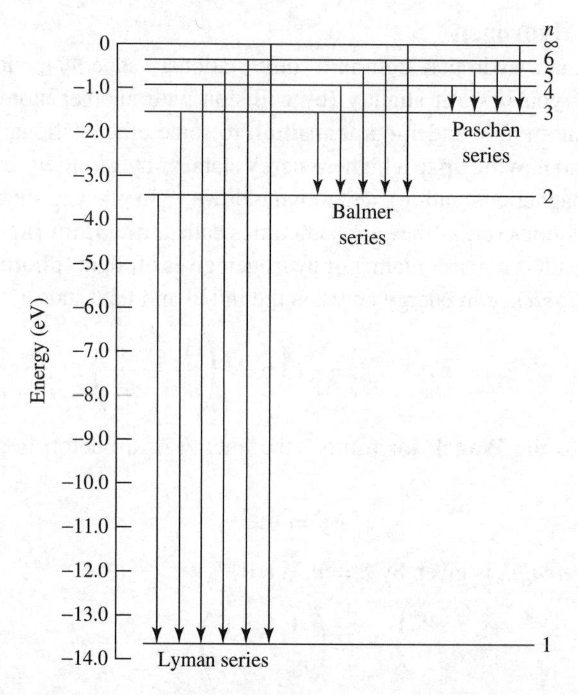

Figure 4.10: Energy levels and transitions in the spectrum of hydrogen.

your answers as appropriate multiples of the hydrogen values.) Where in the electromagnetic spectrum would the Lyman series fall, for $Z = 2$ and $Z = 3$? *Hint:* There's nothing much to *calculate* here—in the potential (Equation 4.52) $e^2 \rightarrow Ze^2$, so all you have to do is make the same substitution in all the final results.

Problem 4.20 Consider the earth–sun system as a gravitational analog to the hydrogen atom.

(a) What is the potential energy function (replacing Equation 4.52)? (Let m_E be the mass of the earth, and M the mass of the sun.)

(b) What is the "Bohr radius," a_g, for this system? Work out the actual number.

(c) Write down the gravitational "Bohr formula," and, by equating E_n to the classical energy of a planet in a circular orbit of radius r_o, show that $n = \sqrt{r_o/a_g}$. From this, estimate the quantum number n of the earth.

(d) Suppose the earth made a transition to the next lower level $(n - 1)$. How much energy (in Joules) would be released? What would the wavelength of the emitted photon (or, more likely, graviton) be? (Express your answer in light years—is the remarkable answer[28] a coincidence?)

[28] Thanks to John Meyer for pointing this out.

4.3 ANGULAR MOMENTUM

As we have seen, the stationary states of the hydrogen atom are labeled by three quantum numbers: n, ℓ, and m. The principal quantum number (n) determines the energy of the state (Equation 4.70); ℓ and m are related to the orbital angular momentum. In the classical theory of central forces, energy and angular momentum are the fundamental conserved quantities, and it is not surprising that angular momentum plays an important role in the quantum theory.

Classically, the angular momentum of a particle (with respect to the origin) is given by the formula

$$\mathbf{L} = \mathbf{r} \times \mathbf{p}, \tag{4.95}$$

which is to say,[29]

$$L_x = y p_z - z p_y, \quad L_y = z p_x - x p_z, \quad L_z = x p_y - y p_x. \tag{4.96}$$

The corresponding quantum operators[30] are obtained by the standard prescription $p_x \to -i\hbar\partial/\partial x$, $p_y \to -i\hbar\partial/\partial y$, $p_z \to -i\hbar\partial/\partial z$. In this section we'll obtain the eigenvalues of the the angular momentum operators by a purely algebraic technique reminiscent of the one we used in Chapter 2 to get the allowed energies of the harmonic oscillator; it is all based on the clever exploitation of commutation relations. After that we will turn to the more difficult problem of determining the eigenfunctions.

4.3.1 Eigenvalues

The operators L_x and L_y do not commute; in fact

$$\begin{aligned}\left[L_x, L_y\right] &= \left[y p_z - z p_y, z p_x - x p_z\right] \\ &= \left[y p_z, z p_x\right] - \left[y p_z, x p_z\right] - \left[z p_y, z p_x\right] + \left[z p_y, x p_z\right].\end{aligned} \tag{4.97}$$

From the canonical commutation relations (Equation 4.10) we know that the only operators here that *fail* to commute are x with p_x, y with p_y, and z with p_z. So the two middle terms drop out, leaving

$$\left[L_x, L_y\right] = y p_x \left[p_z, z\right] + x p_y \left[z, p_z\right] = i\hbar\left(x p_y - y p_x\right) = i\hbar L_z. \tag{4.98}$$

Of course, we could have started out with $\left[L_y, L_z\right]$ or $\left[L_z, L_x\right]$, but there is no need to calculate these separately—we can get them immediately by cyclic permutation of the indices ($x \to y$, $y \to z$, $z \to x$):

$$\boxed{\left[L_x, L_y\right] = i\hbar L_z; \quad \left[L_y, L_z\right] = i\hbar L_x; \quad \left[L_z, L_x\right] = i\hbar L_y.} \tag{4.99}$$

These are the fundamental commutation relations for angular momentum; everything follows from them.

---

[29] Because angular momentum involves the product of position and momentum, you might worry that the ambiguity addressed in Chapter 3 (footnote 15, page 102) would arise. Fortunately, only *different components* of **r** and **p** are multiplied, and they commute (Equation 4.10).

[30] To reduce clutter (and avoid confusion with the unit vectors $\hat{\imath}, \hat{\jmath}, \hat{k}, \hat{r}, \hat{\theta}, \hat{\phi}$) I'm going to take the hats off operators for the rest of the chapter.

Notice that L_x, L_y, and L_z are *incompatible* observables. According to the generalized uncertainty principle (Equation 3.62),

$$\sigma_{L_x}^2 \sigma_{L_y}^2 \geq \left(\frac{1}{2i}\langle i\hbar L_z\rangle\right)^2 = \frac{\hbar^2}{4}\langle L_z\rangle^2,$$

or

$$\sigma_{L_x}\sigma_{L_y} \geq \frac{\hbar}{2}\left|\langle L_z\rangle\right|. \tag{4.100}$$

It would therefore be futile to look for states that are simultaneously eigenfunctions of L_x and L_y. On the other hand, the *square* of the *total* angular momentum,

$$L^2 \equiv L_x^2 + L_y^2 + L_z^2, \tag{4.101}$$

does commute with L_x:

$$\begin{aligned}
\left[L^2, L_x\right] &= \left[L_x^2, L_x\right] + \left[L_y^2, L_x\right] + \left[L_z^2, L_x\right] \\
&= L_y\left[L_y, L_x\right] + \left[L_y, L_x\right]L_y + L_z\left[L_z, L_x\right] + \left[L_z, L_x\right]L_z \\
&= L_y\left(-i\hbar L_z\right) + \left(-i\hbar L_z\right)L_y + L_z\left(i\hbar L_y\right) + \left(i\hbar L_y\right)L_z \\
&= 0.
\end{aligned}$$

(I used Equation 3.65 to reduce the commutators; of course, *any* operator commutes with *itself*.) It follows that L^2 also commutes with L_y and L_z:

$$\left[L^2, L_x\right] = 0, \quad \left[L^2, L_y\right] = 0, \quad \left[L^2, L_z\right] = 0, \tag{4.102}$$

or, more compactly,

$$\left[L^2, \mathbf{L}\right] = 0. \tag{4.103}$$

So L^2 *is* compatible with each component of $\mathbf{L}$, and we *can* hope to find simultaneous eigenstates of L^2 and (say) L_z:

$$L^2 f = \lambda f \quad \text{and} \quad L_z f = \mu f. \tag{4.104}$$

We'll use a **ladder operator** technique, very similar to the one we applied to the harmonic oscillator back in Section 2.3.1. Let

$$L_\pm \equiv L_x \pm i L_y. \tag{4.105}$$

Its commutator with L_z is

$$\left[L_z, L_\pm\right] = \left[L_z, L_x\right] \pm i\left[L_z, L_y\right] = i\hbar L_y \pm i\left(-i\hbar L_x\right) = \pm\hbar\left(L_x \pm i L_y\right),$$

so

$$\left[L_z, L_\pm\right] = \pm\hbar L_\pm. \tag{4.106}$$

Also (from Equation 4.102)

$$\left[L^2, L_\pm\right] = 0. \tag{4.107}$$

I claim that if f is an eigenfunction of L^2 and L_z, so also is $L_\pm f$: Equation 4.107 says

$$L^2\left(L_\pm f\right) = L_\pm\left(L^2 f\right) = L_\pm\left(\lambda f\right) = \lambda\left(L_\pm f\right), \tag{4.108}$$

so $L_\pm f$ is an eigenfunction of L^2, with the same eigenvalue λ, and Equation 4.106 says

$$\begin{aligned}
L_z\left(L_\pm f\right) &= \left(L_z L_\pm - L_\pm L_z\right) f + L_\pm L_z f = \pm\hbar L_\pm f + L_\pm\left(\mu f\right) \\
&= \left(\mu \pm \hbar\right)\left(L_\pm f\right),
\end{aligned} \tag{4.109}$$

Figure 4.11: The "ladder" of angular momentum states.

so $L_\pm f$ is an eigenfunction of L_z with the *new* eigenvalue $\mu \pm \hbar$. We call L_+ the **raising operator**, because it *increases* the eigenvalue of L_z by $\hbar$, and L_- the **lowering operator**, because it *lowers* the eigenvalue by $\hbar$.

For a given value of λ, then, we obtain a "ladder" of states, with each "rung" separated from its neighbors by one unit of $\hbar$ in the eigenvalue of L_z (see Figure 4.11). To ascend the ladder we apply the raising operator, and to descend, the lowering operator. But this process cannot go on forever: Eventually we're going to reach a state for which the z-component exceeds the *total*, and that cannot be.[31] There must exist a "top rung", f_t, such that[32]

$$L_+ f_t = 0. \tag{4.110}$$

Let $\hbar\ell$ be the eigenvalue of L_z at the top rung (the appropriateness of the letter "ℓ" will appear in a moment):

$$L_z f_t = \hbar\ell f_t; \quad L^2 f_t = \lambda f_t. \tag{4.111}$$

[31] Formally, $\langle L^2\rangle = \langle L_x^2\rangle + \langle L_y^2\rangle + \langle L_z^2\rangle$, but $\langle L_x^2\rangle = \langle f | L_x^2 f\rangle = \langle L_x f | L_x f\rangle \geq 0$ (and likewise for L_y), so $\lambda = \langle L_x^2\rangle + \langle L_y^2\rangle + \mu^2 \geq \mu^2$.

[32] Actually, all we can conclude is that $L_+ f_t$ is *not normalizable*—its norm could be *infinite*, instead of zero. Problem 4.21 explores this alternative.

Now,

$$L_{\pm}L_{\mp} = \left(L_x \pm iL_y\right)\left(L_x \mp iL_y\right) = L_x^2 + L_y^2 \mp i\left(L_xL_y - L_yL_x\right)$$
$$= L^2 - L_z^2 \mp i\left(i\hbar L_z\right),$$

or, putting it the other way around,

$$L^2 = L_{\pm}L_{\mp} + L_z^2 \mp \hbar L_z. \tag{4.112}$$

It follows that

$$L^2 f_t = \left(L_-L_+ + L_z^2 + \hbar L_z\right) f_t = \left(0 + \hbar^2\ell^2 + \hbar^2\ell\right) f_t = \hbar^2\ell\left(\ell+1\right) f_t,$$

and hence

$$\lambda = \hbar^2\ell\left(\ell+1\right). \tag{4.113}$$

This tells us the eigenvalue of L^2 in terms of the *maximum* eigenvalue of L_z.

Meanwhile, there is also (for the same reason) a *bottom* rung, f_b, such that

$$L_- f_b = 0. \tag{4.114}$$

Let $\hbar\bar{\ell}$ be the eigenvalue of L_z at this bottom rung:

$$L_z f_b = \hbar\bar{\ell} f_b; \quad L^2 f_b = \lambda f_b. \tag{4.115}$$

Using Equation 4.112, we have

$$L^2 f_b = \left(L_+L_- + L_z^2 - \hbar L_z\right) f_b = \left(0 + \hbar^2\bar{\ell}^2 - \hbar^2\bar{\ell}\right) f_b = \hbar^2\bar{\ell}(\bar{\ell}-1) f_b,$$

and therefore

$$\lambda = \hbar^2\bar{\ell}\left(\bar{\ell}-1\right). \tag{4.116}$$

Comparing Equations 4.113 and 4.116, we see that $\ell\left(\ell+1\right) = \bar{\ell}\left(\bar{\ell}-1\right)$, so either $\bar{\ell} = \ell+1$ (which is absurd—the bottom rung would be higher than the top rung!) or else

$$\bar{\ell} = -\ell. \tag{4.117}$$

So the eigenvalues of L_z are $m\hbar$, where m (the appropriateness of this letter will also be clear in a moment) goes from $-\ell$ to $+\ell$, in N integer steps. In particular, it follows that $\ell = -\ell + N$, and hence $\ell = N/2$, so ℓ must be *an integer or a half-integer*. The eigenfunctions are characterized by the numbers ℓ and m:

$$\boxed{L^2 f_\ell^m = \hbar^2\ell\left(\ell+1\right) f_\ell^m; \quad L_z f_\ell^m = \hbar m f_\ell^m,} \tag{4.118}$$

where

$$\ell = 0,\ 1/2,\ 1,\ 3/2, \ldots; \quad m = -\ell,\ -\ell+1,\ \ldots,\ \ell-1,\ \ell. \tag{4.119}$$

For a given value of ℓ, there are $2\ell+1$ different values of m (i.e. $2\ell+1$ "rungs" on the "ladder").

Some people like to illustrate this with the diagram in Figure 4.12 (drawn for the case $\ell = 2$). The arrows are supposed to represent possible angular momenta (in units of $\hbar$)—they all have the same length $\sqrt{\ell\left(\ell+1\right)}$ (in this case $\sqrt{6} = 2.45$), and their z components are the allowed values of m ($-2, -1, 0, 1, 2$). Notice that the magnitude of the vectors (the radius of the sphere) is *greater* than the maximum z component! (In general, $\sqrt{\ell\left(\ell+1\right)} > \ell$, except for the "trivial" case $\ell = 0$.) Evidently you can't get the angular momentum to point perfectly

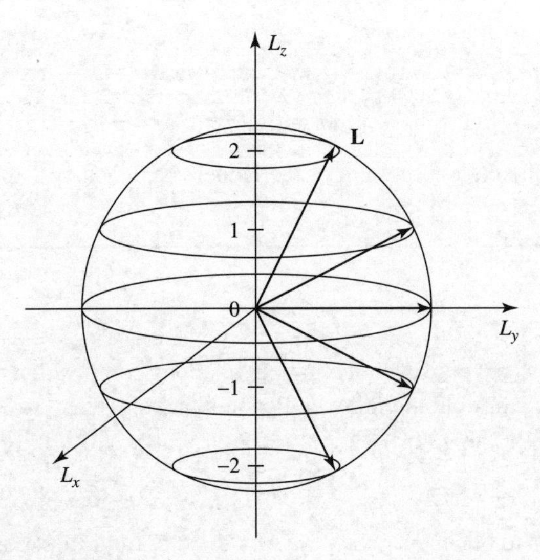

Figure 4.12: Angular momentum states (for $\ell = 2$).

along the z direction. At first, this sounds absurd. "Why can't I just *pick* my axes so that z points along the direction of the angular momentum vector?" Well, to do that you would have to know all three components simultaneously, and the uncertainty principle (Equation 4.100) says that's impossible. "Well, all right, but surely once in a while, by good fortune, I will just *happen* to aim my z axis along the direction of **L**." No, no! You have missed the point. It's not merely that you don't *know* all three components of **L**; there just *aren't* three components—a particle simply cannot *have* a determinate angular momentum vector, any more than it can simultaneously have a determinate position and momentum. If L_z has a well-defined value, then L_x and L_y do *not*. It is misleading even to *draw* the vectors in Figure 4.12—at best they should be smeared out around the latitude lines, to indicate that L_x and L_y are indeterminate.

I hope you're impressed: By *purely algebraic means*, starting with the fundamental commutation relations for angular momentum (Equation 4.99), we have determined the eigenvalues of L^2 and L_z—without ever seeing the eigenfunctions themselves! We turn now to the problem of constructing the eigenfunctions, but I should warn you that this is a much messier business. Just so you know where we're headed, I'll let you in on the punch line: $f_\ell^m = Y_\ell^m$—the eigenfunctions of L^2 and L_z are nothing but the old spherical harmonics, which we came upon by a quite different route in Section 4.1.2 (that's why I chose the same letters ℓ and m, of course). And I can now explain why the spherical harmonics are orthogonal: They are eigenfunctions of hermitian operators $\left(L^2 \text{ and } L_z\right)$ belonging to distinct eigenvalues (Theorem 2, Section 3.3.1).

* **Problem 4.21** The raising and lowering operators change the value of m by one unit:

$$L_+ f_\ell^m = \left(A_\ell^m\right) f_\ell^{m+1}, \quad L_- f_\ell^m = \left(B_\ell^m\right) f_\ell^{m-1} \tag{4.120}$$

where A_ℓ^m and B_ℓ^m are constants. *Question*: What *are* they, if the eigenfunctions are to be normalized? *Hint*: First show that $L_\mp$ is the hermitian conjugate of $L_\pm$ (since L_x and L_y are observables, you may assume they are hermitian... but *prove* it if you like); then use Equation 4.112. *Answer*:

$$A_\ell^m = \hbar\sqrt{\ell(\ell+1) - m(m+1)} = \hbar\sqrt{(\ell-m)(\ell+m+1)},$$
$$B_\ell^m = \hbar\sqrt{\ell(\ell+1) - m(m-1)} = \hbar\sqrt{(\ell+m)(\ell-m+1)}. \tag{4.121}$$

Note what happens at the top and bottom of the ladder (i.e. when you apply L_+ to f_ℓ^ℓ or L_- to $f_\ell^{-\ell}$).

* **Problem 4.22**

(a) Starting with the canonical commutation relations for position and momentum (Equation 4.10), work out the following commutators:

$$\begin{aligned} [L_z, x] &= i\hbar y, & [L_z, y] &= -i\hbar x, & [L_z, z] &= 0 \\ [L_z, p_x] &= i\hbar p_y, & [L_z, p_y] &= -i\hbar p_x, & [L_z, p_z] &= 0. \end{aligned} \tag{4.122}$$

(b) Use these results to obtain $[L_z, L_x] = i\hbar L_y$ directly from Equation 4.96.

(c) Find the commutators $[L_z, r^2]$ and $[L_z, p^2]$ (where, of course, $r^2 = x^2 + y^2 + z^2$ and $p^2 = p_x^2 + p_y^2 + p_z^2$).

(d) Show that the Hamiltonian $H = (p^2/2m) + V$ commutes with all three components of $\mathbf{L}$, provided that V depends only on r. (Thus H, L^2, and L_z are mutually compatible observables.)

** **Problem 4.23**

(a) Prove that for a particle in a potential $V(\mathbf{r})$ the rate of change of the expectation value of the orbital angular momentum $\mathbf{L}$ is equal to the expectation value of the torque:

$$\frac{d}{dt}\langle \mathbf{L}\rangle = \langle \mathbf{N}\rangle,$$

where

$$\mathbf{N} = \mathbf{r} \times (-\nabla V).$$

(This is the rotational analog to Ehrenfest's theorem.)

(b) Show that $d\langle \mathbf{L}\rangle/dt = 0$ for any spherically symmetric potential. (This is one form of the quantum statement of **conservation of angular momentum**.)

4.3.2 Eigenfunctions

First of all we need to rewrite L_x, L_y, and L_z in spherical coordinates. Now, $\mathbf{L} = -i\hbar(\mathbf{r} \times \nabla)$, and the gradient, in spherical coordinates, is:[33]

$$\nabla = \hat{r}\frac{\partial}{\partial r} + \hat{\theta}\frac{1}{r}\frac{\partial}{\partial \theta} + \hat{\phi}\frac{1}{r\sin\theta}\frac{\partial}{\partial \phi}; \tag{4.123}$$

[33] George Arfken and Hans-Jurgen Weber, *Mathematical Methods for Physicists*, 7th edn, Academic Press, Orlando (2013), Section 3.10.

meanwhile, $\mathbf{r} = r\hat{r}$, so

$$\mathbf{L} = -i\hbar\left[r\left(\hat{r}\times\hat{r}\right)\frac{\partial}{\partial r} + \left(\hat{r}\times\hat{\theta}\right)\frac{\partial}{\partial\theta} + \left(\hat{r}\times\hat{\phi}\right)\frac{1}{\sin\theta}\frac{\partial}{\partial\phi}\right].$$

But $\left(\hat{r}\times\hat{r}\right) = 0$, $\left(\hat{r}\times\hat{\theta}\right) = \hat{\phi}$, and $\left(\hat{r}\times\hat{\phi}\right) = -\hat{\theta}$ (see Figure 4.1), and hence

$$\mathbf{L} = -i\hbar\left(\hat{\phi}\frac{\partial}{\partial\theta} - \hat{\theta}\frac{1}{\sin\theta}\frac{\partial}{\partial\phi}\right). \tag{4.124}$$

The unit vectors $\hat{\theta}$ and $\hat{\phi}$ can be resolved into their cartesian components:

$$\hat{\theta} = (\cos\theta\cos\phi)\,\hat{\imath} + (\cos\theta\sin\phi)\,\hat{\jmath} - (\sin\theta)\,\hat{k}; \tag{4.125}$$

$$\hat{\phi} = -(\sin\phi)\,\hat{\imath} + (\cos\phi)\,\hat{\jmath}. \tag{4.126}$$

Thus

$$\mathbf{L} = -i\hbar\left[\left(-\sin\phi\,\hat{\imath} + \cos\phi\,\hat{\jmath}\right)\frac{\partial}{\partial\theta} - \left(\cos\theta\cos\phi\,\hat{\imath} + \cos\theta\sin\phi\,\hat{\jmath} - \sin\theta\,\hat{k}\right)\frac{1}{\sin\theta}\frac{\partial}{\partial\phi}\right].$$

So

$$L_x = -i\hbar\left(-\sin\phi\frac{\partial}{\partial\theta} - \cos\phi\cot\theta\frac{\partial}{\partial\phi}\right), \tag{4.127}$$

$$L_y = -i\hbar\left(+\cos\phi\frac{\partial}{\partial\theta} - \sin\phi\cot\theta\frac{\partial}{\partial\phi}\right), \tag{4.128}$$

and

$$\boxed{L_z = -i\hbar\frac{\partial}{\partial\phi}.} \tag{4.129}$$

We shall also need the raising and lowering operators:

$$L_\pm = L_x \pm iL_y = -i\hbar\left[(-\sin\phi \pm i\cos\phi)\frac{\partial}{\partial\theta} - (\cos\phi \pm i\sin\phi)\cot\theta\frac{\partial}{\partial\phi}\right].$$

But $\cos\phi \pm i\sin\phi = e^{\pm i\phi}$, so

$$L_\pm = \pm\hbar e^{\pm i\phi}\left(\frac{\partial}{\partial\theta} \pm i\cot\theta\frac{\partial}{\partial\phi}\right). \tag{4.130}$$

In particular (Problem 4.24(a)):

$$L_+L_- = -\hbar^2\left(\frac{\partial^2}{\partial\theta^2} + \cot\theta\frac{\partial}{\partial\theta} + \cot^2\theta\frac{\partial^2}{\partial\phi^2} + i\frac{\partial}{\partial\phi}\right), \tag{4.131}$$

and hence (Problem 4.24(b)):

$$\boxed{L^2 = -\hbar^2\left[\frac{1}{\sin\theta}\frac{\partial}{\partial\theta}\left(\sin\theta\frac{\partial}{\partial\theta}\right) + \frac{1}{\sin^2\theta}\frac{\partial^2}{\partial\phi^2}\right].} \tag{4.132}$$

We are now in a position to determine $f_\ell^m(\theta,\phi)$. It's an eigenfunction of L^2, with eigenvalue $\hbar^2\ell(\ell+1)$:

$$L^2 f_\ell^m = -\hbar^2 \left[\frac{1}{\sin\theta} \frac{\partial}{\partial\theta} \left(\sin\theta \frac{\partial}{\partial\theta} \right) + \frac{1}{\sin^2\theta} \frac{\partial^2}{\partial\phi^2} \right] f_\ell^m = \hbar^2 \ell (\ell + 1) f_\ell^m.$$

But this is precisely the "angular equation" (Equation 4.18). And it's also an eigenfunction of L_z, with the eigenvalue $m\hbar$:

$$L_z f_\ell^m = -i\hbar \frac{\partial}{\partial\phi} f_\ell^m = \hbar m f_\ell^m,$$

but this is equivalent to the azimuthal equation (Equation 4.21). We have already solved this system of equations! The result (appropriately normalized) is the spherical harmonic, $Y_\ell^m(\theta, \phi)$. *Conclusion:* Spherical harmonics *are* the eigenfunctions of L^2 and L_z. When we solved the Schrödinger equation by separation of variables, in Section 4.1, we were inadvertently constructing simultaneous eigenfunctions of the three commuting operators H, L^2, and L_z:

$$H\psi = E\psi, \quad L^2\psi = \hbar^2 \ell (\ell + 1) \psi, \quad L_z \psi = \hbar m \psi. \tag{4.133}$$

Incidentally, we can use Equation 4.132 to rewrite the Schrödinger equation (Equation 4.14) more compactly:

$$\frac{1}{2mr^2} \left[-\hbar^2 \frac{\partial}{\partial r} \left(r^2 \frac{\partial}{\partial r} \right) + L^2 \right] \psi + V\psi = E\psi.$$

There is a curious final twist to this story: the *algebraic* theory of angular momentum permits ℓ (and hence also m) to take on *half*-integer values (Equation 4.119), whereas separation of variables yielded eigenfunctions only for *integer* values (Equation 4.29).[34] You might suppose that the half-integer solutions are spurious, but it turns out that they are of profound importance, as we shall see in the following sections.

* **Problem 4.24**

(a) Derive Equation 4.131 from Equation 4.130. *Hint:* Use a test function; otherwise you're likely to drop some terms.

(b) Derive Equation 4.132 from Equations 4.129 and 4.131. *Hint:* Use Equation 4.112.

* **Problem 4.25**

(a) What is $L_+ Y_\ell^\ell$? (No calculation allowed!)

(b) Use the result of (a), together with Equation 4.130 and the fact that $L_z Y_\ell^\ell = \hbar \ell Y_\ell^\ell$, to determine $Y_\ell^\ell(\theta, \phi)$, up to a normalization constant.

(c) Determine the normalization constant by direct integration. Compare your final answer to what you got in Problem 4.7.

Problem 4.26 In Problem 4.4 you showed that

$$Y_2^1(\theta, \phi) = -\sqrt{15/8\pi}\, \sin\theta \cos\theta e^{i\phi}.$$

Apply the raising operator to find $Y_2^2(\theta, \phi)$. Use Equation 4.121 to get the normalization.

[34] For an interesting discussion, see I. R. Gatland, *Am. J. Phys.* **74**, 191 (2006).

** **Problem 4.27** Two particles (masses m_1 and m_2) are attached to the ends of a massless rigid rod of length a. The system is free to rotate in three dimensions about the (fixed) center of mass.

(a) Show that the allowed energies of this **rigid rotor** are

$$E_n = \frac{\hbar^2}{2I}n(n+1), \quad (n = 0,\ 1,\ 2, \ldots), \quad \text{where} \quad I = \frac{m_1 m_2}{(m_1 + m_2)}a^2$$

is the moment of inertia of the system. *Hint*: First express the (classical) energy in terms of the angular momentum.

(b) What are the normalized eigenfunctions for this system? (Let θ and ϕ define the orientation of the rotor axis.) What is the degeneracy of the nth energy level?

(c) What spectrum would you expect for this system? (Give a formula for the frequencies of the spectral lines.) *Answer:* $\nu_j = \hbar j / 2\pi I,\ j = 1, 2, 3, \ldots$.

(d) Figure 4.13 shows a portion of the rotational spectrum of carbon monoxide (CO). What is the frequency separation ($\Delta\nu$) between adjacent lines? Look up the masses of ^{12}C and ^{16}O, and from m_1, m_2, and $\Delta\nu$ determine the distance between the atoms.

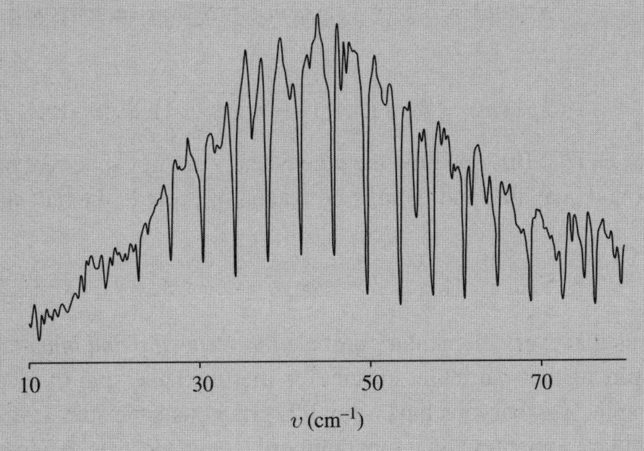

Figure 4.13: Rotation spectrum of CO. Note that the frequencies are in spectroscopist's units: inverse centimeters. To convert to Hertz, multiply by $c = 3.00 \times 10^{10}$ cm/s. Reproduced by permission from John M. Brown and Allan Carrington, *Rotational Spectroscopy of Diatomic Molecules*, Cambridge University Press, 2003, which in turn was adapted from E. V. Loewenstein, *Journal of the Optical Society of America*, **50**, 1163 (1960).

4.4 SPIN

In *classical* mechanics, a rigid object admits two kinds of angular momentum: **orbital** ($\mathbf{L} = \mathbf{r} \times \mathbf{p}$), associated with motion *of* the center of mass, and **spin** ($\mathbf{S} = I\omega$), associated with motion *about* the center of mass. For example, the earth has orbital angular momentum attributable to its annual revolution around the sun, and spin angular momentum coming from its daily rotation about the north–south axis. In the classical context this distinction is largely a matter of convenience, for when you come right down to it, **S** is nothing but the sum total of the "orbital" angular momenta of all the rocks and dirt clods that go to make up the earth, as they circle around the axis. But a similar thing happens in quantum mechanics, and here the distinction is absolutely fundamental. In addition to orbital angular momentum, associated (in the case of hydrogen) with the motion of the electron around the nucleus (and described by the spherical harmonics), the electron also carries *another* form of angular momentum, which has nothing to do with motion in space (and which is not, therefore, described by any function of the position variables r, θ, ϕ) but which is somewhat analogous to classical spin (and for which, therefore, we use the same word). It doesn't pay to press this analogy too far: The electron (as far as we know) is a structureless point, and its spin angular momentum cannot be decomposed into orbital angular momenta of constituent parts (see Problem 4.28).[35] Suffice it to say that elementary particles carry **intrinsic** angular momentum (**S**) in addition to their "extrinsic" angular momentum (**L**).

The *algebraic* theory of spin is a carbon copy of the theory of orbital angular momentum, beginning with the fundamental commutation relations:[36]

$$\left[S_x, S_y\right] = i\hbar S_z, \quad \left[S_y, S_z\right] = i\hbar S_x, \quad \left[S_z, S_x\right] = i\hbar S_y. \tag{4.134}$$

It follows (as before) that the eigenvectors of S^2 and S_z satisfy[37]

$$S^2|s\,m\rangle = \hbar^2 s\,(s+1)\,|s\,m\rangle; \quad S_z|s\,m\rangle = \hbar m|s\,m\rangle; \tag{4.135}$$

and

$$S_{\pm}\,|s\,m\rangle = \hbar\sqrt{s\,(s+1) - m\,(m \pm 1)}\;|s\;(m \pm 1)\rangle\,, \tag{4.136}$$

where $S_{\pm} \equiv S_x \pm iS_y$. But this time the eigenvectors are not spherical harmonics (they're not functions of θ and ϕ at all), and there is no reason to exclude the half-integer values of s and m:

$$s = 0,\ \frac{1}{2},\ 1,\ \frac{3}{2}, \ldots; \qquad m = -s,\ -s+1, \ldots, s-1,\ s. \tag{4.137}$$

It so happens that every elementary particle has a *specific and immutable* value of s, which we call **the spin** of that particular species: π mesons have spin 0; electrons have spin 1/2; photons have spin 1; Δ baryons have spin 3/2; gravitons have spin 2; and so on. By contrast, the *orbital* angular momentum quantum number l (for an electron in a hydrogen atom, say) can take on any (integer) value you please, and will change from one to another when the system is perturbed. But s is *fixed*, for any given particle, and this makes the theory of spin comparatively simple.[38]

---

[35] For a contrary interpretation, see Hans C. Ohanian, "What is Spin?", *Am. J. Phys.* **54**, 500 (1986).

[36] We shall take these as *postulates* for the theory of spin; the analogous formulas for *orbital* angular momentum (Equation 4.99) were *derived* from the known form of the operators (Equation 4.96). Actually, they both follow from rotational invariance in three dimensions, as we shall see in Chapter 6. Indeed, these fundamental commutation relations apply to *all* forms of angular momentum, whether spin, orbital, or the combined angular momentum of a composite system, which could be partly spin and partly orbital.

[37] Because the eigenstates of spin are not *functions*, I will switch now to Dirac notation. By the way, I'm running out of letters, so I'll use m for the eigenvalue of S_z, just as I did for L_z (some authors write m_l and m_s at this stage, just to be absolutely clear).

[38] Indeed, in a mathematical sense, spin 1/2 is the simplest possible nontrivial quantum system, for it admits just two basis states (recall Example 3.8). In place of an infinite-dimensional Hilbert space, with all its subtleties and complications, we find ourselves working in an ordinary two-dimensional vector space; instead of unfamiliar differential equations and fancy functions, we are confronted with 2×2 matrices and two-component vectors. For this reason, some authors *begin* quantum mechanics with the study of spin. (An outstanding example is John

Problem 4.28 If the electron were a classical solid sphere, with radius

$$r_c = \frac{e^2}{4\pi\epsilon_0 mc^2}, \tag{4.138}$$

(the so-called **classical electron radius**, obtained by assuming the electron's mass is attributable to energy stored in its electric field, via the Einstein formula $E = mc^2$), and its angular momentum is $(1/2)\,\hbar$, then how fast (in m/s) would a point on the "equator" be moving? Does this model make sense? (Actually, the radius of the electron is known experimentally to be much less than r_c, but this only makes matters worse.)[39]

4.4.1 Spin 1/2

By far the most important case is $s = 1/2$, for this is the spin of the particles that make up ordinary matter (protons, neutrons, and electrons), as well as all quarks and all leptons. Moreover, once you understand spin 1/2, it is a simple matter to work out the formalism for any higher spin. There are just *two* eigenstates: $\left|\frac{1}{2}\,\frac{1}{2}\right\rangle$, which we call **spin up** (informally, $\uparrow$), and $\left|\frac{1}{2}\left(-\frac{1}{2}\right)\right\rangle$, **spin down** ($\downarrow$). Using these as basis vectors, the general state[40] of a spin-1/2 particle can be represented by a two-element column matrix (or **spinor**):

$$\chi = \begin{pmatrix} a \\ b \end{pmatrix} = a\chi_+ + b\chi_-, \tag{4.139}$$

with

$$\chi_+ = \begin{pmatrix} 1 \\ 0 \end{pmatrix} \tag{4.140}$$

representing spin up, and

$$\chi_- = \begin{pmatrix} 0 \\ 1 \end{pmatrix} \tag{4.141}$$

for spin down.

With respect to this basis the spin operators become 2×2 matrices,[41] which we can work out by noting their effect on χ_+ and χ_-. Equation 4.135 says

---

S. Townsend, *A Modern Approach to Quantum Mechanics,* 2nd edn, University Books, Sausalito, CA, 2012.) But the price of mathematical simplicity is conceptual abstraction, and I prefer not to do it that way.

[39] If it comforts you to picture the electron as a tiny spinning sphere, go ahead; I do, and I don't think it hurts, as long as you don't take it literally.

[40] I'm only talking about the *spin* state, for the moment. If the particle is moving around, we will also need to deal with its *position* state (Ψ), but for the moment let's put that aside.

[41] I hate to be fussy about notation, but perhaps I should reiterate that a ket (such as $|s\,m\rangle$) is a *vector* in Hilbert space (in this case a $(2s+1)$-dimensional vector space), whereas a spinor χ is a set of *components* of a vector, with respect to a particular basis $\left(\left|\frac{1}{2}\,\frac{1}{2}\right\rangle \text{ and } \left|\frac{1}{2}\,-\frac{1}{2}\right\rangle\right.$, in the case of spin $\left.\frac{1}{2}\right)$, displayed as a column. Physicists sometimes write, for instance, $\left|\frac{1}{2}\,\frac{1}{2}\right\rangle = \chi_+$, but technically this confuses a *vector* (which lives "out there" in Hilbert space) with its *components* (a string of numbers). Similarly, S_z (for example) is an operator that acts on kets; it is represented (with respect to the chosen basis) by a matrix S_z (sans serif), which multiplies spinors—but again, $S_z = \mathsf{S}_z$, though perfectly intelligible, is sloppy language.

$$\mathsf{S}^2\chi_+ = \frac{3}{4}\hbar^2\chi_+ \quad \text{and} \quad \mathsf{S}^2\chi_- = \frac{3}{4}\hbar^2\chi_-. \tag{4.142}$$

If we write S^2 as a matrix with (as yet) undetermined elements,

$$\mathsf{S}^2 = \begin{pmatrix} c & d \\ e & f \end{pmatrix},$$

then the first equation says

$$\begin{pmatrix} c & d \\ e & f \end{pmatrix}\begin{pmatrix} 1 \\ 0 \end{pmatrix} = \frac{3}{4}\hbar^2\begin{pmatrix} 1 \\ 0 \end{pmatrix}, \quad \text{or} \quad \begin{pmatrix} c \\ e \end{pmatrix} = \begin{pmatrix} \frac{3}{4}\hbar^2 \\ 0 \end{pmatrix},$$

so $c = (3/4)\hbar^2$ and $e = 0$. The second equation says

$$\begin{pmatrix} c & d \\ e & f \end{pmatrix}\begin{pmatrix} 0 \\ 1 \end{pmatrix} = \frac{3}{4}\hbar^2\begin{pmatrix} 0 \\ 1 \end{pmatrix}, \quad \text{or} \quad \begin{pmatrix} d \\ f \end{pmatrix} = \begin{pmatrix} 0 \\ \frac{3}{4}\hbar^2 \end{pmatrix},$$

so $d = 0$ and $f = (3/4)\,\hbar^2$. *Conclusion:*

$$\mathsf{S}^2 = \frac{3}{4}\hbar^2\begin{pmatrix} 1 & 0 \\ 0 & 1 \end{pmatrix}. \tag{4.143}$$

Similarly,

$$\mathsf{S}_z\chi_+ = \frac{\hbar}{2}\chi_+, \quad \mathsf{S}_z\chi_- = -\frac{\hbar}{2}\chi_-, \tag{4.144}$$

from which it follows that

$$\mathsf{S}_z = \frac{\hbar}{2}\begin{pmatrix} 1 & 0 \\ 0 & -1 \end{pmatrix}. \tag{4.145}$$

Meanwhile, Equation 4.136 says

$$\mathsf{S}_+\chi_- = \hbar\chi_+, \quad \mathsf{S}_-\chi_+ = \hbar\chi_-, \quad \mathsf{S}_+\chi_+ = \mathsf{S}_-\chi_- = 0,$$

so

$$\mathsf{S}_+ = \hbar\begin{pmatrix} 0 & 1 \\ 0 & 0 \end{pmatrix}, \quad \mathsf{S}_- = \hbar\begin{pmatrix} 0 & 0 \\ 1 & 0 \end{pmatrix}. \tag{4.146}$$

Now $S_\pm = S_x \pm i S_y$, so $S_x = (1/2)\,(S_+ + S_-)$ and $S_y = (1/2i)\,(S_+ - S_-)$, and hence

$$\mathsf{S}_x = \frac{\hbar}{2}\begin{pmatrix} 0 & 1 \\ 1 & 0 \end{pmatrix}, \quad \mathsf{S}_y = \frac{\hbar}{2}\begin{pmatrix} 0 & -i \\ i & 0 \end{pmatrix}. \tag{4.147}$$

Since S_x, S_y, and S_z all carry a factor of $\hbar/2$, it is tidier to write $\mathbf{S} = (\hbar/2)\boldsymbol{\sigma}$, where

$$\boxed{\sigma_x \equiv \begin{pmatrix} 0 & 1 \\ 1 & 0 \end{pmatrix}, \ \sigma_y \equiv \begin{pmatrix} 0 & -i \\ i & 0 \end{pmatrix}, \ \sigma_z \equiv \begin{pmatrix} 1 & 0 \\ 0 & -1 \end{pmatrix}.} \tag{4.148}$$

These are the famous **Pauli spin matrices**. Notice that S_x, S_y, S_z, and S^2 are all *hermitian* matrices (as they *should* be, since they represent observables). On the other hand, S_+ and S_- are *not* hermitian—evidently they are not observable.

The eigenspinors of S_z are (or course):

$$\chi_+ = \begin{pmatrix}1\\0\end{pmatrix}, \left(\text{eigenvalue } +\frac{\hbar}{2}\right); \quad \chi_- = \begin{pmatrix}0\\1\end{pmatrix}, \left(\text{eigenvalue } -\frac{\hbar}{2}\right). \tag{4.149}$$

If you measure S_z on a particle in the general state χ (Equation 4.139), you could get $+\hbar/2$, with probability $|a|^2$, or $-\hbar/2$, with probability $|b|^2$. Since these are the *only* possibilities,

$$|a|^2 + |b|^2 = 1 \tag{4.150}$$

(i.e. the spinor must be *normalized*: $\chi^\dagger \chi = 1$).[42]

But what if, instead, you chose to measure S_x? What are the possible results, and what are their respective probabilities? According to the generalized statistical interpretation, we need to know the eigenvalues and eigenspinors of S_x. The characteristic equation is

$$\begin{vmatrix}-\lambda & \hbar/2\\ \hbar/2 & -\lambda\end{vmatrix} = 0 \Rightarrow \lambda^2 = \left(\frac{\hbar}{2}\right)^2 \Rightarrow \lambda = \pm\frac{\hbar}{2}.$$

Not surprisingly (but it's gratifying to see how it works out), the possible values for S_x are the same as those for S_z. The eigenspinors are obtained in the usual way:

$$\frac{\hbar}{2}\begin{pmatrix}0 & 1\\ 1 & 0\end{pmatrix}\begin{pmatrix}\alpha\\ \beta\end{pmatrix} = \pm\frac{\hbar}{2}\begin{pmatrix}\alpha\\ \beta\end{pmatrix} \Rightarrow \begin{pmatrix}\beta\\ \alpha\end{pmatrix} = \pm\begin{pmatrix}\alpha\\ \beta\end{pmatrix},$$

so $\beta = \pm\alpha$. Evidently the (normalized) eigenspinors of S_x are

$$\chi_+^{(x)} = \begin{pmatrix}1/\sqrt{2}\\ 1/\sqrt{2}\end{pmatrix}, \left(\text{eigenvalue } +\frac{\hbar}{2}\right); \ \chi_-^{(x)} = \begin{pmatrix}1/\sqrt{2}\\ -1/\sqrt{2}\end{pmatrix}, \left(\text{eigenvalue } -\frac{\hbar}{2}\right). \tag{4.151}$$

As the eigenvectors of a hermitian matrix, they span the space; the generic spinor χ (Equation 4.139) can be expressed as a linear combination of them:

$$\chi = \left(\frac{a+b}{\sqrt{2}}\right)\chi_+^{(x)} + \left(\frac{a-b}{\sqrt{2}}\right)\chi_-^{(x)}. \tag{4.152}$$

If you measure S_x, the probability of getting $+\hbar/2$ is $(1/2)\,|a+b|^2$, and the probability of getting $-\hbar/2$ is $(1/2)\,|a-b|^2$. (Check for yourself that these probabilities add up to 1.)

Example 4.2

Suppose a spin-1/2 particle is in the state

$$\chi = \frac{1}{\sqrt{6}}\begin{pmatrix}1+i\\ 2\end{pmatrix}.$$

What are the probabilities of getting $+\hbar/2$ and $-\hbar/2$, if you measure S_z and S_x?

Solution: Here $a = (1+i)/\sqrt{6}$ and $b = 2/\sqrt{6}$, so for S_z the probability of getting $+\hbar/2$ is $\left|(1+i)/\sqrt{6}\right|^2 = 1/3$, and the probability of getting $-\hbar/2$ is $\left|2/\sqrt{6}\right|^2 = 2/3$. For

[42] People often say that $|a|^2$ is the "probability that the particle is in the spin-up state," but this is bad language; what they *mean* is that if you *measured* S_z, $|a|^2$ is the probability you'd get $\hbar/2$. See footnote 18, page 103.

S_x the probability of getting $+\hbar/2$ is $(1/2)\left|(3+i)/\sqrt{6}\right|^2 = 5/6$, and the probability of getting $-\hbar/2$ is $(1/2)\left|(-1+i)/\sqrt{6}\right|^2 = 1/6$. Incidentally, the *expectation* value of S_x is

$$\frac{5}{6}\left(+\frac{\hbar}{2}\right)+\frac{1}{6}\left(-\frac{\hbar}{2}\right)=\frac{\hbar}{3},$$

which we could also have obtained more directly:

$$\langle S_x\rangle = \chi^\dagger \mathsf{S}_x \chi = \left((1-i)/\sqrt{6} \quad 2/\sqrt{6}\right)\begin{pmatrix}0 & \hbar/2\\ \hbar/2 & 0\end{pmatrix}\begin{pmatrix}(1+i)/\sqrt{6}\\ 2/\sqrt{6}\end{pmatrix}=\frac{\hbar}{3}.$$

I'd like now to walk you through an imaginary measurement scenario involving spin 1/2, because it serves to illustrate in very concrete terms some of the abstract ideas we discussed back in Chapter 1. Let's say we start out with a particle in the state χ_+. If someone asks, "What is the z-component of that particle's spin angular momentum?", we can answer unambiguously: $+\hbar/2$. For a measurement of S_z is certain to return that value. But if our interrogator asks instead, "What is the x-component of that particle's spin angular momentum?" we are obliged to equivocate: If you measure S_x, the chances are fifty-fifty of getting either $\hbar/2$ or $-\hbar/2$. If the questioner is a classical physicist, or a "realist" (in the sense of Section 1.2), he will regard this as an inadequate—not to say impertinent—response: "Are you telling me that you *don't know* the true state of that particle?" On the contrary; I know *precisely* what the state of the particle is: χ_+. "Well, then, how come you can't tell me what the x-component of its spin is?" Because it simply *does not have* a particular x-component of spin. Indeed, it *cannot*, for if both S_x and S_z were well-defined, the uncertainty principle would be violated.

At this point our challenger grabs the test-tube and *measures* the x-component of the particle's spin; let's say he gets the value $+\hbar/2$. "Aha!" (he shouts in triumph), "You *lied!* This particle has a perfectly well-defined value of S_x: $\hbar/2$." Well, sure—it does *now*, but that doesn't prove it *had* that value, prior to your measurement. "You have obviously been reduced to splitting hairs. And anyway, what happened to your uncertainty principle? I now know both S_x *and* S_z." I'm sorry, but you do *not*: In the course of your measurement, you altered the particle's state; it is now in the state $\chi_+^{(x)}$, and whereas you know the value of S_x, you no longer know the value of S_z. "But I was extremely careful not to disturb the particle when I measured S_x." Very well, if you don't believe me, *check it out*: Measure S_z, and see what you get. (Of course, he *may* get $+\hbar/2$, which will be embarrassing to my case—but if we repeat this whole scenario over and over, half the time he will get $-\hbar/2$.)

To the layman, the philosopher, or the classical physicist, a statement of the form "this particle doesn't have a well-defined position" (or momentum, or x-component of spin angular momentum, or whatever) sounds vague, incompetent, or (worst of all) profound. It is none of these. But its precise meaning is, I think, almost impossible to convey to anyone who has not studied quantum mechanics in some depth. If you find your own comprehension slipping, from time to time (if you *don't*, you probably haven't understood the problem), come back to the spin-1/2 system: It is the simplest and cleanest context for thinking through the conceptual paradoxes of quantum mechanics.

Problem 4.29

(a) Check that the spin matrices (Equations 4.145 and 4.147) obey the fundamental commutation relations for angular momentum, Equation 4.134.

(b) Show that the Pauli spin matrices (Equation 4.148) satisfy the product rule

$$\sigma_j \sigma_k = \delta_{jk} + i \sum_l \epsilon_{jkl} \sigma_l, \tag{4.153}$$

where the indices stand for x, y, or z, and ϵ_{jkl} is the **Levi-Civita** symbol: +1 if $jkl = 123$, 231, or 312; -1 if $jkl = 132$, 213, or 321; 0 otherwise.

* **Problem 4.30** An electron is in the spin state

$$\chi = A \begin{pmatrix} 3i \\ 4 \end{pmatrix}.$$

(a) Determine the normalization constant A.

(b) Find the expectation values of S_x, S_y, and S_z.

(c) Find the "uncertainties" σ_{S_x}, σ_{S_y}, and σ_{S_z}. *Note:* These sigmas are standard deviations, not Pauli matrices!

(d) Confirm that your results are consistent with all three uncertainty principles (Equation 4.100 and its cyclic permutations—only with S in place of L, of course).

* **Problem 4.31** For the most general normalized spinor χ (Equation 4.139), compute $\langle S_x \rangle$, $\langle S_y \rangle$, $\langle S_z \rangle$, $\langle S_x^2 \rangle$, $\langle S_y^2 \rangle$, and $\langle S_z^2 \rangle$. Check that $\langle S_x^2 \rangle + \langle S_y^2 \rangle + \langle S_z^2 \rangle = \langle S^2 \rangle$.

* **Problem 4.32**

(a) Find the eigenvalues and eigenspinors of S_y.

(b) If you measured S_y on a particle in the general state χ (Equation 4.139), what values might you get, and what is the probability of each? Check that the probabilities add up to 1. *Note: a and b need not be real!*

(c) If you measured S_y^2, what values might you get, and with what probabilities?

** **Problem 4.33** Construct the matrix S_r representing the component of spin angular momentum along an arbitrary direction $\hat{r}$. Use spherical coordinates, for which

$$\hat{r} = \sin\theta \cos\phi \, \hat{\imath} + \sin\theta \sin\phi \, \hat{\jmath} + \cos\theta \, \hat{k}. \tag{4.154}$$

Find the eigenvalues and (normalized) eigenspinors of S_r. *Answer:*

$$\chi_+^{(r)} = \begin{pmatrix} \cos(\theta/2) \\ e^{i\phi} \sin(\theta/2) \end{pmatrix}; \quad \chi_-^{(r)} = \begin{pmatrix} e^{-i\phi} \sin(\theta/2) \\ -\cos(\theta/2) \end{pmatrix}. \tag{4.155}$$

Note: You're always free to multiply by an arbitrary phase factor—say, $e^{i\phi}$—so your answer may not *look* exactly the same as mine.

Problem 4.34 Construct the spin matrices (S_x, S_y, and S_z) for a particle of spin 1. *Hint*: How many eigenstates of S_z are there? Determine the action of S_z, S_+, and S_- on each of these states. Follow the procedure used in the text for spin 1/2.

4.4.2 Electron in a Magnetic Field

A spinning charged particle constitutes a magnetic dipole. Its **magnetic dipole moment**, $\boldsymbol{\mu}$, is proportional to its spin angular momentum, $\mathbf{S}$:

$$\boldsymbol{\mu} = \gamma \mathbf{S}; \tag{4.156}$$

the proportionality constant, γ, is called the **gyromagnetic ratio**.[43] When a magnetic dipole is placed in a magnetic field $\mathbf{B}$, it experiences a torque, $\boldsymbol{\mu} \times \mathbf{B}$, which tends to line it up parallel to the field (just like a compass needle). The energy associated with this torque is[44]

$$H = -\boldsymbol{\mu} \cdot \mathbf{B}, \tag{4.157}$$

so the Hamiltonian matrix for a spinning charged particle, at rest[45] in a magnetic field $\mathbf{B}$, is

$$\mathsf{H} = -\gamma \mathbf{B} \cdot \mathsf{S}, \tag{4.158}$$

where S is the appropriate spin matrix (Equations 4.145 and 4.147, in the case of spin 1/2).

Example 4.3

Larmor precession: Imagine a particle of spin 1/2 at rest in a uniform magnetic field, which points in the z-direction:

$$\mathbf{B} = B_0 \hat{k}. \tag{4.159}$$

The Hamiltonian (Equation 4.158) is

$$\mathsf{H} = -\gamma B_0 \mathsf{S}_z = -\frac{\gamma B_0 \hbar}{2} \begin{pmatrix} 1 & 0 \\ 0 & -1 \end{pmatrix}. \tag{4.160}$$

The eigenstates of H are the same as those of S_z:

$$\begin{cases} \chi_+, & \text{with energy } E_+ = -\left(\gamma B_0 \hbar\right)/2, \\ \chi_-, & \text{with energy } E_- = +\left(\gamma B_0 \hbar\right)/2. \end{cases} \tag{4.161}$$

---

[43] See, for example, David J. Griffiths, *Introduction to Electrodynamics*, 4th edn (Pearson, Boston, 2013), Problem 5.58. Classically, the gyromagnetic ratio of an object whose charge and mass are identically distributed is $q/2m$, where q is the charge and m is the mass. For reasons that are fully explained only in relativistic quantum theory, the gyromagnetic ratio of the electron is (almost) exactly *twice* the classical value: $\gamma = -e/m$.

[44] Griffiths (footnote 43), Problem 6.21.

[45] If the particle is allowed to *move*, there will also be kinetic energy to consider; moreover, it will be subject to the Lorentz force ($q\mathbf{v} \times \mathbf{B}$), which is not derivable from a potential energy function, and hence does not fit the Schrödinger equation as we have formulated it so far. I'll show you later on how to handle this (Problem 4.42), but for the moment let's just assume that the particle is free to *rotate*, but otherwise stationary.

The energy is lowest when the dipole moment is parallel to the field—just as it would be classically.

Since the Hamiltonian is time independent, the general solution to the time-dependent Schrödinger equation,

$$i\hbar\frac{\partial\chi}{\partial t} = \mathsf{H}\chi, \tag{4.162}$$

can be expressed in terms of the stationary states:

$$\chi(t) = a\chi_+ e^{-iE_+t/\hbar} + b\chi_- e^{-iE_-t/\hbar} = \begin{pmatrix} ae^{i\gamma B_0 t/2} \\ be^{-i\gamma B_0 t/2} \end{pmatrix}.$$

The constants a and b are determined by the initial conditions:

$$\chi(0) = \begin{pmatrix} a \\ b \end{pmatrix},$$

(of course, $|a|^2 + |b|^2 = 1$). With no essential loss of generality[46] I'll write $a = \cos(\alpha/2)$ and $b = \sin(\alpha/2)$, where α is a fixed angle whose physical significance will appear in a moment. Thus

$$\chi(t) = \begin{pmatrix} \cos(\alpha/2)\ e^{i\gamma B_0 t/2} \\ \sin(\alpha/2)\ e^{-i\gamma B_0 t/2} \end{pmatrix}. \tag{4.163}$$

To get a feel for what is happening here, let's calculate the expectation value of $\mathbf{S}$, as a function of time:

$$\langle S_x \rangle = \chi(t)^\dagger\, \mathsf{S}_x \chi(t)$$

$$= \begin{pmatrix} \cos(\alpha/2)e^{-i\gamma B_0 t/2} & \sin(\alpha/2)e^{i\gamma B_0 t/2} \end{pmatrix} \frac{\hbar}{2}\begin{pmatrix} 0 & 1 \\ 1 & 0 \end{pmatrix} \begin{pmatrix} \cos(\alpha/2)e^{i\gamma B_0 t/2} \\ \sin(\alpha/2)e^{-i\gamma B_0 t/2} \end{pmatrix}$$

$$= \frac{\hbar}{2}\sin\alpha\cos(\gamma B_0 t). \tag{4.164}$$

Similarly,

$$\langle S_y \rangle = \chi(t)^\dagger \mathsf{S}_y \chi(t) = -\frac{\hbar}{2}\sin\alpha\sin(\gamma B_0 t), \tag{4.165}$$

and

$$\langle S_z \rangle = \chi(t)^\dagger \mathsf{S}_z \chi(t) = \frac{\hbar}{2}\cos\alpha. \tag{4.166}$$

Thus $\langle \mathbf{S} \rangle$ is tilted at a constant angle α to the z axis, and precesses about the field at the **Larmor frequency**

$$\omega = \gamma B_0, \tag{4.167}$$

---

[46] This does assume that a and b are *real*; you can work out the general case if you like, but all it does is add a constant to t.

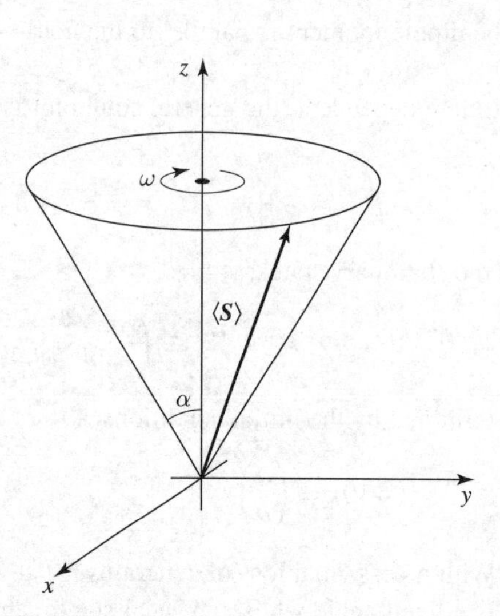

Figure 4.14: Precession of $\langle \mathbf{S} \rangle$ in a uniform magnetic field.

just as it would classically[47] (see Figure 4.14). No surprise here—Ehrenfest's theorem (in the form derived in Problem 4.23) guarantees that $\langle \mathbf{S} \rangle$ evolves according to the classical laws. But it's nice to see how this works out in a specific context.

Example 4.4

The Stern–Gerlach experiment: In an *inhomogeneous* magnetic field, there is not only a *torque*, but also a *force*, on a magnetic dipole:[48]

$$\mathbf{F} = \nabla (\boldsymbol{\mu} \cdot \mathbf{B}). \tag{4.168}$$

This force can be used to separate out particles with a particular spin orientation. Imagine a beam of heavy neutral atoms,[49] traveling in the y direction, which passes through a region of static but inhomogeneous magnetic field (Figure 4.15)—say

$$\mathbf{B}(x, y, z) = -\alpha x \hat{\imath} + (B_0 + \alpha z)\,\hat{k}, \tag{4.169}$$

---

47 See, for instance, Richard P. Feynman and Robert B. Leighton, *The Feynman Lectures on Physics* (Addison-Wesley, Reading, 1964), Volume II, Section 34-3. Of course, in the classical case it is the angular momentum vector itself, not just its expectation value, that precesses around the magnetic field.

48 Griffiths (footnote 43), Section 6.1.2. Note that $\mathbf{F}$ is the negative gradient of the energy (Equation 4.157).

49 We make them neutral so as to avoid the large-scale deflection that would otherwise result from the Lorentz force, and heavy so we can construct localized wave packets and treat the motion in terms of classical particle trajectories. In practice, the Stern–Gerlach experiment doesn't work, for example, with a beam of free electrons. Stern and Gerlach themselves used silver atoms; for the story of their discovery see B. Friedrich and D. Herschbach, *Physics Today* **56**, 53 (2003).

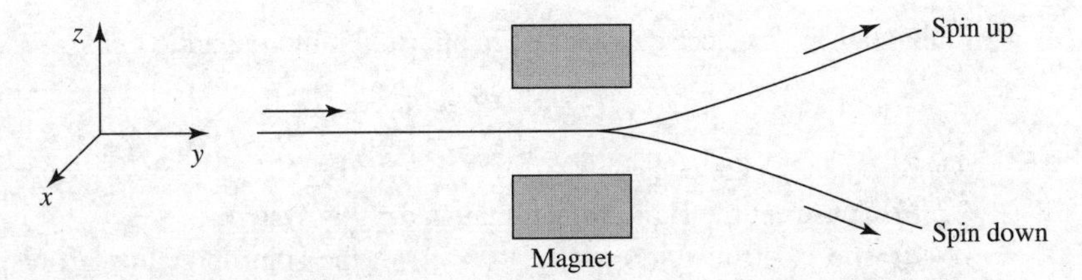

Figure 4.15: The Stern–Gerlach apparatus.

where B_0 is a strong uniform field and the constant α describes a small deviation from homogeneity. (Actually, what we'd *prefer* is just the z component of this field, but unfortunately that's impossible—it would violate the electromagnetic law $\nabla \cdot \mathbf{B} = 0$; like it or not, the x component comes along for the ride.) The force on these atoms is[50]

$$\mathbf{F} = \gamma\alpha\left(-S_x\hat{\imath} + S_z\hat{k}\right).$$

But because of the Larmor precession about $\mathbf{B}_0$, S_x oscillates rapidly, and *averages* to zero; the *net* force is in the z direction:

$$F_z = \gamma\alpha S_z, \tag{4.170}$$

and the beam is deflected up or down, in proportion to the z component of the spin angular momentum. *Classically* we'd expect a *smear* (because S_z would not be quantized), but in fact the beam splits into $2s+1$ separate streams, beautifully demonstrating the quantization of angular momentum. (If you use silver atoms, all the inner electrons are paired, in such a way that their angular momenta cancel. The net spin is simply that of the outermost—unpaired—electron, so in this case $s = 1/2$, and the beam splits in two.)

The Stern–Gerlach experiment has played an important role in the philosophy of quantum mechanics, where it serves both as the prototype for the preparation of a quantum state and as an illuminating model for a certain kind of quantum measurement. We tend casually to assume that the *initial* state of a system is *known* (the Schrödinger equation tells us how it subsequently evolves)—but it is natural to wonder how you get a system into a particular state in the first place. Well, if you want to prepare a beam of atoms in a given spin configuration, you pass an unpolarized beam through a Stern–Gerlach magnet, and select the outgoing stream you are interested in (closing off the others with suitable baffles and shutters). Conversely, if you want to *measure* the z component of an atom's spin, you send it through a Stern–Gerlach apparatus, and record which bin it lands in. I do not claim that this is always the most *practical* way to do the job, but it is *conceptually* very clean, and hence a useful context in which to explore the problems of state preparation and measurement.

Problem 4.35 In Example 4.3:

(a) If you measured the component of spin angular momentum along the x direction, at time t, what is the probability that you would get $+\hbar/2$?

(b) Same question, but for the y component.

(c) Same, for the z component.

[50] For a quantum mechanical justification of this equation see Problem 4.73.

** **Problem 4.36** An electron is at rest in an oscillating magnetic field

$$\mathbf{B} = B_0 \cos(\omega t)\,\hat{k},$$

where B_0 and ω are constants.

(a) Construct the Hamiltonian matrix for this system.

(b) The electron starts out (at $t = 0$) in the spin-up state with respect to the x axis (that is: $\chi(0) = \chi_+^{(x)}$). Determine $\chi(t)$ at any subsequent time. *Beware:* This is a time-*dependent* Hamiltonian, so you cannot get $\chi(t)$ in the usual way from stationary states. Fortunately, in this case you can solve the time-dependent Schrödinger equation (Equation 4.162) directly.

(c) Find the probability of getting $-\hbar/2$, if you measure S_x. *Answer:*

$$\sin^2\left(\frac{\gamma B_0}{2\omega}\sin(\omega t)\right).$$

(d) What is the minimum field (B_0) required to force a complete flip in S_x?

4.4.3 Addition of Angular Momenta

Suppose now that we have *two* particles, with spins s_1 and s_2. Say, the first is in the state $|s_1\, m_1\rangle$ and the second in the state $|s_2\, m_2\rangle$. We denote the composite state by $|s_1\, s_2\, m_1\, m_2\rangle$:

$$\begin{aligned}
S^{(1)^2}\,|s_1\, s_2\, m_1\, m_2\rangle &= s_1\,(s_1+1)\,\hbar^2\,|s_1\, s_2\, m_1\, m_2\rangle\,,\\
S^{(2)^2}\,|s_1\, s_2\, m_1\, m_2\rangle &= s_2\,(s_2+1)\,\hbar^2\,|s_1\, s_2\, m_1\, m_2\rangle\,,\\
S_z^{(1)}\,|s_1\, s_2\, m_1\, m_2\rangle &= m_1\hbar\,|s_1\, s_2\, m_1\, m_2\rangle\,,\\
S_z^{(2)}\,|s_1\, s_2\, m_1\, m_2\rangle &= m_2\hbar\,|s_1\, s_2\, m_1\, m_2\rangle\,.
\end{aligned} \tag{4.171}$$

Question: What is the *total* angular momentum,

$$\mathbf{S} = \mathbf{S}^{(1)} + \mathbf{S}^{(2)}, \tag{4.172}$$

of this system? That is to say: what is the net spin, s, of the combination, and what is the z component, m? The z component is easy:

$$\begin{aligned}
S_z\,|s_1\, s_2\, m_1\, m_2\rangle &= S_z^{(1)}\,|s_1\, s_2\, m_1\, m_2\rangle + S_z^{(2)}\,|s_1\, s_2\, m_1\, m_2\rangle\,, \qquad (4.173)\\
&= \hbar\,(m_1+m_2)\,|s_1\, s_2\, m_1\, m_2\rangle = \hbar m\,|s_1\, s_2\, m_1\, m_2\rangle\,,
\end{aligned}$$

so

$$m = m_1 + m_2; \tag{4.174}$$

it's just the *sum*. But s is much more subtle, so let's begin with the simplest nontrivial example.

Example 4.5

Consider the case of two spin-1/2 particles—say, the electron and the proton in the ground state of hydrogen. Each can have spin up or spin down, so there are four possibilities in all:[51]

[51] More precisely, the composite system is in a *linear combination* of the four states listed. For spin 1/2 I find the arrows more evocative than the four-index kets, but you can always revert to the formal notation if you're worried about it.

$$|\uparrow\uparrow\rangle = \left|\tfrac{1}{2}\,\tfrac{1}{2}\,\tfrac{1}{2}\,\tfrac{1}{2}\right\rangle, \qquad m = 1,$$

$$|\uparrow\downarrow\rangle = \left|\tfrac{1}{2}\,\tfrac{1}{2}\,\tfrac{1}{2}\,\tfrac{-1}{2}\right\rangle, \qquad m = 0,$$

$$|\downarrow\uparrow\rangle = \left|\tfrac{1}{2}\,\tfrac{1}{2}\,\tfrac{-1}{2}\,\tfrac{1}{2}\right\rangle, \qquad m = 0,$$

$$|\downarrow\downarrow\rangle = \left|\tfrac{1}{2}\,\tfrac{1}{2}\,\tfrac{-1}{2}\,\tfrac{-1}{2}\right\rangle, \qquad m = -1.$$

This doesn't look right: m is supposed to advance in integer steps, from $-s$ to $+s$, so it appears that $s = 1$—but there is an "extra" state with $m = 0$.

One way to untangle this problem is to apply the lowering operator, $S_- = S_-^{(1)} + S_-^{(2)}$ to the state $|\uparrow\uparrow\rangle$, using Equation 4.146:

$$S_- |\uparrow\uparrow\rangle = \left(S_-^{(1)} |\uparrow\rangle\right) |\uparrow\rangle + |\uparrow\rangle \left(S_-^{(2)} |\uparrow\rangle\right)$$
$$= (\hbar |\downarrow\rangle) |\uparrow\rangle + |\uparrow\rangle (\hbar |\downarrow\rangle) = \hbar (|\downarrow\uparrow\rangle + |\uparrow\downarrow\rangle).$$

Evidently the three states with $s = 1$ are (in the notation $|s\, m\rangle$):

$$\boxed{\left.\begin{cases} |1\,1\rangle &= |\uparrow\uparrow\rangle \\ |1\,0\rangle &= \frac{1}{\sqrt{2}} (|\uparrow\downarrow\rangle + |\downarrow\uparrow\rangle) \\ |1\,{-1}\rangle &= |\downarrow\downarrow\rangle \end{cases}\right\} \quad s = 1 \ \text{(triplet)}.} \tag{4.175}$$

(As a check, try applying the lowering operator to $|1\,0\rangle$; what *should* you get? See Problem 4.37(a).) This is called the **triplet** combination, for the obvious reason. Meanwhile, the orthogonal state with $m = 0$ carries $s = 0$:

$$\boxed{\left\{ |0\,0\rangle = \frac{1}{\sqrt{2}} (|\uparrow\downarrow\rangle - |\downarrow\uparrow\rangle) \right\} \quad s = 0 \ \text{(singlet)}.} \tag{4.176}$$

(If you apply the raising or lowering operator to *this* state, you'll get *zero*. See Problem 4.37(b).)

I claim, then, that the combination of two spin-1/2 particles can carry a total spin of 1 or 0, depending on whether they occupy the triplet or the singlet configuration. To *confirm* this, I need to prove that the triplet states are eigenvectors of S^2 with eigenvalue $2\hbar^2$, and the singlet is an eigenvector of S^2 with eigenvalue 0. Now,

$$S^2 = \left(\mathbf{S}^{(1)} + \mathbf{S}^{(2)}\right) \cdot \left(\mathbf{S}^{(1)} + \mathbf{S}^{(2)}\right) = \left(S^{(1)}\right)^2 + \left(S^{(2)}\right)^2 + 2\mathbf{S}^{(1)} \cdot \mathbf{S}^{(2)}. \tag{4.177}$$

Using Equations 4.145 and 4.147, we have

$$\mathbf{S}^{(1)} \cdot \mathbf{S}^{(2)} |\uparrow\downarrow\rangle = \left(S_x^{(1)} |\uparrow\rangle\right)\left(S_x^{(2)} |\downarrow\rangle\right) + \left(S_y^{(1)} |\uparrow\rangle\right)\left(S_y^{(2)} |\downarrow\rangle\right) + \left(S_z^{(1)} |\uparrow\rangle\right)\left(S_z^{(2)} |\downarrow\rangle\right)$$
$$= \left(\frac{\hbar}{2} |\downarrow\rangle\right)\left(\frac{\hbar}{2} |\uparrow\rangle\right) + \left(\frac{i\hbar}{2} |\downarrow\rangle\right)\left(\frac{-i\hbar}{2} |\uparrow\rangle\right) + \left(\frac{\hbar}{2} |\uparrow\rangle\right)\left(\frac{-\hbar}{2} |\downarrow\rangle\right)$$
$$= \frac{\hbar^2}{4} (2 |\downarrow\uparrow\rangle - |\uparrow\downarrow\rangle).$$

Similarly,

$$\mathbf{S}^{(1)} \cdot \mathbf{S}^{(2)} (|\downarrow\uparrow\rangle) = \frac{\hbar^2}{4}(2\,|\uparrow\downarrow\rangle - |\downarrow\uparrow\rangle).$$

It follows that

$$\mathbf{S}^{(1)} \cdot \mathbf{S}^{(2)} \,|1\,0\rangle = \frac{\hbar^2}{4}\frac{1}{\sqrt{2}}(2\,|\downarrow\uparrow\rangle - |\uparrow\downarrow\rangle + 2\,|\uparrow\downarrow\rangle - |\downarrow\uparrow\rangle) = \frac{\hbar^2}{4}\,|1\,0\rangle\,, \tag{4.178}$$

and

$$\mathbf{S}^{(1)} \cdot \mathbf{S}^{(2)} \,|0\,0\rangle = \frac{\hbar^2}{4}\frac{1}{\sqrt{2}}(2\,|\downarrow\uparrow\rangle - |\uparrow\downarrow\rangle - 2\,|\uparrow\downarrow\rangle + |\downarrow\uparrow\rangle) = -\frac{3\hbar^2}{4}\,|0\,0\rangle\,. \tag{4.179}$$

Returning to Equation 4.177 (and using Equation 4.142), we conclude that

$$S^2\,|1\,0\rangle = \left(\frac{3\hbar^2}{4} + \frac{3\hbar^2}{4} + 2\frac{\hbar^2}{4}\right)|1\,0\rangle = 2\hbar^2\,|1\,0\rangle\,, \tag{4.180}$$

so $|1\,0\rangle$ is indeed an eigenstate of S^2 with eigenvalue $2\hbar^2$; and

$$S^2\,|0\,0\rangle = \left(\frac{3\hbar^2}{4} + \frac{3\hbar^2}{4} - 2\frac{3\hbar^2}{4}\right)|0\,0\rangle = 0, \tag{4.181}$$

so $|0\,0\rangle$ is an eigenstate of S^2 with eigenvalue 0. (I will leave it for you to confirm that $|1\,1\rangle$ and $|1\,-1\rangle$ are eigenstates of S^2, with the appropriate eigenvalue—see Problem 4.37(c).)

What we have just done (combining spin 1/2 with spin 1/2 to get spin 1 and spin 0) is the simplest example of a larger problem: If you combine spin s_1 with spin s_2, what total spins s can you get?[52] The answer[53] is that you get every spin from $(s_1 + s_2)$ down to $(s_1 - s_2)$—or $(s_2 - s_1)$, if $s_2 > s_1$—in integer steps:

$$s = (s_1 + s_2)\,,\ (s_1 + s_2 - 1)\,,\ (s_1 + s_2 - 2)\,,\ \ldots\,,\ |s_1 - s_2|. \tag{4.182}$$

(Roughly speaking, the highest total spin occurs when the individual spins are aligned parallel to one another, and the lowest occurs when they are antiparallel.) For example, if you package together a particle of spin 3/2 with a particle of spin 2, you could get a total spin of 7/2, 5/2, 3/2, or 1/2, depending on the configuration. Another example: If a hydrogen atom is in the state $\psi_{n\ell m}$, the net angular momentum of the electron (spin plus orbital) is $\ell + 1/2$ or $\ell - 1/2$; if you now throw in spin of the *proton*, the atom's *total* angular momentum quantum number is $\ell + 1$, ℓ, or $\ell - 1$ (and ℓ can be achieved in two distinct ways, depending on whether the electron alone is in the $\ell + 1/2$ configuration or the $\ell - 1/2$ configuration).

The combined state $|s\,m\rangle$ with total spin s and z-component m will be some linear combination of the composite states $|s_1\,s_2\,m_1\,m_2\rangle$:

$$|s\,m\rangle = \sum_{m_1+m_2=m} C^{s_1 s_2 s}_{m_1 m_2 m}\,|s_1\,s_2\,m_1\,m_2\rangle \tag{4.183}$$

(because the z-components *add*, the only composite states that contribute are those for which $m_1 + m_2 = m$). Equations 4.175 and 4.176 are special cases of this general form,

52 I say *spins*, for simplicity, but either one (or both) could just as well be *orbital* angular momentum (for which, however, we would use the letter ℓ).

53 For a proof you must look in a more advanced text; see, for instance, Claude Cohen-Tannoudji, Bernard Diu, and Franck Laloë, *Quantum Mechanics*, Wiley, New York (1977), Vol. 2, Chapter X.

Table 4.8: *Clebsch–Gordan coefficients. (A square root sign is understood for every entry; the minus sign, if present, goes* outside *the radical.)*

1/2 × 1/2

m_1	m_2			
		1, +1		
+1/2	+1/2	1		
		1, 0	**0, 0**	
+1/2	−1/2	1/2	1/2	
−1/2	+1/2	1/2	−1/2	
		1, −1		
−1/2	−1/2	1		

2 × 1/2

m_1	m_2		
		5/2, +5/2	
+2	1/2	1	
		5/2, 3/2	**3/2, +3/2**
+2	−1/2	1/5	4/5
+1	+1/2	4/5	−1/5
		5/2, +1/2	**3/2, +1/2**
+1	−1/2	2/5	3/5
0	+1/2	3/5	−2/5
		5/2, −1/2	**3/2, −1/2**
0	−1/2	3/5	2/5
−1	+1/2	2/5	−3/5
		5/2, −3/2	**3/2, −3/2**
−1	−1/2	4/5	1/5
−2	+1/2	1/5	−4/5
		5/2, −5/2	
−2	−1/2	1	

1 × 1/2

m_1	m_2		
		3/2, +3/2	
+1	+1/2	1	
		3/2, +1/2	**1/2, +1/2**
+1	−1/2	1/3	2/3
0	+1/2	2/3	−1/3
		3/2, −1/2	**1/2, −1/2**
0	−1/2	2/3	1/3
−1	+1/2	1/3	−2/3
		3/2, −3/2	
−1	−1/2	1	

3/2 × 1/2

m_1	m_2		
		2, +2	
+3/2	+1/2	1	
		2, +1	**1, +1**
+3/2	−1/2	1/4	3/4
+1/2	+1/2	3/4	−1/4
		2, 0	**1, 0**
+1/2	−1/2	1/2	1/2
−1/2	+1/2	1/2	−1/2
		2, −1	**1, −1**
−1/2	−1/2	3/4	1/4
−3/2	+1/2	1/4	−3/4
		2, −2	
−3/2	−1/2	1	

2 × 1

m_1	m_2			
		3, +3		
+2	+1	1		
		3, +2	**2, +2**	
+2	0	1/3	2/3	
+1	+1	2/3	−1/3	
		3, +1	**2, +1**	**1, +1**
+2	−1	1/15	1/3	3/5
+1	0	8/15	1/6	−3/10
0	+1	6/15	−1/2	1/10
		3, 0	**2, 0**	**1, 0**
+1	−1	1/5	1/2	3/10
0	0	3/5	0	−2/5
−1	+1	1/5	−1/2	3/10
		3, −1	**2, −1**	**1, −1**
0	−1	6/15	1/2	1/10
−1	0	8/15	−1/6	−3/10
−2	+1	1/15	−1/3	3/5
		3, −2	**2, −2**	
−1	−1	2/3	1/3	
−2	0	1/3	−2/3	
		3, −3		
−2	−1	1		

3/2 × 1

m_1	m_2			
		5/2, +5/2		
+3/2	+1	1		
		5/2, +3/2	**3/2, +3/2**	
+3/2	0	2/5	3/5	
+1/2	+1	3/5	−2/5	
		5/2, +1/2	**3/2, +1/2**	**1/2, +1/2**
+3/2	−1	1/10	2/5	1/2
+1/2	0	3/5	1/15	−1/3
−1/2	+1	3/10	−8/15	1/6
		5/2, −1/2	**3/2, −1/2**	**1/2, −1/2**
+1/2	−1	3/10	8/15	1/6
−1/2	0	3/5	−1/15	−1/3
−3/2	+1	1/10	−2/5	1/2
		5/2, −3/2	**3/2, −3/2**	
−1/2	−1	3/5	2/5	
−3/2	0	2/5	−3/5	
		5/2, −5/2		
−3/2	−1	1		

1 × 1

m_1	m_2			
		2, +2		
+1	+1	1		
		2, +1	**1, +1**	
+1	0	1/2	1/2	
0	+1	1/2	−1/2	
		2, 0	**1, 0**	**0, 0**
+1	−1	1/6	1/2	1/3
0	0	2/3	0	−1/3
−1	+1	1/6	−1/2	1/3
		2, −1	**1, −1**	
0	−1	1/2	1/2	
−1	0	1/2	−1/2	
		2, −2		
−1	−1	1		

with $s_1 = s_2 = 1/2$. The constants $C^{s_1 s_2 s}_{m_1 m_2 m}$ are called **Clebsch–Gordan coefficients**. A few of the simplest cases are listed in Table 4.8.[54] For example, the shaded column of the 2×1 table tells us that

$$|3\,0\rangle = \frac{1}{\sqrt{5}}\,|2\,1\,1\,{-1}\rangle + \sqrt{\frac{3}{5}}\,|2\,1\,0\,0\rangle + \frac{1}{\sqrt{5}}\,|2\,1\,{-1}\,1\rangle .$$

If two particles (of spin 2 and spin 1) are at rest in a box, and the *total* spin is 3, and its z component is 0, then a measurement of $S_z^{(1)}$ could return the value $\hbar$ (with probability 1/5), or 0 (with probability 3/5), or $-\hbar$ (with probability 1/5). Notice that the probabilities add up to 1 (the sum of the squares of any column on the Clebsch–Gordan table is 1).

These tables also work the other way around:

$$|s_1\, s_2\, m_1\, m_2\rangle = \sum_s C^{s_1 s_2 s}_{m_1 m_2 m}\, |s\, m\rangle\,, \qquad (m = m_1 + m_2)\,. \tag{4.184}$$

For example, the shaded *row* in the $3/2 \times 1$ table tells us that

$$\left|\tfrac{3}{2}\, 1\, \tfrac{1}{2}\, 0\right\rangle = \sqrt{\tfrac{3}{5}}\, \left|\tfrac{5}{2}\, \tfrac{1}{2}\right\rangle + \sqrt{\tfrac{1}{15}}\, \left|\tfrac{3}{2}\, \tfrac{1}{2}\right\rangle - \sqrt{\tfrac{1}{3}}\, \left|\tfrac{1}{2}\, \tfrac{1}{2}\right\rangle .$$

If you put particles of spin 3/2 and spin 1 in the box, and you know that the first has $m_1 = 1/2$ and the second has $m_2 = 0$ (so m is necessarily 1/2), and you measured the *total* spin, s, you could get 5/2 (with probability 3/5), or 3/2 (with probability 1/15), or 1/2 (with probability 1/3). Again, the sum of the probabilities is 1 (the sum of the squares of each *row* on the Clebsch–Gordan table is 1).

If you think this is starting to sound like mystical numerology, I don't blame you. We will not be using the Clebsch–Gordan tables much in the rest of the book, but I wanted you to know

---

[54] The general formula is derived in Arno Bohm, *Quantum Mechanics: Foundations and Applications*, 2nd edn, Springer, 1986, p. 172.

where they fit into the scheme of things, in case you encounter them later on. In a mathematical sense this is all applied **group theory**—what we are talking about is the decomposition of the direct product of two irreducible representations of the rotation group into a direct sum of irreducible representations (you can quote that, to impress your friends).

* **Problem 4.37**

(a) Apply S_- to $|1\,0\rangle$ (Equation 4.175), and confirm that you get $\sqrt{2}\hbar\,|1\,-1\rangle$.

(b) Apply $S_\pm$ to $|0\,0\rangle$ (Equation 4.176), and confirm that you get zero.

(c) Show that $|1\,1\rangle$ and $|1\,-1\rangle$ (Equation 4.175) are eigenstates of S^2, with the appropriate eigenvalue.

Problem 4.38 **Quarks** carry spin 1/2. Three quarks bind together to make a **baryon** (such as the proton or neutron); two quarks (or more precisely a quark and an antiquark) bind together to make a **meson** (such as the pion or the kaon). Assume the quarks are in the ground state (so the *orbital* angular momentum is zero).

(a) What spins are possible for baryons?

(b) What spins are possible for mesons?

* **Problem 4.39** Verify Equations 4.175 and 4.176 using the Clebsch–Gordan table.

Problem 4.40

(a) A particle of spin 1 and a particle of spin 2 are at rest in a configuration such that the total spin is 3, and its z component is $\hbar$. If you measured the z-component of the angular momentum of the spin-2 particle, what values might you get, and what is the probability of each one? *Comment:* Using Clebsch–Gordan tables is like driving a stick-shift—scary and frustrating when you start out, but easy once you get the hang of it.

(b) An electron with spin down is in the state ψ_{510} of the hydrogen atom. If you could measure the total angular momentum squared of the electron alone (*not* including the proton spin), what values might you get, and what is the probability of each?

Problem 4.41 Determine the commutator of S^2 with $S_z^{(1)}$ (where $\mathbf{S} \equiv \mathbf{S}^{(1)} + \mathbf{S}^{(2)}$). Generalize your result to show that

$$\left[S^2, \mathbf{S}^{(1)}\right] = 2i\hbar\left(\mathbf{S}^{(1)} \times \mathbf{S}^{(2)}\right). \tag{4.185}$$

Comment: Because $S_z^{(1)}$ does not commute with S^2, we cannot hope to find states that are simultaneous eigenvectors of both. In order to form eigenstates of S^2 we need *linear combinations* of eigenstates of $S_z^{(1)}$. This is precisely what the Clebsch–Gordan coefficients (in Equation 4.183) do for us. On the other hand, it follows by obvious inference from Equation 4.185 that the *sum* $\mathbf{S}^{(1)} + \mathbf{S}^{(2)}$ *does* commute with S^2, which only confirms what we already knew (see Equation 4.103).]

4.5 ELECTROMAGNETIC INTERACTIONS

4.5.1 Minimal Coupling

In classical electrodynamics[55] the force on a particle of charge q moving with velocity $\mathbf{v}$ through electric and magnetic fields $\mathbf{E}$ and $\mathbf{B}$ is given by the **Lorentz force law**:

$$\mathbf{F} = q\left(\mathbf{E} + \mathbf{v} \times \mathbf{B}\right). \tag{4.186}$$

This force cannot be expressed as the gradient of a scalar potential energy function, and therefore the Schrödinger equation in its original form (Equation 1.1) cannot accommodate it. But in the more sophisticated form

$$i\hbar\frac{\partial \Psi}{\partial t} = \hat{H}\Psi \tag{4.187}$$

there is no problem. The classical Hamiltonian for a particle of charge q and momentum $\mathbf{p}$, in the presence of electromagnetic fields is[56]

$$H = \frac{1}{2m}\left(\mathbf{p} - q\mathbf{A}\right)^2 + q\varphi, \tag{4.188}$$

where $\mathbf{A}$ is the vector potential and φ is the scalar potential:

$$\mathbf{E} = -\nabla\varphi - \partial\mathbf{A}/\partial t, \quad \mathbf{B} = \nabla \times \mathbf{A}. \tag{4.189}$$

Making the standard substitution $\mathbf{p} \rightarrow -i\hbar\nabla$, we obtain the Hamiltonian operator[57]

$$\hat{H} = \frac{1}{2m}\left(-i\hbar\nabla - q\mathbf{A}\right)^2 + q\varphi, \tag{4.190}$$

and the Schrödinger equation becomes

$$\boxed{i\hbar\frac{\partial \Psi}{\partial t} = \left[\frac{1}{2m}\left(-i\hbar\nabla - q\mathbf{A}\right)^2 + q\varphi\right]\Psi.} \tag{4.191}$$

This is the quantum implementation of the Lorentz force law; it is sometimes called the **minimal coupling rule**.[58]

*** **Problem 4.42**

(a) Using Equation 4.190 and the generalized Ehrenfest theorem (3.73), show that

$$\frac{d\langle\mathbf{r}\rangle}{dt} = \frac{1}{m}\langle(\mathbf{p} - q\mathbf{A})\rangle. \tag{4.192}$$

Hint: This stands for three equations—one for each component. Work it out for, say, the x component, and then generalize your result.

---

[55] Readers who have not studied electrodynamics may want to skip Section 4.5.

[56] See, for example, Herbert Goldstein, Charles P. Poole, and John Safko, *Classical Mechanics*, 3rd edn, Prentice Hall, Upper Saddle River, NJ, 2002, page 342.

[57] In the case of electro*statics* we can choose $\mathbf{A} = \mathbf{0}$, and $q\varphi$ is the potential energy V.

[58] Note that the potentials are *given*, just like the potential energy V in the regular Schrödinger equation. In quantum electrodynamics (QED) the fields themselves are quantized, but that's an entirely different theory.

(b) As always (see Equation 1.32) we identify $d\langle \mathbf{r} \rangle /dt$ with $\langle \mathbf{v} \rangle$. Show that[59]

$$m\frac{d\langle \mathbf{v} \rangle}{dt} = q\langle \mathbf{E} \rangle + \frac{q}{2m}\langle (\mathbf{p}\times\mathbf{B} - \mathbf{B}\times\mathbf{p}) \rangle - \frac{q^2}{m}\langle (\mathbf{A}\times\mathbf{B}) \rangle . \tag{4.193}$$

(c) In particular, if the fields **E** and **B** are *uniform* over the volume of the wave packet, show that

$$m\frac{d\langle \mathbf{v} \rangle}{dt} = q\left(\mathbf{E} + \langle \mathbf{v} \rangle \times \mathbf{B}\right), \tag{4.194}$$

so the *expectation value* of **v** moves according to the Lorentz force law, as we would expect from Ehrenfest's theorem.

∗∗∗ **Problem 4.43** Suppose

$$\mathbf{A} = \frac{B_0}{2}\left(x\hat{j} - y\hat{i}\right), \quad \text{and} \quad \varphi = Kz^2,$$

where B_0 and K are constants.

(a) Find the fields **E** and **B**.

(b) Find the allowed energies, for a particle of mass m and charge q, in these fields. *Answer:*

$$E(n_1, n_2) = \left(n_1 + \tfrac{1}{2}\right)\hbar\omega_1 + \left(n_2 + \tfrac{1}{2}\right)\hbar\omega_2, \quad (n_1, n_2 = 0, 1, 2, \ldots), \tag{4.195}$$

where $\omega_1 \equiv qB_0/m$ and $\omega_2 \equiv \sqrt{2qK/m}$. *Comment:* In two dimensions (x and y, with $K = 0$) this is the quantum analog to **cyclotron motion**; ω_1 is the classical cyclotron frequency, and ω_2 is zero. The allowed energies, $\left(n_1 + \tfrac{1}{2}\right)\hbar\omega_1$, are called **Landau Levels**.[60]

4.5.2 The Aharonov–Bohm Effect

In classical electrodynamics the potentials **A** and φ are not uniquely determined; the *physical* quantities are the *fields*, **E** and **B**.[61] Specifically, the potentials

$$\varphi' \equiv \varphi - \frac{\partial \Lambda}{\partial t}, \qquad \mathbf{A}' \equiv \mathbf{A} + \nabla\Lambda \tag{4.196}$$

(where Λ is an arbitrary real function of position and time) yield the same fields as φ and **A**. (Check that for yourself, using Equation 4.189.) Equation 4.196 is called a **gauge transformation**, and the theory is said to be **gauge invariant**.

In quantum mechanics the potentials play a more direct role (it is they, not the fields, that appear in the Equation 4.191), and it is of interest to ask whether the theory remains gauge invariant. It is easy to show (Problem 4.44) that

---

59 Note that **p** does not commute with **B**, so $(\mathbf{p}\times\mathbf{B}) \neq -(\mathbf{B}\times\mathbf{p})$, but **A** *does* commute with **B**, so $(\mathbf{A}\times\mathbf{B}) = -(\mathbf{B}\times\mathbf{A})$.

60 For further discussion see Leslie E. Ballentine, *Quantum Mechanics: A Modern Development*, World Scientific, Singapore (1998), Section 11.3.

61 See, for example, Griffiths (footnote 43), Section 10.1.2.

$$\Psi' \equiv e^{iq\Lambda/\hbar}\Psi \tag{4.197}$$

satisfies Equation 4.191 with the gauge-transformed potentials φ' and $\mathbf{A}'$ (Equation 4.196). Since Ψ' differs from Ψ only by a *phase factor*, it represents the same physical state,[62] and in this sense the theory *is* gauge invariant. For a long time it was taken for granted that there could be no electromagnetic influences in regions where **E** and **B** are zero—any more than there can be in the classical theory. But in 1959 Aharonov and Bohm[63] showed that the vector potential *can* affect the quantum behavior of a charged particle, *even when the particle is confined to a region where the field itself is zero.*

Example 4.6

Imagine a particle constrained to move in a circle of radius b (a bead on a wire ring, if you like). Along the axis runs a solenoid of radius $a < b$, carrying a steady electric current I (see Figure 4.16). If the solenoid is extremely long, the magnetic field inside it is uniform, and the field outside is zero. But the vector potential outside the solenoid is *not* zero; in fact (adopting the convenient gauge condition $\nabla \cdot \mathbf{A} = 0$),[64]

$$\mathbf{A} = \frac{\Phi}{2\pi r}\hat{\phi}, \quad (r > a), \tag{4.198}$$

where $\Phi = \pi a^2 B$ is the **magnetic flux** through the solenoid. Meanwhile, the solenoid itself is uncharged, so the scalar potential φ is zero. In this case the Hamiltonian (Equation 4.190) becomes

$$\hat{H} = \frac{1}{2m}\left[-\hbar^2\nabla^2 + q^2A^2 + 2i\hbar q\mathbf{A}\cdot\nabla\right] \tag{4.199}$$

(Problem 4.45(a)). But the wave function depends only on the azimuthal angle ϕ ($\theta = \pi/2$ and $r = b$), so $\nabla \to \left(\hat{\phi}/b\right)(d/d\phi)$, and the Schrödinger equation reads

$$\frac{1}{2m}\left[-\frac{\hbar^2}{b^2}\frac{d^2}{d\phi^2} + \left(\frac{q\Phi}{2\pi b}\right)^2 + i\frac{\hbar q\Phi}{\pi b^2}\frac{d}{d\phi}\right]\psi(\phi) = E\psi(\phi). \tag{4.200}$$

This is a linear differential equation with constant coefficients:

$$\frac{d^2\psi}{d\phi^2} - 2i\beta\frac{d\psi}{d\phi} + \epsilon\psi = 0, \tag{4.201}$$

where

$$\beta \equiv \frac{q\Phi}{2\pi\hbar} \quad \text{and} \quad \epsilon \equiv \frac{2mb^2E}{\hbar^2} - \beta^2. \tag{4.202}$$

---

[62] That is to say, $\langle\mathbf{r}\rangle$, $d\langle\mathbf{r}\rangle/dt$, etc. are unchanged. Because Λ depends on position, $\langle\mathbf{p}\rangle$ (with $\mathbf{p}$ represented by the operator $-i\hbar\nabla$) *does* change, but as you found in Equation 4.192, $\mathbf{p}$ does not represent the mechanical momentum ($m\mathbf{v}$) in this context (in Lagrangian mechanics $\mathbf{p} = m\mathbf{v} + q\mathbf{A}$ is the so-called **canonical momentum**).

[63] Y. Aharonov and D. Bohm, *Phys. Rev.* **115**, 485 (1959). For a significant precursor, see W. Ehrenberg and R. E. Siday, *Proc. Phys. Soc. London* **B62**, 8 (1949).

[64] See, for instance, Griffiths (footnote 43), Equation 5.71.

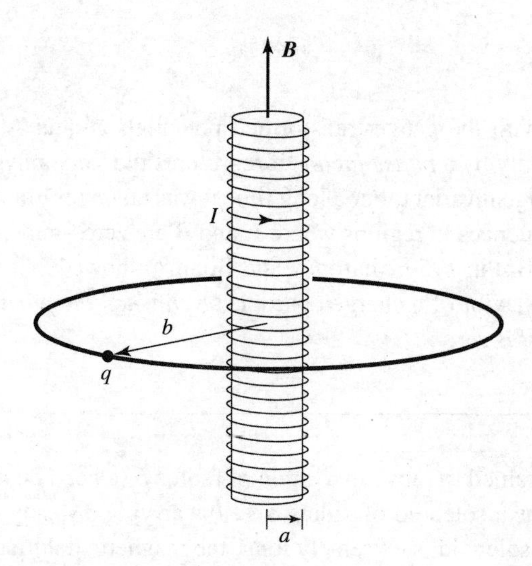

Figure 4.16: Charged bead on a circular ring through which a long solenoid passes.

Solutions are of the form

$$\psi = Ae^{i\lambda\phi}, \tag{4.203}$$

with

$$\lambda = \beta \pm \sqrt{\beta^2 + \epsilon} = \beta \pm \frac{b}{\hbar}\sqrt{2mE}. \tag{4.204}$$

Continuity of $\psi(\phi)$, at $\phi = 2\pi$, requires that λ be an *integer*:

$$\beta \pm \frac{b}{\hbar}\sqrt{2mE} = n, \tag{4.205}$$

and it follows that

$$E_n = \frac{\hbar^2}{2mb^2}\left(n - \frac{q\Phi}{2\pi\hbar}\right)^2, \quad (n = 0, \pm 1, \pm 2, \ldots). \tag{4.206}$$

The solenoid lifts the two-fold degeneracy of the bead-on-a-ring (Problem 2.46): positive n, representing a particle traveling in the *same* direction as the current in the solenoid, has a somewhat *lower* energy (assuming q is positive) than negative n, describing a particle traveling in the *opposite* direction. More important, the allowed energies clearly depend on the field inside the solenoid, *even though the field at the location of the particle is zero*![65]

More generally, suppose a particle is moving through a region where **B** is zero (so $\nabla \times \mathbf{A} = \mathbf{0}$), but **A** itself is *not*. (I'll assume that **A** is static, although the method can be generalized to time-dependent potentials.) The Schrödinger equation,

[65] It is a peculiar property of **superconducting** rings that the enclosed flux is *quantized*: $\Phi = (2\pi\hbar/q)\,n'$, where n' is an integer. In that case the effect is undetectable, since $E_n = \left(\hbar^2/2mb^2\right)\left(n + n'\right)^2$, and $\left(n + n'\right)$ is just another integer. (Incidentally, the charge q here turns out to be *twice* the charge of an electron; the superconducting electrons are locked together in pairs.) However, flux quantization is enforced by the *superconductor* (which induces circulating currents to make up the difference), not by the solenoid or the electromagnetic field, and it does not occur in the (nonsuperconducting) example considered here.

$$\left[\frac{1}{2m}(-i\hbar\nabla - q\mathbf{A})^2\right]\Psi = i\hbar\frac{\partial\Psi}{\partial t}, \tag{4.207}$$

can be simplified by writing

$$\Psi = e^{ig}\Psi', \tag{4.208}$$

where

$$g(\mathbf{r}) \equiv \frac{q}{\hbar}\int_{\mathcal{O}}^{\mathbf{r}} \mathbf{A}(\mathbf{r}')\cdot d\mathbf{r}', \tag{4.209}$$

and $\mathcal{O}$ is some (arbitrarily chosen) reference point. (Note that this definition makes sense *only* when $\nabla\times\mathbf{A} = \mathbf{0}$ throughout the region in question[66]—otherwise the line integral would depend on the *path* taken from $\mathcal{O}$ to $\mathbf{r}$, and hence would not define a function of $\mathbf{r}$.) In terms of Ψ', the gradient of Ψ is

$$\nabla\Psi = e^{ig}\left(i\nabla g\right)\Psi' + e^{ig}\left(\nabla\Psi'\right);$$

but $\nabla g = (q/\hbar)\,\mathbf{A}$, so

$$(-i\hbar\nabla - q\mathbf{A})\,\Psi = -i\hbar e^{ig}\nabla\Psi', \tag{4.210}$$

and it follows that

$$(-i\hbar\nabla - q\mathbf{A})^2\,\Psi = -\hbar^2 e^{ig}\nabla^2\Psi' \tag{4.211}$$

(Problem 4.45(b)). Putting this into Equation 4.207, and cancelling the common factor of e^{ig}, we are left with

$$-\frac{\hbar^2}{2m}\nabla^2\Psi' = i\hbar\frac{\partial\Psi'}{\partial t}. \tag{4.212}$$

Evidently Ψ' satisfies the Schrödinger equation *without* $\mathbf{A}$. If we can solve Equation 4.212, correcting for the presence of a (curl-free) vector potential will be trivial: just tack on the phase factor e^{ig}.

Aharonov and Bohm proposed an experiment in which a beam of electrons is split in two, and they pass either side of a long solenoid before recombining (Figure 4.17). The beams are kept well away from the solenoid itself, so they encounter only regions where $\mathbf{B} = \mathbf{0}$. But $\mathbf{A}$, which is given by Equation 4.198, is *not* zero, and the two beams arrive with *different phases*:[67]

$$g = \frac{q}{\hbar}\int \mathbf{A}\cdot d\mathbf{r} = \frac{q\Phi}{2\pi\hbar}\int\left(\frac{1}{r}\hat{\phi}\right)\cdot\left(r\hat{\phi}\,d\phi\right) = \pm\frac{q\Phi}{2\hbar}. \tag{4.213}$$

[66] The region in question must also be simply connected (no holes). This might seem like a technicality, but in the present example we need to excise the solenoid itself, and that leaves a hole in the space. To get around this we treat each *side* of the solenoid as a separate simply-connected region. If that bothers you, you're not alone; it seems to have bothered Aharanov and Bohm as well, since—in addition to this argument—they provided an alternative solution to confirm their result (Y. Aharonov and D. Bohm, *Phys. Rev.* **115**, 485 (1959)). The Aharonov–Bohm effect can also be cast as an example of **Berry's phase** (see Chapter 11), where this issue does not arise (M. Berry, *Proc. Roy. Soc. Lond.* **A 392**, 45 (1984)).

[67] Use cylindrical coordinates centered on the axis of the solenoid; put $\mathcal{O}$ on the incoming beam, and let ϕ run $0 < \pi$ on one side and $0 > -\pi$ on the other, with $r > a$ always.

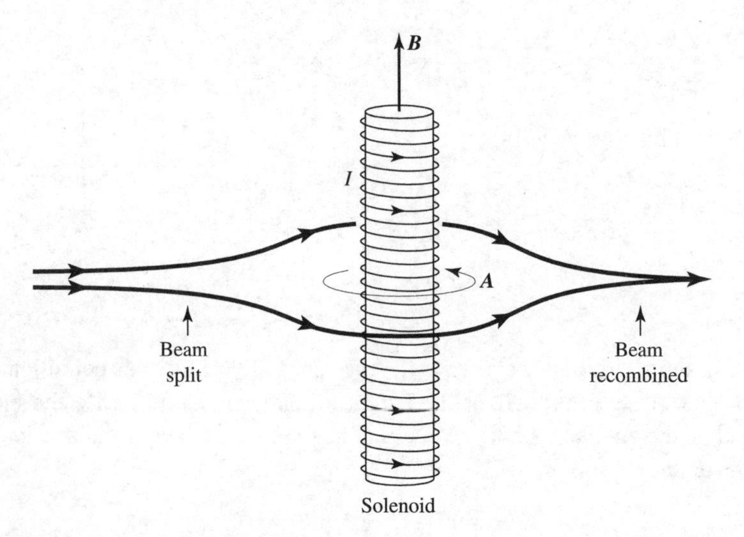

Figure 4.17: The Aharonov–Bohm effect: The electron beam splits, with half passing either side of a long solenoid.

The plus sign applies to the electrons traveling in the same direction as **A**—which is to say, in the same direction as the current in the solenoid. The beams arrive out of phase by an amount proportional to the magnetic flux their paths encircle:

$$\text{phase difference} = \frac{q\Phi}{\hbar}. \tag{4.214}$$

This phase shift leads to measurable interference, which has been confirmed experimentally by Chambers and others.[68]

What are we to make of the Aharonov–Bohm effect? It seems our classical preconceptions are simply *mistaken*: There *can* be electromagnetic effects in regions where the fields are zero. Note, however, that this does not make **A** itself measurable—only the enclosed *flux* comes into the final answer, and the theory remains gauge invariant.[69]

** **Problem 4.44** Show that Ψ' (Equation 4.197) satisfies the Schrödinger equation (Equation 4.191 with the potentials φ' and $\mathbf{A}'$ (Equation 4.196).

Problem 4.45

(a) Derive Equation 4.199 from Equation 4.190.

(b) Derive Equation 4.211, starting with Equation 4.210.

[68] R. G. Chambers, *Phys. Rev. Lett.* **5**, 3 (1960).

[69] Aharonov and Bohm themselves concluded that the vector potential has a physical significance in quantum mechanics that it lacks in classical theory, and most physicists today would agree. For the early history of the Aharonov–Bohm effect see H. Ehrlickson, *Am. J. Phys.* **38**, 162 (1970).

FURTHER PROBLEMS ON CHAPTER 4

* **Problem 4.46** Consider the **three-dimensional harmonic oscillator**, for which the potential is

$$V(r) = \frac{1}{2} m\omega^2 r^2. \tag{4.215}$$

(a) Show that separation of variables in cartesian coordinates turns this into three one-dimensional oscillators, and exploit your knowledge of the latter to determine the allowed energies. *Answer*:

$$E_n = \left(n + \tfrac{3}{2}\right) \hbar\omega. \tag{4.216}$$

(b) Determine the degeneracy $d(n)$ of E_n.

*** **Problem 4.47** Because the three-dimensional harmonic oscillator potential (see Equation 4.215) is spherically symmetrical, the Schrödinger equation can also be handled by separation of variables in *spherical* coordinates. Use the power series method (as in Sections 2.3.2 and 4.2.1) to solve the radial equation. Find the recursion formula for the coefficients, and determine the allowed energies. (Check that your answer is consistent with Equation 4.216.) How is N related to n in this case? Draw the diagram analogous to Figures 4.3 and 4.6, and determine the degeneracy of nth energy level.[70]

** **Problem 4.48**

(a) Prove the **three-dimensional virial theorem**:

$$2\langle T\rangle = \langle \mathbf{r} \cdot \nabla V\rangle \tag{4.217}$$

(for stationary states). *Hint*: refer to Problem 3.37.

(b) Apply the virial theorem to the case of hydrogen, and show that

$$\langle T\rangle = -E_n; \quad \langle V\rangle = 2E_n. \tag{4.218}$$

(c) Apply the virial theorem to the three-dimensional harmonic oscillator (Problem 4.46), and show that in this case

$$\langle T\rangle = \langle V\rangle = E_n/2. \tag{4.219}$$

*** **Problem 4.49** *Warning:* Attempt this problem only if you are familiar with vector calculus. Define the (three-dimensional) **probability current** by generalization of Problem 1.14:

$$\mathbf{J} \equiv \frac{i\hbar}{2m}\left(\Psi\, \nabla\Psi^* - \Psi^*\, \nabla\Psi\right). \tag{4.220}$$

(a) Show that $\mathbf{J}$ satisfies the **continuity equation**

$$\nabla \cdot \mathbf{J} = -\frac{\partial}{\partial t}|\Psi|^2, \tag{4.221}$$

[70] For some damn reason energy levels are traditionally counted starting with $n = 0$, for the harmonic oscillator. That conflicts with good sense and with our explicit convention (footnote 12), but please stick with it for this problem.

which expresses local **conservation of probability**. It follows (from the divergence theorem) that

$$\oint_{\mathcal{S}} \mathbf{J} \cdot d\mathbf{a} = -\frac{d}{dt} \int_{\mathcal{V}} |\Psi|^2 \, d^3\mathbf{r}, \tag{4.222}$$

where $\mathcal{V}$ is a (fixed) volume and $\mathcal{S}$ is its boundary surface. In words: The flow of probability out through the surface is equal to the decrease in probability of finding the particle in the volume.

(b) Find $\mathbf{J}$ for hydrogen in the state $n = 2, l = 1, m = 1$. *Answer:*

$$\frac{\hbar}{64\pi m a^5} r e^{-r/a} \sin\theta \, \hat{\phi}.$$

(c) If we interpret $m\mathbf{J}$ as the flow of *mass*, the angular momentum is

$$\mathbf{L} = m \int (\mathbf{r} \times \mathbf{J}) \, d^3\mathbf{r}.$$

Use this to calculate L_z for the state ψ_{211}, and comment on the result.[71]

∗∗∗ **Problem 4.50** The (time-independent) **momentum space wave function** in three dimensions is defined by the natural generalization of Equation 3.54:

$$\phi(\mathbf{p}) \equiv \frac{1}{(2\pi\hbar)^{3/2}} \int e^{-i(\mathbf{p}\cdot\mathbf{r})/\hbar} \psi(\mathbf{r}) \, d^3\mathbf{r}. \tag{4.223}$$

(a) Find the momentum space wave function for the ground state of hydrogen (Equation 4.80). *Hint*: Use spherical coordinates, setting the polar axis along the direction of $\mathbf{p}$. Do the θ integral first. *Answer*:

$$\phi(\mathbf{p}) = \frac{1}{\pi} \left(\frac{2a}{\hbar}\right)^{3/2} \frac{1}{\left[1 + (ap/\hbar)^2\right]^2}. \tag{4.224}$$

(b) Check that $\phi(\mathbf{p})$ is normalized.

(c) Use $\phi(\mathbf{p})$ to calculate $\langle p^2 \rangle$, in the ground state of hydrogen.

(d) What is the expectation value of the kinetic energy in this state? Express your answer as a multiple of E_1, and check that it is consistent with the virial theorem (Equation 4.218).

∗∗∗ **Problem 4.51** In Section 2.6 we noted that the finite square well (in one dimension) has at least one bound state, no matter how shallow or narrow it may be. In Problem 4.11 you showed that the finite *spherical* well (three dimensions) has *no* bound state, if the potential is sufficiently weak. *Question:* What about the finite *circular* well (two

[71] Schrödinger (*Annalen der Physik* **81**, 109 (1926), Section 7) interpreted $e\mathbf{J}$ as the electric current density (this was *before* Born published his statistical interpretation of the wave function), and noted that it is time-independent (in a stationary state): "we may in a certain sense speak of a *return to electrostatic and magnetostatic atomic models*. In this way the lack of radiation in [a stationary] state would, indeed, find a startlingly simple explanation." (I thank Kirk McDonald for calling this reference to my attention.)

dimensions)? Show that (like the one-dimensional case) there is always at least one bound state. *Hint:* Look up any information you need about Bessel functions, and use a computer to draw the graphs.

Problem 4.52

(a) Construct the spatial wave function (ψ) for hydrogen in the state $n = 3$, $\ell = 2$, $m = 1$. Express your answer as a function of r, θ, ϕ, and a (the Bohr radius) *only*—no other variables (ρ, z, etc.) or functions (Y, v, etc.), or constants (A, c_0, etc.), or derivatives, allowed (π is okay, and e, and 2, etc.).

(b) Check that this wave function is properly normalized, by carrying out the appropriate integrals over r, θ, and ϕ.

(c) Find the expectation value of r^s in this state. For what range of s (positive and negative) is the result finite?

Problem 4.53

(a) Construct the wave function for hydrogen in the state $n = 4$, $\ell = 3$, $m = 3$. Express your answer as a function of the spherical coordinates r, θ, and ϕ.

(b) Find the expectation value of r in this state. (As always, look up any nontrivial integrals.)

(c) If you could somehow measure the observable $L_x^2 + L_y^2$ on an atom in this state, what value (or values) could you get, and what is the probability of each?

Problem 4.54 What is the probability that an electron in the ground state of hydrogen will be found *inside the nucleus*?

(a) First calculate the *exact* answer, assuming the wave function (Equation 4.80) is correct all the way down to $r = 0$. Let b be the radius of the nucleus.

(b) Expand your result as a power series in the small number $\epsilon \equiv 2b/a$, and show that the lowest-order term is the cubic: $P \approx (4/3)(b/a)^3$. This should be a suitable approximation, provided that $b \ll a$ (which it *is*).

(c) Alternatively, we might assume that $\psi(r)$ is essentially constant over the (tiny) volume of the nucleus, so that $P \approx (4/3)\pi b^3 |\psi(0)|^2$. Check that you get the same answer this way.

(d) Use $b \approx 10^{-15}$ m and $a \approx 0.5 \times 10^{-10}$ m to get a numerical estimate for P. Roughly speaking, this represents the "fraction of its time that the electron spends inside the nucleus."

Problem 4.55

(a) Use the recursion formula (Equation 4.76) to confirm that when $\ell = n - 1$ the radial wave function takes the form

$$R_{n(n-1)} = N_n r^{n-1} e^{-r/na},$$

and determine the normalization constant N_n by direct integration.

(b) Calculate $\langle r \rangle$ and $\langle r^2 \rangle$ for states of the form $\psi_{n(n-1)m}$.

(c) Show that the "uncertainty" in r (σ_r) is $\langle r\rangle/\sqrt{2n+1}$ for such states. Note that the fractional spread in r decreases, with increasing n (in this sense the system "begins to look classical," with identifiable circular "orbits," for large n). Sketch the radial wave functions for several values of n, to illustrate this point.

Problem 4.56 Coincident spectral lines.[72] According to the Rydberg formula (Equation 4.93) the wavelength of a line in the hydrogen spectrum is determined by the principal quantum numbers of the initial and final states. Find two distinct pairs $\{n_i, n_f\}$ that yield the *same* λ. For example, $\{6851, 6409\}$ and $\{15283, 11687\}$ will do it, but you're not allowed to use those!

Problem 4.57 Consider the observables $A = x^2$ and $B = L_z$.

(a) Construct the uncertainty principle for $\sigma_A\sigma_B$.
(b) Evaluate σ_B in the hydrogen state $\psi_{n\ell m}$.
(c) What can you conclude about $\langle xy\rangle$ in this state?

Problem 4.58 An electron is in the spin state

$$\chi = A\begin{pmatrix}1-2i\\2\end{pmatrix}.$$

(a) Determine the constant A by normalizing χ.
(b) If you measured S_z on this electron, what values could you get, and what is the probability of each? What is the expectation value of S_z?
(c) If you measured S_x on this electron, what values could you get, and what is the probability of each? What is the expectation value of S_x?
(d) If you measured S_y on this electron, what values could you get, and what is the probability of each? What is the expectation value of S_y?

∗∗∗ **Problem 4.59** Suppose two spin-1/2 particles are known to be in the singlet configuration (Equation 4.176). Let $S_a^{(1)}$ be the component of the spin angular momentum of particle number 1 in the direction defined by the vector **a**. Similarly, let $S_b^{(2)}$ be the component of 2's angular momentum in the direction **b**. Show that

$$\left\langle S_a^{(1)} S_b^{(2)}\right\rangle = -\frac{\hbar^2}{4}\cos\theta, \tag{4.225}$$

where θ is the angle between **a** and **b**.

∗∗∗ **Problem 4.60**

(a) Work out the Clebsch–Gordan coefficients for the case $s_1 = 1/2$, $s_2 =$ anything. *Hint:* You're looking for the coefficients A and B in

$$|s\,m\rangle = A\left|\tfrac{1}{2}\,s_2\,\tfrac{1}{2}\,\left(m-\tfrac{1}{2}\right)\right\rangle + B\left|\tfrac{1}{2}\,s_2\,\tfrac{-1}{2}\,\left(m+\tfrac{1}{2}\right)\right\rangle,$$

[72] Nicholas Wheeler, "Coincident Spectral Lines" (unpublished Reed College report, 2001).

such that $|s\, m\rangle$ is an eigenstate of S^2. Use the method of Equations 4.177 through 4.180. If you can't figure out what $S_x^{(2)}$ (for instance) does to $|s_2\, m_2\rangle$, refer back to Equation 4.136 and the line before Equation 4.147. *Answer*:

$$A = \sqrt{\frac{s_2 \pm m + 1/2}{2s_2 + 1}}; \quad B = \pm\sqrt{\frac{s_2 \mp m + 1/2}{2s_2 + 1}},$$

where the signs are determined by $s = s_2 \pm 1/2$.

(b) Check this general result against three or four entries in Table 4.8.

Problem 4.61 Find the matrix representing S_x for a particle of spin 3/2 (using as your basis the eigenstates of S_z). Solve the characteristic equation to determine the eigenvalues of S_x.

*** **Problem 4.62** Work out the spin matrices for arbitrary spin s, generalizing spin 1/2 (Equations 4.145 and 4.147), spin 1 (Problem 4.34), and spin 3/2 (Problem 4.61). *Answer:*

$$\mathsf{S}_z = \hbar\begin{pmatrix} s & 0 & 0 & \cdots & 0 \\ 0 & s-1 & 0 & \cdots & 0 \\ 0 & 0 & s-2 & \cdots & 0 \\ \vdots & \vdots & \vdots & \cdots & \vdots \\ 0 & 0 & 0 & \cdots & -s \end{pmatrix};$$

$$\mathsf{S}_x = \frac{\hbar}{2}\begin{pmatrix} 0 & b_s & 0 & 0 & \cdots & 0 & 0 \\ b_s & 0 & b_{s-1} & 0 & \cdots & 0 & 0 \\ 0 & b_{s-1} & 0 & b_{s-2} & \cdots & 0 & 0 \\ 0 & 0 & b_{s-2} & 0 & \cdots & 0 & 0 \\ \vdots & \vdots & \vdots & \vdots & \cdots & \vdots & \vdots \\ 0 & 0 & 0 & 0 & \cdots & 0 & b_{-s+1} \\ 0 & 0 & 0 & 0 & \cdots & b_{-s+1} & 0 \end{pmatrix}$$

$$\mathsf{S}_y = \frac{\hbar}{2}\begin{pmatrix} 0 & -ib_s & 0 & 0 & \cdots & 0 & 0 \\ ib_s & 0 & -ib_{s-1} & 0 & \cdots & 0 & 0 \\ 0 & ib_{s-1} & 0 & -ib_{s-2} & \cdots & 0 & 0 \\ 0 & 0 & ib_{s-2} & 0 & \cdots & 0 & 0 \\ \vdots & \vdots & \vdots & \vdots & \cdots & \vdots & \vdots \\ 0 & 0 & 0 & 0 & \cdots & 0 & -ib_{-s+1} \\ 0 & 0 & 0 & 0 & \cdots & ib_{-s+1} & 0 \end{pmatrix}$$

where

$$b_j \equiv \sqrt{(s+j)(s+1-j)}.$$

*** **Problem 4.63** Work out the normalization factor for the spherical harmonics, as follows. From Section 4.1.2 we know that

$$Y_\ell^m = K_\ell^m e^{im\phi} P_\ell^m(\cos\theta);$$

the problem is to determine the factor K_ℓ^m (which I *quoted*, but did not derive, in Equation 4.32). Use Equations 4.120, 4.121, and 4.130 to obtain a recursion relation giving K_ℓ^{m+1} in terms of K_ℓ^m. Solve it by induction on m to get K_ℓ^m up to an overall constant, $C(\ell) \equiv K_\ell^0$. Finally, use the result of Problem 4.25 to fix the constant. You may find the following formula for the derivative of an associated Legendre function useful:

$$\left(1-x^2\right)\frac{dP_\ell^m}{dx} = -\sqrt{1-x^2}\,P_\ell^{m+1} - mx\,P_\ell^m. \tag{4.226}$$

Problem 4.64 The electron in a hydrogen atom occupies the combined spin and position state

$$R_{21}\left(\sqrt{1/3}\,Y_1^0\chi_+ + \sqrt{2/3}\,Y_1^1\chi_-\right).$$

(a) If you measured the orbital angular momentum squared (L^2), what values might you get, and what is the probability of each?
(b) Same for the z component of orbital angular momentum (L_z).
(c) Same for the spin angular momentum squared (S^2).
(d) Same for the z component of spin angular momentum (S_z).

Let $\mathbf{J} \equiv \mathbf{L} + \mathbf{S}$ be the *total* angular momentum.
(e) If you measured J^2, what values might you get, and what is the probability of each?
(f) Same for J_z.
(g) If you measured the *position* of the particle, what is the probability density for finding it at r, θ, ϕ?
(h) If you measured both the z component of the spin *and* the distance from the origin (note that these are compatible observables), what is the probability per unit r for finding the particle with spin up and at radius r?

** **Problem 4.65** If you combine *three* spin-1/2 particles, you can get a total spin of 3/2 or 1/2 (and the latter can be achieved in two distinct ways). Construct the quadruplet and the two doublets, using the notation of Equations 4.175 and 4.176:

$$\left\{\begin{array}{lcl} \left|\frac{3}{2}\,\frac{3}{2}\right\rangle & = & ?? \\ \left|\frac{3}{2}\,\frac{1}{2}\right\rangle & = & ?? \\ \left|\frac{3}{2}\,\frac{-1}{2}\right\rangle & = & ?? \\ \left|\frac{3}{2}\,\frac{-3}{2}\right\rangle & = & ?? \end{array}\right\} \quad s = \frac{3}{2} \;\text{(quadruplet)}$$

$$\left\{\begin{array}{lcl} \left|\frac{1}{2}\,\frac{1}{2}\right\rangle_1 & = & ?? \\ \left|\frac{1}{2}\,\frac{-1}{2}\right\rangle_1 & = & ?? \end{array}\right\} \quad s = \frac{1}{2} \;\text{(doublet 1)}$$

$$\left\{\begin{array}{lcl} \left|\frac{1}{2}\,\frac{1}{2}\right\rangle_2 & = & ?? \\ \left|\frac{1}{2}\,\frac{-1}{2}\right\rangle_2 & = & ?? \end{array}\right\} \quad s = \frac{1}{2} \;\text{(doublet 2)}$$

Hint: The first one is easy: $\left|\frac{3}{2}\,\frac{3}{2}\right\rangle = |\uparrow\uparrow\uparrow\rangle$; apply the lowering operator to get the other states in the quadruplet. For the doublets you might start with the first two in the singlet state, and tack on the third:

$$\left|\tfrac{1}{2}\,\tfrac{1}{2}\right\rangle_1 = \frac{1}{\sqrt{2}}\left(|\uparrow\downarrow\rangle - |\downarrow\uparrow\rangle\right)|\uparrow\rangle\,.$$

Take it from there (make sure $\left|\frac{1}{2}\,\frac{1}{2}\right\rangle_2$ is orthogonal to $\left|\frac{1}{2}\,\frac{1}{2}\right\rangle_1$ and to $\left|\frac{3}{2}\,\frac{1}{2}\right\rangle$). *Note:* the two doublets are not uniquely determined—any linear combination of them would still carry spin 1/2. The point is to construct two *independent* doublets.

Problem 4.66 Deduce the condition for minimum uncertainty in S_x and S_y (that is, *equality* in the expression $\sigma_{S_x}\sigma_{S_y} \geq (\hbar/2)\,|\langle S_z\rangle|$), for a particle of spin 1/2 in the generic state (Equation 4.139). *Answer:* With no loss of generality we can pick a to be real; then the condition for minimum uncertainty is that b is either pure real or else pure imaginary.

∗∗ **Problem 4.67 Magnetic frustration.** Consider three spin-1/2 particles arranged on the corners of a triangle and interacting via the Hamiltonian

$$H = J\left(\mathbf{S}_1 \cdot \mathbf{S}_2 + \mathbf{S}_2 \cdot \mathbf{S}_3 + \mathbf{S}_3 \cdot \mathbf{S}_1\right), \tag{4.227}$$

where J is a positive constant. This interaction favors opposite alignment of neighboring spins (**antiferromagnetism**, if they are magnetic dipoles), but the triangular arrangement means that this condition cannot be satisfied simultaneously for all three pairs (Figure 4.18). This is known as geometrical "frustration."

(a) Show that the Hamiltonian can be written in terms of the square of the total spin, S^2, where $\mathbf{S} = \sum_i \mathbf{S}_i$.

(b) Determine the ground state energy, and its degeneracy.

(c) Now consider *four* spin-1/2 particles arranged on the corners of a square, and interacting with their nearest neighbors:

$$H = J\left(\mathbf{S}_1 \cdot \mathbf{S}_2 + \mathbf{S}_2 \cdot \mathbf{S}_3 + \mathbf{S}_3 \cdot \mathbf{S}_4 + \mathbf{S}_4 \cdot \mathbf{S}_1\right). \tag{4.228}$$

In this case there is a unique ground state. Show that the Hamiltonian in this case can be written

$$H = \frac{1}{2}J\left[S^2 - (\mathbf{S}_1 + \mathbf{S}_3)^2 - (\mathbf{S}_2 + \mathbf{S}_4)^2\right]. \tag{4.229}$$

What is the ground state energy?

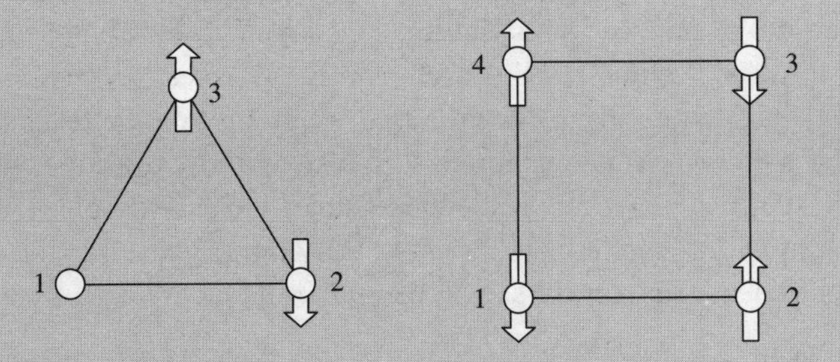

Figure 4.18: The figure shows three spins arranged around a triangle, where there is no way for each spin to be anti-aligned with all of its neighbors. In contrast, there is no such frustration with four spins arranged around a square.

** 

Problem 4.68 Imagine a hydrogen atom at the center of an infinite spherical well of radius b. We will take b to be much greater than the Bohr radius (a), so the low-n states are not much affected by the distant "wall" at $r = b$. But since $u(b) = 0$ we can use the method of Problem 2.61 to solve the radial equation (4.53) numerically.

(a) Show that v_j (in Problem 2.61) takes the form

$$v_j = -\frac{2\beta}{j} + \frac{\ell(\ell+1)}{j^2} \quad \text{where} \quad \beta \equiv \frac{b}{(N+1)\,a}.$$

(b) We want $\Delta r \ll a$ (so as to sample a reasonable number of points within the potential) and $a \ll b$ (so the wall doesn't distort the atom too much). Thus

$$1 \ll \beta^{-1} \ll N.$$

Let's use $\beta = 1/50$ and $N = 1000$. Find the three lowest eigenvalues of H, for $\ell = 0$, $\ell = 1$, and $\ell = 2$, and plot the corresponding eigenfunctions. Compare the known (Bohr) energies (Equation 4.70). *Note:* Unless the wave function drops to zero well before $r = b$, the energies of this system cannot be expected to match those of free hydrogen, but they are of interest in their own right as allowed energies of "compressed" hydrogen.[73]

** 

Problem 4.69 Find a few of the Bohr energies for hydrogen by "wagging the dog" (Problem 2.55), starting with Equation 4.53—or, better yet, Equation 4.56; in fact, why not use Equation 4.68 to set $\rho_0 = 2n$, and tweak n? We know that the correct solutions occur when n is a positive integer, so you might start with $n = 0.9, 1.9, 2.9$, etc., and increase it in small increments—the tail should wag when you pass $1, 2, 3, \ldots$. Find the lowest three ns, to four significant digits, first for $\ell = 0$, and then for $\ell = 1$ and $\ell = 2$. *Warning:* Mathematica doesn't like to divide by zero, so you might change ρ to $(\rho + 0.000001)$ in the denominator. *Note:* $u(0) = 0$ in all cases, but $u'(0) = 0$ only for $\ell \geq 1$ (Equation 4.59). So for $\ell = 0$ you can use $u(0) = 0$, $u'(0) = 1$. For $\ell > 0$ you might be tempted to use $u(0) = 0$ and $u'(0) = 0$, but Mathematica is lazy, and will go for the trivial solution $u(\rho) \equiv 0$; better, therefore, to use (say) $u(1) = 1$ and $u'(0) = 0$.

Problem 4.70 Sequential Spin Measurements.

(a) At time $t = 0$ a large ensemble of spin-1/2 particles is prepared, all of them in the spin-up state (with respect to the z axis).[74] They are not subject to any forces or torques. At time $t_1 > 0$ each spin is measured—some along the z direction and others along the x direction (but we aren't told the results). At time $t_2 > t_1$ their spin is measured again, this time along the x direction, and those with spin up (along x) are saved as a subensemble (those with spin down are discarded). *Question:* Of those remaining (the subensemble), what fraction had spin up (along z or x, depending on which was measured) in the first measurement?

---

[73] For a variety of reasons this system has been much studied in the literature. See, for example, J. M. Ferreyra and C. R. Proetto, *Am. J. Phys.* **81**, 860 (2013).

[74] N. D. Mermin, *Physics Today*, October 2011, page 8.

(b) Part (a) was easy—trivial, really, once you see it. Here's a more pithy generalization: At time $t = 0$ an ensemble of spin-1/2 particles is prepared, all in the spin-up state along direction $\boldsymbol{a}$. At time $t_1 > 0$ their spins are measured along direction $\boldsymbol{b}$ (but we are not told the results), and at time $t_2 > t_1$ their spins are measured along direction $\boldsymbol{c}$. Those with spin up (along $\boldsymbol{c}$) are saved as a subensemble. Of the particles in this subensemble, what fraction had spin up (along $\boldsymbol{b}$) in the first measurement? *Hint:* Use Equation 4.155 to show that the probability of getting spin up (along $\boldsymbol{b}$) in the first measurement is $P_+ = \cos^2(\theta_{ab}/2)$, and (by extension) the probability of getting spin up in *both* measurements is $P_{++} = \cos^2(\theta_{ab}/2)\cos^2(\theta_{bc}/2)$. Find the other three probabilities (P_{+-}, P_{-+}, and P_{--}). *Beware:* If the outcome of the first measurement was spin *down*, the relevant angle is now the *supplement* of θ_{bc}. *Answer:* $\left[1+\tan^2(\theta_{ab}/2)\tan^2(\theta_{bc}/2)\right]^{-1}$.

Problem 4.71 In molecular and solid-state applications, one often uses a basis of orbitals aligned with the cartesian axes rather than the basis $\psi_{n\ell m}$ used throughout this chapter. For example, the orbitals

$$\psi_{2p_x}(r,\theta,\phi) = \frac{1}{\sqrt{32\pi a^3}}\frac{x}{a}e^{-r/2a}$$
$$\psi_{2p_y}(r,\theta,\phi) = \frac{1}{\sqrt{32\pi a^3}}\frac{y}{a}e^{-r/2a}$$
$$\psi_{2p_z}(r,\theta,\phi) = \frac{1}{\sqrt{32\pi a^3}}\frac{z}{a}e^{-r/2a}$$

are a basis for the hydrogen states with $n = 2$ and $\ell = 1$.

(a) Show that each of these orbitals can be written as a linear combination of the orbitals $\psi_{n\ell m}$ with $n = 2$, $\ell = 1$, and $m = -1, 0, 1$.

(b) Show that the states ψ_{2p_i} are eigenstates of the corresponding component of angular momentum: $\hat{L}_i$. What is the eigenvalue in each case.

(c) Make contour plots (as in Figure 4.9) for the three orbitals. In Mathematica use **ContourPlot3D**.

Problem 4.72 Consider a particle with charge q, mass m, and spin s, in a uniform magnetic field $\mathbf{B}_0$. The vector potential can be chosen as

$$\mathbf{A} = -\frac{1}{2}\mathbf{r}\times\mathbf{B}_0 .$$

(a) Verify that this vector potential produces a uniform magnetic field $\mathbf{B}_0$.

(b) Show that the Hamiltonian can be written

$$H = \frac{p^2}{2m} + q\varphi - \mathbf{B}_0\cdot(\gamma_o\mathbf{L}+\gamma\mathbf{S}) + \frac{q^2}{8m}\left[r^2B_0^2 - (\mathbf{r}\cdot\mathbf{B}_0)^2\right], \tag{4.230}$$

where $\gamma_o = q/2m$ is the gyromagnetic ratio for orbital motion. *Note*: The term linear in $\mathbf{B}_0$ makes it energetically favorable for the magnetic moments (orbital and spin) to align with the magnetic field; this is the origin of **paramagnetism** in materials. The term quadratic in $\mathbf{B_0}$ leads to the opposite effect: **diamagnetism**.[75]

Problem 4.73 Example 4.4, couched in terms of forces, was a quasi-classical explanation for the Stern–Gerlach effect. Starting from the Hamiltonian for a neutral, spin-1/2 particle traveling through the magnetic field given by Equation 4.169,

$$H = \frac{p^2}{2m} - \gamma\, \mathbf{B} \cdot \mathbf{S},$$

use the generalized Ehrenfest theorem (Equation 3.73) to show that

$$m \frac{d^2}{dt^2} \langle z \rangle = \gamma\, \alpha\, \langle S_z \rangle .$$

Comment: Equation 4.170 is therefore a correct quantum-mechanical statement, with the understanding that the quantities refer to expectation values.

Problem 4.74 Neither Example 4.4 nor Problem 4.73 actually solved the Schrödinger equation for the Stern–Gerlach experiment. In this problem we will see how to set up that calculation. The Hamiltonian for a neutral, spin-1/2 particle traveling through a Stern–Gerlach device is

$$H = \frac{p^2}{2m} - \gamma\, \mathbf{B} \cdot \mathbf{S}$$

where $\mathbf{B}$ is given by Equation 4.169. The most general wave function for a spin-1/2 particle—including both spatial and spin degrees of freedom—is[76]

$$\mathbf{\Psi}(\mathbf{r}, t) = \Psi_+(\mathbf{r}, t)\, \chi_+ + \Psi_-(\mathbf{r}, t)\, \chi_- .$$

(a) Put $\mathbf{\Psi}(\mathbf{r}, t)$ into the Schrödinger equation

$$H\, \mathbf{\Psi} = i\, \hbar \frac{\partial}{\partial t} \mathbf{\Psi}$$

to obtain a pair of coupled equations for $\Psi_\pm$. *Partial answer:*

$$-\frac{\hbar^2}{2m} \nabla^2 \Psi_+ - \frac{\hbar}{2} \gamma\, (B_0 + \alpha\, z)\, \Psi_+ + \frac{\hbar}{2} \gamma\, \alpha\, x\, \Psi_- = i\, \hbar \frac{\partial}{\partial t} \Psi_+ .$$

(b) We know from Example 4.3 that the spin will precess in a uniform field $B_0\, \hat{k}$. We can factor this behavior out of our solution—with no loss of generality—by writing

$$\Psi_\pm(\mathbf{r}, t) = e^{\pm i\, \gamma\, B_0\, t/2}\, \tilde{\Psi}_\pm(\mathbf{r}, t) .$$

Find the coupled equations for $\tilde{\Psi}_\pm$. *Partial answer:*

$$-\frac{\hbar^2}{2m} \nabla^2 \tilde{\Psi}_+ - \frac{\hbar}{2} \gamma\, \alpha\, z\, \tilde{\Psi}_+ + \frac{\hbar}{2} \gamma\, \alpha\, x\, e^{-i\, \gamma\, B_0\, t}\, \tilde{\Psi}_- = i\, \hbar \frac{\partial}{\partial t} \tilde{\Psi}_+ .$$

[75] That's not obvious but we'll prove it in Chapter 7.

[76] In this notation, $|\Psi_+(\mathbf{r})|^2\, d^3\mathbf{r}$ gives the probability of finding the particle in the vicinity of $\mathbf{r}$ with spin up, and similarly measuring its spin along the z axis to be up, and similarly for $|\Psi_-(\mathbf{r})|^2\, d^3\mathbf{r}$ with spin down.

(c) If one ignores the oscillatory term in the solution to (b)—on the grounds that it averages to zero (see discussion in Example 4.4)—one obtains uncoupled equations of the form

$$-\frac{\hbar^2}{2m}\nabla^2\tilde{\Psi}_\pm + V_\pm\tilde{\Psi}_\pm = i\hbar\frac{\partial}{\partial t}\tilde{\Psi}_\pm.$$

Based upon the motion you would expect for a particle in the "potential" $V_\pm$, explain the Stern–Gerlach experiment.

Problem 4.75 Consider the system of Example 4.6, now with a time-dependent flux $\Phi(t)$ through the solenoid. Show that

$$\Psi(t) = \frac{1}{\sqrt{2\pi}}e^{in\phi}e^{-if(t)}$$

with

$$f(t) = \frac{1}{\hbar}\int_0^t \frac{\hbar^2}{2mb^2}\left(n - \frac{q\Phi(t')}{2\pi\hbar}\right)^2 dt'$$

is a solution to the *time-dependent* Schrödinger equation.

Problem 4.76 The shift in the energy levels in Example 4.6 can be understood from classical electrodynamics. Consider the case where initially no current flows in the solenoid. Now imagine slowly increasing the current.

(a) Calculate (from classical electrodynamics) the emf produced by the changing flux and show that the rate at which work is done on the charge confined to the ring can be written

$$\frac{dW}{d\Phi} = -q\frac{\omega}{2\pi},$$

where ω is the angular velocity of the particle.

(b) Calculate the z component of the mechanical angular momentum,[77]

$$\mathbf{L}_{\text{mechanical}} = \mathbf{r}\times m\mathbf{v} = \mathbf{r}\times(\mathbf{p} - q\mathbf{A}), \tag{4.231}$$

for a particle in the state ψ_n in Example 4.6. Note that the *mechanical* angular momentum is not quantized in integer multiples of $\hbar$![78]

(c) Show that your result from part (a) is precisely equal to the rate at which the stationary state energies change as the flux is increased: $dE_n/d\Phi$.

[77] See footnote 62 for a discussion of the difference between the canonical and mechanical momentum.

[78] However, the electromagnetic fields *also* carry angular momentum, and the *total* (mechanical plus electromagnetic) is quantized in integer multiples of $\hbar$. For a discussion see M. Peshkin, *Physics Reports* **80**, 375 (1981) or Chapter 1 of Frank Wilczek, *Fractional Statistics and Anyon Superconductivity*, World Scientific, New Jersey (1990).

导读 / Guidance

第 5 章　全同粒子

本章以双粒子体系为例，首先介绍玻色子和费米子的概念，给出双粒子体系玻色子和费米子的波函数的构造方法，引出泡利不相容原理，接下来讨论费米子和玻色子的交换作用，给出交换力的物理意义和共价键的本质；其次，以氦原子为例，通过计算，讨论其交换作用的大小；最后，把固体中的电子看成自由电子气模型，讨论固体物理的电子特性，给出费米面、费米能的概念，给出简并压。同时，考虑粒子运动受到周期性晶体势场的作用，求解固体中电子的薛定谔方程，给出其能带结构特征。本章对交换力的计算是本书的特色，它清晰地给出了交换力的物理根源和分子共价键的本质。这一点在国内其他教材上很少涉及。

习题特色

（1）拓展原子物理知识。如对 Li 原子的基态能级的计算，见习题 5.16；对洪特第一定则、洪特第二定则、洪特第三定则进行讨论，见习题 5.18；对相关泡利顺磁性的问题进行讨论，见习题 5.33。

（2）拓展固体物理知识。引入态密度的概念和定义并求解态密度的大小，给出范霍夫奇点的概念，见习题 5.37；对分数量子霍尔效应问题进行讨论，见习题 5.11；引入一维谐振子链模型和一维边界条件讨论相关问题，见习题 5.38、习题 5.39。

（3）拓展天体物理知识。对白矮星相关参数的估算和白矮星性质的讨论见习题 5.35；引入钱德拉塞卡极限的概念，讨论白矮星、中子星的形成条件，见习题 5.36。

5 IDENTICAL PARTICLES

5.1 TWO-PARTICLE SYSTEMS

For a *single* particle, $\Psi(\mathbf{r}, t)$ is a function of the spatial coordinates, $\mathbf{r}$, and the time, t (I'll ignore spin, for the moment). The state of a *two*-particle system is a function of the coordinates of particle one ($\mathbf{r}_1$), the coordinates of particle two ($\mathbf{r}_2$), and the time:

$$\Psi(\mathbf{r}_1, \mathbf{r}_2, t). \tag{5.1}$$

Its time evolution is determined by the Schrödinger equation:

$$i\hbar\frac{\partial \Psi}{\partial t} = \hat{H}\Psi, \tag{5.2}$$

where H is the Hamiltonian for the whole works:

$$\hat{H} = -\frac{\hbar^2}{2m_1}\nabla_1^2 - \frac{\hbar^2}{2m_2}\nabla_2^2 + V(\mathbf{r}_1, \mathbf{r}_2, t) \tag{5.3}$$

(the subscript on ∇ indicates differentiation with respect to the coordinates of particle 1 or particle 2, as the case may be). The statistical interpretation carries over in the obvious way:

$$|\Psi(\mathbf{r}_1, \mathbf{r}_2, t)|^2 \, d^3\mathbf{r}_1 \, d^3\mathbf{r}_2 \tag{5.4}$$

is the probability of finding particle 1 in the volume $d^3\mathbf{r}_1$ *and* particle 2 in the volume $d^3\mathbf{r}_2$; as always, Ψ must be normalized:

$$\int |\Psi(\mathbf{r}_1, \mathbf{r}_2, t)|^2 \, d^3\mathbf{r}_1 \, d^3\mathbf{r}_2 = 1. \tag{5.5}$$

For time-independent potentials, we obtain a complete set of solutions by separation of variables:

$$\Psi(\mathbf{r}_1, \mathbf{r}_2, t) = \psi(\mathbf{r}_1, \mathbf{r}_2)e^{-iEt/\hbar}, \tag{5.6}$$

where the spatial wave function (ψ) satisfies the time-independent Schrödinger equation:

$$-\frac{\hbar^2}{2m_1}\nabla_1^2\psi - \frac{\hbar^2}{2m_2}\nabla_2^2\psi + V\psi = E\psi, \tag{5.7}$$

and E is the total energy of the system. In general, solving Equation 5.7 is difficult, but two special cases can be reduced to one-particle problems:

1. **Noninteracting particles.** Suppose the particles do not interact with one another, but each is subject to some external force. For example, they might be attached to two different springs. In that case the total potential energy is the *sum* of the two:

$$V(\mathbf{r}_1, \mathbf{r}_2) = V_1(\mathbf{r}_1) + V_2(\mathbf{r}_2), \tag{5.8}$$

and Equation 5.7 can be solved by separation of variables:

$$\psi(\mathbf{r}_1, \mathbf{r}_2) = \psi_a(\mathbf{r}_1)\psi_b(\mathbf{r}_2). \tag{5.9}$$

Plugging Equation 5.9 into Equation 5.7, dividing by $\psi(\mathbf{r}_1, \mathbf{r}_2)$, and collecting the terms in $\mathbf{r}_1$ alone and in $\mathbf{r}_2$ alone, we find that $\psi_a(\mathbf{r}_1)$ and $\psi_b(\mathbf{r}_2)$ each satisfy the one-particle Schrödinger equation:

$$\begin{aligned} -\frac{\hbar^2}{2m_1}\nabla_1^2\psi_a(\mathbf{r}_1) + V_1(\mathbf{r}_1)\psi_a(\mathbf{r}_1) &= E_a\psi_a(\mathbf{r}_1), \\ -\frac{\hbar^2}{2m_2}\nabla_2^2\psi_b(\mathbf{r}_2) + V_2(\mathbf{r}_2)\psi_b(\mathbf{r}_2) &= E_b\psi_b(\mathbf{r}_2), \end{aligned} \tag{5.10}$$

and $E = E_a + E_b$. In this case the two-particle wave function is a simple *product* of one-particle wave functions,

$$\begin{aligned} \Psi(\mathbf{r}_1, \mathbf{r}_2, t) &= \psi_a(\mathbf{r}_1)\psi_b(\mathbf{r}_2)e^{-i(E_a+E_b)t/\hbar} \\ &= \left(\psi_a(\mathbf{r}_1)e^{-iE_at/\hbar}\right)\left(\psi_b(\mathbf{r}_2)e^{-iE_bt/\hbar}\right) = \Psi_a(\mathbf{r}_1, t)\Psi_b(\mathbf{r}_2, t), \end{aligned} \tag{5.11}$$

and it makes sense to say that particle 1 is in state a, and particle 2 is in state b. But any linear combination of such solutions will still satisfy the (time-dependent) Schrödinger equation—for instance

$$\Psi(\mathbf{r}_1, \mathbf{r}_2, t) = \frac{3}{5}\Psi_a(\mathbf{r}_1, t)\,\Psi_b(\mathbf{r}_2, t) + \frac{4}{5}\Psi_c(\mathbf{r}_1, t)\,\Psi_d(\mathbf{r}_2, t). \tag{5.12}$$

In this case the state of particle 1 depends on the state of particle 2, and vice versa. If you measured the energy of particle 1, you might get E_a (with probability 9/25), in which case the energy of particle 2 is definitely E_b, or you might get E_c (probability 16/25), in which case the energy of particle 2 is E_d. We say that the two particles are **entangled** (Schrödinger's lovely term). An entangled state is one that *cannot* be written as a product of single-particle states.[1]

2. **Central potentials.** Suppose the particles interact *only* with one another, via a potential that depends on their separation:

$$V(\mathbf{r}_1, \mathbf{r}_2) \to V(|\mathbf{r}_1 - \mathbf{r}_2|). \tag{5.13}$$

The hydrogen atom would be an example, if you include the motion of the proton. In this case the two-body problem reduces to an equivalent one-body problem, just as it does in *classical* mechanics (see Problem 5.1).

In general, though, the two particles will be subject both to external forces *and* to mutual interactions, and this makes the analysis more complicated. For example, think of the two electrons in a helium atom: each feels the Coulomb attraction of the nucleus (charge $2e$), and at the same time they repel one another:

$$V(\mathbf{r}_1, \mathbf{r}_2) = \frac{1}{4\pi\epsilon_0}\left(-\frac{2e^2}{|\mathbf{r}_1|} - \frac{2e^2}{|\mathbf{r}_2|} + \frac{e^2}{|\mathbf{r}_1 - \mathbf{r}_2|}\right). \tag{5.14}$$

We'll take up this problem in later sections.

[1] The classic example of an entangled state is two spin-1/2 particles in the singlet configuration (Equation 4.176).

** **Problem 5.1** A typical interaction potential depends only on the vector displacement $\mathbf{r} \equiv \mathbf{r}_1 - \mathbf{r}_2$ between the two particles: $V(\mathbf{r}_1, \mathbf{r}_2) \rightarrow V(\mathbf{r})$. In that case the Schrödinger equation separates, if we change variables from $\mathbf{r}_1$, $\mathbf{r}_2$ to $\mathbf{r}$ and $\mathbf{R} \equiv (m_1\mathbf{r}_1 + m_2\mathbf{r}_2) / (m_1 + m_2)$ (the center of mass).

(a) Show that $\mathbf{r}_1 = \mathbf{R} + (\mu/m_1)\,\mathbf{r}$, $\mathbf{r}_2 = \mathbf{R} - (\mu/m_2)\,\mathbf{r}$, and $\nabla_1 = (\mu/m_2)\,\nabla_R + \nabla_r$, $\nabla_2 = (\mu/m_1)\,\nabla_R - \nabla_r$, where

$$\mu \equiv \frac{m_1 m_2}{m_1 + m_2} \tag{5.15}$$

is the **reduced mass** of the system.

(b) Show that the (time-independent) Schrödinger equation (5.7) becomes

$$-\frac{\hbar^2}{2(m_1 + m_2)}\nabla_R^2\psi - \frac{\hbar^2}{2\mu}\nabla_r^2\psi + V(\mathbf{r})\psi = E\psi.$$

(c) Separate the variables, letting $\psi(\mathbf{R}, \mathbf{r}) = \psi_R(\mathbf{R})\psi_r(\mathbf{r})$. Note that ψ_R satisfies the one-particle Schrödinger equation, with the *total* mass $(m_1 + m_2)$ in place of m, potential zero, and energy E_R, while ψ_r satisfies the one-particle Schrödinger equation with the *reduced* mass in place of m, potential $V(\mathbf{r})$, and energy E_r. The total energy is the sum: $E = E_R + E_r$. What this tells us is that the center of mass moves like a free particle, and the *relative* motion (that is, the motion of particle 2 with respect to particle 1) is the same as if we had a *single* particle with the *reduced* mass, subject to the potential V. Exactly the same decomposition occurs in *classical* mechanics;[2] it reduces the two-body problem to an equivalent one-body problem.

Problem 5.2 In view of Problem 5.1, we can correct for the motion of the nucleus in hydrogen by simply replacing the electron mass with the reduced mass.

(a) Find (to two significant digits) the percent error in the binding energy of hydrogen (Equation 4.77) introduced by our use of m instead of μ.

(b) Find the separation in wavelength between the red Balmer lines ($n = 3 \rightarrow n = 2$) for hydrogen and deuterium (whose nucleus contains a neutron as well as the proton).

(c) Find the binding energy of **positronium** (in which the proton is replaced by a positron—positrons have the same mass as electrons, but opposite charge).

(d) Suppose you wanted to confirm the existence of **muonic hydrogen**, in which the electron is replaced by a muon (same charge, but 206.77 times heavier). Where (i.e. at what wavelength) would you look for the "Lyman-α" line ($n = 2 \rightarrow n = 1$)?

---

[2] See, for example, Jerry B. Marion and Stephen T. Thornton, *Classical Dynamics of Particles and Systems*, 4th edn, Saunders, Fort Worth, TX (1995), Section 8.2.

Problem 5.3 Chlorine has two naturally occurring isotopes, Cl^{35} and Cl^{37}. Show that the vibrational spectrum of HCl should consist of closely spaced doublets, with a splitting given by $\Delta\nu = 7.51 \times 10^{-4}\nu$, where ν is the frequency of the emitted photon. *Hint:* Think of it as a harmonic oscillator, with $\omega = \sqrt{k/\mu}$, where μ is the reduced mass (Equation 5.15) and k is presumably the same for both isotopes.

5.1.1 Bosons and Fermions

Suppose we have two noninteracting particles, number 1 in the (one-particle) state $\psi_a(\mathbf{r})$, and number 2 in the state $\psi_b(\mathbf{r})$. In that case $\psi(\mathbf{r}_1, \mathbf{r}_2)$ is the product (Equation 5.9):

$$\psi(\mathbf{r}_1, \mathbf{r}_2) = \psi_a(\mathbf{r}_1)\psi_b(\mathbf{r}_2). \tag{5.16}$$

Of course, this assumes that we can tell the particles apart—otherwise it wouldn't make any sense to claim that number 1 is in state ψ_a and number 2 is in state ψ_b; all we could say is that *one* of them is in the state ψ_a and the other is in state ψ_b, but we wouldn't know which is which. If we were talking *classical* mechanics this would be a silly objection: You can *always* tell the particles apart, in principle—just paint one of them red and the other one blue, or stamp identification numbers on them, or hire private detectives to follow them around. But in quantum mechanics the situation is fundamentally different: You can't paint an electron red, or pin a label on it, and a detective's observations will inevitably and unpredictably alter its state, raising the possibility that the two particles might have secretly switched places. The fact is, all electrons are *utterly identical*, in a way that no two classical objects can ever be. It's not just that *we* don't know which electron is which; *God* doesn't know which is which, because there is really *no such thing* as "this" electron, or "that" electron; all we can legitimately speak about is "an" electron.

Quantum mechanics neatly accommodates the existence of particles that are *indistinguishable in principle*: We simply construct a wave function that is *noncommittal* as to which particle is in which state. There are actually two ways to do it:

$$\psi_\pm(\mathbf{r}_1, \mathbf{r}_2) = A\left[\psi_a(\mathbf{r}_1)\psi_b(\mathbf{r}_2) \pm \psi_b(\mathbf{r}_1)\psi_a(\mathbf{r}_2)\right]; \tag{5.17}$$

the theory admits two kinds of identical particles: **bosons** (the plus sign), and **fermions** (the minus sign). Boson states are **symmetric** under interchange, $\psi_+(\mathbf{r}_2, \mathbf{r}_1) = \psi_+(\mathbf{r}_1, \mathbf{r}_2)$; fermion states are **antisymmetric** under interchange, $\psi_-(\mathbf{r}_2, \mathbf{r}_1) = -\psi_-(\mathbf{r}_1, \mathbf{r}_2)$. It so happens that

$$\left\{\begin{array}{l}\text{all particles with } \textit{integer} \text{ spin are bosons, and} \\ \text{all particles with } \textit{half integer} \text{ spin are fermions.}\end{array}\right. \tag{5.18}$$

This **connection between spin and statistics** (bosons and fermions have quite different statistical properties) can be *proved* in *relativistic* quantum mechanics; in the nonrelativistic theory it is simply taken as an axiom.[3]

[3] It seems strange that *relativity* should have anything to do with it, and there has been a lot of discussion as to whether it might be possible to prove the spin-statistics connection in other ways. See, for example, Robert C. Hilborn, *Am. J. Phys.* **63**, 298 (1995); Ian Duck and E. C. G. Sudarshan, *Pauli and the Spin-Statistics Theorem*, World Scientific, Singapore (1997). For a comprehensive bibliography on spin and statistics see C. Curceanu, J. D. Gillaspy, and R. C. Hilborn, *Am. J. Phys.* **80**, 561 (2010).

It follows, in particular, that *two identical fermions* (for example, two electrons) *cannot occupy the same state*. For if $\psi_a = \psi_b$, then

$$\psi_-(\mathbf{r}_1, \mathbf{r}_2) = A\left[\psi_a(\mathbf{r}_1)\psi_a(\mathbf{r}_2) - \psi_a(\mathbf{r}_1)\psi_a(\mathbf{r}_2)\right] = 0,$$

and we are left with no wave function at all.[4] This is the famous **Pauli exclusion principle**. It is not (as you may have been led to believe) a weird ad hoc assumption applying only to electrons, but rather a consequence of the rules for constructing two-particle wave functions, applying to *all* identical fermions.

Example 5.1

Suppose we have two noninteracting (they pass right through one another... never mind how you would set this up in practice!) particles, both of mass m, in the infinite square well (Section 2.2). The one-particle states are

$$\psi_n(x) = \sqrt{\frac{2}{a}}\sin\left(\frac{n\pi}{a}x\right), \quad E_n = n^2 K$$

(where $K \equiv \pi^2\hbar^2/2ma^2$). If the particles are *distinguishable*, with number 1 in state n_1 and number 2 in state n_2, the composite wave function is a simple product:

$$\psi_{n_1 n_2}(x_1, x_2) = \psi_{n_1}(x_1)\psi_{n_2}(x_2), \quad E_{n_1 n_2} = \left(n_1^2 + n_2^2\right) K.$$

For example, the ground state is

$$\psi_{11} = \frac{2}{a}\sin\left(\frac{\pi x_1}{a}\right)\sin\left(\frac{\pi x_2}{a}\right), \quad E_{11} = 2K;$$

the first excited state is doubly degenerate:

$$\psi_{12} = \frac{2}{a}\sin\left(\frac{\pi x_1}{a}\right)\sin\left(\frac{2\pi x_2}{a}\right), \quad E_{12} = 5K,$$

$$\psi_{21} = \frac{2}{a}\sin\left(\frac{2\pi x_1}{a}\right)\sin\left(\frac{\pi x_2}{a}\right), \quad E_{21} = 5K;$$

and so on. If the two particles are identical *bosons*, the ground state is unchanged, but the first excited state is *nondegenerate*:

$$\frac{\sqrt{2}}{a}\left[\sin\left(\frac{\pi x_1}{a}\right)\sin\left(\frac{2\pi x_2}{a}\right) + \sin\left(\frac{2\pi x_1}{a}\right)\sin\left(\frac{\pi x_2}{a}\right)\right]$$

(still with energy $5K$). And if the particles are identical *fermions*, there is *no* state with energy $2K$; the ground state is

$$\frac{\sqrt{2}}{a}\left[\sin\left(\frac{\pi x_1}{a}\right)\sin\left(\frac{2\pi x_2}{a}\right) - \sin\left(\frac{2\pi x_1}{a}\right)\sin\left(\frac{\pi x_2}{a}\right)\right],$$

and its energy is $5K$.

[4] I'm still leaving out the spin, don't forget—if this bothers you (after all, a spinless fermion is an oxymoron), assume they're in the *same* spin state. I'll show you how spin affects the story in Section 5.1.3

∗ **Problem 5.4**

(a) If ψ_a and ψ_b are orthogonal, and both are normalized, what is the constant A in Equation 5.17?

(b) If $\psi_a = \psi_b$ (and it is normalized), what is A? (This case, of course, occurs only for bosons.)

Problem 5.5

(a) Write down the Hamiltonian for two noninteracting identical particles in the infinite square well. Verify that the fermion ground state given in Example 5.1 is an eigenfunction of $\hat{H}$, with the appropriate eigenvalue.

(b) Find the next two excited states (beyond the ones given in the example)—wave functions, energies, and degeneracies—for each of the three cases (distinguishable, identical bosons, identical fermions).

5.1.2 Exchange Forces

To give you some sense of what the symmetrization requirement (Equation 5.17) actually *does*, I'm going to work out a simple one-dimensional example. Suppose one particle is in state $\psi_a(x)$, and the other is in state $\psi_b(x)$, and these two states are orthogonal and normalized. If the two particles are distinguishable, and number 1 is the one in state ψ_a, then the combined wave function is

$$\psi(x_1, x_2) = \psi_a(x_1)\psi_b(x_2); \tag{5.19}$$

if they are identical bosons, the composite wave function is (see Problem 5.4 for the normalization)

$$\psi_+(x_1, x_2) = \frac{1}{\sqrt{2}}\left[\psi_a(x_1)\psi_b(x_2) + \psi_b(x_1)\psi_a(x_2)\right]; \tag{5.20}$$

and if they are identical fermions, it is

$$\psi_-(x_1, x_2) = \frac{1}{\sqrt{2}}\left[\psi_a(x_1)\psi_b(x_2) - \psi_b(x_1)\psi_a(x_2)\right]. \tag{5.21}$$

Let's calculate the expectation value of the square of the separation distance between the two particles,

$$\left\langle (x_1 - x_2)^2 \right\rangle = \langle x_1^2\rangle + \langle x_2^2\rangle - 2\langle x_1 x_2\rangle. \tag{5.22}$$

Case 1: Distinguishable particles. For the wave function in Equation 5.19,

$$\langle x_1^2\rangle = \int x_1^2 \left|\psi_a(x_1)\right|^2 dx_1 \int \left|\psi_b(x_2)\right|^2 dx_2 = \left\langle x^2\right\rangle_a$$

(the expectation value of x^2 in the one-particle state ψ_a),

$$\langle x_2^2\rangle = \int \left|\psi_a(x_1)\right|^2 dx_1 \int x_2^2 \left|\psi_b(x_2)\right|^2 dx_2 = \left\langle x^2\right\rangle_b,$$

and

$$\langle x_1 x_2\rangle = \int x_1 \left|\psi_a(x_1)\right|^2 dx_1 \int x_2 \left|\psi_b(x_2)\right|^2 dx_2 = \langle x\rangle_a \langle x\rangle_b.$$

In this case, then,

$$\left\langle (x_1 - x_2)^2 \right\rangle_d = \left\langle x^2 \right\rangle_a + \left\langle x^2 \right\rangle_b - 2\langle x \rangle_a \langle x \rangle_b. \tag{5.23}$$

(Incidentally, the answer would—of course—be the same if particle 1 had been in state ψ_b, and particle 2 in state ψ_a.)

Case 2: Identical particles. For the wave functions in Equations 5.20 and 5.21,

$$\begin{aligned}
\left\langle x_1^2 \right\rangle = \frac{1}{2}\Bigg[& \int x_1^2 |\psi_a(x_1)|^2 \, dx_1 \int |\psi_b(x_2)|^2 \, dx_2 \\
& + \int x_1^2 |\psi_b(x_1)|^2 \, dx_1 \int |\psi_a(x_2)|^2 \, dx_2 \\
& \pm \int x_1^2 \psi_a(x_1)^* \psi_b(x_1) \, dx_1 \int \psi_b(x_2)^* \psi_a(x_2) \, dx_2 \\
& \pm \int x_1^2 \psi_b(x_1)^* \psi_a(x_1) \, dx_1 \int \psi_a(x_2)^* \psi_b(x_2) \, dx_2 \Bigg] \\
= \frac{1}{2}\Big[& \left\langle x^2 \right\rangle_a + \left\langle x^2 \right\rangle_b \pm 0 \pm 0 \Big] = \frac{1}{2}\left(\left\langle x^2 \right\rangle_a + \left\langle x^2 \right\rangle_b \right).
\end{aligned}$$

Similarly,

$$\left\langle x_2^2 \right\rangle = \frac{1}{2}\left(\left\langle x^2 \right\rangle_b + \left\langle x^2 \right\rangle_a \right).$$

(Naturally, $\left\langle x_2^2 \right\rangle = \left\langle x_1^2 \right\rangle$, since you can't tell them apart.) But

$$\begin{aligned}
\langle x_1 x_2 \rangle = \frac{1}{2}\Bigg[& \int x_1 |\psi_a(x_1)|^2 \, dx_1 \int x_2 |\psi_b(x_2)|^2 \, dx_2 \\
& + \int x_1 |\psi_b(x_1)|^2 \, dx_1 \int x_2 |\psi_a(x_2)|^2 \, dx_2 \\
& \pm \int x_1 \psi_a(x_1)^* \psi_b(x_1) \, dx_1 \int x_2 \psi_b(x_2)^* \psi_a(x_2) \, dx_2 \\
& \pm \int x_1 \psi_b(x_1)^* \psi_a(x_1) \, dx_1 \int x_2 \psi_a(x_2)^* \psi_b(x_2) \, dx_2 \Bigg] \\
= \frac{1}{2} & \left(\langle x \rangle_a \langle x \rangle_b + \langle x \rangle_b \langle x \rangle_a \pm \langle x \rangle_{ab} \langle x \rangle_{ba} \pm \langle x \rangle_{ba} \langle x \rangle_{ab} \right) \\
= \langle x & \rangle_a \langle x \rangle_b \pm |\langle x \rangle_{ab}|^2,
\end{aligned}$$

where

$$\langle x \rangle_{ab} \equiv \int x \psi_a(x)^* \psi_b(x) \, dx. \tag{5.24}$$

Thus

$$\left\langle (x_1 - x_2)^2 \right\rangle_{\pm} = \left\langle x^2 \right\rangle_a + \left\langle x^2 \right\rangle_b - 2\langle x \rangle_a \langle x \rangle_b \mp 2|\langle x \rangle_{ab}|^2. \tag{5.25}$$

Comparing Equations 5.23 and 5.25, we see that the difference resides in the final term:

$$\left\langle (\Delta x)^2 \right\rangle_{\pm} = \left\langle (\Delta x)^2 \right\rangle_d \mp 2 \left| \langle x \rangle_{ab} \right|^2; \tag{5.26}$$

identical bosons (the upper signs) tend to be somewhat closer together, and identical fermions (the lower signs) somewhat farther apart, than distinguishable particles in the same two states. Notice that $\langle x \rangle_{ab}$ *vanishes* unless the two wave functions actually *overlap*: if $\psi_a(x)$ is zero wherever $\psi_b(x)$ is *non*zero, the integral in Equation 5.24 is zero. So if ψ_a represents an electron in an atom in Chicago, and ψ_b represents an electron in an atom in Seattle, it's not going to make any difference whether you antisymmetrize the wave function or not. As a *practical* matter, therefore, it's okay to pretend that electrons with non-overlapping wave functions are distinguishable. (Indeed, this is the only thing that allows chemists to proceed at *all*, for in *principle* every electron in the universe is linked to every other one, via the antisymmetrization of their wave functions, and if this really *mattered*, you wouldn't be able to talk about any *one* unless you were prepared to deal with them *all!*)

The *interesting* case is when the overlap integral (Equation 5.24) is *not* zero. The system behaves as though there were a "force of attraction" between identical bosons, pulling them closer together, and a "force of repulsion" between identical fermions, pushing them apart (remember that we are for the moment ignoring spin). We call it an **exchange force**, although it's not really a force at all[5]—no physical agency is pushing on the particles; rather, it is a purely *geometrical* consequence of the symmetrization requirement. It is also a strictly quantum mechanical phenomenon, with no classical counterpart.

∗ **Problem 5.6** Imagine two noninteracting particles, each of mass m, in the infinite square well. If one is in the state ψ_n (Equation 2.31), and the other in state ψ_l ($l \neq n$), calculate $\langle (x_1 - x_2)^2 \rangle$, assuming (a) they are distinguishable particles, (b) they are identical bosons, and (c) they are identical fermions.

∗∗ **Problem 5.7** Two noninteracting particles (of equal mass) share the same harmonic oscillator potential, one in the ground state and one in the first excited state.

(a) Construct the wave function, $\psi(x_1, x_2)$, assuming (i) they are distinguishable, (ii) they are identical bosons, (iii) they are identical fermions. Plot $|\psi(x_1, x_2)|^2$ in each case (use, for instance, Mathematica's **Plot3D**).

(b) Use Equations 5.23 and 5.25 to determine $\langle (x_1 - x_2)^2 \rangle$ for each case.

(c) Express each $\psi(x_1, x_2)$ in terms of the relative and center-of-mass coordinates $r \equiv x_1 - x_2$ and $R \equiv (x_1 + x_2)/2$, and integrate over R to get the probability of finding the particles a distance $|r|$ apart:

$$P(|r|) = 2 \int |\psi(R, r)|^2 \, dR$$

(the 2 accounts for the fact that r could be positive or negative). Graph $P(r)$ for the three cases.

(d) Define the density operator by

$$n(x) = \sum_{i=1}^{2} \delta(x - x_i);$$

$\langle n(x) \rangle \, dx$ is the expected number of particles in the interval dx. Compute $\langle n(x) \rangle$ for each of the three cases and plot your results. (The result may surprise you.)

[5] For an incisive critique of this terminology see W. J. Mullin and G. Blaylock, *Am. J. Phys.* **71**, 1223 (2003).

Problem 5.8 Suppose you had *three* particles, one in state $\psi_a(x)$, one in state $\psi_b(x)$, and one in state $\psi_c(x)$. Assuming ψ_a, ψ_b, and ψ_c are orthonormal, construct the three-particle states (analogous to Equations 5.19, 5.20, and 5.21) representing (a) distinguishable particles, (b) identical bosons, and (c) identical fermions. Keep in mind that (b) must be completely symmetric, under interchange of *any* pair of particles, and (c) must be completely *anti*-symmetric, in the same sense. *Comment*: There's a cute trick for constructing completely antisymmetric wave functions: Form the **Slater determinant**, whose first row is $\psi_a(x_1)$, $\psi_b(x_1)$, $\psi_c(x_1)$, etc., whose second row is $\psi_a(x_2)$, $\psi_b(x_2)$, $\psi_c(x_2)$, etc., and so on (this device works for any number of particles).[6]

5.1.3 Spin

It is time to bring spin into the story. The *complete* state of an electron (say) includes not only its position wave function, but also a spinor, describing the orientation of its spin:[7]

$$\psi(\mathbf{r})\chi. \tag{5.27}$$

When we put together the two-particle state,[8]

$$\psi(\mathbf{r}_1, \mathbf{r}_2)\chi(1, 2) \tag{5.28}$$

it is the *whole works*, not just the spatial part, that has to be antisymmetric with respect to exchange:

$$\psi(\mathbf{r}_1, \mathbf{r}_2)\chi(1, 2) = -\psi(\mathbf{r}_2, \mathbf{r}_1)\chi(2, 1). \tag{5.29}$$

Now, a glance back at the composite spin states (Equations 4.175 and 4.176) reveals that the singlet combination is antisymmetric (and hence would have to be joined with a *symmetric* spatial function), whereas the three triplet states are all symmetric (and would require an *antisymmetric* spatial function). Thus the Pauli principle actually allows *two* electrons in a given position state, as long as their spins are in the singlet configuration (but they could not be in the same position state *and* in the same spin state—say, both spin up).

** **Problem 5.9** In Example 5.1 and Problem 5.5(b) we ignored spin (or, if you prefer, we assumed the particles are in the same spin state).

(a) Do it now for particles of spin 1/2. Construct the four lowest-energy configurations, and specify their energies and degeneracies. *Suggestion:* Use the notation $\psi_{n_1 n_2}\ |s\ m\rangle$, where $\psi_{n_1 n_2}$ is defined in Example 5.1 and $|s\ m\rangle$ in Section 4.4.3.[9]

---

[6] To construct a completely *symmetric* configuration, use the **permanent** (same as determinant, but without the minus signs).

[7] In the absence of coupling between spin and position, we are free to assume that the state is *separable* in its spin and spatial coordinates. This just says that the probability of getting spin up is independent of the *location* of the particle. In the *presence* of coupling, the general state would take the form of a linear combination: $\psi_+(\mathbf{r})\chi_+ + \psi_-(\mathbf{r})\chi_-$, as in Problem 4.64.

[8] I'll let $\chi(1, 2)$ stand for the combined spin state; in Dirac notation it is some linear combination of the states $|s_1\ s_2\ m_1\ m_2\rangle$. I assume that the state is again a simple product of a position state and a spin state; as you'll see in Problem 5.10, this is not always true when three or more electrons are involved—even in the absence of coupling.

[9] Of course, spin requires three dimensions, whereas we ordinarily think of the infinite square well as existing in one dimension. But it could represent a particle in three dimensions that is confined to a one-dimensional wire.

(b) Do the same for spin 1. *Hint:* First work out the spin-1 analogs to the spin-1/2 singlet and triplet configurations, using the Clebsch–Gordan coefficients; note which of them are symmetric and which antisymmetric.[10]

5.1.4 Generalized Symmetrization Principle

I have assumed, for the sake of simplicity, that the particles are noninteracting, the spin and position are decoupled (with the combined state a product of position and spin factors), and the potential is time-independent. But the fundamental symmetrization/antisymmetrization requirement for identical bosons/fermions is much more general. Let us define the **exchange operator**, $\hat{P}$, which interchanges the two particles:[11]

$$\hat{P}\,|(1,2)\rangle = |(2,1)\rangle. \tag{5.30}$$

Clearly, $\hat{P}^2 = 1$, and it follows (prove it for yourself) that the eigenvalues of $\hat{P}$ are ± 1. Now, if the two particles are identical, the Hamiltonian must treat them the same: $m_1 = m_2$ and $V(\mathbf{r}_1, \mathbf{r}_2, t) = V(\mathbf{r}_2, \mathbf{r}_1, t)$. It follows that $\hat{P}$ and $\hat{H}$ are compatible observables,

$$\left[\hat{P}, \hat{H}\right] = 0, \tag{5.31}$$

and hence (Equation 3.73)

$$\frac{d\left\langle \hat{P}\right\rangle}{dt} = 0. \tag{5.32}$$

If the system starts out in an eigenstate of $\hat{P}$—symmetric $\left(\langle\hat{P}\rangle = 1\right)$, or antisymmetric $\left(\langle\hat{P}\rangle = -1\right)$—then it will stay that way forever. The **symmetrization axiom** says that for identical particles the state is not merely *allowed*, but *required* to satisfy

$$|(1,2)\rangle = \pm|(2,1)\rangle, \tag{5.33}$$

with the plus sign for bosons, and the minus sign for fermions.[12] If you have n identical particles, of course, the state must be symmetric or antisymmetric under the interchange of *any two*:

$$\boxed{|(1,2,\ldots,i,\ldots,j,\ldots,n)\rangle = \pm|(1,2,\ldots,j,\ldots,i,\ldots,n)\rangle,} \tag{5.34}$$

This is the *general* statement, of which Equation 5.17 is a special case.

---

[10] This problem was suggested by Greg Elliott.

[11] $\hat{P}$ switches the particles ($1 \leftrightarrow 2$); this means exchanging their positions, their spins, and any other properties they might possess. If you like, it switches the *labels*, 1 and 2. I claimed (in Chapter 1) that all our operators would involve multiplication or differentiation; that was a lie. The exchange operator is an exception—and for that matter so is the projection operator (Section 3.6.2).

[12] It is sometimes alleged that the symmetrization requirement (Equation 5.33) is *forced* by the fact that $\hat{P}$ and $\hat{H}$ commute. This is false: It is perfectly possible to imagine a system of two *distinguishable* particles (say, an electron and a positron) for which the Hamiltonian is symmetric, and yet there is no requirement that the state be symmetric (or antisymmetric). But *identical* particles *have* to occupy symmetric or antisymmetric states, and this is a *new fundamental law*—on a par, logically, with Schrödinger's equation and the statistical interpretation. Of course, there didn't *have* to be any such things as identical particles; it could have been that every single particle in the universe was distinguishable from every other one. Quantum mechanics allows for the *possibility* of identical particles, and nature (being lazy) seized the opportunity. (But don't complain—this makes matters enormously simpler!)

** **Problem 5.10** For *two* spin-1/2 particles you can construct symmetric and antisymmetric spin states (the triplet and singlet combinations, respectively). For *three* spin-1/2 particles you can construct symmetric combinations (the quadruplet, in Problem 4.65), but no completely *anti*-symmetric configuration is possible.

(a) *Prove* it. *Hint:* The "bulldozer" method is to write down the most general linear combination:

$$\chi(1,2,3) = a\left|\uparrow\uparrow\uparrow\right\rangle + b\left|\uparrow\uparrow\downarrow\right\rangle + c\left|\uparrow\downarrow\uparrow\right\rangle + d\left|\uparrow\downarrow\downarrow\right\rangle + e\left|\downarrow\uparrow\uparrow\right\rangle + f\left|\downarrow\uparrow\downarrow\right\rangle + g\left|\downarrow\downarrow\uparrow\right\rangle + h\left|\downarrow\downarrow\downarrow\right\rangle.$$

What does antisymmetry under $1 \leftrightarrow 2$ tell you about the coefficients? (Note that the eight terms are mutually orthogonal.) Now invoke antisymmetry under $2 \leftrightarrow 3$.

(b) Suppose you put three identical noninteracting spin-1/2 particles in the infinite square well. What is the ground state for this system, what is its energy, and what is its degeneracy? *Note:* You can't put all three in the position state ψ_1 (why not?); you'll need two in ψ_1 and the other in ψ_2. But the symmetric configuration $[\psi_1(x_1)\psi_1(x_2)\psi_2(x_3) + \psi_1(x_1)\psi_2(x_2)\psi_1(x_3) + \psi_2(x_1)\psi_1(x_2)\psi_1(x_3)]$ is no good (because there's no antisymmetric spin combination to go with it), and you can't make a completely antisymmetric combination of those three terms. ... In this case you simply cannot construct an antisymmetric product of a spatial state and a spin state. But you *can* do it with an appropriate *linear combination* of such products. *Hint:* Form the Slater determinant (Problem 5.8) whose top row is $\psi_1(x_1)\left|\uparrow\right\rangle_1$, $\psi_1(x_1)\left|\downarrow\right\rangle_1$, $\psi_2(x_1)\left|\uparrow\right\rangle_1$.

(c) Show that your answer to part (b), properly normalized, can be written in the form

$$\Phi(1,2,3) = \frac{1}{\sqrt{3}}\left[\Phi(1,2)\,\phi(3) - \Phi(1,3)\,\phi(2) + \Phi(2,3)\,\phi(1)\right]$$

where $\Phi(i,j)$ is the wave function of two particles in the $n = 1$ state and the singlet spin configuration,

$$\Phi(i,j) = \psi_1(x_i)\,\psi_1(x_j)\,\frac{\left|\uparrow_i\downarrow_j\right\rangle - \left|\downarrow_i\uparrow_j\right\rangle}{\sqrt{2}}, \tag{5.35}$$

and $\phi(i)$ is the wave function of the ith particle in the $n = 2$ spin up state: $\phi(i) = \psi_2(x_i)\,\left|\uparrow_i\right\rangle$. Noting that $\Phi(i,j)$ is antisymmetric in $i \leftrightarrow j$, check that $\Phi(1,2,3)$ is antisymmetric in all three exchanges $(1 \leftrightarrow 2, 2 \leftrightarrow 3, \text{and } 3 \leftrightarrow 1)$.

** **Problem 5.11** In Section 5.1 we found that for *noninteracting* particles the energy eigenstates can be expressed as products of single-particle states (Equation 5.9)—or, for identical particles, as a symmetrized/antisymmetrized linear combination of such states (Equations 5.20 and 5.21). For *interacting* particles this is no longer the case. A famous example is the **Laughlin wave function**,[13] which is an approximation to the ground state

[13] "Robert B. Laughlin—Nobel Lecture: Fractional Quantization." Nobelprize.org. Nobel Media AB 2014. ⟨http://www.nobelprize.org/nobel_prizes/physics/laureates/1998/laughlin-lecture.html⟩.

of N electrons confined to two dimensions in a perpendicular magnetic field of strength B (the setting for the **fractional quantum Hall effect**). The Laughlin wave function is

$$\psi(z_1, z_2, \ldots, z_N) = A\left[\prod_{j<k}^{N}(z_j - z_k)^q\right]\exp\left[-\frac{1}{2}\sum_{k}^{N}|z_k|^2\right]$$

where q is a positive odd integer and

$$z_j \equiv \sqrt{\frac{e\,B}{2\,\hbar\,c}}\,(x_j + i\,y_j)\,.$$

(Spin is not at issue here; in the ground state all the electrons have spin down with respect to the direction of **B**, and that is a trivially symmetric configuration.)

(a) Show that ψ has the proper antisymmetry for fermions.

(b) For $q = 1$, ψ describes noninteracting particles (by which I mean that it can be written as a single Slater determinant—see Problem 5.8). This is true for any N, but check it explicitly for $N = 3$. What single particle states are occupied in this case?

(c) For values of q greater than 1, ψ *cannot* be written as a single Slater determinant, and describes interacting particles (in practice, Coulomb repulsion of the electrons). It can, however, be written as a *sum* of Slater determinants. Show that, for $q = 3$ and $N = 2$, ψ can be written as a sum of two Slater determinants.

Comment: In the noninteracting case (b) we can describe the wave function as "three particles occupying the three single-particle states ψ_a, ψ_b and ψ_c," but in the interacting case (c), no corresponding statement can be made; in that case, the different Slater determinants that make up ψ correspond to occupation of different sets of single-particle states.

5.2 ATOMS

A neutral atom, of atomic number Z, consists of a heavy nucleus, with electric charge Ze, surrounded by Z electrons (mass m and charge $-e$). The Hamiltonian for this system is[14]

$$\hat{H} = \sum_{j=1}^{Z}\left\{-\frac{\hbar^2}{2m}\nabla_j^2 - \left(\frac{1}{4\pi\epsilon_0}\right)\frac{Ze^2}{r_j}\right\} + \frac{1}{2}\left(\frac{1}{4\pi\epsilon_0}\right)\sum_{j\neq k}^{Z}\frac{e^2}{|\mathbf{r}_j - \mathbf{r}_k|}. \tag{5.36}$$

The term in curly brackets represents the kinetic plus potential energy of the jth electron, in the electric field of the nucleus; the second sum (which runs over all values of j and k except $j = k$) is the potential energy associated with the mutual repulsion of the electrons (the factor

[14] I'm assuming the nucleus is *stationary*. The trick of accounting for nuclear motion by using the reduced mass (Problem 5.1) works only for the *two*-body problem; fortunately, the nucleus is so much heavier than the electrons that the correction is extremely small even in the case of hydrogen (see Problem 5.2(a)), and it is smaller still for other atoms. There are more interesting effects, due to magnetic interactions associated with electron spin, relativistic corrections, and the finite size of the nucleus. We'll look into these in later chapters, but all of them are minute corrections to the "purely coulombic" atom described by Equation 5.36.

of 1/2 in front corrects for the fact that the summation counts each pair twice). The problem is to solve Schrödinger's equation,

$$\hat{H}\psi = E\psi, \tag{5.37}$$

for the wave function $\psi(\mathbf{r}_1, \mathbf{r}_2, \ldots, \mathbf{r}_Z)$.[15]

Unfortunately, the Schrödinger equation with Hamiltonian in Equation 5.36 cannot be solved exactly (at any rate, it *hasn't* been), except for the very simplest case, $Z = 1$ (hydrogen). In practice, one must resort to elaborate approximation methods. Some of these we shall explore in Part II; for now I plan only to sketch some qualitative features of the solutions, obtained by neglecting the electron repulsion term altogether. In Section 5.2.1 we'll study the ground state and excited states of helium, and in Section 5.2.2 we'll examine the ground states of higher atoms.

Problem 5.12

(a) Suppose you could find a solution $\psi(\mathbf{r}_1, \mathbf{r}_2, \ldots, \mathbf{r}_Z)$ to the Schrödinger equation (Equation 5.37), for the Hamiltonian in Equation 5.36. Describe how you would construct from it a completely symmetric function, and a completely antisymmetric function, which also satisfy the Schrödinger equation, with the same energy. What happens to the completely antisymmetric function if $\psi(\mathbf{r}_1, \mathbf{r}_2, \ldots, \mathbf{r}_Z)$ is symmetric in (say) its first two arguments ($\mathbf{r}_1 \leftrightarrow \mathbf{r}_2$)?

(b) By the same logic, show that a completely antisymmetric spin state for Z electrons is impossible, if $Z > 2$ (this generalizes Problem 5.10(a)).

5.2.1 Helium

After hydrogen, the simplest atom is helium ($Z = 2$). The Hamiltonian,

$$\hat{H} = \left\{ -\frac{\hbar^2}{2m}\nabla_1^2 - \frac{1}{4\pi\epsilon_0}\frac{2e^2}{r_1} \right\} + \left\{ -\frac{\hbar^2}{2m}\nabla_2^2 - \frac{1}{4\pi\epsilon_0}\frac{2e^2}{r_2} \right\} + \frac{1}{4\pi\epsilon_0}\frac{e^2}{|\mathbf{r}_1 - \mathbf{r}_2|}, \tag{5.38}$$

consists of two hydrogenic Hamiltonians (with nuclear charge $2e$), one for electron 1 and one for electron 2, together with a final term describing the repulsion of the two electrons. It is this last term that causes all the trouble. If we simply *ignore* it, the Schrödinger equation separates, and the solutions can be written as products of *hydrogen* wave functions:

$$\psi(\mathbf{r}_1, \mathbf{r}_2) = \psi_{n\ell m}(\mathbf{r}_1)\psi_{n'\ell'm'}(\mathbf{r}_2), \tag{5.39}$$

only with half the Bohr radius (Equation 4.72), and four times the Bohr energies (Equation 4.70)—if you don't see why, refer back to Problem 4.19. The total energy would be

$$E = 4(E_n + E_{n'}), \tag{5.40}$$

where $E_n = -13.6/n^2$ eV. In particular, the ground state would be

$$\psi_0(\mathbf{r}_1, \mathbf{r}_2) = \psi_{100}(\mathbf{r}_1)\psi_{100}(\mathbf{r}_2) = \frac{8}{\pi a^3}e^{-2(r_1+r_2)/a}, \tag{5.41}$$

[15] Because the Hamiltonian (5.36) makes no reference to spin, the product $\psi(\mathbf{r}_1, \mathbf{r}_2, \ldots, \mathbf{r}_Z)\chi(\mathbf{s}_1, \mathbf{s}_2, \ldots, \mathbf{s}_Z)$ still satisfies the Schrödinger equation. However, for $Z > 2$ such product states cannot in general meet the (anti-)symmetrization requirement, and it is necessary to construct linear combinations, with permuted indices (see Problem 5.16). But that comes at the end of the story; for the moment we are only concerned with the spatial wave function.

(Equation 4.80), and its energy would be

$$E_0 = 8\,(-13.6\,\text{eV}) = -109\,\text{eV}. \tag{5.42}$$

Because ψ_0 is a symmetric function, the spin state has to be *antisymmetric*, so the ground state of helium should be a *singlet* configuration, with the spins "oppositely aligned." The *actual* ground state of helium is indeed a singlet, but the experimentally determined energy is -78.975 eV, so the agreement is not very good. But this is hardly surprising: We ignored electron–electron repulsion, which is certainly *not* a small contribution. It is clearly *positive* (see Equation 5.38), which is comforting—evidently it brings the total energy up from -109 to -79 eV (see Problem 5.15).

The excited states of helium consist of one electron in the hydrogenic ground state, and the other in an excited state:

$$\psi_{n\ell m}\psi_{100}. \tag{5.43}$$

(If you try to put *both* electrons in excited states, one immediately drops to the ground state, releasing enough energy to knock the other one into the continuum ($E > 0$), leaving you with a helium *ion* (He^+) and a free electron. This is an interesting system in its own right—see Problem 5.13—but it is not our present concern.) We can construct both symmetric and antisymmetric combinations, in the usual way (Equation 5.17); the former go with the *antisymmetric* spin configuration (the singlet)—they are called **parahelium**—while the latter require a *symmetric* spin configuration (the triplet)—they are known as **orthohelium**. The ground state is necessarily parahelium; the excited states come in both forms. Because the symmetric spatial state brings the electrons closer together (as we discovered in Section 5.1.2), we expect a higher interaction energy in parahelium, and indeed, it is experimentally confirmed that the parahelium states have somewhat higher energy than their orthohelium counterparts (see Figure 5.1).

Problem 5.13

(a) Suppose you put both electrons in a helium atom into the $n = 2$ state; what would the energy of the emitted electron be? (Assume no photons are emitted in the process.)

(b) Describe (quantitatively) the spectrum of the helium ion, He^+. That is, state the "Rydberg-like" formula for the emitted wavelengths.

Problem 5.14 Discuss (qualitatively) the energy level scheme for helium if (a) electrons were identical bosons, and (b) if electrons were distinguishable particles (but with the same mass and charge). Pretend these "electrons" still have spin 1/2, so the spin configurations are the singlet and the triplet.

** **Problem 5.15**

(a) Calculate $\langle(1/|\mathbf{r}_1 - \mathbf{r}_2|)\rangle$ for the state ψ_0 (Equation 5.41). *Hint*: Do the $d^3\mathbf{r}_2$ integral first, using spherical coordinates, and setting the polar axis along $\mathbf{r}_1$, so that

$$|\mathbf{r}_1 - \mathbf{r}_2| = \sqrt{r_1^2 + r_2^2 - 2r_1r_2\cos\theta_2}.$$

The θ_2 integral is easy, but be careful to take the *positive root*. You'll have to break the r_2 integral into two pieces, one ranging from 0 to r_1, the other from r_1 to ∞. *Answer:* $5/4a$.

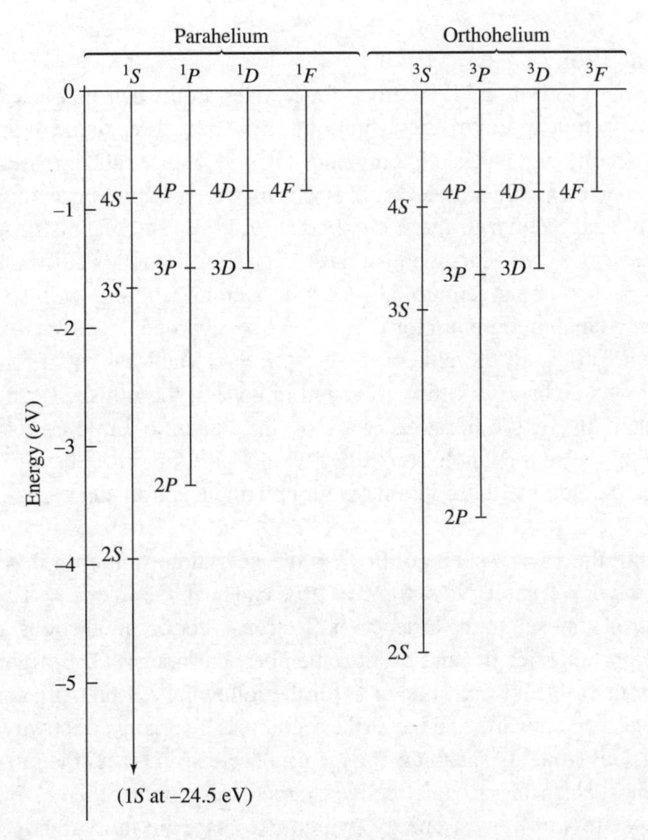

Figure 5.1: Energy level diagram for helium (the notation is explained in Section 5.2.2). Note that parahelium energies are uniformly higher than their orthohelium counterparts. The numerical values on the vertical scale are relative to the ground state of ionized helium (He^+): $4 \times (-13.6)$ eV $= -54.4$ eV; to get the *total* energy of the state, subtract 54.4 eV.

(b) Use your result in (a) to estimate the electron interaction energy in the ground state of helium. Express your answer in electron volts, and add it to E_0 (Equation 5.42) to get a corrected estimate of the ground state energy. Compare the experimental value. (Of course, we're still working with an approximate wave function, so don't expect *perfect* agreement.)

Problem 5.16 The ground state of lithium. Ignoring electron–electron repulsion, construct the ground state of lithium ($Z = 3$). Start with a spatial wave function, analogous to Equation 5.41, but remember that only two electrons can occupy the hydrogenic ground state; the third goes to ψ_{200}.[16] What is the energy of this state? Now tack on the spin, and antisymmetrize (if you get stuck, refer back to Problem 5.10). What's the degeneracy of the ground state?

[16] Actually, $\ell = 1$ would do just as well, but electron–electron repulsion favors $\ell = 0$, as we shall see.

5.2.2 The Periodic Table

The ground state electron configurations for heavier atoms can be pieced together in much the same way. To first approximation (ignoring their mutual repulsion altogether) the individual electrons occupy one-particle hydrogenic states (n, ℓ, m), called **orbitals**, in the Coulomb potential of a nucleus with charge Ze. If electrons were bosons (or distinguishable particles) they would all shake down to the ground state $(1, 0, 0)$, and chemistry would be very dull indeed. But electrons are in fact identical fermions, subject to the Pauli exclusion principle, so only *two* can occupy any given orbital (one with spin up, and one with spin down—or, more precisely, in the singlet configuration). There are n^2 hydrogenic wave functions (all with the same energy E_n) for a given value of n, so the $n = 1$ **shell** has room for two electrons, the $n = 2$ shell holds eight, $n = 3$ takes 18, and in general the nth shell can accommodate $2n^2$ electrons. Qualitatively, the horizonal rows on the **Periodic Table** correspond to filling out each shell (if this were the whole story, they would have lengths 2, 8, 18, 32, 50, etc., instead of 2, 8, 8, 18, 18, etc.; we'll see in a moment how the electron–electron repulsion throws the counting off).

With helium, the $n = 1$ shell is filled, so the next atom, lithium $(Z = 3)$, has to put one electron into the $n = 2$ shell. Now, for $n = 2$ we can have $\ell = 0$ or $\ell = 1$; which of these will the third electron choose? In the absence of electron–electron interactions, they have the same energy (the Bohr energies depend on n, remember, but not on ℓ). But the effect of electron repulsion is to favor the lowest value of ℓ, for the following reason. Angular momentum tends to throw the electron outward, and the farther out it gets, the more effectively the inner electrons **screen** the nucleus (roughly speaking, the innermost electron "sees" the full nuclear charge Ze, but the outermost electron sees an effective charge hardly greater than e). Within a given shell, therefore, the state with lowest energy (which is to say, the most tightly bound electron) is $\ell = 0$, and the energy increases with increasing ℓ. Thus the third electron in lithium occupies the orbital (2,0,0).[17] The next atom (beryllium, with $Z = 4$) also fits into this state (only with "opposite spin"), but boron $(Z = 5)$ has to make use of $\ell = 1$.

Continuing in this way, we reach neon $(Z = 10)$, at which point the $n = 2$ shell is filled, and we advance to the next row of the periodic table and begin to populate the $n = 3$ shell. First there are two atoms (sodium and magnesium) with $\ell = 0$, and then there are six with $\ell = 1$ (aluminum through argon). Following argon there "should" be 10 atoms with $n = 3$ and $\ell = 2$; however, by this time the screening effect is so strong that it overlaps the next shell; potassium $(Z = 19)$ and calcium $(Z = 20)$ choose $n = 4$, $\ell = 0$, in preference to $n = 3$, $\ell = 2$. After that we drop back to pick up the $n = 3$, $\ell = 2$ stragglers (scandium through zinc), followed by $n = 4$, $\ell = 1$ (gallium through krypton), at which point we again make a premature jump to the next row $(n = 5)$, and wait until later to slip in the $\ell = 2$ and $\ell = 3$ orbitals from the $n = 4$ shell. For details of this intricate counterpoint I refer you to any book on atomic physics.[18]

I would be delinquent if I failed to mention the archaic nomenclature for atomic states, because all chemists and most physicists use it (and the people who make up the Graduate Record Exam *love* this sort of thing). For reasons known best to nineteenth-century spectroscopists, $\ell = 0$ is called s (for "sharp"), $\ell = 1$ is p (for "principal"), $\ell = 2$ is d ("diffuse"), and $\ell = 3$ is f ("fundamental"); after that I guess they ran out of imagination, because it now

[17] This standard argument has been called into question by W. Stacey and F. Marsiglio, *EPL*, **100**, 43002 (2012).

[18] See, for example, Ugo Fano and L. Fano, *Basic Physics of Atoms and Molecules*, Wiley, New York (1959), Chapter 18, or the classic by Gerhard Herzberg, *Atomic Spectra and Atomic Structure*, Dover, New York (1944).

continues alphabetically (g, h, i, skip j, just to be utterly perverse, k, l, etc.).[19] The state of a particular electron is represented by the pair $n\ell$, with n (the number) giving the shell, and ℓ (the letter) specifying the orbital angular momentum; the magnetic quantum number m is not listed, but an exponent is used to indicate the number of electrons that occupy the state in question. Thus the configuration

$$(1s)^2\,(2s)^2\,(2p)^2 \tag{5.44}$$

tells us that there are two electrons in the orbital (1,0,0), two in the orbital (2,0,0), and two in some combination of the orbitals (2,1,1), (2,1,0), and (2,1,−1). This happens to be the ground state of carbon.

In that example there are two electrons with orbital angular momentum quantum number 1, so the *total* orbital angular momentum quantum number, L (capital L—not to be confused with the L denoting $n = 2$—instead of ℓ, to indicate that this pertains to the *total*, not to any one particle) could be 2, 1, or 0. Meanwhile, the two $(1s)$ electrons are locked together in the singlet state, with total spin zero, and so are the two $(2s)$ electrons, but the two $(2p)$ electrons could be in the singlet configuration or the triplet configuration. So the *total* spin quantum number S (capital, again, because it's the *total*) could be 1 or 0. Evidently the *grand* total (orbital plus spin), J, could be 3, 2, 1, or 0 (Equation 4.182). There exist rituals, known as **Hund's Rules** (see Problem 5.18) for figuring out what these totals will be, for a particular atom. The result is recorded as the following hieroglyphic:

$$^{2S+1}L_J, \tag{5.45}$$

(where S and J are the numbers, and L the letter—capitalized, because we're talking about the *totals*). The ground state of carbon happens to be $^3P_0$: the total spin is 1 (hence the 3), the total orbital angular momentum is 1 (hence the P), and the *grand* total angular momentum is zero (hence the 0). In Table 5.1 the individual configurations and the total angular momenta (in the notation of Equation 5.45) are listed, for the first four rows of the Periodic Table.[20]

* **Problem 5.17**

(a) Figure out the electron configurations (in the notation of Equation 5.44) for the first two rows of the Periodic Table (up to neon), and check your results against Table 5.1.

Table 5.1 *Ground-state electron configurations for the first four rows of the Periodic Table.*

Z	Element	Configuration	
1	H	$(1s)$	$^2S_{1/2}$
2	He	$(1s)^2$	$^1S_0$
3	Li	(He)$(2s)$	$^2S_{1/2}$
4	Be	(He)$(2s)^2$	$^1S_0$
5	B	(He)$(2s)^2(2p)$	$^2P_{1/2}$
6	C	(He)$(2s)^2(2p)^2$	$^3P_0$

[19] The shells themselves are assigned equally arbitrary nicknames, starting (don't ask me why) with K: The K shell is $n = 1$, the L shell is $n = 2$, M is $n = 3$, and so on (at least they're in alphabetical order).

[20] After krypton—element 36—the situation gets more complicated (fine structure starts to play a significant role in the ordering of the states) so it is not for want of space that the table terminates there.

7	N	$(\text{He})(2s)^2(2p)^3$	$^4S_{3/2}$
8	O	$(\text{He})(2s)^2(2p)^4$	$^3P_2$
9	F	$(\text{He})(2s)^2(2p)^5$	$^2P_{3/2}$
10	Ne	$(\text{He})(2s)^2(2p)^6$	$^1S_0$
11	Na	$(\text{Ne})(3s)$	$^2S_{1/2}$
12	Mg	$(\text{Ne})(3s)^2$	$^1S_0$
13	Al	$(\text{Ne})(3s)^2(3p)$	$^2P_{1/2}$
14	Si	$(\text{Ne})(3s)^2(3p)^2$	$^3P_0$
15	P	$(\text{Ne})(3s)^2(3p)^3$	$^4S_{3/2}$
16	S	$(\text{Ne})(3s)^2(3p)^4$	$^3P_2$
17	Cl	$(\text{Ne})(3s)^2(3p)^5$	$^2P_{3/2}$
18	Ar	$(\text{Ne})(3s)^2(3p)^6$	$^1S_0$
19	K	$(\text{Ar})(4s)$	$^2S_{1/2}$
20	Ca	$(\text{Ar})(4s)^2$	$^1S_0$
21	Sc	$(\text{Ar})(4s)^2(3d)$	$^2D_{3/2}$
22	Ti	$(\text{Ar})(4s)^2(3d)^2$	$^3F_2$
23	V	$(\text{Ar})(4s)^2(3d)^3$	$^4F_{3/2}$
24	Cr	$(\text{Ar})(4s)(3d)^5$	$^7S_3$
25	Mn	$(\text{Ar})(4s)^2(3d)^5$	$^6S_{5/2}$
26	Fe	$(\text{Ar})(4s)^2(3d)^6$	$^5D_4$
27	Co	$(\text{Ar})(4s)^2(3d)^7$	$^4F_{9/2}$
28	Ni	$(\text{Ar})(4s)^2(3d)^8$	$^3F_4$
29	Cu	$(\text{Ar})(4s)(3d)^{10}$	$^2S_{1/2}$
30	Zn	$(\text{Ar})(4s)^2(3d)^{10}$	$^1S_0$
31	Ga	$(\text{Ar})(4s)^2(3d)^{10}(4p)$	$^2P_{1/2}$
32	Ge	$(\text{Ar})(4s)^2(3d)^{10}(4p)^2$	$^3P_0$
33	As	$(\text{Ar})(4s)^2(3d)^{10}(4p)^3$	$^4S_{3/2}$
34	Se	$(\text{Ar})(4s)^2(3d)^{10}(4p)^4$	$^3P_2$
35	Br	$(\text{Ar})(4s)^2(3d)^{10}(4p)^5$	$^2P_{3/2}$
36	Kr	$(\text{Ar})(4s)^2(3d)^{10}(4p)^6$	$^1S_0$

(b) Figure out the corresponding total angular momenta, in the notation of Equation 5.45, for the first four elements. List all the *possibilities* for boron, carbon, and nitrogen.

∗∗ **Problem 5.18**

(a) **Hund's first rule** says that, consistent with the Pauli principle, the state with the highest total spin (S) will have the lowest energy. What would this predict in the case of the excited states of helium?

(b) **Hund's second rule** says that, for a given spin, the state with the highest total orbital angular momentum (L), consistent with overall antisymmetrization, will have the lowest energy. Why doesn't carbon have $L = 2$? *Hint:* Note that the "top of the ladder" ($M_L = L$) is symmetric.

(c) **Hund's third rule** says that if a subshell (n, ℓ) is no more than half filled, then the lowest energy level has $J = |L - S|$; if it is more than half filled, then $J = L + S$ has the lowest energy. Use this to resolve the boron ambiguity in Problem 5.17(b).

(d) Use Hund's rules, together with the fact that a symmetric spin state must go with an antisymmetric position state (and vice versa) to resolve the carbon and nitrogen ambiguities in Problem 5.17(b). *Hint:* Always go to the "top of the ladder" to figure out the symmetry of a state.

Problem 5.19 The ground state of dysprosium (element 66, in the 6th row of the Periodic Table) is listed as $^5I_8$. What are the total spin, total orbital, and grand total angular momentum quantum numbers? Suggest a likely electron configuration for dysprosium.

5.3 SOLIDS

In the solid state, a few of the loosely-bound outermost **valence** electrons in each atom become detached, and roam around throughout the material, no longer subject only to the Coulomb field of a specific "parent" nucleus, but rather to the combined potential of the entire crystal lattice. In this section we will examine two extremely primitive models: first, the "electron gas" theory of Sommerfeld, which ignores *all* forces (except the confining boundaries), treating the wandering electrons as free particles in a box (the three-dimensional analog to an infinite square well); and second, Bloch's theory, which introduces a periodic potential representing the electrical attraction of the regularly spaced, positively charged, nuclei (but still ignores electron–electron repulsion). These models are no more than the first halting steps toward a quantum theory of solids, but already they reveal the critical role of the Pauli exclusion principle in accounting for "solidity," and provide illuminating insight into the remarkable electrical properties of conductors, semi-conductors, and insulators.

5.3.1 The Free Electron Gas

Suppose the object in question is a rectangular solid, with dimensions l_x, l_y, l_z, and imagine that an electron inside experiences no forces at all, except at the impenetrable walls:

$$V(x, y, z) = \begin{cases} 0, & 0 < x < l_x, 0 < y < l_y, 0 < z < l_z; \\ \infty, & \text{otherwise.} \end{cases} \tag{5.46}$$

The Schrödinger equation,

$$-\frac{\hbar^2}{2m}\nabla^2\psi = E\psi,$$

separates, in Cartesian coordinates: $\psi(x, y, z) = X(x)Y(y)Z(z)$, with

$$-\frac{\hbar^2}{2m}\frac{d^2X}{dx^2} = E_x X, \quad -\frac{\hbar^2}{2m}\frac{d^2Y}{dy^2} = E_y Y, \quad -\frac{\hbar^2}{2m}\frac{d^2Z}{dz^2} = E_z Z,$$

and $E = E_x + E_y + E_z$. Letting

$$k_x \equiv \frac{\sqrt{2mE_x}}{\hbar}, \quad k_y \equiv \frac{\sqrt{2mE_y}}{\hbar}, \quad k_z \equiv \frac{\sqrt{2mE_z}}{\hbar},$$

we obtain the general solutions

$$X(x) = A_x \sin(k_x x) + B_x \cos(k_x x), \; Y(y) = A_y \sin(k_y y) + B_y \cos(k_y y),$$

$$Z(z) = A_z \sin(k_z z) + B_z \cos(k_z z).$$

The boundary conditions require that $X(0) = Y(0) = Z(0) = 0$, so $B_x = B_y = B_z = 0$, and $X(l_x) = Y(l_y) = Z(l_z) = 0$, so

$$k_x l_x = n_x \pi, \quad k_y l_y = n_y \pi, \quad k_z l_z = n_z \pi, \tag{5.47}$$

where each n is a positive integer:

$$n_x = 1,\ 2,\ 3,\ \ldots, \quad n_y = 1,\ 2,\ 3,\ \ldots, \quad n_z = 1,\ 2,\ 3,\ \ldots. \tag{5.48}$$

The (normalized) wave functions are

$$\psi_{n_x n_y n_z} = \sqrt{\frac{8}{l_x l_y l_z}} \sin\left(\frac{n_x \pi}{l_x} x\right) \sin\left(\frac{n_y \pi}{l_y} y\right) \sin\left(\frac{n_z \pi}{l_z} z\right), \tag{5.49}$$

and the allowed energies are

$$E_{n_x n_y n_z} = \frac{\hbar^2 \pi^2}{2m}\left(\frac{n_x^2}{l_x^2} + \frac{n_y^2}{l_y^2} + \frac{n_z^2}{l_z^2}\right) = \frac{\hbar^2 k^2}{2m}, \tag{5.50}$$

where k is the magnitude of the **wave vector**, $\mathbf{k} \equiv \left(k_x, k_y, k_z\right)$.

If you imagine a three-dimensional space, with axes k_x, k_y, k_z, and planes drawn in at $k_x = (\pi/l_x), (2\pi/l_x), (3\pi/l_x), \ldots$, at $k_y = \left(\pi/l_y\right), \left(2\pi/l_y\right), \left(3\pi/l_y\right), \ldots$, and at $k_z = (\pi/l_z)$, $(2\pi/l_z), (3\pi/l_z), \ldots$, each intersection point represents a distinct (one-particle) stationary state (Figure 5.2). Each block in this grid, and hence also each state, occupies a volume

$$\frac{\pi^3}{l_x l_y l_z} = \frac{\pi^3}{V} \tag{5.51}$$

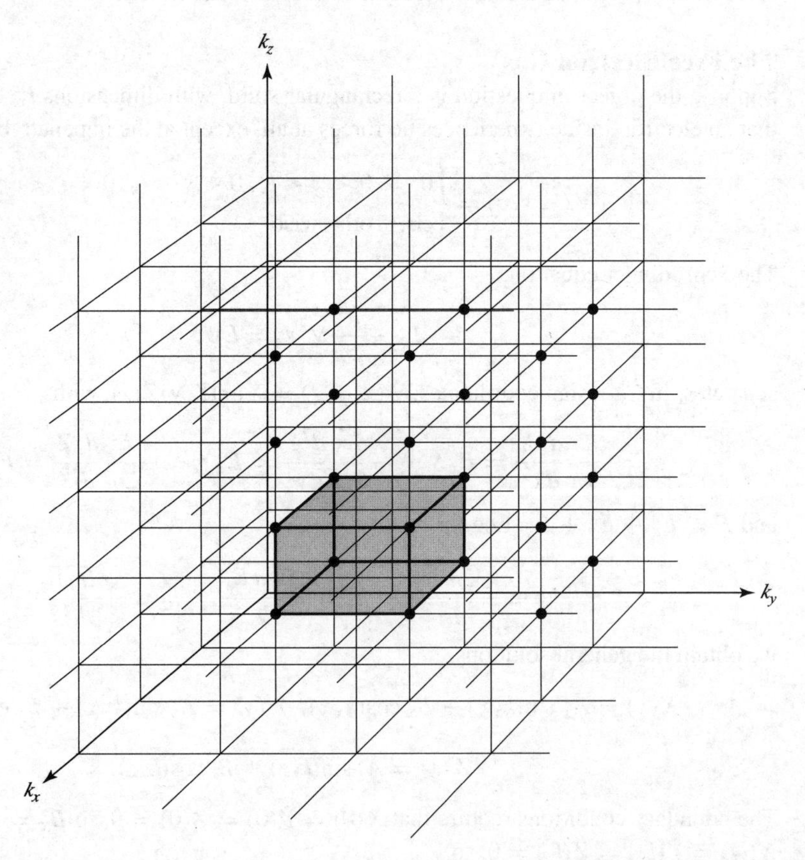

Figure 5.2: Free electron gas. Each intersection on the grid represents a stationary state. The shaded volume is one "block," and there is one state (potentially two electrons) for every block.

of "k-space," where $V \equiv l_x l_y l_z$ is the volume of the object itself. Suppose our sample contains N atoms, and each atom contributes d free electrons. (In practice, N will be enormous—on the order of Avogadro's number, for an object of macroscopic size—whereas d is a small number—1, 2, or 3, typically.) If electrons were bosons (or distinguishable particles), they would all settle down to the ground state, ψ_{111}.[21] But electrons are in fact identical *fermions*, subject to the Pauli exclusion principle, so only two of them can occupy any given state. They will fill up one octant of a *sphere* in k-space,[22] whose radius, k_F, is determined by the fact that each pair of electrons requires a volume π^3/V (Equation 5.51):

$$\frac{1}{8}\left(\frac{4}{3}\pi k_F^3\right) = \frac{Nd}{2}\left(\frac{\pi^3}{V}\right).$$

Thus

$$k_F = \left(3\rho\pi^2\right)^{1/3}, \tag{5.52}$$

where

$$\rho \equiv \frac{Nd}{V} \tag{5.53}$$

is the *free electron density* (the number of free electrons per unit volume).

The boundary separating occupied and unoccupied states, in k-space, is called the **Fermi surface** (hence the subscript F). The corresponding energy is the **Fermi energy**, E_F; for a free electron gas,

$$E_F = \frac{\hbar^2}{2m}\left(3\rho\pi^2\right)^{2/3}. \tag{5.54}$$

The *total* energy of the electron gas can be calculated as follows: A shell of thickness dk (Figure 5.3) contains a volume

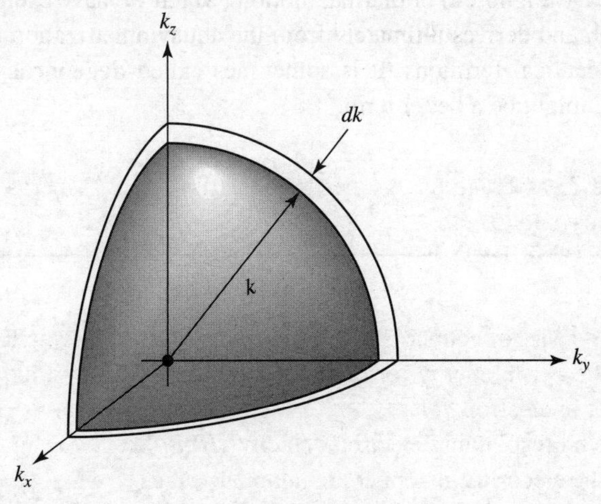

Figure 5.3: One octant of a spherical shell in k-space.

[21] I'm assuming there is no appreciable thermal excitation, or other disturbance, to lift the solid out of its collective ground state. If you like, I'm talking about a "cold" solid, though (as you will see in Problem 5.21(c)), typical solids are still "cold," in this sense, far above room temperature.

[22] Because N is such a huge number, we need not worry about the distinction between the actual jagged edge of the grid and the smooth spherical surface that approximates it.

$$\frac{1}{8}\left(4\pi k^2\right) dk,$$

so the number of electron states in the shell is

$$\frac{2\left[(1/2)\,\pi k^2\, dk\right]}{\left(\pi^3/V\right)} = \frac{V}{\pi^2} k^2\, dk.$$

Each of these states carries an energy $\hbar^2 k^2/2m$ (Equation 5.50), so the energy of the electrons in the shell is

$$dE = \frac{\hbar^2 k^2}{2m} \frac{V}{\pi^2} k^2\, dk, \tag{5.55}$$

and hence the total energy of all the filled states is

$$E_{\text{tot}} = \frac{\hbar^2 V}{2\pi^2 m} \int_0^{k_F} k^4\, dk = \frac{\hbar^2 k_F^5 V}{10\pi^2 m} = \frac{\hbar^2 \left(3\pi^2 N d\right)^{5/3}}{10\pi^2 m} V^{-2/3}. \tag{5.56}$$

This quantum mechanical energy plays a role rather analogous to the internal *thermal* energy (U) of an ordinary gas. In particular, it exerts a *pressure* on the walls, for if the box expands by an amount dV, the total energy decreases:

$$dE_{\text{tot}} = -\frac{2}{3} \frac{\hbar^2 \left(3\pi^2 N d\right)^{5/3}}{10\pi^2 m} V^{-5/3}\, dV = -\frac{2}{3} E_{\text{tot}} \frac{dV}{V},$$

and this shows up as work done on the outside ($dW = P\, dV$) by the quantum pressure P. Evidently

$$P = \frac{2}{3} \frac{E_{\text{tot}}}{V} = \frac{2}{3} \frac{\hbar^2 k_F^5}{10\pi^2 m} = \frac{\left(3\pi^2\right)^{2/3} \hbar^2}{5m} \rho^{5/3}. \tag{5.57}$$

Here, then, is a partial answer to the question of why a cold solid object doesn't simply *collapse*: There is a stabilizing internal pressure, having nothing to do with electron–electron repulsion (which we have ignored) or thermal motion (which we have excluded), but is strictly quantum mechanical, and derives ultimately from the antisymmetrization requirement for the wave functions of identical fermions. It is sometimes called **degeneracy pressure**, though "exclusion pressure" might be a better term.[23]

Problem 5.20 Find the average energy per free electron (E_{tot}/Nd), as a fraction of the Fermi energy. *Answer:* $(3/5)\,E_F$.

Problem 5.21 The density of copper is 8.96 g/cm^3, and its atomic weight is 63.5 g/mole.

(a) Calculate the Fermi energy for copper (Equation 5.54). Assume $d = 1$, and give your answer in electron volts.

(b) What is the corresponding electron velocity? *Hint:* Set $E_F = (1/2)\, mv^2$. Is it safe to assume the electrons in copper are nonrelativistic?

(c) At what temperature would the characteristic thermal energy ($k_B T$, where k_B is the Boltzmann constant and T is the Kelvin temperature) equal the Fermi energy, for copper? *Comment:* This is called the **Fermi temperature**, T_F. As long as the *actual* temperature is substantially below the Fermi temperature, the material can be

[23] We *derived* Equations 5.52, 5.54, 5.56, and 5.57 for the special case of an infinite rectangular well, but they hold for containers of any shape, as long as the number of particles is extremely large.

regarded as "cold," with most of the electrons in the lowest accessible state. Since the melting point of copper is 1356 K, solid copper is *always* cold.

(d) Calculate the degeneracy pressure (Equation 5.57) of copper, in the electron gas model.

Problem 5.22 Helium-3 is fermion with spin 1/2 (unlike the more common isotope helium-4 which is a boson). At low temperatures ($T \ll T_F$), helium-3 can be treated as a Fermi gas (Section 5.3.1). Given a density of 82 kg/m^3, calculate T_F (Problem 5.21(c)) for helium-3.

Problem 5.23 The **bulk modulus** of a substance is the ratio of a small decrease in pressure to the resulting fractional increase in volume:

$$B = -V\frac{dP}{dV}.$$

Show that $B = (5/3)\,P$, in the free electron gas model, and use your result in Problem 5.21(d) to estimate the bulk modulus of copper. *Comment*: The observed value is 13.4×10^{10} N/m^2, but don't expect perfect agreement—after all, we're neglecting all electron–nucleus and electron–electron forces! Actually, it is rather surprising that this calculation comes as close as it *does*.

5.3.2 Band Structure

We're now going to improve on the free electron model, by including the forces exerted on the electrons by the regularly spaced, positively charged, essentially stationary nuclei. The qualitative behavior of solids is dictated to a remarkable degree by the mere fact that this potential is *periodic*—its actual *shape* is relevant only to the finer details. To show you how it goes, I'm going to develop the simplest possible model: a one-dimensional **Dirac comb**, consisting of evenly spaced delta-function spikes (Figure 5.4).[24] But first I need to introduce a powerful theorem that vastly simplifies the analysis of periodic potentials.

A periodic potential is one that repeats itself after some fixed distance a:

$$V(x+a) = V(x). \tag{5.58}$$

Bloch's theorem tells us that for such a potential the solutions to the Schrödinger equation,

$$-\frac{\hbar^2}{2m}\frac{d^2\psi}{dx^2} + V(x)\psi = E\psi, \tag{5.59}$$

can be taken to satisfy the condition

$$\psi(x+a) = e^{iqa}\psi(x), \tag{5.60}$$

[24] It would be more natural to let the delta functions go *down*, so as to represent the attractive force of the nuclei. But then there would be negative energy solutions as well as positive energy solutions, and that makes the calculations more cumbersome (see Problem 5.26). Since all we're trying to do here is explore the consequences of periodicity, it is simpler to adopt this less plausible shape; if it comforts you, think of the nuclei as residing at $\pm a/2$, $\pm 3a/2$, $\pm 5a/2, \ldots$.

for some constant q (by "constant" I mean that it is independent of x; it may well depend on E).[25] In a moment we will discover that q is in fact *real*, so although $\psi(x)$ itself is not periodic, $|\psi(x)|^2$ *is*:

$$|\psi(x+a)|^2 = |\psi(x)|^2 , \tag{5.61}$$

as one would certainly expect.[26]

Of course, no *real* solid goes on forever, and the edges are going to spoil the periodicity of $V(x)$, and render Bloch's theorem inapplicable. However, for any macroscopic crystal, containing something on the order of Avogadro's number of atoms, it is hardly imaginable that edge effects can significantly influence the behavior of electrons deep inside. This suggests the following device to salvage Bloch's theorem: We wrap the x axis around in a circle, and connect it onto its tail, after a large number $N \approx 10^{23}$ of periods; formally, we impose the boundary condition

$$\psi(x+Na) = \psi(x). \tag{5.62}$$

It follows (from Equation 5.60) that

$$e^{iNqa}\psi(x) = \psi(x),$$

so $e^{iNqa} = 1$, or $Nqa = 2\pi n$, and hence

$$q = \frac{2\pi n}{Na}, \quad (n = 0, \pm 1, \pm 2, \ldots). \tag{5.63}$$

In particular, q is necessarily real. The virtue of Bloch's theorem is that we need only solve the Schrödinger equation *within a single cell* (say, on the interval $0 \le x < a$); recursive application of Equation 5.60 generates the solution everywhere else.

Now, suppose the potential consists of a long string of delta-function spikes (the Dirac comb):

$$V(x) = \alpha \sum_{j=0}^{N-1} \delta(x - ja). \tag{5.64}$$

(In Figure 5.4 you must imagine that the x axis has been "wrapped around", so the Nth spike actually appears at $x = -a$.) No one would pretend that this is a *realistic* model, but remember, it is only the effect of *periodicity* that concerns us here; the classic **Kronig–Penney model**[27]

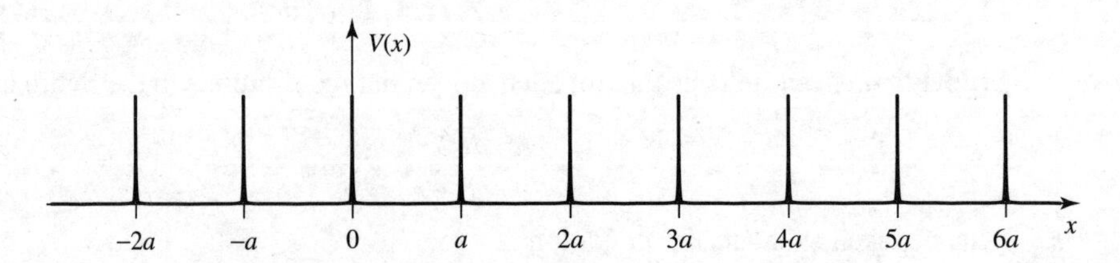

Figure 5.4: The Dirac comb, Equation 5.64.

[25] The proof of Bloch's theorem will come in Chapter 6 (see Section 6.2.2).

[26] Indeed, you might be tempted to reverse the argument, *starting* with Equation 5.61, as a way of proving Bloch's theorem. It doesn't work, for Equation 5.61 alone would allow the phase factor in Equation 5.60 to be a *function of* x.

[27] R. de L. Kronig and W. G. Penney, *Proc. R. Soc. Lond.*, ser. A, **130**, 499 (1930).

used a repeating *rectangular* pattern, and many authors still prefer that one.[28] In the region $0 < x < a$ the potential is zero, so

$$-\frac{\hbar^2}{2m}\frac{d^2\psi}{dx^2} = E\psi,$$

or

$$\frac{d^2\psi}{dx^2} = -k^2\psi,$$

where

$$k \equiv \frac{\sqrt{2mE}}{\hbar}, \tag{5.65}$$

as usual.

The general solution is

$$\psi(x) = A\sin(kx) + B\cos(kx), \quad (0 < x < a). \tag{5.66}$$

According to Bloch's theorem, the wave function in the cell immediately to the *left* of the origin is

$$\psi(x) = e^{-iqa}\{A\sin[k(x+a)] + B\cos[k(x+a)]\}, \qquad (-a < x < 0). \tag{5.67}$$

At $x = 0$, ψ must be continuous, so

$$B = e^{-iqa}[A\sin(ka) + B\cos(ka)]; \tag{5.68}$$

its derivative suffers a discontinuity proportional to the strength of the delta function (Equation 2.128, with the sign of α switched, since these are spikes instead of wells):

$$kA - e^{-iqa}k[A\cos(ka) - B\sin(ka)] = \frac{2m\alpha}{\hbar^2}B. \tag{5.69}$$

Solving Equation 5.68 for $A\sin(ka)$ yields

$$A\sin(ka) = \left[e^{iqa} - \cos(ka)\right]B. \tag{5.70}$$

Substituting this into Equation 5.69, and cancelling kB, we find

$$\left[e^{iqa} - \cos(ka)\right]\left[1 - e^{-iqa}\cos(ka)\right] + e^{-iqa}\sin^2(ka) = \frac{2m\alpha}{\hbar^2 k}\sin(ka),$$

which simplifies to

$$\cos(qa) = \cos(ka) + \frac{m\alpha}{\hbar^2 k}\sin(ka). \tag{5.71}$$

This is the fundamental result, from which all else follows.[29]

Equation 5.71 determines the possible values of k, and hence the allowed energies. To simplify the notation, let

$$z \equiv ka, \quad \text{and} \quad \beta \equiv \frac{m\alpha a}{\hbar^2}, \tag{5.72}$$

so the right side of Equation 5.71 can be written as

$$f(z) \equiv \cos(z) + \beta\frac{\sin(z)}{z}. \tag{5.73}$$

[28] See, for instance, David Park, *Introduction to the Quantum Theory*, 3nd edn, McGraw-Hill, New York (1992).

[29] For the Kronig–Penney potential (footnote 27, page 221), the formula is more complicated, but it shares the qualitative features we are about to explore.

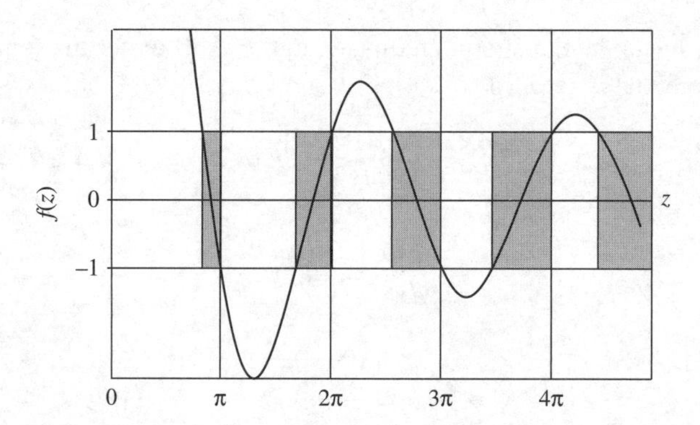

Figure 5.5: Graph of $f(z)$ (Equation 5.73) for $\beta = 10$, showing allowed bands separated by forbidden gaps.

The constant β is a dimensionless measure of the "strength" of the delta function. In Figure 5.5 I have plotted $f(z)$, for the case $\beta = 10$. The important thing to notice is that $f(z)$ strays outside the range $(-1, +1)$, and in such regions there is no hope of solving Equation 5.71, since $|\cos(qa)|$, of course, cannot be greater than 1. These **gaps** represent *forbidden* energies; they are separated by **bands** of *allowed* energies. Within a given band, virtually any energy is allowed, since according to Equation 5.63, $qa = 2\pi n/N$, where N is a huge number, and n can be any integer. You might imagine drawing N horizontal lines on Figure 5.5, at values of $\cos(2\pi n/N)$ ranging from $+1$ $(n = 0)$ down to -1 $(n = N/2)$, and back almost to $+1$ $(n = N - 1)$—at this point the Bloch factor e^{iqa} recycles, so no new solutions are generated by further increasing n. The intersection of each of these lines with $f(z)$ yields an allowed energy. Evidently there are N states in each band, so closely spaced that for most purposes we can regard them as forming a continuum (Figure 5.6).

So far, we've only put *one* electron in our potential. In practice there will be Nd of them, where d is again the number of "free" electrons per atom. Because of the Pauli exclusion principle, only two electrons can occupy a given spatial state, so if $d = 1$, they will half fill the first band, if $d = 2$ they will completely fill the first band, if $d = 3$ they half fill the second band, and so on. (In three dimensions, and with more realistic potentials, the band structure may be more complicated, but the *existence* of allowed bands, separated by forbidden gaps, persists—band structure is the *signature* of a periodic potential.[30])

Now, if the topmost band is only *partly* filled, it takes very little energy to excite an electron to the next allowed level, and such a material will be a **conductor** (a **metal**). On the other hand, if the top band is *completely* filled, it takes a relatively large energy to excite an electron, since it has to jump across the forbidden zone. Such materials are typically **insulators**, though if the gap is rather narrow, and the temperature sufficiently high, then random thermal energy can knock an electron over the hump, and the material is a **semiconductor** (silicon and germanium are examples).[31] In the free electron model *all* solids should be metals, since there are no

[30] Regardless of dimension, if d is an odd integer you are guaranteed to have partially-filled bands and you would expect metallic behavior. If d is an even integer, it depends on the specific band structure whether there will be partially-filled bands or not. Interestingly, some materials, called **Mott insulators**, are nonconductors even though d is odd. In that case it is the interactions between electrons that leads to the insulating behavior, not the presence of gaps in the single-particle energy spectrum.

[31] Semiconductors typically have band gaps of 4 eV or less, small enough that thermal excitation at room temperature ($k_B T \approx 0.025$ eV) produces perceptible conductivity. The conductivity of a semiconductor can be controlled by

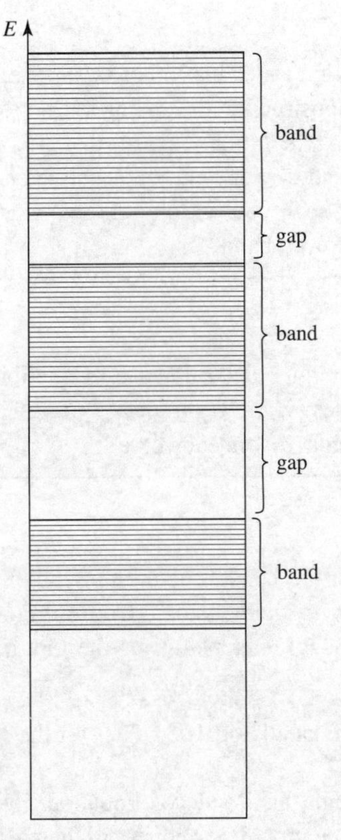

Figure 5.6: The allowed energies for a periodic potential form essentially continuous bands.

large gaps in the spectrum of allowed energies. It takes the band theory to account for the extraordinary range of electrical conductivities exhibited by the solids in nature.

Problem 5.24

(a) Using Equations 5.66 and 5.70, show that the wave function for a particle in the periodic delta function potential can be written in the form

$$\psi(x) = C\left\{\sin(kx) + e^{-iqa}\sin[k(a-x)]\right\}, \qquad (0 \le x \le a).$$

(Don't bother to determine the normalization constant C.)

(b) At the top of a band, where $z = j\pi$, (a) yields $\psi(x) = 0/0$ (indeterminate). Find the correct wave function for this case. Note what happens to ψ at each delta function.

Problem 5.25 Find the energy at the bottom of the first allowed band, for the case $\beta = 10$, correct to three significant digits. For the sake of argument, assume $\alpha/a = 1$ eV.

doping: including a few atoms of larger or smaller d; this puts some "extra" electrons into the next higher band, or creates some **holes** in the previously filled one, allowing in either case for weak electric currents to flow.

∗∗ **Problem 5.26** Suppose we use delta function *wells*, instead of *spikes* (i.e. switch the sign of α in Equation 5.64). Analyze this case, constructing the analog to Figure 5.5. This requires no new calculation, for the positive energy solutions (except that β is now negative; use $\beta = -1.5$ for the graph), but you *do* need to work out the negative energy solutions (let $\kappa \equiv \sqrt{-2mE}/\hbar$ and $z \equiv -\kappa a$, for $E < 0$); your graph will now extend to negative z). How many states are there in the first allowed band?

Problem 5.27 Show that *most* of the energies determined by Equation 5.71 are doubly degenerate. What are the exceptional cases? *Hint*: Try it for $N = 1, 2, 3, 4, \ldots$, to see how it goes. What are the possible values of $\cos(qa)$ in each case?

Problem 5.28 Make a plot of E vs. q for the band structure in Section 5.3.2. Use $\alpha = 1$ (in units where $m = \hbar = a = 1$). *Hint:* In Mathematica, **ContourPlot** will graph $E(q)$ as defined implicitly by Equation 5.71. On other platforms the plot can be obtained as follows:

- Choose a large number (say 30,000) of equally-spaced values for the energy in the range $E = 0$ and $E = 30$.
- For each value of E, compute the right-hand side of Equation 5.71. If the result is between -1 and 1, solve for q from Equation 5.71 and record the pair of values `{q,E}` and `{-q,E}` (there are two solutions for each energy).

You will then have a list of pairs `{{q1,E1},{q2,E2},...}` which you can plot.

FURTHER PROBLEMS ON CHAPTER 5

Problem 5.29 Suppose you have three particles, and three distinct one-particle states ($\psi_a(x)$, $\psi_b(x)$, and $\psi_c(x)$) are available. How many different three-particle states can be constructed (a) if they are distinguishable particles, (b) if they are identical bosons, (c) if they are identical fermions? (The particles need not be in *different* states—$\psi_a(x_1)\psi_a(x_2)\psi_a(x_3)$ would be one possibility, if the particles are distinguishable.)

Problem 5.30 Calculate the Fermi energy for electrons in a *two*-dimensional infinite square well. Let σ be the number of free electrons per unit area.

Problem 5.31 Repeat the analysis of Problem 2.58 to estimate the cohesive energy for a three-dimensional metal, including the effects of spin.

Problem 5.32 Consider a free electron gas (Section 5.3.1) with unequal numbers of spin-up and spin-down particles (N_+ and N_- respectively). Such a gas would have a net **magnetization** (magnetic dipole moment per unit volume)

$$\mathbf{M} = -\frac{(N_+ - N_-)}{V}\mu_B\,\hat{k} = M\hat{k}, \tag{5.74}$$

where $\mu_B = e\hbar/2m_e$ is the **Bohr magneton**. (The minus sign is there, of course, because the charge of the electron is negative.)

(a) Assuming that the electrons occupy the lowest energy levels consistent with the number of particles in each spin orientation, find E_{tot}. Check that your answer reduces to Equation 5.56 when $N_+ = N_-$.

(b) Show that for $M/\mu_B \ll \rho \equiv (N_+ + N_-)/V$ (which is to say, $|N_+ - N_-| \ll (N_+ + N_-)$), the energy density is

$$\frac{1}{V}E_{\text{tot}} = \frac{\hbar^2\left(3\pi^2\rho\right)^{5/3}}{10\pi^2 m}\left[1 + \frac{5}{9}\left(\frac{M}{\rho\,\mu_B}\right)^2\right].$$

The energy is a minimum for $M = 0$, so the ground state will have zero magnetization. However, if the gas is placed in a magnetic field (or in the presence of interactions between the particles) it may be energetically favorable for the gas to magnetize. This is explored in Problems 5.33 and 5.34.

Problem 5.33 Pauli paramagnetism. If the free electron gas (Section 5.3.1) is placed in a uniform magnetic field $\mathbf{B} = B\,\hat{k}$, the energies of the spin-up and spin-down states will be different:[32]

$$E^{\pm}_{n_x n_y n_z} = \frac{\hbar^2\pi^2}{2m}\left(\frac{n_x^2}{l_x^2} + \frac{n_y^2}{l_y^2} + \frac{n_z^2}{l_z^2}\right) \pm \mu_B\,B.$$

There will be more spin-down states occupied than spin-up states (since they are lower in energy), and consequently the system will acquire a magnetization (see Problem 5.32).

(a) In the approximation that $M/\mu_B \ll \rho$, find the magnetization that minimizes the total energy. *Hint:* Use the result of Problem 5.32(b).

(b) The **magnetic susceptibility** is[33]

$$\chi = \mu_0\frac{dM}{dB}.$$

Calculate the magnetic susceptibility for aluminum $\left(\rho = 18.1 \times 10^{22}\,\text{cm}^{-3}\right)$ and compare the experimental value[34] of 22×10^{-6}.

---

32 Here we are considering only the coupling of the spin to the magnetic field, and ignoring any coupling of the orbital motion.

33 Strictly speaking, the susceptibility is dM/dH, but the difference is negligible when, as here, $\chi \ll 1$.

34 For some metals, such as copper, the agreement is not so good—even the sign is wrong: copper is diamagnetic ($\chi < 0$). The explanation for this discrepancy lies in what has been left out of our model. In addition to the

Problem 5.34 The Stoner criterion. The free-electron gas model (Section 5.3.1) ignores the Coulomb repulsion between electrons. Because of the exchange force (Section 5.1.2), Coulomb repulsion has a stronger effect on two electrons with antiparallel spins (which behave in a way like distinguishable particles) than two electrons with parallel spins (whose position wave function must be antisymmetric). As a crude way to take account of Coulomb repulsion, pretend that every pair of electrons with opposite spin carries extra energy U, while electrons with the same spin do not interact at all; this adds $\Delta E = U N_+ N_-$ to the total energy of the electron gas. As you will show, above a critical value of U, it becomes energetically favorable for the gas to spontaneously magnetize ($N_+ \neq N_-$); the material becomes ferromagnetic.

(a) Rewrite ΔE in terms of the density ρ and the magnetization M (Equation 5.74).

(b) Assuming that $M/\mu_B \ll \rho$, for what minimum value of U is a non-zero magnetization energetically favored? *Hint:* Use the result of Problem 5.32(b).

***

Problem 5.35 Certain cold stars (called **white dwarfs**) are stabilized against gravitational collapse by the degeneracy pressure of their electrons (Equation 5.57). Assuming constant density, the radius R of such an object can be calculated as follows:

(a) Write the total electron energy (Equation 5.56) in terms of the radius, the number of nucleons (protons and neutrons) N, the number of electrons per nucleon d, and the mass of the electron m. *Beware:* In this problem we are recycling the letters N and d for a slightly different purpose than in the text.

(b) Look up, or calculate, the gravitational energy of a uniformly dense sphere. Express your answer in terms of G (the constant of universal gravitation), R, N, and M (the mass of a nucleon). Note that the gravitational energy is *negative*.

(c) Find the radius for which the total energy, (a) plus (b), is a minimum. *Answer:*

$$R = \left(\frac{9\pi}{4}\right)^{2/3} \frac{\hbar^2 d^{5/3}}{GmM^2N^{1/3}}.$$

(Note that the radius *decreases* as the total mass *increases!*) Put in the actual numbers, for everything except N, using $d = 1/2$ (actually, d decreases a bit as the atomic number increases, but this is close enough for our purposes). *Answer:* $R = 7.6 \times 10^{25} N^{-1/3}$ m.

(d) Determine the radius, in kilometers, of a white dwarf with the mass of the sun.

(e) Determine the Fermi energy, in electron volts, for the white dwarf in (d), and compare it with the rest energy of an electron. Note that this system is getting dangerously relativistic (see Problem 5.36).

paramagnetic coupling of the spin magnetic moment to an applied field there is a coupling of the orbital magnetic moment to an applied field and this has both paramagnetic and diamagnetic contributions (see Problem 4.72). In addition, the free electron gas model ignores the tightly-bound core electrons and these also couple to the magnetic field. In the case of copper, it is the diamagnetic coupling of the core electrons that dominates.

*** **Problem 5.36** We can extend the theory of a free electron gas (Section 5.3.1) to the relativistic domain by replacing the classical kinetic energy, $E = p^2/2m$, with the relativistic formula, $E = \sqrt{p^2c^2 + m^2c^4} - mc^2$. Momentum is related to the wave vector in the usual way: $\mathbf{p} = \hbar\mathbf{k}$. In particular, in the *extreme* relativistic limit, $E \approx pc = \hbar ck$.

(a) Replace $\hbar^2k^2/2m$ in Equation 5.55 by the ultra-relativistic expression, $\hbar ck$, and calculate E_{tot} in this regime.

(b) Repeat parts (a) and (b) of Problem 5.35 for the ultra-relativistic electron gas. Notice that in this case there is *no* stable minimum, regardless of R; if the total energy is positive, degeneracy forces exceed gravitational forces, and the star will expand, whereas if the total is negative, gravitational forces win out, and the star will collapse. Find the critical number of nucleons, N_c, such that gravitational collapse occurs for $N > N_c$. This is called the **Chandrasekhar limit**. *Answer:* 2.04×10^{57}. What is the corresponding stellar mass (give your answer as a multiple of the sun's mass). Stars heavier than this will not form white dwarfs, but collapse further, becoming (if conditions are right) **neutron stars**.

(c) At extremely high density, **inverse beta decay**, $e^- + p^+ \rightarrow n + \nu$, converts virtually all of the protons and electrons into neutrons (liberating neutrinos, which carry off energy, in the process). Eventually *neutron* degeneracy pressure stabilizes the collapse, just as *electron* degeneracy does for the white dwarf (see Problem 5.35). Calculate the radius of a neutron star with the mass of the sun. Also calculate the (neutron) Fermi energy, and compare it to the rest energy of a neutron. Is it reasonable to treat a neutron star nonrelativistically?

Problem 5.37 An important quantity in many calculations is the **density of states** $G(E)$:

$$G(E)\; dE \equiv \text{number of states with energy between } E \text{ and } E + dE\,.$$

For a one-dimensional band structure,

$$G(E)\; dE = 2\left(\frac{dq}{2\pi/N a}\right),$$

where $dq/(2\pi/N a)$ counts the number of states in the range dq (see Equation 5.63), and the factor of 2 accounts for the fact that states with q and $-q$ have the same energy. Therefore

$$\frac{1}{N a}\, G(E) = \frac{1}{\pi}\, \frac{1}{|dE/dq|}\,.$$

(a) Show that for $\alpha = 0$ (a free particle) the density of states is given by

$$\frac{1}{N a}\, G_{\text{free}}(E) = \frac{1}{\pi\hbar}\sqrt{\frac{m}{2E}}\,.$$

(b) Find the density of states for $\alpha \neq 0$ by differentiating Equation 5.71 with respect to q to determine dE/dq. *Note:* Your answer should be written as a function of E only (well, and α, m, $\hbar$, a, and N) and must not contain q (use k as a shorthand for $\sqrt{2mE}/\hbar$, if you like).

(c) Make a single plot showing $G(E)/Na$ for both $\alpha = 0$ and $\alpha = 1$ (in units where $m = \hbar = a = 1$). *Comment:* The divergences at the band edges are examples of **van Hove singularities.**[35]

∗∗∗ **Problem 5.38** The **harmonic chain** consists of N equal masses arranged along a line and connected to their neighbors by identical springs:

$$\hat{H} = -\frac{\hbar^2}{2m}\sum_{j=1}^{N}\frac{\partial}{\partial x_j^2} + \sum_{j=1}^{N}\frac{1}{2}m\,\omega^2\left(x_{j+1}-x_j\right)^2,$$

where x_j is the displacement of the jth mass from its equilibrium position. This system (and its extension to two or three dimensions—the **harmonic crystal**) can be used to model the vibrations of a solid. For simplicity we will use periodic boundary conditions: $x_{N+1} = x_1$, and introduce the ladder operators[36]

$$\hat{a}_{k\pm} \equiv \frac{1}{\sqrt{N}}\sum_{j=1}^{N} e^{\pm i\,2\pi\,j\,k/N}\left[\sqrt{\frac{m\,\omega_k}{2\,\hbar}}\,x_j \mp \sqrt{\frac{\hbar}{2\,m\,\omega_k}}\frac{\partial}{\partial x_j}\right] \tag{5.75}$$

where $k = 1, \ldots, N-1$ and the frequencies are given by

$$\omega_k = 2\,\omega\,\sin\left(\frac{\pi\,k}{N}\right).$$

(a) Prove that, for integers k and k' between 1 and $N-1$,

$$\frac{1}{N}\sum_{j=1}^{N} e^{i\,2\pi\,j(k-k')/N} = \delta_{k',k}$$

$$\frac{1}{N}\sum_{j=1}^{N} e^{i\,2\pi\,j(k+k')/N} = \delta_{k',N-k}\,.$$

Hint: Sum the geometric series.

(b) Derive the commutation relations for the ladder operators:

$$\left[\hat{a}_{k-},\hat{a}_{k'+}\right] = \delta_{k,k'} \text{ and } \left[\hat{a}_{k-},\hat{a}_{k'-}\right] = \left[\hat{a}_{k+},\hat{a}_{k'+}\right] = 0\,. \tag{5.76}$$

---

[35] These one-dimensional Van Hove singularities have been observed in the spectroscopy of carbon nanotubes; see J. W. G. Wildöer et al., *Nature*, **391**, 59 (1998).

[36] If you are familiar with the *classical* problem of coupled oscillators, these ladder operators are straightforward to construct. Start with the normal mode coordinates you would use to decouple the classical problem, namely

$$q_k = \frac{1}{\sqrt{N}}\sum_{j=1}^{N} e^{-i\,2\pi\,j\,k/N}x_j\,.$$

The frequencies ω_k are the classical normal mode frequencies, and you simply create a pair of ladder operators for each normal mode, by analogy with the single-particle case (Equation 2.48).

(c) Using Equation 5.75, show that

$$x_j = R + \frac{1}{\sqrt{N}} \sum_{k=1}^{N-1} \sqrt{\frac{\hbar}{2\, m\, \omega_k}} \left(\hat{a}_{k-} + \hat{a}_{N-k+}\right) e^{i\, 2\pi\, j\, k/N}$$

$$\frac{\partial}{\partial x_j} = \frac{1}{N} \frac{\partial}{\partial R} + \frac{1}{\sqrt{N}} \sum_{k=1}^{N-1} \sqrt{\frac{m\, \omega_k}{2\, \hbar}} \left(\hat{a}_{k-} - \hat{a}_{N-k+}\right) e^{i\, 2\pi\, j\, k/N}$$

where $R = \sum_j x_j / N$ is the center of mass coordinate.

(d) Finally, show that

$$\hat{H} = -\frac{\hbar^2}{2\,(N\, m)} \frac{\partial^2}{\partial R^2} + \sum_{k=1}^{N-1} \hbar\, \omega_k \left(\hat{a}_{k+}\, \hat{a}_{k-} + \frac{1}{2}\right).$$

Comment: Written in this form above, the Hamiltonian describes $N - 1$ independent oscillators with frequencies ω_k (as well as a center of mass that moves as a free particle of mass $N\, m$). We can immediately write down the allowed energies:

$$E = -\frac{\hbar^2\, K^2}{2\,(N\, m)} + \sum_{k=1}^{N-1} \hbar\, \omega_k \left(n_k + \frac{1}{2}\right)$$

where $\hbar\, K$ is the momentum of the center of mass and $n_k = 0, 1, \ldots$ is the energy level of the kth mode of vibration. It is conventional to call n_k the **number of phonons** in the kth mode. Phonons are the quanta of sound (atomic vibrations), just as photons are the quanta of light. The ladder operators a_{k+} and a_{k-} are called **phonon creation and annihilation operators** since they increase or decrease the number of phonons in the kth mode.

Problem 5.39 In Section 5.3.1 we put the electrons in a box with impenetrable walls. The same results can be obtained using **periodic boundary conditions**. We still imagine the electrons to be confined to a box with sides of length l_x, l_y, and l_z but instead of requiring the wave function to vanish on each wall, we require it to take the same value on opposite walls:

$$\psi(x, y, z) = \psi(x + l_x, y, z) = \psi\left(x, y + l_y, z\right) = \psi(x, y, z + l_z) .$$

In this case we can represent the wave functions as *traveling* waves,

$$\psi = \frac{1}{\sqrt{l_x\, l_y\, l_z}} e^{i\, \mathbf{k} \cdot \mathbf{r}} = \frac{1}{\sqrt{l_x\, l_y\, l_z}} e^{i\left(k_x\, x + k_y\, y + k_z\, z\right)},$$

rather than as *standing* waves (Equation 5.49). Periodic boundary conditions—while certainly not physical—are often easier to work with (to describe something like electrical current a basis of traveling waves is more natural than a basis of standing waves) and if you are computing *bulk* properties of a material it shouldn't matter which you use.

(a) Show that with periodic boundary conditions the wave vector satisfies

$$k_x l_x = 2n_x \pi, \qquad k_y l_y = 2n_y \pi, \qquad k_z l_z = 2n_z \pi,$$

where each n is an integer (*not* necessarily positive). What is the k-space volume occupied by each block on the grid (corresponding to Equation 5.51)?

(b) Compute k_F, E_F, and E_{tot} for the free electron gas with periodic boundary conditions. What compensates for the larger volume occupied by each k-space block (part (a)) to make these all come out the same as in Section 5.3.1?

导读 / Guidance

第 6 章　对称性和守恒律

对经典力学而言，存在一定的对称性就有相应的守恒定律，量子力学也是如此。本章讨论在量子力学系统中的这些对称性及其守恒定律。重点讨论量子力学系统的两个相关特性——能级简并和选择定则。本章首先介绍一些对称操作，如平移算符、宇称算符、旋转算符等。接下来对上述对称操作分别讨论其对称性和相应的守恒律；关于平移对称性，介绍了平移算符的生成元、平移不变性和动量守恒律；关于宇称，讨论了一维和三维情况下宇称算符的性质和宇称选择定则；关于旋转对称性，讨论了一维、三维旋转算符的生成元、旋转不变性和角动量守恒律、标量算符和矢量算符的选择定则（Wigner-Eckart 定理）；关于时间反演对称，讨论了时间反演对称算符的生成元、时间反演不变和能量守恒定律。

习题特色

（1）将平移对称性拓展到研究固体物理的布洛赫定理和第一布里渊区问题，如习题 6.6。

（2）把一些研究前沿知识拓展成一些习题：讨论自旋态的旋转对称性问题，见习题 6.32、习题 6.37;讨论库仑势的对称性变换问题，见习题 6.34、习题 6.35。

（3）以经典物理中的例子，如球体的斜上抛运动，讨论一定质量的带电粒子在磁场中的运动破坏时间反演对称，从经典物理的层面加深学生对对称性的理解，见习题 6.36。

6 SYMMETRIES & CONSERVATION LAWS

6.1 INTRODUCTION

Conservation laws (energy, momentum, and angular momentum) are familiar from your first course in *classical* mechanics. These same conservation laws hold in quantum mechanics; in both contexts they are the result of symmetries. In this chapter we will explain what a symmetry is and what it means for something to be conserved in quantum mechanics—and show how the two are related. Along the way we'll investigate two related properties of quantum systems—energy level degeneracy and the selection rules that distinguish allowed from "forbidden" transitions.

What is a symmetry? It is some transformation that leaves the system unchanged. As an example consider rotating a square piece of paper, as shown in Figure 6.1. If you rotate it by 30° about an axis through its center it will be in a different orientation than the one it started in, but if you rotate it by 90° it will resume its original orientation; you wouldn't even know it had been rotated unless (say) you wrote numbers on the corners (in which case they would be permuted). A square therefore has a *discrete* rotational symmetry: a rotation by $n\pi/2$ for any integer n leaves it unchanged.[1] If you repeated this experiment with a circular piece of paper, a rotation by *any* angle would leave it unchanged; the circle has *continuous* rotational symmetry. We will see that both discrete and continuous symmetries are important in quantum mechanics.

Now imagine that the shapes in Figure 6.1 refer not to pieces of paper, but to the boundaries of a two-dimensional infinite square well. In that case the potential energy would have the same rotational symmetries as the piece of paper and (because the kinetic energy is unchanged by a rotation) the Hamiltonian would also be invariant. In quantum mechanics, when we say that a system has a symmetry, this is what we mean: that the Hamiltonian is unchanged by some transformation, such as a rotation or a translation.

6.1.1 Transformations in Space

In this section, we introduce the quantum mechanical operators that implement translations, inversions, and rotations. We define each of these operators by how it acts on an arbitrary function. The **translation operator** takes a function and shifts it a distance a. The operator that accomplishes this is defined by the relation

[1] A square of course has other symmetries as well, namely mirror symmetries about axes along a diagonal or bisecting two sides. The set of all transformations that leave the square unchanged is called D_4, the "dihedral group" of degree 4.

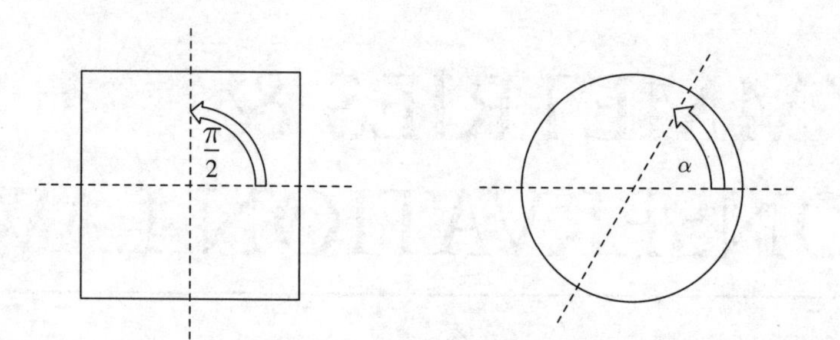

Figure 6.1: A square has a discrete rotational symmetry; it is unchanged when rotated by $\pi/2$ or multiples thereof. A circle has continuous rotational symmetry; it is unchanged when rotated by any angle α.

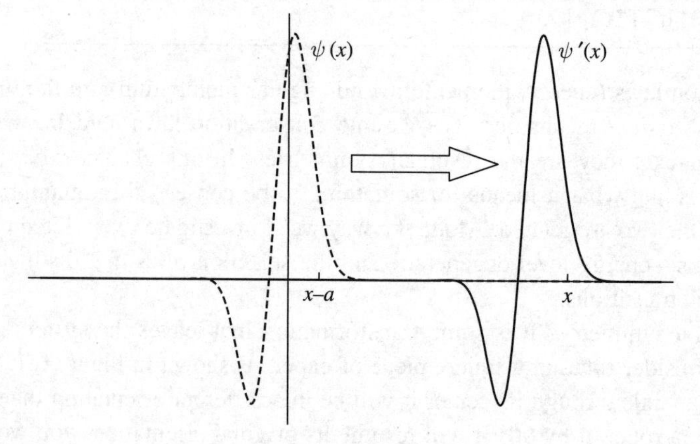

Figure 6.2: A wave function $\psi(x)$ and the translated wave function $\psi'(x) = \hat{T}(a)\,\psi(x)$. Note that the value of ψ' at x is equal to the value of ψ at $x - a$.

$$\hat{T}(a)\,\psi(x) = \psi'(x) = \psi(x-a)\,. \tag{6.1}$$

The sign can be confusing at first; this equation says that the translated function ψ' at x is equal to the untranslated function ψ at $x - a$ (Figure 6.2)—the function itself has been shifted to the *right* by an amount a.

The operator that reflects a function about the origin, the **parity operator** in one dimension, is defined by

$$\hat{\Pi}\,\psi(x) = \psi'(x) = \psi(-x)\,.$$

The effect of parity is shown graphically in Figure 6.3. In three dimensions parity changes the sign of all three coordinates: $\hat{\Pi}\,\psi(x, y, z) = \psi(-x, -y, -z)$.[2]

Finally, the operator that rotates a function about the z axis through an angle φ is most naturally expressed in polar coordinates as

$$\hat{R}_z(\varphi)\,\psi(r, \theta, \phi) = \psi'(r, \theta, \phi) = \psi(r, \theta, \phi - \varphi)\,. \tag{6.2}$$

---

[2] The parity operation in three dimensions can be realized as a mirror reflection followed by a rotation (see Problem 6.1). In two dimensions, the transformation $\psi'(x, y) = \psi(-x, -y)$ is no different from a 180° rotation. We will use the term parity exclusively for spatial inversion, $\hat{\Pi}\,\psi(\mathbf{r}) = \psi(-\mathbf{r})$, in one or three dimensions.

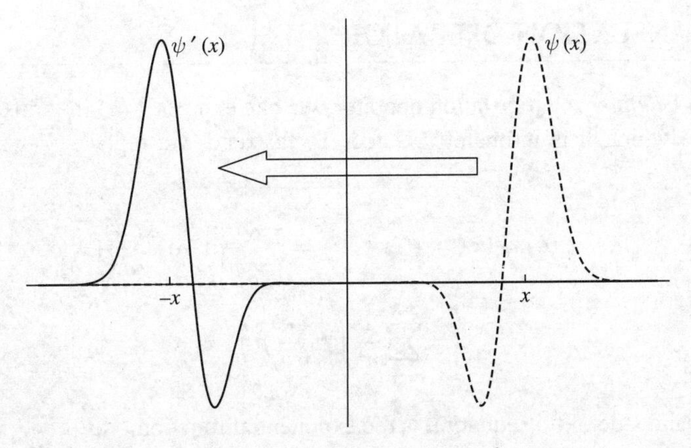

Figure 6.3: A function $\psi(x)$ and the function $\psi'(x) = \hat{\Pi}\,\psi(x) = \psi(-x)$ after a spatial inversion. The value of ψ' at x is equal to the value of ψ at $-x$.

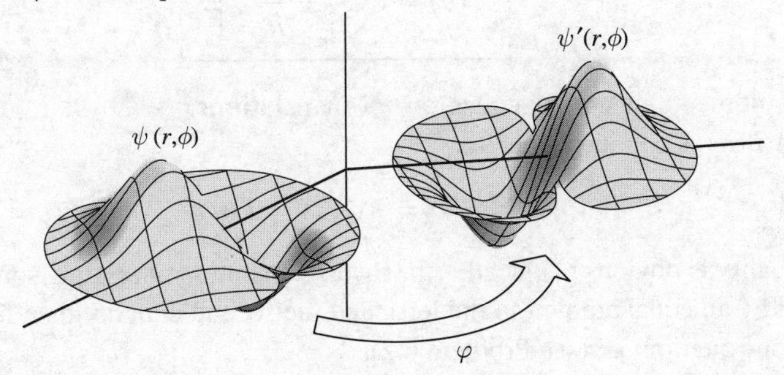

Figure 6.4: A function $\psi(r,\phi)$ and the rotated function $\psi'(r,\phi) = \psi(r,\phi-\varphi)$ after a counter-clockwise rotation about the vertical axis by an angle φ.

When we take up the study of rotations in Section 6.5, we will introduce expressions for rotations about arbitrary axes. The action of the **rotation operator** on a function ψ is illustrated in Figure 6.4.

* **Problem 6.1** Consider the parity operator in three dimensions.

(a) Show that $\hat{\Pi}\,\psi(\mathbf{r}) = \psi'(\mathbf{r}) = \psi(-\mathbf{r})$ is equivalent to a mirror reflection followed by a rotation.

(b) Show that, for ψ expressed in polar coordinates, the action of the parity operator is

$$\hat{\Pi}\,\psi(r,\theta,\phi) = \psi(r,\pi-\theta,\phi+\pi).$$

(c) Show that for the hydrogenic orbitals,

$$\hat{\Pi}\,\psi_{n\ell m}(r,\theta,\phi) = (-1)^{\ell}\,\psi_{n\ell m}(r,\theta,\phi).$$

That is, $\psi_{n\ell m}$ is an eigenstate of the parity operator, with eigenvalue $(-1)^{\ell}$. *Note*: This result actually applies to the stationary states of any central potential $V(\mathbf{r}) = V(r)$. For a central potential, the eigenstates may be written in the separable form $R_{n\ell}(r)\,Y_{\ell}^{m}(\theta,\phi)$ where only the radial function $R_{n\ell}$—which plays no role in determining the parity of the state—depends on the specific functional form of $V(r)$.

6.2 THE TRANSLATION OPERATOR

Equation 6.1 *defines* the translation operator. We can express $\hat{T}(a)$ in terms of the momentum operator, to which it is intimately related. To that end, we replace $\psi(x-a)$ by its Taylor series[3]

$$\hat{T}(a)\,\psi(x) = \psi(x-a) = \sum_{n=0}^{\infty} \frac{1}{n!}(-a)^n \frac{d^n}{dx^n}\psi(x)$$
$$= \sum_{n=0}^{\infty} \frac{1}{n!}\left(\frac{-i\,a}{\hbar}\,\hat{p}\right)^n \psi(x)\,.$$

The right-hand side of this equation is the exponential function,[4] so

$$\boxed{\hat{T}(a) = \exp\left[-\frac{i\,a}{\hbar}\,\hat{p}\right].} \tag{6.3}$$

We say that momentum is the **"generator" of translations.**[5]

Note that $\hat{T}(a)$ is a unitary operator:[6]

$$\hat{T}(a)^{-1} = \hat{T}(-a) = \hat{T}(a)^{\dagger}\,. \tag{6.4}$$

The first equality is obvious physically (the inverse operation of shifting something to the right is shifting it by an equal amount to the left), and the second equality then follows from taking the adjoint of Equation 6.3 (see Problem 6.2).

* **Problem 6.2** Show that, for a Hermitian operator $\hat{Q}$, the operator $\hat{U} = \exp\left[i\,\hat{Q}\right]$ is unitary. *Hint*: First you need to prove that the adjoint is given by $\hat{U}^{\dagger} = \exp\left[-i\,\hat{Q}\right]$; then prove that $\hat{U}^{\dagger}\,\hat{U} = 1$. Problem 3.5 may help.

6.2.1 How Operators Transform

So far I have shown how to translate a *function*; this has an obvious graphical interpretation via Figure 6.2. We can also consider what it means to *translate an operator*. The transformed operator $\hat{Q}'$ is defined to be the operator that gives the same expectation value in the untranslated state ψ as does the operator $\hat{Q}$ in the translated state ψ':

$$\left\langle \psi' \middle| \hat{Q} \middle| \psi' \right\rangle = \left\langle \psi \middle| \hat{Q}' \middle| \psi \right\rangle.$$

---

[3] I'm assuming that our function has a Taylor series expansion, but the final result applies more generally. See Problem 6.31 for the details.

[4] See Section 3.6.2 for the definition of the exponential of an operator.

[5] The term comes from the study of **Lie groups** (the group of translations is an example). If you're interested, an introduction to Lie groups (written for physicists) can be found in George B. Arfken, Hans J. Weber, and Frank E. Harris, *Mathematical Methods for Physicists*, 7th edn, Academic Press, New York (2013), Section 17.7.

[6] Unitary operators are discussed in Problem A.30. A unitary operator is one whose adjoint is also its inverse: $\hat{U}\,\hat{U}^{\dagger} = \hat{U}^{\dagger}\,\hat{U} = 1$.

There are two ways to calculate the effect of a translation on an expectation value. One could actually shift the wave function over some distance (this is called an **active transformation**) *or* one could leave the wave function where it was and shift the origin of our coordinate system by the same amount in the opposite direction (a **passive transformation**). The operator $\hat{Q}'$ is the operator in this shifted coordinate system.

Using Equation 6.1,

$$\left\langle \psi \middle| \hat{T}^{\dagger}\, \hat{Q}\, \hat{T} \middle| \psi \right\rangle = \left\langle \psi \middle| \hat{Q}' \middle| \psi \right\rangle. \tag{6.5}$$

Here I am using the fact that the adjoint of an operator is defined such that, if $\hat{T}\;|f\rangle \equiv |T\,f\rangle$, then $\langle T\,f| = \langle f|\;\hat{T}^{\dagger}$ (see Problem 3.5). Because Equation 6.5 is to hold for *all* ψ, it follows that

$$\boxed{\hat{Q}' = \hat{T}^{\dagger}\, \hat{Q}\, \hat{T}.} \tag{6.6}$$

The transformed operator for the case $\hat{Q} = \hat{x}$ is worked out in Example 6.1. Figure 6.5 illustrates the equivalence of the two ways of carrying out the transformation.

Example 6.1

Find the operator $\hat{x}'$ obtained by applying a translation through a distance a to the operator $\hat{x}$. That is, what is the action of $\hat{x}'$, as defined by Equation 6.6, on an arbitrary $f(x)$?

Solution: Using the definition of $\hat{x}'$ (Equation 6.6) and a test function $f(x)$ we have

$$\hat{x}'\; f(x) = \hat{T}^{\dagger}(a)\; \hat{x}\; \hat{T}(a)\;\; f(x),$$

and since $\hat{T}^{\dagger}(a) = \hat{T}(-a)$ (Equation 6.4),

$$\hat{x}'\; f(x) = \hat{T}(-a)\; \hat{x}\; \hat{T}(a)\;\; f(x).$$

From Equation 6.1

$$\hat{x}'\; f(x) = \hat{T}(-a)\; x\; f(x-a)\,,$$

and from Equation 6.1 again, $\hat{T}(-a)\;\left[x\; f(x-a)\right] = (x+a)\;\; f(x)$, so

$$\hat{x}'\; f(x) = (x+a)\;\; f(x).$$

Finally we may read off the operator

$$\hat{x}' = \hat{x} + a. \tag{6.7}$$

As expected, Equation 6.7 corresponds to shifting the origin of our coordinates to the left by a so that positions in these transformed coordinates are greater by a than in the untransformed coordinates.

In Problem 6.3 you will apply a translation to the momentum operator to show that $\hat{p}' = \hat{T}^{\dagger}\; \hat{p}\; \hat{T} = \hat{p}$: the momentum operator is unchanged by this transformation. Physically, this is because the particle's momentum is independent of where you place the origin of your coordinates, depending only on *differences* in position: $p = m\,dx/dt$. Once you know how the position and momentum operators behave under a translation, you know how *any* operator does, since

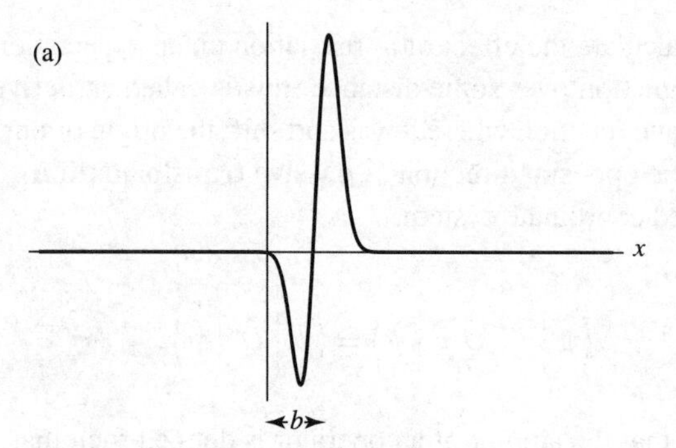

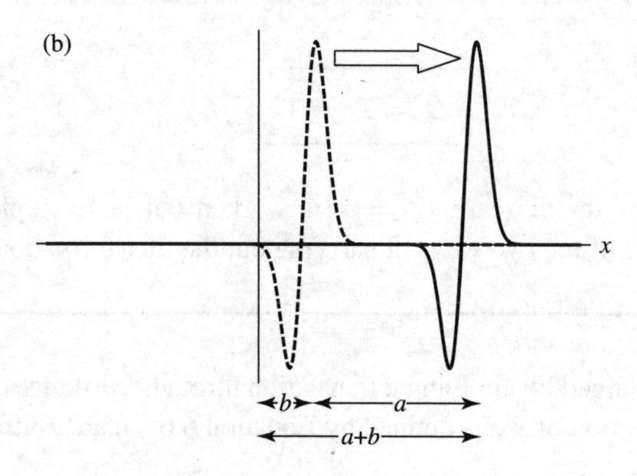

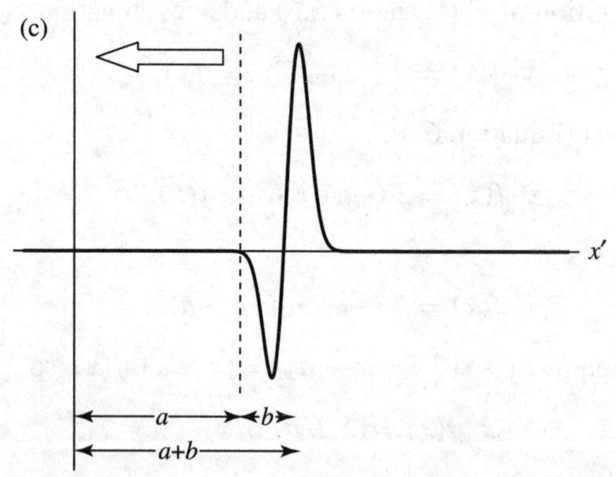

Figure 6.5: Active vs. passive transformations: (a) depicts the original function, (b) illustrates an active transformation in which the function is shifted to the right by an amount a, and (c) illustrates a passive transformation where the axes are shifted to the left by an amount a. A point on the wave a distance b from the origin before the transformation is a distance $a + b$ from the origin after the transformation in either (b) or (c); this is the equivalence of the two pictures.

$$\hat{Q}'(\hat{x}, \hat{p}) = \hat{T}^\dagger\, \hat{Q}(\hat{x}, \hat{p})\, \hat{T} = \hat{Q}(\hat{x}', \hat{p}') = \hat{Q}(\hat{x} + a, \hat{p}). \tag{6.8}$$

Problem 6.4 will walk you through the proof.

Problem 6.3 Show that the operator $\hat{p}'$ obtained by applying a translation to the operator $\hat{p}$ is $\hat{p}' = \hat{T}^\dagger\, \hat{p}\, \hat{T} = \hat{p}$.

Problem 6.4 Prove Equation 6.8. You may assume that $Q(\hat{x}, \hat{p})$ can be written in a power series

$$\hat{Q}(\hat{x}, \hat{p}) = \sum_{m=0}^{\infty} \sum_{n=0}^{\infty} a_{mn} \, \hat{x}^m \, \hat{p}^n$$

for some constants a_{mn}.

6.2.2 Translational Symmetry

So far we have seen how a *function* behaves under a translation and how an *operator* behaves under a translation. I am now in a position to make precise the notion of a symmetry that I mentioned in the introduction. A system is **translationally invariant** (equivalent to saying it has translational symmetry) if the Hamiltonian is unchanged by the transformation:

$$\hat{H}' = \hat{T}^{\dagger} \, \hat{H} \, \hat{T} = \hat{H}.$$

Because $\hat{T}$ is unitary (Equation 6.4) we can multiply both sides of this equation by $\hat{T}$ to get

$$\hat{H} \, \hat{T} = \hat{T} \, \hat{H}.$$

Therefore, a system has translational symmetry if the Hamiltonian commutes with the translation operator:

$$\left[\hat{H}, \hat{T}\right] = 0. \tag{6.9}$$

For a particle of mass m moving in a one-dimensional potential, the Hamiltonian is

$$\hat{H} = \frac{\hat{p}^2}{2m} + V(x).$$

According to Equation 6.8, the transformed Hamiltonian is

$$\hat{H}' = \frac{\hat{p}^2}{2m} + V(x+a)$$

so translational symmetry implies that

$$V(x+a) = V(x). \tag{6.10}$$

Now, there are two very different physical settings where Equation 6.10 might arise. The first is a constant potential, where Equation 6.10 holds for *every* value of a; such a system is said to have **continuous translational symmetry**. The second is a periodic potential, such as an electron might encounter in a crystal, where Equation 6.10 holds only for a discrete set of as; such a system is said to have **discrete translational symmetry**. The two cases are illustrated in Figure 6.6.

Discrete Translational Symmetry and Bloch's Theorem

What are the implications of translational symmetry? For a system with a discrete translational symmetry, the most important consequence is Bloch's theorem; the theorem specifies the form taken by the stationary states. We used this theorem in Section 5.3.2; I will now prove it.

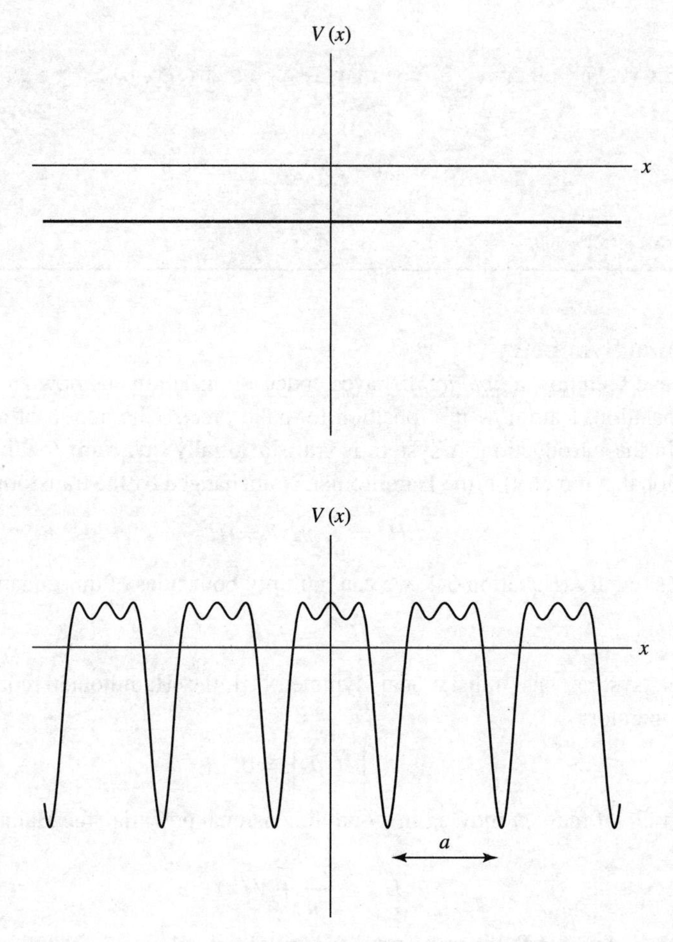

Figure 6.6: Potentials for a system with continuous (top) and discrete (bottom) translational symmetry. In the former case the potential is the same when shifted right or left by *any* amount; in the latter case the potential is the same when shifted right or left by an integer multiple of a.

In Section A.5 it is shown that if two operators commute, then they have a complete set of simultaneous eigenstates. This means that if the Hamiltonian is translationally invariant (which is to say, if it commutes with the translation operator), then the eigenstates $\psi(x)$ of the Hamiltonian can be chosen to be simultaneously eigenstates of $\hat{T}$:

$$\hat{H}\,\psi(x) = E\,\psi(x), \qquad \hat{T}(a)\,\psi(x) = \lambda\,\psi(x),$$

where λ is the eigenvalue associated with $\hat{T}$. Since $\hat{T}$ is unitary, its eigenvalues have magnitude 1 (see Problem A.30), which means that λ can be written as $\lambda = e^{i\phi}$ for some real number ϕ. By convention we write $\phi = -qa$ where $\hbar q$ is called the **crystal momentum**. Therefore, the stationary states of a particle of mass m moving in a periodic potential have the property

$$\psi(x-a) = e^{-i\,qa}\,\psi(x). \tag{6.11}$$

There is a more illuminating way to write Equation 6.11:[7]

$$\boxed{\psi(x) = e^{iqx}\, u(x)} \tag{6.12}$$

where $u(x)$ is a periodic function of x: $u(x+a) = u(x)$ and e^{iqx} is a traveling wave (recall that a traveling wave by itself describes a *free* particle—Section 2.4) with wavelength $2\pi/q$. Equation 6.12 is **Bloch's theorem** and it says that the stationary states of a particle in a periodic potential are periodic functions multiplying traveling waves. Note that just because the Hamiltonian is translationally invariant, that doesn't mean the stationary states *themselves* are translationally invariant, it simply means that they can be chosen to be eigenstates of the translation operator.

Bloch's theorem is truly remarkable. It tells us that the stationary states of a particle in a periodic potential (such as an electron in a crystal) are, apart from a periodic modulation, traveling waves. As such, they have a nonzero velocity.[8] This means that an electron could travel through a perfect crystal without scattering! That has dramatic implications for electronic conduction in solids.

Continuous Translational Symmetry and Momentum Conservation

If a system has continuous translation symmetry then the Hamiltonian commutes with $\hat{T}(a)$ for any choice of a. In this case it is useful to consider an **infinitesimal translation**

$$\hat{T}(\delta) = e^{-i\,\delta\,\hat{p}/\hbar} \approx 1 - i\,\frac{\delta}{\hbar}\,\hat{p},$$

where δ is an infinitesimal length.[9]

If the Hamiltonian has continuous translational symmetry, then it must be unchanged under any translation, including an infinitesimal one; equivalently it commutes with the translation operator, and hence

$$\left[\hat{H}, \hat{T}(\delta)\right] = \left[\hat{H}, 1 - i\,\frac{\delta}{\hbar}\,\hat{p}\right] = 0 \Rightarrow \left[\hat{H}, \hat{p}\right] = 0.$$

So if the Hamiltonian has continuous translational symmetry, it must commute with the momentum operator. And if the Hamiltonian commutes with momentum, then according to the "generalized Ehrenfest's theorem" (Equation 3.73)

$$\frac{d}{dt}\langle p\rangle = \frac{i}{\hbar}\left\langle\left[\hat{H}, \hat{p}\right]\right\rangle = 0. \tag{6.13}$$

---

[7] It is clear that Equation 6.12 satisfies Equation 6.11. In Problem 6.5 you'll prove that they are in fact equivalent statements.

[8] For a delightful proof using perturbation theory, see Neil Ashcroft and N. David Mermin, *Solid State Physics*, Cengage, Belmont, 1976 (p. 765), after you have completed Problem 6.6 and studied Chapter 7.

[9] For the case of continuous symmetries, it is often much easier to work with the infinitesimal form of the transformation; any finite transformation can then be built up as a product of infinitesimal transformations. In particular, the finite translation by a is a sequence of N infinitesimal translations with $\delta = a/N$ in the limit that $N \to \infty$:

$$\lim_{N\to\infty}\left(1 - i\,\frac{a}{N}\,\frac{1}{\hbar}\,\hat{p}\right)^N = \exp\left[-\frac{i\,a}{\hbar}\,\hat{p}\right].$$

For a proof see R. Shankar, *Basic Training in Mathematics: A Fitness Program for Science Students*, Plenum Press, New York, 1995 (p.11).

This is a statement of **momentum conservation** and we have now shown that continuous translational symmetry implies that momentum is conserved. This is our first example of a powerful general principle: *symmetries imply conservation laws.*[10]

Of course, if we're talking about a single particle of mass m moving in a potential $V(x)$, the only potential that has continuous translational symmetry is the constant potential, which is equivalent to the free particle. And it is pretty obvious that momentum is conserved in that case. But the analysis here readily extends to a system of interacting particles (see Problem 6.7). The fact that momentum is conserved in that case as well (so long as the Hamiltonian is translationally invariant) is a highly nontrivial result. In any event, the point to remember is that *conservation of momentum is a consequence of translational symmetry.*

Problem 6.5 Show that Equation 6.12 follows from Equation 6.11. *Hint*: First write $\psi(x) = e^{iqx}\,u(x)$, which is certainly true for *some* $u(x)$, and then show that $u(x)$ is necessarily a periodic function of x.

*** **Problem 6.6** Consider a particle of mass m moving in a potential $V(x)$ with period a. We know from Bloch's theorem that the wave function can be written in the form of Equation 6.12. *Note*: It is conventional to label the states with quantum numbers n and q as $\psi_{nq}(x) = e^{iqx}\,u_{nq}(x)$ where E_{nq} is the nth energy for a given value of q.

(a) Show that u satisfies the equation

$$-\frac{\hbar^2}{2m}\frac{d^2u_{nq}}{dx^2} - \frac{i\hbar^2 q}{m}\frac{du_{nq}}{dx} + V(x)\,u_{nq} = \left(E_{nq} - \frac{\hbar^2 q^2}{2m}\right)u_{nq}. \tag{6.14}$$

(b) Use the technique from Problem 2.61 to solve the differential equation for u_{nq}. You need to use a two-sided difference for the first derivative so that you have a Hermitian matrix to diagonalize: $\frac{d\psi}{dx} \approx \frac{\psi_{j+1}-\psi_{j-1}}{2\Delta x}$. For the potential in the interval 0 to a let

$$V(x) = \begin{cases} -V_0 & a/4 < x < 3a/4 \\ 0 & \text{else} \end{cases}$$

with $V_0 = 20\,\hbar^2/2\,m\,a^2$. (You will need to modify the technique slightly to account for the fact that the function u_{nq} is periodic.) Find the lowest two energies for the following values of the crystal momentum: $q\,a = -\pi, -\pi/2, 0, \pi/2, \pi$. Note that q and $q + 2\pi/a$ describe the same wave function (Equation 6.12), so there is no reason to consider values of $q\,a$ outside of the interval from $-\pi$ to π. In solid state physics, the values of q inside this range constitute the **first Brillouin zone**.

(c) Make a plot of the energies E_{1q} and E_{2q} for values of q between $-\pi/a$ and π/a. If you've automated the code that you used in part (b), you should be able to show a large number of q values in this range. If not, simply plot the values that you computed in (b).

[10] In the case of a discrete translational symmetry, momentum is not conserved, but there is a conserved quantity closely related to the discrete translational symmetry, which is the crystal momentum. For a discussion of crystal momentum see Steven H. Simon, *The Oxford Solid State Basics*, Oxford, 2013, p.84.

Problem 6.7 Consider two particles of mass m_1 and m_2 (in one dimension) that interact via a potential that depends only on the distance between the particles $V(|x_1 - x_2|)$, so that the Hamiltonian is

$$\hat{H} = -\frac{\hbar^2}{2m_1}\frac{\partial^2}{\partial x_1^2} - \frac{\hbar^2}{2m_2}\frac{\partial^2}{\partial x_2^2} + V(|x_1 - x_2|).$$

Acting on a two-particle wave function the translation operator would be

$$\hat{T}(a)\,\psi(x_1, x_2) = \psi(x_1 - a, x_2 - a).$$

(a) Show that the translation operator can be written

$$\hat{T}(a) = e^{-\frac{i a}{\hbar}\hat{P}},$$

where $\hat{P} = \hat{p}_1 + \hat{p}_2$ is the total momentum.

(b) Show that the total momentum is conserved for this system.

6.3 CONSERVATION LAWS

In classical mechanics the meaning of a conservation law is straightforward: the quantity in question is the same before and after some event. Drop a rock, and potential energy is converted into kinetic energy, but the *total* is the same just before it hits the ground as when it was released; collide two billiard balls and momentum is transferred from one to the other, but the total remains unchanged. But in quantum mechanics a system does not in general *have* a definite energy (or momentum) before the process begins (or afterward). What does it *mean*, in that case, to say that the observable Q is (or is not) conserved? Here are two possibilities:

- **First definition**: The *expectation value* $\langle Q\rangle$ is independent of time.
- **Second definition**: The probability of getting any particular value is independent of time.

Under what conditions does each of these conservation laws hold?

Let us stipulate that the observable in question does not depend explicitly on time: $\partial Q/\partial t = 0$. In that case the generalized Ehrenfest theorem (Equation 3.73) tells us that the expectation value of Q is independent of time if *The operator $\hat{Q}$ commutes with the Hamiltonian.* It so happens that the same criterion guarantees conservation by the second definition.

I will now prove this result. Recall that the probability of getting the result q_n in a measurement of Q at time t is (Equation 3.43)

$$P(q_n) = |\langle f_n \mid \Psi(t)\rangle|^2, \tag{6.15}$$

where f_n is the corresponding eigenvector: $\hat{Q}\,|f_n\rangle = q_n\,|f_n\rangle$.[11] We know that the time evolution of the wave function is (Equation 2.17)

[11] If the spectrum of $\hat{Q}$ is degenerate (there are distinct eigenvectors with the same eigenvalue q_n: $\hat{Q}\,|f_n^{(i)}\rangle = q_n\,|f_n^{(i)}\rangle$ for $i = 1, 2, \ldots$), then we need to sum over those states:

$$P(q_n) = \sum_i \left|\left\langle f_n^{(i)} \middle| \Psi(t)\right\rangle\right|^2.$$

Except for the sum over i the proof proceeds unchanged.

$$|\Psi(t)\rangle = \sum_m e^{-i E_m t/\hbar} c_m \, |\psi_m\rangle ,$$

where the $|\psi_n\rangle$ are the eigenstates of $\hat{H}$, and therefore

$$P(q_n) = \left| \sum_m e^{-i E_m t/\hbar} c_m \, \langle f_n | \psi_m \rangle \right|^2 .$$

Now the key point: since $\hat{Q}$ and $\hat{H}$ commute we can find a complete set of simultaneous eigenstates for them (see Section A.5); without loss of generality then $|f_n\rangle = |\psi_n\rangle$. Using the orthonormality of the $|\psi_n\rangle$,

$$P(q_n) = \left| \sum_m e^{-i E_m t/\hbar} c_m \, \langle \psi_n \mid \psi_m \rangle \right|^2 = |c_n|^2 ,$$

which is clearly independent of time.

6.4 PARITY

6.4.1 Parity in One Dimension

A spatial inversion is implemented by the parity operator $\hat{\Pi}$; in one dimension,

$$\hat{\Pi} \, \psi(x) = \psi'(x) = \psi(-x) .$$

Evidently, the parity operator is its own inverse: $\hat{\Pi}^{-1} = \hat{\Pi}$; in Problem 6.8 you will show that it is Hermitian: $\hat{\Pi}^\dagger = \hat{\Pi}$. Putting this together, the parity operator is unitary as well:

$$\hat{\Pi}^{-1} = \hat{\Pi} = \hat{\Pi}^\dagger . \tag{6.16}$$

Operators transform under a spatial inversion as

$$\hat{Q}' = \hat{\Pi}^\dagger \, \hat{Q} \, \hat{\Pi}. \tag{6.17}$$

I won't repeat the argument leading up to Equation 6.17, since it is identical to the one by which we arrived at Equation 6.6 in the case of translations. The position and momentum operators are "odd under parity" (Problem 6.10):

$$\hat{x}' = \hat{\Pi}^\dagger \, \hat{x} \, \hat{\Pi} = -\hat{x}, \tag{6.18}$$

$$\hat{p}' = \hat{\Pi}^\dagger \, \hat{p} \, \hat{\Pi} = -\hat{p}, \tag{6.19}$$

and this tells us how *any* operator transforms (see Problem 6.4):

$$\hat{Q}'(\hat{x}, \hat{p}) = \hat{\Pi}^\dagger \, \hat{Q}(\hat{x}, \hat{p}) \, \hat{\Pi} = \hat{Q}(-\hat{x}, -\hat{p}) .$$

A system has **inversion symmetry** if the Hamiltonian is unchanged by a parity transformation:

$$\hat{H}' = \hat{\Pi}^\dagger \, \hat{H} \, \hat{\Pi} = \hat{H},$$

or, using the unitarity of the parity operator,

$$\left[\hat{H}, \hat{\Pi}\right] = 0. \tag{6.20}$$

If our Hamiltonian describes a particle of mass m in a one-dimensional potential $V(x)$, then inversion symmetry simply means that the potential is an even function of position:

$$V(x) = V(-x).$$

The implications of inversion symmetry are two: First, we can find a complete set of simultaneous eigenstates of $\hat{\Pi}$ and $\hat{H}$. Let such an eigenstate be written ψ_n; it satisfies

$$\hat{\Pi}\,\psi_n(x) = \psi_n(-x) = \pm\,\psi_n(x),$$

since the eigenvalues of the parity operator are restricted to ± 1 (Problem 6.8). So the stationary states of a potential that is an even function of position are themselves even or odd functions (or can be chosen as such, in the case of degeneracy).[12] This property is familiar from the simple harmonic oscillator, the infinite square well (if the origin is placed at the center of the well), and the Dirac delta function potential, and you proved it in general in Problem 2.1.

Second, according to Ehrenfest's theorem, if the Hamiltonian has an inversion symmetry then

$$\frac{d}{dt}\langle \Pi \rangle = \frac{i}{\hbar}\left\langle \left[\hat{H}, \hat{\Pi}\right] \right\rangle = 0$$

so *parity is conserved* for a particle moving in a symmetric potential. And not just the expectation value, but the *probability* of any particular outcome in a measurement, in accord with the theorem of Section 6.3. Parity conservation means, for example, that if the wave function of a particle in a harmonic oscillator potential is even at $t = 0$ then it will be even at any later time t; see Figure 6.7.

* **Problem 6.8**

(a) Show that the parity operator $\hat{\Pi}$ is Hermitian.

(b) Show that the eigenvalues of the parity operator are ± 1.

6.4.2 Parity in Three Dimensions

The spatial inversion generated by the parity operator in three dimensions is

$$\hat{\Pi}\,\psi(\mathbf{r}) = \psi'(\mathbf{r}) = \psi(-\mathbf{r}).$$

The operators $\hat{\mathbf{r}}$ and $\hat{\mathbf{p}}$ transform as

$$\hat{\mathbf{r}}' = \hat{\Pi}^\dagger\,\hat{\mathbf{r}}\,\hat{\Pi} = -\hat{\mathbf{r}}, \tag{6.21}$$

$$\hat{\mathbf{p}}' = \hat{\Pi}^\dagger\,\hat{\mathbf{p}}\,\hat{\Pi} = -\hat{\mathbf{p}}. \tag{6.22}$$

Any other operator transforms as

$$\hat{Q}'(\hat{\mathbf{r}}, \hat{\mathbf{p}}) = \hat{\Pi}^\dagger\,\hat{Q}(\hat{\mathbf{r}}, \hat{\mathbf{p}})\,\hat{\Pi} = \hat{Q}(-\hat{\mathbf{r}}, -\hat{\mathbf{p}}). \tag{6.23}$$

---

[12] For bound (normalizable) states in one dimension, there is no degeneracy and every bound state of a symmetric potential is automatically an eigenstate of parity. (However, see Problem 2.46.) For scattering states, degeneracy does occur.

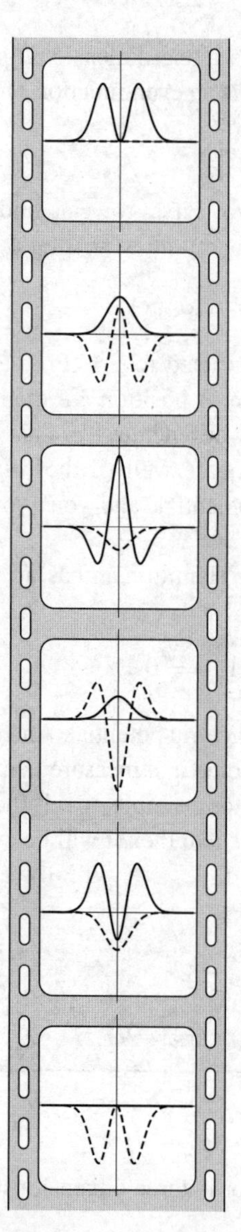

Figure 6.7: This filmstrip shows the time evolution of a particular wave function $\left(\Psi_0(\xi) = A\,\xi^2\,e^{-\xi^2/2}\right)$ for a particle in the harmonic oscillator potential. The solid and dashed curves are the real and imaginary parts of the wave function respectively, and time increases from top to bottom. Since parity is conserved, a wave function which is initially an even function of position (as this one is) remains an even function at all later times.

Example 6.2

Find the parity-transformed angular momentum operator $\hat{\mathbf{L}}' = \hat{\Pi}^\dagger\,\hat{\mathbf{L}}\,\hat{\Pi}$, in terms of $\hat{\mathbf{L}}$.

Solution: Since $\mathbf{L} = \mathbf{r} \times \mathbf{p}$, Equation 6.23 tells us that

$$\hat{\mathbf{L}}' = \hat{\Pi}^\dagger\,\hat{\mathbf{L}}\,\hat{\Pi} = \hat{\mathbf{r}}' \times \hat{\mathbf{p}}' = (-\hat{\mathbf{r}}) \times (-\hat{\mathbf{p}}) = \hat{\mathbf{r}} \times \hat{\mathbf{p}} = \hat{\mathbf{L}}. \tag{6.24}$$

We have a special name for vectors like $\hat{\mathbf{L}}$, that are *even* under parity. We call them **pseudovectors**, since they don't change sign under parity the way "true" vectors, such as $\hat{\mathbf{r}}$ or $\hat{\mathbf{p}}$, do. Similarly, scalars that are *odd* under parity are called **pseudoscalars**, since they do not behave under parity the way that "true" scalars (such as $\hat{\mathbf{r}} \cdot \hat{\mathbf{r}}$ which is even under parity) do. See Problem 6.9. *Note*: The labels **scalar** and **vector** describe how the operators behave under *rotations*; we will define these terms carefully in the next section. "True" vectors and pseudovectors behave the same way under a rotation—they are both vectors.

In three dimensions, the Hamiltonian for a particle of mass m moving in a potential $V(\mathbf{r})$ will have inversion symmetry if $V(-\mathbf{r}) = V(\mathbf{r})$. Importantly, any central potential satisfies this condition. As in the one-dimensional case, parity is conserved for such systems, and the eigenstates of the Hamiltonian may be chosen to be simultaneously eigenstates of parity. In Problem 6.1 you proved that the eigenstates of a particle in a central potential, written $\psi_{n\ell m}(r,\theta,\phi) = R_{n\ell}(r)\, Y_\ell^m(\theta,\phi)$, are eigenstates of parity:[13]

$$\hat{\Pi}\,\psi_{n\ell m}(r,\theta,\phi) = (-1)^\ell\,\psi_{n\ell m}(r,\theta,\phi)\,.$$

* **Problem 6.9**

(a) Under parity, a "true" scalar operator does not change:

$$\hat{\Pi}^\dagger\,\hat{f}\,\hat{\Pi} = \hat{f},$$

whereas a pseudoscalar changes sign. Show therefore that $\left[\hat{\Pi}, \hat{f}\right] = 0$ for a "true" scalar, whereas $\left\{\hat{\Pi}, \hat{f}\right\} = 0$ for a pseudoscalar. *Note*: the **anti-commutator** of two operators $\hat{A}$ and $\hat{B}$ is defined as $\left\{\hat{A}, \hat{B}\right\} \equiv \hat{A}\,\hat{B} + \hat{B}\,\hat{A}$.

(b) Similarly, a "true" vector changes sign

$$\hat{\Pi}^\dagger\,\hat{\mathbf{V}}\,\hat{\Pi} = -\hat{\mathbf{V}},$$

whereas a pseudovector is unchanged. Show therefore that $\left\{\hat{\Pi}, \hat{\mathbf{V}}\right\} = \mathbf{0}$ for a "true" vector and $\left[\hat{\Pi}, \hat{\mathbf{V}}\right] = \mathbf{0}$ for a pseudovector.

6.4.3 Parity Selection Rules

Selection rules tell you when a matrix element is zero based on the symmetry of the situation. Recall that a matrix element is any object of the form $\langle b|\hat{Q}|a\rangle$; an expectation value is a special case of a matrix element with $a = b = \psi$. One operator whose selection rules are physically important is the electric dipole moment operator

$$\hat{\mathbf{p}}_e = q\,\hat{\mathbf{r}}.$$

The selection rules for this operator—the operator itself is nothing more than the charge of the particle times its position—determine which atomic transitions are allowed and which are forbidden (see Chapter 11). It is odd under parity since the position vector $\hat{\mathbf{r}}$ is odd:

[13] Note that Equation 6.24 could equivalently be written as $\left[\hat{\Pi}, \hat{\mathbf{L}}\right] = \mathbf{0}$. The fact that parity commutes with every component of the angular momentum (and therefore also $\hat{L}^2$) is the reason you can find simultaneous eigenstates of $\hat{L}^2$, $\hat{L}_z$, and $\hat{\Pi}$.

$$\hat{\Pi}^{\dagger}\,\hat{\mathbf{p}}_e\,\hat{\Pi} = -\hat{\mathbf{p}}_e. \tag{6.25}$$

Now consider the matrix elements of the electric dipole operator between two states $\psi_{n\ell m}$ and $\psi_{n'\ell'm'}$ (we label the corresponding kets $|n\ell m\rangle$ and $|n'\ell'm'\rangle$). Using Equation 6.25 we have

$$\begin{aligned}\left\langle n'\ell'm' \,\middle|\, \hat{\mathbf{p}}_e \,\middle|\, n\ell m\right\rangle &= -\left\langle n'\ell'm' \,\middle|\, \hat{\Pi}^{\dagger}\,\hat{\mathbf{p}}_e\,\hat{\Pi} \,\middle|\, n\ell m\right\rangle \\ &= -\left\langle n'\ell'm' \,\middle|\, (-1)^{\ell'}\,\hat{\mathbf{p}}_e\,(-1)^{\ell} \,\middle|\, n\ell m\right\rangle \\ &= (-1)^{\ell+\ell'+1}\left\langle n'\ell'm' \,\middle|\, \hat{\mathbf{p}}_e \,\middle|\, n\ell m\right\rangle.\end{aligned}$$

From this we see immediately that

$$\left\langle n'\ell'm' \,\middle|\, \hat{\mathbf{p}}_e \,\middle|\, n\ell m\right\rangle = \mathbf{0} \text{ if } \ell+\ell' \text{ is even.} \tag{6.26}$$

This is called **Laporte's rule**; it says that matrix elements of the dipole moment operator vanish between states with the same parity. The reasoning by which we obtained Equation 6.26 can be generalized to derive selection rules for any operator, as long as you know how that operator transforms under parity. In particular, *Laporte's rule applies to any operator that is odd under parity*. The selection rule for an operator that is *even* under parity, such as $\hat{\mathbf{L}}$, is derived in Problem 6.11.

Problem 6.10 Show that the position and momentum operators are odd under parity. That is, prove Equations 6.18, 6.19, and, by extension, 6.21 and 6.22.

* **Problem 6.11** Consider the matrix elements of $\hat{\mathbf{L}}$ between two definite-parity states: $\langle n'\ell'm'|\hat{\mathbf{L}}|n\ell m\rangle$. Under what conditions is this matrix element guaranteed to vanish? Note that the same selection rule would apply to any pseudovector operator, or any "true" scalar operator.

Problem 6.12 Spin angular momentum, $\hat{\mathbf{S}}$, is even under parity, just like orbital angular momentum $\hat{\mathbf{L}}$:

$$\hat{\Pi}^{\dagger}\,\hat{\mathbf{S}}\,\hat{\Pi} = \hat{\mathbf{S}} \quad \text{or} \quad \left[\hat{\Pi}, \hat{\mathbf{S}}\right] = 0. \tag{6.27}$$

Acting on a spinor written in the standard basis (Equation 4.139), the parity operator becomes a 2×2 matrix. Show that, due to Equation 6.27, this matrix must be a constant times the identity matrix. As such, the parity of a spinor isn't very interesting since both spin states are parity eigenstates with the same eigenvalue. We can arbitrarily choose that parity to be $+1$, so the parity operator has no effect on the spin portion of the wave function.[14]

[14] However, it turns out that *anti*particles of spin $1/2$ have opposite parity. Thus the electron is conventionally assigned parity $+1$, but the positron then has parity -1.

* **Problem 6.13** Consider an electron in a hydrogen atom.

(a) Show that if the electron is in the ground state, then necessarily $\langle \mathbf{p}_e \rangle = 0$. No calculation allowed.

(b) Show that if the electron is in an $n = 2$ state, then $\langle \mathbf{p}_e \rangle$ need not vanish. Give an example of a wave function for the energy level $n = 2$ that has a non-vanishing $\langle \mathbf{p}_e \rangle$ and compute $\langle \mathbf{p}_e \rangle$ for this state.

6.5 ROTATIONAL SYMMETRY

6.5.1 Rotations About the z Axis

The operator that rotates a function about the z axis by an angle φ (Equation 6.2)

$$\hat{R}_z(\varphi)\, \psi(r,\theta,\phi) = \psi'(r,\theta,\phi) = \psi(r,\theta,\phi-\varphi) \tag{6.28}$$

is closely related to the z component of angular momentum (Equation 4.129). By the same reasoning that led to Equation 6.3,

$$\hat{R}_z(\varphi) = \exp\left[-\frac{i\,\varphi}{\hbar}\,\hat{L}_z\right], \tag{6.29}$$

and we say that $\hat{L}_z$ is the **generator of rotations** about the z axis (compare Equation 6.3).

How do the operators $\hat{\mathbf{r}}$ and $\hat{\mathbf{p}}$ transform under rotations? To answer this question we use the infinitesimal form of the operator:

$$\hat{R}_z(\delta) \approx 1 - \frac{i\,\delta}{\hbar}\,\hat{L}_z.$$

Then the operator $\hat{x}$ transforms as

$$\begin{aligned}\hat{x}' = \hat{R}^\dagger\, \hat{x}\, \hat{R} &\approx \left(1 + \frac{i\,\delta}{\hbar}\,\hat{L}_z\right)\hat{x}\left(1 - \frac{i\,\delta}{\hbar}\,\hat{L}_z\right)\\ &\approx \hat{x} + \frac{i\,\delta}{\hbar}\left[\hat{L}_z, \hat{x}\right] \approx \hat{x} - \delta\,\hat{y}\end{aligned}$$

(I used Equation 4.122 for the commutator). Similar calculations show that $\hat{y}' = \hat{y} + \delta\,\hat{x}$ and $\hat{z}' = \hat{z}$. We can combine these results into a matrix equation

$$\begin{pmatrix}\hat{x}'\\ \hat{y}'\\ \hat{z}'\end{pmatrix} = \begin{pmatrix}1 & -\delta & 0\\ \delta & 1 & 0\\ 0 & 0 & 1\end{pmatrix}\begin{pmatrix}\hat{x}\\ \hat{y}\\ \hat{z}\end{pmatrix}. \tag{6.30}$$

That doesn't look quite right for a rotation. Shouldn't it be

$$\begin{pmatrix}\hat{x}'\\ \hat{y}'\\ \hat{z}'\end{pmatrix} = \begin{pmatrix}\cos\varphi & -\sin\varphi & 0\\ \sin\varphi & \cos\varphi & 0\\ 0 & 0 & 1\end{pmatrix}\begin{pmatrix}\hat{x}\\ \hat{y}\\ \hat{z}\end{pmatrix}? \tag{6.31}$$

Yes, but don't forget, we are assuming $\varphi \to \delta$ is infinitesimal, so (dropping terms of order δ^2 and higher) $\cos\varphi \to 1$ and $\sin\varphi \to \delta$.[15]

[15] To go the other way, from infinitesimal to finite, see Problem 6.14.

∗∗ **Problem 6.14** In this problem you will establish the correspondence between Equations 6.30 and 6.31.

(a) Diagonalize the matrix[16]

$$\mathsf{M} = \begin{pmatrix} 1 & -\varphi/N \\ \varphi/N & 1 \end{pmatrix}$$

to obtain the matrix

$$\mathsf{M}' = \mathsf{S}\,\mathsf{M}\,\mathsf{S}^{-1},$$

where S^{-1} is the unitary matrix whose columns are the (normalized) eigenvectors of M.

(b) Use the binomial expansion to show that $\lim_{N\to\infty}(\mathsf{M}')^N$ is a diagonal matrix with entries $e^{-i\varphi}$ and $e^{i\varphi}$ on the diagonal.

(c) Transform back to the original basis to show that

$$\lim_{N\to\infty} \mathsf{M}^N = \mathsf{S}^{-1}\left[\lim_{N\to\infty}(\mathsf{M}')^N\right]\mathsf{S}$$

agrees with the matrix in Equation 6.31.

6.5.2 Rotations in Three Dimensions

Equation 6.29 can be generalized in the obvious way to a rotation about an axis along the unit vector $\mathbf{n}$:

$$\boxed{\hat{R}_{\mathbf{n}}(\varphi) = \exp\left[-\frac{i\,\varphi}{\hbar}\,\mathbf{n}\cdot\hat{\mathbf{L}}\right]} \tag{6.32}$$

Just as linear momentum is the generator of translations, angular momentum is the **generator of rotations.**

Any operator (with three components) that transforms the same way as the position operator under rotations is called a **vector operator**. By "transforms the same way" we mean that $\hat{\mathbf{V}}' = \mathsf{D}\,\hat{\mathbf{V}}$ where D is the same matrix as appears in $\hat{\mathbf{r}}' = \mathsf{D}\,\hat{\mathbf{r}}$. In particular for a rotation about the z axis, we would have (Equation 6.31)

$$\begin{pmatrix} \hat{V}_x' \\ \hat{V}_y' \\ \hat{V}_z' \end{pmatrix} = \begin{pmatrix} \cos\varphi & -\sin\varphi & 0 \\ \sin\varphi & \cos\varphi & 0 \\ 0 & 0 & 1 \end{pmatrix}\begin{pmatrix} \hat{V}_x \\ \hat{V}_y \\ \hat{V}_z \end{pmatrix}.$$

This transformation rule follows from the commutation relations[17]

$$\boxed{\left[\hat{L}_i,\,\hat{V}_j\right] = i\,\hbar\,\epsilon_{ijk}\,\hat{V}_k,} \tag{6.33}$$

[16] See Section A.5.

[17] The Levi-Civita symbol ϵ_{ijk} is defined in Problem 4.29.

Table 6.1: *Operators are classified as vectors or scalars based on their commutation relations with $\hat{\mathbf{L}}$, which encode how they transform under a rotation, and as pseudo- or "true" quantities based on their commutation relations with $\hat{\Pi}$, which encode how they transform under a spatial inversion. The curly brackets in the first column denote the anti-commutator, defined in Problem 6.9. To include the spin $\hat{\mathbf{S}}$ in this table, one simply replaces $\hat{L}_i$ everywhere it appears in the third column with $\hat{J}_i = \hat{L}_i + \hat{S}_i$ (Problems 6.12 and 6.32, respectively, discuss the effect of parity and rotations on spinors). $\hat{\mathbf{S}}$, like $\hat{\mathbf{L}}$, is then a pseudovector and $\hat{\mathbf{p}} \cdot \hat{\mathbf{S}}$ is a pseudoscalar.*

	parity	rotations	examples
true vector $\hat{\mathbf{V}}$	$\left\{\hat{\Pi}, \hat{V}_i\right\} = 0$	$\left[\hat{L}_i, \hat{V}_j\right] = i\,\hbar\epsilon_{ijk}\,\hat{V}_k$	$\hat{\mathbf{r}}, \hat{\mathbf{p}}$
pseudovector $\hat{\mathbf{V}}$	$\left[\hat{\Pi}, \hat{V}_i\right] = 0$	$\left[\hat{L}_i, \hat{V}_j\right] = i\,\hbar\epsilon_{ijk}\,\hat{V}_k$	$\hat{\mathbf{L}}$
true scalar $\hat{f}$	$\left[\hat{\Pi}, \hat{f}\right] = 0$	$\left[\hat{L}_i, \hat{f}\right] = 0$	$\hat{\mathbf{r}} \cdot \hat{\mathbf{r}}$
pseudoscalar $\hat{f}$	$\left\{\hat{\Pi}, \hat{f}\right\} = 0$	$\left[\hat{L}_i, \hat{f}\right] = 0$	

(see Problem 6.16), and we may take Equation 6.33 as the *definition* of a vector operator. So far we have encountered three such operators, $\hat{\mathbf{r}}$, $\hat{\mathbf{p}}$ and $\hat{\mathbf{L}}$:

$$\left[\hat{L}_i, \hat{r}_j\right] = i\,\hbar\epsilon_{ijk}\hat{r}_k, \qquad \left[\hat{L}_i, \hat{p}_j\right] = i\,\hbar\epsilon_{ijk}\,\hat{p}_k, \qquad \left[\hat{L}_i, \hat{L}_j\right] = i\,\hbar\epsilon_{ijk}\,\hat{L}_k$$

(see Equations 4.99 and 4.122).

A **scalar operator** is a single quantity that is unchanged by rotations; this is equivalent to saying that the operator commutes with $\hat{\mathbf{L}}$:

$$\boxed{\left[\hat{L}_i, \hat{f}\right] = 0.} \tag{6.34}$$

We can now classify operators as either scalars or vectors, based on their commutation relations with $\hat{\mathbf{L}}$ (how they transform under a rotation), and as "true" or pseudo-quantities, based on their commutators with $\hat{\Pi}$ (how they transform under parity). These results are summarized in Table 6.1.[18]

CONTINUOUS ROTATIONAL SYMMETRY

For a particle of mass m moving in a potential $V(\mathbf{r})$, the Hamiltonian

$$\hat{H} = \frac{\hat{p}^2}{2m} + V(\mathbf{r})$$

is rotationally invariant if $V(\mathbf{r}) = V(r)$ (the central potentials studied in Section 4.1.1). In this case the Hamiltonian commutes with a rotation by any angle about an arbitrary axis

[18] Of course, not every operator will fit into one of these categories. Scalar and vector operators are simply the first two instances in a hierarchy of tensor operators. Next come second-rank tensors (the inertia tensor from classical mechanics or the quadrupole tensor from electrodynamics are examples), third-rank tensors, and so forth.

$$\left[\hat{H}, \hat{R}_{\mathbf{n}}(\varphi)\right] = 0. \tag{6.35}$$

In particular, Equation 6.35 must hold for an *infinitesimal* rotation

$$R_{\mathbf{n}}(\delta) \approx 1 - i\frac{\delta}{\hbar}\mathbf{n} \cdot \hat{\mathbf{L}},$$

which means that the Hamiltonian commutes with the three components of **L**:

$$\left[\hat{H}, \hat{\mathbf{L}}\right] = \mathbf{0}. \tag{6.36}$$

What, then, are the consequences of rotational invariance?

From Equation 6.36 and Ehrenfest's theorem

$$\frac{d}{dt}\langle\mathbf{L}\rangle = \frac{i}{\hbar}\left\langle\left[\hat{H}, \hat{\mathbf{L}}\right]\right\rangle = \mathbf{0} \tag{6.37}$$

for a central potential. Thus, *angular momentum conservation is a consequence of rotational invariance*. And beyond the statement 6.37, angular momentum conservation means that the probability distributions (for each component of the angular momentum) are independent of time as well—see Section 6.3.

Since the Hamiltonian for a central potential commutes with all three components of angular momentum, it also commutes with $\hat{L}^2$. The operators $\hat{H}$, $\hat{L}_z$, and $\hat{L}^2$ form a **complete set of compatible observables** for the bound states of a central potential. Compatible means that they commute pairwise

$$\begin{aligned} \left[\hat{H}, \hat{L}^2\right] &= 0, \\ \left[\hat{H}, \hat{L}_z\right] &= 0, \\ \left[\hat{L}_z, \hat{L}^2\right] &= 0, \end{aligned} \tag{6.38}$$

so that the eigenstates of $\hat{H}$ can be chosen to be simultaneous eigenstates of $\hat{L}^2$ and $\hat{L}_z$.

$$\begin{aligned} \hat{H}\,\psi_{n\ell m} &= E_n\,\psi_{n\ell m}, \\ \hat{L}_z\,\psi_{n\ell m} &= m\,\hbar\,\psi_{n\ell m}, \\ \hat{L}^2\,\psi_{n\ell m} &= \ell\,(\ell+1)\,\hbar^2\,\psi_{n\ell m}. \end{aligned}$$

Saying they are complete means that the quantum numbers n, ℓ, and m uniquely specify a bound state of the Hamiltonian. This is familiar from our solution to the hydrogen atom, the infinite spherical well, and the three-dimensional harmonic oscillator, but it is true for *any* central potential.[19]

* **Problem 6.15** Show how Equation 6.34 guarantees that a scalar is unchanged by a rotation: $\hat{f}' = \hat{R}^\dagger\,\hat{f}\,\hat{R} = \hat{f}$.

[19] This follows from the fact that the radial Schrödinger equation (Equation 4.35) has at most a single normalizable solution so that, once you have specified ℓ and m, the energy uniquely specifies the state. The principal quantum number n indexes those energy values that lead to normalizable solutions.

* **Problem 6.16** Working from Equation 6.33, find how the vector operator $\hat{\mathbf{V}}$ transforms for an infinitesimal rotation by an angle δ about the y axis. That is, find the matrix D in

$$\hat{\mathbf{V}}' = \mathsf{D}\,\hat{\mathbf{V}}.$$

Problem 6.17 Consider the action of an infinitesimal rotation about the $\mathbf{n}$ axis of an angular momentum eigenstate $\psi_{n\ell m}$. Show that

$$\hat{R}_{\mathbf{n}}(\delta)\ \psi_{n\ell m} = \sum_{m'} D_{m'm}\,\psi_{n\,\ell\,m'},$$

and find the complex numbers $D_{m'm}$ (they will depend on δ, $\mathbf{n}$, and ℓ as well as m and m'). This result makes sense: a rotation doesn't change the magnitude of the angular momentum (specified by ℓ) but does change its projection along the z axis (specified by m).

6.6 DEGENERACY

Symmetry is the source of most[20] degeneracy in quantum mechanics. We have seen that a symmetry implies the existence of an operator $\hat{Q}$ that commutes with the Hamiltonian

$$\left[\hat{H}, \hat{Q}\right] = 0. \tag{6.39}$$

So why does symmetry lead to degeneracy in the energy spectrum? The basic idea is this: if we have a stationary state $|\psi_n\rangle$, then $|\psi_n'\rangle = \hat{Q}\,|\psi_n\rangle$ is a stationary state with the same energy. The proof is straightforward:

$$\hat{H}\,|\psi_n'\rangle = \hat{H}\left(\hat{Q}\,|\psi_n\rangle\right) = \hat{Q}\,\hat{H}\,|\psi_n\rangle = \hat{Q}\,E_n\,|\psi_n\rangle = E_n\left(\hat{Q}\,|\psi_n\rangle\right) = E_n\,|\psi_n'\rangle.$$

For example, if you have an eigenstate of a spherically-symmetric Hamiltonian and you rotate that state about some axis, you must get back another state of the same energy.

You might think that symmetry would *always* lead to degeneracy, and that continuous symmetries would lead to infinite degeneracy, but that is not the case. The reason is that the two states $|\psi_n\rangle$ and $|\psi_n'\rangle$ might be the same.[21] As an example, consider the Hamiltonian for the harmonic oscillator in one dimension; it commutes with parity. All of its stationary states are either even or odd, so when you act on one with the parity operator you get back the same state

20 When we can't identify the symmetry responsible for a particular degeneracy, we call it an **accidental degeneracy**. In most such cases, the degeneracy turns out to be no accident at all, but instead due to symmetry that is more difficult to identify than, say, rotational invariance. The canonical example is the larger symmetry group of the hydrogen atom (Problem 6.34).

21 This is highly non-classical. In classical mechanics, if you take a Keplerian orbit there will always be some axis about which you can rotate it to get a different Keplerian orbit (of the same energy) and in fact there will be an infinite number of such orbits with different orientations. In quantum mechanics, if you rotate the ground state of hydrogen you get back exactly the same state regardless of which axis you choose, and if you rotate one of the states with $n = 2$ and $\ell = 1$, you get back a linear combination of the three orthogonal states with these quantum numbers.

you started with (perhaps multiplied by -1, but that, physically, *is* the same state). There is therefore no degeneracy associated with inversion symmetry in this case.

In fact, if there is only a single symmetry operator $\hat{Q}$ (or if there are multiple symmetry operators that all commute), you do not get degeneracy in the spectrum. The reason is the same theorem we've now quoted many times: since $\hat{Q}$ and $\hat{H}$ commute, we can find simultaneous eigenstates $|\psi_n\rangle$ of $\hat{Q}$ and $\hat{H}$ and these states transform into themselves under the symmetry operation: $\hat{Q}\,|\psi_n\rangle = q_n\,|\psi_n\rangle$.

But what if there are *two* operators that commute with the Hamiltonian (call them $\hat{Q}$ and $\hat{\Lambda}$), but do *not* commute with each other? In this case, degeneracy in the energy spectrum is inevitable. Why?

First, consider a state $|\psi\rangle$ that is an eigenstate of both $\hat{H}$ and $\hat{Q}$ with eigenvalues E_n and q_m respectively. Since $\hat{H}$ and $\hat{\Lambda}$ commute we know that the state $|g\rangle = \hat{\Lambda}\,|\psi\rangle$ is also an eigenstate of $\hat{H}$ with eigenvalue E_n. Since $\hat{Q}$ and $\hat{\Lambda}$ do *not* commute we know (Section A.5) that there cannot exist a complete set of simultaneous eigenstates of all three operators ($\hat{Q}$, $\hat{\Lambda}$ and $\hat{H}$). Therefore, there must be some $|\psi\rangle$ such that $\hat{\Lambda}\,|\psi\rangle$ is distinct from $|\psi\rangle$ (specifically, it is not an eigenstate of $\hat{\Lambda}$) meaning that the energy level E_n is at least doubly degenerate. *The presence of multiple non-commuting symmetry operators guarantees degeneracy of the energy spectrum.*

This is precisely the situation we have encountered in the case of central potentials. Here the Hamiltonian commutes with rotations about any axis (or equivalently with the generators $\hat{L}_x$, $\hat{L}_y$, and $\hat{L}_z$) but those rotations don't commute with each other. So we know that there will be degeneracy in the spectrum of a particle in a central potential. The following example shows exactly how much degeneracy is explained by rotational invariance.

Example 6.3

Consider an eigenstate of a central potential $\psi_{n\ell m}$ with energy E_n. Use the fact that the Hamiltonian for a central potential commutes with any component of $\hat{\mathbf{L}}$, and therefore also with $\hat{L}_+$ and $\hat{L}_-$, to show that $\psi_{n\ell m\pm1}$ are necessarily also eigenstates with the same energy as $\psi_{n\ell m}$.[22]

Solution: Since the Hamiltonian commutes with $\hat{L}_\pm$ we have

$$\left(\hat{H}\,\hat{L}_\pm - \hat{L}_\pm\,\hat{H}\right)\psi_{n\ell m} = 0,$$

so

$$\hat{H}\,\hat{L}_\pm\,\psi_{n\ell m} = \hat{L}_\pm\,\hat{H}\,\psi_{n\ell m} = E_n\,\hat{L}_\pm\,\psi_{n\ell m}$$

or

$$\hat{H}\,\psi_{n\ell m\pm1} = E_n\,\psi_{n\ell m\pm1}$$

(I canceled the constant $\hbar\sqrt{\ell\,(\ell+1) - m\,(m\pm1)}$ from both sides in the last expression). This argument could obviously be repeated to show that $\psi_{n\ell m\pm2}$ has the same energy as

[22] Of course, we already know the energies are equal since the radial equation, Equation 4.35, does not depend on m. This example demonstrates that rotational invariance is behind the degeneracy.

$\psi_{n\ell m\pm1}$, and so on until you've exhausted the ladder of states. Therefore, *rotational invariance explains why states which differ only in the quantum number m have the same energy*, and since there are $2\ell+1$ different values of m, $2\ell+1$ is the "normal" degeneracy for energies in a central potential.

Of course, the degeneracy of hydrogen (neglecting spin) is n^2 $(=1,4,9,\ldots)$ (Equation 4.85) which is *greater* than $2\ell+1$ $(=1,3,5,\ldots)$.[23] Evidently hydrogen has more degeneracy than is explained by rotational invariance alone. The source of the extra degeneracy is an additional symmetry that is unique to the $1/r$ potential; this is explored in Problem 6.34.[24]

In this section we have focused on continuous rotational symmetry, but *discrete* rotational symmetry, as experienced (for instance) by an electron in a crystal, can also be of interest. Problem 6.33 explores one such system.

Problem 6.18 Consider the free particle in one dimension: $\hat{H}=\hat{p}^2/2m$. This Hamiltonian has both translational symmetry and inversion symmetry.

(a) Show that translations and inversion don't commute.

(b) Because of the translational symmetry we know that the eigenstates of $\hat{H}$ can be chosen to be simultaneous eigenstates of momentum, namely $f_p(x)$ (Equation 3.32). Show that the parity operator turns $f_p(x)$ into $f_{-p}(x)$; these two states must therefore have the same energy.

(c) Alternatively, because of the inversion symmetry we know that the eigenstates of $\hat{H}$ can be chosen to be simultaneous eigenstates of parity, namely

$$\frac{1}{\sqrt{\pi\hbar}}\cos\left(\frac{p\,x}{\hbar}\right) \quad\text{and}\quad \frac{1}{\sqrt{\pi\hbar}}\sin\left(\frac{p\,x}{\hbar}\right).$$

Show that the translation operator mixes these two states together; they therefore must be degenerate.

Note: Both parity *and* translational invariance are required to explain the degeneracy in the free-particle spectrum. Without parity, there is no reason for $f_p(x)$ and $f_{-p}(x)$ to have the same energy (I mean no reason based on symmetries discussed thus far ... obviously you can plug them in to the time-independent Schrödinger equation and show it's true).

Problem 6.19 For any vector operator $\hat{\mathbf{V}}$ one can define raising and lowering operators as

$$\hat{V}_\pm=\hat{V}_x\pm i\,\hat{V}_y.$$

(a) Using Equation 6.33, show that

$$\left[\hat{L}_z,\hat{V}_\pm\right]=\pm\hbar\,\hat{V}_\pm.$$

$$\left[\hat{L}^2,\hat{V}_\pm\right]=2\hbar^2\,\hat{V}_\pm\pm2\hbar\,\hat{V}_\pm\,\hat{L}_z\mp2\hbar\,\hat{V}_z\,\hat{L}_\pm.$$

[23] I don't mean that they necessarily occur in this order. Look back at the infinite spherical well (Figure 4.3): starting with the ground state the degeneracies are $=1,3,5,1,7,3,9,5,\ldots$. These are precisely the degrees of degeneracy we expect for rotational invariance ($2\ell+1$ for integer ℓ) but the symmetry considerations don't tell us where in the spectrum each degeneracy will occur.

[24] For the three-dimensional harmonic oscillator the degeneracy is $n\,(n+1)\,/2=1,3,6,10,\ldots$ (Problem 4.46) which again is greater than $2\ell+1$. For a discussion of the additional symmetry in the oscillator problem see D. M. Fradkin, *Am. J. Phys.* **33**, 207 (1965).

(b) Show that, if ψ is an eigenstate of $\hat{L}^2$ and $\hat{L}_z$ with eigenvalues $\ell(\ell+1)\hbar^2$ and $\ell\hbar$ respectively, then either $\hat{V}_+\psi$ is zero or $\hat{V}_+\psi$ is also an eigenstate of $\hat{L}^2$ and $\hat{L}_z$ with eigenvalues $(\ell+1)(\ell+2)\hbar^2$ and $(\ell+1)\hbar$ respectively. This means that, acting on a state with maximal $m_\ell = \ell$, the operator $\hat{V}_+$ either "raises" both the ℓ and m values by 1 or destroys the state.

6.7 ROTATIONAL SELECTION RULES

The most general statement of the rotational selection rules is the **Wigner–Eckart Theorem**; as a practical matter, it is arguably the most important theorem in all of quantum mechanics. Rather than prove the theorem in full generality I will work out the selection rules for the two classes of operators one encounters most often: scalar operators (in Section 6.7.1) and vector operators (in Section 6.7.2). In deriving these selection rules we consider only how the operators behave under a rotation; therefore, the results of this section apply equally well to "true" scalars and pseudoscalars, and those of the next section apply equally well to "true" vectors and pseudeovectors. These selection rules can be combined with the parity selection rules of Section 6.4.3 to obtain a larger set of selection rules for the operator.

6.7.1 Selection Rules for Scalar Operators

The commutation relations for a scalar operator $\hat{f}$ with the three components of angular momentum (Equation 6.34) can be rewritten in terms of the raising and lowering operators as

$$\left[\hat{L}_z, \hat{f}\right] = 0 \tag{6.40}$$

$$\left[\hat{L}_\pm, \hat{f}\right] = 0 \tag{6.41}$$

$$\left[\hat{L}^2, \hat{f}\right] = 0. \tag{6.42}$$

We derive selection rules for $\hat{f}$ by sandwiching these commutators between two states of definite angular momentum, which we will write as $|n\,\ell\,m\rangle$ and $|n'\,\ell'\,m'\rangle$. These might be hydrogenic orbitals, but they need not be (in fact they need not even be eigenstates of any Hamiltonian but I'll leave the quantum number n there so they look familiar); we require only that $|n\,\ell\,m\rangle$ is an eigenstate of $\hat{L}^2$ and $\hat{L}_z$ with quantum numbers ℓ and m respectively.[25]

Sandwiching Equation 6.40 between two such states gives

$$\left\langle n'\,\ell'\,m' \middle| \left[\hat{L}_z, \hat{f}\right] \middle| n\,\ell\,m \right\rangle = 0$$

or

$$\left\langle n'\,\ell'\,m' \middle| \hat{L}_z\,\hat{f} \middle| n\,\ell\,m \right\rangle - \left\langle n'\,\ell'\,m' \middle| \hat{f}\,\hat{L}_z \middle| n\,\ell\,m \right\rangle = 0$$

and therefore

$$(m'-m)\left\langle n'\,\ell'\,m' \middle| \hat{f} \middle| n\,\ell\,m \right\rangle = 0 \tag{6.43}$$

$\left(\text{using the hermiticity of } \hat{L}_z\right)$. Equation 6.43 says that the matrix elements of a scalar operator vanish unless $m' - m \equiv \Delta m = 0$. Repeating this procedure with Equation 6.42 we get

[25] Importantly, they satisfy Equations 4.118 and 4.120.

$$\left\langle n' \ell' m' \middle| \left[\hat{L}^2, \hat{f}\right] \middle| n \ell m \right\rangle = 0$$
$$\left\langle n' \ell' m' \middle| \hat{L}^2 \hat{f} \middle| n \ell m \right\rangle - \left\langle n' \ell' m' \middle| \hat{f} \hat{L}^2 \middle| n \ell m \right\rangle = 0$$
$$\left[\ell' (\ell' + 1) - \ell (\ell + 1)\right] \left\langle n' \ell' m' \middle| \hat{f} \middle| n \ell m \right\rangle = 0. \tag{6.44}$$

This tells us that the matrix elements of a scalar operator vanish unless $\ell' - \ell \equiv \Delta\ell = 0$.[26] These, then, are the selection rules for a scalar operator: $\Delta\ell = 0$ and $\Delta m = 0$.

However, we can get even more information about the matrix elements from the remaining commutators: (I'll just do the $+$ case and leave the $-$ case for Problem 6.20)

$$\left\langle n' \ell' m' \middle| \left[\hat{L}_+, \hat{f}\right] \middle| n \ell m \right\rangle = 0$$
$$\left\langle n' \ell' m' \middle| \hat{L}_+ \hat{f} \middle| n \ell m \right\rangle - \left\langle n' \ell' m' \middle| \hat{f} \hat{L}_+ \middle| n \ell m \right\rangle = 0$$
$$B_{\ell'}^{m'} \left\langle n' \ell' (m' - 1) \middle| \hat{f} \middle| n \ell m \right\rangle - A_\ell^m \left\langle n' \ell' m' \middle| \hat{f} \middle| n \ell (m + 1) \right\rangle = 0, \tag{6.45}$$

where (from Problem 4.21)

$$A_\ell^m = \hbar \sqrt{\ell (\ell + 1) - m (m + 1)}, \qquad \text{and} \qquad B_\ell^m = \hbar \sqrt{\ell (\ell + 1) - m (m - 1)}.$$

(I also used the fact that $\hat{L}_\pm$ is the Hermitian conjugate of $\hat{L}_\mp$: $\langle \psi | \hat{L}_\pm = \langle L_\mp \psi | .)$[27] Both terms in Equation 6.45 are zero unless $m' = m + 1$ and $\ell' = \ell$, as we proved in Equations 6.43 and 6.44. When these conditions are satisfied, the two coefficients are equal $\left(B_\ell^{m+1} = A_\ell^m\right)$ and Equation 6.45 reduces to

$$\left\langle n' \ell m \middle| \hat{f} \middle| n \ell m \right\rangle = \left\langle n' \ell (m + 1) \middle| \hat{f} \middle| n \ell (m + 1) \right\rangle. \tag{6.46}$$

Evidently the matrix elements of a scalar operator are independent of m.

The results of this section can be summarized as follows:

$$\boxed{\left\langle n' \ell' m' \middle| \hat{f} \middle| n \ell m \right\rangle = \delta_{\ell \ell'} \, \delta_{m m'} \, \langle n' \ell \| f \| n \ell \rangle.} \tag{6.47}$$

The funny-looking matrix element on the right, with two bars, is called a **reduced matrix element** and is just shorthand for "a constant that depends on n, ℓ, and n', but not m."

Example 6.4

(a) Find $\langle r^2 \rangle$ for all four of the degenerate $n = 2$ states of a hydrogen atom.

Solution: From Equation 6.47 we have, for the states with $\ell = 1$, the following equality:

$$\left\langle 2 1 1 \middle| r^2 \middle| 2 1 1 \right\rangle = \left\langle 2 1 0 \middle| r^2 \middle| 2 1 0 \right\rangle = \left\langle 2 1 \, -1 \middle| r^2 \middle| 2 1 \, -1 \right\rangle \equiv \left\langle 2 1 \middle\| r^2 \middle\| 2 1 \right\rangle.$$

26 The other root of the quadratic $\ell' (\ell' + 1) - \ell (\ell + 1) = 0$ is $\ell' = -(\ell + 1)$; since ℓ and ℓ' are non-negative integers this isn't possible.

27 Since $\hat{L}_x$ and $\hat{L}_y$ are Hermitian,

$$\hat{L}_\pm^\dagger = \left(\hat{L}_x \pm i \, \hat{L}_y\right)^\dagger = \hat{L}_x^\dagger \pm (-i) \, \hat{L}_y^\dagger = \hat{L}_x \mp i \, \hat{L}_y = \hat{L}_\mp.$$

To calculate the reduced matrix element we simply pick any one of these expectation values:

$$\begin{aligned}\left\langle 2\,1 \left\| r^2 \right\| 2\,1\right\rangle &= \left\langle 2\,1\,0 \middle| r^2 \middle| 2\,1\,0\right\rangle \\ &= \int r^2 \, |\psi_{210}(r)|^2 \, d^3\mathbf{r} \\ &= \int_0^\infty r^4 \, |R_{21}(r)|^2 \, dr \int \left|Y_1^0(\theta,\phi)\right|^2 \, d\Omega.\end{aligned}$$

The spherical harmonics are normalized (Equation 4.31), so the angular integral is 1, and the radial functions $R_{n\ell}$ are listed in Table 4.7, giving

$$\left\langle 2\,1 \left\| r^2 \right\| 2\,1\right\rangle = \int_0^\infty r^4 \, \frac{1}{24\,a^3} \, \frac{r^2}{a^2} \, e^{-r/a} \, dr = 30\,a^2.$$

That determines three of the expectation values. The final expectation value is

$$\begin{aligned}\left\langle 2\,0 \left\| r^2 \right\| 2\,0\right\rangle &= \left\langle 2\,0\,0 \middle| r^2 \middle| 2\,0\,0\right\rangle \\ &= \int r^2 \, |\psi_{200}(r)|^2 \, d^3\mathbf{r} \\ &= \int_0^\infty r^4 \, |R_{20}(r)|^2 \, dr \int \left|Y_0^0(\theta,\phi)\right|^2 \, d\Omega \\ &= \int_0^\infty r^4 \, \frac{1}{2\,a^3} \left(1 - \frac{1}{2}\frac{r}{a}\right)^2 e^{-r/a} \, dr \\ &= 42\,a^2.\end{aligned}$$

Summarizing:

$$\left\langle 200 \middle| r^2 \middle| 200\right\rangle = 42\,a^2, \qquad \left.\begin{array}{r}\langle 211 \,|\, r^2 \,|\, 211\rangle \\ \langle 210 \,|\, r^2 \,|\, 210\rangle \\ \langle 21-1 \,|\, r^2 \,|\, 21-1\rangle\end{array}\right\} = 30\,a^2. \tag{6.48}$$

(b) Find the expectation value of r^2 for an electron in the superposition state

$$|\psi\rangle = \frac{1}{\sqrt{2}} \left(|200\rangle - i\,|211\rangle\right).$$

Solution: We can expand the expectation value as

$$\begin{aligned}\left\langle \psi \middle| r^2 \middle| \psi\right\rangle &= \frac{1}{2} \left(\langle 200| + i\,\langle 211|\right) r^2 \left(|200\rangle - i\,|211\rangle\right) \\ &= \frac{1}{2} \Big(\left\langle 200 \middle| r^2 \middle| 200\right\rangle + i \left\langle 211 \middle| r^2 \middle| 200\right\rangle - i \left\langle 200 \middle| r^2 \middle| 211\right\rangle \\ &\quad + \left\langle 211 \middle| r^2 \middle| 211\right\rangle\Big).\end{aligned}$$

From Equation 6.47 we see that two of these matrix elements vanish, and

$$\left\langle \psi \middle| r^2 \middle| \psi\right\rangle = \frac{1}{2} \left(\left\langle 20 \left\| r^2 \right\| 20\right\rangle + \left\langle 21 \left\| r^2 \right\| 21\right\rangle\right) = 36\,a^2. \tag{6.49}$$

Problem 6.20 Show that the commutator $\left[\hat{L}_-, \hat{f}\right] = 0$ leads to the same rule, Equation 6.46, as does the commutator $\left[\hat{L}_+, \hat{f}\right] = 0$.

* **Problem 6.21** For an electron in the hydrogen state

$$\psi = \frac{1}{\sqrt{2}}\left(\psi_{211} + \psi_{21-1}\right),$$

find $\langle r \rangle$ after first expressing it in terms of a single reduced matrix element.

6.7.2 Selection Rules for Vector Operators

We now move on to the selection rules for a vector operator $\hat{\mathbf{V}}$. This is significantly more work than the scalar case, but the result is central to understanding atomic transitions (Chapter 11). We begin by defining, by analogy with the angular momentum raising and lowering operators, the operators[28]

$$\hat{V}_\pm \equiv \hat{V}_x \pm i\,\hat{V}_y.$$

Written in terms of these operators, Equation 6.33 becomes

$$\left[\hat{L}_z, \hat{V}_z\right] = 0 \tag{6.50}$$

$$\left[\hat{L}_z, \hat{V}_\pm\right] = \pm\hbar\,\hat{V}_\pm \tag{6.51}$$

$$\left[\hat{L}_\pm, \hat{V}_\pm\right] = 0 \tag{6.52}$$

$$\left[\hat{L}_\pm, \hat{V}_z\right] = \mp\hbar\,\hat{V}_\pm \tag{6.53}$$

$$\left[\hat{L}_\pm, \hat{V}_\mp\right] = \pm 2\hbar\,\hat{V}_z \tag{6.54}$$

as you will show in Problem 6.22(a).[29] Just as for the scalar operator in Section 6.7.1, we sandwich each of these commutators between two states of definite angular momentum to derive (a) conditions under which the matrix elements are guaranteed to vanish and (b) relations between matrix elements with differing values of m or different components of $\hat{\mathbf{V}}$.

From Equation 6.51,

$$\left\langle n'\,\ell'\,m'\middle|\,\hat{L}_z\,\hat{V}_\pm\,\middle|\,n\,\ell\,m\right\rangle - \left\langle n'\,\ell'\,m'\middle|\,\hat{V}_\pm\,\hat{L}_z\,\middle|\,n\,\ell\,m\right\rangle = \pm\hbar\left\langle n'\,\ell'\,m'\middle|\,\hat{V}_\pm\,\middle|\,n\,\ell\,m\right\rangle,$$

---

[28] The operators $\hat{V}_\pm$ are, up to constants, components of what are known as **spherical tensor** operators of rank 1, written $\hat{T}_q^{(k)}$ where k is the rank and q the component of the operator:

$$\hat{T}_{\pm 1}^{(1)} = \mp\frac{1}{\sqrt{2}}\,\hat{V}_\pm \qquad \hat{T}_0^{(1)} = \hat{V}_z.$$

Similarly, the scalar operator f treated in Section 6.7.1 is a rank-0 spherical tensor operator:

$$\hat{T}_0^{(0)} = \hat{f}.$$

[29] Equations 6.51–6.54 each stand for *two* equations: read the upper signs all the way across, or the lower signs.

and since our states are eigenstates of $\hat{L}_z$, this simplifies to

$$[m' - (m \pm 1)] \left\langle n' \ell' m' \middle| \hat{V}_{\pm} \middle| n \ell m \right\rangle = 0. \tag{6.55}$$

Equation 6.55 says that *either* $m' = m \pm 1$, or else the matrix element of $\hat{V}_{\pm}$ must vanish. Equation 6.50 works out similarly (see Problem 6.22) and this first set of commutators gives us the selection rules for m:

$$\left\langle n' \ell' m' \middle| \hat{V}_{+} \middle| n \ell m \right\rangle = 0 \qquad \text{unless } m' = m + 1 \tag{6.56}$$

$$\left\langle n' \ell' m' \middle| \hat{V}_{z} \middle| n \ell m \right\rangle = 0 \qquad \text{unless } m' = m \tag{6.57}$$

$$\left\langle n' \ell' m' \middle| \hat{V}_{-} \middle| n \ell m \right\rangle = 0 \qquad \text{unless } m' = m - 1. \tag{6.58}$$

Note that, if desired, these expressions can be turned back into selection rules for the x- and y-components of our operator, since

$$\left\langle n' \ell' m' \middle| \hat{V}_{x} \middle| n \ell m \right\rangle = \frac{1}{2} \left[\left\langle n' \ell' m' \middle| \hat{V}_{-} \middle| n \ell m \right\rangle + \left\langle n' \ell' m' \middle| \hat{V}_{+} \middle| n \ell m \right\rangle \right]$$

$$\left\langle n' \ell' m' \middle| \hat{V}_{y} \middle| n \ell m \right\rangle = \frac{i}{2} \left[\left\langle n' \ell' m' \middle| \hat{V}_{-} \middle| n \ell m \right\rangle - \left\langle n' \ell' m' \middle| \hat{V}_{+} \middle| n \ell m \right\rangle \right].$$

The remaining commutators, Equations 6.52–6.54, yield a selection rule on ℓ and relations among the nonzero matrix elements. As shown in Problem 6.24, the results may be summarized as[30]

$$\left\langle n'\ell'm' \middle| \hat{V}_{+} \middle| n\ell m \right\rangle = -\sqrt{2}\, C^{\ell\,1\,\ell'}_{m 1 m'} \left\langle n'\ell' \,\|\, V \,\|\, n\ell \right\rangle \tag{6.59}$$

$$\left\langle n'\ell'm' \middle| \hat{V}_{-} \middle| n\ell m \right\rangle = \sqrt{2}\, C^{\ell\,1\,\ell'}_{m\,-1\,m'} \left\langle n'\ell' \,\|\, V \,\|\, n\ell \right\rangle \tag{6.60}$$

$$\left\langle n'\ell'm' \middle| \hat{V}_{z} \middle| n\ell m \right\rangle = C^{\ell\,1\,\ell'}_{m 0 m'} \left\langle n'\ell' \,\|\, V \,\|\, n\ell \right\rangle. \tag{6.61}$$

The constants $C^{j_1\, j_2\, J}_{m_1\, m_2\, M}$ in these expressions are precisely the Clebsch–Gordan coefficients that appeared in the addition of angular momenta (Section 4.4.3). The Clebsch–Gordan coefficient $C^{j_1\, j_2\, J}_{m_1\, m_2\, M}$ vanishes unless $M = m_1 + m_2$ (since the z-components of angular momentum add) and unless $J = j_1 + j_2,\ j_1 + j_2 - 1, \ldots, |j_1 - j_2|$ (Equation 4.182). In particular, the matrix elements of any component of a vector operator, $\langle n'\ell'm' | \hat{V}_i | n\ell m \rangle$, are nonzero only if

$$\Delta\ell = 0, \pm 1, \qquad \text{and} \qquad \Delta m = 0, \pm 1. \tag{6.62}$$

---

[30] A warning about notation: In the selection rules for the scalar operator r,

$$\left\langle n'\ell'm' \,|\, r \,|\, n\ell m \right\rangle = \delta_{\ell\ell'}\, \delta_{mm'} \left\langle n'\ell' \,\|\, r \,\|\, n\ell \right\rangle,$$

and for a component (say z) of the vector operator $\mathbf{r}$,

$$\left\langle n'\ell'm' \,|\, z \,|\, n\ell m \right\rangle = C^{\ell\,1\,\ell'}_{m 0 m'} \left\langle n'\ell' \,\|\, r \,\|\, n\ell \right\rangle,$$

the two reduced matrix elements *are not the same*. One is the reduced matrix element for r and one is the reduced matrix element for $\mathbf{r}$, and these are *different operators* that share the same name. You could tack on a subscript ($\langle n'\ell' \,\|\, r \,\|\, n\ell \rangle_s$ and $\langle n'\ell' \,\|\, r \,\|\, n\ell \rangle_v$) to distinguish between the two if that helps keep them straight.

Example 6.5
Find all of the matrix elements of $\hat{\mathbf{r}}$ between the states with $n = 2, \ell = 1$ and $n' = 3, \ell' = 2$:

$$\langle 3\,2\,m' \,|\, r_i \,|\, 2\,1\,m \rangle$$

where $m = -1, 0, 1$, $m' = -2, -1, 0, 1, 2$, and $r_i = x, y, z$.
Solution: With the vector operator $\hat{\mathbf{V}} = \hat{\mathbf{r}}$, our components are $V_z = z$, $V_+ = x + i\,y$, and $V_- = x - i\,y$. We start by calculating *one* of the matrix elements,

$$\begin{aligned}\langle 320 \,|\, z \,|\, 210 \rangle &= \int \psi_{320}(\mathbf{r})\; r\cos\theta\; \psi_{210}(\mathbf{r})\; d^3\mathbf{r} \\ &= \int R_{32}(r)^*\; r\; R_{21}(r)\; r^2\,dr \int Y_2^0(\theta,\phi)^*\; \cos\theta\; Y_1^0(\theta,\phi)\; d\Omega \\ &= \frac{2^{12}\,3^3\,\sqrt{3}}{5^7}\,a.\end{aligned}$$

From Equation 6.61 we can then determine the reduced matrix element

$$\begin{aligned}\langle 320 \,|\, z \,|\, 210 \rangle &= C^{1\,1\,2}_{0\,0\,0}\; \langle 32 \,\|\, V \,\|\, 21 \rangle \\ \frac{2^{12}\,3^3\,\sqrt{3}}{5^7}\,a &= \sqrt{\frac{2}{3}}\; \langle 32 \,\|\, V \,\|\, 21 \rangle\,.\end{aligned}$$

Therefore

$$\langle 32 \,\|\, V \,\|\, 21 \rangle = \frac{2^{12}\,3^4}{5^7\,\sqrt{2}}\,a. \tag{6.63}$$

We can now find all of the remaining matrix elements from Equations 6.59–6.60 with the help of the Clebsch–Gordan table. The relevant coefficients are shown in Figure 6.8. The nonzero matrix elements are

$$\begin{aligned}\left\langle 322 \,\middle|\, \hat{V}_+ \,\middle|\, 211 \right\rangle &= -\sqrt{2}\,C^{112}_{112}\; \langle 32 \,\|\, V \,\|\, 21 \rangle = -\sqrt{2}\; \langle 32 \,\|\, V \,\|\, 21 \rangle \\ \left\langle 321 \,\middle|\, \hat{V}_+ \,\middle|\, 210 \right\rangle &= -\sqrt{2}\,C^{112}_{011}\; \langle 32 \,\|\, V \,\|\, 21 \rangle = -\; \langle 32 \,\|\, V \,\|\, 21 \rangle \\ \left\langle 320 \,\middle|\, \hat{V}_+ \,\middle|\, 21-1 \right\rangle &= -\sqrt{2}\,C^{\;112}_{-110}\; \langle 32 \,\|\, V \,\|\, 21 \rangle = -\frac{1}{\sqrt{3}}\; \langle 32 \,\|\, V \,\|\, 21 \rangle \\ \left\langle 320 \,\middle|\, \hat{V}_- \,\middle|\, 211 \right\rangle &= \sqrt{2}\,C^{1\;\;12}_{1-10}\; \langle 32 \,\|\, V \,\|\, 21 \rangle = \frac{1}{\sqrt{3}}\; \langle 32 \,\|\, V \,\|\, 21 \rangle \\ \left\langle 32-1 \,\middle|\, \hat{V}_- \,\middle|\, 210 \right\rangle &= \sqrt{2}\,C^{1\;\;1\;\;2}_{0-1-1}\; \langle 32 \,\|\, V \,\|\, 21 \rangle = \langle 32 \,\|\, V \,\|\, 21 \rangle \\ \left\langle 32-2 \,\middle|\, \hat{V}_- \,\middle|\, 21-1 \right\rangle &= \sqrt{2}\,C^{\;\;1\;\;1\;\;2}_{-1-1-2}\; \langle 32 \,\|\, V \,\|\, 21 \rangle = \sqrt{2}\; \langle 32 \,\|\, V \,\|\, 21 \rangle \\ \left\langle 321 \,\middle|\, \hat{V}_z \,\middle|\, 211 \right\rangle &= C^{112}_{101}\; \langle 32 \,\|\, V \,\|\, 21 \rangle = \frac{1}{\sqrt{2}}\; \langle 32 \,\|\, V \,\|\, 21 \rangle \\ \left\langle 320 \,\middle|\, \hat{V}_z \,\middle|\, 210 \right\rangle &= C^{112}_{000}\; \langle 32 \,\|\, V \,\|\, 21 \rangle = \sqrt{\frac{2}{3}}\; \langle 32 \,\|\, V \,\|\, 21 \rangle \\ \left\langle 32-1 \,\middle|\, \hat{V}_z \,\middle|\, 21-1 \right\rangle &= C^{\;\;11\;\;2}_{-10-1}\; \langle 32 \,\|\, V \,\|\, 21 \rangle = \frac{1}{\sqrt{2}}\; \langle 32 \,\|\, V \,\|\, 21 \rangle\,,\end{aligned}$$

with the reduced matrix element given by Equation 6.63. The other thirty-six matrix elements vanish due to the selection rules (Equations 6.56–6.58 and 6.62). We have determined all forty-five matrix elements and have only needed to evaluate a *single* integral. I've left the matrix elements in terms of V_+ and V_- but it's straightforward to write them in terms of x and y using the expressions on page 259.

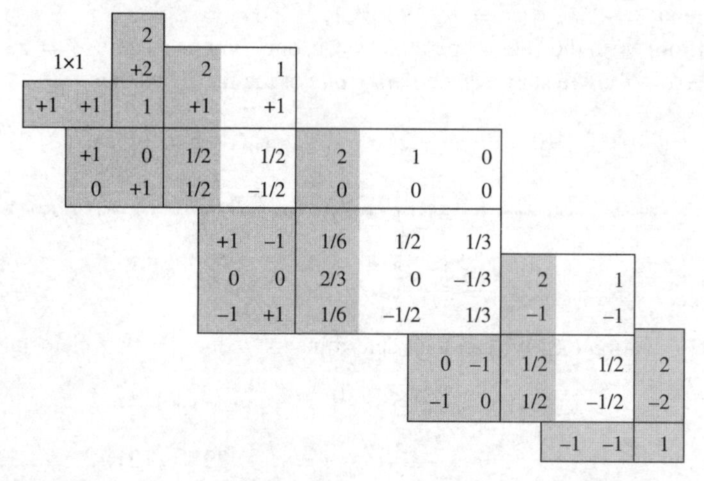

Figure 6.8: The Clebsch–Gordan coefficients for $1 \otimes 1$.

It is no coincidence that the Clebsch–Gordan coefficients appear in Equations 6.59–6.61. States have angular momentum, but *operators* also carry angular momentum. A scalar operator (Equation 6.34) has $\ell = 0$—it is unchanged by a rotation—just as a state of angular momentum 0 is unchanged. A vector operator (Equation 6.33) has $\ell = 1$; its three components transform into each other under a rotation in the same way the triplet of states with angular momentum $\ell = 1$ transform into each other.[31] When we act on a state with an operator, we add together the angular momentum of the state and the operator to obtain the angular momentum of the resultant state; this addition of angular momenta is the source of the Clebsch–Gordan coefficients in Equations 6.59–6.61.[32]

* **Problem 6.22**

(a) Show that the commutation relations, Equations 6.50–6.54, follow from the definition of a vector operator, Equation 6.33. If you did Problem 6.19 you already derived one of these.

(b) Derive Equation 6.57.

---

[31] In the case of the position operator $\hat{\mathbf{r}}$, this correspondence is particularly evident when we rewrite the operator with the help of Table 4.3:

$$x \pm i\,y = r\,\sin\theta\,e^{\pm i\,\phi} = \mp r\sqrt{\frac{8\,\pi}{3}}\,Y_1^{\pm 1}(\theta,\phi)$$

$$z = r\,\cos\theta = r\sqrt{\frac{4\,\pi}{3}}\,Y_1^0(\theta,\phi)\,.$$

[32] Since $C^{\ell\,0\,\ell'}_{m0m'} = \delta_{mm'}\,\delta_{\ell\ell'}$, one could rewrite the selection rules for a scalar operator (Equation 6.47) as

$$\left\langle n'\ell' m' \middle| \hat{f} \middle| n\ell m \right\rangle = C^{\ell\,0\,\ell'}_{m0m'} \left\langle n'\ell' \middle\| f \middle\| n\ell \right\rangle.$$

Problem 6.23 The Clebsch–Gordan coefficients are defined by Equation 4.183. Adding together two states with angular momentum j_1 and j_2 produces a state with total angular momentum J according to

$$|J\,M\rangle = \sum_{m_1,m_2} C^{j_1\,j_2\,J}_{m_1\,m_2\,M}\,|j_1\,j_2\,m_1\,m_2\rangle\,. \tag{6.64}$$

(a) From Equation 6.64, show that the Clebsch–Gordan coefficients satisfy

$$C^{j_1\,j_2\,J}_{m_1\,m_2\,M} = \langle j_1\,j_2\,m_1\,m_2\,|\,J\,M\rangle\,. \tag{6.65}$$

(b) Apply $\hat{J}_\pm = \hat{J}^{(1)}_\pm + \hat{J}^{(2)}_\pm$ to Equation 6.64 to derive the **recursion relations for Clebsch–Gordan coefficients**:

$$\begin{aligned} A^M_J\,C^{j_1\,j_2\,J}_{m_1\,m_2\,M+1} &= B^{m_1}_{j_1}\,C^{j_1\,j_2\,J}_{m_1-1\,m_2\,M} + B^{m_2}_{j_2}\,C^{j_1\,j_2\,J}_{m_1\,m_2-1\,M} \\ B^M_J\,C^{j_1\,j_2\,J}_{m_1\,m_2\,M-1} &= A^{m_1}_{j_1}\,C^{j_1\,j_2\,J}_{m+1\,m_2\,M} + A^{m_2}_{j_2}\,C^{j_1\,j_2\,J}_{m_1\,m_2+1\,M} \end{aligned} \tag{6.66}$$

*** **Problem 6.24**

(a) Sandwich each of the six commutation relations in Equations 6.52–6.54 between $\langle n'\ell'm'|$ and $|n\ell m\rangle$ to obtain relations between matrix elements of $\hat{\mathbf{V}}$. As an example, Equation 6.52 with the upper signs gives

$$B^{m'}_{\ell'}\,\langle n'\,\ell'\,(m'-1)\,|\,V_+\,|\,n\,\ell\,m\rangle = A^m_\ell\,\langle n'\,\ell'\,m'\,|\,V_+\,|\,n\,\ell\,(m+1)\rangle\,.$$

(b) Using the results in Problem 6.23, show that the six expressions you wrote down in part (a) are satisfied by Equations 6.59–6.61.

* **Problem 6.25** Express the expectation value of the dipole moment $\mathbf{p}_e$ for an electron in the hydrogen state

$$\psi = \frac{1}{\sqrt{2}}\left(\psi_{211} + \psi_{200}\right)$$

in terms of a single reduced matrix element, and evaluate the expectation value. *Note*: this is the expectation value of a vector so you need to compute all three components. Don't forget Laporte's rule!

6.8 TRANSLATIONS IN TIME

In this section we study time-translation invariance. Consider a solution $\Psi(x,t)$ to the time-dependent Schrödinger equation

$$\hat{H}\,\Psi(x,t) = i\,\hbar\,\frac{\partial}{\partial t}\,\Psi(x,t)\,.$$

We can define the operator that propagates the wave function forward in time, $\hat{U}(t)$ by

$$\hat{U}(t)\;\Psi(x,0) = \Psi(x,t)\,; \tag{6.67}$$

$\hat{U}(t)$ can be expressed in terms of the Hamiltonian, and doing so is straightforward if the Hamiltonian is not itself a function of time. In that case, expanding the right-hand side of Equation 6.67 in a Taylor series gives[33]

$$\hat{U}(t)\,\Psi(x,0) = \Psi(x,t) = \sum_{n=0}^{\infty} \frac{1}{n!}\,\frac{\partial^n}{\partial t^n}\Psi(x,t)\bigg|_{t=0} t^n \tag{6.68}$$

$$= \sum_{n=0}^{\infty} \frac{1}{n!}\left(-\frac{i}{\hbar}\hat{H}\,t\right)^n \Psi(x,0)\,. \tag{6.69}$$

Therefore, in the case of a time-independent Hamiltonian, the time-evolution operator is[34]

$$\boxed{\hat{U}(t) = \exp\left[-\frac{i\,t}{\hbar}\,\hat{H}\right].} \tag{6.71}$$

We say that the Hamiltonian is the **generator of translations in time**. Note that $\hat{U}(t)$ is a unitary operator (see Problem 6.2).

The time-evolution operator offers a compact way to state the procedure for solving the time-dependent Schrödinger equation. To see the correspondence, write out the wave function at time $t = 0$ as a superposition of stationary states $\left(\hat{H}\,\psi_n = E_n\,\psi_n\right)$:

$$\Psi(x,0) = \sum_n c_n\,\psi_n(x)\,.$$

Then

$$\Psi(x,t) = \hat{U}(t)\,\Psi(x,0) = \sum_n c_n\,\hat{U}(t)\,\psi_n(x)$$

$$= \sum_n c_n\,e^{-i\,\hat{H}\,t/\hbar}\,\psi_n(x) = \sum_n c_n\,e^{-i\,E_n\,t/\hbar}\,\psi_n(x)\,.$$

In this sense Equation 6.71 is shorthand for the process of expanding the initial wave function in terms of stationary states and then tacking on the "wiggle factors" to obtain the wave function at a later time (Section 2.1).

---

[33] Why is this analysis limited to the case where $\hat{H}$ is independent of time? Whether or not $\hat{H}$ depends on time, Schrödinger's equation says $i\,\hbar\,\dot{\Psi} = \hat{H}\,\Psi$. However, if $\hat{H}$ is time dependent then the second derivative of Ψ is given by

$$\frac{\partial^2}{\partial t^2}\Psi = \frac{\partial}{\partial t}\left(\frac{1}{i\,\hbar}\hat{H}\,\Psi\right) = \frac{1}{i\,\hbar}\,\frac{\partial \hat{H}}{\partial t}\,\Psi - \frac{1}{\hbar^2}\,\hat{H}^2\,\Psi$$

and higher derivatives will be even more complicated. Therefore, Equation 6.69 only follows from Equation 6.68 when $\hat{H}$ has no time dependence. See also Problem 11.23.

[34] This derivation assumes that the actual solution to Schrodinger's equation, $\Psi(x,t)$, can be expanded as a Taylor series in t, and nothing guarantees that. B. R. Holstein and A. R. Swift, *A. J. Phys.* **40**, 829 (1989) give an innocent-seeming example where such an expansion does not exist. Nonetheless, Equation 6.71 still holds in such cases as long as we define the exponential function through its spectral decomposition (Equation 3.103):

$$\hat{U}(t) \equiv \sum_n e^{-i\,E_n\,t/\hbar}\,|\psi_n\rangle\,\langle\psi_n|\,. \tag{6.70}$$

See also M. Amaku et al., *Am. J. Phys.* **85**, 692 (2017).

6.8.1 The Heisenberg Picture

Just as for the other transformations studied in this chapter, we can examine the effect of applying time translation to *operators*, as well as to wave functions. The transformed operators are called **Heisenberg-picture** operators and we follow the convention of giving them a subscript H rather than a prime:

$$\boxed{\hat{Q}_H(t) = \hat{U}^\dagger(t)\ \hat{Q}\ \hat{U}(t)\,.} \tag{6.72}$$

Example 6.6

A particle of mass m moves in one dimension in a potential $V(x)$:

$$\hat{H} = \frac{\hat{p}^2}{2m} + V(x)\,.$$

Find the position operator in the Heisenberg picture for an infinitesimal time translation δ.

Solution: From Equation 6.71,

$$\hat{U}(\delta) \approx 1 - i\,\frac{\delta}{\hbar}\,\hat{H}.$$

Applying Equation 6.72, we have

$$\begin{aligned}\hat{x}_H(\delta) &\approx \left(1 + i\,\frac{\delta}{\hbar}\,\hat{H}^\dagger\right)\hat{x}\left(1 - i\,\frac{\delta}{\hbar}\,\hat{H}\right)\\ &\approx \hat{x} - i\frac{\delta}{\hbar}\left[\hat{x}, \hat{H}\right] \approx \hat{x} - i\frac{\delta}{\hbar}\,i\,\hbar\,\frac{\hat{p}}{m}\end{aligned}$$

so

$$\hat{x}_H(\delta) \approx \hat{x}_H(0) + \frac{1}{m}\,\hat{p}_H(0)\,\delta$$

(making use of the fact that the Heisenberg-picture operators at time 0 are just the untransformed operators). This looks exactly like classical mechanics: $x(\delta) \approx x(0) + v(0)\,\delta$. The Heisenberg picture illuminates the connection between classical and quantum mechanics: the quantum operators obey the classical equations of motion (see Problem 6.29).

Example 6.7

A particle of mass m moves in one dimension in a harmonic-oscillator potential:

$$\hat{H} = \frac{\hat{p}^2}{2m} + \frac{1}{2}\,m\,\omega^2\,x^2.$$

Find the position operator in the Heisenberg picture at time t.

Solution: Consider the action of $\hat{x}_H$ on a stationary state ψ_n. (Introducing ψ_n allows us to replace the *operator* $e^{-i\hat{H}t/\hbar}$ with the *number* $e^{-iE_nt/\hbar}$, since $e^{-i\hat{H}t/\hbar}\,\psi_n = e^{-iE_nt/\hbar}\,\psi_n$.) Writing $\hat{x}$ in terms of raising and lowering operators we have (using Equations 2.62, 2.67, and 2.70)

$$\begin{aligned}\hat{x}_H(t)\,\psi_n(x) &= \hat{U}^\dagger(t)\,\hat{x}\,\hat{U}(t)\,\psi_n(x)\\ &= e^{i\hat{H}t/\hbar}\sqrt{\frac{\hbar}{2m\omega}}\,(\hat{a}_+ + \hat{a}_-)\,e^{-i\hat{H}t/\hbar}\,\psi_n(x)\end{aligned}$$

$$= \sqrt{\frac{\hbar}{2m\omega}}\, e^{-i E_n t/\hbar}\, e^{i \hat{H} t/\hbar} \left(\hat{a}_+ + \hat{a}_-\right) \psi_n(x)$$
$$= \sqrt{\frac{\hbar}{2m\omega}}\, e^{-i E_n t/\hbar}\, e^{i \hat{H} t/\hbar} \left[\sqrt{n+1}\, \psi_{n+1}(x) + \sqrt{n}\, \psi_{n-1}(x)\right]$$
$$= \sqrt{\frac{\hbar}{2m\omega}}\, e^{-i E_n t/\hbar} \left[\sqrt{n+1}\, e^{i E_{n+1} t/\hbar}\, \psi_{n+1}(x) + \sqrt{n}\, e^{i E_{n-1} t/\hbar}\, \psi_{n-1}(x)\right]$$
$$= \sqrt{\frac{\hbar}{2m\omega}} \left[\sqrt{n+1}\, e^{i\omega t}\, \psi_{n+1}(x) + \sqrt{n}\, e^{-i\omega t}\, \psi_{n-1}(x)\right]. \tag{6.73}$$

Thus[35]

$$\hat{x}_H(t) = \sqrt{\frac{\hbar}{2m\omega}} \left[e^{i\omega t} \hat{a}_+ + e^{-i\omega t}\, \hat{a}_-\right].$$

Or, using Equation 2.48 to express $\hat{a}_\pm$ in terms of $\hat{x}$ and $\hat{p}$,

$$\hat{x}_H(t) = \hat{x}_H(0)\, \cos(\omega t) + \frac{1}{m\omega}\, \hat{p}_H(0)\, \sin(\omega t). \tag{6.74}$$

As in Example 6.6 we see that the Heisenberg-picture operator satisfies the *classical* equation of motion for a mass on a spring.

In this book we have been working in the **Schrödinger picture**, so-named by Dirac because it was the picture that Schrödinger himself had in mind. In the Schrödinger picture, the wave function evolves in time according to the Schrödinger equation

$$\hat{H}\, \Psi(x,t) = i\hbar \frac{\partial}{\partial t} \Psi(x,t).$$

The operators $\hat{x} = x$ and $\hat{p} = -i\hbar\,\partial_x$ have no time dependence of their own, and the time dependence of expectation values (or, more generally, matrix elements) comes from the time dependence of the wave function:[36]

$$\langle \hat{Q} \rangle = \langle \Psi(t) | \hat{Q} | \Psi(t) \rangle.$$

In the Heisenberg picture, the wave function is constant in time, $\Psi_H(x) = \Psi(x,0)$, and the *operators* evolve in time according to Equation 6.72. In the Heisenberg picture, the time dependence of expectation values (or matrix elements) is carried by the *operators*.

$$\langle \hat{Q} \rangle = \langle \Psi_H | \hat{Q}_H(t) | \Psi_H \rangle.$$

Of course, the two pictures are entirely equivalent since

$$\langle \Psi(t) | \hat{Q} | \Psi(t) \rangle = \langle \Psi(0) | \hat{U}^\dagger \hat{Q}\, \hat{U} | \Psi(0) \rangle = \langle \Psi_H | \hat{Q}_H(t) | \Psi_H \rangle.$$

[35] Since Equation 6.73 holds for *any* stationary state ψ_n and since the ψ_n constitute a complete set of states, the operators must in fact be identical.

[36] I am assuming that $\hat{Q}$, like $\hat{x}$ or $\hat{p}$, has no *explicit* time dependence.

A nice analogy for the two pictures runs as follows. On an ordinary clock, the hands move in a clockwise direction while the numbers stay fixed. But one could equally well design a clock where the hands are stationary and the numbers move in the counter-clockwise direction. The correspondence between these two clocks is roughly the correspondence between the Schrödinger and Heisenberg pictures, the hands representing the wave function and the numbers representing the operator. Other pictures could be introduced as well, in which both the hands of the clock and the numbers on the dial move at intermediate rates such that the clock still tells the correct time.[37]

* **Problem 6.26** Work out $\hat{p}_H(t)$ for the system in Example 6.7 and comment on the correspondence with the classical equation of motion.

** **Problem 6.27** Consider a free particle of mass m. Show that the position and momentum operators in the Heisenberg picture are given by

$$\hat{x}_H(t) = \hat{x}_H(0) + \frac{1}{m}\,\hat{p}_H(0)\,t$$
$$\hat{p}_H(t) = \hat{p}_H(0)\,.$$

Comment on the relationship between these equations and the classical equations of motion. *Hint*: you will first need to evaluate the commutator $\left[\hat{x}, \hat{H}^n\right]$; this will allow you to evaluate the commutator $\left[\hat{x}, \hat{U}\right]$.

6.8.2 Time-Translation Invariance

If the Hamiltonian is time-*de*pendent one can still write the formal solution to the Schrödinger equation in terms of the time-translation operator, $\hat{U}$:

$$\Psi(x,t) = \hat{U}(t,t_0)\;\Psi(x,t_0)\,, \tag{6.75}$$

but $\hat{U}$ no longer takes the simple form 6.71.[38] (See Problem 11.23 for the general case.) For an infinitesimal time interval δ (see Problem 6.28)

$$\hat{U}(t_0+\delta,t_0) \approx 1 - \frac{i}{\hbar}\,\hat{H}(t_0)\;\delta. \tag{6.76}$$

Time-translation invariance means that the time evolution is independent of which time interval we are considering. In other words

$$\hat{U}(t_1+\delta,t_1) = \hat{U}(t_2+\delta,t_2) \tag{6.77}$$

for any choice of t_1 and t_2. This ensures that if the system starts in state $|\alpha\rangle$ at time t_1 and evolves for a time δ then it will end up in the same state $|\beta\rangle$ as if the system started in the same state $|\alpha\rangle$ at time t_2 and evolved for the same amount of time δ; i.e. the experiment proceeds the same on Thursday as it did on Tuesday, assuming identical conditions. Plugging Equation 6.76 into Equation 6.77 we see that the requirement for this to be true is $\hat{H}(t_1) = \hat{H}(t_2)$, and since

[37] Of these other possible pictures the most important is the **interaction picture** (or Dirac picture) which is often employed in time-dependent perturbation theory.

[38] And is a function of *both* the initial time t_0 and the final time t, not simply the amount of time for which the wave function has evolved.

this must hold true for all t_1 and t_2, it must be that the Hamiltonian is in fact time-independent after all (for time-translation invariance to hold):

$$\frac{\partial \hat{H}}{\partial t} = 0.$$

In that case the generalized Ehrenfest theorem says

$$\frac{d}{dt}\langle \hat{H} \rangle = \frac{i}{\hbar}\left\langle \left[\hat{H}, \hat{H}\right] \right\rangle + \left\langle \frac{\partial \hat{H}}{\partial t} \right\rangle = 0.$$

Therefore, **energy conservation is a consequence of time-translation invariance.**

We have now recovered all the classical conservation laws: conservation of momentum, energy, and angular momentum, and seen that they are each related to a continuous symmetry of the Hamiltonian (spatial translation, time translation, and rotation, respectively). And in quantum mechanics, discrete symmetries (such as parity) can also lead to conservation laws.

Problem 6.28 Show that Equations 6.75 and 6.76 are the solution to the Schrödinger equation for an infinitesimal time δ. *Hint*: expand $\Psi(x, t)$ in a Taylor series.

* **Problem 6.29** Differentiate Equation 6.72 to obtain the **Heisenberg equations of motion**

$$i\hbar \frac{d}{dt}\hat{Q}_H(t) = \left[\hat{Q}_H(t), \hat{H}\right] \tag{6.78}$$

(for $\hat{Q}$ and $\hat{H}$ independent of time).[39] Plug in $\hat{Q} = \hat{x}$ and $\hat{Q} = \hat{p}$ to obtain the differential equations for $\hat{x}_H$ and $\hat{p}_H$ in the Heisenberg picture for a single particle of mass m moving in a potential $V(x)$.

*** **Problem 6.30** Consider a time-independent Hamiltonian for a particle moving in one dimension that has stationary states $\psi_n(x)$ with energies E_n.

(a) Show that the solution to the time-dependent Schrödinger equation can be written

$$\Psi(x, t) = \hat{U}(t)\, \Psi(x, 0) = \int K\left(x, x', t\right)\, \Psi\left(x', 0\right)\, dx',$$

where $K\left(x, x', t\right)$, known as the **propagator**, is

$$K\left(x, x', t\right) = \sum_n \psi_n^*\left(x'\right)\, e^{-i E_n t/\hbar}\, \psi_n(x)\,. \tag{6.79}$$

Here $\left|K\left(x, x', t\right)\right|^2$ is the probability for a quantum mechanical particle to travel from position x' to position x in time t.

(b) Find K for a particle of mass m in a simple harmonic oscillator potential of frequency ω. You will need the identity

[39] For time-dependent $\hat{Q}$ and $\hat{H}$ the generalization is

$$i\hbar \frac{d}{dt}\hat{Q}_H(t) = \left[\hat{Q}_H(t), \hat{H}_H(t)\right] + \hat{U}^\dagger \frac{\partial \hat{Q}}{\partial t}\hat{U}.$$

$$\frac{1}{\sqrt{1-z^2}} \exp\left[-\frac{\xi^2+\eta^2-2\xi\,\eta\,z}{1-z^2}\right] = e^{-\xi^2} e^{-\eta^2} \sum_{n=0}^{\infty} \frac{z^n}{2^n\, n!} H_n(\xi)\, H_n(\eta).$$

(c) Find $\Psi(x,t)$ if the particle from part (a) is initially in the state[40]

$$\Psi(x,0) = \left(\frac{2a}{\pi}\right)^{1/4} e^{-a(x-x_0)^2}.$$

Compare your answer with Problem 2.49. *Note*: Problem 2.49 is a special case with $a = m\omega/2\hbar$.

(d) Find K for a free particle of mass m. In this case the stationary states are continuous, not discrete, and one must make the replacement

$$\sum_n \rightarrow \int_{-\infty}^{\infty} dp$$

in Equation 6.79.

(e) Find $\Psi(x,t)$ for a free particle that starts out in the state

$$\Psi(x,0) = \left(\frac{2a}{\pi}\right)^{1/4} e^{-a\,x^2}.$$

Compare your answer with Problem 2.21.

FURTHER PROBLEMS ON CHAPTER 6

Problem 6.31 In deriving Equation 6.3 we assumed that our function had a Taylor series. The result holds more generally if we define the exponential of an operator by its spectral decomposition,

$$\hat{T}(a) = \int e^{-i\,a\,p/\hbar}\, |p\rangle\, \langle p|\, dp, \tag{6.80}$$

rather than its power series. Here I've given the operator in Dirac notation; acting on a position-space function (see the discussion on page 123) this means

$$\hat{T}(a)\; \psi(x) = \int_{-\infty}^{\infty} e^{-i\,a\,p/\hbar}\, f_p(x)\; \Phi(p)\; dp, \tag{6.81}$$

where $\Phi(p)$ is the momentum space wave function corresponding to $\psi(x)$ and $f_p(x)$ is defined in Equation 3.32. Show that the operator $\hat{T}(a)$, as given by Equation 6.81, applied to the function

$$\psi(x) = \sqrt{\lambda}\, e^{-\lambda\,|x|}$$

(whose first derivative is undefined at $x = 0$) gives the correct result.

[40] The integrals in (c)–(e) can *all* be done with the following identity:

$$\int_{-\infty}^{\infty} e^{-ax^2+bx}\, dx = \sqrt{\frac{\pi}{a}}\, e^{b^2/4a},$$

which was derived in Problem 2.21.

** **Problem 6.32** Rotations on spin states are given by an expression identical to Equation 6.32, with the spin angular momentum replacing the orbital angular momentum:

$$\mathsf{R}_{\mathbf{n}}(\varphi) = \exp\left[-i\,\frac{\varphi}{\hbar}\,\mathbf{n}\cdot\mathsf{S}\right].$$

In this problem we will consider rotations of a spin-1/2 state.

(a) Show that

$$(\mathbf{a}\cdot\boldsymbol{\sigma})\,(\mathbf{b}\cdot\boldsymbol{\sigma}) = \mathbf{a}\cdot\mathbf{b} + i\,(\mathbf{a}\times\mathbf{b})\cdot\boldsymbol{\sigma},$$

where the σ_i are the Pauli spin matrices and **a** and **b** are ordinary vectors. Use the result of Problem 4.29.

(b) Use your result from part (a) to show that

$$\exp\left[-i\,\frac{\varphi}{\hbar}\,\mathbf{n}\cdot\mathsf{S}\right] = \cos\left(\frac{\varphi}{2}\right) - i\,\sin\left(\frac{\varphi}{2}\right)\,\mathbf{n}\cdot\boldsymbol{\sigma}.$$

Recall that $\mathsf{S} = (\hbar/2)\,\boldsymbol{\sigma}$.

(c) Show that your result from part (b) becomes, in the standard basis of spin up and spin down along the z axis, the matrix

$$\mathsf{R}_{\mathbf{n}} = \cos\left(\frac{\varphi}{2}\right)\begin{pmatrix}1 & 0\\ 0 & 1\end{pmatrix} - i\,\sin\left(\frac{\varphi}{2}\right)\begin{pmatrix}\cos\theta & \sin\theta\, e^{-i\phi}\\ \sin\theta\, e^{i\phi} & -\cos\theta\end{pmatrix}$$

where θ and ϕ are the polar coordinates of the unit vector **n** that describes the axis of rotation.

(d) Verify that the matrix $\mathsf{R}_{\mathbf{n}}$ in part (c) is unitary.

(e) Compute explicitly the matrix $\mathsf{S}'_x = \mathsf{R}^\dagger\,\mathsf{S}_x\,\mathsf{R}$ where R is a rotation by an angle φ about the z axis and verify that it returns the expected result. *Hint*: rewrite your result for S'_x in terms of S_x and S_y.

(f) Construct the matrix for a π rotation about the x axis and verify that it turns an up spin into a down spin.

(g) Find the matrix describing a 2π rotation about the z axis. Why is this answer surprising?[41]

Problem 6.33 Consider a particle of mass m in a two-dimensional infinite square well with sides of length L. With the origin placed at the center of the well, the stationary states can be written as

$$\psi_{n_x n_y}(x, y) = \frac{2}{L}\,\sin\left[\frac{n_x\pi}{L}\left(x - \frac{L}{2}\right)\right]\,\sin\left[\frac{n_y\pi}{L}\left(y - \frac{L}{2}\right)\right],$$

with energies

$$E_{n_x n_y} = \frac{\pi^2\hbar^2}{2m\,L^2}\left(n_x^2 + n_y^2\right),$$

for positive integers n_x and n_y.

[41] For a discussion of how this sign change is actually *measured*, see S. A. Werner et al., *Phys. Rev. Lett.* **35**, 1053 (1975).

(a) The two states ψ_{ab} and ψ_{ba} for $a \neq b$ are clearly degenerate. Show that a rotation by 90° counterclockwise about the center of the square carries one into the other,

$$\hat{R}\,\psi_{ab} \propto \psi_{ba},$$

and determine the constant of proportionality. *Hint*: write ψ_{ab} in polar coordinates.

(b) Suppose that instead of ψ_{ab} and ψ_{ba} we choose the basis ψ_+ and ψ_- for our two degenerate states:

$$\psi_{\pm} = \frac{\psi_{ab} \pm \psi_{ba}}{\sqrt{2}}.$$

Show that if a and b are both even or both odd, then ψ_+ and ψ_- are eigenstates of the rotation operator.

(c) Make a contour plot of the state ψ_- for $a = 5$ and $b = 7$ and verify (visually) that it is an eigenstate of every symmetry operation of the square (rotation by an integer multiple of $\pi/2$, reflection across a diagonal, or reflection along a line bisecting two sides). The fact that ψ_+ and ψ_- are not connected to each other by any symmetry of the square means that there must be additional symmetry explaining the degeneracy of these two states.[42]

∗∗∗ **Problem 6.34** The Coulomb potential has *more symmetry* than simply rotational invariance. This additional symmetry is manifest in an additional conserved quantity, the **Laplace–Runge–Lenz vector**

$$\hat{\mathbf{M}} = \frac{\hat{\mathbf{p}} \times \hat{\mathbf{L}} - \hat{\mathbf{L}} \times \hat{\mathbf{p}}}{2m} + V(r)\,\mathbf{r},$$

where $V(\mathbf{r})$ is the potential energy, $V(r) = -e^2/4\pi\epsilon_0 r$.[43] The complete set of commutators for the conserved quantities in the hydrogen atom is

$$\text{(i)} \quad \left[\hat{H}, \hat{M}_i\right] = 0$$

$$\text{(ii)} \quad \left[\hat{H}, \hat{L}_i\right] = 0$$

$$\text{(iii)} \quad \left[\hat{L}_i, \hat{L}_j\right] = i\,\hbar\,\epsilon_{ijk}\,\hat{L}_k$$

$$\text{(iv)} \quad \left[\hat{L}_i, \hat{M}_j\right] = i\,\hbar\,\epsilon_{ijk}\,\hat{M}_k$$

$$\text{(v)} \quad \left[\hat{M}_i, \hat{M}_j\right] = \frac{\hbar}{i}\,\epsilon_{ijk}\,L_k\,\frac{2}{m}\,\hat{H}.$$

The *physical* content of these equations is that (i) **M** is a conserved quantity, (ii) **L** is a conserved quantity, (iii) **L** is a vector, and (iv) **M** is a vector ((v) has no obvious interpretation). There are two additional relations between the quantities $\hat{\mathbf{L}}$, $\hat{\mathbf{M}}$, and $\hat{H}$. They are

---

[42] See F. Leyvraz, et al., *Am. J. Phys.* **65**, 1087 (1997) for a discussion of this "accidental" degeneracy.

[43] The full symmetry of the Coulomb Hamiltonian is not just the obvious three-dimensional rotation group (known to mathematicians as SO(3)), but the four-dimensional rotation group (SO(4)), which has six generators (**L** and **M**). (If the four axes are w, x, y, and z, the generators correspond to rotations in each of the six orthogonal planes, wx, wy, wz (that's **M**) and yz, zx, xy (that's **L**).

(vi) $\hat{M}^2 = \left(\frac{e^2}{4\pi\,\epsilon_0}\right)^2 + \frac{2}{m}\hat{H}\left(\hat{L}^2 + \hbar^2\right)$

(vii) $\hat{\mathbf{M}} \cdot \hat{\mathbf{L}} = 0.$

(a) From the result of Problem 6.19, and the fact that $\hat{\mathbf{M}}$ is a conserved quantity, we know that $\hat{M}_+ \,\psi_{n\ell\ell} = c_{n\ell}\,\psi_{n(\ell+1)(\ell+1)}$ for some constants $c_{n\ell}$. Apply (vii) to the state $\psi_{n\ell\ell}$ to show that

$$\hat{M}_z\,\psi_{n\ell\ell} = -\frac{1}{\sqrt{2}}\frac{1}{\sqrt{\ell+1}}\,c_{n\ell}\,\psi_{n(\ell+1)\ell}.$$

(b) Use (vi) to show that

$$\hat{M}_-\,\hat{M}_+\,\psi_{n\ell\ell} = \left(\frac{e^2}{4\pi\,\epsilon_0}\right)^2\left[1-\left(\frac{\ell+1}{n}\right)^2\right]\psi_{n\ell\ell} - \hat{M}_z^2\,\psi_{n\ell\ell}.$$

(c) From your results to parts (a) and (b), obtain the constants $c_{n\ell}$. You should find that $c_{n\ell}$ is nonzero unless $\ell = n-1$. *Hint*: Consider $\int |M_+\psi_{n\ell m}|^2 d^3\mathbf{r}$ and use the fact that $M_\pm$ are Hermitian conjugates. Figure 6.9 shows how the degenerate states of hydrogen are related by the generators $\hat{\mathbf{L}}$ and $\hat{\mathbf{M}}$.

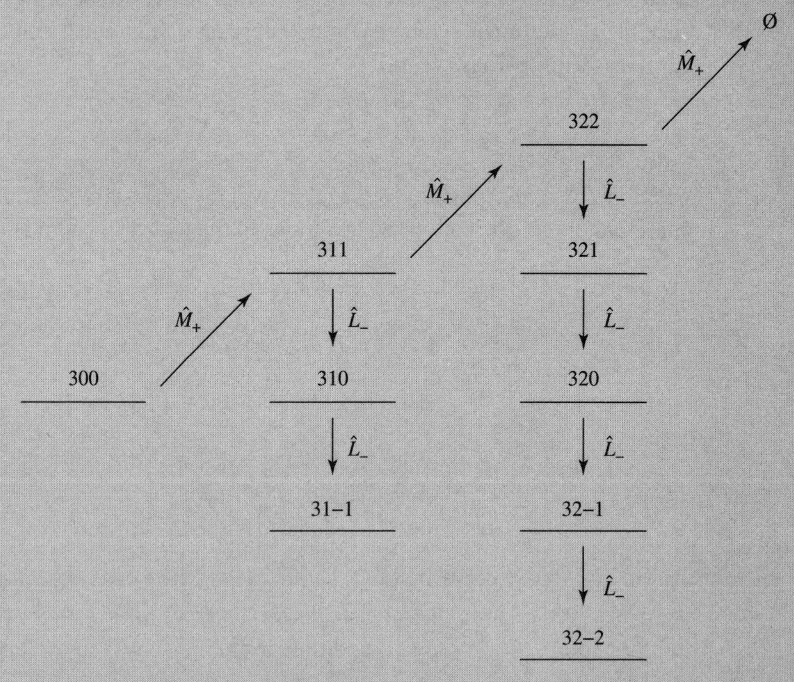

Figure 6.9: The degenerate $n = 3$ states of the hydrogen atom, and the symmetry operations that connect them.

** **Problem 6.35** A **Galilean transformation** performs a boost from a reference frame $\mathcal{S}$ to a reference frame $\mathcal{S}'$ moving with velocity $-v$ with respect to $\mathcal{S}$ (the origins of the two frames coincide at $t = 0$). The unitary operator that carries out a Galilean transformation at time t is

$$\hat{\Gamma}(v,t) = \exp\left[-\frac{i}{\hbar}\, v\left(t\,\hat{p} - m\,\hat{x}\right)\right].$$

(a) Find $\hat{x}' = \hat{\Gamma}^{\dagger}\,\hat{x}\,\hat{\Gamma}$ and $\hat{p}' = \hat{\Gamma}^{\dagger}\,\hat{p}\,\hat{\Gamma}$ for an infinitesimal transformation with velocity δ. What is the physical meaning of your result?

(b) Show that

$$\hat{\Gamma}(v,t) = \exp\left[\frac{i}{\hbar}\left(m\,x\,v - \frac{1}{2}\,m\,v^2\,t\right)\right]\hat{T}(vt)$$
$$= \hat{T}(vt)\,\exp\left[\frac{i}{\hbar}\left(m\,x\,v + \frac{1}{2}\,m\,v^2\,t\right)\right].$$

where $\hat{T}$ is the spatial translation operator (Equation 6.3). You will need to use the Baker–Campbell–Hausdorff formula (Problem 3.29).

(c) Show that if Ψ is a solution to the time-dependent Schrödinger equation with Hamiltonian

$$\hat{H} = \frac{\hat{p}^2}{2\,m} + V(x)$$

then the boosted wave function $\Psi' = \hat{\Gamma}(v,t)\,\Psi$ is a solution to the time-dependent Schrödinger equation with the potential $V(x)$ in motion:

$$\hat{H} = \frac{\hat{p}^2}{2\,m} + V(x - v\,t).$$

Note: $\left(d/dt\right) e^{\hat{A}} = e^{\hat{A}}\left(d\hat{A}/dt\right)$ only if $\left[\hat{A}, \left(d\hat{A}/dt\right)\right] = 0$.

(d) Show that the result of Problem 2.50(a) is an example of this result.

Problem 6.36 A ball thrown through the air leaves your hand at position $\mathbf{r}_0$ with a velocity of $\mathbf{v}_0$ and arrives a time t later at position $\mathbf{r}_1$ traveling with a velocity $\mathbf{v}_1$ (Figure 6.10). Suppose we could instantaneously reverse the ball's velocity when it reaches $\mathbf{r}_1$. Neglecting air resistance, it would retrace the path that took it from $\mathbf{r}_0$ to $\mathbf{r}_1$ and arrive back at $\mathbf{r}_0$ after another time t had passed, traveling with a velocity $-\mathbf{v}_0$. This is an example of **time-reversal invariance**—reverse the motion of a particle at any point along its trajectory and it will retrace its path with an equal and opposite velocity at all positions.

Why is this called time reversal? After all, it was the velocity that was reversed, *not* time. Well, if we showed you a movie of the ball traveling from $\mathbf{r}_1$ to $\mathbf{r}_0$, there would be no way to tell if you were watching a movie of the ball after the reversal playing forward, or a movie of the ball before the reversal playing backward. In a time-reversal invariant system, playing the movie backwards represents another possible motion.

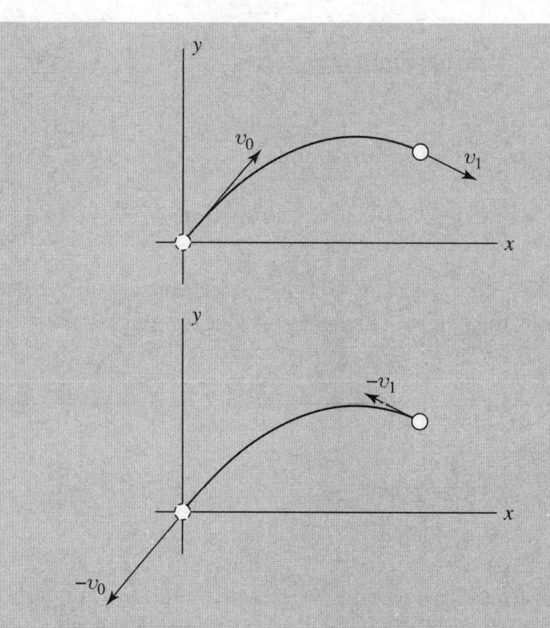

Figure 6.10: A ball thrown through the air (ignore air resistance) is an example of a system with time-reversal symmetry. If we flip the velocity of the particle at any point along its trajectory, it will retrace its path.

A familiar example of a system that does *not* exhibit time-reversal symmetry is a charged particle moving in an external magnetic field.[44] In that case, when you reverse the velocity of the particle, the Lorentz force will also change sign and the particle will not retrace its path; this is illustrated in Figure 6.11.

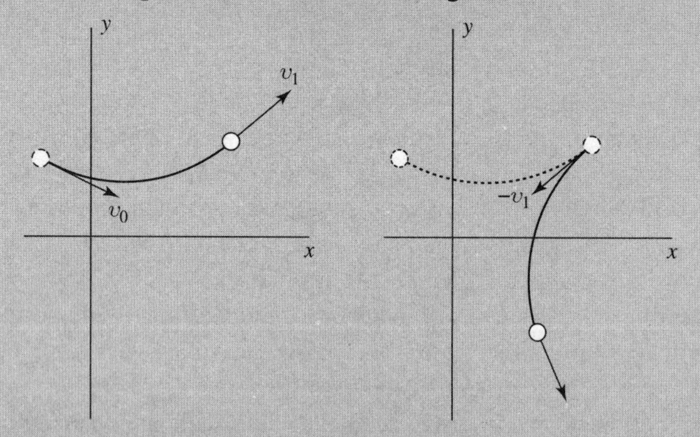

Figure 6.11: An external magnetic field breaks time-reversal symmetry. Shown is the trajectory of a particle of charge $+q$ traveling in a uniform magnetic field pointing into the page. If we flip the particle's velocity from $\mathbf{v}_1$ to $-\mathbf{v}_1$ at the point shown, the particle does not retrace its path, but instead moves onto a new circular orbit.

[44] By external magnetic field, I mean that we we only reverse the velocity of our charge q, and not the velocities of the charges producing the magnetic field. If we reversed those velocities as well, the magnetic field would also switch directions, the Lorentz force on the charge q would be unchanged by the reversal, and the system would in fact be time-reversal invariant.

The **time-reversal operator** $\hat{\Theta}$ is the operator that reverses the momentum of the particle ($\mathbf{p} \to -\mathbf{p}$), leaving its position unchanged. A better name would really be the "reversal of the direction of motion" operator.[45] For a spinless particle, the time-reversal operator $\hat{\Theta}$ simply complex conjugates the position-space wave function[46]

$$\hat{\Theta}\,\Psi(x,t) = \Psi^*(x,t)\,. \tag{6.82}$$

(a) Show that the operators $\hat{x}$ and $\hat{p}$ transform under time reversal as

$$\hat{x}' = \hat{\Theta}^{-1}\,\hat{x}\,\hat{\Theta} = \hat{x}$$
$$\hat{p}' = \hat{\Theta}^{-1}\,\hat{p}\,\hat{\Theta} = -\hat{p}.$$

Hint: Do this by calculating the action of $\hat{x}'$ and $\hat{p}'$ on an arbitrary test function $f(x)$.

(b) We can write down a mathematical statement of time-reversal invariance from our discussion above. We take a system, evolve it for a time t, reverse its momentum, and evolve it for time t again. If the system is time-reversal invariant it will be back where it started, albeit with its momentum reversed (Figure 6.10). As an operator statement this says

$$\hat{U}(t)\;\hat{\Theta}\,\hat{U}(t) = \hat{\Theta}.$$

If this is to hold for any time interval, it must hold in particular for an infinitesimal time interval δ. Show that time-reversal invariance requires

$$\left[\hat{\Theta}, \hat{H}\right] = 0. \tag{6.83}$$

(c) Show that, for a time-reversal invariant Hamiltonian, if $\psi_n(x)$ is a stationary state with energy E_n, then $\psi_n^*(x)$ is also a stationary state with the same energy E_n. If the energy is nondegenerate, this means that the stationary state can be chosen as real.

(d) What do you get by time-reversing a momentum eigenfunction $f_p(x)$ (Equation 3.32)? How about a hydrogen wave function $\psi_{n\ell m}(r,\theta,\phi)$? Comment on each state's relation to the untransformed state and verify that the transformed and untransformed states share the same energy, as guaranteed by (c).

Problem 6.37 As an angular momentum, a particle's spin must flip under time reversal (Problem 6.36). The action of time-reversal on a spinor (Section 4.4.1) is in fact

---

[45] See Eugene P. Wigner, *Group Theory and its Applications to Quantum Mechanics and Atomic Spectra* (Academic Press, New York, 1959), p. 325.

[46] Time reversal is an **anti-unitary operator**. An anti-unitary operator satisfies

$$\langle \Theta\, f \mid \Theta\, g \rangle = \langle f \mid g \rangle^*$$
$$\hat{\Theta}\,(a\,|\alpha\rangle + b\,|\beta\rangle) = a^*\,\hat{\Theta}\,|\alpha\rangle + b^*\,\hat{\Theta}\,|\beta\rangle$$

whereas a unitary operator satisfies the same two equations without the complex conjugates. I won't define the adjoint of an anti-unitary operator; instead I use $\hat{\Theta}^{-1}$ for an anti-unitary operator where we might have used $\hat{U}^\dagger$ or $\hat{U}^{-1}$ interchangeably for a unitary operator.

$$\hat{\Theta}\begin{pmatrix} a \\ b \end{pmatrix} = \begin{pmatrix} -b^* \\ a^* \end{pmatrix} \tag{6.84}$$

so that, in addition to the complex conjugation, the up and down components are interchanged.[47]

(a) Show that $\hat{\Theta}^2 = -1$ for a spin-1/2 particle.

(b) Consider an eigenstate $|\psi_n\rangle$ of a time-reversal invariant Hamiltonian (Equation 6.83) with energy E_n. We know that $|\psi'_n\rangle = \hat{\Theta}\,|\psi_n\rangle$ is also an eigenstate of $\hat{H}$ with the same energy E_n. There two possibilities: either $|\psi'_n\rangle$ and $|\psi_n\rangle$ are the *same* state (meaning $|\psi'_n\rangle = c\,|\psi_n\rangle$ for some complex constant c) or they are distinct states. Show that the first case leads to a contradiction in the case of a spin-1/2 particle, meaning the energy level must be (at least) two-fold degenerate in that case.

Comment: What you have proved is a special case of **Kramer's degeneracy**: for an odd number of spin-1/2 particles (or any half-integer spin for that matter), every energy level (of a time-reversal-invariant Hamiltonian) is at least two-fold degenerate. This is because—as you just showed—for half-integer spin a state and its time-reversed state are necessarily distinct.[48]

---

[47] For arbitrary spin,

$$\hat{\Theta} = e^{-i\pi \hat{S}_y/\hbar}\,\hat{K} \tag{6.85}$$

where the first term is a rotation by π about the y axis and $\hat{K}$ is the operator that performs the complex conjugation.

[48] What about in the case of a spin-0 particle—does time-reversal symmetry tell us anything interesting? Actually it does. For one thing, the stationary states can be chosen as real; you proved this back in Problem 2.2 but we now see that it is a consequence of time-reversal symmetry. Another example is the degeneracy of the energy levels in a periodical potential (Section 5.3.2 and Problem 6.6) for states with crystal momentum q and $-q$. This can be ascribed to inversion symmetry if the potential is symmetric, but the degeneracy persists even when inversion symmetry is absent (try it out!); that is a result of time-reversal symmetry.

第Ⅱ部分

应　用

II

APPLICATIONS

导读 / Guidance

第 7 章　定态微扰理论

本章第 1 节讨论非简并微扰理论，给出了非简并微扰体系的能级一级修正、二级修正和波函数的一级修正；第 2 节讨论简并微扰理论，以简单的一个二重简并体系，给出二重简并情况下的能级修正公式。作者采用的这个讲法和目前国内大多数教材不一样，国内大多数教材是从一般的多重简并出发，给出一个普遍的能级修正公式；第 3 节是把非简并微扰理论应用到氢原子来讨论其精细结构，给出相对论效应、自旋轨道耦合两种情况下对氢原子能级的修正；在第 4 节中，作者着重介绍塞曼效应，讨论弱场、强场和一般情况下的塞曼效应对氢原子能级的修正，强调微扰理论在解决氢原子问题中起的关键作用。最后讨论由自旋 - 自旋耦合引起的氢原子的超精细分裂。

习题特色

（1）量子力学的综合题目：把求解一维无限方势阱的求解能级和全同粒子结合起来，讨论两个全同玻色子在一维无限方势阱中的运动问题，如习题 7.3。

（2）拓展性知识和挑战型性题目：所给的题目本身就可能不存在确切的答案，需要学生做深入思考和探讨，从语言上告诉你如果你能解决这个问题，你可以得诺贝尔奖，如习题 7.14、习题 7.18、习题 7.22 和习题 7.30 等。

（3）引入一些新的物理概念，如 μ 子氢、正电子素、反 μ 子素（见习题 7.32）、范德瓦耳斯作用（见习题 7.37）、克拉默斯关系（见习题 7.43）、Crandall 之谜（见习题 7.55）等。

（4）拓展利用计算机进行数值模拟的题目，如习题 7.7、习题 7.53。

7 TIME-INDEPENDENT PERTURBATION THEORY

7.1 NONDEGENERATE PERTURBATION THEORY

7.1.1 General Formulation

Suppose we have solved the (time-independent) Schrödinger equation for some potential (say, the one-dimensional infinite square well):

$$H^0 \psi_n^0 = E_n^0 \psi_n^0, \tag{7.1}$$

obtaining a complete set of orthonormal eigenfunctions, ψ_n^0,

$$\left\langle \psi_n^0 \middle| \psi_m^0 \right\rangle = \delta_{nm}, \tag{7.2}$$

and the corresponding eigenvalues E_n^0. Now we perturb the potential slightly (say, by putting a little bump in the bottom of the well—Figure 7.1). We'd *like* to find the new eigenfunctions and eigenvalues:

$$H \psi_n = E_n \psi_n, \tag{7.3}$$

but unless we are very lucky, we're not going to be able to solve the Schrödinger equation exactly, for this more complicated potential. **Perturbation theory** is a systematic procedure for obtaining *approximate* solutions to the perturbed problem, by building on the known exact solutions to the *unperturbed* case.

To begin with we write the new Hamiltonian as the sum of two terms:

$$H = H^0 + \lambda H', \tag{7.4}$$

where H' is the perturbation (the superscript 0 always identifies the *un*perturbed quantity). For the moment we'll take λ to be a small number; later we'll crank it up to 1, and H will be the true Hamiltonian. Next we write ψ_n and E_n as power series in λ:

$$\psi_n = \psi_n^0 + \lambda \psi_n^1 + \lambda^2 \psi_n^2 + \cdots ; \tag{7.5}$$

$$E_n = E_n^0 + \lambda E_n^1 + \lambda^2 E_n^2 + \cdots . \tag{7.6}$$

Here E_n^1 is the **first-order correction** to the nth eigenvalue, and ψ_n^1 is the first-order correction to the nth eigenfunction; E_n^2 and ψ_n^2 are the **second-order corrections**, and so on. Plugging Equations 7.5 and 7.6 into Equation 7.3, we have:

$$\begin{aligned} &\left(H^0 + \lambda H'\right)\left[\psi_n^0 + \lambda \psi_n^1 + \lambda^2 \psi_n^2 + \cdots\right] \\ &= \left(E_n^0 + \lambda E_n^1 + \lambda^2 E_n^2 + \cdots\right)\left[\psi_n^0 + \lambda \psi_n^1 + \lambda^2 \psi_n^2 + \cdots\right], \end{aligned}$$

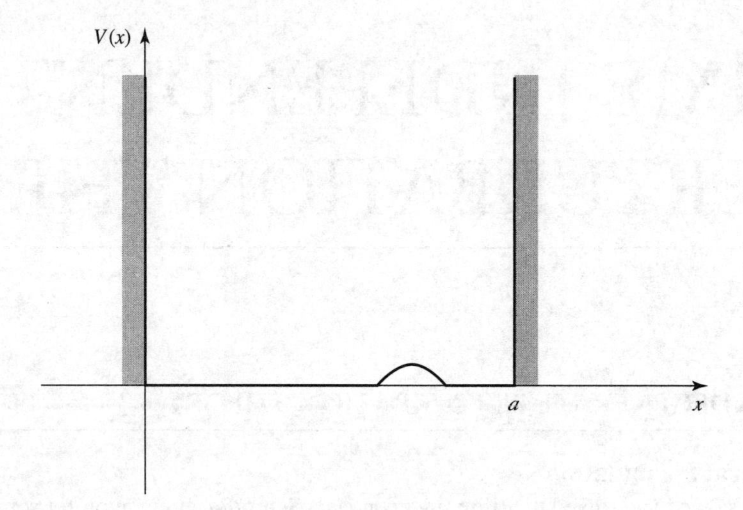

Figure 7.1: Infinite square well with small perturbation.

or (collecting like powers of λ):

$$H^0\psi_n^0 + \lambda\left(H^0\psi_n^1 + H'\psi_n^0\right) + \lambda^2\left(H^0\psi_n^2 + H'\psi_n^1\right) + \cdots$$
$$= E_n^0\psi_n^0 + \lambda\left(E_n^0\psi_n^1 + E_n^1\psi_n^0\right) + \lambda^2\left(E_n^0\psi_n^2 + E_n^1\psi_n^1 + E_n^2\psi_n^0\right) + \cdots.$$

To lowest order[1] $\left(\lambda^0\right)$ this yields $H^0\psi_n^0 = E_n^0\psi_n^0$, which is nothing new (Equation 7.1). To first order $\left(\lambda^1\right)$,

$$H^0\psi_n^1 + H'\psi_n^0 = E_n^0\psi_n^1 + E_n^1\psi_n^0. \tag{7.7}$$

To second order $\left(\lambda^2\right)$,

$$H^0\psi_n^2 + H'\psi_n^1 = E_n^0\psi_n^2 + E_n^1\psi_n^1 + E_n^2\psi_n^0, \tag{7.8}$$

and so on. (I'm done with λ, now—it was just a device to keep track of the different orders—so crank it up to 1.)

7.1.2 First-Order Theory

Taking the inner product of Equation 7.7 with ψ_n^0 (that is, multiplying by $\left(\psi_n^0\right)^*$ and integrating),

$$\left\langle\psi_n^0\middle|H^0\psi_n^1\right\rangle + \left\langle\psi_n^0\middle|H'\psi_n^0\right\rangle = E_n^0\left\langle\psi_n^0\middle|\psi_n^1\right\rangle + E_n^1\left\langle\psi_n^0\middle|\psi_n^0\right\rangle.$$

But H^0 is hermitian, so

$$\left\langle\psi_n^0\middle|H^0\psi_n^1\right\rangle = \left\langle H^0\psi_n^0\middle|\psi_n^1\right\rangle = \left\langle E_n^0\psi_n^0\middle|\psi_n^1\right\rangle = E_n^0\left\langle\psi_n^0\middle|\psi_n^1\right\rangle,$$

---

[1] As always (footnote 34, page 49) the uniqueness of power series expansions guarantees that the coefficients of like powers are equal.

and this cancels the first term on the right. Moreover, $\langle\psi_n^0|\psi_n^0\rangle = 1$, so[2]

$$\boxed{E_n^1 = \left\langle \psi_n^0 \middle| H' \middle| \psi_n^0 \right\rangle.} \tag{7.9}$$

This is the fundamental result of first-order perturbation theory; as a *practical* matter, it may well be the most frequently used equation in quantum mechanics. It says that the first-order correction to the energy is the *expectation value* of the perturbation, in the *unperturbed* state.

Example 7.1

The unperturbed wave functions for the infinite square well are (Equation 2.31)

$$\psi_n^0(x) = \sqrt{\frac{2}{a}} \sin\left(\frac{n\pi}{a}x\right).$$

Suppose we perturb the system by simply raising the "floor" of the well a constant amount V_0 (Figure 7.2). Find the first-order correction to the energies.

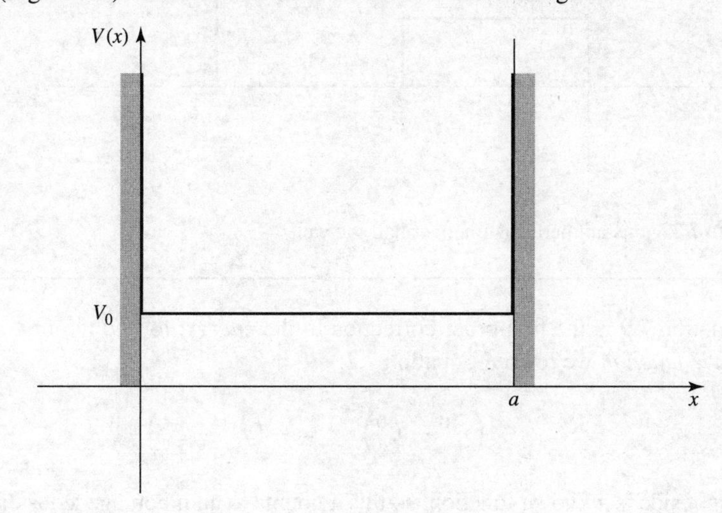

Figure 7.2: Constant perturbation over the whole well.

Solution: In this case $H' = V_0$, and the first-order correction to the energy of the nth state is

$$E_n^1 = \left\langle \psi_n^0 \middle| V_0 \middle| \psi_n^0 \right\rangle = V_0 \left\langle \psi_n^0 \middle| \psi_n^0 \right\rangle = V_0.$$

The corrected energy levels, then, are $E_n \approx E_n^0 + V_0$; they are simply lifted by the amount V_0. Of *course!* The only surprising thing is that in this case the first-order theory yields the *exact* answer. Evidently for a *constant* perturbation all the higher corrections vanish.[3]

---

[2] In this context it doesn't matter whether we write $\langle\psi_n^0|H'\psi_n^0\rangle$ or $\langle\psi_n^0|H'|\psi_n^0\rangle$ (with the extra vertical bar), because we are using the wave function itself to label the state. But the latter notation is preferable, because it frees us from this convention. For instance, if we used $|n\rangle$ to denote the nth state of the harmonic oscillator (Equation 2.86), $H'\,|n\rangle$ makes sense, but $|H'n\rangle$ is unintelligible (operators act on *vectors*/functions, not on *numbers*).

[3] Incidentally, nothing here depends on the specific nature of the infinite square well—the same holds for *any* potential, when the perturbation is a constant.

On the other hand, if the perturbation extends only half-way across the well (Figure 7.3), then

$$E_n^1 = \frac{2V_0}{a}\int_0^{a/2} \sin^2\left(\frac{n\pi}{a}x\right) dx = \frac{V_0}{2}.$$

In this case every energy level is lifted by $V_0/2$. That's not the *exact* result, presumably, but it does seem reasonable, as a first-order approximation.

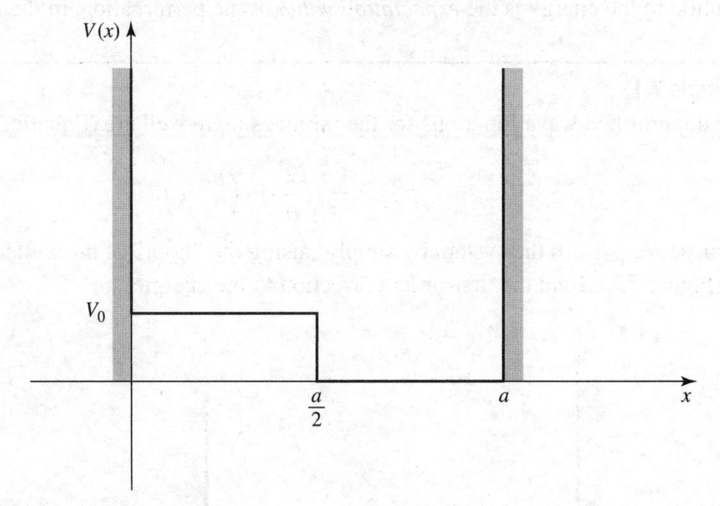

Figure 7.3: Constant perturbation over half the well.

Equation 7.9 is the first-order correction to the *energy*; to find the first-order correction to the *wave function* we rewrite Equation 7.7:

$$\left(H^0 - E_n^0\right)\psi_n^1 = -\left(H' - E_n^1\right)\psi_n^0. \tag{7.10}$$

The right side is a known function, so this amounts to an inhomogeneous differential equation for ψ_n^1. Now, the unperturbed wave functions constitute a complete set, so ψ_n^1 (like any other function) can be expressed as a linear combination of them:

$$\psi_n^1 = \sum_{m \neq n} c_m^{(n)} \psi_m^0. \tag{7.11}$$

(There is no need to include $m = n$ in the sum, for if ψ_n^1 satisfies Equation 7.10, so too does $(\psi_n^1 + \alpha\psi_n^0)$, for any constant α, and we can use this freedom to subtract off the ψ_n^0 term.[4]) If we could determine the coefficients $c_m^{(n)}$, we'd be done.

[4] Alternatively, a glance at Equation 7.5 reveals that any ψ_n^0 component in ψ_n^1 might as well be pulled out and combined with the first term. In fact, the choice $c_n^{(n)} = 0$ ensures that ψ_n—with 1 as the coefficient of ψ_n^0 in Equation 7.5—is *normalized* (to first order in λ): $\langle\psi_n|\psi_n\rangle = \langle\psi_n^0|\psi_n^0\rangle + \lambda\left(\langle\psi_n^1|\psi_n^0\rangle + \langle\psi_n^0|\psi_n^1\rangle\right) + \lambda^2(\cdots) + \cdots$, but the orthonormality of the unperturbed states means that the first term is 1 and $\langle\psi_n^1|\psi_n^0\rangle = \langle\psi_n^0|\psi_n^1\rangle = 0$, as long as ψ_n^1 has no ψ_n^0 component.

Well, putting Equation 7.11 into Equation 7.10, and using the fact that ψ_m^0 satisfies the unperturbed Schrödinger equation (Equation 7.1), we have

$$\sum_{m\neq n}\left(E_m^0 - E_n^0\right) c_m^{(n)} \psi_m^0 = -\left(H' - E_n^1\right)\psi_n^0.$$

Taking the inner product with ψ_l^0,

$$\sum_{m\neq n}\left(E_m^0 - E_n^0\right) c_m^{(n)} \left\langle \psi_l^0 \middle| \psi_m^0 \right\rangle = -\left\langle \psi_l^0 \middle| H' \middle| \psi_n^0 \right\rangle + E_n^1 \left\langle \psi_l^0 \middle| \psi_n^0 \right\rangle.$$

If $l = n$, the left side is zero, and we recover Equation 7.9; if $l \neq n$, we get

$$\left(E_l^0 - E_n^0\right) c_l^{(n)} = -\left\langle \psi_l^0 \middle| H' \middle| \psi_n^0 \right\rangle,$$

or

$$c_m^{(n)} = \frac{\left\langle \psi_m^0 \middle| H' \middle| \psi_n^0 \right\rangle}{E_n^0 - E_m^0}, \tag{7.12}$$

so

$$\boxed{\psi_n^1 = \sum_{m\neq n} \frac{\left\langle \psi_m^0 \middle| H' \middle| \psi_n^0 \right\rangle}{\left(E_n^0 - E_m^0\right)} \psi_m^0.} \tag{7.13}$$

Notice that the denominator is safe (since there is no coefficient with $m = n$) *as long as the unperturbed energy spectrum is nondegenerate*. But if two different unperturbed states share the same energy, we're in serious trouble (we divided by zero to get Equation 7.12); in that case we need **degenerate perturbation theory**, which I'll come to in Section 7.2.

That completes first-order perturbation theory: The first-order correction to the energy, E_n^1, is given by Equation 7.9, and the first-order correction to the wave function, ψ_n^1, is given by Equation 7.13.

* **Problem 7.1** Suppose we put a delta-function bump in the center of the infinite square well:

$$H' = \alpha\delta(x - a/2),$$

where α is a constant.

(a) Find the first-order correction to the allowed energies. Explain why the energies are not perturbed for even n.

(b) Find the first three nonzero terms in the expansion (Equation 7.13) of the correction to the ground state, ψ_1^1.

* **Problem 7.2** For the harmonic oscillator $\left[V(x) = (1/2)kx^2\right]$, the allowed energies are

$$E_n = (n + 1/2)\hbar\omega, \quad (n = 0, 1, 2, \ldots),$$

where $\omega = \sqrt{k/m}$ is the classical frequency. Now suppose the spring constant increases slightly: $k \to (1+\epsilon)\,k$. (Perhaps we cool the spring, so it becomes less flexible.)

(a) Find the *exact* new energies (trivial, in this case). Expand your formula as a power series in ϵ, up to second order.

(b) Now calculate the first-order perturbation in the energy, using Equation 7.9. What is H' here? Compare your result with part (a). *Hint:* It is not necessary—in fact, it is not *permitted*—to calculate a single integral in doing this problem.

Problem 7.3 Two identical spin-zero bosons are placed in an infinite square well (Equation 2.22). They interact weakly with one another, via the potential

$$V(x_1, x_2) = -aV_0\delta(x_1 - x_2)$$

(where V_0 is a constant with the dimensions of energy, and a is the width of the well).

(a) First, ignoring the interaction between the particles, find the ground state and the first excited state—both the wave functions and the associated energies.

(b) Use first-order perturbation theory to estimate the effect of the particle–particle interaction on the energies of the ground state and the first excited state.

7.1.3 Second-Order Energies

Proceeding as before, we take the inner product of the *second*-order equation (Equation 7.8) with ψ_n^0:

$$\left\langle \psi_n^0 \middle| H^0\psi_n^2 \right\rangle + \left\langle \psi_n^0 \middle| H'\psi_n^1 \right\rangle = E_n^0 \left\langle \psi_n^0 \middle| \psi_n^2 \right\rangle + E_n^1 \left\langle \psi_n^0 \middle| \psi_n^1 \right\rangle + E_n^2 \left\langle \psi_n^0 \middle| \psi_n^0 \right\rangle.$$

Again, we exploit the hermiticity of H^0:

$$\left\langle \psi_n^0 \middle| H^0\psi_n^2 \right\rangle = \left\langle H^0\psi_n^0 \middle| \psi_n^2 \right\rangle = E_n^0 \left\langle \psi_n^0 \middle| \psi_n^2 \right\rangle,$$

so the first term on the left cancels the first term on the right. Meanwhile, $\langle \psi_n^0 | \psi_n^0 \rangle = 1$, and we are left with a formula for E_n^2:

$$E_n^2 = \left\langle \psi_n^0 \middle| H' \middle| \psi_n^1 \right\rangle - E_n^1 \left\langle \psi_n^0 \middle| \psi_n^1 \right\rangle. \tag{7.14}$$

But

$$\langle \psi_n^0 | \psi_n^1 \rangle = \sum_{m \neq n} c_m^{(n)} \langle \psi_n^0 | \psi_m^0 \rangle = 0$$

(because the sum excludes $m = n$, and all the others are orthogonal), so

$$E_n^2 = \left\langle \psi_n^0 \middle| H' \middle| \psi_n^1 \right\rangle = \sum_{m \neq n} c_m^{(n)} \left\langle \psi_n^0 \middle| H' \middle| \psi_m^0 \right\rangle = \sum_{m \neq n} \frac{\langle \psi_m^0 | H' | \psi_n^0 \rangle \langle \psi_n^0 | H' | \psi_m^0 \rangle}{E_n^0 - E_m^0},$$

or, finally,

$$\boxed{E_n^2 = \sum_{m \neq n} \frac{\left| \langle \psi_m^0 | H' | \psi_n^0 \rangle \right|^2}{E_n^0 - E_m^0}.} \tag{7.15}$$

This is the fundamental result of second-order perturbation theory.

We could go on to calculate the second-order correction to the wave function (ψ_n^2), the third-order correction to the energy, and so on, but in practice Equation 7.15 is ordinarily as far as it is useful to pursue this method.[5]

** **Problem 7.4** Apply perturbation theory to the most general two-level system. The unperturbed Hamiltonian is

$$\mathsf{H}^0 = \begin{pmatrix} E_a^0 & 0 \\ 0 & E_b^0 \end{pmatrix}$$

and the perturbation is

$$\mathsf{H}' = \lambda \begin{pmatrix} V_{aa} & V_{ab} \\ V_{ba} & V_{bb} \end{pmatrix}$$

with $V_{ba} = V_{ab}^*$, V_{aa} and V_{bb} real, so that H is hermitian. As in Section 7.1.1, λ is a constant that will later be set to 1.

(a) Find the exact energies for this two-level system.

(b) Expand your result from (a) to second order in λ (and then set λ to 1). Verify that the terms in the series agree with the results from perturbation theory in Sections 7.1.2 and 7.1.3. Assume that $E_b > E_a$.

(c) Setting $V_{aa} = V_{bb} = 0$, show that the series in (b) only converges if

$$\left| \frac{V_{ab}}{E_b^0 - E_a^0} \right| < \frac{1}{2}.$$

Comment: In general, perturbation theory is only valid if the matrix elements of the perturbation are small compared to the energy level spacings. Otherwise, the first few terms (which are all we ever calculate) will give a poor approximation to the quantity of interest and, as shown here, the series may fail to converge at all, in which case the first few terms tell us nothing.

* **Problem 7.5**

(a) Find the second-order correction to the energies (E_n^2) for the potential in Problem 7.1. *Comment:* You can sum the series explicitly, obtaining $-2m(\alpha/\pi\hbar n)^2$ for odd n.

[5] In the short-hand notation $V_{mn} \equiv \langle\psi_m^0|H'|\psi_n^0\rangle$, $\Delta_{mn} \equiv E_m^0 - E_n^0$, the first three corrections to the nth energy are

$$E_n^1 = V_{nn}, \quad E_n^2 = \sum_{m\neq n} \frac{|V_{nm}|^2}{\Delta_{nm}}, \quad E_n^3 = \sum_{l,m\neq n} \frac{V_{nl}V_{lm}V_{mn}}{\Delta_{nl}\Delta_{nm}} - V_{nn}\sum_{m\neq n}\frac{|V_{nm}|^2}{\Delta_{nm}^2}.$$

The third-order correction is given in Landau and Lifschitz, *Quantum Mechanics: Non-Relativistic Theory*, 3rd edn, Pergamon, Oxford (1977), page 136; the fourth and fifth orders (together with a powerful general technique for obtaining the higher orders) are developed by Nicholas Wheeler, *Higher-Order Spectral Perturbation* (unpublished Reed College report, 2000). Illuminating alternative formulations of time-independent perturbation theory include the Dalgarno–Lewis method and the closely related "logarithmic" perturbation theory (see, for example, T. Imbo and U. Sukhatme, *Am. J. Phys.* **52**, 140 (1984), for LPT, and H. Mavromatis, *Am. J. Phys.* **59**, 738 (1991), for Dalgarno–Lewis).

(b) Calculate the second-order correction to the ground state energy $\left(E_0^2\right)$ for the potential in Problem 7.2. Check that your result is consistent with the exact solution.

** **Problem 7.6** Consider a charged particle in the one-dimensional harmonic oscillator potential. Suppose we turn on a weak electric field (E), so that the potential energy is shifted by an amount $H' = -qEx$.

(a) Show that there is no first-order change in the energy levels, and calculate the second-order correction. *Hint:* See Problem 3.39.

(b) The Schrödinger equation can be solved directly in this case, by a change of variables: $x' \equiv x - \left(qE/m\omega^2\right)$. Find the exact energies, and show that they are consistent with the perturbation theory approximation.

**  **Problem 7.7** Consider a particle in the potential shown in Figure 7.3.

(a) Find the first-order correction to the ground-state wave function. The first three nonzero terms in the sum will suffice.

(b) Using the method of Problem 2.61 find (numerically) the ground-state wave function and energy. Use $V_0 = 4\,\hbar^2/m\,a^2$ and $N = 100$. Compare the energy obtained numerically to the result from first-order perturbation theory (see Example 7.1).

(c) Make a single plot showing (i) the unperturbed ground-state wave function, (ii) the numerical ground-state wave function, and (ii) the first-order approximation to the ground-state wave function. *Note*: Make sure you've properly normalized your numerical result,

$$1 = \int |\psi(x)|^2\,dx \approx \sum_{i=1}^{N} |\psi_i|^2\,\Delta x.$$

7.2 DEGENERATE PERTURBATION THEORY

If the unperturbed states are degenerate—that is, if two (or more) distinct states (ψ_a^0 and ψ_b^0) share the same energy—then ordinary perturbation theory fails: $c_a^{(b)}$ (Equation 7.12) and E_a^2 (Equation 7.15) blow up (unless, perhaps, the numerator vanishes, $\left\langle\psi_a^0\left|H'\right|\psi_b^0\right\rangle = 0$—a loophole that will be important to us later on). In the degenerate case, therefore, there is no reason to trust even the *first*-order correction to the energy (Equation 7.9), and we must look for some other way to handle the problem. Note this is *not* a minor problem; almost all applications of perturbation theory involve degeneracy.

7.2.1 Two-Fold Degeneracy

Suppose that

$$H^0\psi_a^0 = E^0\psi_a^0, \quad H^0\psi_b^0 = E^0\psi_b^0, \quad \left\langle\psi_a^0\middle|\psi_b^0\right\rangle = 0, \tag{7.16}$$

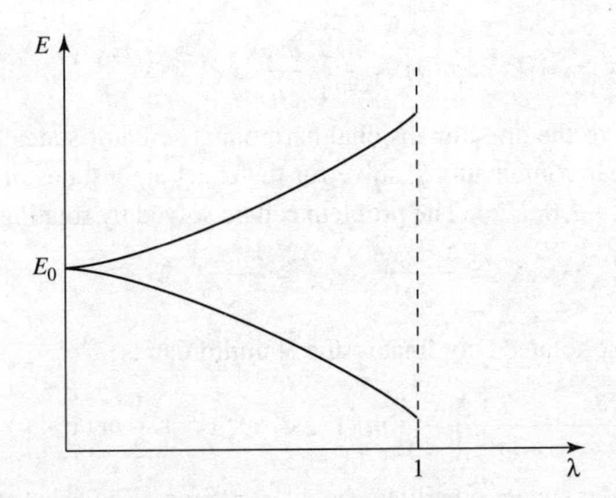

Figure 7.4: "Lifting" of a degeneracy by a perturbation.

with ψ_a^0 and ψ_b^0 both normalized. Note that any linear combination of these states,

$$\psi^0 = \alpha\psi_a^0 + \beta\psi_b^0, \tag{7.17}$$

is still an eigenstate of H^0, with the same eigenvalue E^0:

$$H^0\psi^0 = E^0\psi^0. \tag{7.18}$$

Typically, the perturbation $\left(H'\right)$ will "break" (or "lift") the degeneracy: As we increase λ (from 0 to 1), the common unperturbed energy E^0 splits into two (Figure 7.4). Going the other direction, when we turn *off* the perturbation, the "upper" state reduces down to *one* linear combination of ψ_a^0 and ψ_b^0, and the "lower" state reduces to some (orthogonal) linear combination, but we don't know a priori what these "**good**" linear combinations will be. For this reason we can't even calculate the *first*-order energy (Equation 7.9)—we don't know what unperturbed states to use.

The "good" states are *defined* as the limit of the true eigenstates as the perturbation is switched off ($\lambda \to 0$) but that isn't how you find them in realistic situations (if you *knew* the exact eigenstates you wouldn't need perturbation theory). Before I show you the practical techniques for calculating them, we'll look at an example where we can take the $\lambda \to 0$ limit of the exact eigenstates.

Example 7.2

Consider a particle of mass m in a two-dimensional oscillator potential

$$H^0 = \frac{p^2}{2m} + \frac{1}{2} m\,\omega^2 \left(x^2 + y^2\right)$$

to which is added a perturbation

$$H' = \epsilon\, m\,\omega^2\, x\, y.$$

The unperturbed first-excited state (with $E^0 = 2\,\hbar\,\omega$) is two-fold degenerate, and *one* basis for those two degenerate states is

$$\psi_a^0 = \psi_0(x)\ \psi_1(y) = \sqrt{\frac{2}{\pi}}\,\frac{m\,\omega}{\hbar}\, y\, e^{-\frac{m\omega}{2\hbar}(x^2+y^2)}$$

$$\psi_b^0 = \psi_1(x)\ \psi_0(y) = \sqrt{\frac{2}{\pi}\ \frac{m\,\omega}{\hbar}}\ x\, e^{-\frac{m\omega}{2\hbar}(x^2+y^2)}, \tag{7.19}$$

where ψ_0 and ψ_1 refer to the one-dimensional harmonic oscillator states (Equation 2.86). To find the "good" linear combinations, solve for the exact eigenstates of $H = H^0 + H'$ and take their limit as $\epsilon \to 0$. *Hint*: The problem can be solved by rotating coordinates

$$x' = \frac{x+y}{\sqrt{2}} \qquad y' = \frac{x-y}{\sqrt{2}}. \tag{7.20}$$

Solution: In terms of the rotated coordinates, the Hamiltonian is

$$H = -\frac{\hbar^2}{2m}\left(\frac{\partial^2}{\partial x'^2} + \frac{\partial^2}{\partial y'^2}\right) + \frac{1}{2} m\,(1+\epsilon)\,\omega^2\, x'^2 + \frac{1}{2} m\,(1-\epsilon)\,\omega^2\, y'^2.$$

This amounts to two independent one-dimensional oscillators. The exact solutions are

$$\psi_{mn} = \psi_m^+(x')\ \psi_n^-(y'),$$

where $\psi_m^\pm$ are one-dimensional oscillator states with frequencies $\omega_\pm = \sqrt{1\pm\epsilon}\,\omega$ respectively. The first few exact energies,

$$E_{mn} = \left(m + \frac{1}{2}\right)\hbar\omega_+ + \left(n + \frac{1}{2}\right)\hbar\omega_-, \tag{7.21}$$

are shown in Figure 7.5.

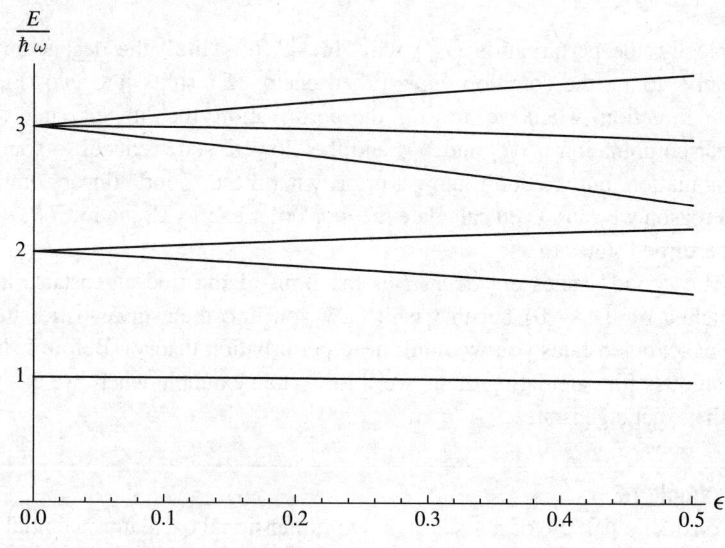

Figure 7.5: Exact energy levels as a function of ϵ for Example 7.2. Level (1) is E_{00}, (2) are E_{01} and E_{10}, (3) are E_{20}, E_{11}, and E_{02}. The lines are not straight (except the first).

The two states which grow out of the degenerate first-excited states as ϵ is increased have $m = 0, n = 1$ (lower state) and $m = 1, n = 0$ (upper state). If we track these states back to $\epsilon = 0$ (in that limit $\omega_+ = \omega_- = \omega$) we get

$$\lim_{\epsilon\to 0}\psi_{01} = \lim_{\epsilon\to 0}\psi_0^+(x')\ \psi_1^-(y') = \psi_0\left(\frac{x+y}{\sqrt{2}}\right)\ \psi_1\left(\frac{x-y}{\sqrt{2}}\right)$$

$$= \sqrt{\frac{2}{\pi}\ \frac{m\,\omega}{\hbar}}\ \frac{x-y}{\sqrt{2}}\ e^{-\frac{m\omega}{2\hbar}(x^2+y^2)} = \frac{-\psi_a^0 + \psi_b^0}{\sqrt{2}}.$$

$$\lim_{\epsilon \to 0} \psi_{10} = \frac{\psi_a^0 + \psi_b^0}{\sqrt{2}}. \tag{7.22}$$

Therefore the "good" states for this problem are

$$\boxed{\psi_\pm^0 \equiv \frac{1}{\sqrt{2}}\left(\psi_b^0 \pm \psi_a^0\right).} \tag{7.23}$$

In this example we were able to find the exact eigenstates of H and then turn off the perturbation to see what states they evolve from. But how do we find the "good" states when we *can't* solve the system exactly?

For the moment let's just write the "good" unperturbed states in generic form (Equation 7.17), keeping α and β adjustable. We want to solve the Schrödinger equation,

$$H\psi = E\psi, \tag{7.24}$$

with $H = H^0 + \lambda H'$ and

$$E = E^0 + \lambda E^1 + \lambda^2 E^2 + \cdots, \quad \psi = \psi^0 + \lambda\psi^1 + \lambda^2\psi^2 + \cdots. \tag{7.25}$$

Plugging these into Equation 7.24, and collecting like powers of λ (as before) we find

$$H^0\psi^0 + \lambda\left(H'\psi^0 + H^0\psi^1\right) + \cdots = E^0\psi^0 + \lambda\left(E^1\psi^0 + E^0\psi^1\right) + \cdots.$$

But $H^0\psi^0 = E^0\psi^0$ (Equation 7.18), so the first terms cancel; at order λ^1 we have

$$H^0\psi^1 + H'\psi^0 = E^0\psi^1 + E^1\psi^0. \tag{7.26}$$

Taking the inner product with ψ_a^0:

$$\left\langle\psi_a^0 \middle| H^0\psi^1\right\rangle + \left\langle\psi_a^0 \middle| H'\psi^0\right\rangle = E^0\left\langle\psi_a^0 \middle| \psi^1\right\rangle + E^1\left\langle\psi_a^0 \middle| \psi^0\right\rangle.$$

Because H^0 is hermitian, the first term on the left cancels the first term on the right. Putting in Equation 7.17, and exploiting the orthonormality condition (Equation 7.16), we obtain

$$\alpha\left\langle\psi_a^0 \middle| H' \middle| \psi_a^0\right\rangle + \beta\left\langle\psi_a^0 \middle| H' \middle| \psi_b^0\right\rangle = \alpha E^1,$$

or, more compactly,

$$\alpha W_{aa} + \beta W_{ab} = \alpha E^1, \tag{7.27}$$

where

$$W_{ij} \equiv \left\langle\psi_i^0 \middle| H' \middle| \psi_j^0\right\rangle. \quad (i, j = a, b). \tag{7.28}$$

Similarly, the inner product with ψ_b^0 yields

$$\alpha W_{ba} + \beta W_{bb} = \beta E^1. \tag{7.29}$$

Notice that the Ws are (in principle) *known*—they are just the "matrix elements" of H', with respect to the unperturbed wave functions ψ_a^0 and ψ_b^0. Written in matrix form, Equations 7.27 and 7.29 are

$$\underbrace{\begin{pmatrix} W_{aa} & W_{ab} \\ W_{ba} & W_{bb} \end{pmatrix}}_{\mathsf{W}} \begin{pmatrix} \alpha \\ \beta \end{pmatrix} = E^1 \begin{pmatrix} \alpha \\ \beta \end{pmatrix}. \tag{7.30}$$

The eigenvalues of the matrix W give the first-order corrections to the energy $\left(E^1\right)$ and the corresponding eigenvectors tell us the coefficients α and β that determine the "good" states.[6]

The Appendix (Section A.5) shows how to obtain the eigenvalues of a matrix; I'll reproduce those steps here to find a general solution for E^1. First, move all the terms in Equation 7.30 to the left-hand side.

$$\begin{pmatrix} W_{aa} - E^1 & W_{ab} \\ W_{ba} & W_{bb} - E^1 \end{pmatrix} \begin{pmatrix} \alpha \\ \beta \end{pmatrix} = 0 \tag{7.31}$$

This equation only has non-trivial solutions if the matrix on the left is non-invertible—that is to say, if its determinant vanishes:

$$\begin{vmatrix} W_{aa} - E^1 & W_{ab} \\ W_{ba} & W_{bb} - E^1 \end{vmatrix} = \left(W_{aa} - E^1\right)\left(W_{bb} - E^1\right) - |W_{ab}|^2 = 0, \tag{7.32}$$

where we used the fact that $W_{ba} = W_{ab}^*$. Solving the quadratic,

$$\boxed{E_{\pm}^1 = \frac{1}{2}\left[W_{aa} + W_{bb} \pm \sqrt{(W_{aa} - W_{bb})^2 + 4\,|W_{ab}|^2}\right].} \tag{7.33}$$

This is the fundamental result of degenerate perturbation theory; the two roots correspond to the two perturbed energies.

Example 7.3

Returning to Example 7.2, show that diagonalizing the matrix W gives the same "good" states we found by solving the problem exactly.

Solution: We need to calculate the matrix elements of W. First,

$$\begin{aligned} \mathsf{W}_{aa} &= \int\!\!\int \psi_a^0(x, y)\; H'\; \psi_a^0(x, y)\; dx\, dy \\ &= \epsilon\, m\, \omega^2 \int |\psi_0(x)|^2\; x\, dx \int |\psi_1(y)|^2\; y\, dy = 0 \end{aligned}$$

(the integrands are both odd functions). Similarly, $\mathsf{W}_{bb} = 0$, and we need only compute

$$\begin{aligned} \mathsf{W}_{ab} &= \int\!\!\int \psi_a^0(x, y)\; H'\; \psi_b^0(x, y)\; dx\, dy \\ &= \epsilon\, m\, \omega^2 \int \psi_0(x)\; x\; \psi_1(x)\; dx \int \psi_1(y)\; y\; \psi_0(y)\; dy. \end{aligned}$$

These two integrals are equal, and recalling (Equation 2.70)

$$x = \sqrt{\frac{\hbar}{2m\,\omega}}\,(a_+ + a_-)$$

[6] This assumes that the eigenvalues of W are *distinct* so that the degeneracy lifts at first order. If not, any choice of α and β satisfies Equation 7.30; you *still* don't know what the good states are. The first-order energies are correctly given by Equation 7.33 when this happens, and in many cases that's all you require. But if you need to know the "good" states—for example to calculate higher-order corrections—you will have to use *second-order* degenerate perturbation theory (see Problems 7.39, 7.40, and 7.41) or employ the theorem of Section 7.2.2.

we have

$$\mathsf{W}_{ab} = \epsilon\, m\, \omega^2 \left[\int \psi_0(x) \sqrt{\frac{\hbar}{2\, m\, \omega}}\, (a_+ + a_-)\, \psi_1(x)\, dx \right]^2$$

$$= \epsilon\, \frac{\hbar\, \omega}{2} \left[\int \psi_0(x)\, \psi_0(x)\, dx \right]^2 = \epsilon\, \frac{\hbar\, \omega}{2}.$$

Therefore, the matrix W is

$$\mathsf{W} = \epsilon\, \frac{\hbar\, \omega}{2} \begin{pmatrix} 0 & 1 \\ 1 & 0 \end{pmatrix}.$$

The (normalized) eigenvectors of this matrix are

$$\frac{1}{\sqrt{2}} \begin{pmatrix} 1 \\ 1 \end{pmatrix} \text{ and } \frac{1}{\sqrt{2}} \begin{pmatrix} -1 \\ 1 \end{pmatrix}.$$

These eigenvectors tell us which linear combination of ψ_a^0 and ψ_b^0 are the good states:

$$\boxed{\psi_\pm^0 = \frac{1}{\sqrt{2}} \left(\psi_b^0 \pm \psi_a^0 \right),}$$

just as in Equation 7.23. The eigenvalues of the matrix W,

$$E^1 = \pm\, \epsilon\, \frac{\hbar\, \omega}{2}, \tag{7.34}$$

give the first-order corrections to the energy (compare 7.33).

If it happens that $W_{ab} = 0$ in Equation 7.30 then the two eigenvectors are

$$\begin{pmatrix} 1 \\ 0 \end{pmatrix} \text{ and } \begin{pmatrix} 0 \\ 1 \end{pmatrix}$$

and the energies,

$$E_+^1 = W_{aa} = \left\langle \psi_a^0 \middle| H' \middle| \psi_a^0 \right\rangle, \quad E_-^1 = W_{bb} = \left\langle \psi_b^0 \middle| H' \middle| \psi_b^0 \right\rangle, \tag{7.35}$$

are precisely what we would have obtained using *non*degenerate perturbation theory (Equation 7.9). We have simply been *lucky*: The states ψ_a^0 and ψ_b^0 were *already* the "good" linear combinations. Obviously, it would be greatly to our advantage if we could somehow *guess* the "good" states right from the start—then we could go ahead and use *non*degenerate perturbation theory. As it turns out, we can very often do this by exploiting the theorem in the following section.

7.2.2 "Good" States

Theorem: Let A be a hermitian operator that commutes with H^0 and H'. If ψ_a^0 and ψ_b^0 (the degenerate eigenfunctions of H^0) are also eigenfunctions of A, with distinct eigenvalues,

$$A\psi_a^0 = \mu\psi_a^0, \quad A\psi_b^0 = \nu\psi_b^0, \quad \text{and } \mu \neq \nu,$$

then ψ_a^0 and ψ_b^0 are the "good" states to use in perturbation theory.

Proof: Since $H(\lambda) = H^0 + \lambda H'$ and A commute, there exist simultaneous eigenstates $\psi_\gamma(\lambda)$ where

$$H(\lambda)\ \psi_\gamma(\lambda) = E(\lambda)\ \psi_\gamma(\lambda) \text{ and } A\,\psi_\gamma(\lambda) = \gamma\,\psi_\gamma(\lambda)\,. \tag{7.36}$$

The fact that A is hermitian means

$$\left\langle \psi_a^0 \middle| A\psi_\gamma(\lambda)\right\rangle = \left\langle A\psi_a^0 \middle| \psi_\gamma(\lambda)\right\rangle$$

$$\gamma \left\langle \psi_a^0 \middle| \psi_\gamma(\lambda)\right\rangle = \mu^* \left\langle \psi_a^0 \middle| \psi_\gamma(\lambda)\right\rangle \tag{7.37}$$

$$(\gamma - \mu) \left\langle \psi_a^0 \middle| \psi_\gamma(\lambda)\right\rangle = 0, \tag{7.38}$$

(making use of the fact that μ is real). This holds true for any value of λ and taking the limit as $\lambda \to 0$ we have

$$\left\langle \psi_a^0 \middle| \psi_\gamma(0)\right\rangle = 0 \text{ unless } \gamma = \mu,$$

and similarly

$$\left\langle \psi_b^0 \middle| \psi_\gamma(0)\right\rangle = 0 \text{ unless } \gamma = \nu.$$

Now the good states are linear combinations of ψ_a^0 and ψ_b^0: $\psi_\gamma(0) = \alpha\,\psi_a^0 + \beta\,\psi_b^0$. From above it follows that either $\gamma = \mu$, in which case $\beta = \langle \psi_b^0 \mid \psi_\gamma(0)\rangle = 0$ and the good state is simply ψ_a^0, or $\gamma = \nu$ and the good state is ψ_b^0. QED

Once we identify the "good" states, either by solving Equation 7.30 or by applying this theorem, we can use these "good" states as our unperturbed states and apply ordinary non-degenerate perturbation theory.[7] In most cases, the operator A will be suggested by symmetry; as you saw in Chapter 6, symmetries are associated with operators that commute with H—precisely what are required to identify the good states.

Example 7.4

Find an operator A that satisfies the requirements of the preceding theorem to construct the "good" states in Examples 7.2 and 7.3.

Solution: The perturbation H' has less symmetry than H^0. H^0 had continuous rotational symmetry, but $H = H^0 + H'$ is only invariant under rotations by integer multiples of π. For A, take the operator $R(\pi)$ that rotates a function counterclockwise by an angle π. Acting on our states ψ_a and ψ_b we have

$$R(\pi)\ \psi_a^0(x,y) = \psi_a^0(-x,-y) = -\psi_a^0(x,y)\,,$$
$$R(\pi)\ \psi_b^0(x,y) = \psi_b^0(-x,-y) = -\psi_b^0(x,y)\,.$$

That's no good; we need an operator with distinct eigenvalues. How about the operator that interchanges x and y? This is a reflection about a 45° diagonal of the well. Call this operator D. D commutes with both H^0 and H', since they are unchanged when you switch x and y. Now,

[7] Note that the theorem is more general than Equation 7.30. In order to identify the good states from Equation 7.30, the energies $E_\pm^1$ need to be different. In some cases they are the same and the energies of the degenerate states split at second, third, or higher order in perturbation theory. But the *theorem* allows you to identify the good states in every case.

$$D\,\psi_a^0(x,y) = \psi_a^0(y,x) = \psi_b^0(x,y)\,,$$
$$D\,\psi_b^0(x,y) = \psi_b^0(y,x) = \psi_a^0(x,y)\,.$$

So our degenerate eigenstates are not eigenstates of D. But we can construct linear combinations that *are*:

$$\psi_\pm^0 \equiv \pm\psi_a^0 + \psi_b^0. \tag{7.39}$$

Then

$$D\left(\pm\psi_a^0 + \psi_b^0\right) = \pm\,D\,\psi_a^0 + D\,\psi_b^0 = \pm\psi_b^0 + \psi_a^0 = \pm\left(\pm\psi_a^0 + \psi_b^0\right).$$

These are "good" states, since they are eigenstates of an operator D with distinct eigenvalues (± 1), and D commutes with both H^0 and H'.

Moral: If you're faced with degenerate states, look around for some hermitian operator A that commutes with H^0 and H'; pick as your unperturbed states ones that are simultaneously eigenfunctions of H^0 and A (with *distinct* eigenvalues). Then use *ordinary* first-order perturbation theory. If you can't find such an operator, you'll have to resort to Equation 7.33, but in practice this is seldom necessary.

Problem 7.8 Let the two "good" unperturbed states be

$$\psi_\pm^0 = \alpha_\pm\psi_a^0 + \beta_\pm\psi_b^0,$$

where $\alpha_\pm$ and $\beta_\pm$ are determined (up to normalization) by Equation 7.27 (or Equation 7.29). Show explicitly that

(a) $\psi_\pm^0$ are orthogonal $\left(\langle\psi_+^0|\psi_-^0\rangle = 0\right)$;

(b) $\left\langle\psi_+^0\left|H'\right|\psi_-^0\right\rangle = 0$;

(c) $\left\langle\psi_\pm^0\left|H'\right|\psi_\pm^0\right\rangle = E_\pm^1$, with $E_\pm^1$ given by Equation 7.33.

Problem 7.9 Consider a particle of mass m that is free to move in a one-dimensional region of length L that closes on itself (for instance, a bead that slides frictionlessly on a circular wire of circumference L, as in Problem 2.46).

(a) Show that the stationary states can be written in the form

$$\psi_n(x) = \frac{1}{\sqrt{L}}e^{2\pi i n x/L}, \quad (-L/2 < x < L/2)\,,$$

where $n = 0, \pm 1, \pm 2, \ldots$, and the allowed energies are

$$E_n = \frac{2}{m}\left(\frac{n\pi\hbar}{L}\right)^2.$$

Notice that—with the exception of the ground state ($n = 0$)—these are all doubly degenerate.

(b) Now suppose we introduce the perturbation

$$H' = -V_0 e^{-x^2/a^2},$$

where $a \ll L$. (This puts a little "dimple" in the potential at $x = 0$, as though we bent the wire slightly to make a "trap".) Find the first-order correction to E_n, using Equation 7.33. *Hint:* To evaluate the integrals, exploit the fact that $a \ll L$ to extend the limits from $\pm L/2$ to $\pm\infty$; after all, H' is essentially zero outside $-a < x < a$.

(c) What are the "good" linear combinations of ψ_n and ψ_{-n}, for this problem? (*Hint:* use Eq. 7.27.) Show that with these states you get the first-order correction using Equation 7.9.

(d) Find a hermitian operator A that fits the requirements of the theorem, and show that the simultaneous eigenstates of H^0 and A are precisely the ones you used in (c).

7.2.3 Higher-Order Degeneracy

In the previous section I assumed the degeneracy was two-fold, but it is easy to see how the method generalizes. In the case of n-fold degeneracy, we look for the eigenvalues of the $n \times n$ matrix

$$W_{ij} = \left\langle \psi_i^0 \middle| H' \middle| \psi_j^0 \right\rangle. \tag{7.40}$$

For three-fold degeneracy (with degenerate states ψ_a^0, ψ_b^0, and ψ_c^0) the first-order corrections to the energies E^1 are the eigenvalues of W, determined by solving

$$\begin{pmatrix} W_{aa} & W_{ab} & W_{ac} \\ W_{ba} & W_{bb} & W_{bc} \\ W_{ca} & W_{cb} & W_{cc} \end{pmatrix} \begin{pmatrix} \alpha \\ \beta \\ \gamma \end{pmatrix} = E^1 \begin{pmatrix} \alpha \\ \beta \\ \gamma \end{pmatrix}, \tag{7.41}$$

and the "good" states are the corresponding eigenvectors:[8]

$$\psi^0 = \alpha\,\psi_a^0 + \beta\,\psi_b^0 + \gamma\,\psi_c^0. \tag{7.42}$$

Once again, if you can think of an operator A that *commutes* with H^0 and H', and use the simultaneous eigenfunctions of A and H^0, then the W matrix will *automatically* be diagonal, and you won't have to fuss with calculating the off-diagonal elements of W or solving the characteristic equation.[9] (If you're nervous about my generalization from two-fold degeneracy to n-fold degeneracy, work Problem 7.13.)

Problem 7.10 Show that the first-order energy corrections computed in Example 7.3 (Equation 7.34) agree with an expansion of the exact solution (Equation 7.21) to first order in ϵ.

Problem 7.11 Suppose we perturb the infinite cubical well (Problem 4.2) by putting a delta function "bump" at the point $(a/4, a/2, 3a/4)$:

$$H' = a^3 V_0 \delta(x - a/4)\delta(y - a/2)\delta(z - 3a/4).$$

Find the first-order corrections to the energy of the ground state and the (triply degenerate) first excited states.

[8] If the eigenvalues of W are degenerate, see footnote 6.

[9] Degenerate perturbation theory amounts to diagonalization of the *degenerate part* of the Hamiltonian; see Problems 7.34 and 7.35.

* **Problem 7.12** Consider a quantum system with just three linearly independent states. Suppose the Hamiltonian, in matrix form, is

$$\mathsf{H} = V_0 \begin{pmatrix} (1-\epsilon) & 0 & 0 \\ 0 & 1 & \epsilon \\ 0 & \epsilon & 2 \end{pmatrix},$$

where V_0 is a constant, and ϵ is some small number ($\epsilon \ll 1$).

(a) Write down the eigenvectors and eigenvalues of the *unperturbed* Hamiltonian ($\epsilon = 0$).

(b) Solve for the *exact* eigenvalues of H. Expand each of them as a power series in ϵ, up to second order.

(c) Use first- and second-order *non*-degenerate perturbation theory to find the approximate eigenvalue for the state that grows out of the nondegenerate eigenvector of H^0. Compare the exact result, from (b).

(d) Use *degenerate* perturbation theory to find the first-order correction to the two initially degenerate eigenvalues. Compare the exact results.

Problem 7.13 In the text I asserted that the first-order corrections to an n-fold degenerate energy are the eigenvalues of the W matrix, and I justified this claim as the "natural" generalization of the case $n = 2$. *Prove* it, by reproducing the steps in Section 7.2.1, starting with

$$\psi^0 = \sum_{j=1}^{n} \alpha_j \psi_j^0$$

(generalizing Equation 7.17), and ending by showing that the analog to Equation 7.27 can be interpreted as the eigenvalue equation for the matrix W.

7.3 THE FINE STRUCTURE OF HYDROGEN

In our study of the hydrogen atom (Section 4.2) we took the Hamiltonian—called the Bohr Hamiltonian—to be

$$H_{\text{Bohr}} = -\frac{\hbar^2}{2m}\nabla^2 - \frac{e^2}{4\pi\epsilon_0}\frac{1}{r} \tag{7.43}$$

(electron kinetic energy plus Coulombic potential energy). But this is not quite the whole story. We have already learned how to correct for the motion of the nucleus: Just replace m by the reduced mass (Problem 5.1). More significant is the so-called **fine structure**, which is actually due to two distinct mechanisms: a **relativistic correction**, and **spin-orbit coupling**. Compared to the Bohr energies (Equation 4.70), fine structure is a tiny perturbation—smaller by a factor of α^2, where

$$\alpha \equiv \frac{e^2}{4\pi\epsilon_0\hbar c} \approx \frac{1}{137.036} \tag{7.44}$$

is the famous **fine structure constant**. Smaller still (by another factor of α) is the **Lamb shift**, associated with the quantization of the electric field, and smaller by yet another order of magnitude is the **hyperfine structure**, which is due to the interaction between the magnetic

Table 7.1: *Hierarchy of corrections to the Bohr energies of hydrogen.*

Bohr energies:	of order	$\alpha^2 mc^2$
Fine structure:	of order	$\alpha^4 mc^2$
Lamb shift:	of order	$\alpha^5 mc^2$
Hyperfine splitting:	of order	$(m/m_p)\alpha^4 mc^2$

dipole moments of the electron and the proton. This hierarchy is summarized in Table 7.1. In the present section we will analyze the fine structure of hydrogen, as an application of time-independent perturbation theory.

Problem 7.14

(a) Express the Bohr energies in terms of the fine structure constant and the rest energy (mc^2) of the electron.

(b) Calculate the fine structure constant from first principles (i.e., without recourse to the empirical values of ϵ_0, e, $\hbar$, and c). *Comment:* The fine structure constant is undoubtedly the most fundamental pure (dimensionless) number in all of physics. It relates the basic constants of electromagnetism (the charge of the electron), relativity (the speed of light), and quantum mechanics (Planck's constant). If you can solve part (b), you have the most certain Nobel Prize in history waiting for you. But I wouldn't recommend spending a lot of time on it right now; many smart people have tried, and all (so far) have failed.

7.3.1 The Relativistic Correction

The first term in the Hamiltonian is supposed to represent kinetic energy:

$$T = \frac{1}{2}mv^2 = \frac{p^2}{2m}, \tag{7.45}$$

and the canonical substitution $\mathbf{p} \rightarrow -i\hbar\nabla$ yields the operator

$$T = -\frac{\hbar^2}{2m}\nabla^2. \tag{7.46}$$

But Equation 7.45 is the *classical* expression for kinetic energy; the *relativistic* formula is

$$T = \frac{mc^2}{\sqrt{1-(v/c)^2}} - mc^2. \tag{7.47}$$

The first term is the *total* relativistic energy (not counting *potential* energy, which we aren't concerned with at the moment), and the second term is the *rest* energy—the *difference* is the energy attributable to motion.

We need to express T in terms of the (relativistic) momentum,

$$p = \frac{mv}{\sqrt{1-(v/c)^2}}, \tag{7.48}$$

instead of velocity. Notice that

$$p^2c^2 + m^2c^4 = \frac{m^2v^2c^2 + m^2c^4\left[1 - (v/c)^2\right]}{1 - (v/c)^2} = \frac{m^2c^4}{1 - (v/c)^2} = \left(T + mc^2\right)^2,$$

so

$$T = \sqrt{p^2c^2 + m^2c^4} - mc^2. \tag{7.49}$$

This relativistic equation for kinetic energy reduces (of course) to the classical result (Equation 7.45), in the nonrelativistic limit $p \ll mc$; expanding in powers of the small number (p/mc), we have

$$T = mc^2\left[\sqrt{1 + \left(\frac{p}{mc}\right)^2} - 1\right] = mc^2\left[1 + \frac{1}{2}\left(\frac{p}{mc}\right)^2 - \frac{1}{8}\left(\frac{p}{mc}\right)^4 \cdots - 1\right]$$

$$= \frac{p^2}{2m} - \frac{p^4}{8m^3c^2} + \cdots. \tag{7.50}$$

The lowest-order[10] relativistic correction to the Hamiltonian is therefore

$$H_r' = -\frac{p^4}{8m^3c^2}. \tag{7.51}$$

In first-order perturbation theory, the correction to E_n is given by the expectation value of H' in the unperturbed state (Equation 7.9):

$$E_r^1 = \langle H_r' \rangle = -\frac{1}{8m^3c^2}\left\langle \psi \,\middle|\, p^4\psi \right\rangle = -\frac{1}{8m^3c^2}\left\langle p^2\psi \,\middle|\, p^2\psi \right\rangle. \tag{7.52}$$

Now, the Schrödinger equation (for the unperturbed states) says

$$p^2\psi = 2m(E - V)\psi, \tag{7.53}$$

and hence[11]

$$E_r^1 = -\frac{1}{2mc^2}\left\langle (E - V)^2 \right\rangle = -\frac{1}{2mc^2}\left[E^2 - 2E\langle V \rangle + \left\langle V^2 \right\rangle\right]. \tag{7.54}$$

So far this is entirely general; but we're interested in hydrogen, for which $V(r) = -(1/4\pi\epsilon_0)\,e^2/r$:

$$E_r^1 = -\frac{1}{2mc^2}\left[E_n^2 + 2E_n\left(\frac{e^2}{4\pi\epsilon_0}\right)\left\langle \frac{1}{r} \right\rangle + \left(\frac{e^2}{4\pi\epsilon_0}\right)^2\left\langle \frac{1}{r^2} \right\rangle\right], \tag{7.55}$$

where E_n is the Bohr energy of the state in question.

To complete the job, we need the expectation values of $1/r$ and $1/r^2$, in the (unperturbed) state $\psi_{n\ell m}$ (Equation 4.89). The first is easy (see Problem 7.15):

$$\left\langle \frac{1}{r} \right\rangle = \frac{1}{n^2 a}, \tag{7.56}$$

---

[10] The kinetic energy of the electron in hydrogen is on the order of 10 eV, which is minuscule compared to its rest energy (511,000 eV), so the hydrogen atom is basically nonrelativistic, and we can afford to keep only the lowest-order correction. In Equation 7.50, p is the *relativistic* momentum (Equation 7.48), *not* the classical momentum mv. It is the former that we now associate with the quantum operator $-i\hbar\nabla$, in Equation 7.51.

[11] An earlier edition of this book claimed that p^4 is not hermitian for states with $\ell = 0$ (calling into question the maneuver leading to Equation 7.54). That was incorrect—p^4 *is* hermitian, for all ℓ (see Problem 7.18).

where a is the Bohr radius (Equation 4.72). The second is not so simple to derive (see Problem 7.42), but the answer is[12]

$$\left\langle \frac{1}{r^2} \right\rangle = \frac{1}{(\ell + 1/2)\, n^3 a^2}. \tag{7.57}$$

It follows that

$$E_r^1 = -\frac{1}{2mc^2}\left[E_n^2 + 2E_n\left(\frac{e^2}{4\pi\epsilon_0}\right)\frac{1}{n^2 a} + \left(\frac{e^2}{4\pi\epsilon_0}\right)^2 \frac{1}{(\ell + 1/2)\, n^3 a^2}\right],$$

or, eliminating a (using Equation 4.72) and expressing everything in terms of E_n (using Equation 4.70):

$$E_r^1 = -\frac{(E_n)^2}{2mc^2}\left[\frac{4n}{\ell + 1/2} - 3\right]. \tag{7.58}$$

Evidently the relativistic correction is smaller than E_n, by a factor of about $E_n/mc^2 = 2 \times 10^{-5}$.

You might have noticed that I used *non*-degenerate perturbation theory in this calculation (Equation 7.52), in spite of the fact that the hydrogen atom is highly degenerate. But the perturbation is spherically symmetric, so it commutes with L^2 and L_z. Moreover, the eigenfunctions of these operators (taken together) have distinct eigenvalues for the n^2 states with a given E_n. Luckily, then, the wave functions $\psi_{n\ell m}$ *are* the "good" states for this problem (or, as we say, n, ℓ, and m are the **good quantum numbers**), so as it happens the use of nondegenerate perturbation theory was legitimate (see the "Moral" to Section 7.2.1).

From Equation 7.58 we see that *some* of the degeneracy of the nth energy level has lifted. The $(2\ell + 1)$-fold degeneracy in m remains; as we saw in Example 6.3 it is due to rotational symmetry, a symmetry that remains intact with this perturbation. On the other hand, the "accidental" degeneracy in ℓ has disappeared; since its source is an additional symmetry unique to the $1/r$ potential (see Problem 6.34), we expect that degeneracy to be broken by practically *any* perturbation.

∗ **Problem 7.15** Use the virial theorem (Problem 4.48) to prove Equation 7.56.

Problem 7.16 In Problem 4.52 you calculated the expectation value of r^s in the state ψ_{321}. Check your answer for the special cases $s = 0$ (trivial), $s = -1$ (Equation 7.56), $s = -2$ (Equation 7.57), and $s = -3$ (Equation 7.66). Comment on the case $s = -7$.

∗∗ **Problem 7.17** Find the (lowest-order) relativistic correction to the energy levels of the one-dimensional harmonic oscillator. *Hint:* Use the technique of Problem 2.12.

[12] The general formula for the expectation value of *any* power of r is given in Hans A. Bethe and Edwin E. Salpeter, *Quantum Mechanics of One- and Two-Electron Atoms*, Plenum, New York (1977), p. 17.

∗∗∗ **Problem 7.18** Show that p^2 is hermitian, for hydrogen states with $\ell = 0$. *Hint:* For such states ψ is independent of θ and ϕ, so

$$p^2 = -\frac{\hbar^2}{r^2}\frac{d}{dr}\left(r^2\frac{d}{dr}\right)$$

(Equation 4.13). Using integration by parts, show that

$$\left\langle f \,\middle|\, p^2 g\right\rangle = -4\pi\hbar^2\left(r^2 f\frac{dg}{dr} - r^2 g\frac{df}{dr}\right)\Big|_0^\infty + \left\langle p^2 f \,\middle|\, g\right\rangle.$$

Check that the boundary term vanishes for ψ_{n00}, which goes like

$$\psi_{n00} \sim \frac{1}{\sqrt{\pi}\,(na)^{3/2}}\exp\left(-r/na\right)$$

near the origin.

The case of p^4 is more subtle. The Laplacian of $1/r$ picks up a delta function (see, for example, D. J. Griffiths, *Introduction to Electrodynamics,* 4th edn, Eq. 1.102). Show that

$$\nabla^4\left[e^{-kr}\right] = \left(-\frac{4k^3}{r} + k^4\right)e^{-kr} + 8\pi k\,\delta^3(\mathbf{r}),$$

and confirm that p^4 is hermitian.[13]

7.3.2 Spin-Orbit Coupling

Imagine the electron in orbit around the nucleus; from the *electron's* point of view, the proton is circling around *it* (Figure 7.6). This orbiting positive charge sets up a magnetic field $\mathbf{B}$, in the electron frame, which exerts a torque on the spinning electron, tending to align its magnetic moment ($\boldsymbol{\mu}$) along the direction of the field. The Hamiltonian (Equation 4.157) is

$$H = -\boldsymbol{\mu}\cdot\mathbf{B}. \tag{7.59}$$

To begin with, we need to figure out the magnetic field of the proton ($\mathbf{B}$) and the dipole moment of the electron ($\boldsymbol{\mu}$).

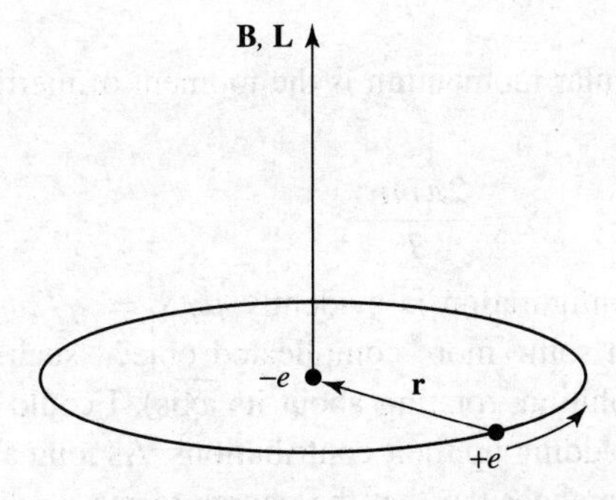

Figure 7.6: Hydrogen atom, from the electron's perspective.

[13] Thanks to Edward Ross and Li Yi-ding for fixing this problem.

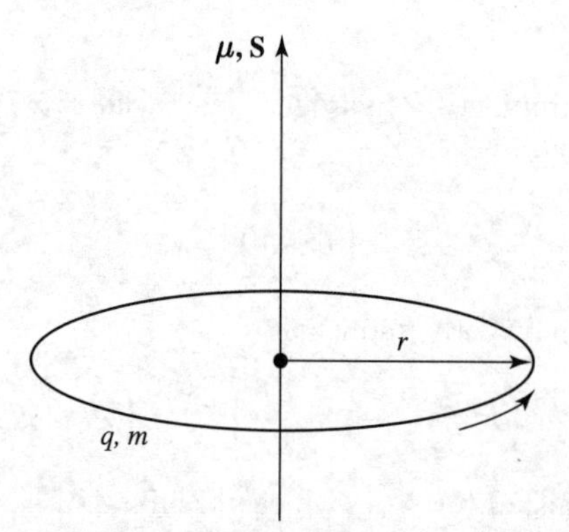

Figure 7.7: A ring of charge, rotating about its axis.

The Magnetic Field of the Proton. If we picture the proton (from the electron's perspective) as a continuous current loop (Figure 7.6), its magnetic field can be calculated from the Biot–Savart law:

$$B = \frac{\mu_0 I}{2r},$$

with an effective current $I = e/T$, where e is the charge of the proton and T is the period of the orbit. On the other hand, the orbital angular momentum of the *electron* (in the rest frame of the *nucleus*) is $L = rmv = 2\pi mr^2/T$. Moreover, **B** and **L** point in the same direction (up, in Figure 7.6), so

$$\mathbf{B} = \frac{1}{4\pi\epsilon_0}\frac{e}{mc^2r^3}\mathbf{L}. \tag{7.60}$$

(I used $c = 1/\sqrt{\epsilon_0\mu_0}$ to eliminate μ_0 in favor of ϵ_0.)

The Magnetic Dipole Moment of the Electron. The magnetic dipole moment of a spinning charge is related to its (spin) angular momentum; the proportionality factor is the gyromagnetic ratio (which we already encountered in Section 4.4.2). Let's derive it, this time, using classical electrodynamics. Consider first a charge q smeared out around a ring of radius r, which rotates about the axis with period T (Figure 7.7). The magnetic dipole moment of the ring is defined as the current (q/T) times the area $\left(\pi r^2\right)$:

$$\mu = \frac{q\pi r^2}{T}.$$

If the mass of the ring is m, its angular momentum is the moment of inertia $\left(mr^2\right)$ times the angular velocity $(2\pi/T)$:

$$S = \frac{2\pi mr^2}{T}.$$

The gyromagnetic ratio for this configuration is evidently $\mu/S = q/2m$. Notice that it is independent of r (and T). If I had some more complicated object, such as a sphere (all I require is that it be a figure of revolution, rotating about its axis), I could calculate $\boldsymbol{\mu}$ and **S** by chopping it into little rings, and adding up their contributions. As long as the mass and the charge are distributed in the same manner (so that the charge-to-mass ratio is uniform), the gyromagnetic ratio will be the same for each ring, and hence also for the object as a whole. Moreover, the directions of $\boldsymbol{\mu}$ and **S** are the same (or opposite, if the charge is negative), so

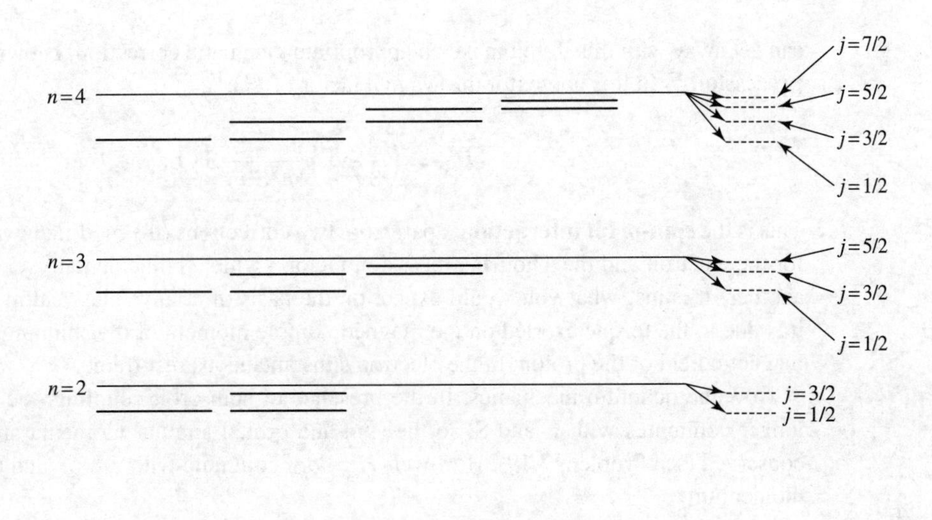

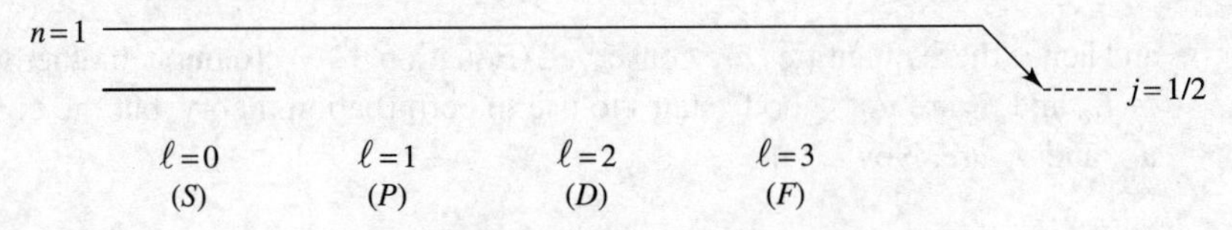

Figure 7.8: Energy levels of hydrogen, including fine structure (not to scale).

$$\boldsymbol{\mu} = \left(\frac{q}{2m}\right)\mathbf{S}. \tag{7.61}$$

That was a purely *classical* calculation, however; as it turns out the electron's magnetic moment is *twice* the classical value:

$$\boldsymbol{\mu}_e = -\frac{e}{m}\mathbf{S}. \tag{7.62}$$

The "extra" factor of 2 was explained by Dirac, in his relativistic theory of the electron.[14]

Putting all this together, we have

$$H = \left(\frac{e^2}{4\pi\epsilon_0}\right)\frac{1}{m^2c^2r^3}\mathbf{S}\cdot\mathbf{L}.$$

But there is a serious fraud in this calculation: I did the analysis in the rest frame of the electron, but that's *not an inertial system*—it *accelerates*, as the electron orbits around the nucleus. You

[14] We have already noted that it can be dangerous to picture the electron as a spinning sphere (see Problem 4.28), and it is not too surprising that the naive classical model gets the gyromagnetic ratio wrong. The *deviation* from the classical expectation is known as the **g-factor**: $\boldsymbol{\mu} = g\,(q/2m)\,\mathbf{S}$. Thus the g-factor of the electron, in Dirac's theory, is exactly 2. But **quantum electrodynamics** reveals tiny corrections to this: g_e is actually $2 + (\alpha/\pi) + \cdots = 2.002\ldots$. The calculation and measurement (which agree to exquisite precision) of the so-called **anomalous magnetic moment** of the electron were among the greatest achievements of twentieth-century physics.

can get away with this if you make an appropriate kinematic correction, known as the **Thomas precession.**[15] In this context it throws in a factor of 1/2:[16]

$$H'_{\mathrm{so}} = \left(\frac{e^2}{8\pi\epsilon_0}\right)\frac{1}{m^2c^2r^3}\mathbf{S}\cdot\mathbf{L}. \tag{7.63}$$

This is the **spin-orbit interaction**; apart from two corrections (the modified gyromagnetic ratio for the electron and the Thomas precession factor—which, coincidentally, exactly cancel one another) it is just what you would expect on the basis of a naive classical model. Physically, it is due to the torque exerted on the magnetic dipole moment of the spinning electron, by the magnetic field of the proton, in the electron's instantaneous rest frame.

Now the quantum mechanics. In the presence of spin-orbit coupling, the Hamiltonian no longer commutes with **L** and **S**, so the spin and orbital angular momenta are not separately conserved (see Problem 7.19). However, H'_{so} *does* commute with L^2, S^2 and the *total* angular momentum

$$\mathbf{J} \equiv \mathbf{L} + \mathbf{S}, \tag{7.64}$$

and hence these quantities *are* conserved (Equation 3.73). To put it another way, the eigenstates of L_z and S_z are not "good" states to use in perturbation theory, but the eigenstates of L^2, S^2, J^2, and J_z *are*. Now

$$J^2 = (\mathbf{L}+\mathbf{S})\cdot(\mathbf{L}+\mathbf{S}) = L^2 + S^2 + 2\mathbf{L}\cdot\mathbf{S},$$

so

$$\mathbf{L}\cdot\mathbf{S} = \frac{1}{2}\left(J^2 - L^2 - S^2\right), \tag{7.65}$$

and therefore the eigenvalues of $\mathbf{L}\cdot\mathbf{S}$ are

$$\frac{\hbar^2}{2}\left[j(j+1) - \ell(\ell+1) - s(s+1)\right].$$

In this case, of course, $s = 1/2$. Meanwhile, the expectation value of $1/r^3$ (see Problem 7.43)[17] is

$$\left\langle\frac{1}{r^3}\right\rangle = \frac{1}{\ell(\ell+1/2)(\ell+1)n^3a^3}, \tag{7.66}$$

---

[15] One way of thinking of it is that the electron is continually stepping from one inertial system to another; Thomas precession amounts to the cumulative effect of all these Lorentz transformations. We could avoid the whole problem, of course, by staying in the *lab* frame, in which the nucleus is at rest. In that case the field of the proton is purely *electric*, and you might wonder why it exerts any torque on the electron. Well, the fact is that a moving *magnetic* dipole acquires an *electric* dipole moment, and in the lab frame the spin-orbit coupling is due to the interaction of the *electric* field of the nucleus with the *electric* dipole moment of the electron. Because this analysis requires more sophisticated electrodynamics, it seems best to adopt the electron's perspective, where the physical mechanism is more transparent.

[16] More precisely, Thomas precession subtracts 1 from the g-factor (see R. R. Haar and L. J. Curtis, *Am. J. Phys.*, **55**, 1044 (1987)).

[17] In Problem 7.43 the expectation values are calculated using the hydrogen wave functions $\psi(n\ell m)$—that is, eigenstates of L_z—whereas we now want eigenstates of J_z—which are linear combinations of $m = m_j + \frac{1}{2}$ and $m = m_j - \frac{1}{2}$. But since $\langle r^s\rangle$ is independent of m, it doesn't matter.

and we conclude that

$$E_{\text{so}}^1 = \langle H'_{\text{so}} \rangle = \frac{e^2}{8\pi\epsilon_0} \frac{1}{m^2c^2} \frac{(\hbar^2/2)\left[j(j+1) - \ell(\ell+1) - 3/4\right]}{\ell(\ell+1/2)(\ell+1)n^3a^3},$$

or, expressing it all in terms of E_n:[18]

$$E_{\text{so}}^1 = \frac{(E_n)^2}{mc^2}\left\{\frac{n\left[j(j+1) - \ell(\ell+1) - 3/4\right]}{\ell(\ell+1/2)(\ell+1)}\right\}. \tag{7.67}$$

It is remarkable, considering the totally different physical mechanisms involved, that the relativistic correction and the spin-orbit coupling are of the same order $\left(E_n^2/mc^2\right)$. Adding them together, we get the complete fine-structure formula (see Problem 7.20):

$$E_{\text{fs}}^1 = \frac{(E_n)^2}{2mc^2}\left(3 - \frac{4n}{j+1/2}\right). \tag{7.68}$$

Combining this with the Bohr formula, we obtain the grand result for the energy levels of hydrogen, including fine structure:

$$\boxed{E_{nj} = -\frac{13.6\,\text{eV}}{n^2}\left[1 + \frac{\alpha^2}{n^2}\left(\frac{n}{j+1/2} - \frac{3}{4}\right)\right].} \tag{7.69}$$

Fine structure breaks the degeneracy in ℓ (that is, for a given n, the different allowed values of ℓ do not all carry the same energy), but it still preserves degeneracy in j (see Figure 7.8). The z-component eigenvalues for orbital and spin angular momentum (m_l and m_s) are no longer "good" quantum numbers—the stationary states are linear combinations of states with different values of these quantities; the "good" quantum numbers are n, ℓ, s, j, and m_j.[19]

Problem 7.19 Evaluate the following commutators: (a) $[\mathbf{L}\cdot\mathbf{S}, \mathbf{L}]$, (b) $[\mathbf{L}\cdot\mathbf{S}, \mathbf{S}]$, (c) $[\mathbf{L}\cdot\mathbf{S}, \mathbf{J}]$, (d) $\left[\mathbf{L}\cdot\mathbf{S}, L^2\right]$, (e) $\left[\mathbf{L}\cdot\mathbf{S}, S^2\right]$, (f) $\left[\mathbf{L}\cdot\mathbf{S}, J^2\right]$. *Hint:* **L** and **S** satisfy the fundamental commutation relations for angular momentum (Equations 4.99 and 4.134), but they commute with each other.

---

[18] The case $\ell = 0$ looks problematic, since we are ostensibly dividing by zero. On the other hand, the numerator is *also* zero, since in this case $j = s$; so Equation 7.67 is indeterminate. On physical grounds there shouldn't be any spin-orbit coupling when $\ell = 0$. In any event, the problem disappears when the spin-orbit coupling is added to the relativistic correction, and their *sum* (Equation 7.68) is correct for *all* ℓ. If you're feeling uneasy about this whole calculation, I don't blame you; take comfort in the fact that the *exact* solution can be obtained by using the (relativistic) Dirac equation in place of the (nonrelativistic) Schrödinger equation, and it confirms the results we obtain here by less rigorous means (see Problem 7.22.)

[19] To write $|j m_j\rangle$ (for given ℓ and s) as a linear combination of $|\ell\, s\, m_\ell\, m_s\rangle$ we would use the appropriate Clebsch–Gordan coefficients (Equation 4.183).

* **Problem 7.20** Derive the fine structure formula (Equation 7.68) from the relativistic correction (Equation 7.58) and the spin-orbit coupling (Equation 7.67). *Hint:* Note that $j = \ell \pm 1/2$ (except for $\ell = 0$, where only the plus sign occurs); treat the plus sign and the minus sign separately, and you'll find that you get the same final answer either way.

** **Problem 7.21** The most prominent feature of the hydrogen spectrum in the visible region is the red Balmer line, coming from the transition $n = 3$ to $n = 2$. First of all, determine the wavelength and frequency of this line according to the Bohr theory. Fine structure splits this line into several closely-spaced lines; the question is: *How many*, and *what is their spacing*? *Hint:* First determine how many sublevels the $n = 2$ level splits into, and find E^1_{fs} for each of these, in eV. Then do the same for $n = 3$. Draw an energy level diagram showing all possible transitions from $n = 3$ to $n = 2$. The energy released (in the form of a photon) is $(E_3 - E_2) + \Delta E$, the first part being common to all of them, and ΔE (due to fine structure) varying from one transition to the next. Find ΔE (in eV) for each transition. Finally, convert to photon frequency, and determine the spacing between adjacent spectral lines (in Hz)—*not* the frequency interval between each line and the *unperturbed* line (which is, of course, unobservable), but the frequency interval between each line and the *next* one. Your final answer should take the form: "The red Balmer line splits into (???) lines. In order of increasing frequency, they come from the transitions (1) $j =$ (???) to $j =$ (???), (2) $j =$ (???) to $j =$ (???), ... The frequency spacing between line (1) and line (2) is (???) Hz, the spacing between line (2) and line (3) is (???) Hz, ..."

Problem 7.22 The *exact* fine-structure formula for hydrogen (obtained from the Dirac equation without recourse to perturbation theory) is[20]

$$E_{nj} = mc^2\left\{\left[1 + \left(\frac{\alpha}{n - (j+1/2) + \sqrt{(j+1/2)^2 - \alpha^2}}\right)^2\right]^{-1/2} - 1\right\}.$$

Expand to order α^4 (noting that $\alpha \ll 1$), and show that you recover Equation 7.69.

7.4 THE ZEEMAN EFFECT

When an atom is placed in a uniform external magnetic field $\mathbf{B}_{\text{ext}}$, the energy levels are shifted. This phenomenon is known as the **Zeeman effect**. For a single electron, the perturbation is[21]

$$H'_Z = -\left(\boldsymbol{\mu}_l + \boldsymbol{\mu}_s\right) \cdot \mathbf{B}_{\text{ext}}, \tag{7.70}$$

[20] Bethe and Salpeter (footnote 12, page 298), page 238.

[21] This is correct to first order in B. We are ignoring a term of order B^2 in the Hamiltonian (the exact result was calculated in Problem 4.72). In addition, the orbital magnetic moment (Equation 7.72) is proportional to the mechanical angular momentum, not the canonical angular momentum (see Problem 7.49). These neglected terms give corrections of order B^2, comparable to the second-order corrections from H'_Z. Since we're working to first order, they are safe to ignore in this context.

where

$$\boldsymbol{\mu}_s = -\frac{e}{m}\mathbf{S} \tag{7.71}$$

(Equation 7.62) is the magnetic dipole moment associated with electron spin, and

$$\boldsymbol{\mu}_l = -\frac{e}{2m}\mathbf{L} \tag{7.72}$$

(Equation 7.61) is the dipole moment associated with orbital motion.[22] Thus

$$H'_Z = \frac{e}{2m}(\mathbf{L}+2\mathbf{S})\cdot\mathbf{B}_{\text{ext}}. \tag{7.73}$$

The nature of the Zeeman splitting depends critically on the strength of the external field in comparison with the *internal* field (Equation 7.60) that gives rise to spin-orbit coupling. If $B_{\text{ext}} \ll B_{\text{int}}$, then fine structure dominates, and H'_Z can be treated as a small perturbation, whereas if $B_{\text{ext}} \gg B_{\text{int}}$, then the Zeeman effect dominates, and fine structure becomes the perturbation. In the intermediate zone, where the two fields are comparable, we need the full machinery of degenerate perturbation theory, and it is necessary to diagonalize the relevant portion of the Hamiltonian "by hand." In the following sections we shall explore each of these regimes briefly, for the case of hydrogen.

Problem 7.23 Use Equation 7.60 to estimate the internal field in hydrogen, and characterize quantitatively a "strong" and "weak" Zeeman field.

7.4.1 Weak-Field Zeeman Effect

If $B_{\text{ext}} \ll B_{\text{int}}$, fine structure dominates; we treat $H_{\text{Bohr}} + H'_{fs}$ as the "unperturbed" Hamiltonian and H'_Z as the perturbation. Our "unperturbed" eigenstates are then those appropriate to fine structure: $|n\ell j m_j\rangle$ and the "unperturbed" energies are E_{nj} (Equation 7.69). Even though fine structure has lifted some of the degeneracy in the Bohr model, these states are still degenerate, since the energy does not depend on m_j or ℓ. Luckily the states $|n\ell j m_j\rangle$ are the "good" states for treating the perturbation H'_Z (meaning we don't have to write down the W matrix for H'_Z—it's already diagonal) since H'_Z commutes with J_z (so long as we align $\mathbf{B}_{\text{ext}}$ with the z axis) and L^2, and each of the degenerate states is *uniquely* labeled by the two quantum numbers m_j and ℓ.

In first-order perturbation theory, the Zeeman correction to the energy is

$$E^1_Z = \langle n\,\ell\,j\,m_j \,|H'_Z|\, n\,\ell\,j\,m_j\rangle = \frac{e}{2\,m} B_{\text{ext}}\,\hat{k}\cdot\langle\mathbf{L}+2\mathbf{S}\rangle, \tag{7.74}$$

where, as mentioned above, we align $\mathbf{B}_{\text{ext}}$ with the z axis to eliminate the off-diagonal elements of W. Now $\mathbf{L}+2\mathbf{S} = \mathbf{J}+\mathbf{S}$. Unfortunately, we do not immediately know the expectation value of $\mathbf{S}$. But we can figure it out, as follows: The total angular momentum $\mathbf{J} = \mathbf{L}+\mathbf{S}$ is constant (Figure 7.9); $\mathbf{L}$ and $\mathbf{S}$ precess rapidly about this fixed vector. In particular, the (time) *average* value of $\mathbf{S}$ is just its projection along $\mathbf{J}$:

$$\mathbf{S}_{\text{ave}} = \frac{(\mathbf{S}\cdot\mathbf{J})}{J^2}\mathbf{J}. \tag{7.75}$$

---

[22] The gyromagnetic ratio for *orbital* motion is just the classical value $(q/2m)$—it is only for *spin* that there is an "extra" factor of 2.

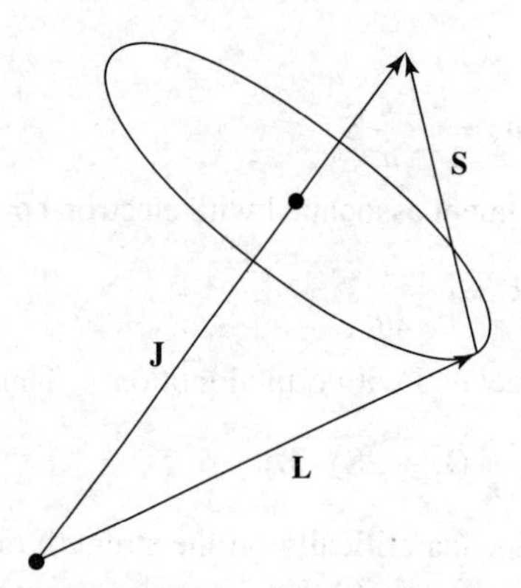

Figure 7.9: In the presence of spin-orbit coupling, **L** and **S** are not separately conserved; they precess about the fixed total angular momentum, **J**.

But $\mathbf{L} = \mathbf{J} - \mathbf{S}$, so $L^2 = J^2 + S^2 - 2\mathbf{J}\cdot\mathbf{S}$, and hence

$$\mathbf{S}\cdot\mathbf{J} = \frac{1}{2}\left(J^2 + S^2 - L^2\right) = \frac{\hbar^2}{2}\left[j(j+1) + s(s+1) - \ell(\ell+1)\right], \tag{7.76}$$

from which it follows that[23]

$$\langle \mathbf{L} + 2\mathbf{S}\rangle = \left\langle\left(1 + \frac{\mathbf{S}\cdot\mathbf{J}}{J^2}\right)\mathbf{J}\right\rangle = \left[1 + \frac{j(j+1) - \ell(\ell+1) + s(s+1)}{2j(j+1)}\right]\langle\mathbf{J}\rangle. \tag{7.78}$$

The term in square brackets is known as the **Landé g-factor**, g_J.[24]

The energy corrections are then

$$E_Z^1 = \mu_B g_J B_{\text{ext}} m_j, \tag{7.79}$$

where

$$\mu_B \equiv \frac{e\hbar}{2m} = 5.788 \times 10^{-5}\ \text{eV/T} \tag{7.80}$$

is the so-called **Bohr magneton**. Recall (Example 6.3) that degeneracy in the quantum number m is a consequence of rotational invariance.[25] The perturbation H_Z' picks out a specific direction in space (the direction of **B**) which breaks the rotational symmetry and lifts the degeneracy in m.

---

[23] While Equation 7.78 was derived by replacing **S** by its average value, the result is not an approximation; $\mathbf{L} + 2\mathbf{S}$ and **J** are both vector operators and the states are angular-momentum eigenstates. Therefore, the matrix elements can be evaluated by use of the Wigner–Eckart theorem (Equations 6.59–6.61). It follows (Problem 7.25) that the matrix elements are proportional:

$$\left\langle n\ell j m_j \middle| \mathbf{L} + 2\mathbf{S} \middle| n\ell j m_j' \right\rangle = g_J \left\langle n\ell j m_j \middle| \mathbf{J} \middle| n\ell j m_j' \right\rangle \tag{7.77}$$

and the constant of proportionality g_J is the ratio of reduced matrix elements. All that remains is to evaluate g_J: see Claude Cohen-Tannoudji, Bernard Diu, and Franck Laloë, *Quantum Mechanics*, Wiley, New York (1977), Vol. 2, Chapter X.

[24] In the case of a single electron, where $j = \ell \pm \frac{1}{2}$, $g_J = (2j+1)/(2\ell+1)$.

[25] That example specifically treated orbital angular momentum, but the same argument holds for the total angular momentum.

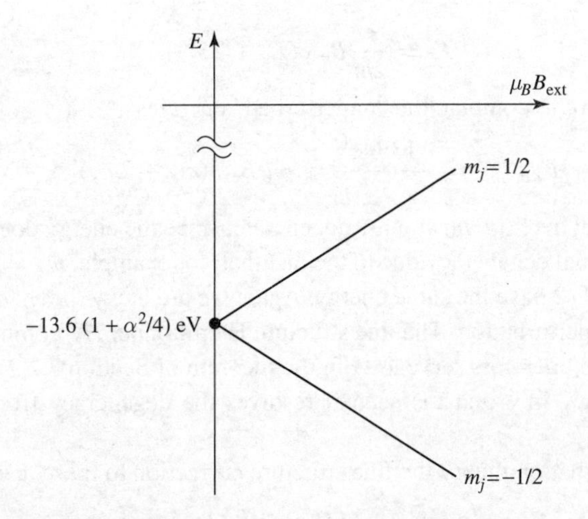

Figure 7.10: Weak-field Zeeman splitting of the ground state of hydrogen; the upper line $(m_j = 1/2)$ has slope 1, the lower line $(m_j = -1/2)$ has slope -1.

The *total* energy is the sum of the fine-structure part (Equation 7.69) and the Zeeman contribution (Equation 7.79). For example, the ground state ($n = 1$, $\ell = 0$, $j = 1/2$, and therefore $g_J = 2$) splits into two levels:

$$-13.6\,\text{eV}\left(1 + \alpha^2/4\right) \pm \mu_B B_{\rm ext}, \tag{7.81}$$

with the plus sign for $m_j = 1/2$, and minus for $m_j = -1/2$. These energies are plotted (as functions of $B_{\rm ext}$) in Figure 7.10.

* **Problem 7.24** Consider the (eight) $n = 2$ states, $|2\ell\, j\, m_j\rangle$. Find the energy of each state, under weak-field Zeeman splitting, and construct a diagram like Figure 7.10 to show how the energies evolve as $B_{\rm ext}$ increases. Label each line clearly, and indicate its slope.

Problem 7.25 Use the Wigner–Eckart theorem (Equations 6.59–6.61) to prove that the matrix elements of any two vector operators, **V** and **W**, are proportional in a basis of angular-momentum eigenstates:

$$\langle n\,\ell'\,m' \,|\, \mathbf{V} \,|\, n\,\ell\,m\rangle = \alpha\,\langle n\,\ell'\,m' \,|\, \mathbf{W} \,|\, n\,\ell\,m\rangle. \tag{7.82}$$

Comment: With ℓ replaced by j (the theorem holds regardless of whether the states are eigenstates of orbital, spin, or total angular momentum), $\mathbf{V} = \mathbf{L} + 2\,\mathbf{S}$ and $\mathbf{W} = \mathbf{J}$, this proves Equation 7.77.

7.4.2 Strong-Field Zeeman Effect

If $B_{\rm ext} \gg B_{\rm int}$, the Zeeman effect dominates[26] and we take the "unperturbed" Hamiltonian to be $H_{\rm Bohr} + H'_Z$ and the perturbation to be $H'_{\rm fs}$. The Zeeman Hamiltonian is

[26] In this regime the Zeeman effect is also known as the **Paschen–Back effect**.

$$H_Z' = \frac{e}{2m} B_{\text{ext}} \left(L_z + 2S_z \right),$$

and it is straightforward to compute the "unperturbed" energies:

$$E_{nm_\ell m_s} = -\frac{13.6 \text{ eV}}{n^2} + \mu_B B_{\text{ext}} \left(m_\ell + 2m_s \right). \tag{7.83}$$

The states we are using here: $|n \ell m_\ell m_s\rangle$ are degenerate, since the energy does not depend on ℓ, and there is an additional degeneracy due to the fact that, for example, $m_\ell = 3$ and $m_s = -1/2$ or $m_\ell = 1$ and $m_s = 1/2$ have the same energy. Again we are lucky; $|n \ell m_\ell m_s\rangle$ are the "good" states for treating the perturbation. The fine structure Hamiltonian H_{fs}' commutes with both L^2 and with J_z (these two operators serve as A in the theorem of Section 7.2.2); the first operator resolves the degeneracy in ℓ and the second resolves the degeneracy from coincidences in $m_\ell + 2\, m_s = m_j + m_s$.

In first-order perturbation theory the fine structure correction to these levels is

$$E_{\text{fs}}^1 = \left\langle n\, \ell\, m_\ell\, m_s \,\middle|\, \left(H_r' + H_{\text{so}}' \right) \,\middle|\, n\, \ell\, m_\ell\, m_s \right\rangle. \tag{7.84}$$

The relativistic contribution is the same as before (Equation 7.58); for the spin-orbit term (Equation 7.63) we need

$$\langle \mathbf{S} \cdot \mathbf{L} \rangle = \langle S_x \rangle \langle L_x \rangle + \langle S_y \rangle \langle L_y \rangle + \langle S_z \rangle \langle L_z \rangle = \hbar^2 m_l m_s \tag{7.85}$$

(note that $\langle S_x \rangle = \langle S_y \rangle = \langle L_x \rangle = \langle L_y \rangle = 0$ for eigenstates of S_z and L_z). Putting all this together (Problem 7.26), we conclude that

$$E_{\text{fs}}^1 = \frac{13.6 \text{ eV}}{n^3} \alpha^2 \left\{ \frac{3}{4n} - \left[\frac{\ell(\ell+1) - m_\ell m_s}{\ell(\ell+1/2)(\ell+1)} \right] \right\}. \tag{7.86}$$

(The term in square brackets is indeterminate for $\ell = 0$; its correct value in this case is 1—see Problem 7.28.) The *total* energy is the sum of the Zeeman part (Equation 7.83) and the fine structure contribution (Equation 7.86).

Problem 7.26 Starting with Equation 7.84, and using Equations 7.58, 7.63, 7.66, and 7.85, derive Equation 7.86.

** **Problem 7.27** Consider the (eight) $n = 2$ states, $|2\, \ell\, m_\ell\, m_s\rangle$. Find the energy of each state, under strong-field Zeeman splitting. Express each answer as the sum of three terms: the Bohr energy, the fine-structure (proportional to α^2), and the Zeeman contribution (proportional to $\mu_B B_{\text{ext}}$). If you ignore fine structure altogether, how many distinct levels are there, and what are their degeneracies?

Problem 7.28 If $\ell = 0$, then $j = s$, $m_j = m_s$, and the "good" states are the same ($|n\, m_s\rangle$) for weak *and* strong fields. Determine E_Z^1 (from Equation 7.74) and the fine structure energies (Equation 7.69), and write down the general result for the $\ell = 0$ Zeeman effect—*regardless* of the strength of the field. Show that the strong-field formula (Equation 7.86) reproduces this result, provided that we interpret the indeterminate term in square brackets as 1.

7.4.3 Intermediate-Field Zeeman Effect

In the intermediate regime neither H'_Z nor H'_{fs} dominates, and we must treat the two on an equal footing, as perturbations to the Bohr Hamiltonian (Equation 7.43):

$$H' = H'_Z + H'_{\text{fs}}. \tag{7.87}$$

I'll confine my attention here to the case $n = 2$ (you get to do $n = 3$ in Problem 7.30). It's not obvious what the "good" states are, so we'll have to resort to the full machinery of degenerate perturbation theory. I'll choose basis states characterized by ℓ, j, and m_j.[27] Using the Clebsch–Gordan coefficients (Problem 4.60 or Table 4.8) to express $|j\, m_j\rangle$ as linear combinations of $|\ell\, s\, m_l\, m_s\rangle$,[28] we have:

$\underline{\ell = 0:}$

$$\psi_1 \equiv \left|\tfrac{1}{2}\,\tfrac{1}{2}\right\rangle = \left|0\,\tfrac{1}{2}\,0\,\tfrac{1}{2}\right\rangle,$$

$$\psi_2 \equiv \left|\tfrac{1}{2}\,\tfrac{-1}{2}\right\rangle = \left|0\,\tfrac{1}{2}\,0\,\tfrac{-1}{2}\right\rangle,$$

$\underline{\ell = 1:}$

$$\psi_3 \equiv \left|\tfrac{3}{2}\,\tfrac{3}{2}\right\rangle = \left|1\,\tfrac{1}{2}\,1\,\tfrac{1}{2}\right\rangle,$$

$$\psi_4 \equiv \left|\tfrac{3}{2}\,\tfrac{-3}{2}\right\rangle = \left|1\,\tfrac{1}{2}\,{-1}\,\tfrac{-1}{2}\right\rangle,$$

$$\psi_5 \equiv \left|\tfrac{3}{2}\,\tfrac{1}{2}\right\rangle = \sqrt{2/3}\left|1\,\tfrac{1}{2}\,0\,\tfrac{1}{2}\right\rangle + \sqrt{1/3}\left|1\,\tfrac{1}{2}\,1\,\tfrac{-1}{2}\right\rangle,$$

$$\psi_6 \equiv \left|\tfrac{1}{2}\,\tfrac{1}{2}\right\rangle = -\sqrt{1/3}\left|1\,\tfrac{1}{2}\,0\,\tfrac{1}{2}\right\rangle + \sqrt{2/3}\left|1\,\tfrac{1}{2}\,1\,\tfrac{-1}{2}\right\rangle,$$

$$\psi_7 \equiv \left|\tfrac{3}{2}\,\tfrac{-1}{2}\right\rangle = \sqrt{1/3}\left|1\,\tfrac{1}{2}\,{-1}\,\tfrac{1}{2}\right\rangle + \sqrt{2/3}\left|1\,\tfrac{1}{2}\,0\,\tfrac{-1}{2}\right\rangle,$$

$$\psi_8 \equiv \left|\tfrac{1}{2}\,\tfrac{-1}{2}\right\rangle = -\sqrt{2/3}\left|1\,\tfrac{1}{2}\,{-1}\,\tfrac{1}{2}\right\rangle + \sqrt{1/3}\left|1\,\tfrac{1}{2}\,0\,\tfrac{-1}{2}\right\rangle.$$

In this basis the nonzero matrix elements of H'_{fs} are all on the diagonal, and given by Equation 7.68; H'_Z has four off-diagonal elements, and the complete matrix $-\mathsf{W}$ is (see Problem 7.29):

$$\begin{pmatrix}
5\gamma-\beta & 0 & 0 & 0 & 0 & 0 & 0 & 0\\
0 & 5\gamma+\beta & 0 & 0 & 0 & 0 & 0 & 0\\
0 & 0 & \gamma-2\beta & 0 & 0 & 0 & 0 & 0\\
0 & 0 & 0 & \gamma+2\beta & 0 & 0 & 0 & 0\\
0 & 0 & 0 & 0 & \gamma-\frac{2}{3}\beta & \frac{\sqrt{2}}{3}\beta & 0 & 0\\
0 & 0 & 0 & 0 & \frac{\sqrt{2}}{3}\beta & 5\gamma-\frac{1}{3}\beta & 0 & 0\\
0 & 0 & 0 & 0 & 0 & 0 & \gamma+\frac{2}{3}\beta & \frac{\sqrt{2}}{3}\beta\\
0 & 0 & 0 & 0 & 0 & 0 & \frac{\sqrt{2}}{3}\beta & 5\gamma+\frac{1}{3}\beta
\end{pmatrix}$$

[27] You can use ℓ, m_ℓ, m_s states if you prefer—this makes the matrix elements of H'_Z easier, but those of H'_{fs} more difficult; the W matrix will be more complicated, but its eigenvalues (which are independent of basis) are the same either way.

[28] Don't confuse the notation $|\ell\, s\, m_\ell\, m_s\rangle$ in the Clebsch–Gordan tables with $|n\,\ell\, j\, m_j\rangle$ (in Section 7.4.1) or $|n\,\ell\, m_\ell\, m_s\rangle$ (in Section 7.4.2); here n is always 2, and s (of course) is always 1/2.

Table 7.2: *Energy levels for the $n = 2$ states of hydrogen, with fine structure and Zeeman splitting.*

ϵ_1	=	$E_2 - 5\gamma + \beta$
ϵ_2	=	$E_2 - 5\gamma - \beta$
ϵ_3	=	$E_2 - \gamma + 2\beta$
ϵ_4	=	$E_2 - \gamma - 2\beta$
ϵ_5	=	$E_2 - 3\gamma + \beta/2 + \sqrt{4\gamma^2 + (2/3)\gamma\beta + \beta^2/4}$
ϵ_6	=	$E_2 - 3\gamma + \beta/2 - \sqrt{4\gamma^2 + (2/3)\gamma\beta + \beta^2/4}$
ϵ_7	=	$E_2 - 3\gamma - \beta/2 + \sqrt{4\gamma^2 - (2/3)\gamma\beta + \beta^2/4}$
ϵ_8	=	$E_2 - 3\gamma - \beta/2 - \sqrt{4\gamma^2 - (2/3)\gamma\beta + \beta^2/4}$

where

$$\gamma \equiv (\alpha/8)^2 \; 13.6 \text{ eV} \quad \text{and} \quad \beta \equiv \mu_B B_{\text{ext}}.$$

The first four eigenvalues are already displayed along the diagonal; it remains only to find the eigenvalues of the two 2×2 blocks. The characteristic equation for the first of these is

$$\lambda^2 - \lambda\,(6\gamma - \beta) + \left(5\gamma^2 - \frac{11}{3}\gamma\beta\right) = 0,$$

and the quadratic formula gives the eigenvalues:

$$\lambda_\pm = 3\gamma - (\beta/2) \pm \sqrt{4\gamma^2 + (2/3)\,\gamma\beta + \left(\beta^2/4\right)}. \tag{7.88}$$

The eigenvalues of the second block are the same, but with the sign of β reversed. The eight energies are listed in Table 7.2, and plotted against B_{ext} in Figure 7.11. In the zero-field limit ($\beta = 0$) they reduce to the fine structure values; for weak fields ($\beta \ll \gamma$) they reproduce what you got in Problem 7.24; for strong fields ($\beta \gg \gamma$) we recover the results of Problem 7.27 (note the convergence to five distinct energy levels, at very high fields, as predicted in Problem 7.27).

Problem 7.29 Work out the matrix elements of H'_Z and H'_{fs}, and construct the W matrix given in the text, for $n = 2$.

*** **Problem 7.30** Analyze the Zeeman effect for the $n = 3$ states of hydrogen, in the weak, strong, and intermediate field regimes. Construct a table of energies (analogous to Table 7.2), plot them as functions of the external field (as in Figure 7.11), and check that the intermediate-field results reduce properly in the two limiting cases. *Hint*: The Wigner–Eckart theorem comes in handy here. In Chapter 6 we wrote the theorem in terms of the orbital angular momentum ℓ but it also holds for states of total angular momentum j. In particular,

$$\left\langle j' m'_j \middle| V^z \middle| j m_j \right\rangle = C^{j\;1\;j}_{m_j 0 m'_j} \left\langle j' \,\|\, V \,\|\, j \right\rangle$$

for any vector operator $\mathbf{V}$ (and $\mathbf{L} + 2\mathbf{S}$ is a vector operator).

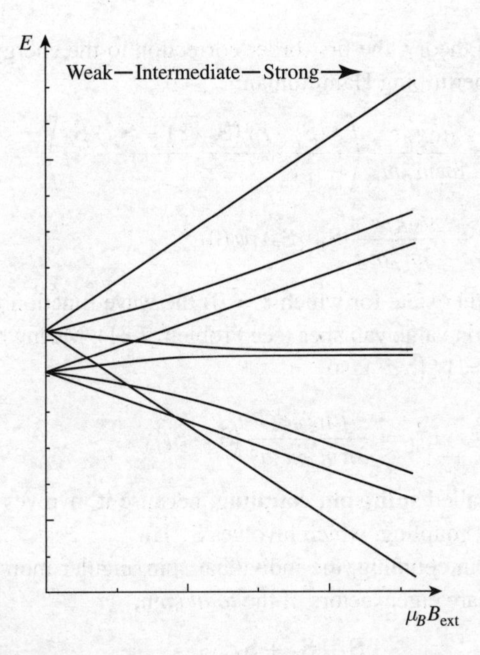

Figure 7.11: Zeeman splitting of the $n = 2$ states of hydrogen, in the weak, intermediate, and strong field regimes.

7.5 HYPERFINE SPLITTING IN HYDROGEN

The proton itself constitutes a magnetic dipole, though its dipole moment is much smaller than the electron's because of the mass in the denominator (Equation 7.62):

$$\boldsymbol{\mu}_p = \frac{g_p e}{2m_p}\mathbf{S}_p, \quad \boldsymbol{\mu}_e = -\frac{e}{m_e}\mathbf{S}_e. \tag{7.89}$$

(The proton is a composite structure, made up of three quarks, and its gyromagnetic ratio is not as simple as the electron's—hence the explicit g-factor (g_p), whose measured value is 5.59 as opposed to 2.00 for the electron.) According to classical electrodynamics, a dipole $\boldsymbol{\mu}$ sets up a magnetic field[29]

$$\mathbf{B} = \frac{\mu_0}{4\pi r^3}\left[3\left(\boldsymbol{\mu}\cdot\hat{r}\right)\hat{r} - \boldsymbol{\mu}\right] + \frac{2\mu_0}{3}\boldsymbol{\mu}\delta^3(\mathbf{r}). \tag{7.90}$$

So the Hamiltonian of the electron, in the magnetic field due to the proton's magnetic dipole moment, is (Equation 7.59)

$$H'_{\rm hf} = \frac{\mu_0 g_p e^2}{8\pi m_p m_e}\frac{\left[3\left(\mathbf{S}_p\cdot\hat{r}\right)\left(\mathbf{S}_e\cdot\hat{r}\right) - \mathbf{S}_p\cdot\mathbf{S}_e\right]}{r^3} + \frac{\mu_0 g_p e^2}{3m_p m_e}\mathbf{S}_p\cdot\mathbf{S}_e\delta^3(\mathbf{r}). \tag{7.91}$$

[29] If you are unfamiliar with the delta function term in Equation 7.90, you can derive it by treating the dipole as a spinning charged spherical shell, in the limit as the radius goes to zero and the charge goes to infinity (with $\boldsymbol{\mu}$ held constant). See D. J. Griffiths, *Am. J. Phys.*, **50**, 698 (1982).

According to perturbation theory, the first-order correction to the energy (Equation 7.9) is the expectation value of the perturbing Hamiltonian:

$$E_{\mathrm{hf}}^{1}=\frac{\mu_0 g_p e^2}{8\pi m_p m_e}\left\langle\frac{3\left(\mathbf{S}_p\cdot\hat{r}\right)\left(\mathbf{S}_e\cdot\hat{r}\right)-\mathbf{S}_p\cdot\mathbf{S}_e}{r^3}\right\rangle$$
$$+\frac{\mu_0 g_p e^2}{3m_p m_e}\left\langle\mathbf{S}_p\cdot\mathbf{S}_e\right\rangle|\psi(0)|^2. \tag{7.92}$$

In the ground state (or any other state for which $\ell=0$) the wave function is spherically symmetric, and the first expectation value vanishes (see Problem 7.31). Meanwhile, from Equation 4.80 we find that $|\psi_{100}(0)|^2=1/\left(\pi a^3\right)$, so

$$E_{\mathrm{hf}}^{1}=\frac{\mu_0 g_p e^2}{3\pi m_p m_e a^3}\left\langle\mathbf{S}_p\cdot\mathbf{S}_e\right\rangle, \tag{7.93}$$

in the ground state. This is called **spin-spin coupling**, because it involves the dot product of two spins (contrast spin-orbit coupling, which involves $\mathbf{S}\cdot\mathbf{L}$).

In the presence of spin-spin coupling, the individual spin angular momenta are no longer conserved; the "good" states are eigenvectors of the *total* spin,

$$\mathbf{S}\equiv\mathbf{S}_e+\mathbf{S}_p. \tag{7.94}$$

As before, we square this out to get

$$\mathbf{S}_p\cdot\mathbf{S}_e=\frac{1}{2}\left(S^2-S_e^2-S_p^2\right). \tag{7.95}$$

But the electron and proton both have spin 1/2, so $S_e^2=S_p^2=(3/4)\,\hbar^2$. In the triplet state (spins "parallel") the total spin is 1, and hence $S^2=2\hbar^2$; in the singlet state the total spin is 0, and $S^2=0$. Thus

$$E_{\mathrm{hf}}^{1}=\frac{4g_p\hbar^4}{3m_p m_e^2c^2a^4}\begin{cases}+1/4, & \text{(triplet)};\\ -3/4, & \text{(singlet)}.\end{cases} \tag{7.96}$$

Spin-spin coupling breaks the spin degeneracy of the ground state, lifting the triplet configuration and depressing the singlet (see Figure 7.12). The energy gap is

$$\Delta E=\frac{4g_p\hbar^4}{3m_p m_e^2c^2a^4}=5.88\times10^{-6}\mathrm{eV}. \tag{7.97}$$

The frequency of the photon emitted in a transition from the triplet to the singlet state is

$$\nu=\frac{\Delta E}{h}=1420\ \mathrm{MHz}, \tag{7.98}$$

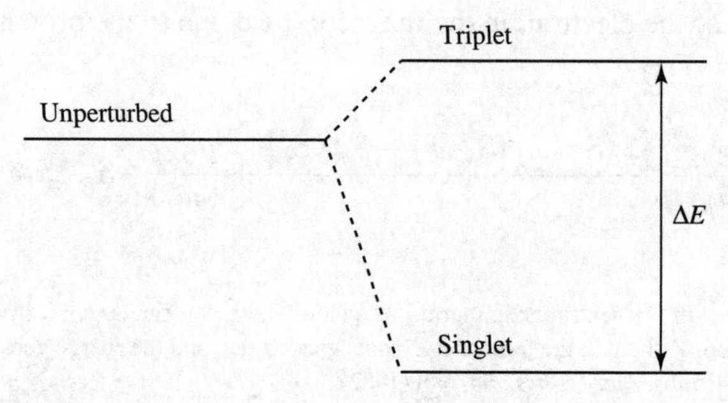

Figure 7.12: Hyperfine splitting in the ground state of hydrogen.

and the corresponding wavelength is $c/\nu = 21$ cm, which falls in the microwave region. This famous **21-centimeter line** is among the most pervasive forms of radiation in the universe.

Problem 7.31 Let **a** and **b** be two constant vectors. Show that

$$\int (\mathbf{a} \cdot \hat{r}) (\mathbf{b} \cdot \hat{r}) \sin\theta \, d\theta \, d\phi = \frac{4\pi}{3} (\mathbf{a} \cdot \mathbf{b}) \tag{7.99}$$

(the integration is over the usual range: $0 < \theta < \pi, 0 < \phi < 2\pi$). Use this result to demonstrate that

$$\left\langle \frac{3 (\mathbf{S}_p \cdot \hat{r}) (\mathbf{S}_e \cdot \hat{r}) - \mathbf{S}_p \cdot \mathbf{S}_e}{r^3} \right\rangle = 0,$$

for states with $\ell = 0$. *Hint*: $\hat{r} = \sin\theta \cos\phi \, \hat{\imath} + \sin\theta \sin\phi \, \hat{\jmath} + \cos\theta \, \hat{k}$. Do the angular integrals first.

Problem 7.32 By appropriate modification of the hydrogen formula, determine the hyperfine splitting in the ground state of (a) **muonic hydrogen** (in which a muon—same charge and g-factor as the electron, but 207 times the mass—substitutes for the electron), (b) **positronium** (in which a positron—same mass and g-factor as the electron, but opposite charge—substitutes for the proton), and (c) **muonium** (in which an anti-muon—same mass and g-factor as a muon, but opposite charge—substitutes for the proton). *Hint*: Don't forget to use the reduced mass (Problem 5.1) in calculating the "Bohr radius" of these exotic "atoms," but use the *actual* masses in the gyromagnetic ratios. Incidentally, the answer you get for positronium (4.82×10^{-4} eV) is quite far from the experimental value (8.41×10^{-4} eV); the large discrepancy is due to pair annihilation $(e^+ + e^- \to \gamma + \gamma)$, which contributes an extra $(3/4)\,\Delta E$, and does not occur (of course) in ordinary hydrogen, muonic hydrogen, or muonium.[30]

FURTHER PROBLEMS ON CHAPTER 7

Problem 7.33 Estimate the correction to the ground state energy of hydrogen due to the finite size of the nucleus. Treat the proton as a uniformly charged spherical shell of radius b, so the potential energy of an electron inside the shell is *constant:* $-e^2/(4\pi\epsilon_0 b)$; this isn't very realistic, but it is the simplest model, and it will give us the right order of magnitude. Expand your result in powers of the small parameter (b/a), where a is the Bohr radius, and keep only the leading term, so your final answer takes the form

$$\frac{\Delta E}{E} = A\,(b/a)^n .$$

Your business is to determine the constant A and the power n. Finally, put in $b \approx 10^{-15}$ m (roughly the radius of the proton) and work out the actual number. How does it compare with fine structure and hyperfine structure?

[30] For details see Griffiths, footnote 29, page 311.

Problem 7.34 In this problem you will develop an alternative approach to degenerate perturbation theory. Consider an unperturbed Hamiltonian H^0 with two degenerate states ψ_a^0 and ψ_b^0 (energy E^0), and a perturbation H'. Define the operator that projects[31] onto the degenerate subspace:

$$P_D = \left|\psi_a^0\right\rangle\left\langle\psi_a^0\right| + \left|\psi_b^0\right\rangle\left\langle\psi_b^0\right|. \tag{7.100}$$

The Hamiltonian can be written

$$H = H^0 + H' = \tilde{H}^0 + \tilde{H}' \tag{7.101}$$

where

$$\tilde{H}^0 = H^0 + P_D\, H'\, P_D, \qquad \tilde{H}' = H' - P_D\, H'\, P_D. \tag{7.102}$$

The idea is to treat $\tilde{H}^0$ as the "unperturbed" Hamiltonian and $\tilde{H}'$ as the perturbation; as you'll soon discover, $\tilde{H}^0$ is nondegenerate, so we can use ordinary nondegenerate perturbation theory.

(a) First we need to find the eigenstates of $\tilde{H}^0$.

i. Show that any eigenstate ψ_n^0 (other than ψ_a^0 or ψ_b^0) of H^0 is also an eigenstate of $\tilde{H}^0$ with the same eigenvalue.

ii. Show that the "good" states $\psi^0 = \alpha\,\psi_a^0 + \beta\,\psi_b^0$ (with α and β determined by solving Equation 7.30) are eigenstates of $\tilde{H}^0$ with energies $E^0 + E_\pm^1$.

(b) Assuming that E_+^1 and E_-^1 are distinct, you now have a *nondegenerate* unperturbed Hamiltonian $\tilde{H}^0$ and you can do nondegenerate perturbation theory using the perturbation $\tilde{H}'$. Find an expression for the energy to second order for the states $\psi_\pm^0$ in (ii).

Comment: One advantage of this approach is that it also handles the case where the unperturbed energies are not *exactly* equal, but very close:[32] $E_a^0 \approx E_b^0$. In this case one must still use degenerate perturbation theory; an important example of this occurs in the **nearly-free electron approximation** for calculating band structure.[33]

Problem 7.35 Here is an application of the technique developed in Problem 7.34. Consider the Hamiltonian

$$\mathsf{H}^0 = \begin{pmatrix} \epsilon & 0 & 0 \\ 0 & \epsilon & 0 \\ 0 & 0 & \epsilon' \end{pmatrix}, \quad \mathsf{H}' = \begin{pmatrix} 0 & a & b \\ a^* & 0 & c \\ b^* & c^* & 0 \end{pmatrix}.$$

(a) Find the projection operator P_D (it's a 3×3 matrix) that projects onto the subspace spanned by

$$\left|\psi_a^0\right\rangle = \begin{pmatrix} 1 \\ 0 \\ 0 \end{pmatrix} \quad\text{and}\quad \left|\psi_b^0\right\rangle = \begin{pmatrix} 0 \\ 1 \\ 0 \end{pmatrix}.$$

Then construct the matrices $\tilde{\mathsf{H}}^0$ and $\tilde{\mathsf{H}}'$.

---

[31] See page 118 for a discussion of projection operators.

[32] See Problem 7.4 for a discussion of what close means in this context.

[33] See, for example, Steven H. Simon, *The Oxford Solid State Basics* (Oxford University Press, 2013), Section 15.1.

(b) Solve for the eigenstates of $\tilde{\mathsf{H}}^0$ and verify...

i. that its spectrum is nondegenerate,

ii. that the nondegenerate eigenstate of H^0

$$\left|\psi_c^0\right\rangle = \begin{pmatrix} 0 \\ 0 \\ 1 \end{pmatrix}$$

is also an eigenstate of $\tilde{\mathsf{H}}^0$ with the same eigenvalue.

(c) What are the "good" states, and what are their energies, to first order in the perturbation?

Problem 7.36 Consider the isotropic three-dimensional harmonic oscillator (Problem 4.46). Discuss the effect (in first order) of the perturbation

$$H' = \lambda x^2 yz$$

(for some constant λ) on

(a) the ground state;

(b) the (triply degenerate) first excited state. *Hint:* Use the answers to Problems 2.12 and 3.39.

∗∗∗ **Problem 7.37 Van der Waals interaction.** Consider two atoms a distance R apart. Because they are electrically neutral you might suppose there would be no force between them, but if they are polarizable there is in fact a weak attraction. To model this system, picture each atom as an electron (mass m, charge $-e$) attached by a spring (spring constant k) to the nucleus (charge $+e$), as in Figure 7.13. We'll assume the nuclei are heavy, and essentially motionless. The Hamiltonian for the unperturbed system is

$$H^0 = \frac{1}{2m}p_1^2 + \frac{1}{2}kx_1^2 + \frac{1}{2m}p_2^2 + \frac{1}{2}kx_2^2. \tag{7.103}$$

The Coulomb interaction between the atoms is

$$H' = \frac{1}{4\pi\epsilon_0}\left(\frac{e^2}{R} - \frac{e^2}{R - x_1} - \frac{e^2}{R + x_2} + \frac{e^2}{R - x_1 + x_2}\right). \tag{7.104}$$

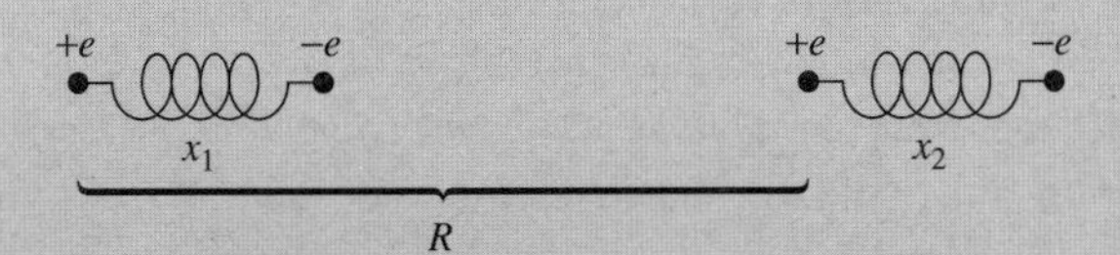

Figure 7.13: Two nearby polarizable atoms (Problem 7.37).

(a) Explain Equation 7.104. Assuming that $|x_1|$ and $|x_2|$ are both much less than R, show that

$$H' \cong -\frac{e^2 x_1 x_2}{2\pi\epsilon_0 R^3}. \tag{7.105}$$

(b) Show that the total Hamiltonian (H^0 plus Equation 7.105) separates into two harmonic oscillator Hamiltonians:

$$H = \left[\frac{1}{2m}p_+^2 + \frac{1}{2}\left(k - \frac{e^2}{2\pi\epsilon_0 R^3}\right)x_+^2\right] + \left[\frac{1}{2m}p_-^2 + \frac{1}{2}\left(k + \frac{e^2}{2\pi\epsilon_0 R^3}\right)x_-^2\right], \tag{7.106}$$

under the change variables

$$x_\pm \equiv \frac{1}{\sqrt{2}}(x_1 \pm x_2), \quad \text{which entails} \quad p_\pm = \frac{1}{\sqrt{2}}(p_1 \pm p_2). \tag{7.107}$$

(c) The ground state energy for this Hamiltonian is evidently

$$E = \frac{1}{2}\hbar(\omega_+ + \omega_-), \quad \text{where} \quad \omega_\pm = \sqrt{\frac{k \mp (e^2/2\pi\epsilon_0 R^3)}{m}}. \tag{7.108}$$

Without the Coulomb interaction it would have been $E_0 = \hbar\omega_0$, where $\omega_0 = \sqrt{k/m}$. Assuming that $k \gg \left(e^2/2\pi\epsilon_0 R^3\right)$, show that

$$\Delta V \equiv E - E_0 \cong -\frac{\hbar}{8m^2\omega_0^3}\left(\frac{e^2}{2\pi\epsilon_0}\right)^2 \frac{1}{R^6}. \tag{7.109}$$

Conclusion: There is an attractive potential between the atoms, proportional to the inverse sixth power of their separation. This is the **van der Waals interaction** between two neutral atoms.

(d) Now do the same calculation using second-order perturbation theory. *Hint:* The unperturbed states are of the form $\psi_{n_1}(x_1)\,\psi_{n_2}(x_2)$, where $\psi_n(x)$ is a one-particle oscillator wave function with mass m and spring constant k; ΔV is the second-order correction to the ground state energy, for the perturbation in Equation 7.105 (notice that the *first*-order correction is zero).[34]

∗∗ **Problem 7.38** Suppose the Hamiltonian H, for a particular quantum system, is a function of some parameter λ; let $E_n(\lambda)$ and $\psi_n(\lambda)$ be the eigenvalues and eigenfunctions of $H(\lambda)$. The **Feynman–Hellmann theorem**[35] states that

$$\frac{\partial E_n}{\partial \lambda} = \left\langle \psi_n \middle| \frac{\partial H}{\partial \lambda} \middle| \psi_n \right\rangle \tag{7.110}$$

(assuming either that E_n is nondegenerate, or—if degenerate—that the ψ_ns are the "good" linear combinations of the degenerate eigenfunctions).

(a) Prove the Feynman–Hellmann theorem.

(b) Apply it to the one-dimensional harmonic oscillator, (i) using $\lambda = \omega$ (this yields a formula for the expectation value of V), (ii) using $\lambda = \hbar$ (this yields $\langle T\rangle$), and (iii) using $\lambda = m$ (this yields a relation between $\langle T\rangle$ and $\langle V\rangle$). Compare your answers to Problem 2.12, and the virial theorem predictions (Problem 3.37).

[34] There is an interesting fraud in this well-known problem. If you expand H' to order $1/R^5$, the extra term has a nonzero expectation value in the ground state of H^0, so there is a nonzero first-order perturbation, and the dominant contribution goes like $1/R^5$, not $1/R^6$. The model gets the power "right" in three dimensions (where the expectation value is zero), but not in one. See A. C. Ipsen and K. Splittorff, *Am. J. Phys.* **83**, 150 (2015).

[35] Feynman obtained Equation 7.110 while working on his undergraduate thesis at MIT (R. P. Feynman, *Phys. Rev.* **56**, 340, 1939); Hellmann's work was published four years earlier in an obscure Russian journal.

Problem 7.39 Consider a three-level system with the unperturbed Hamiltonian

$$\mathsf{H}^0 = \begin{pmatrix} \epsilon_a & 0 & 0 \\ 0 & \epsilon_a & 0 \\ 0 & 0 & \epsilon_c \end{pmatrix} \tag{7.111}$$

($\epsilon_a > \epsilon_c$) and the perturbation

$$\mathsf{H}' = \begin{pmatrix} 0 & 0 & V \\ 0 & 0 & V \\ V^* & V^* & 0 \end{pmatrix}. \tag{7.112}$$

Since the (2×2) matrix W is diagonal (and in fact identically 0) in the basis of states $(1, 0, 0)$ and $(0, 1, 0)$, you might assume they are the good states, but they're not. To see this:

(a) Obtain the exact eigenvalues for the perturbed Hamiltonian $\mathsf{H} = \mathsf{H}^0 + \mathsf{H}'$.

(b) Expand your results from part (a) as a power series in $|V|$ up to second order.

(c) What do you obtain by applying *non*degenerate perturbation theory to find the energies of all three states (up to second order)? This would work if the assumption about the good states above were correct.

Moral: If any of the eigenvalues of W are equal, the states that diagonalize W are not unique, and diagonalizing W does not determine the "good" states. When this happens (and it's not uncommon), you need to use *second*-order degenerate perturbation theory (see Problem 7.40).

Problem 7.40 If it happens that the square root in Equation 7.33 vanishes, then $E_+^1 = E_-^1$; the degeneracy is not lifted at first order. In this case, diagonalizing the W matrix puts no restriction on α and β and you still don't know what the "good" states are. If you need to determine the "good" states—for example to calculate higher-order corrections—you need to use **second-order degenerate perturbation theory**.

(a) Show that, for the two-fold degeneracy studied in Section 7.2.1, the first-order correction to the wave function in degenerate perturbation theory is

$$\psi^1 = \sum_{m \neq a,b} \frac{\alpha\, V_{ma} + \beta\, V_{mb}}{E^0 - E_m^0}\, \psi_m^0.$$

(b) Consider the terms of order λ^2 (corresponding to Equation 7.8 in the nondegenerate case) to show that α and β are determined by finding the eigenvectors of the matrix W^2 (the superscript denotes second order, not W squared) where

$$\left[\mathsf{W}^2\right]_{ij} = \sum_{m \neq a,b} \frac{\langle \psi_i^0 | H' | \psi_m^0 \rangle \langle \psi_m^0 | H' | \psi_j^0 \rangle}{E^0 - E_m^0}$$

and that the eigenvalues of this matrix correspond to the second-order energies E^2.

(c) Show that second-order degenerate perturbation theory, developed in (*b*), gives the correct energies to second order for the three-state Hamiltonian in Problem 7.39.

** **Problem 7.41** A free particle of mass m is confined to a ring of circumference L such that $\psi(x + L) = \psi(x)$. The unperturbed Hamiltonian is

$$H^0 = -\frac{\hbar^2}{2m}\frac{d^2}{dx^2},$$

to which we add a perturbation

$$H' = V_0 \cos\left(2\pi \frac{x}{L}\right).$$

(a) Show that the unperturbed states may be written

$$\psi_n^0(x) = \frac{1}{\sqrt{L}} e^{i 2\pi n x/L}$$

for $n = 0, \pm 1, \pm 2$ and that, apart from $n = 0$, all of these states are two-fold degenerate.

(b) Find a general expression for the matrix elements of the perturbation:

$$H'_{mn} = \left\langle \psi_m^0 \middle| H' \middle| \psi_n^0 \right\rangle.$$

(c) Consider the degenerate pair of states with $n = \pm 1$. Construct the matrix W and calculate the first-order energy corrections, E^1. Note that the degeneracy *does not lift at first order*. Therefore, diagonalizing W does not tell us what the "good" states are.

(d) Construct the matrix W^2 (Problem 7.40) for the states $n = \pm 1$, and show that the degeneracy lifts at second order. What are the good linear combinations of the states with $n = \pm 1$?

(e) What are the energies, accurate to second order, for these states?[36]

** **Problem 7.42** The Feynman–Hellmann theorem (Problem 7.38) can be used to determine the expectation values of $1/r$ and $1/r^2$ for hydrogen.[37] The effective Hamiltonian for the radial wave functions is (Equation 4.53)

$$H = -\frac{\hbar^2}{2m}\frac{d^2}{dr^2} + \frac{\hbar^2}{2m}\frac{\ell(\ell+1)}{r^2} - \frac{e^2}{4\pi\epsilon_0}\frac{1}{r},$$

and the eigenvalues (expressed in terms of ℓ)[38] are (Equation 4.70)

$$E_n = -\frac{me^4}{32\pi^2\epsilon_0^2\hbar^2(N+\ell)^2}.$$

(a) Use $\lambda = e$ in the Feynman–Hellmann theorem to obtain $\langle 1/r \rangle$. Check your result against Equation 7.56.

(b) Use $\lambda = \ell$ to obtain $\left\langle 1/r^2 \right\rangle$. Check your answer with Equation 7.57.

[36] See D. Kiang, *Am. J. Phys.* **46** (11), 1978 and L.-K. Chen, Am. J. Phys. **72** (7), 2004 for further discussion of this problem. It is shown that each degenerate energy level, $E_{\pm n}^0$, splits at order $2n$ in perturbation theory. The exact solution to the problem can also be obtained as the time-independent Schrödinger equation for $H = H^0 + H^1$ reduces to the **Mathieu equation**.

[37] C. Sánchez del Rio, *Am. J. Phys.*, **50**, 556 (1982); H. S. Valk, *Am. J. Phys.*, **54**, 921 (1986).

[38] In part (b) we treat ℓ as a continuous variable; n becomes a function of ℓ, according to Equation 4.67, because N, which must be an integer, is fixed. To avoid confusion, I have eliminated n, to reveal the dependence on ℓ explicitly.

*** **Problem 7.43** Prove **Kramers' relation**:[39]

$$\frac{s+1}{n^2}\langle r^s\rangle-(2s+1)\,a\langle r^{s-1}\rangle+\frac{s}{4}\left[(2\ell+1)^2-s^2\right]a^2\langle r^{s-2}\rangle=0, \qquad (7.113)$$

which relates the expectation values of r to three different powers (s, $s-1$, and $s-2$), for an electron in the state $\psi_{n\ell m}$ of hydrogen. *Hint:* Rewrite the radial equation (Equation 4.53) in the form

$$u''=\left[\frac{\ell(\ell+1)}{r^2}-\frac{2}{ar}+\frac{1}{n^2a^2}\right]u,$$

and use it to express $\int(ur^su'')\,dr$ in terms of $\langle r^s\rangle$, $\langle r^{s-1}\rangle$, and $\langle r^{s-2}\rangle$. Then use integration by parts to reduce the second derivative. Show that $\int(ur^su')\,dr=-(s/2)\langle r^{s-1}\rangle$, and $\int(u'r^su')\,dr=-\left[2/(s+1)\right]\int(u''r^{s+1}u')\,dr$. Take it from there.

Problem 7.44

(a) Plug $s=0$, $s=1$, $s=2$, and $s=3$ into Kramers' relation (Equation 7.113) to obtain formulas for $\langle r^{-1}\rangle$, $\langle r\rangle$, $\langle r^2\rangle$, and $\langle r^3\rangle$. Note that you could continue indefinitely, to find *any* positive power.

(b) In the *other* direction, however, you hit a snag. Put in $s=-1$, and show that all you get is a relation between $\langle r^{-2}\rangle$ and $\langle r^{-3}\rangle$.

(c) But if you can get $\langle r^{-2}\rangle$ by some *other* means, you can apply the Kramers relation to obtain the rest of the negative powers. Use Equation 7.57 (which is derived in Problem 7.42) to determine $\langle r^{-3}\rangle$, and check your answer against Equation 7.66.

*** **Problem 7.45** When an atom is placed in a uniform external electric field $\mathbf{E}_{\text{ext}}$, the energy levels are shifted—a phenomenon known as the **Stark effect** (it is the electrical analog to the Zeeman effect). In this problem we analyze the Stark effect for the $n=1$ and $n=2$ states of hydrogen. Let the field point in the z direction, so the potential energy of the electron is

$$H_S'=eE_{\text{ext}}z=eE_{\text{ext}}r\cos\theta.$$

Treat this as a perturbation on the Bohr Hamiltonian (Equation 7.43). (Spin is irrelevant to this problem, so ignore it, and neglect the fine structure.)

(a) Show that the ground state energy is not affected by this perturbation, in first order.

(b) The first excited state is four-fold degenerate: ψ_{200}, ψ_{211}, ψ_{210}, ψ_{21-1}. Using degenerate perturbation theory, determine the first-order corrections to the energy. Into how many levels does E_2 split?

(c) What are the "good" wave functions for part (b)? Find the expectation value of the electric dipole moment ($\mathbf{p}_e=-e\mathbf{r}$), in each of these "good" states. Notice that the results are independent of the applied field—evidently hydrogen in its first excited state can carry a *permanent* electric dipole moment.

[39] This is also known as the (second) **Pasternack relation**. See H. Beker, *Am. J. Phys.* **65**, 1118 (1997). For a proof based on the Feynman–Hellmann theorem (Problem 7.38) see S. Balasubramanian, *Am. J. Phys.* **68**, 959 (2000).

Hint: There are lots of integrals in this problem, but almost all of them are zero. So study each one carefully, before you do any calculations: If the ϕ integral vanishes, there's not much point in doing the r and θ integrals! You can avoid those integrals altogether if you use the selection rules of Sections 6.4.3 and 6.7.2. *Partial answer:* $W_{13} = W_{31} = -3eaE_{\text{ext}}$; all other elements are zero.

∗∗∗ **Problem 7.46** Consider the Stark effect (Problem 7.45) for the $n = 3$ states of hydrogen. There are initially nine degenerate states, $\psi_{3\ell m}$ (neglecting spin, as before), and we turn on an electric field in the z direction.

(a) Construct the 9×9 matrix representing the perturbing Hamiltonian. *Partial answer:* $\langle 3\,0\,0|z|3\,1\,0\rangle = -3\sqrt{6}a$, $\langle 3\,1\,0|z|3\,2\,0\rangle = -3\sqrt{3}a$, $\langle 3\,1\,\pm1\,|z|\,3\,2\,\pm1\rangle = -(9/2)\,a$.

(b) Find the eigenvalues, and their degeneracies.

Problem 7.47 Calculate the wavelength, in centimeters, of the photon emitted under a hyperfine transition in the ground state ($n = 1$) of **deuterium**. Deuterium is "heavy" hydrogen, with an extra neutron in the nucleus; the proton and neutron bind together to form a **deuteron**, with spin 1 and magnetic moment

$$\boldsymbol{\mu}_d = \frac{g_d e}{2m_d}\mathbf{S}_d;$$

the deuteron g-factor is 1.71.

∗∗∗ **Problem 7.48** In a crystal, the electric field of neighboring ions perturbs the energy levels of an atom. As a crude model, imagine that a hydrogen atom is surrounded by three pairs of point charges, as shown in Figure 7.14. (Spin is irrelevant to this problem, so ignore it.)

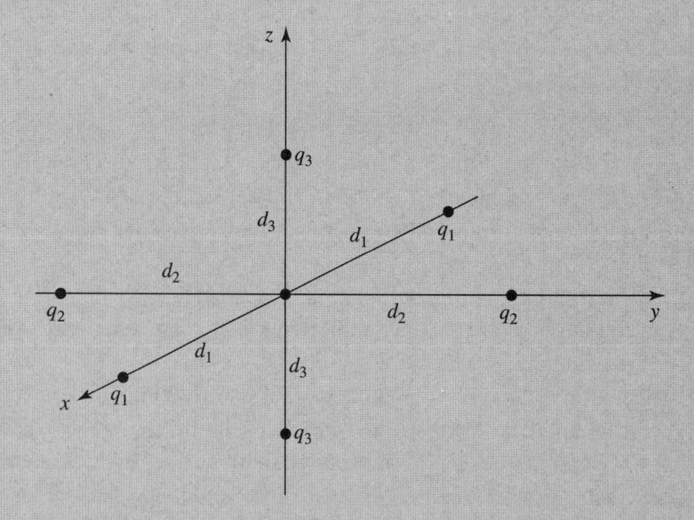

Figure 7.14: Hydrogen atom surrounded by six point charges (crude model for a crystal lattice); Problem 7.48.

(a) Assuming that $r \ll d_1$, $r \ll d_2$, and $r \ll d_3$, show that

$$H' = V_0 + 3\left(\beta_1 x^2 + \beta_2 y^2 + \beta_3 z^2\right) - (\beta_1 + \beta_2 + \beta_3)\, r^2,$$

where

$$\beta_i \equiv -\frac{e}{4\pi\epsilon_0}\frac{q_i}{d_i^3}, \quad \text{and } V_0 = 2\left(\beta_1 d_1^2 + \beta_2 d_2^2 + \beta_3 d_3^2\right).$$

(b) Find the lowest-order correction to the ground state energy.

(c) Calculate the first-order corrections to the energy of the first excited states ($n = 2$). Into how many levels does this four-fold degenerate system split, (i) in the case of **cubic symmetry**, $\beta_1 = \beta_2 = \beta_3$; (ii) in the case of **tetragonal symmetry**, $\beta_1 = \beta_2 \neq \beta_3$; (iii) in the general case of **orthorhombic symmetry** (all three different)? *Note*: you might recognize the "good" states from Problem 4.71.

Problem 7.49 A hydrogen atom is placed in a uniform magnetic field $\mathbf{B}_0 = B_0\,\hat{z}$ (the Hamiltonian can be written as in Equation 4.230). Use the Feynman–Hellman theorem (Problem 7.38) to show that

$$\frac{\partial E_n}{\partial B_0} = -\left\langle \psi_n \left| \mu_z \right| \psi_n \right\rangle \tag{7.114}$$

where the electron's magnetic dipole moment[40] (orbital plus spin) is

$$\boldsymbol{\mu} = \gamma_o\, \mathbf{L}_{\text{mechanical}} + \gamma\, \mathbf{S}.$$

The mechanical angular momentum is defined in Equation 4.231.

Note: From Equation 7.114 it follows that the magnetic susceptibility of N atoms in a volume V and at 0 K (when they're all in the ground state) is[41]

$$\chi = \mu_0 \frac{\partial M}{\partial B_0} = -\frac{N}{V}\mu_0 \frac{\partial^2 E_0}{\partial B_0^2}, \tag{7.115}$$

where E_0 is the ground-state energy. Although we derived Equation 7.114 for a hydrogen atom, the expression applies to multi-electron atoms as well—even when electron–electron interactions are included.

Problem 7.50 For an atom in a uniform magnetic field $\mathbf{B}_0 = B_0\,\hat{z}$, Equation 4.230 gives

$$H = H_{\text{atom}} - B_0\left(\gamma_o L_z + \gamma S_z\right) + \frac{e^2}{8m} B_0^2 \sum_{i=1}^{Z}\left(x_i^2 + y_i^2\right),$$

where L_z and S_z refer to the total orbital and spin angular momentum of all the electrons.

---

40 For most purposes we can take this to be the magnetic moment of the atom as well. The proton's larger mass means that its contribution to the dipole moment is orders of magnitude smaller than the electron's contribution.

41 See Problem 5.33 for the definition of magnetic susceptibility. This formula does not apply when the ground state is degenerate (see Neil W. Ashcroft and N. David Mermin, *Solid State Physics* (Belmont: Cengage, 1976), p. 655); atoms with non-degenerate ground states have $J = 0$ (see Table 5.1).

(a) Treating the terms involving B_0 as a perturbation, compute the shift of the ground state energy of a helium atom to second order in B_0. Assume that the helium ground state is given by

$$\psi_0 = \psi_{100}(r_1)\,\psi_{100}(r_2)\,\frac{|\uparrow\downarrow\rangle - |\downarrow\uparrow\rangle}{\sqrt{2}}$$

where ψ_{100} refers to the hydrogenic ground state (with $Z = 2$).

(b) Use the results of Problem 7.49 to calculate the magnetic susceptibility of helium. Given a density of $0.166\,\mathrm{kg/m^3}$, obtain a numerical value for the susceptibility. *Note*: The experimental result is -1.0×10^{-9} (the negative sign means that helium is a **diamagnet**). The results can be brought closer by taking account of screening, which increases the orbital radius (see Section 8.2).

*** **Problem 7.51** Sometimes it is possible to solve Equation 7.10 directly, without having to expand ψ_n^1 in terms of the unperturbed wave functions (Equation 7.11). Here are two particularly nice examples.

(a) Stark effect in the ground state of hydrogen.

(i) Find the first-order correction to the ground state of hydrogen in the presence of a uniform external electric field E_{ext} (see Problem 7.45). *Hint:* Try a solution of the form

$$\left(A + Br + Cr^2\right) e^{-r/a} \cos\theta;$$

your problem is to find the constants A, B, and C that solve Equation 7.10.

(ii) Use Equation 7.14 to determine the second-order correction to the ground state energy (the first-order correction is zero, as you found in Problem 7.45(a)). *Answer:* $-m\left(3a^2 e E_{\text{ext}}/2\hbar\right)^2$.

(b) If the proton had an *electric* dipole moment p, the potential energy of the electron in hydrogen would be perturbed in the amount

$$H' = -\frac{ep\cos\theta}{4\pi\epsilon_0 r^2}.$$

(i) Solve Equation 7.10 for the first-order correction to the ground state wave function.

(ii) Show that the *total* electric dipole moment of the atom is (surprisingly) *zero*, to this order.

(iii) Use Equation 7.14 to determine the second-order correction to the ground state energy. What is the *first*-order correction?

Problem 7.52 Consider a spinless particle of charge q and mass m constrained to move in the xy plane under the influence of the two-dimensional harmonic oscillator potential

$$V(x, y) = \frac{1}{2} m\omega^2 \left(x^2 + y^2\right).$$

(a) Construct the ground state wave function, $\psi_0(x, y)$, and write down its energy. Do the same for the (degenerate) first excited states.

(b) Now imagine that we turn on a weak magnetic field of magnitude B_0 pointing in the z-direction, so that (to first order in B_0) the Hamiltonian acquires an extra term

$$H' = -\boldsymbol{\mu} \cdot \mathbf{B} = -\frac{q}{2m} (\mathbf{L} \cdot \mathbf{B}) = -\frac{q B_0}{2m} \left(x\, p_y - y\, p_x \right).$$

Treating this as a perturbation, find the first-order corrections to the energies of the ground state and first excited states.

Problem 7.53 Imagine an infinite square well (Equation 2.22) into which we introduce a delta-function perturbation,

$$H'(x) = \lambda\, \delta(x - x_0),$$

where λ is a positive constant, and $0 < x_0 < a$ (to simplify matters, let $x_0 = pa$, where $0 < p < 1$).[42]

(a) Find the first-order correction to the nth allowed energy (Equation 2.30), assuming λ is small. (What does "small" mean, in this context?)

(b) Find the second-order correction to the allowed energies. (Leave your answer as a sum.)

(c) Now solve the Schrödinger equation exactly, treating separately the regions $0 \le x < x_0$ and $x_0 < x < a$, and imposing the boundary conditions at x_0. Derive the transcendental equation for the energies:

$$u_n \sin(u_n) + \Lambda \sin(p u_n) \sin\left[(1 - p)\, u_n\right] = 0 \qquad (E > 0). \qquad (7.116)$$

Here $\Lambda \equiv 2ma\lambda/\hbar^2$, $u_n \equiv k_n a$, and $k_n \equiv \sqrt{2mE_n}/\hbar$. Check that Equation 7.116 reproduces your result from part (a), in the appropriate limit.

(d) Everything so far holds just as well if λ is *negative*, but in that case there may be an additional solution with negative energy. Derive the transcendental equation for a negative-energy state:

$$v \sinh(v) + \Lambda \sinh(pv) \sinh\left[(1 - p)\, v\right] = 0 \qquad (E < 0), \qquad (7.117)$$

where $v \equiv \kappa a$ and $\kappa \equiv \sqrt{-2mE}/\hbar$. Specialize to the symmetrical case $p = 1/2$, and show that you recover the energy of the delta-function well (Equation 2.132), in the appropriate regime.

(e) There is in fact exactly *one* negative-energy solution, provided that $|\Lambda| > 1/\left[p\,(1 - p)\right]$. First, prove this (graphically), for the case $p = 1/2$. (Below that critical value there is no negative-energy solution.) Next, by computer, plot the solution v, as a function of p, for $\Lambda = -4.1, -5$, and -10. Verify that the solution only exists within the predicted range of p.

(f) For $p = 1/2$, plot the ground state wave function, $\psi_1(x)$, for $\Lambda = 0, -2, -3, -3.5, -4.5, -5$, and -10, to show how the sinusoidal shape (Figure 2.2) evolves into the exponential shape (Figure 2.13), as the delta function well "deepens."[43]

*** **Problem 7.54** Suppose you want to calculate the expectation value of some observable Ω, in the nth energy eigenstate of a system that is perturbed by H':

$$\langle \Omega \rangle = \left\langle \psi_n \middle| \hat{\Omega} \middle| \psi_n \right\rangle.$$

---

[42] We adopt the notation of Y. N. Joglekar, *Am. J. Phys.* **77**, 734 (2009), from which this problem is drawn.

[43] For the corresponding analysis of the delta function *barrier* (positive λ) see Problem 11.34.

Replacing ψ_n by its perturbation expansion, Equation 7.5,[44]

$$\langle \Omega \rangle = \left\langle \psi_n^0 \middle| \hat{\Omega} \middle| \psi_n^0 \right\rangle + \lambda \left[\left\langle \psi_n^1 \middle| \hat{\Omega} \middle| \psi_n^0 \right\rangle + \left\langle \psi_n^0 \middle| \hat{\Omega} \middle| \psi_n^1 \right\rangle \right] + \lambda^2 (\cdots) + \cdots .$$

The first-order correction to $\langle \Omega \rangle$ is therefore

$$\langle \Omega \rangle^1 = 2\mathrm{Re} \left[\left\langle \psi_n^0 \middle| \hat{\Omega} \middle| \psi_n^1 \right\rangle \right],$$

or, using Equation 7.13,

$$\langle \Omega \rangle^1 = 2\,\mathrm{Re} \sum_{m \neq n} \frac{\left\langle \psi_n^0 \middle| \hat{\Omega} \middle| \psi_m^0 \right\rangle \left\langle \psi_m^0 \middle| H' \middle| \psi_n^0 \right\rangle}{E_n^0 - E_m^0} \tag{7.118}$$

(assuming the unperturbed energies are nondegenerate, or that we are using the "good" basis states).

(a) Suppose $\Omega = H'$ (the perturbation itself). What does Equation 7.118 tell us in this case? Explain (carefully) why this is consistent with Equation 7.15.

(b) Consider a particle of charge q (maybe an electron in a hydrogen atom, or a pith ball connected to a spring), that is placed in a weak electric field E_{ext} pointing in the x direction, so that

$$H' = -qE_{\text{ext}}x.$$

The field will induce an electric dipole moment, $p_e = qx$, in the "atom." The expectation value of p_e is proportional to the applied field, and the proportionality factor is called the **polarizability**, α. Show that

$$\alpha = -2q^2 \sum_{m \neq n} \frac{\left| \left\langle \psi_n^0 \mid x \mid \psi_m^0 \right\rangle \right|^2}{E_n^0 - E_m^0}. \tag{7.119}$$

Find the polarizability of the ground state of a one-dimensional harmonic oscillator. Compare the classical answer.

(c) Now imagine a particle of mass m in a one-dimensional harmonic oscillator with a small anharmonic perturbation[45]

$$H' = -\frac{1}{6}\kappa x^3. \tag{7.120}$$

Find $\langle x \rangle$ (to first order), in the nth energy eigenstate. *Answer:* $\left(n + \frac{1}{2}\right) \hbar\kappa / \left(2m^2\omega^3\right)$. *Comment:* As the temperature increases, higher-energy states are populated, and the particles move farther (on average) from their equilibrium positions; that's why most solids expand with rising temperature.

Problem 7.55 Crandall's Puzzle.[46] Stationary states of the one-dimensional Schrödinger equation ordinarily respect three "rules of thumb": (1) the energies are nondegenerate,

---

[44] In general, Equation 7.5 does not deliver a *normalized* wave function, but the choice $c_n^{(n)} = 0$ in Equation 7.11 guarantees normalization to first order in λ, which is all we require here (see footnote 4, page 282).

[45] This is just a generic tweak to the simple harmonic oscillator potential, $\frac{1}{2}kx^2$; κ is some constant, and the factor of $-1/6$ is for convenience.

[46] Richard Crandall introduced me to this problem.

(2) the ground state has no nodes, the first excited state has one node, the second has two, and so on, and (3) if the potential is an even function of x, the ground state is even, the first excited state is odd, the second is even, and so on. We have already seen that the "bead-on-a-ring" (Problem 2.46) violates the first of these; now suppose we introduce a "nick" in at the origin:

$$H' = -\alpha\delta(x).$$

(If you don't like the delta function, make it a gaussian, as in Problem 7.9.) This lifts the degeneracy, but what is the sequence of even and odd wave functions, and what is the sequence of node numbers? *Hint:* You don't really need to do any calculations, here, and you're welcome to assume that α is small, but by all means solve the Schrödinger equation exactly if you prefer.

*** **Problem 7.56** In this problem we treat the electron–electron repulsion term in the helium Hamiltonian (Equation 5.38) as a perturbation,

$$H' = \frac{1}{4\pi\epsilon_0}\frac{e^2}{|\mathbf{r}_1 - \mathbf{r}_2|}.$$

(This will not be very accurate, because the perturbation is not small, in comparison to the Coulomb attraction of the nucleus ... but it's a start.)

(a) Find the first-order correction to the ground state,

$$\psi_0(\mathbf{r}_1, \mathbf{r}_2) = \psi_{100}(\mathbf{r}_1)\psi_{100}(\mathbf{r}_2).$$

(You have already done this calculation, if you worked Problem 5.15—only we didn't call it perturbation theory back then.)

(b) Now treat the first excited state, in which one electron is in the hydrogenic ground state, ψ_{100}, and the other is in the state ψ_{200}. Actually, there are *two* such states, depending on whether the electron spins occupy the singlet configuration (parahelium) or the triplet (orthohelium):[47]

$$\psi_\pm(\mathbf{r}_1, \mathbf{r}_2) = \frac{1}{\sqrt{2}}\left[\psi_{100}(\mathbf{r}_1)\psi_{200}(\mathbf{r}_2) \pm \psi_{200}(\mathbf{r}_1)\psi_{100}(\mathbf{r}_2)\right].$$

Show that

$$E^1_\pm = \frac{1}{2}(K \pm J),$$

where

$$K \equiv 2\int \psi_{100}(\mathbf{r}_1)\psi_{200}(\mathbf{r}_2)H'\psi_{100}(\mathbf{r}_1)\psi_{200}(\mathbf{r}_2)\, d^3\mathbf{r}_1\, d^3\mathbf{r}_2,$$

$$J \equiv 2\int \psi_{100}(\mathbf{r}_1)\psi_{200}(\mathbf{r}_2)H'\psi_{200}(\mathbf{r}_1)\psi_{100}(\mathbf{r}_2)\, d^3\mathbf{r}_1\, d^3\mathbf{r}_2.$$

Evaluate these two integrals, put in the actual numbers, and compare your results with Figure 5.2 (the measured energies are $-59.2\,\text{eV}$ and $-58.4\,\text{eV}$).[48]

---

[47] It seems strange, at first glance, that spin has anything to do with it, since the perturbation itself doesn't involve spin (and I'm not even bothering to include the spin state explicitly). The point, of course, is that an antisymmetric spin state forces a symmetric (position) wave function, and vice versa, and this *does* affect the result.

[48] If you want to pursue this problem further, see R. C. Massé and T. G. Walker, *Am. J. Phys.* **83**, 730 (2015).

Problem 7.57 The Hamiltonian for the Bloch functions (Equation 6.12) can be analyzed with perturbation theory by defining H^0 and H' such that

$$H^0\, u_{n0} = E_{n0}\, u_{n0},$$
$$\left(H^0 + H'\right) u_{nq} = E_{nq}\, u_{nq}.$$

In this problem, don't assume anything about the form of $V(x)$.

(a) Determine the operators H^0 and H' (express them in terms of $\hat{p}$).

(b) Find E_{nq} to second order in q. That is, find expressions for A_n, B_n, and C_n (in terms of the E_{n0} and matrix elements of $\hat{p}$ in the "unperturbed" states u_{n0})

$$E_{nq} \approx A_n + B_n\, q + C_n\, q^2 .$$

(c) Show that the constants B_n are all zero. *Hint*: See Problem 2.1(b) to get started. Remember that $u_{n0}(x)$ is periodic.

Comment: It is conventional to write $C_n = \hbar^2/2\,m_n^*$ where m_n^* is the **effective mass** of particles in the nth band since then, as you've just shown,

$$E_{nq} \approx \text{constant} + \frac{\hbar^2\, q^2}{2\, m_n^*}$$

just like the free particle (Equation 2.92) with $k \to q$.

导读 / Guidance

第 8 章　变分原理

本章讨论的内容是变分原理，作者开门见山，直接给出变分原理的公式，接下来用较多的例子介绍变分原理求解量子力学体系的基态能级的过程，让学生从例题中感受尝试波函数的确定和变分法对求解量子力学问题的有效性。其次，介绍用变分原理处理在氦原子的基态、氢分子离子、氢分子基态能级等 3 个体系中，如何选择合理尝试波函数的思想。这部分内容较多使用了一些较复杂的积分公式，如两体积分等。

习题特色

（1）新知识拓展：补充了汤川势模型（见习题 8.21、习题 8.28）、量子点模型求束缚态能量的阈值（见习题 8.27）。μ 子催化问题（见习题 8.26）等知识。

（2）给出一些利用计算机的数值模拟的题目，见习题 8.15、习题 8.30。

8 THE VARIATIONAL PRINCIPLE

8.1 THEORY

Suppose you want to calculate the ground state energy, E_{gs}, for a system described by the Hamiltonian H, but you are unable to solve the (time-independent) Schrödinger equation. The **variational principle** will get you an *upper bound* for E_{gs}, which is sometimes all you need, and often, if you're clever about it, very close to the exact value. Here's how it works: Pick *any normalized function* ψ *whatsoever*; I claim that

$$\boxed{E_{gs} \leq \langle\psi|H|\psi\rangle \equiv \langle H\rangle .} \tag{8.1}$$

That is, the expectation value of H, in the (presumably incorrect) state ψ is certain to *overestimate* the ground state energy. Of course, if ψ just happens to be one of the *excited* states, then *obviously* $\langle H\rangle$ exceeds E_{gs}; the point is that the same holds for any ψ whatsoever.

Proof: Since the (unknown) eigenfunctions of H form a complete set, we can express ψ as a linear combination of them:[1]

$$\psi = \sum_n c_n\psi_n, \quad \text{with } H\psi_n = E_n\psi_n.$$

Since ψ is normalized,

$$1 = \langle\psi|\psi\rangle = \left\langle \sum_m c_m\psi_m \middle| \sum_n c_n\psi_n \right\rangle = \sum_m\sum_n c_m^* c_n \langle\psi_m|\psi_n\rangle = \sum_n |c_n|^2,$$

(assuming the eigenfunctions themselves have been orthonormalized: $\langle\psi_m|\psi_n\rangle = \delta_{mn}$). Meanwhile,

$$\langle H\rangle = \left\langle \sum_m c_m\psi_m \middle| H\sum_n c_n\psi_n \right\rangle = \sum_m\sum_n c_m^* E_n c_n \langle\psi_m|\psi_n\rangle = \sum_n E_n|c_n|^2.$$

But the ground state energy is, by definition, the *smallest* eigenvalue, $E_{gs} \leq E_n$, and hence

$$\langle H\rangle \geq E_{gs}\sum_n |c_n|^2 = E_{gs},$$

which is what we were trying to prove.

[1] If the Hamiltonian admits scattering states, as well as bound states, then we'll need an integral as well as a sum, but the argument is unchanged.

This is hardly surprising. After all, ψ might be the *actual* wave function (at, say, $t = 0$). If you measured the particle's energy you'd be certain to get one of the eigenvalues of H, the smallest of which is E_{gs}, so the *average* of multiple measurements ($\langle H \rangle$) cannot be lower than E_{gs}.

Example 8.1

Suppose we want to find the ground state energy for the one-dimensional harmonic oscillator:

$$H = -\frac{\hbar^2}{2m}\frac{d^2}{dx^2} + \frac{1}{2}m\omega^2 x^2.$$

Of course, we already know the *exact* answer in this case (Equation 2.62): $E_{gs} = (1/2)\,\hbar\omega$; but this makes it a good test of the method. We might pick as our "trial" wave function the gaussian,

$$\psi(x) = Ae^{-bx^2}, \tag{8.2}$$

where b is a constant, and A is determined by normalization:

$$1 = |A|^2 \int_{-\infty}^{\infty} e^{-2bx^2}\,dx = |A|^2\sqrt{\frac{\pi}{2b}} \Rightarrow A = \left(\frac{2b}{\pi}\right)^{1/4}. \tag{8.3}$$

Now

$$\langle H \rangle = \langle T \rangle + \langle V \rangle, \tag{8.4}$$

where, in this case,

$$\langle T \rangle = -\frac{\hbar^2}{2m}|A|^2 \int_{-\infty}^{\infty} e^{-bx^2}\frac{d^2}{dx^2}\left(e^{-bx^2}\right)dx = \frac{\hbar^2 b}{2m}, \tag{8.5}$$

and

$$\langle V \rangle = \frac{1}{2}m\omega^2|A|^2 \int_{-\infty}^{\infty} e^{-2bx^2}x^2\,dx = \frac{m\omega^2}{8b},$$

so

$$\langle H \rangle = \frac{\hbar^2 b}{2m} + \frac{m\omega^2}{8b}. \tag{8.6}$$

According to Equation 8.1, this exceeds E_{gs} *for any* b; to get the *tightest* bound, let's *minimize* $\langle H \rangle$:

$$\frac{d}{db}\langle H \rangle = \frac{\hbar^2}{2m} - \frac{m\omega^2}{8b^2} = 0 \Rightarrow b = \frac{m\omega}{2\hbar}.$$

Putting this back into $\langle H \rangle$, we find

$$\langle H \rangle_{\min} = \frac{1}{2}\hbar\omega. \tag{8.7}$$

In this case we hit the ground state energy right on the nose—because (obviously) I "just happened" to pick a trial function with precisely the form of the *actual* ground state (Equation 2.60). But the gaussian is very easy to work with, so it's a popular trial function, even when it bears little resemblance to the true ground state.

Example 8.2

Suppose we're looking for the ground state energy of the delta function potential:

$$H = -\frac{\hbar^2}{2m}\frac{d^2}{dx^2} - \alpha\delta(x).$$

Again, we already know the exact answer (Equation 2.132): $E_{gs} = -m\alpha^2/2\hbar^2$. As before, we'll use a gaussian trial function (Equation 8.2). We've already determined the normalization, and calculated $\langle T\rangle$; all we need is

$$\langle V\rangle = -\alpha|A|^2\int_{-\infty}^{\infty} e^{-2bx^2}\delta(x)\,dx = -\alpha\sqrt{\frac{2b}{\pi}}.$$

Evidently

$$\langle H\rangle = \frac{\hbar^2 b}{2m} - \alpha\sqrt{\frac{2b}{\pi}}, \tag{8.8}$$

and we know that this exceeds E_{gs} for all b. Minimizing it,

$$\frac{d}{db}\langle H\rangle = \frac{\hbar^2}{2m} - \frac{\alpha}{\sqrt{2\pi b}} = 0 \Rightarrow b = \frac{2m^2\alpha^2}{\pi\hbar^4}.$$

So

$$\langle H\rangle_{\min} = -\frac{m\alpha^2}{\pi\hbar^2}, \tag{8.9}$$

which is indeed somewhat higher than E_{gs}, since $\pi > 2$.

I said you can use *any* (normalized) trial function ψ whatsoever, and this is true in a sense. However, for *discontinuous* functions it takes some fancy footwork to assign a sensible meaning to the second derivative (which you need, in order to calculate $\langle T\rangle$). Continuous functions with kinks in them are fair game, however, as long as you are careful; the next example shows how to handle them.[2]

Example 8.3

Find an upper bound on the ground state energy of the one-dimensional infinite square well (Equation 2.22), using the "triangular" trial wave function (Figure 8.1):[3]

$$\psi(x) = \begin{cases} Ax, & 0 \le x \le a/2, \\ A(a-x), & a/2 \le x \le a, \\ 0, & \text{otherwise,} \end{cases} \tag{8.10}$$

where A is determined by normalization:

$$1 = |A|^2\left[\int_0^{a/2} x^2\,dx + \int_{a/2}^{a}(a-x)^2\,dx\right] = |A|^2\frac{a^3}{12} \Rightarrow A = \frac{2}{a}\sqrt{\frac{3}{a}}. \tag{8.11}$$

[2] For a collection of interesting examples see W. N. Mei, *Int. J. Math. Educ. Sci. Tech.* **30**, 513 (1999).

[3] There is no point in trying a function (such as the gaussian) that extends outside the well, because you'll get $\langle V\rangle = \infty$, and Equation 8.1 tells you nothing.

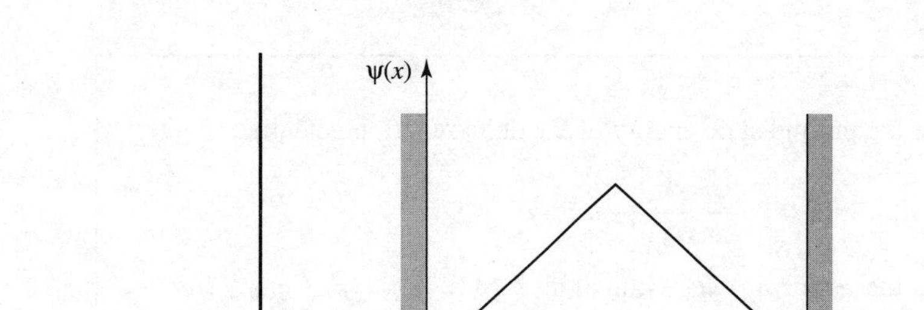

Figure 8.1: Triangular trial wave function for the infinite square well (Equation 8.10).

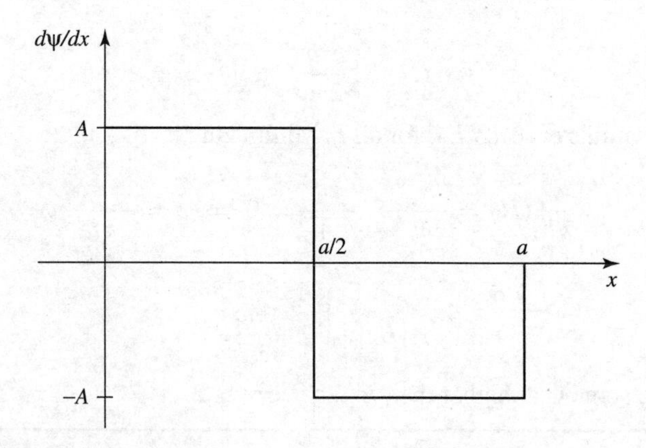

Figure 8.2: Derivative of the wave function in Figure 8.1.

In this case

$$\frac{d\psi}{dx} = \begin{cases} A, & 0 \le x \le a/2, \\ -A, & a/2 \le x \le a, \\ 0, & \text{otherwise}, \end{cases}$$

as indicated in Figure 8.2. Now, the derivative of a step function is a delta function (see Problem 2.23(b)):

$$\frac{d^2\psi}{dx^2} = A\delta(x) - 2A\delta(x - a/2) + A\delta(x - a), \tag{8.12}$$

and hence

$$\begin{aligned} \langle H \rangle &= -\frac{\hbar^2 A}{2m} \int \left[\delta(x) - 2\delta(x - a/2) + \delta(x - a)\right] \psi(x)\, dx \\ &= -\frac{\hbar^2 A}{2m} \left[\psi(0) - 2\psi(a/2) + \psi(a)\right] = \frac{\hbar^2 A^2 a}{2m} = \frac{12\hbar^2}{2ma^2}. \end{aligned} \tag{8.13}$$

The exact ground state energy is $E_{\text{gs}} = \pi^2\hbar^2/2ma^2$ (Equation 2.30), so the theorem works $\left(12 > \pi^2\right)$.

Alternatively, you can exploit the hermiticity of $\hat{p}$:

$$\langle H\rangle = \frac{1}{2m}\left\langle \hat{p}^2\right\rangle = \frac{1}{2m}\langle \hat{p}\psi \mid \hat{p}\psi\rangle = \frac{1}{2m}\int_0^a \left(-i\hbar\frac{d\psi}{dx}\right)^* \left(-i\hbar\frac{d\psi}{dx}\right) dx$$

$$= \frac{\hbar^2}{2m}\left[\int_0^{a/2} (A)^2\, dx + \int_{a/2}^{a} (-A)^2\, dx\right] = \frac{\hbar^2}{2m}A^2 a = \frac{12\hbar^2}{2ma^2}. \qquad (8.14)$$

The variational principle is extraordinarily powerful, and embarrassingly easy to use. What a physical chemist does, to find the ground state energy of some complicated molecule, is write down a trial wave function with a large number of adjustable parameters, calculate $\langle H\rangle$, and tweak the parameters to get the lowest possible value. Even if ψ has little resemblance to the true wave function, you often get miraculously accurate values for E_{gs}. Naturally, if you have some way of guessing a *realistic* ψ, so much the better. The only *trouble* with the method is that you never know for sure how close you are to the target—all you can be *certain* of is that you've got an *upper bound*.[4] Moreover, as it stands the technique applies only to the ground state (see, however, Problem 8.4).[5]

* **Problem 8.1** Use a gaussian trial function (Equation 8.2) to obtain the lowest upper bound you can on the ground state energy of (a) the linear potential: $V(x) = \alpha|x|$; (b) the quartic potential: $V(x) = \alpha x^4$.

** **Problem 8.2** Find the best bound on E_{gs} for the one-dimensional harmonic oscillator using a trial wave function of the form

$$\psi(x) = \frac{A}{x^2 + b^2},$$

where A is determined by normalization and b is an adjustable parameter.

Problem 8.3 Find the best bound on E_{gs} for the delta function potential $V(x) = -\alpha\delta(x)$, using a triangular trial function (Equation 8.10, only centered at the origin). This time a is an adjustable parameter.

Problem 8.4

(a) Prove the following corollary to the variational principle: If $\langle\psi|\psi_{\text{gs}}\rangle = 0$, then $\langle H\rangle \geq E_{\text{fe}}$, where E_{fe} is the energy of the first excited state. *Comment:* If we can find a trial function that is orthogonal to the exact ground state, we can get

[4] In practice this isn't much of a limitation, and there are sometimes ways of estimating the accuracy. The binding energy of helium has been calculated to many significant digits in this way (see for example G. W. Drake *et al.*, *Phys. Rev. A* **65**, 054501 (2002), or Vladimir I. Korobov, *Phys. Rev. A* **66**, 024501 (2002).

[5] For a systematic extension of the variational principle to the calculation of excited state energies see, for example, Linus Pauling and E. Bright Wilson, *Introduction to Quantum Mechanics, With Applications to Chemistry*, McGraw-Hill, New York (1935, paperback edition 1985), Section 26.

an upper bound on the *first excited state*. In general, it's difficult to be sure that ψ is orthogonal to ψ_{gs}, since (presumably) we don't *know* the latter. However, if the potential $V(x)$ is an *even* function of x, then the ground state is likewise even, and hence any *odd* trial function will automatically meet the condition for the corollary.[6]

(b) Find the best bound on the first excited state of the one-dimensional harmonic oscillator using the trial function

$$\psi(x) = Axe^{-bx^2}.$$

Problem 8.5 Using a trial function of your own devising, obtain an upper bound on the ground state energy for the "bouncing ball" potential (Equation 2.185), and compare it with the exact answer (Problem 2.59): $E_{gs} = 2.33811\left(mg^2\hbar^2/2\right)^{1/3}$.

Problem 8.6

(a) Use the variational principle to prove that first-order non-degenerate perturbation theory always *overestimates* (or at any rate never *underestimates*) the ground state energy.

(b) In view of (a), you would expect that the *second*-order correction to the ground state is always negative. Confirm that this is indeed the case, by examining Equation 7.15.

8.2 THE GROUND STATE OF HELIUM

The helium atom (Figure 8.3) consists of two electrons in orbit around a nucleus containing two protons (also some neutrons, which are irrelevant to our purpose). The Hamiltonian for this system (ignoring fine structure and smaller corrections) is:

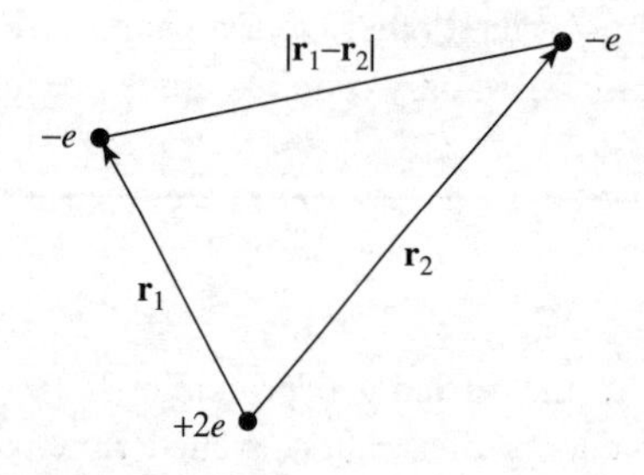

Figure 8.3: The helium atom.

[6] You can extend this trick to other symmetries. Suppose there is a Hermitian operator A such that $[A, H] = 0$. The ground state (assuming it is nondegenerate) must be an eigenstate of A; call the eigenvalue λ: $A\,\psi_{gs} = \lambda\,\psi_{gs}$. If you choose a variational function ψ that is an eigenstate of A with a *different* eigenvalue: $A\,\psi = \nu\,\psi$ with $\lambda \neq \nu$, you can be certain that ψ and ψ_{gs} are orthogonal (see Section 3.3). For an application see Problem 8.20.

$$H = -\frac{\hbar^2}{2m}\left(\nabla_1^2 + \nabla_2^2\right) - \frac{e^2}{4\pi\epsilon_0}\left(\frac{2}{r_1} + \frac{2}{r_2} - \frac{1}{|\mathbf{r}_1 - \mathbf{r}_2|}\right). \tag{8.15}$$

Our problem is to calculate the ground state energy, E_{gs}. Physically, this represents the amount of energy it would take to strip off both electrons. (Given E_{gs} it is easy to figure out the "ionization energy" required to remove a *single* electron—see Problem 8.7.) The ground state energy of helium has been measured to great precision in the laboratory:

$$E_{gs} = -78.975 \text{ eV} \quad \text{(experimental)}. \tag{8.16}$$

This is the number we would like to reproduce theoretically.

It is curious that such a simple and important problem has no known exact solution.[7] The trouble comes from the electron–electron repulsion,

$$V_{ee} = \frac{e^2}{4\pi\epsilon_0}\frac{1}{|\mathbf{r}_1 - \mathbf{r}_2|}. \tag{8.17}$$

If we ignore this term altogether, H splits into two independent hydrogen Hamiltonians (only with a nuclear charge of $2e$, instead of e); the exact solution is just the product of hydrogenic wave functions:

$$\psi_0(\mathbf{r}_1, \mathbf{r}_2) \equiv \psi_{100}(\mathbf{r}_1)\psi_{100}(\mathbf{r}_2) = \frac{8}{\pi a^3}e^{-2(r_1+r_2)/a}, \tag{8.18}$$

and the energy is $8E_1 = -109$ eV (Equation 5.42).[8] This is a long way from -79 eV, but it's a start.

To get a better approximation for E_{gs} we'll apply the variational principle, using ψ_0 as the trial wave function. This is a particularly convenient choice because it's an eigenfunction of *most* of the Hamiltonian:

$$H\psi_0 = (8E_1 + V_{ee})\,\psi_0. \tag{8.19}$$

Thus

$$\langle H\rangle = 8E_1 + \langle V_{ee}\rangle, \tag{8.20}$$

where[9]

$$\langle V_{ee}\rangle = \left(\frac{e^2}{4\pi\epsilon_0}\right)\left(\frac{8}{\pi a^3}\right)^2 \int \frac{e^{-4(r_1+r_2)/a}}{|\mathbf{r}_1 - \mathbf{r}_2|}\, d^3\mathbf{r}_1\, d^3\mathbf{r}_2. \tag{8.21}$$

I'll do the $\mathbf{r}_2$ integral first; for this purpose $\mathbf{r}_1$ is fixed, and we may as well orient the $\mathbf{r}_2$ coordinate system so that the polar axis lies along $\mathbf{r}_1$ (see Figure 8.4). By the law of cosines,

$$|\mathbf{r}_1 - \mathbf{r}_2| = \sqrt{r_1^2 + r_2^2 - 2r_1 r_2\cos\theta_2}, \tag{8.22}$$

[7] There do exist exactly soluble three-body problems with many of the qualitative features of helium, but using non-Coulombic potentials (see Problem 8.24).

[8] Here a is the ordinary Bohr radius and $E_n = -13.6/n^2$ eV is the nth Bohr energy; recall that for a nucleus with atomic number Z, $E_n \to Z^2E_n$ and $a \to a/Z$ (Problem 4.19). The spin configuration associated with Equation 8.18 will be antisymmetric (the singlet).

[9] You can, if you like, interpret Equation 8.21 as first-order perturbation theory, with $H' = V_{ee}$ (Problem 7.56(a)). However, I regard this as a misuse of the method, since the perturbation is comparable in size to the unperturbed potential. I prefer, therefore, to think of it as a variational calculation, in which we are looking for a rigorous upper bound on E_{gs}.

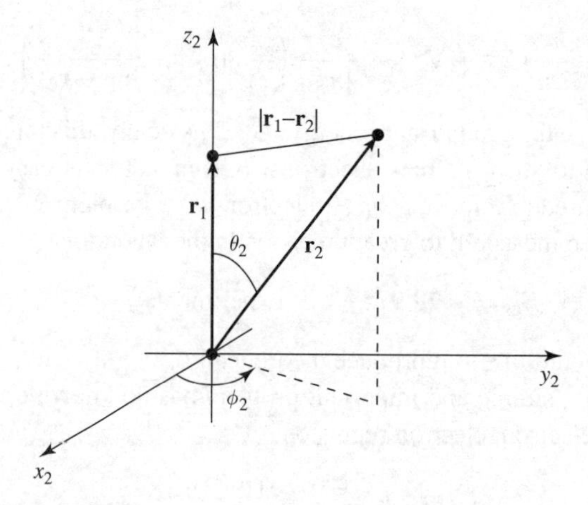

Figure 8.4: Choice of coordinates for the $\mathbf{r}_2$-integral (Equation 8.21).

and hence

$$I_2 \equiv \int \frac{e^{-4r_2/a}}{|\mathbf{r}_1-\mathbf{r}_2|}\, d^3\mathbf{r}_2 = \int \frac{e^{-4r_2/a}}{\sqrt{r_1^2+r_2^2-2r_1r_2\cos\theta_2}} r_2^2 \sin\theta_2\, dr_2\, d\theta_2\, d\phi_2. \tag{8.23}$$

The ϕ_2 integral is trivial (2π); the θ_2 integral is

$$\int_0^\pi \frac{\sin\theta_2}{\sqrt{r_1^2+r_2^2-2r_1r_2\cos\theta_2}}\, d\theta_2 = \left.\frac{\sqrt{r_1^2+r_2^2-2r_1r_2\cos\theta_2}}{r_1r_2}\right|_0^\pi$$

$$\begin{aligned} &= \frac{1}{r_1r_2}\left(\sqrt{r_1^2+r_2^2+2r_1r_2} - \sqrt{r_1^2+r_2^2-2r_1r_2}\right) \\ &= \frac{1}{r_1r_2}\left[(r_1+r_2) - |r_1-r_2|\right] = \begin{cases} 2/r_1, & r_2 < r_1, \\ 2/r_2, & r_2 > r_1. \end{cases} \end{aligned} \tag{8.24}$$

Thus

$$\begin{aligned} I_2 &= 4\pi\left(\frac{1}{r_1}\int_0^{r_1} e^{-4r_2/a} r_2^2\, dr_2 + \int_{r_1}^\infty e^{-4r_2/a} r_2\, dr_2\right) \\ &= \frac{\pi a^3}{8r_1}\left[1 - \left(1+\frac{2r_1}{a}\right)e^{-4r_1/a}\right]. \end{aligned} \tag{8.25}$$

It follows that $\langle V_{ee}\rangle$ is equal to

$$\left(\frac{e^2}{4\pi\epsilon_0}\right)\left(\frac{8}{\pi a^3}\right)\int \left[1 - \left(1+\frac{2r_1}{a}\right)e^{-4r_1/a}\right] e^{-4r_1/a} r_1 \sin\theta_1\, dr_1 d\theta_1 d\phi_1.$$

The angular integrals are easy (4π), and the r_1 integral becomes

$$\int_0^\infty \left[r e^{-4r/a} - \left(r + \frac{2r^2}{a}\right)e^{-8r/a}\right] dr = \frac{5a^2}{128}.$$

Finally, then,

$$\langle V_{ee}\rangle = \frac{5}{4a}\left(\frac{e^2}{4\pi\epsilon_0}\right) = -\frac{5}{2}E_1 = 34\text{ eV}, \tag{8.26}$$

and therefore

$$\langle H\rangle = -109\text{ eV} + 34\text{ eV} = -75\text{ eV}. \tag{8.27}$$

Not bad (remember, the experimental value is -79 eV). But we can do better.

We need to think up a more realistic trial function than ψ_0 (which treats the two electrons as though they did not interact at all). Rather than completely *ignoring* the influence of the other electron, let us say that, on the average, each electron represents a cloud of negative charge which partially *shields* the nucleus, so that the other electron actually sees an *effective* nuclear charge (Z) that is somewhat *less* than 2. This suggests that we use a trial function of the form

$$\psi_1(\mathbf{r}_1, \mathbf{r}_2) \equiv \frac{Z^3}{\pi a^3}e^{-Z(r_1+r_2)/a}. \tag{8.28}$$

We'll treat Z as a variational parameter, picking the value that minimizes $\langle H\rangle$. (Please note that in the variational method we *never touch the Hamiltonian itself*—the Hamiltonian for helium is, and remains, Equation 8.15. But it's fine to *think* about approximating the Hamiltonian *as a way of motivating the choice of the trial wave function.*)

This wave function is an eigenstate of the "unperturbed" Hamiltonian (neglecting electron repulsion), only with Z, instead of 2, in the Coulomb terms. With this in mind, we rewrite H (Equation 8.15) as follows:

$$H = -\frac{\hbar^2}{2m}\left(\nabla_1^2 + \nabla_2^2\right) - \frac{e^2}{4\pi\epsilon_0}\left(\frac{Z}{r_1} + \frac{Z}{r_2}\right) + \frac{e^2}{4\pi\epsilon_0}\left(\frac{(Z-2)}{r_1} + \frac{(Z-2)}{r_2} + \frac{1}{|\mathbf{r}_1 - \mathbf{r}_2|}\right). \tag{8.29}$$

The expectation value of H is evidently

$$\langle H\rangle = 2Z^2E_1 + 2(Z-2)\left(\frac{e^2}{4\pi\epsilon_0}\right)\left\langle\frac{1}{r}\right\rangle + \langle V_{ee}\rangle. \tag{8.30}$$

Here $\langle 1/r\rangle$ is the expectation value of $1/r$ in the (one-particle) hydrogenic ground state ψ_{100} (with nuclear charge Z); according to Equation 7.56,

$$\left\langle\frac{1}{r}\right\rangle = \frac{Z}{a}. \tag{8.31}$$

The expectation value of V_{ee} is the same as before (Equation 8.26), except that instead of $Z = 2$ we now want *arbitrary* Z—so we multiply a by $2/Z$:

$$\langle V_{ee}\rangle = \frac{5Z}{8a}\left(\frac{e^2}{4\pi\epsilon_0}\right) = -\frac{5Z}{4}E_1. \tag{8.32}$$

Putting all this together, we find

$$\langle H\rangle = \left[2Z^2 - 4Z(Z-2) - (5/4)Z\right]E_1 = \left[-2Z^2 + (27/4)Z\right]E_1. \tag{8.33}$$

According to the variational principle, this quantity exceeds E_{gs} for *any* value of Z. The *lowest* upper bound occurs when $\langle H \rangle$ is minimized:

$$\frac{d}{dZ}\langle H \rangle = \left[-4Z + (27/4)\right] E_1 = 0,$$

from which it follows that

$$Z = \frac{27}{16} = 1.69. \tag{8.34}$$

This seems reasonable; it tells us that the other electron partially screens the nucleus, reducing its effective charge from 2 down to about 1.69. Putting in this value for Z, we find

$$\langle H \rangle = \frac{1}{2}\left(\frac{3}{2}\right)^6 E_1 = -77.5 \text{ eV}. \tag{8.35}$$

The ground state of helium has been calculated with great precision in this way, using increasingly complicated trial wave functions, with more and more adjustable parameters.[10] But we're within 2% of the correct answer, and, frankly, at this point my own interest in the problem begins to wane.[11]

Problem 8.7 Using $E_{gs} = -79.0$ eV for the ground state energy of helium, calculate the ionization energy (the energy required to remove just *one* electron). *Hint:* First calculate the ground state energy of the helium ion, He^+, with a single electron orbiting the nucleus; then subtract the two energies.

* **Problem 8.8** Apply the techniques of this Section to the H^- and Li^+ ions (each has two electrons, like helium, but nuclear charges $Z = 1$ and $Z = 3$, respectively). Find the effective (partially shielded) nuclear charge, and determine the best upper bound on E_{gs}, for each case. *Comment:* In the case of H^- you should find that $\langle H \rangle > -13.6$ eV, which would appear to indicate that there is no bound state at all, since it would be energetically favorable for one electron to fly off, leaving behind a neutral hydrogen atom. This is not entirely surprising, since the electrons are less strongly attracted to the nucleus than they are in helium, and the electron repulsion tends to break the atom apart. However, it turns out to be incorrect. With a more sophisticated trial wave function (see Problem 8.25) it can be shown that $E_{gs} < -13.6$ eV, and hence that a bound state *does* exist. It's only *barely* bound, however, and there are no excited bound states,[12] so H^- has no discrete spectrum (all transitions are to and from the continuum). As a result, it is difficult to study in the laboratory, although it exists in great abundance on the surface of the sun.[13]

[10] The classic studies are E. A. Hylleraas, *Z. Phys.* **65**, 209 (1930); C. L. Pekeris, *Phys. Rev.* **115**, 1216 (1959). For more recent work, see footnote 4.

[11] The first excited state of helium can be calculated in much the same way, using a trial wave function orthogonal to the ground state. See Phillip J. E. Peebles, *Quantum Mechanics*, Princeton U.P., Princeton, NJ (1992), Section 40.

[12] Robert N. Hill, *J. Math. Phys.* **18**, 2316 (1977).

[13] For further discussion see Hans A. Bethe and Edwin E. Salpeter, *Quantum Mechanics of One- and Two-Electron Atoms*, Plenum, New York (1977), Section 34.

8.3 THE HYDROGEN MOLECULE ION

Another classic application of the variational principle is to the hydrogen molecule ion, H_2^+, consisting of a single electron in the Coulomb field of two protons (Figure 8.5). I shall assume for the moment that the protons are fixed in position, a specified distance R apart, although one of the most interesting byproducts of the calculation is going to be the actual *value* of R. The Hamiltonian is

$$H = -\frac{\hbar^2}{2m}\nabla^2 - \frac{e^2}{4\pi\epsilon_0}\left(\frac{1}{r} + \frac{1}{r'}\right), \tag{8.36}$$

where r and r' are the distances to the electron from the respective protons. As always, our strategy will be to guess a reasonable trial wave function, and invoke the variational principle to get a bound on the ground state energy. (Actually, our main interest is in finding out whether this system bonds at *all*—that is, whether its energy is less than that of a neutral hydrogen atom plus a free proton. If our trial wave function indicates that there *is* a bound state, a *better* trial function can only make the bonding even stronger.)

To construct the trial wave function, imagine that the ion is formed by taking a hydrogen atom in its ground state (Equation 4.80),

$$\psi_0(\mathbf{r}) = \frac{1}{\sqrt{\pi a^3}} e^{-r/a}, \tag{8.37}$$

bringing the second proton in from "infinity," and nailing it down a distance R away. If R is substantially greater than the Bohr radius, the electron's wave function probably isn't changed very much. But we would like to treat the two protons on an equal footing, so that the electron has the same probability of being associated with either one. This suggests that we consider a trial function of the form

$$\psi = A\left[\psi_0(r) + \psi_0(r')\right]. \tag{8.38}$$

(Quantum chemists call this the **LCAO** technique, because we are expressing the *molecular* wave function as a **l**inear **c**ombination of **a**tomic **o**rbitals.)

Our first task is to *normalize* the trial function:

$$1 = \int |\psi|^2 d^3\mathbf{r} = |A|^2 \left[\int \psi_0(r)^2 d^3\mathbf{r} + \int \psi_0(r')^2 d^3\mathbf{r} + 2\int \psi_0(r)\psi_0(r')\, d^3\mathbf{r}\right]. \tag{8.39}$$

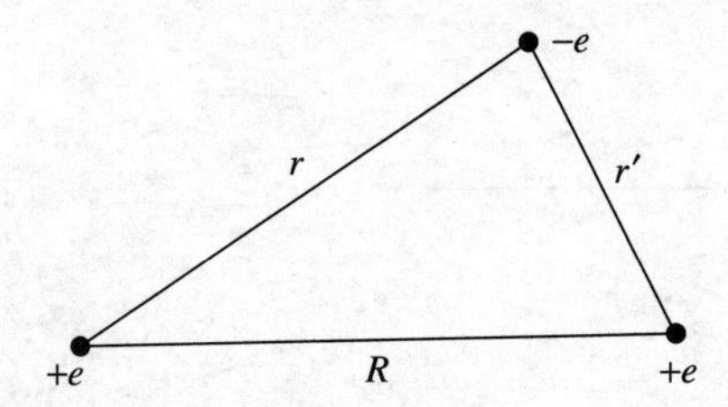

Figure 8.5: The hydrogen molecule ion, H_2^+.

The first two integrals are 1 (since ψ_0 itself is normalized); the third is more tricky. Let

$$I \equiv \langle \psi_0(r) \mid \psi_0(r') \rangle = \frac{1}{\pi a^3} \int e^{-(r+r')/a} \, d^3\mathbf{r}. \tag{8.40}$$

Picking coordinates so that proton 1 is at the origin and proton 2 is on the z axis at the point R (Figure 8.6), we have

$$r' = \sqrt{r^2 + R^2 - 2rR\cos\theta}, \tag{8.41}$$

and therefore

$$I = \frac{1}{\pi a^3} \int e^{-r/a} e^{-\sqrt{r^2+R^2-2rR\cos\theta}/a} r^2 \sin\theta \, dr \, d\theta \, d\phi. \tag{8.42}$$

The ϕ integral is trivial (2π). To do the θ integral, let $y \equiv \sqrt{r^2 + R^2 - 2rR\cos\theta}$, so that $d(y^2) = 2y\,dy = 2rR\sin\theta\,d\theta$. Then

$$\int_0^{\pi} e^{-\sqrt{r^2+R^2-2rR\cos\theta}/a} \sin\theta \, d\theta = \frac{1}{rR} \int_{|r-R|}^{r+R} e^{-y/a} y \, dy$$

$$= -\frac{a}{rR} \left[e^{-(r+R)/a} (r + R + a) - e^{-|r-R|/a} (|r - R| + a) \right].$$

The r integral is now straightforward:

$$I = \frac{2}{a^2 R} \left[-e^{-R/a} \int_0^{\infty} (r + R + a) e^{-2r/a} r \, dr + e^{-R/a} \int_0^{R} (R - r + a) \, r \, dr \right.$$
$$\left. + e^{R/a} \int_R^{\infty} (r - R + a) \, e^{-2r/a} r \, dr \right].$$

Evaluating the integrals, we find (after some algebraic simplification),

$$I = e^{-R/a} \left[1 + \left(\frac{R}{a} \right) + \frac{1}{3} \left(\frac{R}{a} \right)^2 \right]. \tag{8.43}$$

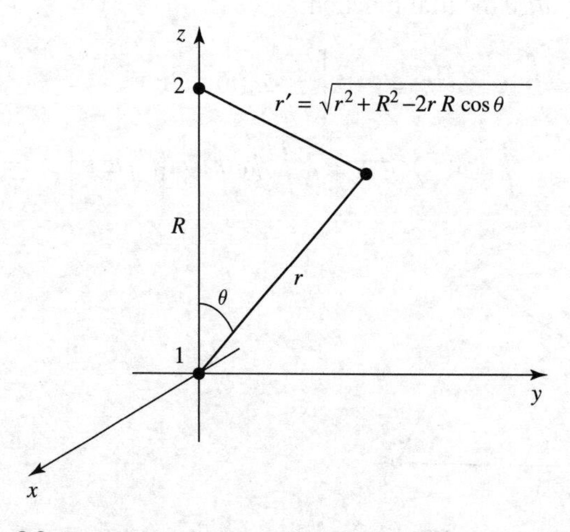

Figure 8.6: Coordinates for the calculation of I (Equation 8.40).

I is called an **overlap** integral; it measures the amount by which $\psi_0(r)$ overlaps $\psi_0(r')$ (notice that it goes to 1 as $R \to 0$, and to 0 as $R \to \infty$). In terms of I, the normalization factor (Equation 8.39) is

$$|A|^2 = \frac{1}{2(1+I)}. \tag{8.44}$$

Next we must calculate the expectation value of H in the trial state ψ. Noting that

$$\left(-\frac{\hbar^2}{2m}\nabla^2 - \frac{e^2}{4\pi\epsilon_0}\frac{1}{r}\right)\psi_0(r) = E_1\psi_0(r)$$

(where $E_1 = -13.6$ eV is the ground state energy of atomic hydrogen)—and the same with r' in place of r—we have

$$\begin{aligned} H\psi &= A\left[-\frac{\hbar^2}{2m}\nabla^2 - \frac{e^2}{4\pi\epsilon_0}\left(\frac{1}{r}+\frac{1}{r'}\right)\right]\left[\psi_0(r)+\psi_0(r')\right] \\ &= E_1\psi - A\left(\frac{e^2}{4\pi\epsilon_0}\right)\left[\frac{1}{r'}\psi_0(r) + \frac{1}{r}\psi_0(r')\right]. \end{aligned}$$

It follows that

$$\langle H\rangle = E_1 - 2|A|^2\left(\frac{e^2}{4\pi\epsilon_0}\right)\left[\left\langle\psi_0(r)\left|\frac{1}{r'}\right|\psi_0(r)\right\rangle + \left\langle\psi_0(r)\left|\frac{1}{r}\right|\psi_0(r')\right\rangle\right]. \tag{8.45}$$

I'll let you calculate the two remaining quantities, the so-called **direct integral**,

$$D \equiv a\left\langle\psi_0(r)\left|\frac{1}{r'}\right|\psi_0(r)\right\rangle, \tag{8.46}$$

and the **exchange integral**,

$$X \equiv a\left\langle\psi_0(r)\left|\frac{1}{r}\right|\psi_0(r')\right\rangle. \tag{8.47}$$

The results (see Problem 8.9) are

$$D = \frac{a}{R} - \left(1+\frac{a}{R}\right)e^{-2R/a}, \tag{8.48}$$

and

$$X = \left(1+\frac{R}{a}\right)e^{-R/a}. \tag{8.49}$$

Putting all this together, and recalling (Equations 4.70 and 4.72) that $E_1 = -\left(e^2/4\pi\epsilon_0\right)(1/2a)$, we conclude:

$$\langle H\rangle = \left[1 + 2\frac{(D+X)}{(1+I)}\right]E_1. \tag{8.50}$$

According to the variational principle, the ground state energy is *less* than $\langle H\rangle$. Of course, this is only the *electron's* energy—there is also potential energy associated with the proton–proton repulsion:

$$V_{pp} = \frac{e^2}{4\pi\epsilon_0}\frac{1}{R} = -\frac{2a}{R}E_1. \tag{8.51}$$

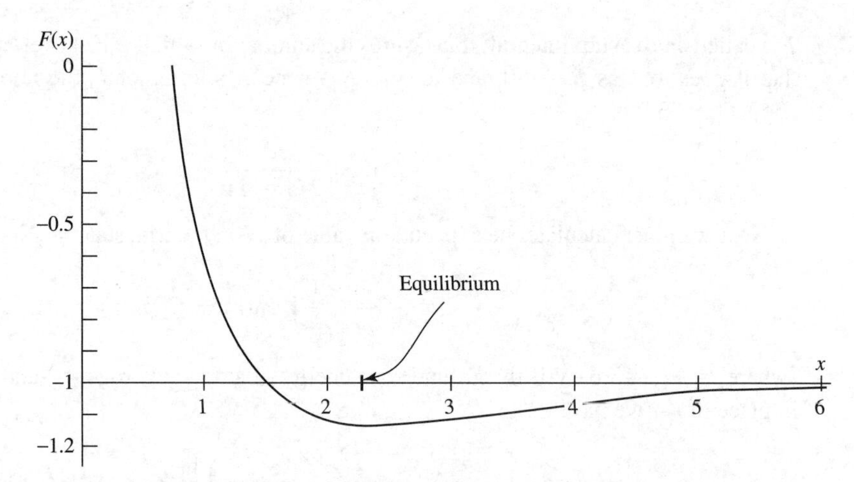

Figure 8.7: Plot of the function $F(x)$, Equation 8.52, showing existence of a bound state.

Thus the *total* energy of the system, in units of $-E_1$, and expressed as a function of $x \equiv R/a$, is less than

$$F(x) = -1 + \frac{2}{x}\left\{\frac{\left(1-(2/3)x^2\right)e^{-x} + (1+x)\,e^{-2x}}{1+\left(1+x+(1/3)x^2\right)e^{-x}}\right\}. \tag{8.52}$$

This function is plotted in Figure 8.7. Evidently bonding *does* occur, for there exists a region in which the graph goes below -1, indicating that the energy is less than that of a neutral atom plus a free proton (-13.6 eV). It's a **covalent bond**, with the electron shared equally by the two protons. The equilibrium separation of the protons is about 2.4 Bohr radii, or 1.3 Å (the experimental value is 1.06 Å). The calculated binding energy is 1.8 eV, whereas the experimental value is 2.8 eV (the variational principle, as always, *over*estimates the ground state energy—and hence *under*estimates the strength of the bond—but never mind: The essential point was to see whether binding occurs at all; a better variational function can only make the potential well even deeper.

* **Problem 8.9** Evaluate D and X (Equations 8.46 and 8.47). Check your answers against Equations 8.48 and 8.49.

** **Problem 8.10** Suppose we used a *minus* sign in our trial wave function (Equation 8.38):

$$\psi = A\left[\psi_0(r) - \psi_0(r')\right]. \tag{8.53}$$

Without doing any new integrals, find $F(x)$ (the analog to Equation 8.52) for this case, and construct the graph. Show that there is no evidence of bonding.[14] (Since the variational principle only gives an *upper bound*, this doesn't *prove* that bonding cannot occur for such a state, but it certainly doesn't look promising.)

[14] The wave function with the plus sign (Equation 8.38) is called the **bonding orbital**. Bonding is associated with a buildup of electron probability in between the two nuclei. The odd linear combination (Equation 8.53) has a *node* at the center, so it's not surprising that this configuration doesn't lead to bonding; it is called the **anti-bonding orbital**.

Problem 8.11 The second derivative of $F(x)$, at the equilibrium point, can be used to estimate the natural frequency of vibration (ω) of the two protons in the hydrogen molecule ion (see Section 2.3). If the ground state energy ($\hbar\omega/2$) of this oscillator exceeds the binding energy of the system, it will fly apart. Show that in fact the oscillator energy is small enough that this will *not* happen, and estimate how many bound vibrational levels there are. *Note*: You're not going to be able to obtain the position of the minimum—still less the second derivative at that point—analytically. Do it numerically, on a computer.

8.4 THE HYDROGEN MOLECULE

Now consider the hydrogen molecule itself, adding a second electron to the hydrogen molecule ion we studied in Section 8.3. Taking the two protons to be at rest, the Hamiltonian is

$$\hat{H} = -\frac{\hbar^2}{2m}\left(\nabla_1^2 + \nabla_2^2\right) + \frac{e^2}{4\pi\epsilon_0}\left(\frac{1}{r_{12}} + \frac{1}{R} - \frac{1}{r_1} - \frac{1}{r_1'} - \frac{1}{r_2} - \frac{1}{r_2'}\right) \tag{8.54}$$

where r_1 and r_1' are the distances of electron 1 from each proton and r_2 and r_2' are the distances of electron 2 from each proton; as shown in Figure 8.8. The six potential energy terms describe the repulsion between the two electrons, the repulsion between the two protons, and the attraction of each electron to each proton.

For the variational wave function, associate one electron with each proton, and symmetrize:

$$\psi_+(\mathbf{r}_1, \mathbf{r}_2) = A_+\left[\psi_0(r_1)\,\psi_0(r_2') + \psi_0(r_1')\,\psi_0(r_2)\right]. \tag{8.55}$$

We'll calculate the normalization A_+ in a moment. Since this spatial wave function is symmetric under interchange, the electrons must occupy the antisymmetric (singlet) spin state. Of course, we could also choose the trial wave function

$$\psi_-(\mathbf{r}_1, \mathbf{r}_2) = A_-\left[\psi_0(r_1)\,\psi_0(r_2') - \psi_0(r_1')\,\psi_0(r_2)\right] \tag{8.56}$$

in which case the electrons would be in a symmetric (triplet) spin state. These two variational wave functions constitute the **Heitler–London approximation**.[15] It is not obvious which of

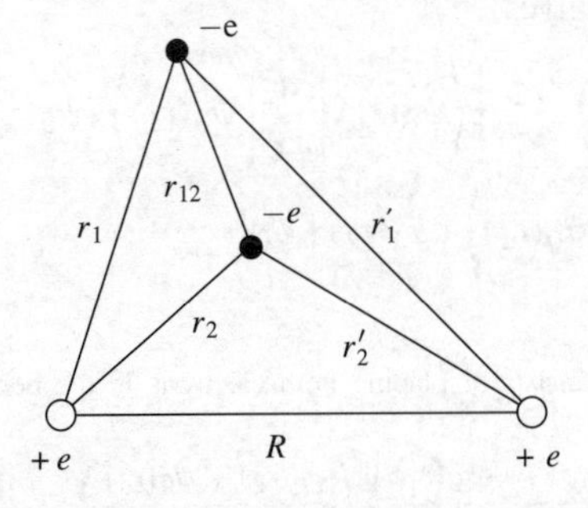

Figure 8.8: Diagram of H_2 showing the distances on which the potential energy depends.

[15] W. Heitler and F. London, *Z. Phys.* **44**, 455 (1928). For an English translation see Hinne Hettema, *Quantum Chemistry: Classic Scientific Papers*, World Scientific, New Jersey, PA, 2000.

Equations 8.55 or 8.56 would be energetically favored, so let's calculate the energy of each one, and find out.[16]

First we need to normalize the wave functions. Note that

$$|\psi_\pm(\mathbf{r}_1, \mathbf{r}_2)|^2 = A_\pm^2 \Big[\psi_0(r_1)^2\, \psi_0(r_2')^2 + \psi_0(r_1')^2\, \psi_0(r_2)^2 \pm 2\,\psi_0(r_1)\, \psi_0(r_2')\, \psi_0(r_1')\, \psi_0(r_2)\Big]. \tag{8.58}$$

Normalization requires

$$\begin{aligned} 1 &= \int\!\!\int |\psi_\pm(\mathbf{r}_1, \mathbf{r}_2)|^2\, d^3\mathbf{r}_1\, d^3\mathbf{r}_2 \\ &= A_\pm^2 \Big[\int \psi_0(r_1)^2\, d^3\mathbf{r}_1 \int \psi_0(r_2')^2\, d^3\mathbf{r}_2 + \int \psi_0(r_1')^2\, d^3\mathbf{r}_1 \int \psi_0(r_2)^2\, d^3\mathbf{r}_2 \\ &\quad \pm 2 \int \psi_0(r_1)\, \psi_0(r_1')\, d^3\mathbf{r}_1 \int \psi_0(r_2')\, \psi_0(r_2)\, d^3\mathbf{r}_2\Big]. \end{aligned} \tag{8.59}$$

The individual orbitals are normalized and the overlap integral was given the symbol I and calculated in Equation 8.43. Thus

$$A_\pm = \frac{1}{\sqrt{2\,(1 \pm I^2)}}. \tag{8.60}$$

To calculate the expectation value of the energy, we will start with the kinetic energy of particle 1. Since ψ_0 is the ground state of the hydrogen Hamiltonian, the same trick that brought us to Equation 8.45 gives

$$\begin{aligned} -\frac{\hbar^2}{2m}\nabla_1^2\,\psi_\pm &= A_\pm \left[\left(-\frac{\hbar^2}{2m}\nabla_1^2\,\psi_0(r_1)\right)\psi_0(r_2') \pm \left(-\frac{\hbar^2}{2m}\nabla_1^2\,\psi_0(r_1')\right)\psi_0(r_2)\right] \\ &= A_\pm \left[\left(E_1 + \frac{e^2}{4\pi\epsilon_0 r_1}\right)\psi_0(r_1)\,\psi_0(r_2') \pm \left(E_1 + \frac{e^2}{4\pi\epsilon_0 r_1'}\right)\psi_0(r_1')\,\psi_0(r_2)\right] \\ &= E_1\,\psi_\pm + \frac{e^2}{4\pi\epsilon_0 a} A_\pm \left(\frac{a}{r_1}\psi_0(r_1)\,\psi_0(r_2') \pm \frac{a}{r_1'}\psi_0(r_1')\,\psi_0(r_2)\right). \end{aligned}$$

Taking the inner product with $\psi_\pm$ then gives

$$\begin{aligned} \left\langle -\frac{\hbar^2}{2m}\nabla_1^2 \right\rangle &= E_1 + \left(\frac{e^2}{4\pi\epsilon_0 a}\right) A_\pm^2 \Big[\left\langle \psi_0(r_1) \middle| \frac{a}{r_1} \middle| \psi_0(r_1)\right\rangle \langle \psi_0(r_2') \,|\, \psi_0(r_2')\rangle \\ &\quad \pm \left\langle \psi_0(r_1') \middle| \frac{a}{r_1} \middle| \psi_0(r_1)\right\rangle \langle \psi_0(r_2) \,|\, \psi_0(r_2')\rangle \end{aligned}$$

---

[16] Another natural variational wave function consists of placing both electrons in the bonding orbital studied in Section 8.3:

$$\psi(\mathbf{r}_1, \mathbf{r}_2) = \left(\frac{\psi_0(r_1) + \psi_0(r_1')}{\sqrt{2(1+I)}}\right)\left(\frac{\psi_0(r_2) + \psi_0(r_2')}{\sqrt{2(1+I)}}\right), \tag{8.57}$$

also paired with a singlet spin state. If you expand this function you'll see that half the terms—such as $\psi_0(r_1)\,\psi_0(r_2)$—involve attaching two electrons to the same proton, which is energetically costly because of the electron–electron repulsion in Equation 8.54. The Heitler–London approximation, Equation 8.55, amounts to dropping the offending terms from Equation 8.57.

$$\pm\left\langle\psi_0(r_1)\left|\frac{a}{r_1'}\right|\psi_0(r_1')\right\rangle\langle\psi_0(r_2')\,|\,\psi_0(r_2)\rangle$$
$$+\left\langle\psi_0(r_1')\left|\frac{a}{r_1'}\right|\psi_0(r_1')\right\rangle\langle\psi_0(r_2)\,|\,\psi_0(r_2)\rangle\Bigg]. \tag{8.61}$$

These inner products were calculated in Section 8.3 and the kinetic energy of particle 1 is

$$\left\langle-\frac{\hbar^2}{2m}\nabla_1^2\right\rangle = E_1+\left(\frac{e^2}{4\pi\,\epsilon_0\,a}\right)\frac{1\pm I\,X}{1\pm I^2}. \tag{8.62}$$

The kinetic energy of particle 2 is of course the same, so the total kinetic energy is simply twice Equation 8.62. The calculation of the electron–proton potential energy is similar; you will show in Problem 8.13 that

$$\left\langle-\frac{e^2}{4\pi\,\epsilon_0\,r_1}\right\rangle = -\frac{1}{2}\left(\frac{e^2}{4\pi\,\epsilon_0\,a}\right)\frac{1+D\pm 2\,I\,X}{1\pm I^2} \tag{8.63}$$

and the total electron–proton potential energy is four times this amount.

The electron–electron potential energy is given by

$$\langle V_{ee}\rangle = \left(\frac{e^2}{4\pi\,\epsilon_0\,a}\right)\int\!\!\int|\psi_\pm(\mathbf{r}_1,\mathbf{r}_2)|^2\,\frac{a}{r_{12}}\,d^3\mathbf{r}_1\,d^3\mathbf{r}_2$$
$$= \left(\frac{e^2}{4\pi\,\epsilon_0\,a}\right)A_\pm^2\left[\int\!\!\int\psi_0(r_1)^2\,\frac{a}{r_{12}}\,\psi_0(r_2')^2\,d^3\mathbf{r}_1\,d^3\mathbf{r}_2\right.$$
$$+\int\!\!\int\psi_0(r_1')^2\,\frac{a}{r_{12}}\,\psi_0(r_2)^2\,d^3\mathbf{r}_1\,d^3\mathbf{r}_2$$
$$\left.\pm 2\int\!\!\int\psi_0(r_1)\,\psi_0(r_1')\,\frac{a}{r_{12}}\,\psi_0(r_2)\,\psi_0(r_2')\,d^3\mathbf{r}_1\,d^3\mathbf{r}_2\right]. \tag{8.64}$$

The first two integrals in Equation 8.64 are equal, as you can see by interchanging the labels 1 and 2. We will give the two remaining integrals the names

$$D_2 = \int\!\!\int|\psi_0(r_1)|^2\,\frac{a}{r_{12}}\,|\psi_0(r_2')|^2\,d^3\mathbf{r}_1\,d^3\mathbf{r}_2 \tag{8.65}$$

$$X_2 = \int\!\!\int\psi_0(r_1)\,\psi_0(r_1')\,\frac{a}{r_{12}}\,\psi_0(r_2)\,\psi_0(r_2')\,d^3\mathbf{r}_1\,d^3\mathbf{r}_2 \tag{8.66}$$

so that

$$\langle V_{ee}\rangle = \left(\frac{e^2}{4\pi\,\epsilon_0\,a}\right)\frac{D_2\pm X_2}{1\pm I^2}. \tag{8.67}$$

The evaluation of these integrals is discussed in Problem 8.14. Note that the integral D_2 is just the electrostatic potential energy of two charge distributions $\rho_1 = |\psi_0(r_1)|^2$ and $\rho_2 = |\psi_0(r_2)|^2$. The exchange term X_2 has no such classical counterpart.

When we add all of the contributions to the energy—the kinetic energy, the electron–proton potential energy, the electron–electron potential energy, and the proton–proton potential energy (which is a constant, $e^2/4\pi\,\epsilon_0\,R$)—we get

$$\boxed{\langle H\rangle_{\pm}=2E_1\left[1-\frac{a}{R}+\frac{2D-D_2\pm(2IX-X_2)}{1\pm I^2}\right].}\tag{8.68}$$

A plot of $\langle H\rangle_+$ and $\langle H\rangle_-$ is shown in Figure 8.9. Recall that the state ψ_+ requires placing the two electrons in the singlet spin configuration, whereas ψ_- means putting them in a triplet spin configuration. According to the figure, bonding only occurs if the two electrons are in a singlet configuration—something that is confirmed experimentally. Again, it's a covalent bond.

Locating the minimum on the plot, our calculation predicts a bond length of 1.64 Bohr radii (the experimental value is 1.40 Bohr radii), and suggests a binding energy of 3.15 eV (whereas the experimental value is 4.75 eV). The trends here follow those of the Hydrogen molecule ion: the calculation overestimates the bond length and underestimates the binding energy, but the agreement is surprisingly good for a variational calculation with no adjustable parameters.

The difference between the singlet and triplet energies is called the **exchange splitting** J. In the Heitler–London approximation it is

$$J=\langle H\rangle_+-\langle H\rangle_-=4E_1\frac{(D_2-2D)\,I^2-(X_2-2IX)}{1-I^4},\tag{8.69}$$

which is roughly $-10\,\text{eV}$ (negative because the singlet is lower in energy) at the equilibrium separation. This means a strong preference for having the electron spins anti-aligned. But in this treatment of H_2 we've left out completely the (magnetic) spin–spin interaction between the electrons—remember that the spin–spin interaction between the *proton* and the electron is what leads to hyperfine splitting (Section 7.5). Were we right to ignore it here? Absolutely: applying Equation 7.92 to two electrons a distance R apart, the energy of the spin–spin interaction is something like 10^{-4} eV in this system, five orders of magnitude smaller than the exchange splitting.

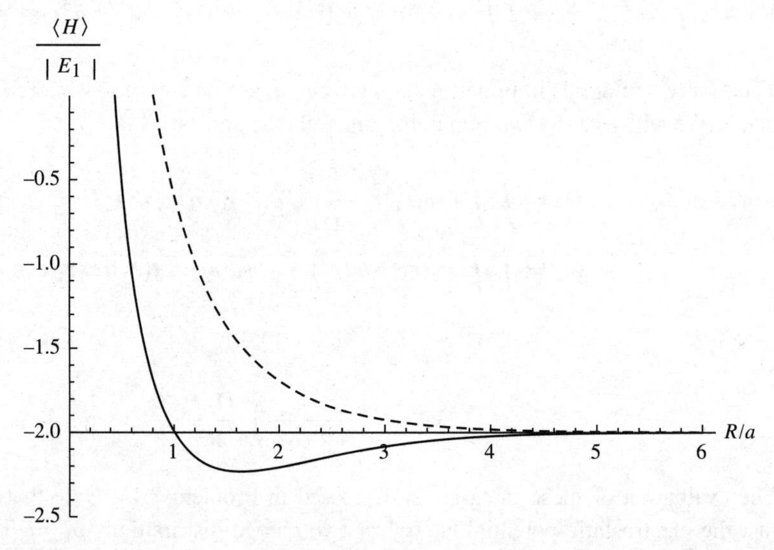

Figure 8.9: The total energy of the singlet (solid curve) and triplet (dashed curve) states for H_2, as a function of the separation R between the protons. The singlet state has a minimum at around 1.6 Bohr radii, representing a stable bond. The triplet state is unstable and will dissociate, as the energy is minimized for $R\to\infty$.

This calculation shows us that different spin configurations can have very different energies, even when the interaction between the spins is negligible. And that helps us understand ferromagnetism (where the spins in a material align) and anti-ferromagnetism (where the spins alternate). As we've just seen, the spin–spin interaction is *way* too weak to account for this—but the exchange splitting isn't. Counterintuitively, it's not a *magnetic* interaction that accounts for ferromagnetism, but an electrostatic one! H_2 is a sort of inchoate anti-ferromagnet where the Hamiltonian, which is independent of the spin, selects a certain *spatial* ground state and the *spin* state comes along for the ride, to satisfy the Fermi statistics.

Problem 8.12 Show that the antisymmetric state (Equation 8.56) can be expressed in terms of the molecular orbitals of Section 8.3—specifically, by placing one electron in the bonding orbital (Equation 8.38) and one in the anti-bonding orbital (Equation 8.53).

Problem 8.13 Verify Equation 8.63 for the electron–proton potential energy.

∗∗∗ **Problem 8.14** The two-body integrals D_2 and X_2 are defined in Equations 8.65 and 8.66. To evaluate D_2 we write

$$D_2 = \int \left|\psi_0(r_2')\right|^2 \Phi(r_2)\, d^3\mathbf{r}_2$$

$$= \iiint \frac{e^{-2\sqrt{R^2+r_2^2-2Rr_2\cos\theta_2}/a}}{\pi a^3} \Phi(r_2)\, r_2^2\, dr_2\, \sin\theta_2\, d\theta_2\, d\phi_2$$

where θ_2 is the angle between $\boldsymbol{R}$ and $\boldsymbol{r_2}$ (Figure 8.8), and

$$\Phi(r_2) \equiv \int |\psi_0(r_1)|^2 \frac{a}{|\mathbf{r}_1 - \mathbf{r}_2|} d^3\mathbf{r}_1 .$$

(a) Consider first the integral over $\mathbf{r}_1$. Align the z axis with $\mathbf{r}_2$ (which is a constant vector for the purposes of this first integral) so that

$$\Phi(r_2) = \frac{1}{\pi a^3} \iiint \frac{a e^{-2r_1/a}}{\sqrt{r_1^2 + r_2^2 - 2r_1 r_2 \cos\theta_1}} r_1^2\, dr_1\, \sin\theta_1\, d\theta_1\, d\phi_1 .$$

Do the angular integration first and show that

$$\Phi(r_2) = \frac{a}{r_2} - \left(1 + \frac{a}{r_2}\right) e^{-2r_2/a} .$$

(b) Plug your result from part (a) back into the relation for D_2, and show that

$$D_2 = \frac{a}{R} - e^{-2R/a} \left[\frac{1}{6}\left(\frac{R}{a}\right)^2 + \frac{3}{4}\left(\frac{R}{a}\right) + \frac{11}{8} + \frac{a}{R}\right]. \tag{8.70}$$

Again, do the angular integration first.

Comment: The integral X_2 *can* also be evaluated in closed form, but the procedure is rather involved.[17] We will simply quote the result,

[17] The calculation was done by Y. Sugiura, *Z. Phys.* **44**, 455 (1927).

$$X_2 = e^{-2R/a}\left[\frac{5}{8} - \frac{23}{20}\frac{R}{a} - \frac{3}{5}\left(\frac{R}{a}\right)^2 - \frac{1}{15}\left(\frac{R}{a}\right)^3\right]$$
$$+ \frac{6}{5}\frac{a}{R}I^2\left[\gamma + \log\left(\frac{R}{a}\right) + \left(\frac{\tilde{I}}{I}\right)^2 \mathrm{Ei}\left(-\frac{4R}{a}\right) - 2\frac{\tilde{I}}{I}\mathrm{Ei}\left(-\frac{2R}{a}\right)\right], \quad (8.71)$$

where $\gamma = 0.5772\ldots$ is Euler's constant, $\mathrm{Ei}(x)$ is the exponential integral

$$\mathrm{Ei}(x) = -\int_{-x}^{\infty} \frac{e^{-t}}{t}\,dt,$$

and $\tilde{I}$ is obtained from I by switching the sign of R:

$$\tilde{I} = e^{R/a}\left[1 - \frac{R}{a} + \frac{1}{3}\left(\frac{R}{a}\right)^2\right]. \quad (8.72)$$

* 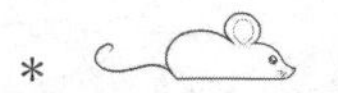

Problem 8.15 Make a plot of the kinetic energy for both the singlet and triplet states of H_2, as a function of R/a. Do the same for the electron-proton potential energy and for the electron–electron potential energy. You should find that the triplet state has lower potential energy than the singlet state for all values of R. However, the singlet state's *kinetic* energy is *so* much smaller that its total energy comes out lower. *Comment*: In situations where there is not a large kinetic energy cost to aligning the spins, such as two electrons in a partially filled orbital in an atom, the triplet state can come out lower in energy. This is the physics behind Hund's first rule.

FURTHER PROBLEMS ON CHAPTER 8

Problem 8.16

(a) Use the function $\psi(x) = Ax(a-x)$ (for $0 < x < a$, otherwise 0) to get an upper bound on the ground state of the infinite square well.

(b) Generalize to a function of the form $\psi(x) = A[x(a-x)]^p$, for some real number p. What is the optimal value of p, and what is the best bound on the ground state energy? Compare the exact value. *Answer:* $\left(5 + 2\sqrt{6}\right)\hbar^2/2ma^2$.

Problem 8.17

(a) Use a trial wave function of the form

$$\psi(x) = \begin{cases} A\cos(\pi x/a), & -a/2 < x < a/2, \\ 0, & \text{otherwise}, \end{cases}$$

to obtain a bound on the ground state energy of the one-dimensional harmonic oscillator. What is the "best" value of a? Compare $\langle H\rangle_{\min}$ with the exact energy. *Note:* This trial function has a "kink" in it (a discontinuous derivative) at $\pm a/2$; do you need to take account of this, as I did in Example 8.3?

(b) Use $\psi(x) = B\sin(\pi x/a)$ on the interval $(-a, a)$ to obtain a bound on the first excited state. Compare the exact answer.

** **Problem 8.18**

(a) Generalize Problem 8.2, using the trial wave function[18]

$$\psi(x) = \frac{A}{\left(x^2 + b^2\right)^n},$$

for arbitrary n. *Partial answer:* The best value of b is given by

$$b^2 = \frac{\hbar}{m\omega}\left[\frac{n(4n-1)(4n-3)}{2(2n+1)}\right]^{1/2}.$$

(b) Find the least upper bound on the first excited state of the harmonic oscillator using a trial function of the form

$$\psi(x) = \frac{Bx}{\left(x^2 + b^2\right)^n}.$$

Partial answer: The best value of b is given by

$$b^2 = \frac{\hbar}{m\omega}\left[\frac{n(4n-5)(4n-3)}{2(2n+1)}\right]^{1/2}.$$

(c) Notice that the bounds approach the exact energies as $n \to \infty$. Why is that? *Hint:* Plot the trial wave functions for $n = 2$, $n = 3$, and $n = 4$, and compare them with the true wave functions (Equations 2.60 and 2.63). To do it analytically, start with the identity

$$e^z = \lim_{n\to\infty}\left(1 + \frac{z}{n}\right)^n.$$

Problem 8.19 Find the lowest bound on the ground state of hydrogen you can get using a gaussian trial wave function

$$\psi(\mathbf{r}) = Ae^{-br^2},$$

where A is determined by normalization and b is an adjustable parameter. *Answer:* -11.5 eV.

Problem 8.20 Find an upper bound on the energy of the first excited state of the hydrogen atom. A trial function with $\ell = 1$ will automatically be orthogonal to the ground state (see footnote 6); for the radial part of ψ you can use the same function as in Problem 8.19.

** **Problem 8.21** If the photon had a nonzero mass $\left(m_\gamma \neq 0\right)$, the Coulomb potential would be replaced by the **Yukawa potential**,

$$V(\mathbf{r}) = -\frac{e^2}{4\pi\epsilon_0}\frac{e^{-\mu r}}{r}, \tag{8.73}$$

where $\mu = m_\gamma c/\hbar$. With a trial wave function of your own devising, estimate the binding energy of a "hydrogen" atom with this potential. Assume $\mu a \ll 1$, and give your answer correct to order $(\mu a)^2$.

[18] W. N. Mei, *Int. J. Educ. Sci. Tech.* **27**, 285 (1996).

Problem 8.22 Suppose you're given a two-level quantum system whose (time-independent) Hamiltonian H^0 admits just two eigenstates, ψ_a (with energy E_a), and ψ_b (with energy E_b). They are orthogonal, normalized, and nondegenerate (assume E_a is the smaller of the two energies). Now we turn on a perturbation H', with the following matrix elements:

$$\langle \psi_a | H' | \psi_a \rangle = \langle \psi_b | H' | \psi_b \rangle = 0; \quad \langle \psi_a | H' | \psi_b \rangle = \langle \psi_b | H' | \psi_a \rangle = h, \tag{8.74}$$

where h is some specified constant.

(a) Find the exact eigenvalues of the perturbed Hamiltonian.

(b) Estimate the energies of the perturbed system using second-order perturbation theory.

(c) Estimate the ground state energy of the perturbed system using the variational principle, with a trial function of the form

$$\psi = (\cos\phi)\,\psi_a + (\sin\phi)\,\psi_b, \tag{8.75}$$

where ϕ is an adjustable parameter. *Note:* Writing the linear combination in this way is just a neat way to guarantee that ψ is normalized.

(d) Compare your answers to (a), (b), and (c). Why is the variational principle so accurate, in this case?

Problem 8.23 As an explicit example of the method developed in Problem 8.22, consider an electron at rest in a uniform magnetic field $\mathbf{B} = B_z\hat{k}$, for which the Hamiltonian is (Equation 4.158):

$$H^0 = \frac{eB_z}{m} S_z. \tag{8.76}$$

The eigenspinors, χ_a and χ_b, and the corresponding energies, E_a and E_b, are given in Equation 4.161. Now we turn on a perturbation, in the form of a uniform field in the x direction:

$$H' = \frac{eB_x}{m} S_x. \tag{8.77}$$

(a) Find the matrix elements of H', and confirm that they have the structure of Equation 8.74. What is h?

(b) Using your result in Problem 8.22(b), find the new ground state energy, in second-order perturbation theory.

(c) Using your result in Problem 8.22(c), find the variational principle bound on the ground state energy.

∗∗∗ **Problem 8.24** Although the Schrödinger equation for helium itself cannot be solved exactly, there exist "helium-like" systems that do admit exact solutions. A simple example[19] is "rubber-band helium," in which the Coulomb forces are replaced by Hooke's law forces:

[19] For a more sophisticated model, see R. Crandall, R. Whitnell, and R. Bettega, *Am. J. Phys* **52**, 438 (1984).

$$H = -\frac{\hbar^2}{2m}\left(\nabla_1^2 + \nabla_2^2\right) + \frac{1}{2}m\omega^2\left(r_1^2 + r_2^2\right) - \frac{\lambda}{4}m\omega^2|\mathbf{r}_1 - \mathbf{r}_2|^2. \tag{8.78}$$

(a) Show that the change of variables from $\mathbf{r}_1$, $\mathbf{r}_2$, to

$$\mathbf{u} \equiv \frac{1}{\sqrt{2}}(\mathbf{r}_1 + \mathbf{r}_2), \quad \mathbf{v} \equiv \frac{1}{\sqrt{2}}(\mathbf{r}_1 - \mathbf{r}_2), \tag{8.79}$$

turns the Hamiltonian into two independent three-dimensional harmonic oscillators:

$$H = \left[-\frac{\hbar^2}{2m}\nabla_u^2 + \frac{1}{2}m\omega^2u^2\right] + \left[-\frac{\hbar^2}{2m}\nabla_v^2 + \frac{1}{2}(1-\lambda)\,m\omega^2v^2\right]. \tag{8.80}$$

(b) What is the *exact* ground state energy for this system?

(c) If we didn't know the exact solution, we might be inclined to apply the method of Section 8.2 to the Hamiltonian in its original form (Equation 8.78). Do so (but don't bother with shielding). How does your result compare with the exact answer? *Answer:* $\langle H\rangle = 3\hbar\omega(1 - \lambda/4)$.

*** **Problem 8.25** In Problem 8.8 we found that the trial wave function with shielding (Equation 8.28), which worked well for helium, is inadequate to confirm the existence of a bound state for the negative hydrogen ion. Chandrasekhar[20] used a trial wave function of the form

$$\psi(\mathbf{r}_1, \mathbf{r}_2) \equiv A\left[\psi_1(r_1)\psi_2(r_2) + \psi_2(r_1)\psi_1(r_2)\right], \tag{8.81}$$

where

$$\psi_1(r) \equiv \sqrt{\frac{Z_1^3}{\pi a^3}}e^{-Z_1 r/a}, \quad \text{and} \quad \psi_2(r) \equiv \sqrt{\frac{Z_2^3}{\pi a^3}}e^{-Z_2 r/a}. \tag{8.82}$$

In effect, he allowed two *different* shielding factors, suggesting that one electron is relatively close to the nucleus, and the other is farther out. (Because electrons are identical particles, the spatial wave function must be symmetrized with respect to interchange. The *spin* state—which is irrelevant to the calculation—is evidently antisymmetric.) Show that by astute choice of the adjustable parameters Z_1 and Z_2 you can get $\langle H\rangle$ less than -13.6 eV. *Answer:*

$$\langle H\rangle = \frac{E_1}{x^6 + y^6}\left(-x^8 + 2x^7 + \frac{1}{2}x^6y^2 - \frac{1}{2}x^5y^2 - \frac{1}{8}x^3y^4 + \frac{11}{8}xy^6 - \frac{1}{2}y^8\right),$$

where $x \equiv Z_1 + Z_2$ and $y \equiv 2\sqrt{Z_1Z_2}$. Chandrasekhar used $Z_1 = 1.039$ (since this is larger than 1, the motivating interpretation as an effective nuclear charge cannot be sustained, but never mind—it's still an acceptable trial wave function) and $Z_2 = 0.283$.

Problem 8.26 The fundamental problem in harnessing nuclear fusion is getting the two particles (say, two deuterons) close enough together for the attractive (but short-range) nuclear force to overcome the Coulomb repulsion. The "bulldozer" method is to heat

[20] S. Chandrasekhar, *Astrophys. J.* **100**, 176 (1944).

the particles up to fantastic temperatures, and allow the random collisions to bring them together. A more exotic proposal is **muon catalysis**, in which we construct a "hydrogen molecule ion," only with deuterons in place of protons, and a *muon* in place of the electron. Predict the equilibrium separation distance between the deuterons in such a structure, and explain why muons are superior to electrons for this purpose.[21]

∗∗∗ **Problem 8.27 Quantum dots.** Consider a particle constrained to move in two dimensions in the cross-shaped region shown in Figure 8.10. The "arms" of the cross continue out to infinity. The potential is zero within the cross, and infinite in the shaded areas outside. Surprisingly, this configuration admits a positive-energy bound state.[22]

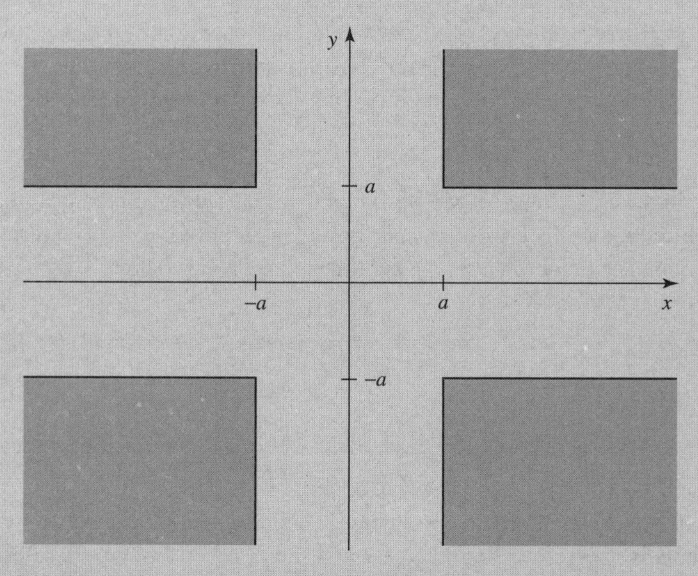

Figure 8.10: The cross-shaped region for Problem 8.27.

(a) Show that the lowest energy that can propagate off to infinity is

$$E_{\text{threshold}} = \frac{\pi^2\hbar^2}{8ma^2};$$

any solution with energy *less* than that has to be a bound state. *Hint:* Go way out one arm (say, $x \gg a$), and solve the Schrödinger equation by separation of variables; if the wave function propagates out to infinity, the dependence on x must take the form $\exp(ik_x x)$ with $k_x > 0$.

---

21 The classic paper on muon-catalyzed fusion is J. D. Jackson, *Phys. Rev.* **106**, 330 (1957); for a more recent popular review, see J. Rafelski and S. Jones, *Scientific American*, November 1987, page 84.

22 This model is taken from R. L. Schult *et al.*, *Phys. Rev. B* **39**, 5476 (1989). For further discussion see J. T. Londergan and D. P. Murdock, *Am. J. Phys.* **80**, 1085 (2012). In the presence of quantum tunneling a classically bound state can become unbound; this is the reverse: A classically *un*bound state is quantum mechanically *bound*.

(b) Now use the variational principle to show that the ground state has energy less than $E_{\text{threshold}}$. Use the following trial wave function (suggested by Jim McTavish):

$$\psi(x,y) = A\begin{cases} \left[\cos(\pi x/2a) + \cos(\pi y/2a)\right]e^{-\alpha}, & |x| \le a \text{ and } |y| \le a \\ \cos(\pi x/2a)e^{-\alpha|y|/a}, & |x| \le a \text{ and } |y| > a \\ \cos(\pi y/2a)e^{-\alpha|x|/a}, & |x| > a \text{ and } |y| \le a \\ 0, & \text{elsewhere.} \end{cases}$$

Normalize it to determine A, and calculate the expectation value of H. *Answer:*

$$\langle H\rangle = \frac{\hbar^2}{ma^2}\left[\frac{\pi^2}{8} - \left(\frac{1-(\alpha/4)}{1+\left(8/\pi^2\right)+(1/2\alpha)}\right)\right].$$

Now minimize with respect to α, and show that the result is less than $E_{\text{threshold}}$. *Hint:* Take full advantage of the symmetry of the problem—you only need to integrate over 1/8 of the open region, since the other seven integrals will be the same. Note however that whereas the trial wave function is continuous, its *derivatives* are *not*—there are "roof-lines" at the joins, and you will need to exploit the technique of Example 8.3.[23]

Problem 8.28 In Yukawa's original theory (1934), which remains a useful approximation in nuclear physics, the "strong" force between protons and neutrons is mediated by the exchange of π-mesons. The potential energy is

$$V(r) = -r_0 V_0 \frac{e^{-r/r_0}}{r}, \tag{8.83}$$

where r is the distance between the nucleons, and the range r_0 is related to the mass of the meson: $r_0 = \hbar/m_\pi c$. *Question:* Does this theory account for the existence of the **deuteron** (a bound state of the proton and the neutron)?

The Schrödinger equation for the proton/neutron system is (see Problem 5.1):

$$-\frac{\hbar^2}{2\mu}\nabla^2\psi(\mathbf{r}) + V(r)\psi(\mathbf{r}) = E\psi(\mathbf{r}), \tag{8.84}$$

where μ is the reduced mass (the proton and neutron have almost identical masses, so call them both m), and $\mathbf{r}$ is the position of the neutron (say) relative to the proton: $\mathbf{r} = \mathbf{r}_n - \mathbf{r}_p$. Your task is to show that there exists a solution with negative energy (a bound state), using a variational trial wave function of the form

$$\psi_\beta(\mathbf{r}) = Ae^{-\beta r/r_0}. \tag{8.85}$$

---

[23] W.-N. Mei gets a somewhat better bound (and avoids the roof-lines) using

$$\psi(x,y) = Ae^{-\alpha\left(x^2+y^2\right)/a^2}\begin{cases}\left(1-x^2y^2/a^4\right), \\ \left(1-x^2/a^2\right), \\ \left(1-y^2/a^2\right),\end{cases}$$

but the integrals have to be done numerically.

(a) Determine A, by normalizing $\psi_\beta(\mathbf{r})$.

(b) Find the expectation value of the Hamiltonian $\left(H = -\frac{\hbar^2}{2\mu}\nabla^2 + V\right)$ in the state ψ_β. *Answer:*

$$E(\beta) = \frac{\hbar^2}{2\mu r_0^2}\beta^2\left[1 - \frac{4\gamma\beta}{(1+2\beta)^2}\right], \quad \text{where } \gamma \equiv \frac{2\mu r_0^2}{\hbar^2}V_0. \tag{8.86}$$

(c) Optimize your trial wave function, by setting $dE(\beta)/d\beta = 0$. This tells you β as a function of γ (and hence—everything else being constant—of V_0), but let's use it instead to eliminate γ in favor of β:

$$E_{\min} = \frac{\hbar^2}{2\mu r_0^2}\frac{\beta^2(1-2\beta)}{(3+2\beta)}. \tag{8.87}$$

(d) Setting $\hbar^2/2\mu r_0^2 = 1$, plot $E_{\min}$ as a function of β, for $0 \le \beta \le 1$. What does this tell you about the binding of the deuteron? What is the minimum value of V_0 for which you can be confident there is a bound state (look up the necessary masses)? The experimental value is 52 MeV.

Problem 8.29 Existence of Bound States. A potential "well" (in one dimension) is a function $V(x)$ that is never positive ($V(x) \le 0$ for all x), and goes to zero at infinity ($V(x) \to 0$ as $x \to \pm\infty$).[24]

(a) Prove the following **Theorem:** If a potential well $V_1(x)$ supports at least one bound state, then any deeper/wider well ($V_2(x) \le V_1(x)$ for all x) will also support at least one bound state. *Hint:* Use the ground state of V_1, $\psi_1(x)$, as a variational test function.

(b) Prove the following **Corollary:** Every potential well in one dimension has a bound state.[25] *Hint:* Use a finite square well (Section 2.6) for V_1.

(c) Does the Theorem generalize to two and three dimensions? How about the Corollary? *Hint:* You might want to review Problems 4.11 and 4.51.

** 

Problem 8.30 Performing a variational calculation requires finding the minimum of the energy, as a function of the variational parameters. This is, in general, a very hard problem. However, if we choose the form of our trial wave function judiciously, we can develop an efficient algorithm. In particular, suppose we use a *linear* combination of functions $\phi_n(x)$:

$$\psi(x) = \sum_{n=1}^{N} c_n\,\phi_n(x)\,, \tag{8.88}$$

[24] To exclude trivial cases, we also assume it has nonzero area $\left(\int V(x)\,dx \neq 0\right)$. Notice that for the purposes of this problem neither the infinite square well nor the harmonic oscillator is a "potential well," though both of them, of course, have bound states.

[25] K. R. Brownstein, *Am. J. Phys.* **68**, 160 (2000) proves that any one-dimensional potential satisfying $\int_{-\infty}^{\infty} V(x)\,dx \le 0$ admits a bound state (as long as $V(x)$ is not identically zero)—even if it runs positive in some places.

where the c_n are the variational parameters. If the ϕ_n are an orthonormal set ($\langle\phi_m|\phi_n\rangle = \delta_{mn}$), but $\psi(x)$ is not necessarily normalized, then $\langle H\rangle$ is

$$\varepsilon = \frac{\langle\psi \mid H \mid \psi\rangle}{\langle\psi \mid \psi\rangle} = \frac{\sum_{mn} c_m^* H_{mn} c_n}{\sum_n |c_n|^2} \tag{8.89}$$

where $H_{mn} = \langle\phi_m|H|\phi_n\rangle$. Taking the derivative with respect to c_j^* (and setting the result equal to 0) gives[26]

$$\sum_n H_{jn} c_n = \varepsilon c_j , \tag{8.90}$$

recognizable as the jth row in an eigenvalue problem:

$$\begin{pmatrix} H_{11} & H_{12} & \cdots & H_{1N} \\ H_{21} & H_{22} & \cdots & H_{2N} \\ \vdots & \vdots & \ddots & \vdots \\ H_{N1} & H_{N2} & \cdots & H_{NN} \end{pmatrix} \begin{pmatrix} c_1 \\ c_2 \\ \vdots \\ c_N \end{pmatrix} = \varepsilon \begin{pmatrix} c_1 \\ c_2 \\ \vdots \\ c_N \end{pmatrix} . \tag{8.91}$$

The smallest eigenvalue of this matrix H gives a bound on the ground state energy and the corresponding eigenvector determines the best variational wave function of the form 8.88.

(a) Verify Equation 8.90.

(b) Now take the derivative of Equation 8.89 with respect to c_j and show that you get a result redundant with Equation 8.90.

(c) Consider a particle in an infinite square well of width a, with a sloping floor:

$$V(x) = \begin{cases} \infty & x < 0, \\ V_0\, x/a & 0 \le x \le a, \\ \infty & x > a \end{cases} .$$

Using a linear combination of the first ten stationary states of the infinite square well as the basis functions,

$$\phi_n = \sqrt{\frac{2}{a}} \sin\left(\frac{n\pi x}{a}\right) ,$$

determine a bound for the ground state energy in the case $V_0 = 100\,\hbar^2/m\,a^2$. Make a plot of the optimized variational wave function. [*Note*: The exact result is $39.9819\,\hbar^2/m\,a^2$.]

---

[26] Each c_j, being complex, stands for two independent parameters (its real and imaginary parts). One *could* take derivatives with respect to the real and imaginary parts,

$$\frac{\partial}{\partial\,\mathrm{Re}\left[c_j\right]} E = 0 \quad \text{and} \quad \frac{\partial}{\partial\,\mathrm{Im}\left[c_j\right]} E = 0,$$

but it is also legitimate (and simpler) to treat c_j and c_j^* as the independent parameters:

$$\frac{\partial}{\partial c_j} E = 0 \quad \text{and} \quad \frac{\partial}{\partial c_j^*} E = 0 .$$

You get the same result either way.

导读 / Guidance

第 9 章　WKB 近似

WKB 近似是一种近似求解一维定态薛定谔方程的准经典方法，此法对计算束缚态能量和势垒穿透率都是非常有用的。本章首先讨论在经典区域和非经典区域的波函数形式，对非经典区域的讨论给出有效的计算量子隧穿系数的方法；其次引入艾里函数，给出经典和非经典区域的连接公式，为早期量子论中的角动量量子化条件提供了量子力学的依据，指出了它所适用的条件。

习题特色

（1）拓展高能物理中一些衰变现象问题（见例题 9.2），对加莫夫的 α 衰变理论进行讨论，如习题 9.4，要求计算 U^{238} 和 Po^{212} 的寿命等。

（2）联系固体物理的齐纳隧穿效应，讨论齐纳二极管，见习题 9.5。

（3）结合原子物理的斯塔克效应，设计斯塔克效应隧穿问题，计算粒子隧穿所需要的时间，见习题 9.18。

（4）设计一些和经典物理联系的题目。通过计算，加深对什么时候使用经典物理、什么时候使用量子物理处理条件的理解，如习题 9.6、习题 9.7 对地板上一个质量为 *m* 的球的弹性反弹的一些计算；习题 9.19 用量子隧穿讨论在室温下的一罐啤酒大概需要多长时间能自发地倾倒。

9 THE WKB APPROXIMATION

The **WKB** (Wentzel, Kramers, Brillouin)[1] method is a technique for obtaining approximate solutions to the time-independent Schrödinger equation in one dimension (the same basic idea can be applied to many other differential equations, and to the radial part of the Schrödinger equation in three dimensions). It is particularly useful in calculating bound state energies and tunneling rates through potential barriers.

The essential idea is as follows: Imagine a particle of energy E moving through a region where the potential $V(x)$ is *constant*. If $E > V$, the wave function is of the form

$$\psi(x) = Ae^{\pm ikx}, \quad \text{with} \quad k \equiv \sqrt{2m(E - V)}/\hbar.$$

The plus sign indicates that the particle is traveling to the right, and the minus sign means it is going to the left (the general solution, of course, is a linear combination of the two). The wave function is oscillatory, with fixed wavelength ($\lambda = 2\pi/k$) and unchanging amplitude (A). Now suppose that $V(x)$ is *not* constant, but varies rather slowly in comparison to λ, so that over a region containing many full wavelengths the potential is *essentially* constant. Then it is reasonable to suppose that ψ remains *practically* sinusoidal, except that the wavelength and the amplitude change slowly with x. This is the inspiration behind the WKB approximation. In effect, it identifies two different levels of x-dependence: rapid oscillations, *modulated* by gradual variation in amplitude and wavelength.

By the same token, if $E < V$ (and V is constant), then ψ is exponential:

$$\psi(x) = Ae^{\pm\kappa x}, \quad \text{with} \quad \kappa \equiv \sqrt{2m(V - E)}/\hbar.$$

And if $V(x)$ is *not* constant, but varies slowly in comparison with $1/\kappa$, the solution remains *practically* exponential, except that A and κ are now slowly-varying functions of x.

Now, there is one place where this whole program is bound to fail, and that is in the immediate vicinity of a classical **turning point**, where $E \approx V$. For here λ (or $1/\kappa$) goes to infinity, and $V(x)$ can hardly be said to vary "slowly" in comparison. As we shall see, a proper handling of the turning points is the most difficult aspect of the WKB approximation, though the final results are simple to state and easy to implement.

9.1 THE "CLASSICAL" REGION

The Schrödinger equation,

$$-\frac{\hbar^2}{2m}\frac{d^2\psi}{dx^2} + V(x)\psi = E\psi,$$

[1] In Holland it's KWB, in France it's BWK, and in England it's JWKB (for Jeffreys).

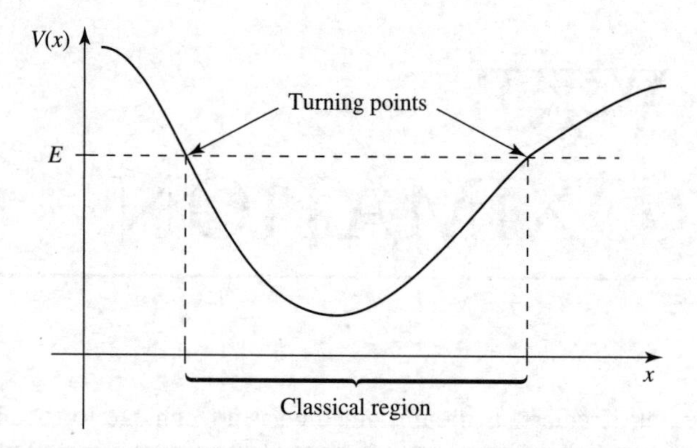

Figure 9.1: Classically, the particle is confined to the region where $E \geq V(x)$.

can be rewritten in the following way:

$$\frac{d^2\psi}{dx^2} = -\frac{p^2}{\hbar^2}\psi, \tag{9.1}$$

where

$$p(x) \equiv \sqrt{2m\left[E - V(x)\right]} \tag{9.2}$$

is the classical formula for the (magnitude of the) momentum of a particle with total energy E and potential energy $V(x)$. For the moment, I'll assume that $E > V(x)$, so that $p(x)$ is *real*; we call this the "classical" region, for obvious reasons—classically the particle is *confined* to this range of x (see Figure 9.1). In general, ψ is some complex function; we can express it in terms of its *amplitude*, $A(x)$, and its *phase*, $\phi(x)$—both of which are *real*:

$$\psi(x) = A(x)e^{i\phi(x)}. \tag{9.3}$$

Using a prime to denote the derivative with respect to x,

$$\frac{d\psi}{dx} = \left(A' + iA\phi'\right)e^{i\phi},$$

and

$$\frac{d^2\psi}{dx^2} = \left[A'' + 2iA'\phi' + iA\phi'' - A\left(\phi'\right)^2\right]e^{i\phi}. \tag{9.4}$$

Putting this into Equation 9.1:

$$A'' + 2iA'\phi' + iA\phi'' - A\left(\phi'\right)^2 = -\frac{p^2}{\hbar^2}A. \tag{9.5}$$

This is equivalent to two *real* equations, one for the real part and one for the imaginary part:

$$A'' - A\left(\phi'\right)^2 = -\frac{p^2}{\hbar^2}A, \quad \text{or} \quad A'' = A\left[\left(\phi'\right)^2 - \frac{p^2}{\hbar^2}\right], \tag{9.6}$$

and

$$2A'\phi' + A\phi'' = 0, \quad \text{or} \quad \left(A^2\phi'\right)' = 0. \tag{9.7}$$

Equations 9.6 and 9.7 are entirely equivalent to the original Schrödinger equation. The second one is easily solved:

$$A^2\phi' = C^2, \quad \text{or} \quad A = \frac{C}{\sqrt{|\phi'|}}, \tag{9.8}$$

where C is a (real) constant. The first one (Equation 9.6) cannot be solved in general—so here comes the approximation: *We assume that the amplitude A varies slowly*, so the A'' term is negligible. (More precisely, we assume that A''/A is much less than both $(\phi')^2$ and $p^2/\hbar^2$.) In that case we can drop the left side of Equation 9.6, and we are left with

$$(\phi')^2 = \frac{p^2}{\hbar^2}, \quad \text{or} \quad \frac{d\phi}{dx} = \pm\frac{p}{\hbar},$$

and therefore

$$\phi(x) = \pm\frac{1}{\hbar}\int p(x)\,dx. \tag{9.9}$$

(I'll write this as an *indefinite* integral, for now—any constant of integration can be absorbed into C, which thereby becomes complex. I'll also absorb a factor of $\sqrt{\hbar}$.) Then

$$\boxed{\psi(x) \approx \frac{C}{\sqrt{p(x)}}e^{\pm\frac{i}{\hbar}\int p(x)\,dx}.} \tag{9.10}$$

Notice that

$$|\psi(x)|^2 \approx \frac{|C|^2}{p(x)}, \tag{9.11}$$

which says that the probability of finding the particle at point x is inversely proportional to its (classical) momentum (and hence its velocity) at that point. This is exactly what you would expect—the particle doesn't spend long in the places where it is moving rapidly, so the probability of getting caught there is small. In fact, the WKB approximation is sometimes *derived* by starting with this "semi-classical" observation, instead of by dropping the A'' term in the differential equation. The latter approach is cleaner mathematically, but the former offers a more illuminating physical rationale. The general (approximate) solution, of course, will be a linear combination the two solutions in Equation 9.10, one with each sign.

Example 9.1

Potential well with two vertical walls. Suppose we have an infinite square well with a bumpy bottom (Figure 9.2):

$$V(x) = \begin{cases} \text{some specified function}, & (0 < x < a), \\ \infty, & \text{(otherwise)}. \end{cases} \tag{9.12}$$

Inside the well (assuming $E > V(x)$ throughout) we have

$$\psi(x) \approx \frac{1}{\sqrt{p(x)}}\left[C_+e^{i\phi(x)} + C_-e^{-i\phi(x)}\right], \tag{9.13}$$

or, more conveniently,

$$\psi(x) \approx \frac{1}{\sqrt{p(x)}}\left[C_1\sin\phi(x) + C_2\cos\phi(x)\right], \tag{9.14}$$

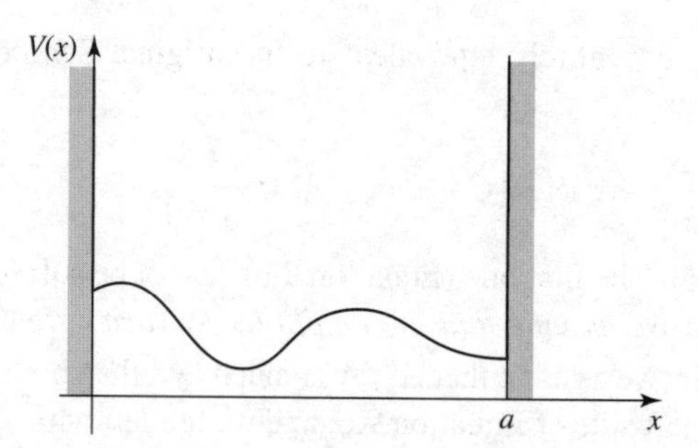

Figure 9.2: Infinite square well with a bumpy bottom.

where (exploiting the freedom noted earlier to impose a convenient lower limit on the integral)[2]

$$\phi(x) = \frac{1}{\hbar}\int_0^x p(x')\,dx'. \tag{9.15}$$

Now, $\psi(x)$ must go to zero at $x = 0$, and therefore (since $\phi(0) = 0$) $C_2 = 0$. Also, $\psi(x)$ goes to zero at $x = a$, so

$$\phi(a) = n\pi \quad (n = 1, 2, 3, \ldots). \tag{9.16}$$

Conclusion:

$$\boxed{\int_0^a p(x)\,dx = n\pi\hbar.} \tag{9.17}$$

This quantization condition determines the (approximate) allowed energies.

For instance, if the well has a *flat* bottom ($V(x) = 0$), then $p(x) = \sqrt{2mE}$ (a constant), and Equation 9.17 says $pa = n\pi\hbar$, or

$$E_n = \frac{n^2\pi^2\hbar^2}{2ma^2},$$

which is the old formula for the energy levels of the infinite square well (Equation 2.30). In this case the WKB approximation yields the *exact* answer (the amplitude of the true wave function is *constant*, so dropping A'' cost us nothing).

* **Problem 9.1** Use the WKB approximation to find the allowed energies (E_n) of an infinite square well with a "shelf," of height V_0, extending half-way across (Figure 7.3):

$$V(x) = \begin{cases} V_0, & (0 < x < a/2), \\ 0, & (a/2 < x < a), \\ \infty, & \text{(otherwise).} \end{cases}$$

[2] We might as well take the positive sign, since both are covered by Equation 9.13.

Express your answer in terms of V_0 and $E_n^0 \equiv (n\pi\hbar)^2/2ma^2$ (the nth allowed energy for the infinite square well with *no* shelf). Assume that $E_1^0 > V_0$, but do *not* assume that $E_n \gg V_0$. Compare your result with what we got in Section 7.1.2, using first-order perturbation theory. Note that they are in agreement if either V_0 is very small (the perturbation theory regime) or n is very large (the WKB—semi-classical—regime).

** **Problem 9.2** An alternative derivation of the WKB formula (Equation 9.10) is based on an expansion in powers of $\hbar$. Motivated by the free-particle wave function, $\psi = A\exp(\pm ipx/\hbar)$, we write

$$\psi(x) = e^{if(x)/\hbar},$$

where $f(x)$ is some complex function. (Note that there is no loss of generality here—*any* nonzero function can be written in this way.)

(a) Put this into Schrödinger's equation (in the form of Equation 9.1), and show that

$$i\hbar f'' - (f')^2 + p^2 = 0.$$

(b) Write $f(x)$ as a power series in $\hbar$:

$$f(x) = f_0(x) + \hbar f_1(x) + \hbar^2 f_2(x) + \cdots,$$

and, collecting like powers of $\hbar$, show that

$$(f_0')^2 = p^2, \quad if_0'' = 2f_0'f_1', \quad if_1'' = 2f_0'f_2' + (f_1')^2, \quad \text{etc.}$$

(c) Solve for $f_0(x)$ and $f_1(x)$, and show that—to first order in $\hbar$—you recover Equation 9.10.

Note: The logarithm of a negative number is defined by $\ln(-z) = \ln(z) + in\pi$, where n is an odd integer. If this formula is new to you, try exponentiating both sides, and you'll see where it comes from.

9.2 TUNNELING

So far, I have assumed that $E > V$, so $p(x)$ is real. But we can easily write down the corresponding result in the *non*-classical region ($E < V$)—it's the same as before (Equation 9.10), only now $p(x)$ is *imaginary*:[3]

$$\boxed{\psi(x) \approx \frac{C}{\sqrt{|p(x)|}} e^{\pm\frac{1}{\hbar}\int |p(x)|\,dx}.} \tag{9.18}$$

Consider, for example, the problem of scattering from a rectangular barrier with a bumpy top (Figure 9.3). To the left of the barrier ($x < 0$),

$$\psi(x) = Ae^{ikx} + Be^{-ikx}, \tag{9.19}$$

[3] In this case the wave function is *real*, and the analogs to Equations 9.6 and 9.7 do not follow *necessarily* from Equation 9.5, although they are still *sufficient*. If this bothers you, study the alternative derivation in Problem 9.2.

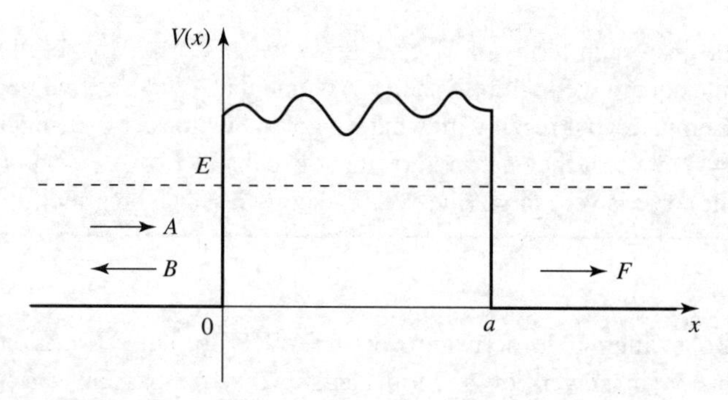

Figure 9.3: Scattering from a rectangular barrier with a bumpy top.

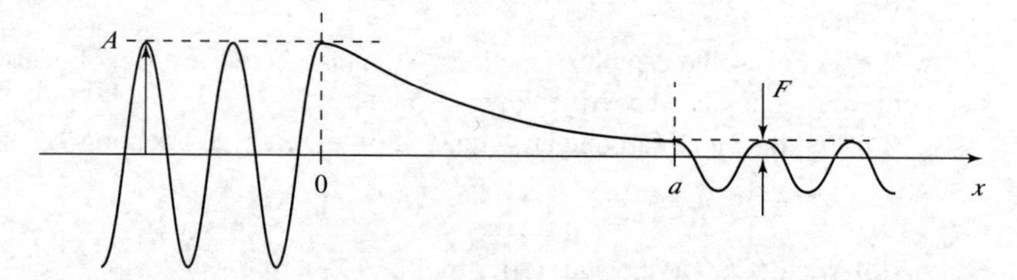

Figure 9.4: Qualitative structure of the wave function, for scattering from a high, broad barrier.

where A is the incident amplitude, B is the reflected amplitude, and $k \equiv \sqrt{2mE}/\hbar$ (see Section 2.5). To the right of the barrier ($x > a$),

$$\psi(x) = Fe^{ikx}, \tag{9.20}$$

where F is the transmitted amplitude. The transmission probability is

$$T = \frac{|F|^2}{|A|^2}. \tag{9.21}$$

In the tunneling region ($0 \leq x \leq a$), the WKB approximation gives

$$\psi(x) \approx \frac{C}{\sqrt{|p(x)|}} e^{\frac{1}{\hbar}\int_0^x |p(x')|\,dx'} + \frac{D}{\sqrt{|p(x)|}} e^{-\frac{1}{\hbar}\int_0^x |p(x')|\,dx'}. \tag{9.22}$$

If the barrier is very high and/or very wide (which is to say, if the probability of tunneling is small), then the coefficient of the exponentially *increasing* term (C) must be small (in fact, it would be *zero* if the barrier were *infinitely* broad), and the wave function looks something like[4] Figure 9.4. The relative amplitudes of the incident and transmitted waves are determined essentially by the total decrease of the exponential over the nonclassical region:

$$\frac{|F|}{|A|} \sim e^{-\frac{1}{\hbar}\int_0^a |p(x')|\,dx'},$$

so

$$\boxed{T \sim e^{-2\gamma}, \quad \text{with} \quad \gamma \equiv \frac{1}{\hbar}\int_0^a |p(x)|\,dx.} \tag{9.23}$$

[4] This heuristic argument can be made more rigorous—see Problem 9.11.

Example 9.2

Gamow's theory of alpha decay.[5] In 1928, George Gamow (and, independently, Condon and Gurney) used Equation 9.23 to provide the first successful explanation of alpha decay (the spontaneous emission of an alpha particle—two protons and two neutrons—by certain radioactive nuclei).[6] Since the alpha particle carries a positive charge ($2e$), it will be electrically repelled by the leftover nucleus (charge Ze), as soon as it gets far enough away to escape the nuclear binding force. But first it has to negotiate a potential barrier that was already known (in the case of uranium) to be more than twice the energy of the emitted alpha particle. Gamow approximated the potential energy by a finite square well (representing the attractive nuclear force), extending out to r_1 (the radius of the nucleus), joined to a repulsive Coulombic tail (Figure 9.5), and identified the escape mechanism as quantum tunneling (this was, by the way, the first time that quantum mechanics had been applied to nuclear physics).

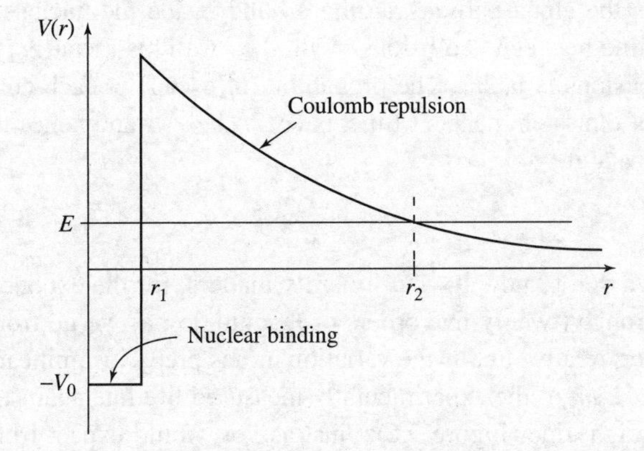

Figure 9.5: Gamow's model for the potential energy of an alpha particle in a radioactive nucleus.

If E is the energy of the emitted alpha particle, the outer turning point (r_2) is determined by

$$\frac{1}{4\pi\epsilon_0}\frac{2Ze^2}{r_2} = E. \tag{9.24}$$

The exponent γ (Equation 9.23) is evidently[7]

$$\gamma = \frac{1}{\hbar}\int_{r_1}^{r_2}\sqrt{2m\left(\frac{1}{4\pi\epsilon_0}\frac{2Ze^2}{r} - E\right)}\,dr = \frac{\sqrt{2mE}}{\hbar}\int_{r_1}^{r_2}\sqrt{\frac{r_2}{r} - 1}\,dr.$$

The integral can be done by substitution $\left(\text{let} r \equiv r_2 \sin^2 u\right)$, and the result is

$$\gamma = \frac{\sqrt{2mE}}{\hbar}\left[r_2\left(\frac{\pi}{2} - \sin^{-1}\sqrt{\frac{r_1}{r_2}}\right) - \sqrt{r_1(r_2 - r_1)}\right]. \tag{9.25}$$

[5] For a more complete discussion, and alternative formulations, see B. R. Holstein, *Am. J. Phys.* **64**, 1061 (1996).

[6] For an interesting brief history see E. Merzbacher, "The Early History of Quantum Tunneling," *Physics Today*, August 2002, p. 44.

[7] In this case the potential does not drop to zero on the left side of the barrier (moreover, this is really a three-dimensional problem), but the essential idea, contained in Equation 9.23, is all we really need.

Typically, $r_1 \ll r_2$, and we can simplify this result using the small angle approximation ($\sin\epsilon \approx \epsilon$):

$$\gamma \approx \frac{\sqrt{2mE}}{\hbar}\left[\frac{\pi}{2}r_2 - 2\sqrt{r_1 r_2}\right] = K_1\frac{Z}{\sqrt{E}} - K_2\sqrt{Z r_1}, \tag{9.26}$$

where

$$K_1 \equiv \left(\frac{e^2}{4\pi\epsilon_0}\right)\frac{\pi\sqrt{2m}}{\hbar} = 1.980\,\mathrm{MeV}^{1/2}, \tag{9.27}$$

and

$$K_2 \equiv \left(\frac{e^2}{4\pi\epsilon_0}\right)^{1/2}\frac{4\sqrt{m}}{\hbar} = 1.485\,\mathrm{fm}^{-1/2}. \tag{9.28}$$

(One fermi (fm) is 10^{-15} m, which is about the size of a typical nucleus.)

If we imagine the alpha particle rattling around inside the nucleus, with an average velocity v, the time between "collisions" with the "wall" is about $2r_1/v$, and hence the *frequency* of collisions is $v/2r_1$. The probability of escape at each collision is $e^{-2\gamma}$, so the probability of emission, per unit time, is $(v/2r_1)\,e^{-2\gamma}$, and hence the **lifetime** of the parent nucleus is about

$$\tau = \frac{2r_1}{v}e^{2\gamma}. \tag{9.29}$$

Unfortunately, we don't know v—but it hardly matters, for the exponential factor varies over a *fantastic* range (twenty-five orders of magnitude), as we go from one radioactive nucleus to another; relative to this the variation in v is pretty insignificant. In particular, if you plot the *logarithm* of the experimentally measured lifetime against $1/\sqrt{E}$, the result is a beautiful straight line (Figure 9.6),[8] just as you would expect from Equations 9.26 and 9.29.

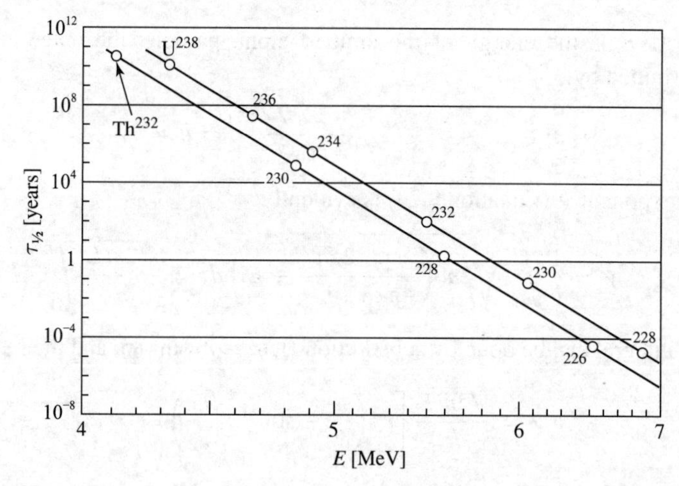

Figure 9.6: Graph of the (base 10) logarithm of the half-life $\left(\tau_{1/2} = \tau \ln 2\right)$ versus $1/\sqrt{E}$ (where E is the energy of the emitted alpha particle), for isotopes of uranium and thorium.

[8] This figure is reprinted by permission from David Park, *Introduction to the Quantum Theory*, 3rd edn, Dover Publications, New York (2005); it was adapted from I. Perlman and J. O. Rasmussen, "Alpha Radioactivity," *Encyclopedia of Physics*, Vol. **42**, Springer (1957). It is known in the literature as a **Geiger-Nuttall** plot (note that E increases to the right, so $1/\sqrt{E}$ increases to the left).

* **Problem 9.3** Use Equation 9.23 to calculate the approximate transmission probability for a particle of energy E that encounters a finite square barrier of height $V_0 > E$ and width $2a$. Compare your answer with the exact result (Problem 2.33), to which it should reduce in the WKB regime $T \ll 1$.

** **Problem 9.4** Calculate the lifetimes of U^{238} and Po^{212}, using Equations 9.26 and 9.29. *Hint*: The density of nuclear matter is relatively constant (i.e. the same for all nuclei), so $(r_1)^3$ is proportional to A (the number of neutrons plus protons). Empirically,

$$r_1 \approx (1.25\,\text{fm})\,A^{1/3}. \tag{9.30}$$

The energy of the emitted alpha particle can be deduced by using Einstein's formula $(E = mc^2)$:

$$E = m_p c^2 - m_d c^2 - m_\alpha c^2, \tag{9.31}$$

where m_p is the mass of the parent nucleus, m_d is the mass of the daughter nucleus, and m_α is the mass of the alpha particle (which is to say, the He^4 nucleus). To figure out what the daughter nucleus is, note that the alpha particle carries off two protons and two neutrons, so Z decreases by 2 and A by 4. Look up the relevant nuclear masses. To estimate v, use $E = (1/2)\,m_\alpha v^2$; this ignores the (negative) potential energy inside the nucleus, and surely *underestimates* v, but it's about the best we can do at this stage. Incidentally, the experimental lifetimes are 6×10^9 yrs and $0.5\,\mu\text{s}$, respectively.

Problem 9.5 Zener Tunneling. In a semiconductor, an electric field (if it's large enough) can produce transitions between energy bands—a phenomenon known as Zener tunneling. A uniform electric field $\mathbf{E} = -E_0\,\hat{\imath}$, for which

$$H' = -e\,E_0\,x\,,$$

makes the energy bands position dependent, as shown in Figure 9.7. It is then possible for an electron to tunnel from the valence (lower) band to the conduction (upper) band; this phenomenon is the basis for the **Zener diode**. Treating the gap as a potential barrier through which the electron may tunnel, find the tunneling probability in terms of E_g and E_0 (as well as m, $\hbar$, e).

9.3 THE CONNECTION FORMULAS

In the discussion so far I have assumed that the "walls" of the potential well (or the barrier) are *vertical*, so that the "exterior" solution is simple, and the boundary conditions trivial. As it turns out, our main results (Equations 9.17 and 9.23) are reasonably accurate even when the edges are not so abrupt (indeed, in Gamow's theory they were applied to just such a case). Nevertheless, it is of some interest to study more closely what happens to the wave function at a turning point ($E = V$), where the "classical" region joins the "nonclassical" region, and

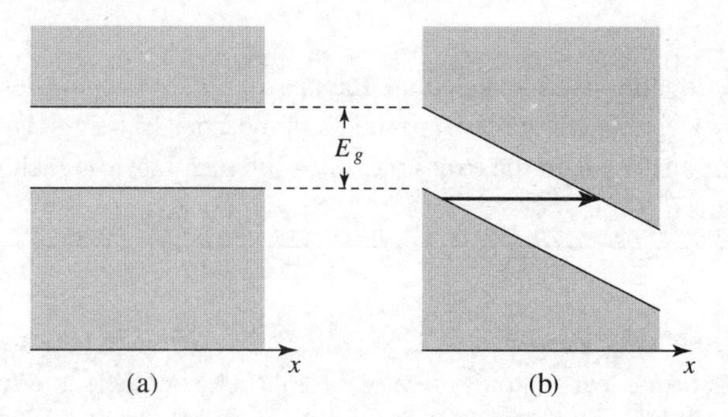

Figure 9.7: (a) The energy bands in the absence of an electric field. (b) In the presence of an electric field an electron can tunnel between the energy bands.

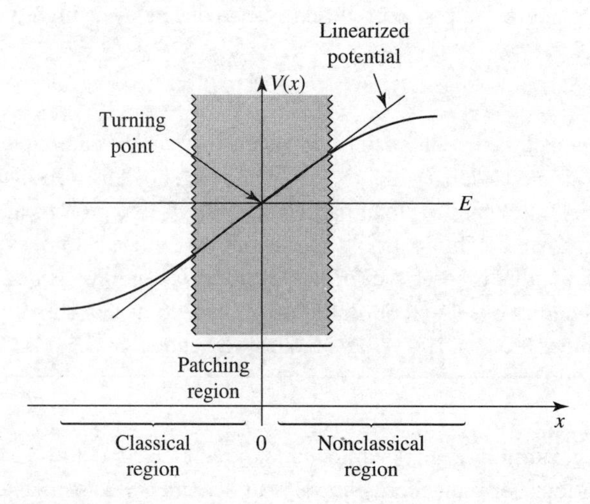

Figure 9.8: Enlarged view of the right-hand turning point.

the WKB approximation itself breaks down. In this section I'll treat the bound state problem (Figure 9.1); you get to do the scattering problem for yourself (Problem 9.11).[9]

For simplicity, let's shift the axes over so that the right hand turning point occurs at $x = 0$ (Figure 9.8). In the WKB approximation, we have

$$\psi(x) \approx \begin{cases} \dfrac{1}{\sqrt{p(x)}}\left[Be^{\frac{i}{\hbar}\int_x^0 p(x')\,dx'} + Ce^{-\frac{i}{\hbar}\int_x^0 p(x')\,dx'}\right], & x < 0, \\ \dfrac{1}{\sqrt{|p(x)|}}De^{-\frac{1}{\hbar}\int_0^x |p(x')|\,dx'}, & x > 0. \end{cases} \tag{9.32}$$

(Assuming $V(x)$ remains greater than E for *all* $x > 0$, we can exclude the positive exponent in this region, because it blows up as $x \to \infty$.) Our task is to join the two solutions at the boundary. But there is a serious difficulty here: In the WKB approximation, ψ goes to *infinity* at the turning point (where $p(x) \to 0$). The *true* wave function, of course, has no such wild behavior—as anticipated, the WKB method simply fails in the vicinity of a turning point. And yet, it is precisely the boundary conditions at the turning points that determine the

[9] *Warning:* The following argument is quite technical, and you may wish to skip it on a first reading.

allowed energies. What we need to do, then, is *splice* the two WKB solutions together, using a "patching" wave function that straddles the turning point.

Since we only need the patching wave function (ψ_p) in the neighborhood of the origin, we'll *approximate the potential by a straight line*:

$$V(x) \approx E + V'(0)x, \tag{9.33}$$

and solve the Schrödinger equation for this linearized V:

$$-\frac{\hbar^2}{2m}\frac{d^2\psi_p}{dx^2} + (E + V'(0)x)\psi_p = E\psi_p,$$

or

$$\frac{d^2\psi_p}{dx^2} = \alpha^3 x\psi_p, \tag{9.34}$$

where

$$\alpha \equiv \left[\frac{2m}{\hbar^2}V'(0)\right]^{1/3}. \tag{9.35}$$

The αs can be absorbed into the independent variable by defining

$$z \equiv \alpha x, \tag{9.36}$$

so that

$$\frac{d^2\psi_p}{dz^2} = z\psi_p. \tag{9.37}$$

This is **Airy's equation**, and the solutions are called **Airy functions**.[10] Since the Airy equation is a second-order differential equation, there are two linearly independent Airy functions, $\mathrm{Ai}(z)$ and $\mathrm{Bi}(z)$. They are related to Bessel functions of order 1/3; some of their properties are listed in Table 9.1 and they are plotted in Figure 9.9. Evidently the patching wave function is a linear combination of $\mathrm{Ai}(z)$ and $\mathrm{Bi}(z)$:

$$\psi_p(x) = a\mathrm{Ai}(\alpha x) + b\mathrm{Bi}(\alpha x), \tag{9.38}$$

for appropriate constants a and b.

Now, ψ_p is the (approximate) wave function in the neighborhood of the origin; our job is to match it to the WKB solutions in the overlap regions on either side (see Figure 9.10). These overlap zones are close enough to the turning point that the linearized potential is reasonably accurate (so that ψ_p is a good approximation to the true wave function), and yet far enough away from the turning point that the WKB approximation is reliable.[11] In the overlap regions Equation 9.33 holds, and therefore (in the notation of Equation 9.35)

$$p(x) \approx \sqrt{2m(E - E - V'(0)x)} = \hbar\alpha^{3/2}\sqrt{-x}. \tag{9.39}$$

[10] *Classically*, a linear potential means a constant force, and hence a constant acceleration—the simplest nontrivial motion possible, and the *starting* point for elementary mechanics. It is ironic that the same potential in *quantum* mechanics yields stationary states that are unfamiliar transcendental functions, and plays only a peripheral role in the theory. Still, wave *packets* can be reasonably simple—see Problem 2.51 and especially footnote 61, page 81.

[11] This is a delicate double constraint, and it is possible to concoct potentials so pathological that no such overlap region exists. However, in practical applications this seldom occurs. See Problem 9.9.

Table 9.1: *Some properties of the Airy functions.*

Differential Equation: $\dfrac{d^2y}{dz^2} = zy.$

Solutions: Linear combinations of Airy functions, Ai(z) and Bi(z).

Integral Representation:

$$\text{Ai}(z) = \frac{1}{\pi}\int_0^\infty \cos\left(\frac{s^3}{3} + sz\right) ds,$$

$$\text{Bi}(z) = \frac{1}{\pi}\int_0^\infty \left[e^{-\frac{s^3}{3} + sz} + \sin\left(\frac{s^3}{3} + sz\right)\right] ds.$$

Asymptotic Forms:

$$\left.\begin{aligned} \text{Ai}(z) &\sim \frac{1}{2\sqrt{\pi}z^{1/4}} e^{-\frac{2}{3}z^{3/2}} \\ \text{Bi}(z) &\sim \frac{1}{\sqrt{\pi}z^{1/4}} e^{\frac{2}{3}z^{3/2}} \end{aligned}\right\} z \gg 0; \qquad \left.\begin{aligned} \text{Ai}(z) &\sim \frac{1}{\sqrt{\pi}(-z)^{1/4}} \sin\left[\frac{2}{3}(-z)^{3/2} + \frac{\pi}{4}\right] \\ \text{Bi}(z) &\sim \frac{1}{\sqrt{\pi}(-z)^{1/4}} \cos\left[\frac{2}{3}(-z)^{3/2} + \frac{\pi}{4}\right] \end{aligned}\right\} z \ll 0.$$

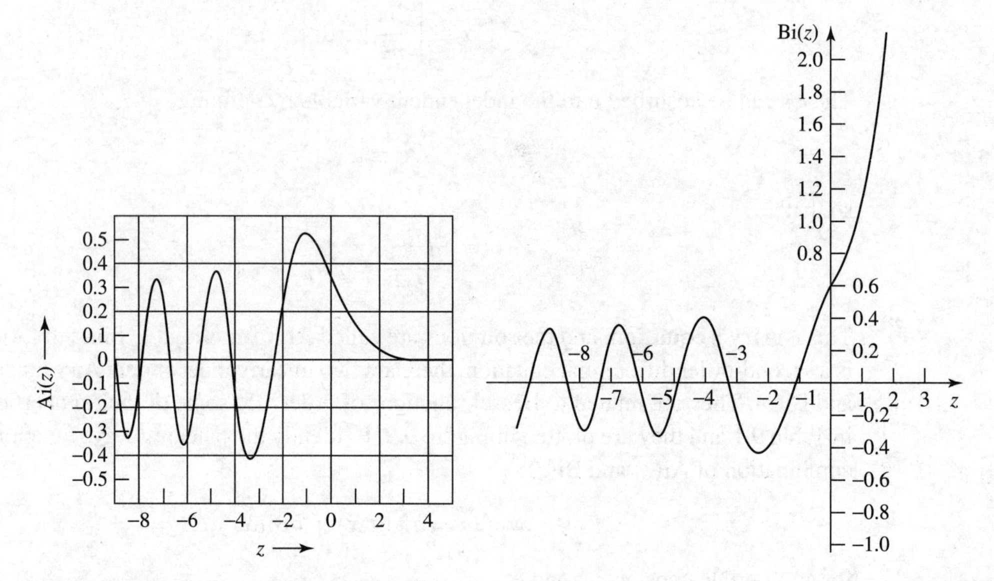

Figure 9.9: Graph of the Airy functions.

In particular, in overlap region 2,

$$\int_0^x \left|p\left(x'\right)\right| dx' \approx \hbar\alpha^{3/2} \int_0^x \sqrt{x'}\, dx' = \frac{2}{3}\hbar(\alpha x)^{3/2},$$

and therefore the WKB wave function (Equation 9.32) can be written as

$$\psi(x) \approx \frac{D}{\sqrt{\hbar}\alpha^{3/4} x^{1/4}} e^{-\frac{2}{3}(\alpha x)^{3/2}}. \tag{9.40}$$

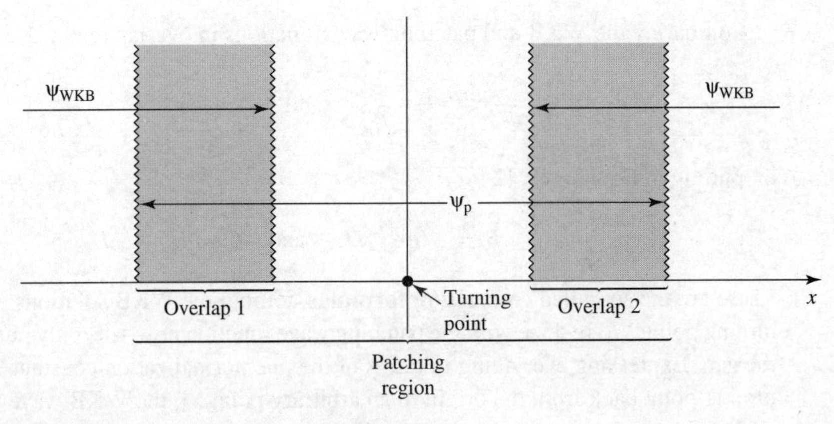

Figure 9.10: Patching region and the two overlap zones.

Meanwhile, using the large-z asymptotic forms[12] of the Airy functions (from Table 9.1), the patching wave function (Equation 9.38) in overlap region 2 becomes

$$\psi_p(x) \approx \frac{a}{2\sqrt{\pi}(\alpha x)^{1/4}} e^{-\frac{2}{3}(\alpha x)^{3/2}} + \frac{b}{\sqrt{\pi}(\alpha x)^{1/4}} e^{\frac{2}{3}(\alpha x)^{3/2}}. \tag{9.41}$$

Comparing the two solutions, we see that

$$a = \sqrt{\frac{4\pi}{\alpha\hbar}} D, \quad \text{and} \quad b = 0. \tag{9.42}$$

Now we go back and repeat the procedure for overlap region 1. Once again, $p(x)$ is given by Equation 9.39, but this time x is *negative*, so

$$\int_x^0 p(x')\, dx' \approx \frac{2}{3}\hbar(-\alpha x)^{3/2} \tag{9.43}$$

and the WKB wave function (Equation 9.32) is

$$\psi(x) \approx \frac{1}{\sqrt{\hbar}\alpha^{3/4}(-x)^{1/4}} \left[B e^{i\frac{2}{3}(-\alpha x)^{3/2}} + C e^{-i\frac{2}{3}(-\alpha x)^{3/2}} \right]. \tag{9.44}$$

Meanwhile, using the asymptotic form of the Airy function for large *negative* z (Table 9.1), the patching function (Equation 9.38, with $b = 0$) reads

$$\begin{aligned} \psi_p(x) &\approx \frac{a}{\sqrt{\pi}(-\alpha x)^{1/4}} \sin\left[\frac{2}{3}(-\alpha x)^{3/2} + \frac{\pi}{4}\right] \\ &= \frac{a}{\sqrt{\pi}(-\alpha x)^{1/4}} \frac{1}{2i} \left[e^{i\pi/4} e^{i\frac{2}{3}(-\alpha x)^{3/2}} - e^{-i\pi/4} e^{-i\frac{2}{3}(-\alpha x)^{3/2}} \right]. \end{aligned} \tag{9.45}$$

[12] At first glance it seems absurd to use a *large-z* approximation in this region, which after all is supposed to be reasonably close to the turning point at $z = 0$ (so that the linear approximation to the potential is valid). But notice that the argument here is αx, and if you study the matter carefully (see Problem 9.9) you will find that there *is* (typically) a region in which αx is large, but at the same time it is reasonable to approximate $V(x)$ by a straight line. Indeed, the asymptotic forms of the Airy functions *are* precisely the WKB solutions to Airy's equation, and since we are already using $\psi_{\rm WKB}$ in the overlap region (Figure 9.10) it is not really a *new* approximation to do the same for ψ_p.

Comparing the WKB and patching wave functions in overlap region 1, we find

$$\frac{a}{2i\sqrt{\pi}}e^{i\pi/4} = \frac{B}{\sqrt{\hbar\alpha}} \quad \text{and} \quad \frac{-a}{2i\sqrt{\pi}}e^{-i\pi/4} = \frac{C}{\sqrt{\hbar\alpha}},$$

or, putting in Equation 9.42 for a:

$$B = -ie^{i\pi/4}D, \quad \text{and} \quad C = ie^{-i\pi/4}D. \tag{9.46}$$

These are the so-called **connection formulas**, joining the WKB solutions at either side of the turning point. We're done with the patching wave function now—its only purpose was to bridge the gap. Expressing everything in terms of the one normalization constant D, and shifting the turning point back from the origin to an arbitrary point x_2, the WKB wave function (Equation 9.32) becomes

$$\psi(x) \approx \begin{cases} \frac{2D}{\sqrt{p(x)}} \sin\left[\frac{1}{\hbar}\int_x^{x_2} p(x')\,dx' + \frac{\pi}{4}\right], & x < x_2; \\ \frac{D}{\sqrt{|p(x)|}} \exp\left[-\frac{1}{\hbar}\int_{x_2}^{x} |p(x')|\,dx'\right], & x > x_2. \end{cases} \tag{9.47}$$

Example 9.3

Potential well with one vertical wall. Imagine a potential well that has one vertical side (at $x = 0$) and one sloping side (Figure 9.11). In this case $\psi(0) = 0$, so Equation 9.47 says

$$\frac{1}{\hbar}\int_0^{x_2} p(x)\,dx + \frac{\pi}{4} = n\pi, \quad (n = 1, 2, 3, \ldots),$$

or

$$\boxed{\int_0^{x_2} p(x)\,dx = \left(n - \frac{1}{4}\right)\pi\hbar.} \tag{9.48}$$

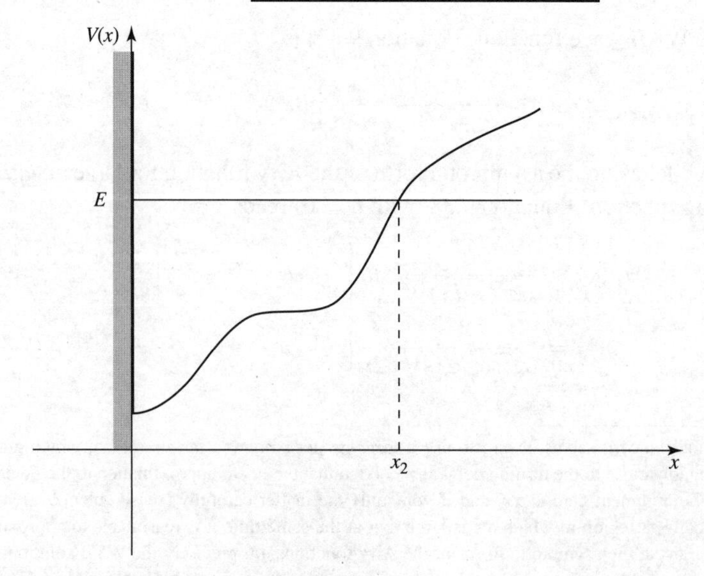

Figure 9.11: Potential well with one vertical wall.

For instance, consider the "half-harmonic oscillator",

$$V(x) = \begin{cases} \frac{1}{2}m\omega^2 x^2, & (x > 0), \\ \infty, & (\text{otherwise}). \end{cases} \tag{9.49}$$

In this case

$$p(x) = \sqrt{2m\left[E - (1/2)m\omega^2 x^2\right]} = m\omega\sqrt{x_2^2 - x^2},$$

where

$$x_2 = \frac{1}{\omega}\sqrt{\frac{2E}{m}}$$

is the turning point. So

$$\int_0^{x_2} p(x)\,dx = m\omega \int_0^{x_2} \sqrt{x_2^2 - x^2}\,dx = \frac{\pi}{4} m\omega x_2^2 = \frac{\pi E}{2\omega},$$

and the quantization condition (Equation 9.48) yields

$$E_n = \left(2n - \frac{1}{2}\right)\hbar\omega = \left(\frac{3}{2}, \frac{7}{2}, \frac{11}{2}, \ldots\right)\hbar\omega. \tag{9.50}$$

In this particular case the WKB approximation actually delivers the *exact* allowed energies (which are precisely the *odd* energies of the *full* harmonic oscillator—see Problem 2.41).

Example 9.4

Potential well with no vertical walls. Equation 9.47 connects the WKB wave functions at a turning point where the potential slopes *upward* (Figure 9.12(a)); the same reasoning, applied to a *downward*-sloping turning point (Figure 9.12(b)), yields (Problem 9.10)

$$\psi(x) \approx \begin{cases} \frac{D'}{\sqrt{|p(x)|}} \exp\left[-\frac{1}{\hbar}\int_x^{x_1} \left|p\left(x'\right)\right| dx'\right], & x < x_1; \\ \frac{2D'}{\sqrt{p(x)}} \sin\left[\frac{1}{\hbar}\int_{x_1}^{x} p\left(x'\right) dx' + \frac{\pi}{4}\right], & x > x_1. \end{cases} \tag{9.51}$$

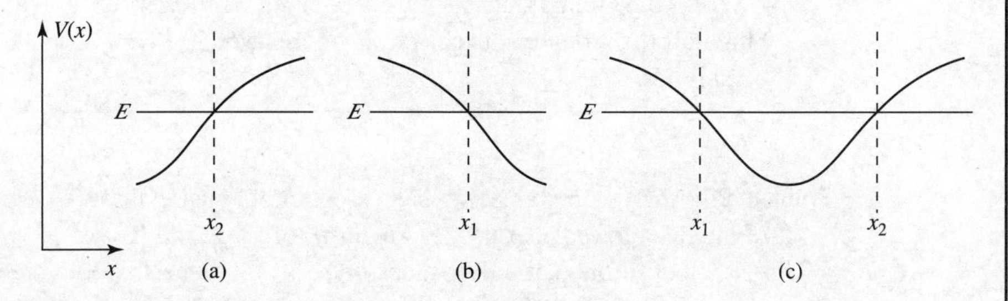

Figure 9.12: Upward-sloping and downward-sloping turning points.

In particular, if we're talking about a potential *well* (Figure 9.12(c)), the wave function in the "interior" region ($x_1 < x < x_2$) can be written *either* as

$$\psi(x) \approx \frac{2D}{\sqrt{p(x)}} \sin\theta_2(x), \quad \text{where} \quad \theta_2(x) \equiv \frac{1}{\hbar}\int_x^{x_2} p\left(x'\right) dx' + \frac{\pi}{4},$$

(Equation 9.47), *or* as

$$\psi(x) \approx \frac{-2D'}{\sqrt{p(x)}} \sin\theta_1(x), \quad \text{where} \quad \theta_1(x) \equiv -\frac{1}{\hbar}\int_{x_1}^{x} p(x')\,dx' - \frac{\pi}{4},$$

(Equation 9.51). Evidently the arguments of the sine functions must be equal, modulo π:[13] $\theta_2 = \theta_1 + n\pi$, from which it follows that

$$\boxed{\int_{x_1}^{x_2} p(x)\,dx = \left(n - \frac{1}{2}\right)\pi\hbar, \quad \text{with} \quad n = 1, 2, 3, \ldots.} \tag{9.52}$$

This quantization condition determines the allowed energies for the "typical" case of a potential well with two sloping sides. Notice that it differs from the formulas for two vertical walls (Equation 9.17) or one vertical wall (Equation 9.48) only in the number that is subtracted from n (0, 1/4, or 1/2). Since the WKB approximation works best in the semi-classical (large n) regime, the distinction is more in appearance than in substance. In any event, the result is extraordinarily powerful, for it enables us to calculate (approximate) allowed energies *without ever solving the Schrödinger equation*, by simply evaluating one integral. The wave function itself has dropped out of sight.

** **Problem 9.6 The "bouncing ball" revisited.** Consider the quantum mechanical analog to the classical problem of a ball (mass m) bouncing elastically on the floor.[14]

(a) What is the potential energy, as a function of height x above the floor? (For negative x, the potential is *infinite*—the ball can't get there at all.)

(b) Solve the Schrödinger equation for this potential, expressing your answer in terms of the appropriate Airy function (note that $\text{Bi}(z)$ blows up for large z, and must therefore be rejected). Don't bother to normalize $\psi(x)$.

(c) Using $g = 9.80\ \text{m/s}^2$ and $m = 0.100$ kg, find the first four allowed energies, in joules, correct to three significant digits. *Hint:* See Milton Abramowitz and Irene A. Stegun, *Handbook of Mathematical Functions*, Dover, New York (1970), page 478; the notation is defined on page 450.

(d) What is the ground state energy, in eV, of an *electron* in this gravitational field? How high off the ground is this electron, on the average? *Hint:* Use the virial theorem to determine $\langle x\rangle$.

* **Problem 9.7** Analyze the bouncing ball (Problem 9.6) using the WKB approximation.

(a) Find the allowed energies, E_n, in terms of m, g, and $\hbar$.

(b) Now put in the particular values given in Problem 9.6(c), and compare the WKB approximation to the first four energies with the "exact" results.

(c) About how large would the quantum number n have to be to give the ball an average height of, say, 1 meter above the ground?

[13] *Not* 2π—an overall minus sign can be absorbed into the normalization factors D and D'.

[14] For more on the quantum bouncing ball see Problem 2.59, J. Gea-Banacloche, *Am. J. Phys.* **67**, 776 (1999), and N. Wheeler, "Classical/quantum dynamics in a uniform gravitational field", unpublished Reed College report (2002). This may sound like an awfully artificial problem, but the experiment has actually been done, using neutrons (V. V. Nesvizhevsky *et al.*, *Nature* **415**, 297 (2002)).

* **Problem 9.8** Use the WKB approximation to find the allowed energies of the harmonic oscillator.

Problem 9.9 Consider a particle of mass m in the nth stationary state of the harmonic oscillator (angular frequency ω).

(a) Find the turning point, x_2.

(b) How far (d) could you go *above* the turning point before the error in the linearized potential (Equation 9.33, but with the turning point at x_2) reaches 1%? That is, if

$$\frac{V(x_2+d)-V_{\text{lin}}(x_2+d)}{V(x_2)}=0.01,$$

what is d?

(c) The asymptotic form of Ai(z) is accurate to 1% as long as $z \geq 5$. For the d in part (b), determine the smallest n such that $\alpha d \geq 5$. (For any n larger than this there exists an overlap region in which the linearized potential is good to 1% *and* the large-z form of the Airy function is good to 1%.)

** **Problem 9.10** Derive the connection formulas at a downward-sloping turning point, and confirm Equation 9.51.

*** **Problem 9.11** Use appropriate connection formulas to analyze the problem of scattering from a barrier with sloping walls (Figure 9.13). *Hint:* Begin by writing the WKB wave function in the form

$$\psi(x)\approx\begin{cases}\frac{1}{\sqrt{p(x)}}\left[Ae^{-\frac{i}{\hbar}\int_x^{x_1}p(x')\,dx'}+Be^{\frac{i}{\hbar}\int_x^{x_1}p(x')\,dx'}\right], & x<x_1;\\ \frac{1}{\sqrt{|p(x)|}}\left[Ce^{\frac{1}{\hbar}\int_{x_1}^{x}|p(x')|\,dx'}+De^{-\frac{1}{\hbar}\int_{x_1}^{x}|p(x')|\,dx'}\right], & x_1<x<x_2;\\ \frac{1}{\sqrt{p(x)}}\left[Fe^{\frac{i}{\hbar}\int_{x_2}^{x}p(x')\,dx'}\right], & x>x_2.\end{cases} \tag{9.53}$$

Do *not* assume $C=0$. Calculate the tunneling probability, $T=|F|^2/|A|^2$, and show that your result reduces to Equation 9.23 in the case of a broad, high barrier.

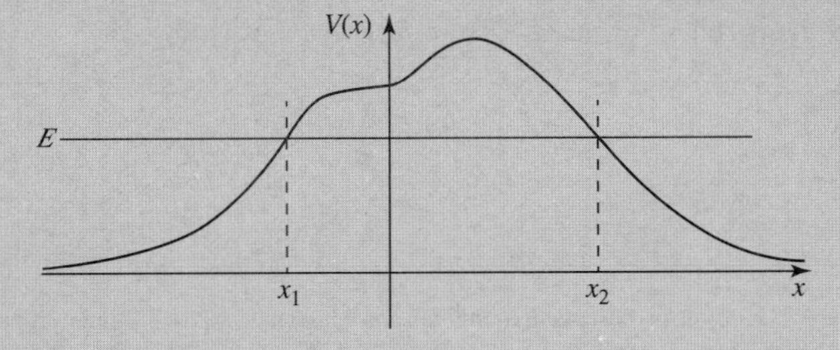

Figure 9.13: Barrier with sloping walls.

** 

Problem 9.12 For the "half-harmonic oscillator" (Example 9.3), make a plot comparing the normalized WKB wave function for $n = 3$ to the exact solution. You'll have to experiment to determine how wide to make the patching region. *Note*: You *can* do the integrals of $p(x)$ by hand, but feel free to do them numerically. You'll need to do the integral of $|\psi_{WKB}|^2$ numerically to normalize the wave function.

FURTHER PROBLEMS ON CHAPTER 9

** **Problem 9.13** Use the WKB approximation to find the allowed energies of the general power-law potential:

$$V(x) = \alpha |x|^{\nu},$$

where ν is a positive number. Check your result for the case $\nu = 2$. *Answer:*[15]

$$E_n = \alpha \left[(n - 1/2)\, \hbar \sqrt{\frac{\pi}{2m\alpha}} \frac{\Gamma\left(\frac{1}{\nu} + \frac{3}{2}\right)}{\Gamma\left(\frac{1}{\nu} + 1\right)} \right]^{\left(\frac{2\nu}{\nu+2}\right)}. \tag{9.54}$$

** **Problem 9.14** Use the WKB approximation to find the bound state energy for the potential in Problem 2.52. Compare the exact answer: $-\left[(9/8) - \left(1/\sqrt{2}\right)\right] \hbar^2 a^2/m$.

Problem 9.15 For spherically symmetrical potentials we can apply the WKB approximation to the radial part (Equation 4.37). In the case $l = 0$ it is reasonable[16] to use Equation 9.48 in the form

$$\int_0^{r_0} p(r)\, dr = (n - 1/4)\, \pi \hbar, \tag{9.55}$$

where r_0 is the turning point (in effect, we treat $r = 0$ as an infinite wall). Exploit this formula to estimate the allowed energies of a particle in the logarithmic potential

$$V(r) = V_0 \ln (r/a)$$

(for constants V_0 and a). Treat only the case $l = 0$. Show that the spacing between the levels is independent of mass. *Partial answer:*

$$E_{n+1} - E_n = V_0 \ln \left(\frac{n + 3/4}{n - 1/4} \right).$$

[15] As always, the WKB result is most accurate in the semi-classical (large n) regime. In particular, Equation 9.54 is not very good for the ground state ($n = 1$). See W. N. Mei, *Am. J. Phys.* **66**, 541 (1998).

[16] Application of the WKB approximation to the radial equation raises some delicate and subtle problems, which I will not go into here. The classic paper on the subject is R. Langer, *Phys. Rev.* **51**, 669 (1937).

** **Problem 9.16** Use the WKB approximation in the form of Equation 9.52,

$$\int_{r_1}^{r_2} p(r)\,dr = \left(n' - 1/2\right)\pi\hbar \tag{9.56}$$

to estimate the bound state energies for hydrogen. Don't forget the centrifugal term in the effective potential (Equation 4.38). The following integral may help:

$$\int_a^b \frac{1}{x}\sqrt{(x-a)(b-x)}\,dx = \frac{\pi}{2}\left(\sqrt{b}-\sqrt{a}\right)^2. \tag{9.57}$$

Answer:

$$E_{n'\ell} \approx \frac{-13.6\text{ eV}}{\left[n' - (1/2) + \sqrt{\ell(\ell+1)}\right]^2}. \tag{9.58}$$

I put a prime on n', because there is no reason to suppose it corresponds to the n in the Bohr formula. Rather, it orders the states *for a given* ℓ, counting the number of nodes in the radial wave function.[17] In the notation of Chapter 4, $n' = N = n - \ell$ (Equation 4.67). Put this in, expand the square root $\left(\sqrt{1+\epsilon} = 1 + \frac{1}{2}\epsilon - \frac{1}{8}\epsilon^2 + \cdots\right)$, and compare your result to the Bohr formula.

*** **Problem 9.17** Consider the case of a symmetrical double well, such as the one pictured in Figure 9.14. We are interested in bound states with $E < V(0)$.

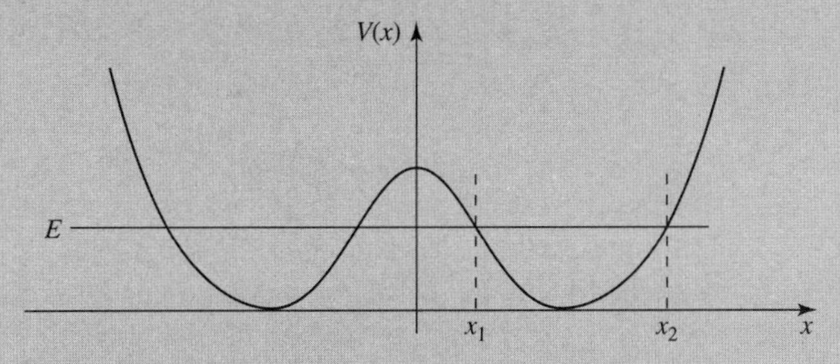

Figure 9.14: Symmetric double well; Problem 9.17.

(a) Write down the WKB wave functions in regions (i) $x > x_2$, (ii) $x_1 < x < x_2$, and (iii) $0 < x < x_1$. Impose the appropriate connection formulas at x_1 and x_2 (this has already been done, in Equation 9.47, for x_2; you will have to work out x_1 for yourself), to show that

$$\psi(x) \approx \begin{cases} \frac{D}{\sqrt{|p(x)|}} \exp\left[-\frac{1}{\hbar}\int_{x_2}^{x} |p(x')|\,dx'\right], & \text{(i)} \\ \frac{2D}{\sqrt{p(x)}} \sin\left[\frac{1}{\hbar}\int_{x}^{x_2} p(x')\,dx' + \frac{\pi}{4}\right], & \text{(ii)} \\ \frac{D}{\sqrt{|p(x)|}} \left[2\cos\theta\, e^{\frac{1}{\hbar}\int_x^{x_1}|p(x')|dx'} + \sin\theta\, e^{-\frac{1}{\hbar}\int_x^{x_1}|p(x')|dx'}\right], & \text{(iii)} \end{cases}$$

where

$$\theta \equiv \frac{1}{\hbar}\int_{x_1}^{x_2} p(x)\,dx. \tag{9.59}$$

---

[17] I thank Ian Gatland and Owen Vajk for pointing this out.

(b) Because $V(x)$ is symmetric, we need only consider even (+) and odd (−) wave functions. In the former case $\psi'(0) = 0$, and in the latter case $\psi(0) = 0$. Show that this leads to the following quantization condition:

$$\tan\theta = \pm 2e^{\phi}, \tag{9.60}$$

where

$$\phi \equiv \frac{1}{\hbar}\int_{-x_1}^{x_1} \left|p(x')\right| dx'. \tag{9.61}$$

Equation 9.60 determines the (approximate) allowed energies (note that E comes into x_1 and x_2, so θ and ϕ are both functions of E).

(c) We are particularly interested in a high and/or broad central barrier, in which case ϕ is large, and e^{ϕ} is *huge*. Equation 9.60 then tells us that θ must be very close to a half-integer multiple of π. With this in mind, write $\theta = (n + 1/2)\pi + \epsilon$, where $|\epsilon| \ll 1$, and show that the quantization condition becomes

$$\theta \approx \left(n + \frac{1}{2}\right)\pi \mp \frac{1}{2}e^{-\phi}. \tag{9.62}$$

(d) Suppose each well is a parabola:[18]

$$V(x) = \begin{cases} \frac{1}{2}m\omega^2 (x+a)^2, & x < 0, \\ \frac{1}{2}m\omega^2 (x-a)^2, & x > 0. \end{cases} \tag{9.63}$$

Sketch this potential, find θ (Equation 9.59), and show that

$$E_n^{\pm} \approx \left(n + \frac{1}{2}\right)\hbar\omega \mp \frac{\hbar\omega}{2\pi}e^{-\phi}. \tag{9.64}$$

Comment: If the central barrier were *impenetrable* ($\phi \to \infty$), we would simply have two detached harmonic oscillators, and the energies, $E_n = (n + 1/2)\hbar\omega$, would be doubly degenerate, since the particle could be in the left well or in the right one. When the barrier becomes *finite* (putting the two wells into "communication"), the degeneracy is lifted. The even states $\left(\psi_n^+\right)$ have slightly *lower* energy, and the odd ones $\left(\psi_n^-\right)$ have slightly higher energy.

(e) Suppose the particle starts out in the *right* well—or, more precisely, in a state of the form

$$\Psi(x,0) = \frac{1}{\sqrt{2}}\left(\psi_n^+ + \psi_n^-\right),$$

which, assuming the phases are picked in the "natural" way, will be concentrated in the right well. Show that it oscillates back and forth between the wells, with a period

$$\tau = \frac{2\pi^2}{\omega}e^{\phi}. \tag{9.65}$$

[18] Even if $V(x)$ is not strictly parabolic in each well, this calculation of θ, and hence the result (Equation 9.64) will be *approximately* correct, in the sense discussed in Section 2.3, with $\omega \equiv \sqrt{V''(x_0)/m}$, where x_0 is the position of the minimum.

(f) Calculate ϕ, for the specific potential in part (d), and show that for $V(0) \gg E$, $\phi \sim m\omega a^2/\hbar$.

Problem 9.18 Tunneling in the Stark Effect. When you turn on an external electric field, the electron in an atom can, in principle, tunnel out, ionizing the atom. *Question:* Is this likely to happen in a typical Stark effect experiment? We can estimate the probability using a crude one-dimensional model, as follows. Imagine a particle in a very deep finite square well (Section 2.6).

(a) What is the energy of the ground state, measured up from the bottom of the well? Assume $V_0 \gg \hbar^2/ma^2$. *Hint:* This is just the ground state energy of the *infinite* square well (of width $2a$).

(b) Now introduce a perturbation $H' = -\alpha x$ (for an electron in an electric field $\mathbf{E} = -E_{\text{ext}}\hat{\imath}$ we would have $\alpha = eE_{\text{ext}}$). Assume it is relatively weak $\left(\alpha a \ll \hbar^2/ma^2\right)$. Sketch the total potential, and note that the particle can now tunnel out, in the direction of positive x.

(c) Calculate the tunneling factor γ (Equation 9.23), and estimate the time it would take for the particle to escape (Equation 9.29). *Answer:* $\gamma = \sqrt{8mV_0^3}/3\alpha\hbar$, $\tau = \left(8ma^2/\pi\hbar\right)e^{2\gamma}$.

(d) Put in some reasonable numbers: $V_0 = 20$ eV (typical binding energy for an outer electron), $a = 10^{-10}$ m (typical atomic radius), $E_{\text{ext}} = 7 \times 10^6$ V/m (strong laboratory field), e and m the charge and mass of the electron. Calculate τ, and compare it to the age of the universe.

Problem 9.19 About how long would it take for a (full) can of beer at room temperature to topple over spontaneously, as a result of quantum tunneling? *Hint:* Treat it as a uniform cylinder of mass m, radius R, and height h. As the can tips, let x be the height of the center above its equilibrium position ($h/2$). The potential energy is mgx, and it topples when x reaches the critical value $x_0 = \sqrt{R^2 + (h/2)^2} - h/2$. Calculate the tunneling probability (Equation 9.23), for $E = 0$. Use Equation 9.29, with the thermal energy $((1/2)\,mv^2 = (1/2)\,k_B T)$ to estimate the velocity. Put in reasonable numbers, and give your final answer in years.[19]

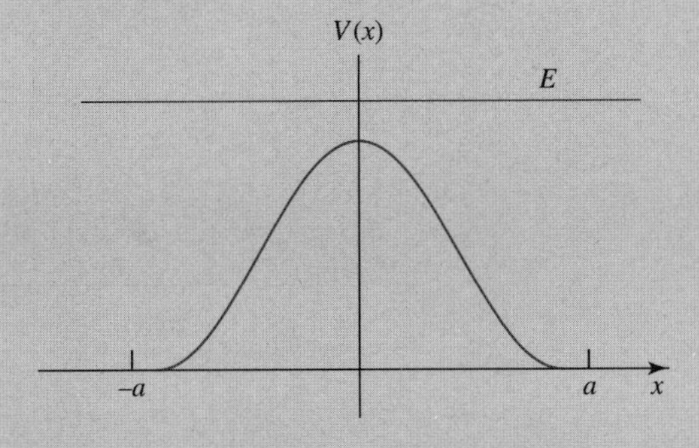

Figure 9.15: Reflection from a barrier (Problem 9.20).

[19] R. E. Crandall, *Scientific American*, February, 1997, p. 74.

Problem 9.20 Equation 9.23 tells us the (approximate) transmission probability for tunneling through a barrier, when $E < V_{\max}$—a classically forbidden process. In this problem we explore the complementary phenomenon: reflection from a barrier when $E > V_{\max}$ (again, a classically forbidden process). We'll assume that $V(x)$ is an even analytic function, that goes to zero as $x \to \pm\infty$ (Figure 9.15). *Question:* What is the analog to Equation 9.23?

(a) Try the obvious approach: assume the potential vanishes for $|x| \geq a$, and use the WKB approximation (Equation 9.13) in the scattering region:

$$\psi(x) \begin{cases} = Ae^{ikx} + Be^{-ikx}, & x < a, \\ \approx \frac{1}{\sqrt{p(x)}}\left[C_+ e^{i\phi(x)} + C_- e^{-i\phi(x)}\right], & -a < x < a, \\ = Ce^{ikx}, & x > a. \end{cases} \tag{9.66}$$

Impose the usual boundary conditions at $\pm a$, and solve for the reflection probability, $R = |B|^2/|A|^2$.

Unfortunately, the result ($R = 0$) is uninformative. It's true that the R is exponentially small (just as the transmission coefficient is, for $E < V_{\max}$), but we've thrown the baby out with the bath water—this approximation is simply too drastic. The correct formula is

$$R = e^{-2\lambda}, \quad \text{where} \quad \lambda \equiv \frac{2}{\hbar}\int_0^{y_0} p(iy)\,dy \tag{9.67}$$

and y_0 is defined by $p(iy_0) = 0$. Notice that λ (like γ in Equation 9.23) goes like $1/\hbar$; it is in fact the leading term in an expansion in powers of $\hbar$: $\lambda = c_1/\hbar + c_2 + c_3\hbar + c_4\hbar^2 + \cdots$. In the classical limit ($\hbar \to 0$), λ and γ go to infinity, so R and T go to zero, as expected. It is not easy to derive Equation 9.67,[20] but let's look at some examples.

(b) Suppose $V(x) = V_0\,\text{sech}^2(x/a)$, for some positive constants V_0 and a. Plot $V(x)$, plot $p(iy)$ for $0 \leq y \leq y_0$, and show that $\lambda = (\pi a/\hbar)\left(\sqrt{2mE} - \sqrt{2mV_0}\right)$. Plot R as a function of E, for fixed V_0.

(c) Suppose $V(x) = V_0/\left[1 + (x/a)^2\right]$. Plot $V(x)$, and express λ in terms of an elliptic integral. Plot R as a function of E.

[20] L. D. Landau and E. M. Lifshitz, *Quantum Mechanics: Non-Relativistic Theory*, Pergamon Press, Oxford (1958), pages 190–191. R. L. Jaffe, *Am. J. Phys.* **78**, 620 (2010) shows that *reflection* (for $E > V_{\max}$) can be regarded as *tunneling* in momentum space, and obtains Equation 9.67 by a clever analog to the argument yielding Equation 9.23.

导读 / Guidance

第 10 章　散射

散射态是一种非束缚态，和处理束缚态不同，处理散射问题所感兴趣的不再是能量本征值，而是散射粒子角分布以及散射过程中各种粒子的性质的变化。本章与前面所讨论的内容都有很大区别，首先以硬球散射为例介绍经典的散射理论，对比讨论量子散射理论，给出散射振幅、微分散射截面的概念。接下来，讨论分波分析和玻恩近似两种计算散射截面的方法。分波分析非常适合处理中心力场作用情况下的低能散射情况，计算时并不是把所有的分波都考虑进去，而是只考虑一些重要的分波。对于弹性散射，可以认为振幅不变，只有相位的改变，并给出相移计算公式。玻恩近似适合处理高能散射的情况，类同束缚态的微扰近似，把入射粒子和靶子的相互作用看成微扰，逐级近似求解，还讨论了一级玻恩近似和玻恩级数。

习题特色

（1）和固体物理联系，拓展晶体中中子散射模型，利用计算机计算散射截面，如习题 10.21。

（2）拓展二维散射理论，如习题 10.22；拓展两个全同粒子的散射理论，如习题 10.23。

10 SCATTERING

10.1 INTRODUCTION

10.1.1 Classical Scattering Theory

Imagine a particle incident on some scattering center (say, a marble bouncing off a bowling ball, or a proton fired at a heavy nucleus). It comes in with energy E and **impact parameter** b, and it emerges at some **scattering angle** θ—see Figure 10.1. (I'll assume for simplicity that the target is symmetrical about the z axis, so the trajectory remains in one plane, and that the target is very heavy, so its recoil is negligible.) The essential problem of classical scattering theory is this: *Given the impact parameter, calculate the scattering angle.* Ordinarily, of course, the smaller the impact parameter, the greater the scattering angle.

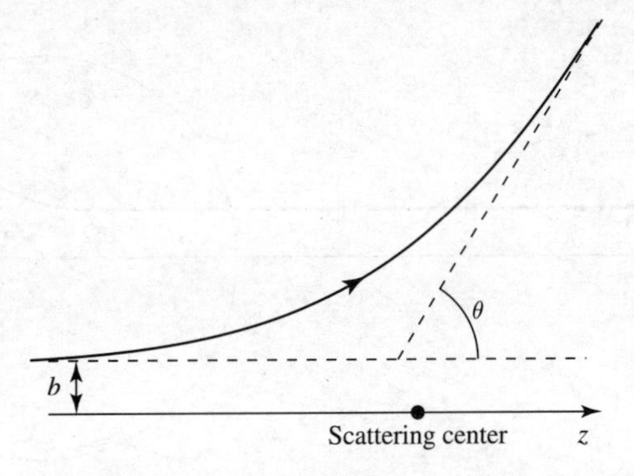

Figure 10.1: The classical scattering problem, showing the impact parameter b and the scattering angle θ.

Example 10.1

Hard-sphere scattering. Suppose the target is a billiard ball, of radius R, and the incident particle is a BB, which bounces off elastically (Figure 10.2). In terms of the angle α, the impact parameter is $b = R\sin\alpha$, and the scattering angle is $\theta = \pi - 2\alpha$, so

$$b = R\sin\left(\frac{\pi}{2} - \frac{\theta}{2}\right) = R\cos\left(\frac{\theta}{2}\right). \tag{10.1}$$

Evidently

$$\theta = \begin{cases} 2\cos^{-1}(b/R), & (b \le R), \\ 0, & (b \ge R). \end{cases} \tag{10.2}$$

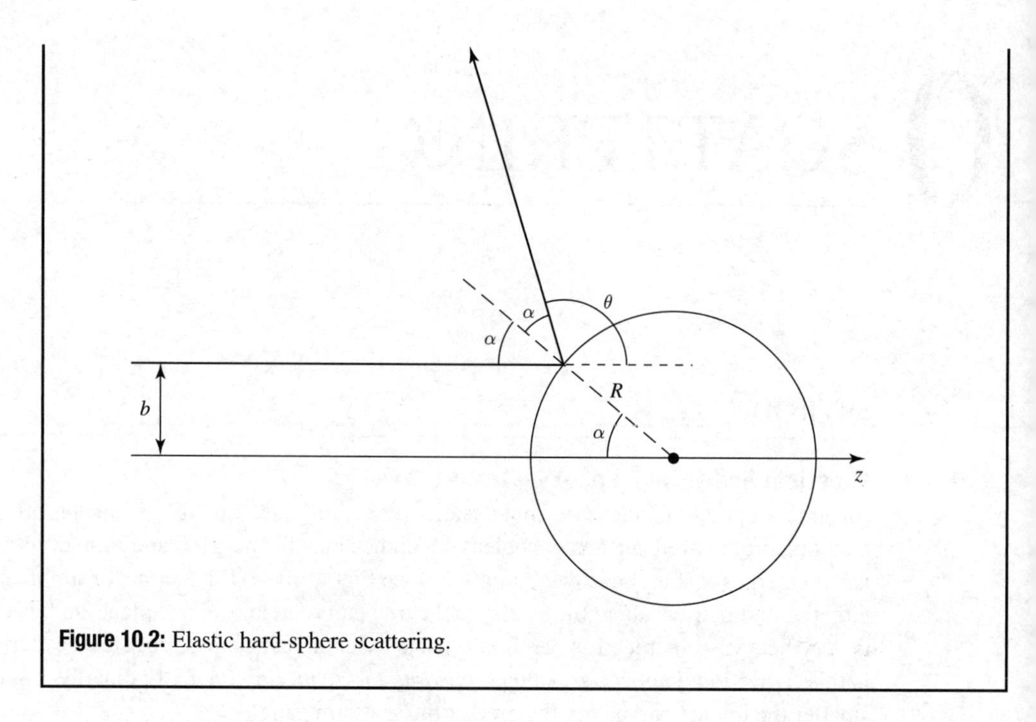

Figure 10.2: Elastic hard-sphere scattering.

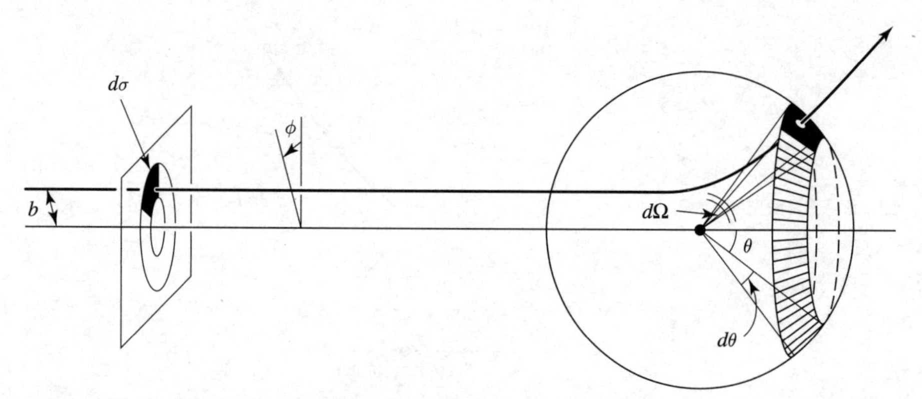

Figure 10.3: Particles incident in the area $d\sigma$ scatter into the solid angle $d\Omega$.

More generally, particles incident within an infinitesimal patch of cross-sectional area $d\sigma$ will scatter into a corresponding infinitesimal solid angle $d\Omega$ (Figure 10.3). The larger $d\sigma$ is, the bigger $d\Omega$ will be; the proportionality factor, $D(\theta) \equiv d\sigma/d\Omega$, is called the **differential (scattering) cross-section**:[1]

$$\boxed{d\sigma = D(\theta)\, d\Omega.} \tag{10.3}$$

[1] This is terrible language: D isn't a *differential*, and it isn't a cross-section. To my ear, the words "differential cross-section" would attach more naturally to $d\sigma$. But I'm afraid we're stuck with this terminology. I should also warn you that the notation $D(\theta)$ is nonstandard—most people just call it $d\sigma/d\Omega$ (which makes Equation 10.3 look like a tautology). I think it will be less confusing if we give the differential cross-section its own symbol.

In terms of the impact parameter and the azimuthal angle ϕ, $d\sigma = b\,db\,d\phi$ and $d\Omega = \sin\theta\,d\theta\,d\phi$, so

$$D(\theta) = \frac{b}{\sin\theta}\left|\frac{db}{d\theta}\right|. \tag{10.4}$$

(Since θ is typically a *decreasing* function of b, the derivative is actually negative—hence the absolute value sign.)

Example 10.2

Hard-sphere scattering (continued). In the case of hard-sphere scattering (Example 10.1)

$$\frac{db}{d\theta} = -\frac{1}{2}R\sin\left(\frac{\theta}{2}\right), \tag{10.5}$$

so

$$D(\theta) = \frac{R\cos(\theta/2)}{\sin\theta}\left(\frac{R\sin(\theta/2)}{2}\right) = \frac{R^2}{4}. \tag{10.6}$$

This example is unusual, in that the differential cross-section is independent of θ.

The **total cross-section** is the *integral* of $D(\theta)$, over all solid angles:

$$\boxed{\sigma \equiv \int D(\theta)\,d\Omega;} \tag{10.7}$$

roughly speaking, it is the total area of incident beam that is scattered by the target. For example, in the case of hard-sphere scattering,

$$\sigma = \left(R^2/4\right)\int d\Omega = \pi R^2, \tag{10.8}$$

which is just what we would expect: It's the cross-sectional area of the sphere; BB's incident within this area will hit the target, and those farther out will miss it completely. But the virtue of the formalism developed here is that it applies just as well to "soft" targets (such as the Coulomb field of a nucleus) that are *not* simply "hit-or-miss".

Finally, suppose we have a *beam* of incident particles, with uniform intensity (or **luminosity**, as particle physicists call it)

$$\mathcal{L} \equiv \text{number of incident particles per unit area, per unit time.} \tag{10.9}$$

The number of particles entering area $d\sigma$ (and hence scattering into solid angle $d\Omega$), per unit time, is $dN = \mathcal{L}\,d\sigma = \mathcal{L}\,D(\theta)\,d\Omega$, so

$$D(\theta) = \frac{1}{\mathcal{L}}\frac{dN}{d\Omega}. \tag{10.10}$$

This is sometimes taken as the *definition* of the differential cross-section, because it makes reference only to quantities easily measured in the laboratory: If the detector subtends a solid angle $d\Omega$, we simply count the *number* recorded per unit time (the **event rate**, dN), divide by $d\Omega$, and normalize to the luminosity of the incident beam.

∗∗∗ **Problem 10.1 Rutherford scattering.** An incident particle of charge q_1 and kinetic energy E scatters off a heavy stationary particle of charge q_2.

(a) Derive the formula relating the impact parameter to the scattering angle.[2] *Answer:* $b = (q_1 q_2/8\pi\epsilon_0 E)\cot(\theta/2)$.

(b) Determine the differential scattering cross-section. *Answer:*

$$D(\theta) = \left[\frac{q_1 q_2}{16\pi\epsilon_0 E\sin^2(\theta/2)}\right]^2. \tag{10.11}$$

(c) Show that the total cross-section for Rutherford scattering is *infinite*.

10.1.2 Quantum Scattering Theory

In the quantum theory of scattering, we imagine an incident plane wave, $\psi(z) = Ae^{ikz}$, traveling in the z direction, which encounters a scattering potential, producing an outgoing *spherical* wave (Figure 10.4).[3] That is, we look for solutions to the Schrödinger equation of the generic form

$$\psi(r,\theta) \approx A\left\{e^{ikz} + f(\theta)\frac{e^{ikr}}{r}\right\}, \quad \text{for large } r. \tag{10.12}$$

(The spherical wave carries a factor of $1/r$, because this portion of $|\psi|^2$ must go like $1/r^2$ to conserve probability.) The **wave number** k is related to the energy of the incident particles in the usual way:

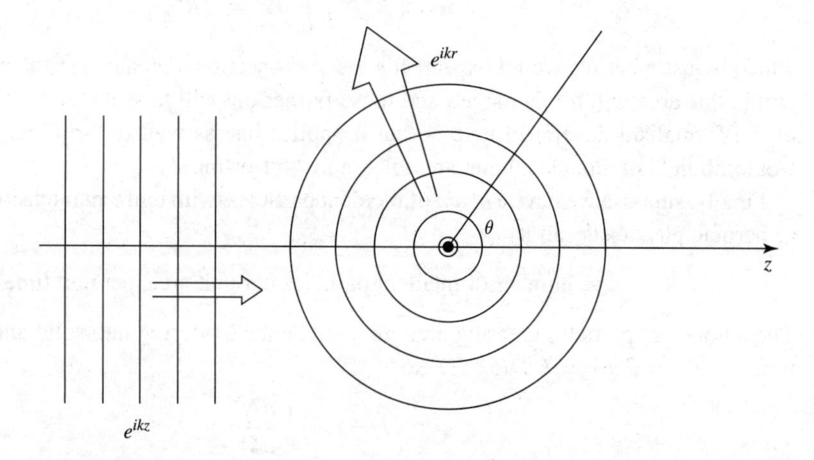

Figure 10.4: Scattering of waves; an incoming plane wave generates an outgoing spherical wave.

[2] This isn't easy, and you might want to refer to a book on classical mechanics, such as Jerry B. Marion and Stephen T. Thornton, *Classical Dynamics of Particles and Systems*, 4th edn, Saunders, Fort Worth, TX (1995), Section 9.10.

[3] For the moment, there's not much *quantum* mechanics in this; what we're really talking about is the scattering of *waves*, as opposed to *particles*, and you could even think of Figure 10.4 as a picture of water waves encountering a rock, or (better, since we're interested in three-dimensional scattering) sound waves bouncing off a basketball.

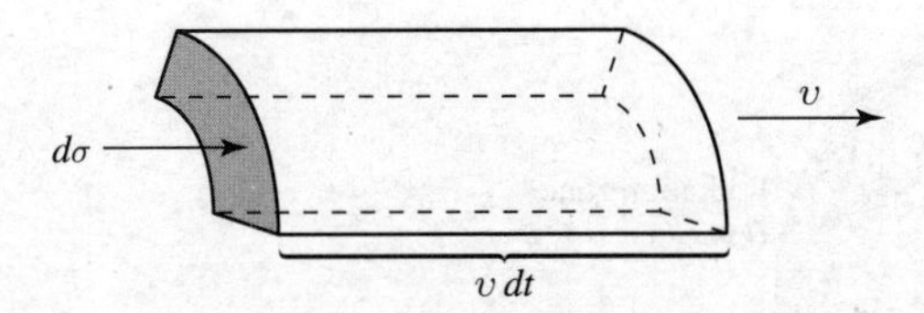

Figure 10.5: The volume dV of incident beam that passes through area $d\sigma$ in time dt.

$$k \equiv \frac{\sqrt{2mE}}{\hbar}. \tag{10.13}$$

(As before, I assume the target is azimuthally symmetrical; in the more general case f would depend on ϕ as well as θ.)

The whole problem is to determine the **scattering amplitude** $f(\theta)$; it tells you the *probability of scattering in a given direction* θ, and hence is related to the differential cross-section. Indeed, the probability that the incident particle, traveling at speed v, passes through the infinitesimal area $d\sigma$, in time dt, is (see Figure 10.5)

$$dP = |\psi_{\text{incident}}|^2 \, dV = |A|^2 \, (v\,dt)\, d\sigma.$$

But this is equal to the probability that the particle scatters into the corresponding solid angle $d\Omega$:

$$dP = |\psi_{\text{scattered}}|^2 \, dV = \frac{|A|^2 \, |f|^2}{r^2} (v\,dt)\, r^2 \, d\Omega,$$

from which it follows that $d\sigma = |f|^2 \, d\Omega$, and hence

$$\boxed{D(\theta) = \frac{d\sigma}{d\Omega} = |f(\theta)|^2 .} \tag{10.14}$$

Evidently the differential cross-section (which is the quantity of interest to the experimentalist) is equal to the absolute square of the scattering amplitude (which is obtained by solving the Schrödinger equation). In the following sections we will study two techniques for calculating the scattering amplitude: **partial wave analysis** and the **Born approximation**.

Problem 10.2 Construct the analogs to Equation 10.12 for one-dimensional and two-dimensional scattering.

10.2 PARTIAL WAVE ANALYSIS

10.2.1 Formalism

As we found in Chapter 4, the Schrödinger equation for a spherically symmetrical potential $V(r)$ admits the separable solutions

$$\psi(r, \theta, \phi) = R(r) Y_\ell^m(\theta, \phi), \tag{10.15}$$

where Y_ℓ^m is a spherical harmonic (Equation 4.32), and $u(r) = rR(r)$ satisfies the radial equation (Equation 4.37):

$$-\frac{\hbar^2}{2m}\frac{d^2u}{dr^2} + \left[V(r) + \frac{\hbar^2}{2m}\frac{\ell(\ell+1)}{r^2}\right] u = Eu. \tag{10.16}$$

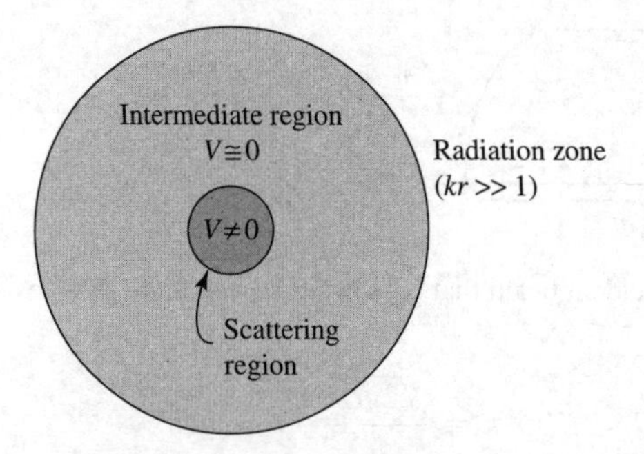

Figure 10.6: Scattering from a localized potential: the scattering region (dark), the intermediate region, where $V = 0$ (shaded), and the radiation zone (where $kr \gg 1$).

At *very* large r the potential goes to zero, and the centrifugal contribution is negligible, so

$$\frac{d^2u}{dr^2} \approx -k^2u.$$

The general solution is

$$u(r) = Ce^{ikr} + De^{-ikr};$$

the first term represents an *outgoing* spherical wave, and the second an *incoming* one—for the scattered wave we want $D = 0$. At very large r, then,

$$R(r) \sim \frac{e^{ikr}}{r},$$

as we already deduced (on physical grounds) in the previous section (Equation 10.12).

That's for *very* large r (more precisely, for $kr \gg 1$; in optics it would be called the **radiation zone**). As in one-dimensional scattering theory, we assume that the potential is "localized," in the sense that exterior to some finite scattering region it is essentially zero (Figure 10.6). In the intermediate region (where V can be ignored but the centrifugal term cannot),[4] the radial equation becomes

$$\frac{d^2u}{dr^2} - \frac{\ell(\ell+1)}{r^2}u = -k^2u, \tag{10.17}$$

and the general solution (Equation 4.45) is a linear combination of spherical Bessel functions:

$$u(r) = Arj_\ell(kr) + Brn_\ell(kr). \tag{10.18}$$

However, neither j_ℓ (which is somewhat like a sine function) nor n_ℓ (which is a sort of generalized cosine function) represents an outgoing (or an incoming) wave. What we need are the linear combinations analogous to e^{ikr} and e^{-ikr}; these are known as **spherical Hankel functions**:

$$h_\ell^{(1)}(x) \equiv j_\ell(x) + in_\ell(x); \quad h_\ell^{(2)}(x) \equiv j_\ell(x) - in_\ell(x). \tag{10.19}$$

[4] What follows does not apply to the Coulomb potential, since $1/r$ goes to zero more slowly than $1/r^2$, as $r \to \infty$, and the centrifugal term does *not* dominate in this region. In this sense the Coulomb potential is not localized, and partial wave analysis is inapplicable.

Table 10.1: *Spherical Hankel functions, $h_\ell^{(1)}(x)$ and $h_\ell^{(2)}(x)$.*

$$h_0^{(1)} = -i\frac{e^{ix}}{x} \qquad h_0^{(2)} = i\frac{e^{-ix}}{x}$$

$$h_1^{(1)} = \left(-\frac{i}{x^2} - \frac{1}{x}\right)e^{ix} \qquad h_1^{(2)} = \left(\frac{i}{x^2} - \frac{1}{x}\right)e^{-ix}$$

$$h_2^{(1)} = \left(-\frac{3i}{x^3} - \frac{3}{x^2} + \frac{i}{x}\right)e^{ix} \qquad h_2^{(2)} = \left(\frac{3i}{x^3} - \frac{3}{x^2} - \frac{i}{x}\right)e^{-ix}$$

$$\left.\begin{aligned} h_\ell^{(1)} &\to \frac{1}{x}(-i)^{\ell+1}e^{ix} \\ h_\ell^{(2)} &\to \frac{1}{x}(i)^{\ell+1}e^{-ix} \end{aligned}\right\} \text{for } x \gg 1$$

The first few spherical Hankel functions are listed in Table 10.1. At large r, $h_\ell^{(1)}(kr)$ (the **Hankel function of the first kind**) goes like e^{ikr}/r, whereas $h_\ell^{(2)}(kr)$ (the **Hankel function of the second kind**) goes like e^{-ikr}/r; for outgoing waves, then, we need spherical Hankel functions of the first kind:

$$R(r) \sim h_\ell^{(1)}(kr). \tag{10.20}$$

The exact wave function, in the exterior region (where $V(r) = 0$), is

$$\psi(r,\theta,\phi) = A\left\{e^{ikz} + \sum_{\ell,m} C_{\ell,m}\, h_\ell^{(1)}(kr)\, Y_\ell^m(\theta,\phi)\right\}. \tag{10.21}$$

The first term is the incident plane wave, and the sum (with expansion coefficients $C_{\ell,m}$) is the scattered wave. But since we are assuming the potential is spherically symmetric, the wave function cannot depend on ϕ.[5] So only terms with $m = 0$ survive (remember, $Y_\ell^m \sim e^{im\phi}$). Now (from Equations 4.27 and 4.32)

$$Y_\ell^0(\theta,\phi) = \sqrt{\frac{2\ell+1}{4\pi}}\, P_\ell(\cos\theta), \tag{10.22}$$

where P_ℓ is the ℓth Legendre polynomial. It is customary to redefine the expansion coefficients $\left(C_{\ell,0} \equiv i^{\ell+1}k\sqrt{4\pi\,(2\ell+1)}\,a_\ell\right)$:

$$\psi(r,\theta) = A\left\{e^{ikz} + k\sum_{\ell=0}^{\infty} i^{\ell+1}\,(2\ell+1)\;a_\ell\, h_\ell^{(1)}\,(kr)\;P_\ell(\cos\theta)\right\}. \tag{10.23}$$

You'll see in a moment why this peculiar notation is convenient; a_ℓ is called the ℓth **partial wave amplitude**.

For *very large* r, the Hankel function goes like $(-i)^{\ell+1}\,e^{ikr}/kr$ (Table 10.1), so

$$\psi(r,\theta) \approx A\left\{e^{ikz} + f(\theta)\frac{e^{ikr}}{r}\right\}, \tag{10.24}$$

[5] There's nothing wrong with θ dependence, of course, because the incoming plane wave defines a z direction, breaking the spherical symmetry. But the *azimuthal* symmetry remains; the incident plane wave has no ϕ dependence, and there is nothing in the scattering process that could introduce any ϕ dependence in the outgoing wave.

where

$$f(\theta)=\sum_{\ell=0}^{\infty}(2\ell+1)\,a_\ell\,P_\ell(\cos\theta). \tag{10.25}$$

This confirms more rigorously the general structure postulated in Equation 10.12, and tells us how to compute the scattering amplitude, $f(\theta)$, in terms of the partial wave amplitudes (a_ℓ). The differential cross-section is

$$D(\theta)=|f(\theta)|^2=\sum_{\ell}\sum_{\ell'}(2\ell+1)\left(2\ell'+1\right)\,a_\ell^*\,a_{\ell'}\,P_\ell(\cos\theta)\,P_{\ell'}(\cos\theta), \tag{10.26}$$

and the total cross-section is

$$\sigma=4\pi\sum_{\ell=0}^{\infty}(2\ell+1)\,|a_\ell|^2\,. \tag{10.27}$$

(I used the orthogonality of the Legendre polynomials, Equation 4.34, to do the angular integration.)

10.2.2 Strategy

All that remains is to determine the partial wave amplitudes, a_ℓ, for the potential in question. This is accomplished by solving the Schrödinger equation in the *interior* region (where $V(r)$ is *not* zero), and matching it to the exterior solution (Equation 10.23), using the appropriate boundary conditions. The only problem is that as it stands my notation is hybrid: I used *spherical* coordinates for the scattered wave, but *cartesian* coordinates for the incident wave. We need to rewrite the wave function in a more consistent notation.

Of course, e^{ikz} satisfies the Schrödinger equation with $V=0$. On the other hand, I just argued that the *general* solution to the Schrödinger equation with $V=0$ can be written in the form

$$\sum_{\ell,m}\left[A_{\ell,m}\,j_\ell(kr)+B_{\ell,m}\,n_\ell(kr)\right]Y_\ell^m(\theta,\phi).$$

In particular, then, it must be possible to express e^{ikz} in this way. But e^{ikz} is finite at the origin, so no Neumann functions are allowed in the sum ($n_\ell(kr)$ blows up at $r=0$), and since $z=r\cos\theta$ has no ϕ dependence, only $m=0$ terms occur. The resulting expansion of a plane wave in terms of spherical waves is known as **Rayleigh's formula**:[6]

$$e^{ikz}=\sum_{\ell=0}^{\infty}i^\ell\,(2\ell+1)\,j_\ell(kr)P_\ell(\cos\theta). \tag{10.28}$$

Using this, the wave function in the exterior region (Equation 10.23) can be expressed entirely in terms of r and θ:

$$\psi(r,\theta)=A\sum_{\ell=0}^{\infty}i^\ell\,(2\ell+1)\left[j_\ell(kr)+ik\,a_\ell\,h_\ell^{(1)}(kr)\right]P_\ell(\cos\theta). \tag{10.29}$$

[6] For a guide to the proof, see George Arfken and Hans-Jurgen Weber, *Mathematical Methods for Physicists*, 7th edn, Academic Press, Orlando (2013), Exercises 15.2.24 and 15.2.25.

Example 10.3

Quantum hard-sphere scattering. Suppose

$$V(r) = \begin{cases} \infty, & (r \le a), \\ 0, & (r > a). \end{cases} \tag{10.30}$$

The boundary condition, then, is

$$\psi(a, \theta) = 0, \tag{10.31}$$

so

$$\sum_{\ell=0}^{\infty} i^{\ell}(2\ell+1)\left[j_{\ell}(ka) + ik\,a_{\ell}\,h_{\ell}^{(1)}(ka)\right] P_{\ell}(\cos\theta) = 0 \tag{10.32}$$

for all θ, from which it follows (Problem 10.3) that

$$a_{\ell} = i\,\frac{j_{\ell}(ka)}{k\,h_{\ell}^{(1)}(ka)}. \tag{10.33}$$

In particular, the total cross-section (Equation 10.27) is

$$\sigma = \frac{4\pi}{k^2}\sum_{\ell=0}^{\infty}(2\ell+1)\left|\frac{j_{\ell}(ka)}{h_{\ell}^{(1)}(ka)}\right|^2. \tag{10.34}$$

That's the *exact* answer, but it's not terribly illuminating, so let's consider the limiting case of *low-energy scattering*: $ka \ll 1$. (Since $k = 2\pi/\lambda$, this amounts to saying that the wavelength is much greater than the radius of the sphere.) Referring to Table 4.4, we note that $n_{\ell}(z)$ is much larger than $j_{\ell}(z)$, for small z, so

$$\begin{aligned} \frac{j_{\ell}(z)}{h_{\ell}^{(1)}(z)} &= \frac{j_{\ell}(z)}{j_{\ell}(z) + i n_{\ell}(z)} \approx -i\,\frac{j_{\ell}(z)}{n_{\ell}(z)} \\ &\approx -i\,\frac{2^{\ell}\ell! z^{\ell}/(2\ell+1)!}{-(2\ell)! z^{-\ell-1}/2^{\ell}\ell!} = \frac{i}{2\ell+1}\left[\frac{2^{\ell}\ell!}{(2\ell)!}\right]^2 z^{2\ell+1}, \end{aligned} \tag{10.35}$$

and hence

$$\sigma \approx \frac{4\pi}{k^2}\sum_{\ell=0}^{\infty}\frac{1}{2\ell+1}\left[\frac{2^{\ell}\ell!}{(2\ell)!}\right]^4 (ka)^{4\ell+2}.$$

But we're assuming $ka \ll 1$, so the higher powers are negligible—in the low-energy approximation the scattering is dominated by the $\ell = 0$ term. (This means that the differential cross-section is independent of θ, just as it was in the classical case.) Evidently

$$\sigma \approx 4\pi a^2, \tag{10.36}$$

for low energy hard-sphere scattering. Surprisingly, the scattering cross-section is *four times* the geometrical cross-section—in fact, σ is the *total surface area of the sphere*. This "larger effective size" is characteristic of long-wavelength scattering (it would be true in optics, as well); in a sense, these waves "feel" their way around the whole sphere, whereas classical *particles* only see the head-on cross-section (Equation 10.8).

Problem 10.3 Prove Equation 10.33, starting with Equation 10.32. *Hint:* Exploit the orthogonality of the Legendre polynomials to show that the coefficients with different values of ℓ must separately vanish.

** **Problem 10.4** Consider the case of low-energy scattering from a spherical delta-function shell:

$$V(r) = \alpha\delta(r - a),$$

where α and a are positive constants. Calculate the scattering amplitude, $f(\theta)$, the differential cross-section, $D(\theta)$, and the total cross-section, σ. Assume $ka \ll 1$, so that only the $\ell = 0$ term contributes significantly. (To simplify matters, throw out all $\ell \neq 0$ terms right from the start.) The main problem, of course, is to determine C_0. Express your answer in terms of the dimensionless quantity $\beta \equiv 2ma\alpha/\hbar^2$. *Answer:* $\sigma = 4\pi a^2\beta^2/(1+\beta)^2$.

10.3 PHASE SHIFTS

Consider first the problem of *one*-dimensional scattering from a localized potential $V(x)$ on the half-line $x < 0$ (Figure 10.7). I'll put a "brick wall" at $x = 0$, so a wave incident from the left,

$$\psi_i(x) = Ae^{ikx} \qquad (x < -a) \tag{10.37}$$

is entirely reflected

$$\psi_r(x) = Be^{-ikx} \qquad (x < -a). \tag{10.38}$$

Whatever happens in the interaction region ($-a < x < 0$), the amplitude of the reflected wave has *got* to be the same as that of the incident wave ($|B| = |A|$), by conservation of probability. But it need not have the same *phase*. If there were no potential at all (just the wall at $x = 0$), then $B = -A$, since the total wave function (incident plus reflected) must vanish at the origin:

$$\psi(x) = A\left(e^{ikx} - e^{-ikx}\right) \qquad (V(x) = 0). \tag{10.39}$$

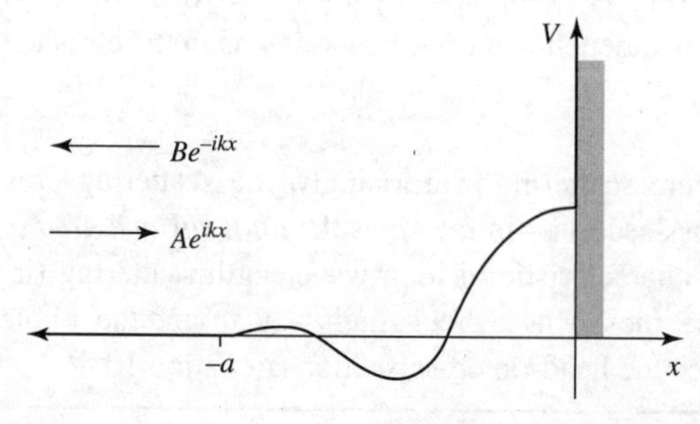

Figure 10.7: One-dimensional scattering from a localized potential bounded on the right by an infinite wall.

If the potential is *not* zero, the wave function (for $x < -a$) takes the form

$$\psi(x) = A\left(e^{ikx} - e^{i(2\delta - kx)}\right) \qquad (V(x) \neq 0). \tag{10.40}$$

The whole theory of scattering reduces to the problem of calculating the **phase shift**[7] δ (as a function of k, and hence of the energy $E = \hbar^2k^2/2m$), for a specified potential. We do this, of course, by solving the Schrödinger equation in the scattering region ($-a < x < 0$), and imposing appropriate boundary conditions (see Problem 10.5). The advantage of working with the phase shift (as opposed to the complex number B) is that it exploits the physics to simplify the mathematics (trading a complex quantity—two real numbers—for a single real quantity).

Now let's return to the three-dimensional case. The incident plane wave $\left(Ae^{ikz}\right)$ carries no angular momentum in the z direction (Rayleigh's formula contains no terms with $m \neq 0$), but it includes all values of the *total* angular momentum ($\ell = 0, 1, 2, \ldots$). Because angular momentum is conserved (by a spherically symmetric potential), each **partial wave** (labelled by a particular ℓ) scatters independently, with (again) no change in amplitude[8]—only in phase.

If there is no potential at all, then $\psi = Ae^{ikz}$, and the ℓth partial wave is (Equation 10.28)

$$\psi^{(\ell)} = Ai^{\ell}\,(2\ell + 1)\; j_\ell(kr)\, P_\ell(\cos\theta) \qquad (V(r) = 0). \tag{10.41}$$

But (from Equation 10.19 and Table 10.1)

$$j_\ell(x) = \frac{1}{2}\left[h_\ell^{(1)}(x) + h_\ell^{(2)}(x)\right] \approx \frac{1}{2x}\left[(-i)^{\ell+1}\, e^{ix} + i^{\ell+1} e^{-ix}\right] \quad (x \gg 1). \tag{10.42}$$

So for large r

$$\psi^{(\ell)} \approx A\frac{(2\ell + 1)}{2ikr}\left[e^{ikr} - (-1)^{\ell}\, e^{-ikr}\right] P_\ell(\cos\theta) \qquad (V(r) = 0). \tag{10.43}$$

The second term inside the square brackets represents an incoming spherical wave; it comes from the incident plane wave, and is unchanged when we now introduce a potential. The first term is the outgoing wave; it picks up a phase shift (due to the scattering potential):

$$\psi^{(\ell)} \approx A\frac{(2\ell + 1)}{2ikr}\left[e^{i(kr + 2\delta_\ell)} - (-1)^{\ell}\, e^{-ikr}\right] P_\ell(\cos\theta) \qquad (V(r) \neq 0). \tag{10.44}$$

Think of it as a converging spherical wave (the e^{-ikr} term, due exclusively to the $h_\ell^{(2)}$ component in e^{ikz}), which is phase shifted an amount δ_ℓ on the way in, and again δ_ℓ on the way out (hence the 2), emerging as an outgoing spherical wave (the e^{ikr} term, due to the $h_\ell^{(1)}$ part of e^{ikz} plus the scattered wave).

In Section 10.2.1 the whole theory was expressed in terms of the partial wave amplitudes a_ℓ; now we have formulated it in terms of the phase shifts δ_ℓ. There must be a connection between the two. Indeed, comparing the asymptotic (large r) form of Equation 10.23

$$\psi^{(\ell)} \approx A\left\{\frac{(2\ell + 1)}{2ikr}\left[e^{ikr} - (-1)^{\ell}\, e^{-ikr}\right] + \frac{(2\ell + 1)}{r}\, a_\ell\, e^{ikr}\right\} P_\ell(\cos\theta) \tag{10.45}$$

---

[7] The 2 in front of δ is conventional. We think of the incident wave as being phase shifted once on the way in, and again on the way out; δ is the "one way" phase shift, and the *total* is 2δ.

[8] One reason this subject can be so confusing is that practically everything is called an "amplitude": $f(\theta)$ is the "scattering amplitude", a_ℓ is the "partial wave amplitude", but the first is a function of θ, and both are complex numbers. I'm *now* talking about "amplitude" in the original sense: the (*real*, of course) height of a sinusoidal wave.

with the generic expression in terms of δ_ℓ (Equation 10.44), we find[9]

$$a_\ell = \frac{1}{2ik}\left(e^{2i\delta_\ell} - 1\right) = \frac{1}{k}e^{i\delta_\ell}\sin(\delta_\ell). \tag{10.46}$$

It follows in particular (Equation 10.25) that

$$f(\theta) = \frac{1}{k}\sum_{\ell=0}^{\infty}(2\ell+1)\,e^{i\delta_\ell}\sin(\delta_\ell)\,P_\ell(\cos\theta) \tag{10.47}$$

and (Equation 10.27)

$$\sigma = \frac{4\pi}{k^2}\sum_{\ell=0}^{\infty}(2\ell+1)\sin^2(\delta_\ell). \tag{10.48}$$

Again, the advantage of working with phase shifts (as opposed to partial wave amplitudes) is that they are easier to interpret physically, and simpler mathematically—the phase shift formalism exploits conservation of angular momentum to reduce a complex quantity a_ℓ (two real numbers) to a single real one δ_ℓ.

Problem 10.5 A particle of mass m and energy E is incident from the left on the potential

$$V(x) = \begin{cases} 0, & (x < -a), \\ -V_0, & (-a \le x \le 0), \\ \infty, & (x > 0). \end{cases}$$

(a) If the incoming wave is $Ae^{i\kappa x}$ (where $\kappa = \sqrt{2mE}/\hbar$ and $E < V_0$), find the reflected wave.
Answer:

$$Ae^{-2i\kappa a}\left[\frac{\kappa - ik\cot(ka)}{\kappa + ik\cot(ka)}\right]e^{-i\kappa x}, \quad \text{where } k = \sqrt{2m(E+V_0)}/\hbar.$$

(b) Confirm that the reflected wave has the same amplitude as the incident wave.

(c) Find the phase shift δ (Equation 10.40) for a very deep well ($E \ll V_0$). *Answer:* $\delta = -ka$.

Problem 10.6 What are the partial wave phase shifts (δ_ℓ) for hard-sphere scattering (Example 10.3)?

Problem 10.7 Find the S-wave ($\ell = 0$) partial wave phase shift $\delta_0(k)$ for scattering from a delta-function shell (Problem 10.4). Assume that the radial wave function $u(r)$ goes to 0 as $r \to 0$. *Answer:*

$$-\cot^{-1}\left[\cot(ka) + \frac{ka}{\beta\sin^2(ka)}\right], \quad \text{where} \quad \beta \equiv \frac{2m\alpha a}{\hbar^2}.$$

[9] Although I used the asymptotic form of the wave function to draw the connection between a_ℓ and δ_ℓ, there is nothing approximate about the result (Equation 10.46). Both of them are *constants* (independent of r), and δ_ℓ *means* the phase shift in the asymptotic region (where the Hankel functions have settled down to $e^{\pm ikr}/kr$).

10.4 THE BORN APPROXIMATION

10.4.1 Integral Form of the Schrödinger Equation

The time-independent Schrödinger equation,

$$-\frac{\hbar^2}{2m}\nabla^2\psi + V\psi = E\psi, \tag{10.49}$$

can be written more succinctly as

$$\left(\nabla^2 + k^2\right)\psi = Q, \tag{10.50}$$

where

$$k \equiv \frac{\sqrt{2mE}}{\hbar} \quad \text{and} \quad Q \equiv \frac{2m}{\hbar^2}V\psi. \tag{10.51}$$

This has the superficial appearance of the **Helmholtz equation**; note, however, that the "inhomogeneous" term (Q) *itself* depends on ψ. Suppose we could find a function $G(\mathbf{r})$ that solves the Helmholtz equation with a *delta function* "source":

$$\left(\nabla^2 + k^2\right)G(\mathbf{r}) = \delta^3(\mathbf{r}). \tag{10.52}$$

Then we could express ψ as an integral:

$$\psi(\mathbf{r}) = \int G(\mathbf{r}-\mathbf{r}_0)Q(\mathbf{r}_0)\,d^3\mathbf{r}_0. \tag{10.53}$$

For it is easy to show that this satisfies Schrödinger's equation, in the form of Equation 10.50:

$$\begin{aligned}\left(\nabla^2 + k^2\right)\psi(\mathbf{r}) &= \int\left[\left(\nabla^2 + k^2\right)G(\mathbf{r}-\mathbf{r}_0)\right]Q(\mathbf{r}_0)\,d^3\mathbf{r}_0\\ &= \int \delta^3(\mathbf{r}-\mathbf{r}_0)Q(\mathbf{r}_0)\,d^3\mathbf{r}_0 = Q(\mathbf{r}).\end{aligned}$$

$G(\mathbf{r})$ is called the **Green's function** for the Helmholtz equation. (In general, the Green's function for a linear differential equation represents the "response" to a delta-function source.)

Our first task[10] is to solve Equation 10.52 for $G(\mathbf{r})$. This is most easily accomplished by taking the Fourier transform, which turns the *differential* equation into an *algebraic* equation. Let

$$G(\mathbf{r}) = \frac{1}{(2\pi)^{3/2}}\int e^{i\mathbf{s}\cdot\mathbf{r}}g(\mathbf{s})\,d^3\mathbf{s}. \tag{10.54}$$

Then

$$\left(\nabla^2 + k^2\right)G(\mathbf{r}) = \frac{1}{(2\pi)^{3/2}}\int\left[\left(\nabla^2 + k^2\right)e^{i\mathbf{s}\cdot\mathbf{r}}\right]g(\mathbf{s})\,d^3\mathbf{s}.$$

But

$$\nabla^2 e^{i\mathbf{s}\cdot\mathbf{r}} = -s^2 e^{i\mathbf{s}\cdot\mathbf{r}}, \tag{10.55}$$

and (see Equation 2.147)

$$\delta^3(\mathbf{r}) = \frac{1}{(2\pi)^3}\int e^{i\mathbf{s}\cdot\mathbf{r}}\,d^3\mathbf{s}, \tag{10.56}$$

10 *Warning:* You are approaching two pages of heavy analysis, including contour integration; if you wish, skip straight to the answer, Equation 10.65.

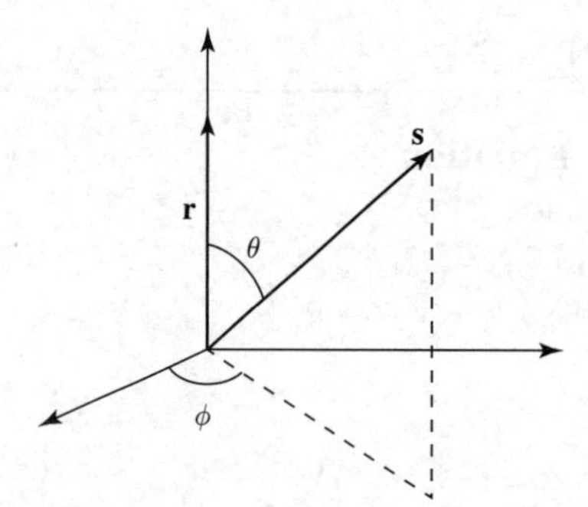

Figure 10.8: Convenient coordinates for the integral in Equation 10.58.

so Equation 10.52 says

$$\frac{1}{(2\pi)^{3/2}}\int\left(-s^2+k^2\right)e^{i\mathbf{s}\cdot\mathbf{r}}g\left(\mathbf{s}\right)d^3\mathbf{s}=\frac{1}{(2\pi)^3}\int e^{i\mathbf{s}\cdot\mathbf{r}}\,d^3\mathbf{s}.$$

It follows[11] that

$$g(\mathbf{s})=\frac{1}{(2\pi)^{3/2}\left(k^2-s^2\right)}. \tag{10.57}$$

Putting this back into Equation 10.54, we find:

$$G(\mathbf{r})=\frac{1}{(2\pi)^3}\int e^{i\mathbf{s}\cdot\mathbf{r}}\frac{1}{\left(k^2-s^2\right)}\,d^3\mathbf{s}. \tag{10.58}$$

Now, $\mathbf{r}$ is *fixed*, as far as the $\mathbf{s}$ integration is concerned, so we may as well choose spherical coordinates (s,θ,ϕ) with the polar axis along $\mathbf{r}$ (Figure 10.8). Then $\mathbf{s}\cdot\mathbf{r}=sr\cos\theta$, the ϕ integral is trivial (2π), and the θ integral is

$$\int_0^\pi e^{isr\cos\theta}\sin\theta\,d\theta=-\left.\frac{e^{isr\cos\theta}}{isr}\right|_0^\pi=\frac{2\sin(sr)}{sr}. \tag{10.59}$$

Thus

$$G(\mathbf{r})=\frac{1}{(2\pi)^2}\frac{2}{r}\int_0^\infty\frac{s\sin(sr)}{k^2-s^2}ds=\frac{1}{4\pi^2 r}\int_{-\infty}^\infty\frac{s\sin(sr)}{k^2-s^2}ds. \tag{10.60}$$

The remaining integral is not so simple. It pays to revert to exponential notation, and factor the denominator:

$$\begin{aligned}G(\mathbf{r})&=\frac{i}{8\pi^2 r}\left\{\int_{-\infty}^\infty\frac{se^{isr}}{(s-k)\,(s+k)}ds-\int_{-\infty}^\infty\frac{se^{-isr}}{(s-k)\,(s+k)}ds\right\}\\&=\frac{i}{8\pi^2 r}\left(I_1-I_2\right).\end{aligned} \tag{10.61}$$

These two integrals can be evaluated using **Cauchy's integral formula**:

$$\oint\frac{f(z)}{(z-z_0)}dz=2\pi i f(z_0), \tag{10.62}$$

[11] This is clearly *sufficient*, but it is also *necessary*, as you can easily show by combining the two terms into a single integral, and using Plancherel's theorem, Equation 2.103.

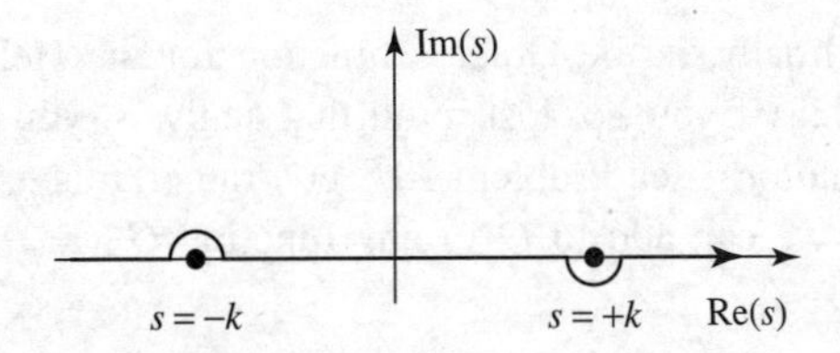

Figure 10.9: Skirting the poles in the contour integral (Equation 10.61).

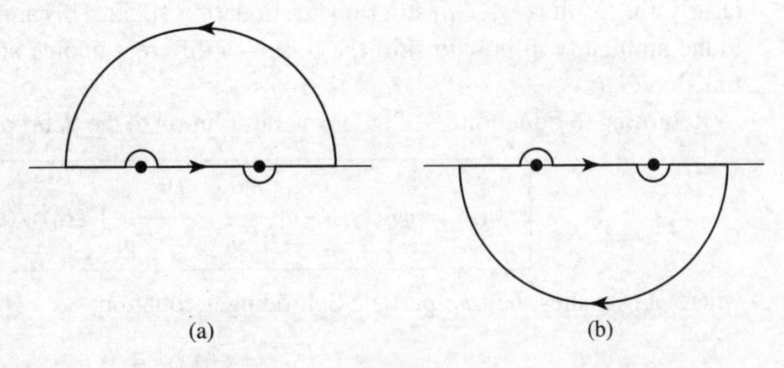

Figure 10.10: Closing the contour in Equations 10.63 and 10.64.

if z_0 lies within the contour (otherwise the integral is zero). In the present case the integration is along the real axis, and it passes *right over* the pole singularities at $\pm k$. We have to decide how to skirt the poles—I'll go *over* the one at $-k$ and *under* the one at $+k$ (Figure 10.9). (You're welcome to choose some *other* convention if you like—even winding seven times around each pole—you'll get a different Green's function, but, as I'll show you in a minute, they're all equally acceptable.)[12]

For each integral in Equation 10.61 I must "close the contour" in such a way that the semi-circle at infinity contributes nothing. In the case of I_1, the factor e^{isr} goes to zero when s has a large *positive* imaginary part; for this one I close *above* (Figure 10.10(a)). The contour encloses only the singularity at $s = +k$, so

$$I_1 = \oint \left[\frac{se^{isr}}{s+k}\right]\frac{1}{s-k}\,ds = 2\pi i\left[\frac{se^{isr}}{s+k}\right]\bigg|_{s=k} = i\pi e^{ikr}. \tag{10.63}$$

In the case of I_2, the factor e^{-isr} goes to zero when s has a large *negative* imaginary part, so we close *below* (Figure 10.10(b)); this time the contour encloses the singularity at $s = -k$ (and it goes around in the *clockwise* direction, so we pick up a minus sign):

$$I_2 = -\oint \left[\frac{se^{-isr}}{s-k}\right]\frac{1}{s+k}\,ds = -2\pi i\left[\frac{se^{-isr}}{s-k}\right]\bigg|_{s=-k} = -i\pi e^{ikr}. \tag{10.64}$$

Conclusion:

$$G(\mathbf{r}) = \frac{i}{8\pi^2 r}\left[\left(i\pi e^{ikr}\right) - \left(-i\pi e^{ikr}\right)\right] = -\frac{e^{ikr}}{4\pi r}. \tag{10.65}$$

[12] If you are unfamiliar with this technique you have every right to be suspicious. In truth, the integral in Equation 10.60 is simply ill-defined—it does not converge, and it's something of a miracle that we can make sense of it at all. The root of the problem is that $G(\mathbf{r})$ doesn't really have a legitimate Fourier transform; we're exceeding the speed limit, here, and just hoping we won't get caught.

This, finally, is the Green's function for the Helmholtz equation—the solution to Equation 10.52. (If you got lost in all that analysis, you might want to *check* the result by direct differentiation—see Problem 10.8.) Or rather, it is *a* Green's function for the Helmholtz equation, for we can add to $G(\mathbf{r})$ any function $G_0(\mathbf{r})$ that satisfies the *homogeneous* Helmholtz equation:

$$\left(\nabla^2 + k^2\right) G_0(\mathbf{r}) = 0; \tag{10.66}$$

clearly, the result $(G + G_0)$ still satisfies Equation 10.52. This ambiguity corresponds precisely to the ambiguity in how to skirt the poles—a different choice amounts to picking a different function $G_0(\mathbf{r})$.

Returning to Equation 10.53, the general solution to the Schrödinger equation takes the form

$$\boxed{\psi(\mathbf{r}) = \psi_0(\mathbf{r}) - \frac{m}{2\pi\hbar^2}\int \frac{e^{ik|\mathbf{r}-\mathbf{r}_0|}}{|\mathbf{r}-\mathbf{r}_0|} V(\mathbf{r}_0)\psi(\mathbf{r}_0)\, d^3\mathbf{r}_0,} \tag{10.67}$$

where ψ_0 satisfies the *free*-particle Schrödinger equation,

$$\left(\nabla^2 + k^2\right) \psi_0 = 0. \tag{10.68}$$

Equation 10.67 is the **integral form of the Schrödinger equation**; it is entirely equivalent to the more familiar differential form. At first glance it *looks* like an explicit *solution* to the Schrödinger equation (for any potential)—which is too good to be true. Don't be deceived: There's a ψ under the integral sign on the right hand side, so you can't do the integral unless you already know the solution! Nevertheless, the integral form can be very powerful, and it is particularly well suited to scattering problems, as we'll see in the following section.

Problem 10.8 Check that Equation 10.65 satisfies Equation 10.52, by direct substitution. *Hint:* $\nabla^2(1/r) = -4\pi\delta^3(\mathbf{r})$.[13]

** **Problem 10.9** Show that the ground state of hydrogen (Equation 4.80) satisfies the integral form of the Schrödinger equation, for the appropriate V and E (note that E is *negative*, so $k = i\kappa$, where $\kappa \equiv \sqrt{-2mE}/\hbar$).

10.4.2 The First Born Approximation

Suppose $V(\mathbf{r}_0)$ is localized about $\mathbf{r}_0 = 0$—that is, the potential drops to zero outside some finite region (as is typical for a scattering problem), and we want to calculate $\psi(\mathbf{r})$ at points *far away* from the scattering center. Then $|\mathbf{r}| \gg |\mathbf{r}_0|$ for all points that contribute to the integral in Equation 10.67, so

$$|\mathbf{r}-\mathbf{r}_0|^2 = r^2 + r_0^2 - 2\mathbf{r}\cdot\mathbf{r}_0 \approx r^2\left(1 - 2\frac{\mathbf{r}\cdot\mathbf{r}_0}{r^2}\right), \tag{10.69}$$

and hence

[13] See, for example, D. Griffiths, *Introduction to Electrodynamics*, 4th edn (Cambridge University Press, Cambridge, UK, 2017), Section 1.5.3.

$$|\mathbf{r}-\mathbf{r}_0| \approx r - \hat{r}\cdot\mathbf{r}_0. \tag{10.70}$$

Let

$$\mathbf{k} \equiv k\hat{r}; \tag{10.71}$$

then

$$e^{ik|\mathbf{r}-\mathbf{r}_0|} \approx e^{ikr}e^{-i\mathbf{k}\cdot\mathbf{r}_0}, \tag{10.72}$$

and therefore

$$\frac{e^{ik|\mathbf{r}-\mathbf{r}_0|}}{|\mathbf{r}-\mathbf{r}_0|} \approx \frac{e^{ikr}}{r}e^{-i\mathbf{k}\cdot\mathbf{r}_0}. \tag{10.73}$$

(In the *denominator* we can afford to make the more radical approximation $|\mathbf{r}-\mathbf{r}_0| \approx r$; in the *exponent* we need to keep the next term. If this puzzles you, try including the next term in the expansion of the denominator. What we are doing is expanding in powers of the small quantity (r_0/r), and dropping all but the lowest order.)

In the case of scattering, we want

$$\psi_0(\mathbf{r}) = Ae^{ikz}, \tag{10.74}$$

representing an incident plane wave. For large r, then,

$$\psi(\mathbf{r}) \approx Ae^{ikz} - \frac{m}{2\pi\hbar^2}\frac{e^{ikr}}{r}\int e^{-i\mathbf{k}\cdot\mathbf{r}_0}V(\mathbf{r}_0)\psi(\mathbf{r}_0)\,d^3\mathbf{r}_0. \tag{10.75}$$

This is in the standard form (Equation 10.12), and we can read off the scattering amplitude:

$$f(\theta,\phi) = -\frac{m}{2\pi\hbar^2 A}\int e^{-i\mathbf{k}\cdot\mathbf{r}_0}V(\mathbf{r}_0)\psi(\mathbf{r}_0)\,d^3\mathbf{r}_0. \tag{10.76}$$

This is *exact*.[14] Now we invoke the **Born approximation**: Suppose the incoming plane wave is *not substantially altered in the scattering region* (where V is nonzero); then it makes sense to use

$$\psi(\mathbf{r}_0) \approx \psi_0(\mathbf{r}_0) = Ae^{ikz_0} = Ae^{i\mathbf{k}'\cdot\mathbf{r}_0}, \tag{10.77}$$

where

$$\mathbf{k}' \equiv k\hat{z}, \tag{10.78}$$

inside the integral. (This would be the *exact* wave function, if V were zero; it is essentially a *weak potential* approximation.[15]) In the Born approximation, then,

$$\boxed{f(\theta,\phi) \approx -\frac{m}{2\pi\hbar^2}\int e^{i(\mathbf{k}'-\mathbf{k})\cdot\mathbf{r}_0}V(\mathbf{r}_0)\,d^3\mathbf{r}_0.} \tag{10.79}$$

(In case you have lost track of the definitions of $\mathbf{k}'$ and $\mathbf{k}$, they both have magnitude k, but the former points in the direction of the incident beam, while the latter points toward the detector—see Figure 10.11; $\hbar\left(\mathbf{k}-\mathbf{k}'\right)$ is the **momentum transfer** in the process.)

---

[14] Remember, $f(\theta,\phi)$ is by definition the coefficient of Ae^{ikr}/r at *large* r.

[15] Either the potential is intrinsically weak, or the incident energy is high. Typically, partial wave analysis is useful when the incident particle has low energy, for then only the first few terms in the series contribute significantly; the Born approximation is more useful at *high* energy, when the deflection is relatively small.

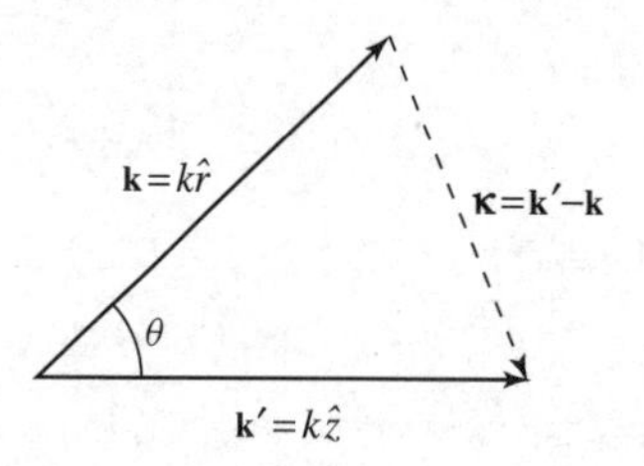

Figure 10.11: Two wave vectors in the Born approximation: $\mathbf{k}'$ points in the *incident* direction, $\mathbf{k}$ in the *scattered* direction.

In particular, for **low energy** (long wavelength) **scattering,** the exponential factor is essentially constant over the scattering region, and the Born approximation simplifies to

$$\boxed{f(\theta, \phi) \approx -\frac{m}{2\pi\hbar^2}\int V(\mathbf{r})\, d^3\mathbf{r}, \quad \text{(low energy).}} \tag{10.80}$$

(I dropped the subscript on $\mathbf{r}$, since there is no likelihood of confusion at this point.)

Example 10.4

Low-energy soft-sphere scattering.[16] Suppose

$$V(\mathbf{r}) = \begin{cases} V_0, & (r \le a), \\ 0, & (r > a). \end{cases} \tag{10.81}$$

In this case the low-energy scattering amplitude is

$$f(\theta, \phi) \approx -\frac{m}{2\pi\hbar^2} V_0 \left(\frac{4}{3}\pi a^3\right), \tag{10.82}$$

(independent of θ and ϕ), the differential cross-section is

$$\frac{d\sigma}{d\Omega} = |f|^2 \approx \left(\frac{2m V_0 a^3}{3\hbar^2}\right)^2, \tag{10.83}$$

and the total cross-section is

$$\sigma \approx 4\pi \left(\frac{2m V_0 a^3}{3\hbar^2}\right)^2. \tag{10.84}$$

For a **spherically symmetrical potential,** $V(\mathbf{r}) = V(r)$—but *not* necessarily at low energy—the Born approximation again reduces to a simpler form. Define

$$\boldsymbol{\kappa} \equiv \mathbf{k}' - \mathbf{k}, \tag{10.85}$$

[16] You can't apply the Born approximation to *hard*-sphere scattering ($V_0 = \infty$)—the integral blows up. The point is that we assumed the potential is *weak*, and doesn't change the wave function much in the scattering region. But a *hard* sphere changes it *radically*—from Ae^{ikz} to *zero*.

and let the polar axis for the $\mathbf{r}_0$ integral lie along $\boldsymbol{\kappa}$, so that

$$\left(\mathbf{k}' - \mathbf{k}\right) \cdot \mathbf{r}_0 = \kappa r_0 \cos\theta_0. \tag{10.86}$$

Then

$$f(\theta) \approx -\frac{m}{2\pi\hbar^2} \int e^{i\kappa r_0 \cos\theta_0} V(r_0) r_0^2 \sin\theta_0 \, dr_0 \, d\theta_0 \, d\phi_0. \tag{10.87}$$

The ϕ_0 integral is trivial (2π), and the θ_0 integral is one we have encountered before (see Equation 10.59). Dropping the subscript on r, we are left with

$$\boxed{f(\theta) \approx -\frac{2m}{\hbar^2 \kappa} \int_0^\infty r V(r) \sin(\kappa r) \, dr, \text{ (spherical symmetry).}} \tag{10.88}$$

The angular dependence of f is carried by κ; in Figure 10.11 we see that

$$\kappa = 2k \sin(\theta/2). \tag{10.89}$$

Example 10.5

Yukawa scattering. The **Yukawa potential** (which is a crude model for the binding force in an atomic nucleus) has the form

$$V(r) = \beta \frac{e^{-\mu r}}{r}, \tag{10.90}$$

where β and μ are constants. The Born approximation gives

$$f(\theta) \approx -\frac{2m\beta}{\hbar^2 \kappa} \int_0^\infty e^{-\mu r} \sin(\kappa r) \, dr = -\frac{2m\beta}{\hbar^2 \left(\mu^2 + \kappa^2\right)}. \tag{10.91}$$

(You get to work out the integral for yourself, in Problem 10.11.)

Example 10.6

Rutherford scattering. If we put in $\beta = q_1 q_2 / 4\pi\epsilon_0$, $\mu = 0$, the Yukawa potential reduces to the Coulomb potential, describing the electrical interaction of two point charges. Evidently the scattering amplitude is

$$f(\theta) \approx -\frac{2m q_1 q_2}{4\pi\epsilon_0 \hbar^2 \kappa^2}, \tag{10.92}$$

or (using Equations 10.89 and 10.51):

$$f(\theta) \approx -\frac{q_1 q_2}{16\pi\epsilon_0 E \sin^2(\theta/2)}. \tag{10.93}$$

The differential cross-section is the square of this:

$$\frac{d\sigma}{d\Omega} = \left[\frac{q_1 q_2}{16\pi\epsilon_0 E \sin^2(\theta/2)}\right]^2, \tag{10.94}$$

which is precisely the Rutherford formula (Equation 10.11). It happens that for the Coulomb potential classical mechanics, the Born approximation, and quantum field theory all yield the same result. As they say in the computer business, the Rutherford formula is amazingly "robust."

* **Problem 10.10** Find the scattering amplitude, in the Born approximation, for soft-sphere scattering at arbitrary energy. Show that your formula reduces to Equation 10.82 in the low-energy limit.

Problem 10.11 Evaluate the integral in Equation 10.91, to confirm the expression on the right.

** **Problem 10.12** Calculate the total cross-section for scattering from a Yukawa potential, in the Born approximation. Express your answer as a function of E.

* **Problem 10.13** For the potential in Problem 10.4,

(a) calculate $f(\theta)$, $D(\theta)$, and σ, in the low-energy Born approximation;

(b) calculate $f(\theta)$ for arbitrary energies, in the Born approximation;

(c) show that your results are consistent with the answer to Problem 10.4, in the appropriate regime.

10.4.3 The Born Series

The Born approximation is similar in spirit to the **impulse approximation** in classical scattering theory. In the impulse approximation we begin by pretending that the particle keeps going in a straight line (Figure 10.12), and compute the transverse impulse that would be delivered to it in that case:

$$I = \int F_{\perp}\, dt. \tag{10.95}$$

If the deflection is relatively small, this should be a good approximation to the transverse momentum imparted to the particle, and hence the scattering angle is

$$\theta \approx \tan^{-1}(I/p), \tag{10.96}$$

where p is the incident momentum. This is, if you like, the "first-order" impulse approximation (the *zeroth*-order is what we *started* with: no deflection at all). Likewise, in the zeroth-order Born approximation the incident plane wave passes by with no modification, and what we

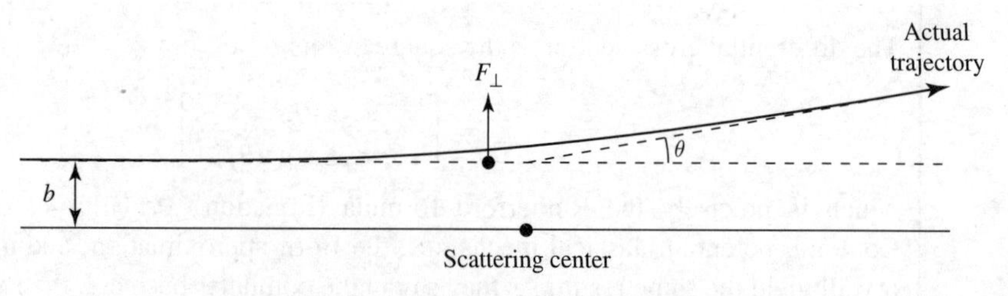

Figure 10.12: The impulse approximation assumes that the particle continues undeflected, and calculates the transverse momentum delivered.

explored in the previous section is really the first-order correction to this. But the same idea can be iterated to generate a series of higher-order corrections, which presumably converge to the exact answer.

The integral form of the Schrödinger equation reads

$$\psi(\mathbf{r}) = \psi_0(\mathbf{r}) + \int g(\mathbf{r} - \mathbf{r}_0) V(\mathbf{r}_0) \psi(\mathbf{r}_0)\, d^3\mathbf{r}_0, \tag{10.97}$$

where ψ_0 is the incident wave,

$$g(\mathbf{r}) \equiv -\frac{m}{2\pi\hbar^2} \frac{e^{ikr}}{r} \tag{10.98}$$

is the Green's function (into which I have now incorporated the factor $2m/\hbar^2$, for convenience), and V is the scattering potential. Schematically,

$$\psi = \psi_0 + \int gV\psi. \tag{10.99}$$

Suppose we take this expression for ψ, and plug it in under the integral sign:

$$\psi = \psi_0 + \int gV\psi_0 + \int\int gVgV\psi. \tag{10.100}$$

Iterating this procedure, we obtain a formal series for ψ:

$$\psi = \psi_0 + \int gV\psi_0 + \int\int gVgV\psi_0 + \int\int\int gVgVgV\psi_0 + \cdots. \tag{10.101}$$

In each integrand only the *incident* wave function (ψ_0) appears, together with more and more powers of gV. The *first* Born approximation truncates the series after the second term, but it is pretty clear how one generates the higher-order corrections.

The Born series can be represented diagrammatically as shown in Figure 10.13. In zeroth order ψ is untouched by the potential; in first order it is "kicked" once, and then "propagates" out in some new direction; in second order it is kicked, propagates to a new location, is kicked again, and then propagates out; and so on. In this context the Green's function is sometimes called the **propagator**—it tells you how the disturbance propagates between one interaction and the next. The Born series was the inspiration for Feynman's formulation of relativistic quantum mechanics, which is expressed entirely in terms of **vertex factors** (V) and propagators (g), connected together in **Feynman diagrams**.

Problem 10.14 Calculate θ (as a function of the impact parameter) for Rutherford scattering, in the impulse approximation. Show that your result is consistent with the exact expression (Problem 10.1(a)), in the appropriate limit.

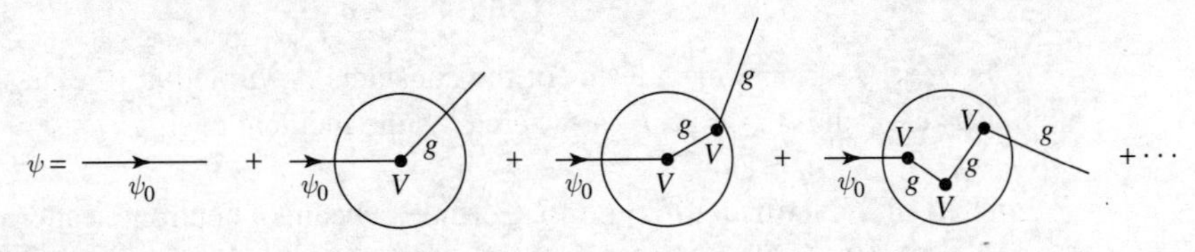

Figure 10.13: Diagrammatic interpretation of the Born series (Equation 10.101).

∗∗∗ **Problem 10.15** Find the scattering amplitude for low-energy soft-sphere scattering in the *second* Born approximation. *Answer:*

$$-\left(2mV_0a^3/3\hbar^2\right)\left[1-\left(4mV_0a^2/5\hbar^2\right)\right].$$

FURTHER PROBLEMS ON CHAPTER 10

∗∗∗ **Problem 10.16** Find the Green's function for the *one*-dimensional Schrödinger equation, and use it to construct the integral form (analogous to Equation 10.66). *Answer:*

$$\psi(x)=\psi_0(x)-\frac{im}{\hbar^2 k}\int_{-\infty}^{\infty} e^{ik|x-x_0|}V(x_0)\psi(x_0)\,dx_0. \tag{10.102}$$

∗∗ **Problem 10.17** Use your result in Problem 10.16 to develop the Born approximation for one-dimensional scattering (on the interval $-\infty<x<\infty$, with no "brick wall" at the origin). That is, choose $\psi_0(x)=Ae^{ikx}$, and assume $\psi(x_0)\approx\psi_0(x_0)$ to evaluate the integral. Show that the reflection coefficient takes the form:

$$R\approx\left(\frac{m}{\hbar^2 k}\right)^2\left|\int_{-\infty}^{\infty} e^{2ikx}V(x)\,dx\right|^2. \tag{10.103}$$

Problem 10.18 Use the one-dimensional Born approximation (Problem 10.17) to compute the transmission coefficient ($T=1-R$) for scattering from a delta function (Equation 2.117) and from a finite square well (Equation 2.148). Compare your results with the exact answers (Equations 2.144 and 2.172).

Problem 10.19 Prove the **optical theorem**, which relates the total cross-section to the imaginary part of the forward scattering amplitude:

$$\sigma=\frac{4\pi}{k}\text{Im}\left[f(0)\right]. \tag{10.104}$$

Hint: Use Equations 10.47 and 10.48.

Problem 10.20 Use the Born approximation to determine the total cross-section for scattering from a gaussian potential

$$V(\mathbf{r})=A\,e^{-\mu r^2}.$$

Express your answer in terms of the constants A, μ, and m (the mass of the incident particle), and $k\equiv\sqrt{2mE}/\hbar$, where E is the incident energy.

Problem 10.21 Neutron diffraction. Consider a beam of neutrons scattering from a crystal (Figure 10.14). The interaction between neutrons and the nuclei in the crystal is short ranged, and can be approximated as

$$V(\mathbf{r}) = \frac{2\pi\hbar^2 b}{m} \sum_i \delta^3(\mathbf{r} - \mathbf{r}_i) ,$$

where the $\mathbf{r}_i$ are the locations of the nuclei and the strength of the potential is expressed in terms of the **nuclear scattering length** b.

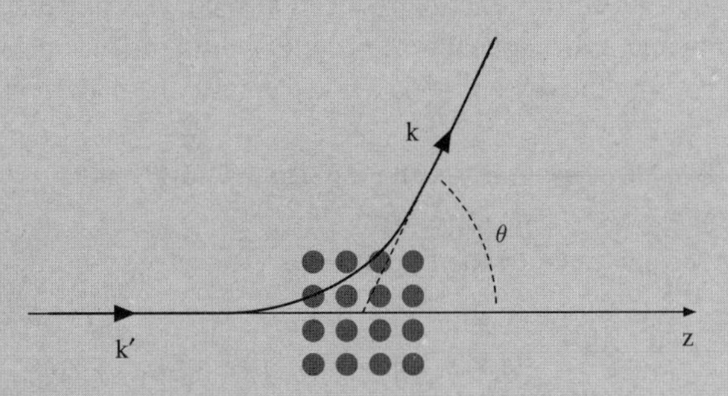

Figure 10.14: Neutron scattering from a crystal.

(a) In the first Born approximation, show that

$$\frac{d\sigma}{d\Omega} = b^2 \left| \sum_i e^{-i\,\mathbf{q}\cdot\mathbf{r}_i} \right|^2$$

where $\mathbf{q} \equiv \mathbf{k} - \mathbf{k}'$.

(b) Now consider the case where the nuclei are arranged on a cubic lattice with spacing a. Take the positions to be

$$\mathbf{r}_i = l\,a\,\hat{\imath} + m\,a\,\hat{\jmath} + n\,a\,\hat{k}$$

where l, m, and n all range from 0 to $N-1$, so there are a total of N^3 nuclei.[17] Show that

$$\frac{d\sigma}{d\Omega} = b^2 \frac{\sin^2(N\,q_x\,a/2)}{\sin^2(q_x\,a/2)} \frac{\sin^2\left(N\,q_y\,a/2\right)}{\sin^2\left(q_y\,a/2\right)} \frac{\sin^2(N\,q_z\,a/2)}{\sin^2(q_z\,a/2)} .$$

(c) Plot

$$\frac{1}{N} \frac{\sin^2(N\,q_x\,a/2)}{\sin^2(q_x\,a/2)}$$

as a function of $q_x\,a$ for several values of N ($N = 1, 5, 10$) to show that the function describes a series of peaks that become progressively sharper as N increases.

(d) In light of (c), in the limit of large N the differential scattering cross section is negligibly small except at one of these peaks:

$$\mathbf{q} = \mathbf{G}_{\ell mn} = \frac{2\pi}{a}\left(l\,\hat{\imath} + m\,\hat{\jmath} + n\,\hat{k}\right)$$

[17] It makes no difference that this crystal isn't "centered" at the origin: shifting the crystal by $\mathbf{R}$ amounts to adding $\mathbf{R}$ to each of the $\mathbf{r}_i$, and that doesn't affect $d\sigma/d\Omega$. After all, we're assuming an incident plane wave, which extends to $\pm\infty$ in the x and y directions.

for integer l, m, and n. The vectors $\mathbf{G}_{lmn}$ are called **reciprocal lattice vectors**. Find the scattering angles (θ) at which peaks occur. If the neutron's wavelength is equal to the crystal spacing a, what are the three smallest (nonzero) angles?

Comment: Neutron diffraction is one method used, to determine crystal structures (electrons and x-rays can also be used and the same expression for the locations of the peaks holds). In this problem we looked at a cubic arrangement of atoms, but a different arrangement (hexagonal for example) would produce peaks at a different set of angles. Thus from the scattering data one can infer the underlying crystal structure.

∗∗∗ **Problem 10.22 Two-dimensional scattering theory.** By analogy with Section 10.2, develop partial wave analysis for two dimensions.

(a) In polar coordinates (r, θ) the Laplacian is

$$\nabla^2 = \frac{\partial^2}{\partial x^2} + \frac{\partial^2}{\partial y^2} = \frac{\partial^2}{\partial r^2} + \frac{1}{r}\frac{\partial}{\partial r} + \frac{1}{r^2}\frac{\partial^2}{\partial \theta^2}. \tag{10.105}$$

Find the separable solutions to the (time-independent) Schrödinger equation, for a potential with azimuthal symmetry ($V(r, \theta) \to V(r)$). *Answer:*

$$\psi(r, \theta) = R(r)\, e^{ij\theta}, \tag{10.106}$$

where j is an integer, and $u \equiv \sqrt{r}\, R$ satisfies the radial equation

$$-\frac{\hbar^2}{2m}\frac{d^2 u}{dr^2} + \left[V(r) + \frac{\hbar^2}{2m}\frac{\left(j^2 - 1/4\right)}{r^2}\right] u = Eu. \tag{10.107}$$

(b) By solving the radial equation for very large r (where both $V(r)$ and the centrifugal term go to zero), show that an outgoing radial wave has the asymptotic form

$$R(r) \sim \frac{e^{ikr}}{\sqrt{r}}, \tag{10.108}$$

where $k \equiv \sqrt{2mE}/\hbar$. Check that an incident wave of the form Ae^{ikx} satisfies the Schrödinger equation, for $V(r) = 0$ (this is trivial, if you use cartesian coordinates). Write down the two-dimensional analog to Equation 10.12, and compare your result to Problem 10.2. *Answer:*

$$\psi(r, \theta) \approx A\left[e^{ikx} + f(\theta)\frac{e^{ikr}}{\sqrt{r}}\right], \quad \text{for large } r. \tag{10.109}$$

(c) Construct the analog to Equation 10.21 (the wave function in the region where $V(r) = 0$ but the centrifugal term *cannot* be ignored). *Answer:*

$$\psi(r, \theta) = A\left\{e^{ikx} + \sum_{j=-\infty}^{\infty} c_j H_j^{(1)}(kr) e^{ij\theta}\right\}, \tag{10.110}$$

where $H_j^{(1)}$ is the Hankel function (*not* the *spherical* Hankel function!) of order j.[18]

[18] See Mary Boas, *Mathematical Methods in the Physical Sciences,* 3rd edn (Wiley, New York, 2006), Section 12.17.

(d) For large z,

$$H_j^{(1)}(z) \sim \sqrt{2/\pi}\, e^{-i\pi/4}\,(-i)^j\,\frac{e^{iz}}{\sqrt{z}}. \tag{10.111}$$

Use this to show that

$$f(\theta) = \sqrt{2/\pi k}\, e^{-i\pi/4} \sum_{j=-\infty}^{\infty} (-i)^j\, c_j e^{ij\theta}. \tag{10.112}$$

(e) Adapt the argument of Section 10.1.2 to this two-dimensional geometry. Instead of the *area* $d\sigma$, we have a *length*, db, and in place of the solid angle $d\Omega$ we have the increment of scattering angle $|d\theta|$; the role of the differential cross-section is played by

$$D(\theta) \equiv \left|\frac{db}{d\theta}\right|, \tag{10.113}$$

and the effective "width" of the target (analogous to the total cross-section) is

$$B \equiv \int_0^{2\pi} D(\theta)\, d\theta. \tag{10.114}$$

Show that

$$D(\theta) = |f(\theta)|^2, \quad \text{and} \quad B = \frac{4}{k}\sum_{j=-\infty}^{\infty} |c_j|^2. \tag{10.115}$$

(f) Consider the case of scattering from a hard disk (or, in three dimensions, an infinite cylinder[19]) of radius a:

$$V(r) = \begin{cases} \infty, & (r \leq a), \\ 0, & (r > a). \end{cases} \tag{10.116}$$

By imposing appropriate boundary conditions at $r = a$, determine B. You'll need the analog to Rayleigh's formula:

$$e^{ikx} = \sum_{j=-\infty}^{\infty} (i)^j\, J_j(kr)\, e^{ij\theta} \tag{10.117}$$

(where J_j is the Bessel function of order J). Plot B as a function of ka, for $0 < ka < 2$.

Problem 10.23 Scattering of identical particles. The results for scattering of a particle from a fixed target also apply to the scattering of two particles in the center of mass frame. With $\psi(\mathbf{R}, \mathbf{r}) = \psi_R(\mathbf{R})\, \psi_r(\mathbf{r})$, $\psi_r(\mathbf{r})$ satisfies

$$-\frac{\hbar}{2\mu}\nabla^2\psi_r + V(r)\,\psi_r = E_r\,\psi_r \tag{10.118}$$

19 S. McAlinden and J. Shertzer, *Am. J. Phys.* **84**, 764 (2016).

(see Problem 5.1) where $V(r)$ is the interaction between the particles (assumed here to depend only on their separation distance). This is the *one*-particle Schrödinger equation (with the reduced mass μ in place of m).

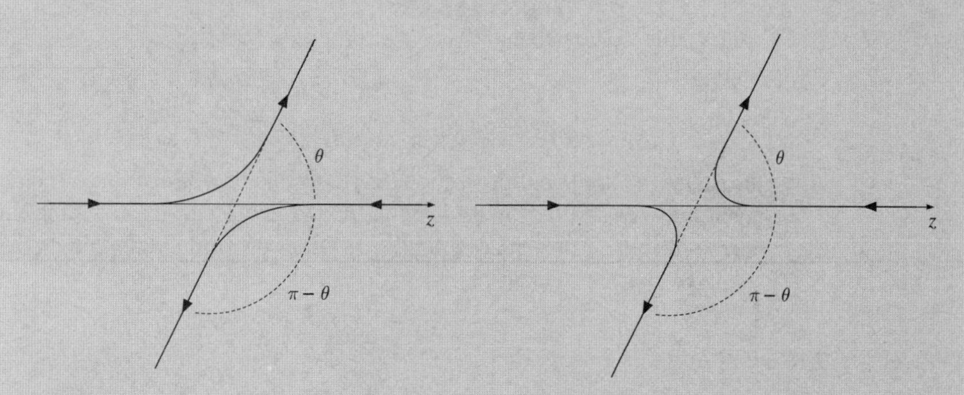

Figure 10.15: Scattering of identical particles.

(a) Show that if the two particles are identical (spinless) bosons, then $\psi_r(\mathbf{r})$ must be an even function of $\mathbf{r}$ (Figure 10.15).

(b) By symmetrizing Equation 10.12 (why is this allowed?), show that the scattering amplitude in this case is

$$f_B(\theta) = f(\theta) + f(\pi - \theta)$$

where $f(\theta)$ is the scattering amplitude of a single particle of mass μ from a fixed target $V(r)$.

(c) Show that the partial wave amplitudes of f_B vanish for all odd powers of ℓ.

(d) How are the results of (a)–(c) different if the particles are identical fermions (in a triplet spin state).

(e) Show that the scattering amplitude for identical fermions vanishes at $\pi/2$.

(f) Plot the logarithm of the differential scattering cross section for fermions and for bosons in Rutherford scattering (Equation 10.93).[20]

[20] Equation 10.93 was derived by taking the limit of Yukawa scattering (Example 10.5) and the result for $f(\theta)$ is missing a phase factor (see Albert Messiah, *Quantum Mechanics*, Dover, New York, NY (1999), Section XI.7). That factor drops out of the cross-section for scattering from a fixed potential—giving the correct answer in Example 10.6—but would show up in the cross-section for scattering of identical particles.

导读 / Guidance

第 11 章　量子动力学

量子力学中处理量子态的问题可以分成两类：一类是量子力学体系可能状态的问题，也就是力学量本征值和本征态的问题，本书前 10 章的内容所介绍的属于此类问题；另一类为本章所讨论的量子力学体系受到随时间变化的微扰作用，体系状态如何随时间演化的问题。首先，与国内教材不同，本书仍用两能级系统为例，介绍含时微扰理论，给出量子力学体系两个能级状态间的跃迁几率。其次，在此基础上讨论电磁波的吸收、受激辐射和自发辐射过程；对自发辐射过程，介绍自发辐射的爱因斯坦理论，给出爱因斯坦自发辐射速率系数 A、吸收系数 B、激发态寿命、选择定则和费米黄金规则等。最后，讨论量子力学体系的绝热定理，引出贝瑞相概念。作为小结，告诉读者本书所讨论的内容都是以绝热近似为基础的。

习题特色

（1）拓展课题：光电效应问题（光电截面），见习题 11.17；戴森方程问题，见习题 11.23；磁共振、核磁共振问题的求解，见习题 11.29；量子芝诺悖论，见习题 11.37。

（2）利用计算机计算跃迁几率问题，如习题 11.34、习题 11.38、习题 11.39。

11 QUANTUM DYNAMICS

So far, practically everything we have done belongs to the subject that might properly be called **quantum statics**, in which the *potential energy function is independent of time*: $V(\mathbf{r}, t) = V(\mathbf{r})$. In that case the (time-dependent) Schrödinger equation,

$$i\hbar \frac{\partial \Psi}{\partial t} = \hat{H}\Psi, \tag{11.1}$$

can be solved by separation of variables:

$$\Psi(\mathbf{r}, t) = \psi(\mathbf{r}) e^{-iEt/\hbar}, \tag{11.2}$$

where $\psi(\mathbf{r})$ satisfies the time-*in*dependent Schrödinger equation,

$$\hat{H}\psi = E\psi. \tag{11.3}$$

Because the time dependence of separable solutions is carried by the exponential factor $\left(e^{-iEt/\hbar}\right)$, which cancels out when we construct the physically relevant quantity $|\Psi|^2$, all probabilities and expectation values (for such states) are constant in time. By forming *linear combinations* of these stationary states we obtain wave functions with more interesting time dependence,

$$\Psi(\mathbf{r}, t) = \sum c_n \psi_n(\mathbf{r}) e^{-iE_n t/\hbar}, \tag{11.4}$$

but even then the possible values of the energy (E_n), and their respective probabilities $\left(|c_n|^2\right)$, are constant.

If we want to allow for **transitions** (**quantum jumps**, as they are sometimes called) between one energy level and another, we must introduce a *time-dependent* potential (**quantum dynamics**). There are precious few exactly solvable problems in quantum dynamics. However, if the time-dependent part of the Hamiltonian is *small* (compared to the time-independent part), it can be treated as a perturbation. The main purpose of this chapter is to develop **time-dependent perturbation theory**, and study its most important application: the emission or absorption of radiation by an atom.

Problem 11.1 Why isn't it *trivial* to solve the time-dependent Schrödinger equation (11.1), in its dependence on t? After all, it's a first-order differential equation.

(a) How would you solve the equation

$$\frac{df}{dt} = kf$$

(for $f(t)$), if k were a constant?

(b) What if k is itself a function of t? (Here $k(t)$ and $f(t)$ might also depend on other variables, such as $\mathbf{r}$—it doesn't matter.)

(c) Why not do the same thing for the Schrödinger equation (with a time-dependent Hamiltonian)? To see that this doesn't work, consider the simple case

$$\hat{H}(t) = \begin{cases} \hat{H}_1, & (0 < t < \tau), \\ \hat{H}_2, & (t > \tau), \end{cases}$$

where $\hat{H}_1$ and $\hat{H}_2$ are themselves time-independent. If the solution in part (b) held for the Schrödinger equation, the wave function at time $t > \tau$ would be

$$\Psi(t) = e^{-i\left[\hat{H}_1\tau + \hat{H}_2(t-\tau)\right]/\hbar}\Psi(0),$$

but of course we could also write

$$\Psi(t) = e^{-i\hat{H}_2(t-\tau)/\hbar}\Psi(\tau) = e^{-i\hat{H}_2(t-\tau)/\hbar}e^{-i\hat{H}_1\tau/\hbar}\Psi(0).$$

Why are these generally *not* the same? [This is a subtle matter; if you want to pursue it further, see Problem 11.23.]

11.1 TWO-LEVEL SYSTEMS

To begin with, let us suppose that there are just *two* states of the (unperturbed) system, ψ_a and ψ_b. They are eigenstates of the unperturbed Hamiltonian, $\hat{H}^0$:

$$\hat{H}^0\psi_a = E_a\psi_a, \quad \text{and} \quad \hat{H}^0\psi_b = E_b\psi_b, \tag{11.5}$$

and they are orthonormal:

$$\langle\psi_i|\psi_j\rangle = \delta_{ij}, \quad (i, j = a, b). \tag{11.6}$$

Any state can be expressed as a linear combination of them; in particular,

$$\Psi(0) = c_a\psi_a + c_b\psi_b. \tag{11.7}$$

The states ψ_a and ψ_b might be position-space wave functions, or spinors, or something more exotic—it doesn't matter. It is the *time* dependence that concerns us here, so when I write $\Psi(t)$, I simply mean the state of the system at time t. In the absence of any perturbation, each component evolves with its characteristic wiggle factor:

$$\Psi(t) = c_a\psi_a e^{-iE_a t/\hbar} + c_b\psi_b e^{-iE_b t/\hbar}. \tag{11.8}$$

Informally, we say that $|c_a|^2$ is the "probability that the particle is in state ψ_a"—by which we *really* mean the probability that a measurement of the energy would yield the value E_a. Normalization of Ψ requires, of course, that

$$|c_a|^2 + |c_b|^2 = 1. \tag{11.9}$$

11.1.1 The Perturbed System

Now suppose we turn on a time-dependent perturbation, $\hat{H}'(t)$. Since ψ_a and ψ_b constitute a complete set, the wave function $\Psi(t)$ can still be expressed as a linear combination of them. The only difference is that c_a and c_b are now *functions of t*:

$$\Psi(t) = c_a(t)\psi_a e^{-iE_a t/\hbar} + c_b(t)\psi_b e^{-iE_b t/\hbar}. \tag{11.10}$$

(I could absorb the exponential factors into $c_a(t)$ and $c_b(t)$, and some people prefer to do it this way, but I think it is nicer to keep visible the part of the time dependence that would be present even *without* the perturbation.) The whole problem is to determine c_a and c_b, as functions of time. If, for example, the particle started out in the state ψ_a ($c_a(0) = 1, c_b(0) = 0$), and at some later time t_1 we find that $c_a(t_1) = 0$, $c_b(t_1) = 1$, we shall report that the system underwent a transition from ψ_a to ψ_b.

We solve for $c_a(t)$ and $c_b(t)$ by demanding that $\Psi(t)$ satisfy the time-dependent Schrödinger equation,

$$\hat{H}\Psi = i\hbar\frac{\partial\Psi}{\partial t}, \quad \text{where} \quad \hat{H} = \hat{H}^0 + \hat{H}'(t). \tag{11.11}$$

From Equations 11.10 and 11.11, we find:

$$c_a\left(\hat{H}^0\psi_a\right)e^{-iE_a t/\hbar} + c_b\left(\hat{H}^0\psi_b\right)e^{-iE_b t/\hbar} + c_a\left(\hat{H}'\psi_a\right)e^{-iE_a t/\hbar}$$
$$+ c_b\left(\hat{H}'\psi_b\right)e^{-iE_b t/\hbar} = i\hbar\Big[\dot{c}_a\psi_a e^{-iE_a t/\hbar} + \dot{c}_b\psi_b e^{-iE_b t/\hbar}$$
$$+ c_a\psi_a\left(-\frac{iE_a}{\hbar}\right)e^{-iE_a t/\hbar} + c_b\psi_b\left(-\frac{iE_b}{\hbar}\right)e^{-iE_b t/\hbar}\Big].$$

In view of Equation 11.5, the first two terms on the left cancel the last two terms on the right, and hence

$$c_a\left(\hat{H}'\psi_a\right)e^{-iE_a t/\hbar} + c_b\left(\hat{H}'\psi_b\right)e^{-iE_b t/\hbar} = i\hbar\left(\dot{c}_a\psi_a e^{-iE_a t/\hbar} + \dot{c}_b\psi_b e^{-iE_b t/\hbar}\right). \tag{11.12}$$

To isolate $\dot{c}_a$, we use the standard trick: Take the inner product with ψ_a, and exploit the orthogonality of ψ_a and ψ_b (Equation 11.6):

$$c_a\left\langle\psi_a\left|\hat{H}'\right|\psi_a\right\rangle e^{-iE_a t/\hbar} + c_b\left\langle\psi_a\left|\hat{H}'\right|\psi_b\right\rangle e^{-iE_b t/\hbar} = i\hbar\dot{c}_a e^{-iE_a t/\hbar}.$$

For short, we define

$$H'_{ij} \equiv \left\langle\psi_i\left|\hat{H}'\right|\psi_j\right\rangle; \tag{11.13}$$

note that the hermiticity of $\hat{H}'$ entails $H'_{ji} = \left(H'_{ij}\right)^*$. Multiplying through by $-(i/\hbar)\,e^{iE_a t/\hbar}$, we conclude that:

$$\dot{c}_a = -\frac{i}{\hbar}\left[c_a H'_{aa} + c_b H'_{ab} e^{-i(E_b - E_a)t/\hbar}\right]. \tag{11.14}$$

Similarly, the inner product with ψ_b picks out $\dot{c}_b$:

$$c_a\left\langle\psi_b\left|\hat{H}'\right|\psi_a\right\rangle e^{-iE_a t/\hbar} + c_b\left\langle\psi_b\left|\hat{H}'\right|\psi_b\right\rangle e^{-iE_b t/\hbar} = i\hbar\dot{c}_b e^{-iE_b t/\hbar},$$

and hence

$$\dot{c}_b = -\frac{i}{\hbar}\left[c_b H'_{bb} + c_a H'_{ba} e^{i(E_b - E_a)t/\hbar}\right]. \tag{11.15}$$

Equations 11.14 and 11.15 determine $c_a(t)$ and $c_b(t)$; taken together, they are completely equivalent to the (time-dependent) Schrödinger equation, for a two-level system. Typically, the diagonal matrix elements of $\hat{H}'$ vanish (see Problem 11.5 for the general case):

$$H'_{aa} = H'_{bb} = 0. \tag{11.16}$$

If so, the equations simplify:

$$\dot{c}_a = -\frac{i}{\hbar} H'_{ab} e^{-i\omega_0 t} c_b, \quad \dot{c}_b = -\frac{i}{\hbar} H'_{ba} e^{i\omega_0 t} c_a, \tag{11.17}$$

where

$$\omega_0 \equiv \frac{E_b - E_a}{\hbar}. \tag{11.18}$$

(I'll assume that $E_b \geq E_a$, so $\omega_0 \geq 0$.)

* **Problem 11.2** A hydrogen atom is placed in a (time-dependent) electric field $\mathbf{E} = E(t)\hat{k}$. Calculate all four matrix elements H'_{ij} of the perturbation $\hat{H}' = eEz$ between the ground state ($n = 1$) and the (quadruply degenerate) first excited states ($n = 2$). Also show that $H'_{ii} = 0$ for all five states. *Note:* There is only one integral to be done here, if you exploit oddness with respect to z; only one of the $n = 2$ states is "accessible" from the ground state by a perturbation of this form, and therefore the system functions as a two-state configuration—assuming transitions to higher excited states can be ignored.

* **Problem 11.3** Solve Equation 11.17 for the case of a *time-independent* perturbation, assuming that $c_a(0) = 1$ and $c_b(0) = 0$. Check that $|c_a(t)|^2 + |c_b(t)|^2 = 1$. *Comment:* Ostensibly, this system oscillates between "pure ψ_a" and "some ψ_b." Doesn't this contradict my general assertion that no transitions occur for time-independent perturbations? No, but the reason is rather subtle: In this case ψ_a and ψ_b are not, and never were, eigenstates of the Hamiltonian—a measurement of the energy *never* yields E_a or E_b. In time-dependent perturbation theory we typically contemplate turning *on* the perturbation for a while, and then turning it *off* again, in order to examine the system. At the beginning, and at the end, ψ_a and ψ_b are eigenstates of the exact Hamiltonian, and only in this context does it make sense to say that the system underwent a transition from one to the other. For the present problem, then, assume that the perturbation was turned on at time $t = 0$, and off again at time T—this doesn't affect the *calculations*, but it allows for a more sensible interpretation of the result.

** **Problem 11.4** Suppose the perturbation takes the form of a delta function (in time):

$$\hat{H}' = \hat{U}\delta(t);$$

assume that $U_{aa} = U_{bb} = 0$, and let $U_{ab} = U^*_{ba} \equiv \alpha$. If $c_a(-\infty) = 1$ and $c_b(-\infty) = 0$, find $c_a(t)$ and $c_b(t)$, and check that $|c_a(t)|^2 + |c_b(t)|^2 = 1$. What is the net probability ($P_{a\to b}$ for $t \to \infty$) that a transition occurs? *Hint:* You might want to treat the delta function as the limit of a sequence of rectangles. *Answer:* $P_{a\to b} = \sin^2(|\alpha|/\hbar)$.

11.1.2 Time-Dependent Perturbation Theory

So far, everything is *exact*: We have made no assumption about the *size* of the perturbation. But if $\hat{H}'$ is "small," we can solve Equation 11.17 by a process of successive approximations, as follows. Suppose the particle starts out in the lower state:

$$c_a(0) = 1, \quad c_b(0) = 0. \tag{11.19}$$

If there were *no perturbation at all*, they would stay this way forever:

Zeroth Order:

$$c_a^{(0)}(t) = 1, \quad c_b^{(0)}(t) = 0. \tag{11.20}$$

(I'll use a superscript in parentheses to indicate the order of the approximation.)

To calculate the first-order approximation, we insert the zeroth-order values on the right side of Equation 11.17:

First Order:

$$\frac{dc_a^{(1)}}{dt} = 0 \Rightarrow c_a^{(1)}(t) = 1;$$
$$\frac{dc_b^{(1)}}{dt} = -\frac{i}{\hbar} H'_{ba} e^{i\omega_0 t} \Rightarrow c_b^{(1)} = -\frac{i}{\hbar} \int_0^t H'_{ba}(t') e^{i\omega_0 t'}\, dt'. \tag{11.21}$$

Now we insert *these* expressions on the right side of Equation 11.17 to obtain the *second*-order approximation:

Second Order:

$$\frac{dc_a^{(2)}}{dt} = -\frac{i}{\hbar} H'_{ab} e^{-i\omega_0 t} \left(-\frac{i}{\hbar}\right) \int_0^t H'_{ba}(t') e^{i\omega_0 t'}\, dt' \quad \Rightarrow$$
$$c_a^{(2)}(t) = 1 - \frac{1}{\hbar^2} \int_0^t H'_{ab}(t') e^{-i\omega_0 t'} \left[\int_0^{t'} H'_{ba}(t'') e^{i\omega_0 t''}\, dt'' \right] dt', \tag{11.22}$$

while c_b is unchanged $(c_b^{(2)}(t) = c_b^{(1)}(t))$. (Notice that $c_a^{(2)}(t)$ *includes* the zeroth-order term; the second-order *correction* would be the integral part alone.)

In principle, we could continue this ritual indefinitely, always inserting the nth-order approximation into the right side of Equation 11.17, and solving for the $(n+1)$th order. The zeroth order contains *no* factors of $\hat{H}'$, the first-order correction contains *one* factor of $\hat{H}'$, the second-order correction has *two* factors of $\hat{H}'$, and so on.[1] The error in the first-order approximation is evident in the fact that $\left|c_a^{(1)}(t)\right|^2 + \left|c_b^{(1)}(t)\right|^2 \neq 1$ (the *exact* coefficients must, of course, obey Equation 11.9). However, $\left|c_a^{(1)}(t)\right|^2 + \left|c_b^{(1)}(t)\right|^2$ *is* equal to 1 *to first order in* $\hat{H}'$, which is all we can expect from a first-order approximation. And the same goes for the higher orders.

Equation 11.21 can be written in the form

$$c_b^{(1)}(t)\, e^{-i E_b t/\hbar} = -\frac{i}{\hbar} \int_0^t e^{-i E_b (t-t')/\hbar}\, H'_{ba}(t')\, e^{-i E_a t'/\hbar} dt' \tag{11.23}$$

(where I've restored the exponential we factored out in Equation 11.10). This suggests a nice pictorial interpretation: reading from right to left, the system remains in state a from time 0 to time t' (picking up the "wiggle factor" $e^{-i E_a t'/\hbar}$), makes a transition from state a to state b at time t', and then remains in state b until time t (picking up the "wiggle factor" $e^{-i E_b(t-t')/\hbar}$). This process is represented in Figure 11.1. (Don't take the picture too literally: there is no

[1] Notice that c_a is modified in every *even* order, and c_b in every *odd* order; this would not be true if the perturbation included diagonal terms, or if the system started out in a linear combination of the two states.

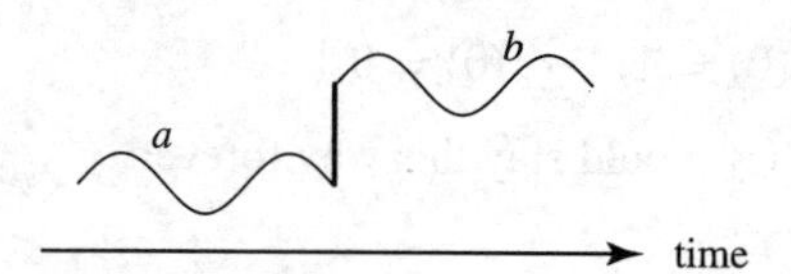

Figure 11.1: Pictorial representation of Equation 11.23.

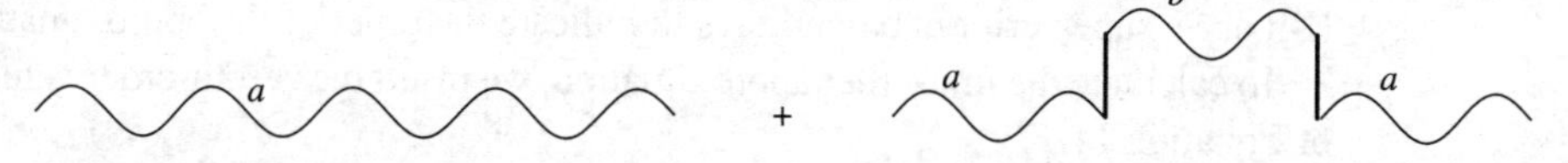

Figure 11.2: Pictorial representation of Equation 11.24.

sharp transition between these states; in fact, you integrate over all the times t' at which this transition can occur.)

This interpretation of the perturbation series is especially illuminating at higher orders and for multi-level systems, where the expressions become complicated. Consider Equation 11.22, which can be written

$$c_a^{(2)}(t)\, e^{-i E_a t/\hbar} = e^{-i E_a t/\hbar} + \left(-\frac{i}{\hbar}\right)^2 \int_0^t \int_0^{t'} e^{-i E_a (t-t')/\hbar} \times H'_{ab}\left(t'\right) e^{-i E_b (t'-t'')/\hbar} H'_{ba}\left(t''\right) e^{-i E_a t''/\hbar}\, dt''\, dt' . \tag{11.24}$$

The two terms here describe a process where the system remains in state a for the entire time, and a second process where the system transitions from a to b at time t'' and then back to a at time t'. Graphically, this is shown in Figure 11.2.

With the insight provided by these pictures, it is easy to write down the general result for a multi-level system:[2]

$$\begin{aligned} c_n^{(2)}(t)\, e^{-i E_n t/\hbar} &= \delta_{ni}\, e^{-i E_i t/\hbar} + \left(\frac{-i}{\hbar}\right) \int_0^t e^{-i E_n (t-t')/\hbar} H'_{ni}\left(t'\right) e^{-i E_i t'/\hbar}\, dt' \\ &+ \sum_m \left(-\frac{i}{\hbar}\right)^2 \int_0^t \int_0^{t'} e^{-i E_n (t-t')} H'_{nm}\left(t'\right) e^{-i E_m (t'-t'')/\hbar} \\ &\times H'_{mi}\left(t''\right) e^{-i E_i t''/\hbar}\, dt''\, dt' . \end{aligned} \tag{11.25}$$

For $n \neq i$, this is represented by the diagram in Figure 11.3. The first-order term describes a direct transition from i to n, and the second-order term describes a process where the transition occurs via an intermediate (or "virtual") state m.

** **Problem 11.5** Suppose you *don't* assume $H'_{aa} = H'_{bb} = 0$.

(a) Find $c_a(t)$ and $c_b(t)$ in first-order perturbation theory, for the case $c_a(0) = 1$, $c_b(0) = 0$. Show that $\left|c_a^{(1)}(t)\right|^2 + \left|c_b^{(1)}(t)\right|^2 = 1$, to first order in $\hat{H}'$.

(b) There is a nicer way to handle this problem. Let

$$d_a \equiv e^{\frac{i}{\hbar} \int_0^t H'_{aa}(t')dt'} c_a, \quad d_b \equiv e^{\frac{i}{\hbar} \int_0^t H'_{bb}(t')dt'} c_b. \tag{11.26}$$

[2] Perturbation theory for multi-level systems is treated in Problem 11.24.

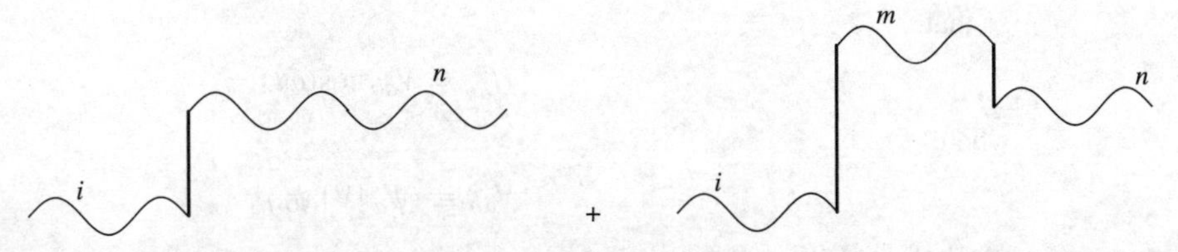

Figure 11.3: Pictorial representation of Equation 11.25 for $n \neq i$.

Show that

$$\dot{d}_a = -\frac{i}{\hbar} e^{i\phi} H'_{ab} e^{-i\omega_0 t} d_b; \quad \dot{d}_b = -\frac{i}{\hbar} e^{-i\phi} H'_{ba} e^{i\omega_0 t} d_a, \tag{11.27}$$

where

$$\phi(t) \equiv \frac{1}{\hbar} \int_0^t \left[H'_{aa}(t') - H'_{bb}(t')\right] dt'. \tag{11.28}$$

So the equations for d_a and d_b are identical in structure to Equation 11.17 (with an extra factor $e^{i\phi}$ tacked onto $\hat{H}'$).

(c) Use the method in part (b) to obtain $c_a(t)$ and $c_b(t)$ in first-order perturbation theory, and compare your answer to (a). Comment on any discrepancies.

* **Problem 11.6** Solve Equation 11.17 to second order in perturbation theory, for the general case $c_a(0) = a,\ c_b(0) = b$.

** **Problem 11.7** Calculate $c_a(t)$ and $c_b(t)$, to second order, for the perturbation in Problem 11.3. Compare your answer with the exact result.

* **Problem 11.8** Consider a perturbation to a two-level system with matrix elements

$$H'_{ab} = H'_{ba} = \frac{\alpha}{\sqrt{\pi}\,\tau} e^{-(t/\tau)^2}, \qquad H'_{aa} = H'_{bb} = 0.$$

where τ and α are positive constants with the appropriate units.

(a) According to first-order perturbation theory, if the system starts off in the state $c_a = 1, c_b = 0$ at $t = -\infty$, what is the probability that it will be found in the state b at $t = \infty$?

(b) In the limit that $\tau \to 0$, $H'_{ab} = \alpha\,\delta(t)$. Compute the $\tau \to 0$ limit of your expression from part (a) and compare the result of Problem 11.4.

(c) Now consider the opposite extreme: $\omega_0 \tau \gg 1$. What is the limit of your expression from part (a)? *Comment*: This is an example of the adiabatic theorem (Section 11.5.2).

11.1.3 Sinusoidal Perturbations

Suppose the perturbation has sinusoidal time dependence:

$$\hat{H}'(\mathbf{r}, t) = V(\mathbf{r}) \cos(\omega t), \tag{11.29}$$

so that

$$H'_{ab} = V_{ab}\cos(\omega t), \tag{11.30}$$

where

$$V_{ab} \equiv \langle \psi_a | V | \psi_b \rangle . \tag{11.31}$$

(As before, I'll assume the diagonal matrix elements vanish, since this is almost always the case in practice.) To first order (from now on we'll work *exclusively* in first order, and I'll dispense with the superscripts) we have (Equation 11.21):

$$\begin{aligned} c_b(t) &\approx -\frac{i}{\hbar} V_{ba} \int_0^t \cos(\omega t') e^{i\omega_0 t'}\, dt' = -\frac{i V_{ba}}{2\hbar} \int_0^t \left[e^{i(\omega_0+\omega)t'} + e^{i(\omega_0-\omega)t'} \right] dt' \\ &= -\frac{V_{ba}}{2\hbar} \left[\frac{e^{i(\omega_0+\omega)t} - 1}{\omega_0+\omega} + \frac{e^{i(\omega_0-\omega)t} - 1}{\omega_0-\omega} \right]. \end{aligned} \tag{11.32}$$

That's the *answer*, but it's a little cumbersome to work with. Things simplify substantially if we restrict our attention to driving frequencies (ω) that are very close to the transition frequency (ω_0), so that the second term in the square brackets dominates; specifically, we assume:

$$\omega_0 + \omega \gg |\omega_0 - \omega| . \tag{11.33}$$

This is not much of a limitation, since perturbations at *other* frequencies have a negligible probability of causing a transition anyway. Dropping the first term, we have

$$\begin{aligned} c_b(t) &\approx -\frac{V_{ba}}{2\hbar} \frac{e^{i(\omega_0-\omega)t/2}}{\omega_0-\omega} \left[e^{i(\omega_0-\omega)t/2} - e^{-i(\omega_0-\omega)t/2} \right] \\ &= -i \frac{V_{ba}}{\hbar} \frac{\sin[(\omega_0-\omega)t/2]}{\omega_0-\omega} e^{i(\omega_0-\omega)t/2}. \end{aligned} \tag{11.34}$$

The **transition probability**—the probability that a particle which started out in the state ψ_a will be found, at time t, in the state ψ_b—is

$$\boxed{P_{a\to b}(t) = |c_b(t)|^2 \approx \frac{|V_{ab}|^2}{\hbar^2} \frac{\sin^2[(\omega_0-\omega)t/2]}{(\omega_0-\omega)^2}.} \tag{11.35}$$

The most remarkable feature of this result is that, as a function of time, the transition probability *oscillates* sinusoidally (Figure 11.4). After rising to a maximum of $|V_{ab}|^2/\hbar^2(\omega_0-\omega)^2$—necessarily much less than 1, else the assumption that the perturbation is "small" would be invalid—it drops back down to zero! At times $t_n = 2n\pi/|\omega_0-\omega|$, where $n = 1, 2, 3, \ldots$, the particle is *certain* to be back in the lower state. If you want to maximize your chances of provoking a transition, you should *not* keep the perturbation on for a long period; you do better to *turn it off* after a time $\pi/|\omega_0-\omega|$, and hope to "catch" the system in the upper state. In Problem 11.9 it is shown that this "flopping" is not an artifact of perturbation theory—it occurs also in the exact solution, though the flopping *frequency* is modified somewhat.

As I noted earlier, the probability of a transition is greatest when the driving frequency is close to the "natural" frequency, ω_0.[3] This is illustrated in Figure 11.5, where $P_{a\to b}$ is plotted as a function of ω. The peak has a height of $(|V_{ab}|t/2\hbar)^2$ and a width $4\pi/t$; evidently it

[3] For *very* small t, $P_{a\to b}(t)$ is independent of ω; it takes a couple of cycles for the system to "realize" that the perturbation is periodic.

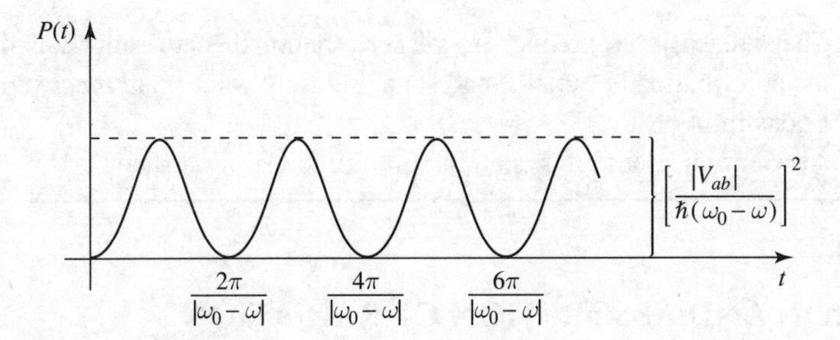

Figure 11.4: Transition probability as a function of time, for a sinusoidal perturbation (Equation 11.35).

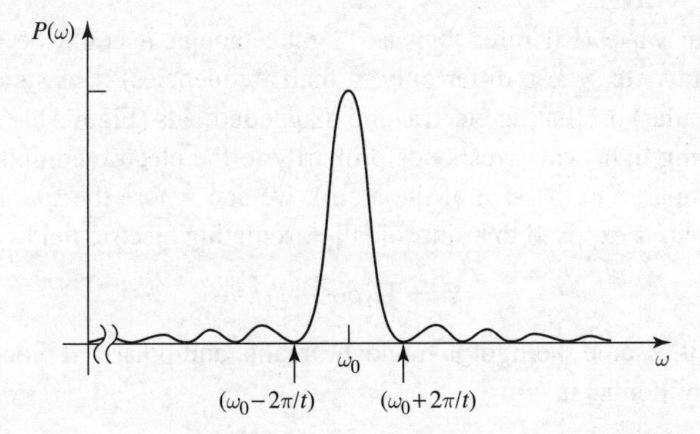

Figure 11.5: Transition probability as a function of driving frequency (Equation 11.35).

gets higher and narrower as time goes on. (Ostensibly, the maximum increases without limit. However, the perturbation assumption breaks down before it gets close to 1, so we can believe the result only for relatively small t. In Problem 11.9 you will see that the *exact* result never exceeds 1.)

** **Problem 11.9** The first term in Equation 11.32 comes from the $e^{i\omega t}/2$ part of $\cos(\omega t)$, and the second from $e^{-i\omega t}/2$. Thus dropping the first term is formally equivalent to writing $\hat{H}' = (V/2)\, e^{-i\omega t}$, which is to say,

$$H'_{ba} = \frac{V_{ba}}{2} e^{-i\omega t}, \quad H'_{ab} = \frac{V_{ab}}{2} e^{i\omega t}. \tag{11.36}$$

(The latter is required to make the Hamiltonian matrix hermitian—or, if you prefer, to pick out the dominant term in the formula analogous to Equation 11.32 for $c_a(t)$.) Rabi noticed that if you make this so-called **rotating wave approximation** at the *beginning* of the calculation, Equation 11.17 can be solved exactly, with no need for perturbation theory, and no assumption about the strength of the field.

(a) Solve Equation 11.17 in the rotating wave approximation (Equation 11.36), for the usual initial conditions: $c_a(0) = 1$, $c_b(0) = 0$. Express your results ($c_a(t)$ and $c_b(t)$) in terms of the **Rabi flopping frequency**,

$$\omega_r \equiv \frac{1}{2}\sqrt{(\omega - \omega_0)^2 + (|V_{ab}|/\hbar)^2}. \tag{11.37}$$

(b) Determine the transition probability, $P_{a\to b}(t)$, and show that it never exceeds 1. Confirm that $|c_a(t)|^2 + |c_b(t)|^2 = 1$.

(c) Check that $P_{a\to b}(t)$ reduces to the perturbation theory result (Equation 11.35) when the perturbation is "small," and state precisely what small *means* in this context, as a constraint on V.

(d) At what time does the system first return to its initial state?

11.2 EMISSION AND ABSORPTION OF RADIATION

11.2.1 Electromagnetic Waves

An electromagnetic wave (I'll refer to it as "light", though it could be infrared, ultraviolet, microwave, x-ray, etc.; these differ only in their frequencies) consists of transverse (and mutually perpendicular) oscillating electric and magnetic fields (Figure 11.6). An atom, in the presence of a passing light wave, responds primarily to the electric component. If the wavelength is long (compared to the size of the atom), we can ignore the *spatial* variation in the field;[4] the atom, then, is exposed to a sinusoidally oscillating electric field

$$\mathbf{E} = E_0 \cos(\omega t)\,\hat{k} \tag{11.38}$$

(for the moment I'll assume the light is monochromatic, and polarized along the z direction). The perturbing Hamiltonian is[5]

$$H' = -qE_0 z \cos(\omega t), \tag{11.39}$$

where q is the charge of the electron.[6] Evidently[7]

$$H'_{ba} = -\wp E_0 \cos(\omega t), \quad \text{where} \quad \wp \equiv q\langle\psi_b\,|\,z\,|\,\psi_a\rangle. \tag{11.40}$$

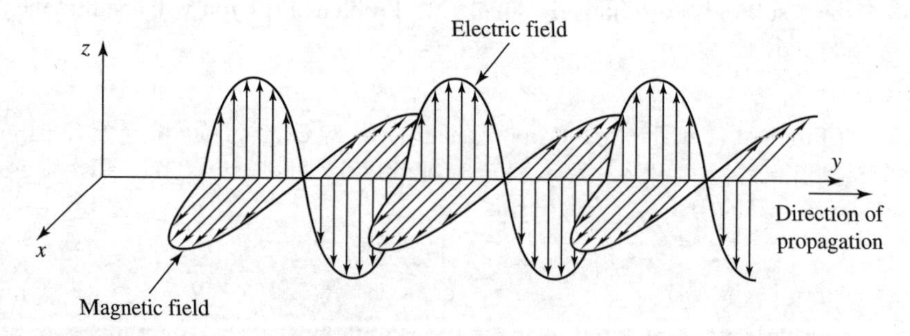

Figure 11.6: An electromagnetic wave.

[4] For visible light $\lambda \sim 5000$ Å, while the diameter of an atom is around 1 Å, so this approximation is reasonable; but it would *not* be for x-rays. Problem 11.31 explores the effect of spatial variation of the field.

[5] The energy of a charge q in a static field $\mathbf{E}$ is $-q\int \mathbf{E}\cdot d\mathbf{r}$. You may well object to the use of an electro*static* formula for a manifestly time-dependent field. I am implicitly assuming that the period of oscillation is long compared to the time it takes the charge to move around (within the atom).

[6] As usual, we assume the nucleus is heavy and stationary; it is the wave function of the *electron* that concerns us.

[7] The letter $\wp$ is supposed to remind you of **electric dipole moment** (for which, in electrodynamics, the letter p is customarily used—in this context it is rendered as a squiggly $\wp$ to avoid confusion with momentum). Actually, $\wp$ is the off-diagonal matrix element of the z component of the dipole moment operator, $q\mathbf{r}$. Because of its association with electric dipole moments, radiation governed by Equation 11.40 is called **electric dipole radiation**; it is overwhelmingly the dominant kind, at least in the visible region. See Problem 11.31 for generalizations and terminology.

Typically, ψ is an even or odd function of z; in either case $z\,|\psi|^2$ is odd, and integrates to zero (this is Laporte's rule, Section 6.4.3; for some examples see Problem 11.2). This licenses our usual assumption that the diagonal matrix elements of $\hat{H}'$ vanish. Thus the interaction of light with matter is governed by precisely the kind of oscillatory perturbation we studied in Section 11.1.3, with

$$V_{ba} = -\wp E_0. \tag{11.41}$$

11.2.2 Absorption, Stimulated Emission, and Spontaneous Emission

If an atom starts out in the "lower" state ψ_a, and you shine a polarized monochromatic beam of light on it, the probability of a transition to the "upper" state ψ_b is given by Equation 11.35, which (in view of Equation 11.41) takes the form

$$P_{a\to b}(t) = \left(\frac{|\wp|\,E_0}{\hbar}\right)^2 \frac{\sin^2[(\omega_0-\omega)\,t/2]}{(\omega_0-\omega)^2}. \tag{11.42}$$

In this process, the atom absorbs energy $E_b - E_a = \hbar\omega_0$ from the electromagnetic field, so it's called **absorption**. (Informally, we say that the atom has "absorbed a photon" (Figure 11.7(a).) Technically, the word "photon" belongs to **quantum electrodynamics**—the quantum theory of the electromagnetic field—whereas we are treating the field itself *classically*. But this language is convenient, as long as you don't read too much into it.)

I could, of course, go back and run the whole derivation for a system that starts off in the *upper* state ($c_a(0) = 0$, $c_b(0) = 1$). Do it for yourself, if you like; it comes out *exactly the same*—except that this time we're calculating $P_{b\to a} = |c_a(t)|^2$, the probability of a transition *down* to the *lower* level:

$$P_{b\to a}(t) = \left(\frac{|\wp|\,E_0}{\hbar}\right)^2 \frac{\sin^2[(\omega_0-\omega)\,t/2]}{(\omega_0-\omega)^2}. \tag{11.43}$$

(It *has* to come out this way—all we're doing is switching $a \leftrightarrow b$, which substitutes $-\omega_0$ for ω_0. When we get to Equation 11.32 we now keep the *first* term, with $-\omega_0 + \omega$ in the denominator, and the rest is the same as before.) But when you stop to think of it, this is an absolutely *astonishing* result: If the particle is in the *upper* state, and you shine light on it, it can make a transition to the *lower* state, and in fact the probability of such a transition is exactly the same as for a transition *upward* from the *lower* state. This process, which was first predicted by Einstein, is called **stimulated emission**.

In the case of stimulated emission the electromagnetic field *gains* energy $\hbar\omega_0$ from the atom; we say that one photon went in and *two* photons came out—the original one that caused the transition plus another one from the transition itself (Figure 11.7(b)). This raises the possibility of *amplification*, for if I had a bottle of atoms, all in the upper state, and triggered it with a single incident photon, a chain reaction would occur, with the first photon producing two, these two producing four, and so on. We'd have an enormous number of photons coming out, all with the same frequency and at virtually the same instant. This is the principle behind the **laser** (light

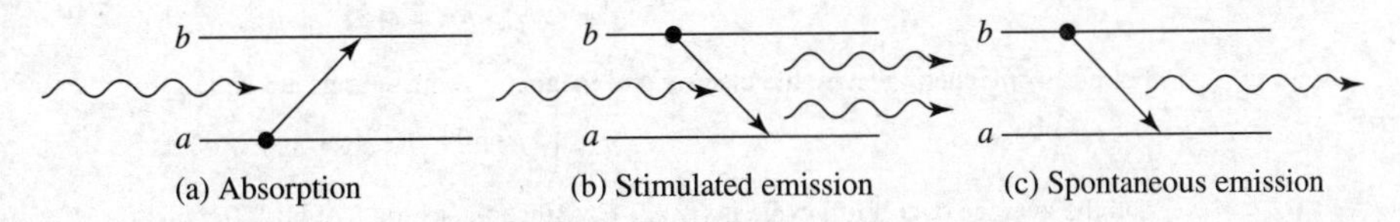

Figure 11.7: Three ways in which light interacts with atoms: (a) absorption, (b) stimulated emission, (c) spontaneous emission.

amplification by stimulated emission of radiation). Note that it is essential (for laser action) to get a majority of the atoms into the upper state (a so-called **population inversion**), because *absorption* (which *costs* one photon) competes with stimulated emission (which *creates* one); if you started with an even mixture of the two states, you'd get no amplification at all.

There is a *third* mechanism (in addition to absorption and stimulated emission) by which radiation interacts with matter; it is called **spontaneous emission**. Here an atom in the excited state makes a transition downward, with the release of a photon, but *without* any applied electromagnetic field to initiate the process (Figure 11.7(c)). This is the mechanism that accounts for the normal decay of an atomic excited state. At first sight it is far from clear why spontaneous emission should occur at *all*. If the atom is in a stationary state (albeit an excited one), and there is no external perturbation, it should just sit there forever. And so it *would*, if it were *really* free of all external perturbations. However, in quantum electrodynamics the fields are nonzero *even in the ground state*—just as the harmonic oscillator (for example) has nonzero energy (to wit: $\hbar\omega/2$) in its ground state. You can turn out all the lights, and cool the room down to absolute zero, but there is still some electromagnetic radiation present, and it is this "zero point" radiation that serves to catalyze spontaneous emission. When you come right down to it, there is really no such thing as *truly* spontaneous emission; it's *all* stimulated emission. The only distinction to be made is whether the field that does the stimulating is one that *you* put there, or one that *God* put there. In this sense it is exactly the reverse of the classical radiative process, in which it's *all* spontaneous, and there is no such thing as *stimulated* emission.

Quantum electrodynamics is beyond the scope of this book,[8] but there is a lovely argument, due to Einstein,[9] which interrelates the three processes (absorption, stimulated emission, and spontaneous emission). Einstein did not identify the *mechanism* responsible for spontaneous emission (perturbation by the ground-state electromagnetic field), but his results nevertheless enable us to calculate the spontaneous emission rate, and from that the natural lifetime of an excited atomic state.[10] Before we turn to that, however, we need to consider the response of an atom to non-monochromatic, unpolarized, incoherent electromagnetic waves coming in from all directions—such as it would encounter, for instance, if it were immersed in thermal radiation.

11.2.3 Incoherent Perturbations

The energy density in an electromagnetic wave is[11]

$$u = \frac{\epsilon_0}{2} E_0^2, \tag{11.44}$$

---

[8] For an accessible treatment see Rodney Loudon, *The Quantum Theory of Light*, 2nd edn (Clarendon Press, Oxford, 1983).

[9] Einstein's paper was published in 1917, well before the Schrödinger equation. Quantum electrodynamics comes into the argument via the Planck blackbody formula, which dates from 1900.

[10] For an alternative derivation using "seat-of-the-pants" quantum electrodynamics, see Problem 11.11.

[11] David J. Griffiths, *Introduction to Electrodynamics*, 4th edn, (Cambridge University Press, Cambridge, UK, 2017), Section 9.2.3. In general, the energy per unit volume in electromagnetic fields is

$$u = (\epsilon_0/2)\,E^2 + (1/2\mu_0)\,B^2.$$

For electromagnetic waves, the electric and magnetic contributions are equal, so

$$u = \epsilon_0 E^2 = \epsilon_0 E_0^2 \cos^2(\omega t),$$

and the average over a full cycle is $(\epsilon_0/2)\,E_0^2$, since the average of $\cos^2$ (or $\sin^2$) is 1/2.

where E_0 is (as before) the amplitude of the electric field. So the transition probability (Equation 11.43) is (not surprisingly) proportional to the energy density of the fields:

$$P_{b\to a}(t) = \frac{2u}{\epsilon_0 \hbar^2} |\wp|^2 \frac{\sin^2[(\omega_0 - \omega)t/2]}{(\omega_0 - \omega)^2}. \tag{11.45}$$

But this is for a **monochromatic** wave, at a single frequency ω. In many applications the system is exposed to electromagnetic waves at a whole *range* of frequencies; in that case $u \to \rho(\omega)d\omega$, where $\rho(\omega)d\omega$ is the energy density in the frequency range $d\omega$, and the net transition probability takes the form of an integral:[12]

$$P_{b\to a}(t) = \frac{2}{\epsilon_0 \hbar^2} |\wp|^2 \int_0^\infty \rho(\omega) \left\{ \frac{\sin^2[(\omega_0 - \omega)t/2]}{(\omega_0 - \omega)^2} \right\} d\omega. \tag{11.46}$$

The term in curly brackets is sharply peaked about ω_0 (Figure 11.5), whereas $\rho(\omega)$ is ordinarily quite broad, so we may as well replace $\rho(\omega)$ by $\rho(\omega_0)$, and take it outside the integral:

$$P_{b\to a}(t) \approx \frac{2|\wp|^2}{\epsilon_0 \hbar^2} \rho(\omega_0) \int_0^\infty \frac{\sin^2[(\omega_0 - \omega)t/2]}{(\omega_0 - \omega)^2} d\omega. \tag{11.47}$$

Changing variables to $x \equiv (\omega_0 - \omega)t/2$, extending the limits of integration to $x = \pm\infty$ (since the integrand is essentially zero out there anyway), and looking up the definite integral

$$\int_{-\infty}^{\infty} \frac{\sin^2 x}{x^2} dx = \pi, \tag{11.48}$$

we find

$$P_{b\to a}(t) \approx \frac{\pi |\wp|^2}{\epsilon_0 \hbar^2} \rho(\omega_0) t. \tag{11.49}$$

This time the transition probability is proportional to t. The bizarre "flopping" phenomenon characteristic of a monochromatic perturbation gets "washed out" when we hit the system with an incoherent spread of frequencies. In particular, the **transition rate** ($R \equiv dP/dt$) is now a *constant*:

$$R_{b\to a} = \frac{\pi}{\epsilon_0 \hbar^2} |\wp|^2 \rho(\omega_0). \tag{11.50}$$

Up to now, we have assumed that the perturbing wave is coming in along the y direction (Figure 11.6), and polarized in the z direction. But we are interested in the case of an atom bathed in radiation coming from *all* directions, and with all possible polarizations; the energy in the fields ($\rho(\omega)$) is shared equally among these different modes. What we need, in place of $|\wp|^2$, is the *average* of $|\boldsymbol{\wp} \cdot \hat{n}|^2$, where

$$\boldsymbol{\wp} \equiv q\langle \psi_b | \mathbf{r} | \psi_a \rangle \tag{11.51}$$

(generalizing Equation 11.40), and the average is over all polarizations and all incident directions.

[12] Equation 11.46 assumes that the perturbations at different frequencies are *independent*, so that the total transition probability is a sum of the individual probabilities. If the different components are **coherent** (phase-correlated), then we should add *amplitudes* ($c_b(t)$), not *probabilities* ($|c_b(t)|^2$), and there will be cross-terms. For the applications we will consider the perturbations are always incoherent.

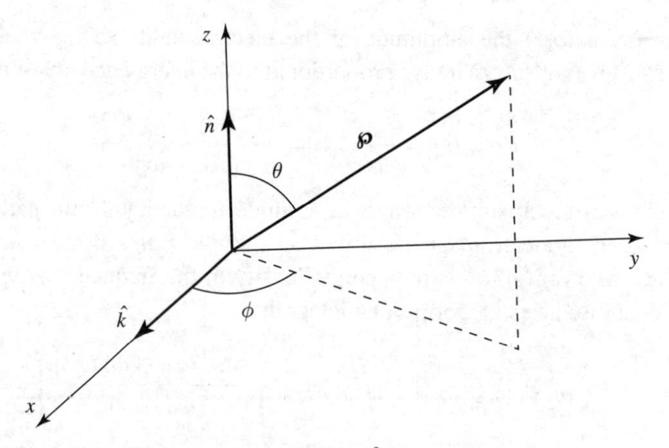

Figure 11.8: Axes for the averaging of $|\wp \cdot \hat{n}|^2$.

The averaging can be carried out as follows: Choose spherical coordinates such that the direction of propagation $(\hat{k})$ is along x, the polarization $(\hat{n})$ is along z, and the vector $\wp$ defines the spherical angles θ and ϕ (Figure 11.8).[13] (Actually, $\wp$ is *fixed*, here, and we're averaging over all $\hat{k}$ and $\hat{n}$ consistent with $\hat{k} \perp \hat{n}$—which is to say, over all θ and ϕ. But we might as well integrate over all directions of $\wp$, keeping $\hat{k}$ and $\hat{n}$ fixed—it amounts to the same thing.) Then

$$\wp \cdot \hat{n} = \wp \cos\theta, \tag{11.52}$$

and

$$\begin{aligned} |\wp \cdot \hat{n}|^2_{\text{ave}} &= \frac{1}{4\pi} \int |\wp|^2 \cos^2\theta \sin\theta \, d\theta \, d\phi \\ &= \frac{|\wp|^2}{4\pi} \left(-\frac{\cos^3\theta}{3} \right) \Big|_0^\pi (2\pi) = \frac{1}{3} |\wp|^2 . \end{aligned} \tag{11.53}$$

Conclusion: The transition rate for stimulated emission from state b to state a, under the influence of incoherent, unpolarized light incident from all directions, is

$$\boxed{R_{b\to a} = \frac{\pi}{3\epsilon_0 \hbar^2} |\wp|^2 \rho(\omega_0),} \tag{11.54}$$

where $\wp$ is the matrix element of the electric dipole moment between the two states (Equation 11.51), and $\rho(\omega_0)$ is the energy density in the fields, per unit frequency, evaluated at $\omega_0 = (E_b - E_a)/\hbar$.

---

[13] I'll treat $\wp$ as though it were *real*, even though in general it will be complex. Since

$$|\wp \cdot \hat{n}|^2 = |\text{Re}(\wp) \cdot \hat{n} + i\text{Im}(\wp) \cdot \hat{n}|^2 = |\text{Re}(\wp) \cdot \hat{n}|^2 + |\text{Im}(\wp) \cdot \hat{n}|^2$$

we can do the whole calculation for the real and imaginary parts separately, and simply add the results. In Equation 11.54 the absolute value signs denote *both* the vector magnitude *and* the complex amplitude:

$$|\wp|^2 = |\wp_x|^2 + |\wp_y|^2 + |\wp_z|^2 .$$

11.3 SPONTANEOUS EMISSION

11.3.1 Einstein's A and B Coefficients

Picture a container of atoms, N_a of them in the lower state (ψ_a), and N_b of them in the upper state (ψ_b). Let A be the spontaneous emission rate,[14] so that the number of particles leaving the upper state by this process, per unit time, is $N_b A$.[15] The transition rate for stimulated emission, as we have seen (Equation 11.54), is proportional to the energy density of the electromagnetic field: $B_{ba}\rho(\omega_0)$, where $B_{ba} = \pi\,|\boldsymbol{\wp}|^2/3\epsilon_0\hbar^2$; the number of particles leaving the upper state by this mechanism, per unit time, is $N_b B_{ba}\rho(\omega_0)$. The absorption rate is likewise proportional to $\rho(\omega_0)$—call it $B_{ab}\rho(\omega_0)$; the number of particles per unit time *joining* the upper level is therefore $N_a B_{ab}\rho(\omega_0)$. All told, then,

$$\frac{dN_b}{dt} = -N_b A - N_b B_{ba}\rho(\omega_0) + N_a B_{ab}\rho(\omega_0). \tag{11.55}$$

Suppose these atoms are in thermal equilibrium with the ambient field, so that the number of particles in each level is *constant*. In that case $dN_b/dt = 0$, and it follows that

$$\rho(\omega_0) = \frac{A}{(N_a/N_b)\,B_{ab} - B_{ba}}. \tag{11.56}$$

On the other hand, we know from statistical mechanics[16] that the number of particles with energy E, in thermal equilibrium at temperature T, is proportional to the **Boltzmann factor**, $\exp(-E/k_B T)$, so

$$\frac{N_a}{N_b} = \frac{e^{-E_a/k_B T}}{e^{-E_b/k_B T}} = e^{\hbar\omega_0/k_B T}, \tag{11.57}$$

and hence

$$\rho(\omega_0) = \frac{A}{e^{\hbar\omega_0/k_B T}\,B_{ab} - B_{ba}}. \tag{11.58}$$

But Planck's blackbody formula[17] tells us the energy density of thermal radiation:

$$\rho(\omega) = \frac{\hbar}{\pi^2 c^3}\frac{\omega^3}{e^{\hbar\omega/k_B T} - 1}; \tag{11.59}$$

comparing the two expressions, we conclude that

$$B_{ab} = B_{ba} \tag{11.60}$$

and

$$A = \frac{\omega_0^3\hbar}{\pi^2 c^3} B_{ba}. \tag{11.61}$$

---

[14] Normally I'd use R for a transition rate, but out of deference to *der Alte* everyone follows Einstein's notation in this context.

[15] Assume that N_a and N_b are very large, so we can treat them as continuous functions of time and ignore statistical fluctuations.

[16] See, for example, Daniel Schroeder, *An Introduction to Thermal Physics* (Pearson, Upper Saddle River, NJ, 2000), Section 6.1.

[17] Schroeder, footnote 16, Section 7.4.

Equation 11.60 confirms what we already knew: the transition rate for stimulated emission is the same as for absorption. But it was an astonishing result in 1917—indeed, Einstein was forced to "invent" stimulated emission in order to reproduce Planck's formula. Our present attention, however, focuses on Equation 11.61, for this tells us the spontaneous emission rate (A)—which is what we are looking for—in terms of the stimulated emission rate ($B_{ba}\rho(\omega_0)$)—which we already know. From Equation 11.54 we read off

$$B_{ba} = \frac{\pi}{3\epsilon_0 \hbar^2} |\wp|^2, \tag{11.62}$$

and it follows that the spontaneous emission rate is

$$\boxed{A = \frac{\omega_0^3 |\wp|^2}{3\pi \epsilon_0 \hbar c^3}.} \tag{11.63}$$

Problem 11.10 As a mechanism for downward transitions, spontaneous emission competes with thermally stimulated emission (stimulated emission for which blackbody radiation is the source). Show that at room temperature ($T = 300$ K) thermal stimulation dominates for frequencies well below 5×10^{12} Hz, whereas spontaneous emission dominates for frequencies well above 5×10^{12} Hz. Which mechanism dominates for visible light?

Problem 11.11 You could derive the spontaneous emission rate (Equation 11.63) without the detour through Einstein's A and B coefficients if you knew the ground state energy density of the electromagnetic field, $\rho_0(\omega)$, for then it would simply be a case of stimulated emission (Equation 11.54). To do this honestly would require quantum electrodynamics, but if you are prepared to believe that the ground state consists of *one photon in each classical mode*, then the derivation is fairly simple:

(a) To obtain the classical modes, consider an empty cubical box, of side l, with one corner at the origin. Electromagnetic fields (in vacuum) satisfy the classical wave equation[18]

$$\left(\frac{1}{c^2}\frac{\partial^2}{\partial t^2} - \nabla^2\right) f(x, y, z, t) = 0,$$

where f stands for any component of **E** or of **B**. Show that separation of variables, and the imposition of the boundary condition $f = 0$ on all six surfaces yields the standing wave patterns

$$f_{n_x,n_y,n_z} = A\cos(\omega t)\sin\left(\frac{n_x\pi}{l}x\right)\sin\left(\frac{n_y\pi}{l}y\right)\sin\left(\frac{n_z\pi}{l}z\right),$$

with

$$\omega = \frac{\pi c}{l}\sqrt{n_x^2 + n_y^2 + n_z^2}.$$

There are two modes for each triplet of positive integers $(n_x, n_y, n_z = 1, 2, 3, \ldots)$, corresponding to the two polarization states.

[18] Griffiths, footnote 11, Section 9.2.1.

(b) The energy of a photon is $E = h\nu = \hbar\omega$ (Equation 4.92), so the energy in the mode (n_x, n_y, n_z) is

$$E_{n_x,n_y,n_z} = 2\frac{\pi\hbar c}{l}\sqrt{n_x^2 + n_y^2 + n_z^2}.$$

What, then, is the *total* energy per unit volume in the frequency range $d\omega$, if each mode gets one photon? Express your answer in the form

$$\frac{1}{l^3}dE = \rho_0(\omega)\,d\omega$$

and read off $\rho_0(\omega)$. *Hint:* refer to Figure 5.3.

(c) Use your result, together with Equation 11.54, to obtain the spontaneous emission rate. Compare Equation 11.63.

11.3.2 The Lifetime of an Excited State

Equation 11.63 is our fundamental result; it gives the transition rate for spontaneous emission. Suppose, now, that you have somehow pumped a large number of atoms into the excited state. As a result of spontaneous emission, this number will decrease as time goes on; specifically, in a time interval dt you will lose a fraction $A\,dt$ of them:

$$dN_b = -AN_b\,dt, \tag{11.64}$$

(assuming there is no mechanism to replenish the supply).[19] Solving for $N_b(t)$, we find:

$$N_b(t) = N_b(0)e^{-At}; \tag{11.65}$$

evidently the number remaining in the excited state decreases exponentially, with a time constant

$$\tau = \frac{1}{A}. \tag{11.66}$$

We call this the **lifetime** of the state—technically, it is the time it takes for $N_b(t)$ to reach $1/e \approx 0.368$ of its initial value.

I have assumed all along that there are only *two* states for the system, but this was just for notational simplicity—the spontaneous emission formula (Equation 11.63) gives the transition rate for $\psi_b \to \psi_a$ regardless of what other states may be accessible (see Problem 11.24). Typically, an excited atom has many different **decay modes** (that is: ψ_b can decay to a large number of different lower-energy states, $\psi_{a_1}, \psi_{a_2}, \psi_{a_3}, \ldots$). In that case the transition rates *add*, and the net lifetime is

$$\tau = \frac{1}{A_1 + A_2 + A_3 + \cdots}. \tag{11.67}$$

[19] This situation is not to be confused with the case of thermal equilibrium, which we considered in the previous section. We assume here that the atoms have been lifted *out* of equilibrium, and are in the process of cascading back down to their equilibrium levels.

Example 11.1

Suppose a charge q is attached to a spring and constrained to oscillate along the x axis. Say it starts out in the state $|n\rangle$ (Equation 2.68), and decays by spontaneous emission to state $|n'\rangle$. From Equation 11.51 we have

$$\wp = q\,\langle n\,|\,x\,|\,n'\rangle\,\hat{\imath}.$$

You calculated the matrix elements of x back in Problem 3.39:

$$\langle n\,|\,x\,|\,n'\rangle = \sqrt{\frac{\hbar}{2m\omega}}\left(\sqrt{n'}\,\delta_{n,n'-1} + \sqrt{n}\,\delta_{n',n-1}\right),$$

where ω is the natural frequency of the oscillator (I no longer need this letter for the frequency of the stimulating radiation). But we're talking about *emission*, so n' must be *lower* than n; for our purposes, then,

$$\wp = q\sqrt{\frac{n\hbar}{2m\omega}}\,\delta_{n',n-1}\,\hat{\imath}. \tag{11.68}$$

Evidently transitions occur only to states one step lower on the "ladder", and the frequency of the photon emitted is

$$\omega_0 = \frac{E_n - E_{n'}}{\hbar} = \frac{(n+1/2)\,\hbar\omega - \left(n'+1/2\right)\hbar\omega}{\hbar} = \left(n-n'\right)\omega = \omega. \tag{11.69}$$

Not surprisingly, the system radiates at the classical oscillator frequency. The transition rate (Equation 11.63) is

$$A = \frac{nq^2\omega^2}{6\pi\epsilon_0 mc^3}, \tag{11.70}$$

and the lifetime of the nth stationary state is

$$\tau_n = \frac{6\pi\epsilon_0 mc^3}{nq^2\omega^2}. \tag{11.71}$$

Meanwhile, each radiated photon carries an energy $\hbar\omega$, so the *power* radiated is $A\hbar\omega$:

$$P = \frac{q^2\omega^2}{6\pi\epsilon_0 mc^3}\,(n\hbar\omega),$$

or, since the energy of an oscillator in the nth state is $E = (n+1/2)\,\hbar\omega$,

$$P = \frac{q^2\omega^2}{6\pi\epsilon_0 mc^3}\left(E - \frac{1}{2}\hbar\omega\right). \tag{11.72}$$

This is the average power radiated by a quantum oscillator with (initial) energy E.

For comparison, let's determine the average power radiated by a *classical* oscillator with the same energy. According to classical electrodynamics, the power radiated by an accelerating charge q is given by the **Larmor formula**:[20]

$$P = \frac{q^2a^2}{6\pi\epsilon_0 c^3}. \tag{11.73}$$

[20] See, for example, Griffiths, footnote 11, Section 11.2.1.

For a harmonic oscillator with amplitude x_0, $x(t) = x_0 \cos(\omega t)$, and the acceleration is $a = -x_0\omega^2 \cos(\omega t)$. Averaging over a full cycle, then,

$$P = \frac{q^2 x_0^2 \omega^4}{12\pi\epsilon_0 c^3}.$$

But the *energy* of the oscillator is $E = (1/2)\, m\omega^2 x_0^2$, so $x_0^2 = 2E/m\omega^2$, and hence

$$P = \frac{q^2\omega^2}{6\pi\epsilon_0 m c^3} E. \tag{11.74}$$

This is the average power radiated by a *classical* oscillator with energy E. In the classical limit ($\hbar \to 0$) the classical and quantum formulas agree;[21] however, the quantum formula (Equation 11.72) protects the ground state: If $E = (1/2)\,\hbar\omega$ the oscillator does not radiate.

Problem 11.12 The **half-life** $(t_{1/2})$ of an excited state is the time it would take for half the atoms in a large sample to make a transition. Find the relation between $t_{1/2}$ and τ (the "lifetime" of the state).

* **Problem 11.13** Calculate the lifetime (in seconds) for each of the four $n = 2$ states of hydrogen. *Hint:* You'll need to evaluate matrix elements of the form $\langle\psi_{100}\,|x|\,\psi_{200}\rangle$, $\langle\psi_{100}\,|y|\,\psi_{211}\rangle$, and so on. Remember that $x = r\sin\theta\cos\phi$, $y = r\sin\theta\sin\phi$, and $z = r\cos\theta$. Most of these integrals are zero, so inspect them closely before you start calculating. *Answer:* 1.60×10^{-9} seconds for all except ψ_{200}, which is infinite.

11.3.3 Selection Rules

The calculation of spontaneous emission rates has been reduced to a matter of evaluating matrix elements of the form

$$\langle\psi_b\,|\,\mathbf{r}\,|\,\psi_a\rangle.$$

As you will have discovered if you worked Problem 11.13, (if you *didn't*, go back right now and *do* so!) these quantities are very often *zero*, and it would be helpful to know in advance when this is going to happen, so we don't waste a lot of time evaluating unnecessary integrals. Suppose we are interested in systems like hydrogen, for which the Hamiltonian is spherically symmetrical. In that case we can specify the states with the usual quantum numbers n, ℓ, and m, and the matrix elements are

$$\langle n'\ell'm'\,|\,\mathbf{r}\,|\,n\ell m\rangle.$$

Now, $\mathbf{r}$ is a vector operator, and we can invoke the results of Chapter 6 to obtain the **selection rules**[22]

$$\boxed{\Delta\ell \equiv \ell' - \ell = \pm 1, \qquad \Delta m \equiv m' - m = 0 \text{ or } \pm 1.} \tag{11.75}$$

[21] This is an example of Bohr's **Correspondence Principle**. In fact, if we express P in terms of the energy *above the ground state*, the two formulas are identical.

[22] See Equation 6.62 (Equation 6.26 eliminates $\Delta\ell = 0$), or derive them from scratch using Problems 11.14 and 11.15.

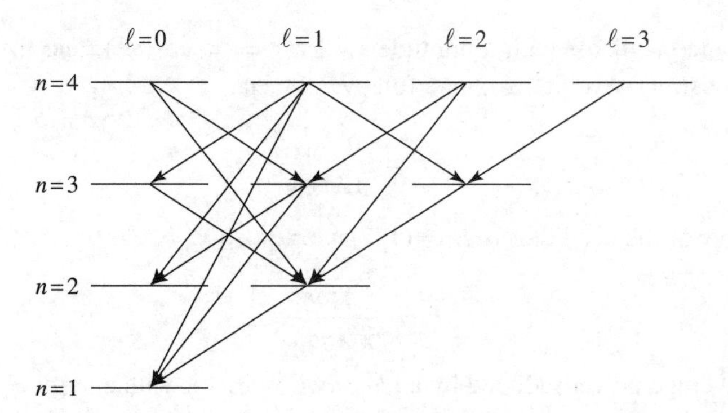

Figure 11.9: Allowed decays for the first four Bohr levels in hydrogen.

These conditions follow from symmetry alone. If they are *not* met, then the matrix element is zero, and the transition is said to be **forbidden**. Moreover, it follows from Equations 6.56–6.58 that

$$\begin{cases} \text{if } m' = m, & \text{then } \langle n'\ell'm' \,|\, x \,|\, n\ell m\rangle = \langle n'\ell'm' \,|\, y \,|\, n\ell m\rangle = 0, \\ \text{if } m' = m \pm 1, & \text{then } \langle n'\ell'm' \,|\, x \,|\, n\ell m\rangle = \pm i \langle n'\ell'm' \,|\, y \,|\, n\ell m\rangle \\ & \qquad \text{and } \langle n'\ell'm' \,|\, z \,|\, n\ell m\rangle = 0. \end{cases} \tag{11.76}$$

So it is never necessary to compute the matrix elements of both x *and* y; you can always get one from the other.

Evidently not all transitions to lower-energy states can proceed by electric dipole radiation; most are forbidden by the selection rules. The scheme of allowed transitions for the first four Bohr levels in hydrogen is shown in Figure 11.9. Notice that the $2S$ state (ψ_{200}) is "stuck": it cannot decay, because there is no lower-energy state with $\ell = 1$. It is called a **metastable** state, and its lifetime is indeed much longer than that of, for example, the $2P$ states (ψ_{211}, ψ_{210}, and ψ_{21-1}). Metastable states do eventually decay, by collisions, or by "forbidden" transitions (Problem 11.31), or by multiphoton emission.

Problem 11.14 From the commutators of L_z with x, y, and z (Equation 4.122):

$$\left[L_z, x\right] = i\hbar y, \quad \left[L_z, y\right] = -i\hbar x, \quad \left[L_z, z\right] = 0, \tag{11.77}$$

obtain the selection rule for Δm and Equation 11.76. *Hint:* Sandwich each commutator between $\langle n'\ell'm'|$ and $|n\ell m\rangle$.

** **Problem 11.15** Obtain the selection rule for $\Delta\ell$ as follows:

(a) Derive the commutation relation

$$\left[L^2, \left[L^2, \mathbf{r}\right]\right] = 2\hbar^2\left(\mathbf{r}L^2 + L^2\mathbf{r}\right). \tag{11.78}$$

Hint: First show that

$$\left[L^2, z\right] = 2i\hbar\left(xL_y - yL_x - i\hbar z\right).$$

Use this, and (in the final step) the fact that $\mathbf{r}\cdot\mathbf{L} = \mathbf{r}\cdot(\mathbf{r}\times\mathbf{p}) = \mathbf{0}$, to demonstrate that

$$\left[L^2, \left[L^2, z\right]\right] = 2\hbar^2\left(zL^2 + L^2 z\right).$$

The generalization from z to $\mathbf{r}$ is trivial.

(b) Sandwich this commutator between $\langle n'\ell'm'|$ and $|n\ell m\rangle$, and work out the implications.

** **Problem 11.16** An electron in the $n = 3$, $\ell = 0$, $m = 0$ state of hydrogen decays by a sequence of (electric dipole) transitions to the ground state.

(a) What decay routes are open to it? Specify them in the following way:

$$|300\rangle \to |n\ell m\rangle \to |n'\ell'm'\rangle \to \cdots \to |100\rangle .$$

(b) If you had a bottle full of atoms in this state, what fraction of them would decay via each route?

(c) What is the lifetime of this state? *Hint:* Once it's made the first transition, it's no longer in the state $|300\rangle$, so only the first step in each sequence is relevant in computing the lifetime.

11.4 FERMI'S GOLDEN RULE

In the previous sections we considered transitions between two *discrete* energy states, such as two bound states of an atom. We saw that such a transition was most likely when the final energy satisfied the resonance condition: $E_f = E_i + \hbar\omega$, where ω is the frequency associated with the perturbation. I now want to look at the case where E_f falls in a *continuum* of states (Figure 11.10). To stick close to the example of Section 11.2, if the radiation is energetic enough it can ionize the atom—the **photoelectric effect**—exciting the electron from a bound state into the continuum of scattering states.

We can't talk about a transition to a *precise* state in that continuum (any more than we can talk about someone being *precisely* 16 years old), but we can compute the probability that the system makes a transition to a state with an energy in some finite range ΔE about E_f. That is given by the integral of Equation 11.35 over all the final states:

$$P = \int_{E_f-\Delta E/2}^{E_f+\Delta E/2} \frac{|V_{in}|^2}{\hbar^2} \left\{ \frac{\sin^2[(\omega_0-\omega)\,t/2]}{(\omega_0-\omega)^2} \right\} \rho(E_n)\, dE_n\,, \tag{11.79}$$

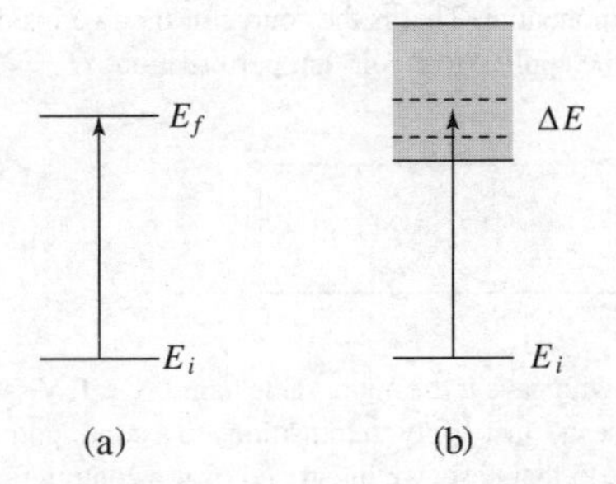

Figure 11.10: A transition (a) between two discrete states and (b) between a discrete state and a continuum of states.

where $\omega_0 = (E_n - E_i)/\hbar$. The quantity $\rho(E)\,dE$ is the number of states with energy between E and $E + dE$; $\rho(E)$ is called the **density of states**, and I'll show you how it's calculated in Example 11.2.

At short times, Equation 11.79 leads to a transition probability proportional to t^2, just as for a transition between discrete states. On the other hand, at long times the quantity in curly brackets in Equation 11.79 is sharply peaked: as a function of E_n its maximum occurs at $E_f = E_i + \hbar\omega$ and the central peak has a width of $4\pi\,\hbar/t$. For sufficiently large t, we can therefore approximate Equation 11.79 as[23]

$$P = \frac{|V_{if}|^2}{\hbar^2}\,\rho(E_f)\int_{-\infty}^{\infty}\frac{\sin^2\left[(\omega_0-\omega)\,t/2\right]}{(\omega_0-\omega)^2}\,dE_n\,.$$

The remaining integral was already evaluated in Section 11.2.3:

$$P = \frac{2\pi}{\hbar}\left|\frac{V_{if}}{2}\right|^2\rho(E_f)\,t\,. \tag{11.80}$$

The oscillatory behavior of P has again been "washed out," giving a constant transition rate:[24]

$$\boxed{R = \frac{2\pi}{\hbar}\left|\frac{V_{if}}{2}\right|^2\rho(E_f)\,.} \tag{11.81}$$

Equation 11.81 is known as **Fermi's Golden Rule**.[25] Apart from the factor of $2\pi/\hbar$, it says that the transition rate is the square of the matrix element (this encapsulates all the relevant information about the *dynamics* of the process) times the density of states (how many final states are accessible, given the energy supplied by the perturbation—the more roads are open, the faster the traffic will flow). It makes sense.

[23] This is the same set of approximations we made in Equations 11.46–11.48.

[24] In deriving Equation 11.35, our perturbation was

$$\hat{H}' = V\cos(\omega t) \to \frac{V}{2}e^{-i\omega t}$$

since we dropped the other (off-resonance) exponential. That is the source of the two inside the absolute value in Equation 11.81. Fermi's Golden rule can also be applied to a constant perturbation, $\hat{H}' = \hat{V}$, if we set $\omega = 0$ and drop the 2:

$$\boxed{R = \frac{2\pi}{\hbar}\,|V_{if}|^2\,\rho(E_f)\,.}$$

[25] It is actually due to Dirac, but Fermi is the one who gave it the memorable name. See T. Visser, *Am. J. Phys.* **77**, 487 (2009) for the history. Fermi's Golden Rule doesn't just apply to transitions to a continuum of states. For instance, Equation 11.54 can be considered an example. In that case, we integrated over a continuous range of *perturbation frequencies*—not a continuum of final states—but the end result is the same.

Example 11.2

Use Fermi's Golden Rule to obtain the differential scattering cross-section for a particle of mass m and incident wave vector $\mathbf{k}'$ scattering from a potential $V(\mathbf{r})$ (Figure 11.11).

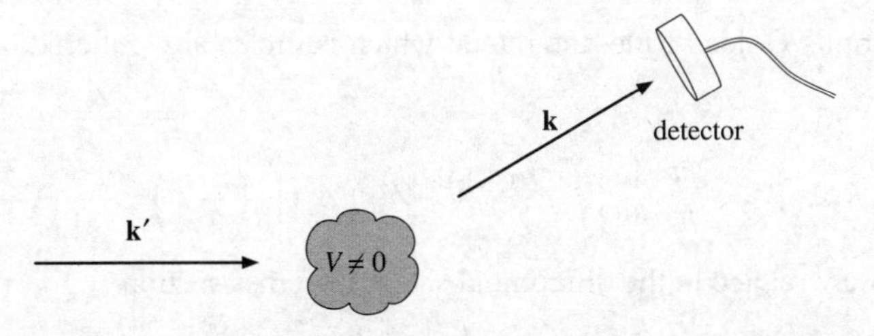

Figure 11.11: A particle with incident wave vector $\mathbf{k}'$ is scattered into a state with wave vector $\mathbf{k}$.

Solution:

We take our initial and final states to be plane waves:

$$\psi_i = \frac{1}{\sqrt{l^3}}\, e^{i\,\mathbf{k}'\cdot\mathbf{r}} \text{ and } \psi_f = \frac{1}{\sqrt{l^3}}\, e^{i\,\mathbf{k}\cdot\mathbf{r}}. \tag{11.82}$$

Here I've used a technique called **box normalization**; I place the whole setup inside a box of length l on a side. This makes the free-particle states normalizable and countable. Formally, we want the limit $l \to \infty$; in practice l will drop out of our final expression. Using periodic boundary conditions,[26] the allowed values of $\mathbf{k}$ are

$$\mathbf{k} = \frac{2\pi}{l}\left(n_x\,\hat{\imath} + n_y\,\hat{\jmath} + n_z\,\hat{k}\right) \tag{11.83}$$

for integers n_x, n_y, and n_z. Our perturbation is the scattering potential, $\hat{H}' = V(\mathbf{r})$, and the relevant matrix element is

$$V_{fi} = \int \psi_f^*(r)\, V(\mathbf{r})\, \psi_i(\mathbf{r})\, d^3\mathbf{r} = \frac{1}{l^3}\int e^{i(\mathbf{k}'-\mathbf{k})\cdot\mathbf{r}}\, V(\mathbf{r})\, d^3\mathbf{r}. \tag{11.84}$$

We need to determine the density of states. In a scattering experiment we measure the number of particles scatterred into a solid angle $d\Omega$. We want to count the number of states with energies between E and $E + dE$, with wave vectors $\mathbf{k}$ lying inside $d\Omega$. In k space these states occupy a section of a spherical shell of radius k and thickness dk that subtends a solid angle $d\Omega$; it has a volume

$$k^2\, dk\, d\Omega$$

and contains a number of states[27]

$$\rho(E)\, dE = \frac{k^2\, dk\, d\Omega}{(2\pi/l)^3} = \left(\frac{l}{2\pi}\right)^3 k^2 \frac{dk}{dE}\, dE\, d\Omega.$$

---

[26] Periodic boundary conditions are discussed in Problem 5.39. In the present context we use periodic boundary conditions—as opposed to impenetrable walls—because they admit traveling-wave solutions.

[27] Each state in k-space "occupies" a volume of $(2\pi/l)^3$, as shown in Problem 5.39.

Since $E = \hbar^2 k^2/2m$ this gives

$$\rho(E) = \left(\frac{l}{2\pi}\right)^3 \frac{\sqrt{2m^3 E}}{\hbar^3} \, d\Omega . \tag{11.85}$$

From Fermi's Golden Rule, the rate at which particles are scattered into the solid angle $d\Omega$ is[28]

$$R_{i\to d\Omega} = \frac{2\pi}{\hbar}\frac{1}{l^6}\left|\int e^{i(\mathbf{k}'-\mathbf{k})\cdot\mathbf{r}}\, V(\mathbf{r})\, d^3\mathbf{r}\right|^2 \left(\frac{l}{2\pi}\right)^3 \frac{\sqrt{2m^3 E_f}}{\hbar^3}\, d\Omega .$$

This is closely related to the differential scattering cross section:

$$\frac{d\sigma}{d\Omega} = \frac{R_{i\to d\Omega}}{J_i\, d\Omega} \tag{11.86}$$

where J_i is the flux (or probability current) of incident particles. For an incident wave of the form $\psi_i = A\, e^{i\,\mathbf{k}'\cdot\mathbf{r}}$, the probability current is (Equation 4.220).

$$J_i = |A|^2\, v = \frac{1}{l^3}\frac{\hbar k'}{m} \tag{11.87}$$

and

$$\frac{d\sigma}{d\Omega} = \left| -\frac{m}{2\pi\hbar^2}\int e^{i(\mathbf{k}-\mathbf{k}')\cdot\mathbf{r}}\, V(\mathbf{r})\, d^3\mathbf{r}\right|^2 . \tag{11.88}$$

This is exactly what we got from the first Born approximation (Equation 10.79).

∗∗∗ **Problem 11.17** In the photoelectric effect, light can ionize an atom if its energy ($\hbar\omega$) exceeds the binding energy of the electron. Consider the photoelectric effect for the ground state of hydrogen, where the electron is kicked out with momentum $\hbar\mathbf{k}$. The initial state of the electron is $\psi_0(r)$ (Equation 4.80) and its final state is[29]

$$\psi_f = \frac{1}{\sqrt{l^3}} e^{i\,\mathbf{k}\cdot\mathbf{r}},$$

as in Example 11.2.

(a) For light polarized along the z direction, use Fermi's Golden Rule to compute the rate at which electrons are ejected into the solid angle $d\Omega$ in the dipole approximation.[30]

$$\left[\textit{Answer}: R_{i\to d\Omega} = 256\alpha \frac{\epsilon_0 E_0^2 c}{2\hbar\omega}\frac{k^3 a^5}{[1+(ka)^2]^5}\cos^2\theta\, d\Omega . \right]$$

[28] See footnote 24.

[29] This is an approximation; we really should be using a scattering state of hydrogen. For an extended discussion of the photoelectric effect, including comparison to experiment and the validity of this approximation, see W. Heitler, *The Quantum Theory of Radiation*, 3rd edn, Oxford University Press, London (1954), Section 21.

[30] The result here is too large by a factor of four; correcting this requires a more careful derivation of the matrix element for radiative transitions (see Problem 11.30). Only the overall factor is affected though; the more interesting features (the dependence on k and θ) are correct.

Hint: To evaluate the matrix element, use the following trick. Write

$$z\, e^{i\,\mathbf{k}\cdot\mathbf{r}} = -i\,\frac{d}{dk_z}\,e^{i\,\mathbf{k}\cdot\mathbf{r}},$$

pull d/dk_z outside the integral, and what remains is straightforward to compute.

(b) The **photoelectric cross section** is defined as

$$\sigma(k) = \frac{R_{i\to\text{all}}\,\hbar\,\omega}{\frac{1}{2}\,\epsilon_0\,E_0^2\,c}$$

where the quantity in the numerator is the rate at which *energy* is absorbed ($\hbar\omega = \frac{\hbar^2k^2}{2m} - E_1$ per photoelectron) and the quantity in the denominator is the intensity of the incident light. Integrate your result from (a) over all angles to obtain $R_{i\to\text{all}}$, and compute the photoelectric cross section.

(c) Obtain a numerical value for the photoelectric cross section for ultraviolet light of wavelength 220 Å (n.b. this is the wavelength of the incident light, not the scattered electron). Express your answer in mega-barns $\left(\text{Mb} = 10^{-22}\,\text{m}^2\right)$.

11.5 THE ADIABATIC APPROXIMATION

11.5.1 Adiabatic Processes

Imagine a perfect pendulum, with no friction or air resistance, oscillating back and forth in a vertical plane. If you grab the support and shake it in a jerky manner the bob will swing around chaotically. But if you *very gently* move the support (Figure 11.12), the pendulum will continue to swing in a nice smooth way, in the same plane (or one parallel to it), with the same amplitude. This *gradual change of the external conditions* defines an **adiabatic** process. Notice that there are two characteristic times involved: T_i, the "internal" time, representing the motion of the system itself (in this case the period of the pendulum's oscillations), and T_e, the "external" time, over which the parameters of the system change appreciably (if the pendulum were mounted on a rotating platform, for example, T_e would be the period of the *platform's* motion). An adiabatic process is one for which $T_e \gg T_i$ (the pendulum executes many oscillations before the platform has moved appreciably).[31]

What if I took this pendulum up to the North Pole, and set it swinging—say, in the direction of Portland (Figure 11.13). For the moment, pretend the earth is not rotating. Very gently (that is, *adiabatically*), I carry it down the longitude line passing through Portland, to the equator. At this point it is swinging north-south. Now I carry it (still swinging north–south) part way around the equator. And finally, I take it back up to the North Pole, along the new longitude line. The pendulum will no longer be swinging in the same plane as it was when I set out—indeed, the new plane makes an angle Θ with the old one, where Θ is the angle between the southbound and the northbound longitude lines. More generally, if you transport the pendulum around a closed loop on the surface of the earth, the angular deviation (between the initial plane of the swing and the final plane) is equal to the solid angle subtended by the path with respect to the center of the sphere, as you can prove for yourself if you are interested.

Incidentally, the **Foucault pendulum** is an example of precisely this sort of adiabatic transport around a closed loop on a sphere—only this time instead of *me* carrying the pendulum

[31] For an interesting discussion of classical adiabatic processes, see Frank S. Crawford, *Am. J. Phys.* **58**, 337 (1990).

Figure 11.12: Adiabatic motion: If the case is transported very gradually, the pendulum inside keeps swinging with the same amplitude, in a plane parallel to the original one.

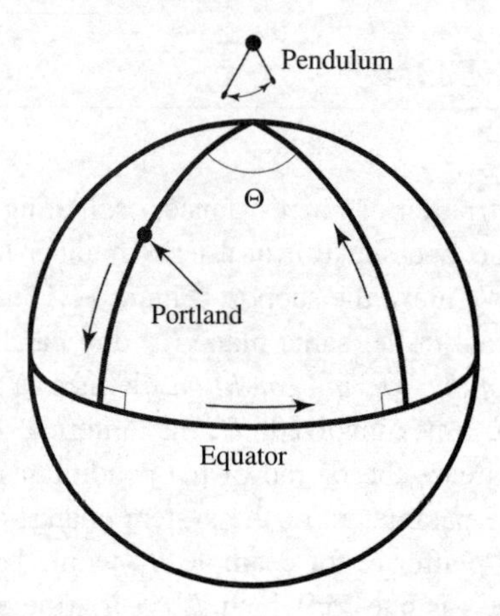

Figure 11.13: Itinerary for adiabatic transport of a pendulum on the surface of the earth.

around, I let the *rotation of the earth* do the job. The solid angle subtended by a latitude line θ_0 (Figure 11.14) is

$$\Omega = \int \sin\theta \, d\theta \, d\phi = 2\pi \left(-\cos\theta\right)\Big|_0^{\theta_0} = 2\pi \left(1 - \cos\theta_0\right). \tag{11.89}$$

Relative to the earth (which has meanwhile turned through an angle of 2π), the daily precession of the Foucault pendulum is $2\pi \cos\theta_0$—a result that is ordinarily obtained by appeal to Coriolis forces in the rotating reference frame,[32] but is seen in this context to admit a purely *geometrical* interpretation.

[32] See, for example, Jerry B. Marion and Stephen T. Thornton, *Classical Dynamics of Particles and Systems*, 4th edn, Saunders, Fort Worth, TX (1995), Example 10.5. Geographers measure latitude (λ) up from the equator, rather than down from the pole, so $\cos\theta_0 = \sin\lambda$.

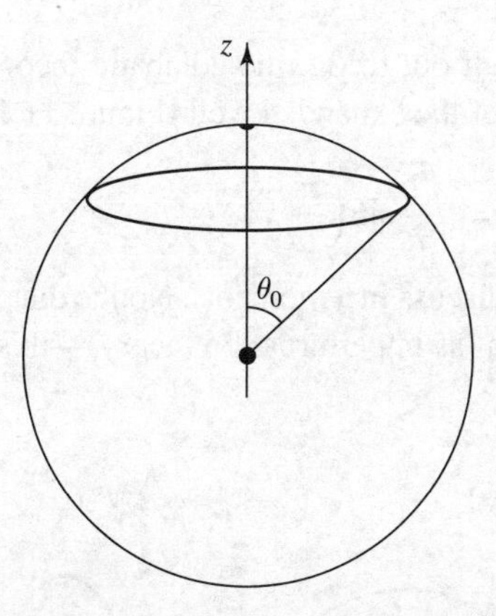

Figure 11.14: Path of a Foucault pendulum, in the course of one day.

The basic strategy for analyzing an adiabatic process is first to solve the problem with the external parameters held *constant*, and only at the *end* of the calculation allow them to vary (slowly) with time. For example, the classical period of a pendulum of (fixed) length L is $2\pi\sqrt{L/g}$; if the length is now gradually *changing*, the period will be $2\pi\sqrt{L(t)/g}$. A more subtle example occurred in our discussion of the hydrogen molecule ion (Section 8.3). We began by assuming that the nuclei were *at rest*, a fixed distance R apart, and we solved for the motion of the electron. Once we had found the ground state energy of the system as a function of R, we located the equilibrium separation and from the curvature of the graph we obtained the frequency of vibration of the nuclei (Problem 8.11). In molecular physics this technique (beginning with nuclei at rest, calculating electronic wave functions, and using these to obtain information about the positions and—relatively sluggish—motion of the nuclei) is known as the **Born–Oppenheimer approximation**.

11.5.2 The Adiabatic Theorem

In quantum mechanics, the essential content of the **adiabatic approximation** can be cast in the form of a theorem. Suppose the Hamiltonian changes *gradually* from some initial form $\hat{H}(0)$ to some final form $\hat{H}(T)$. The **adiabatic theorem**[33] states that if the particle was initially in the nth eigenstate of $\hat{H}(0)$, it will be carried (under the Schrödinger equation) into the nth eigenstate of $\hat{H}(T)$. (I assume that the spectrum is discrete and nondegenerate throughout the transition, so there is no ambiguity about the ordering of the states; these conditions can be relaxed, given a suitable procedure for "tracking" the eigenfunctions, but I'm not going to pursue that here.)

Example 11.3

Suppose we prepare a particle in the ground state of the infinite square well (Figure 11.15(a)):

$$\psi^i(x) = \sqrt{\frac{2}{a}}\sin\left(\frac{\pi}{a}x\right). \tag{11.90}$$

[33] The adiabatic theorem, which is usually attributed to Ehrenfest, is simple to state, and it *sounds* plausible, but it is not easy to prove. The argument will be found in earlier editions of this book, Section 10.1.2.

If we now gradually move the right wall out to $2a$, the adiabatic theorem says that the particle will end up in the ground state of the expanded well (Figure 11.15(b)):

$$\psi^f(x) = \sqrt{\frac{1}{a}} \sin\left(\frac{\pi}{2a}x\right), \tag{11.91}$$

(apart from a phase factor, which we'll discuss in a moment). Notice that we're not talking about a *small* change in the Hamiltonian (as in perturbation theory)—this one is *huge*. All we require is that it happen *slowly*.

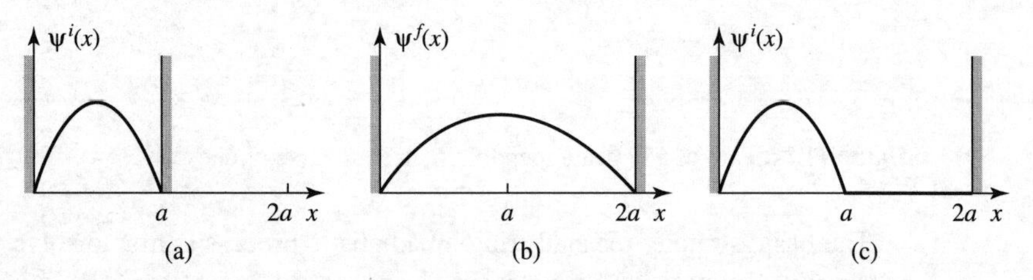

Figure 11.15: (a) Particle starts out in the ground state of the infinite square well. (b) If the wall moves *slowly*, the particle remains in the ground state. (c) If the wall moves *rapidly*, the particle is left (momentarily) in its initial state.

Energy is *not* conserved here—of *course* not: Whoever is moving the wall is extracting energy from the system, just like the piston on a slowly expanding cylinder of gas. By contrast, if the well expands *suddenly*, the resulting state is still $\psi^i(x)$ (Figure 11.15c), which is a complicated linear combination of eigenstates of the new Hamiltonian (Problem 11.18). In this case energy *is* conserved (at least, its *expectation value* is); as in the *free* expansion of a gas (into a vacuum) when the barrier is suddenly removed; no work is done.

According to the adiabatic theorem, a system that starts out in the nth eigenstate of the initial Hamiltonian ($\hat{H}(0)$) will evolve as the nth eigenstate of the instantaneous Hamiltonian ($\hat{H}(t)$), as the Hamiltonian gradually changes. However, this doesn't tell us what happens to the *phase* of the wave function. For a *constant* Hamiltonian it would pick up the standard "wiggle factor"

$$e^{-iE_n t/\hbar},$$

but the eigenvalue E_n may now itself be a function of time, so the wiggle factor naturally generalizes to

$$e^{i\theta_n(t)}, \quad \text{where} \quad \theta_n(t) \equiv -\frac{1}{\hbar}\int_0^t E_n(t')\,dt'. \tag{11.92}$$

This is called the **dynamic phase**. But it may not be the end of the story; for all we know there may be an additional phase factor, $\gamma_n(t)$, the so-called **geometric phase**. In the adiabatic limit, then, the wave function at time t takes the form[34]

$$\Psi_n(t) = e^{i\theta_n(t)} e^{i\gamma_n(t)} \psi_n(t), \tag{11.93}$$

[34] I'm suppressing the dependence on other variables; only the time dependence is at issue here.

where $\psi_n(t)$ is the nth eigenstate of the instantaneous Hamiltonian,

$$\hat{H}(t)\psi_n(t) = E_n(t)\psi_n(t). \tag{11.94}$$

Equation 11.93 is the formal statement of the adiabatic theorem.

Of course, the phase of $\psi_n(t)$ is itself arbitrary (it's still an eigenfunction, with the same eigenvalue, whatever phase you choose), so the geometric phase itself carries no physical significance. But what if we carry the system around a *closed cycle* (like the pendulum we hauled down to the equator, around, and back to the north pole), so that the Hamiltonian at the end is identical to the Hamiltonian at the beginning? Then the *net* phase change is a measurable quantity. The dynamic phase depends on the elapsed time, but the geometric phase, around an adiabatic closed cycle, depends only on the path taken.[35] It is called **Berry's phase**:[36]

$$\gamma_B \equiv \gamma(T) - \gamma(0). \tag{11.95}$$

Example 11.4

Imagine an electron (charge $-e$, mass m) at rest at the origin, in the presence of a magnetic field whose *magnitude* (B_0) is constant, but whose *direction* sweeps out a cone, of opening angle α, at constant angular velocity ω (Figure 11.16):

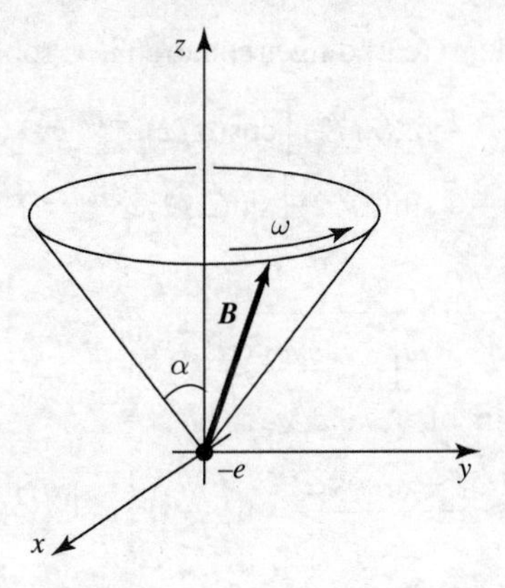

Figure 11.16: The magnetic field sweeps around in a cone, at angular velocity ω (Equation 11.96).

$$\mathbf{B}(t) = B_0\left[\sin\alpha\cos(\omega t)\hat{\imath} + \sin\alpha\sin(\omega t)\hat{\jmath} + \cos\alpha\hat{k}\right]. \tag{11.96}$$

The Hamiltonian (Equation 4.158) is

$$\begin{aligned}\hat{H}(t) &= \frac{e}{m}\mathbf{B}\cdot\mathbf{S} = \frac{e\hbar B_0}{2m}\left[\sin\alpha\cos(\omega t)\sigma_x + \sin\alpha\sin(\omega t)\sigma_y + \cos\alpha\sigma_z\right] \\ &= \frac{\hbar\omega_1}{2}\begin{pmatrix}\cos\alpha & e^{-i\omega t}\sin\alpha \\ e^{i\omega t}\sin\alpha & -\cos\alpha\end{pmatrix},\end{aligned} \tag{11.97}$$

[35] As Michael Berry puts it, the dynamic phase answers the question "How long did your trip take?" and the geometric phase, "Where have you been?"

[36] For more on this subject see Alfred Shapere and Frank Wilczek, eds., *Geometric Phases in Physics*, World Scientific, Singapore (1989); Andrei Bernevig and Taylor Hughes, *Topological Insulators and Topological Superconductors*, Princeton University Press, Princeton, NJ (2013), Chapter 2.

where

$$\omega_1 \equiv \frac{eB_0}{m}. \tag{11.98}$$

The normalized eigenspinors of $\hat{H}(t)$ are

$$\chi_+(t) = \begin{pmatrix} \cos(\alpha/2) \\ e^{i\omega t}\sin(\alpha/2) \end{pmatrix}, \tag{11.99}$$

and

$$\chi_-(t) = \begin{pmatrix} e^{-i\omega t}\sin(\alpha/2) \\ -\cos(\alpha/2) \end{pmatrix}; \tag{11.100}$$

they represent spin up and spin down, respectively, *along the instantaneous direction of* $\boldsymbol{B}(t)$ (see Problem 4.33). The corresponding eigenvalues are

$$E_\pm = \pm\frac{\hbar\omega_1}{2}. \tag{11.101}$$

Suppose the electron starts out with spin up, along $\mathbf{B}(0)$:

$$\chi(0) = \begin{pmatrix} \cos(\alpha/2) \\ \sin(\alpha/2) \end{pmatrix}. \tag{11.102}$$

The exact solution to the time-dependent Schrödinger equation is (Problem 11.20):

$$\chi(t) = \begin{pmatrix} \left[\cos(\lambda t/2) - i\frac{(\omega_1-\omega)}{\lambda}\sin(\lambda t/2)\right]\cos(\alpha/2)e^{-i\omega t/2} \\ \left[\cos(\lambda t/2) - i\frac{(\omega_1+\omega)}{\lambda}\sin(\lambda t/2)\right]\sin(\alpha/2)e^{+i\omega t/2} \end{pmatrix}, \tag{11.103}$$

where

$$\lambda \equiv \sqrt{\omega^2 + \omega_1^2 - 2\omega\omega_1\cos\alpha}, \tag{11.104}$$

or, expressing it as a linear combination of χ_+ and χ_-:

$$\chi(t) = \left[\cos\left(\frac{\lambda t}{2}\right) - i\frac{(\omega_1 - \omega\cos\alpha)}{\lambda}\sin\left(\frac{\lambda t}{2}\right)\right]e^{-i\omega t/2}\chi_+(t) + i\left[\frac{\omega}{\lambda}\sin\alpha\sin\left(\frac{\lambda t}{2}\right)\right]e^{+i\omega t/2}\chi_-(t). \tag{11.105}$$

Evidently the (exact) probability of a transition to spin down (along the current direction of $\mathbf{B}$) is

$$|\langle\chi(t)\,|\,\chi_-(t)\rangle|^2 = \left[\frac{\omega}{\lambda}\sin\alpha\sin\left(\frac{\lambda t}{2}\right)\right]^2. \tag{11.106}$$

The adiabatic theorem says that this transition probability should vanish in the limit $T_e \gg T_i$, where T_e is the characteristic time for changes in the Hamiltonian (in this case, $1/\omega$) and T_i is the characteristic time for changes in the wave function (in this case, $\hbar/(E_+ - E_-) = 1/\omega_1$). Thus the adiabatic approximation means $\omega \ll \omega_1$: the field rotates slowly, in comparison with the phase of the (unperturbed) wave functions. In the adiabatic regime $\lambda \approx \omega_1$ (Equation 11.104), and therefore

$$|\langle\chi(t)\,|\,\chi_-(t)\rangle|^2 \approx \left[\frac{\omega}{\omega_1}\sin\alpha\sin\left(\frac{\lambda t}{2}\right)\right]^2 \to 0, \tag{11.107}$$

as advertised. The magnetic field leads the electron around by its nose, with the spin always pointing in the direction of **B**. By contrast, if $\omega \gg \omega_1$ then $\lambda \approx \omega$, and the system bounces back and forth between spin up and spin down (Figure 11.17).

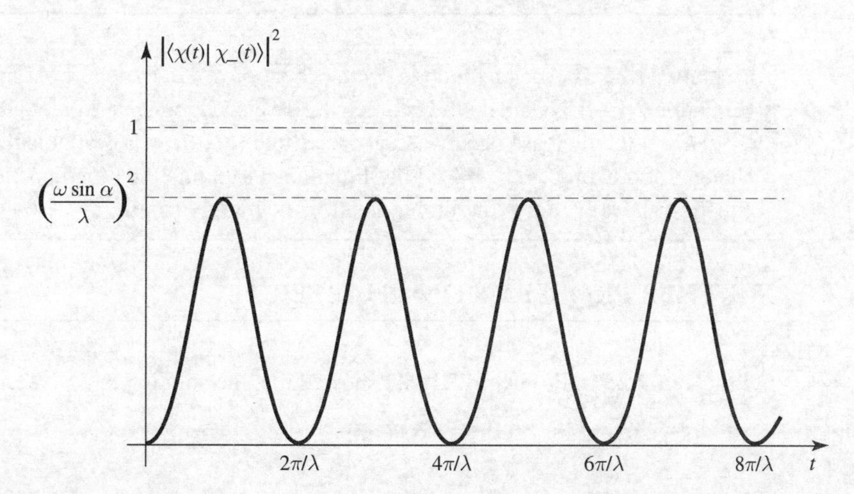

Figure 11.17: Plot of the transition probability, Equation 11.106, in the *non*-adiabatic regime ($\omega \gg \omega_1$).

* **Problem 11.18** A particle of mass m is in the ground state of the infinite square well (Equation 2.22). Suddenly the well expands to twice its original size—the right wall moving from a to $2a$—leaving the wave function (momentarily) undisturbed. The energy of the particle is now measured.

(a) What is the most probable result? What is the probability of getting that result?

(b) What is the *next* most probable result, and what is its probability? Suppose your measurement returned this value; what would you conclude about conservation of energy?

(c) What is the *expectation value* of the energy? *Hint:* If you find yourself confronted with an infinite series, try another method.

Problem 11.19 A particle is in the ground state of the harmonic oscillator with classical frequency ω, when suddenly the spring constant quadruples, so $\omega' = 2\omega$, without initially changing the wave function (of course, Ψ will now *evolve* differently, because the Hamiltonian has changed). What is the probability that a measurement of the energy would still return the value $\hbar\omega/2$? What is the probability of getting $\hbar\omega$? *Answer:* 0.943.

** **Problem 11.20** Check that Equation 11.103 satisfies the time-dependent Schrödinger equation for the Hamiltonian in Equation 11.97. Also confirm Equation 11.105, and show that the sum of the squares of the coefficients is 1, as required for normalization.

* **Problem 11.21** Find Berry's phase for one cycle of the process in Example 11.4. *Hint:* Use Equation 11.105 to determine the *total* phase change, and subtract off the dynamical part. You'll need to expand λ (Equation 11.104) to first order in ω/ω_1.

Problem 11.22 The delta function well (Equation 2.117) supports a single bound state (Equation 2.132). Calculate the geometric phase change when α gradually increases from α_1 to α_2. If the increase occurs at a constant rate ($d\alpha/dt = c$), what is the dynamic phase change for this process?[37] *Hint*: Plug Equation 11.93 into the time-dependent Schrödinger equation and solve for $\dot{\gamma}$, assuming that $\dot{\alpha}$ is negligibly small.

FURTHER PROBLEMS ON CHAPTER 11

*** **Problem 11.23** In Problem 11.1 you showed that the solution to

$$\frac{df}{dt} = k(t)\, f(t)$$

(where $k(t)$ is a function of t) is

$$f(t) = e^{K(t)} f(0), \quad \text{where} \quad K(t) \equiv \int_0^t k(t')\, dt'.$$

This suggests that the solution to the Schrödinger equation (11.1) might be

$$\Psi(t) = e^{\hat{G}(t)}\Psi(0), \quad \text{where} \quad \hat{G}(t) \equiv -\frac{i}{\hbar}\int_0^t \hat{H}(t')\, dt'. \tag{11.108}$$

It doesn't work, because $\hat{H}(t)$ is an *operator*, not a function, and $\hat{H}(t_1)$ does not (in general) commute with $\hat{H}(t_2)$.

(a) Try calculating $i\hbar\, \partial\Psi/\partial t$, using Equation 11.108. *Note:* as always, the exponentiated operator is to be interpreted as a power series:

$$e^{\hat{G}} \equiv 1 + \hat{G} + \frac{1}{2}\hat{G}\hat{G} + \frac{1}{3!}\hat{G}\hat{G}\hat{G} + \cdots.$$

Show that *if* $\left[\hat{G}, \hat{H}\right] = 0$, then Ψ satisfies the Schrödinger equation.

(b) Check that the correct solution in the general case $\left(\left[\hat{G}, \hat{H}\right] \neq 0\right)$ is

$$\Psi(t) = \left\{1 + \left(-\frac{i}{\hbar}\right)\int_0^t \hat{H}(t_1)\, dt_1 + \left(-\frac{i}{\hbar}\right)^2 \int_0^t \hat{H}(t_1)\left[\int_0^{t_1} \hat{H}(t_2)\, dt_2\right] dt_1 \right.$$
$$\left. + \left(-\frac{i}{\hbar}\right)^3 \int_0^t \hat{H}(t_1)\left[\int_0^{t_1} \hat{H}(t_2)\left(\int_0^{t_2} \hat{H}(t_3)\, dt_3\right) dt_2\right] dt_1 + \cdots\right\}\Psi(0). \tag{11.109}$$

[37] If $\psi_n(t)$ is *real*, the geometric phase vanishes. You might try to beat the rap by tacking an unnecessary (but perfectly legal) phase factor onto the eigenfunctions: $\psi_n'(t) \equiv e^{i\phi_n}\psi_n(t)$, where ϕ_n is an arbitrary (real) function. Try it. You'll get a nonzero geometric phase, all right, but note what happens when you put it back into Equation 11.93. And for a *closed* loop it gives *zero*.

UGLY! Notice that the operators in each term are "time-ordered," in the sense that the *latest* $\hat{H}$ appears at the far left, followed by the next latest, and so on $(t \geq t_1 \geq t_2 \geq t_3 \cdots)$. Dyson introduced the **time-ordered product** of two operators:

$$\mathbf{T}\left[\hat{H}(t_i)\hat{H}(t_j)\right] \equiv \begin{cases} \hat{H}(t_i)\hat{H}(t_j), & t_i \geq t_j \\ \hat{H}(t_j)\hat{H}(t_i), & t_j \geq t_i \end{cases} \tag{11.110}$$

or, more generally,

$$\mathbf{T}\left[\hat{H}(t_1)\hat{H}(t_2)\cdots\hat{H}(t_n)\right] \equiv \hat{H}(t_{j_1})\hat{H}(t_{j_2})\cdots\hat{H}(t_{j_n}), \tag{11.111}$$

where $t_{j_1} \geq t_{j_2} \geq \cdots \geq t_{j_n}$.

(c) Show that

$$\mathbf{T}\left[\hat{G}\hat{G}\right] = -\frac{2}{\hbar^2}\int_0^t \hat{H}(t_1)\left[\int_0^{t_1} \hat{H}(t_2)\,dt_2\right]dt_1,$$

and generalize to higher powers of $\hat{G}$. In place of $\hat{G}^n$, in equation 11.108, we really want $\mathbf{T}\left[\hat{G}^n\right]$:

$$\Psi(t) = \mathbf{T}\left[e^{-\frac{i}{\hbar}\int_0^t \hat{H}(t')\,dt'}\right]\Psi(0). \tag{11.112}$$

This is **Dyson's formula**; it's a compact way of writing Equation 11.109, the formal solution to Schrödinger's equation. Dyson's formula plays a fundamental role in quantum field theory.[38]

** **Problem 11.24** In this problem we develop time-dependent perturbation theory for a multi-level system, starting with the generalization of Equations 11.5 and 11.6:

$$\hat{H}_0\psi_n = E_n\psi_n, \quad \langle\psi_n|\psi_m\rangle = \delta_{nm}. \tag{11.113}$$

At time $t = 0$ we turn on a perturbation $H'(t)$, so that the total Hamiltonian is

$$\hat{H} = \hat{H}_0 + \hat{H}'(t). \tag{11.114}$$

[38] The **interaction picture** is intermediate between the Heisenberg and Schrödinger pictures (see Section 6.8.1). In the interaction picture, the wave function satisfies the "Schrödinger equation"

$$i\hbar\frac{d}{dt}|\Psi_I(t)\rangle = \hat{H}'_I(t)\,|\Psi_I(t)\rangle\,,$$

where the interaction- and Schrödinger-picture operators are related by

$$\hat{H}'_I(t) = e^{i\hat{H}_0 t/\hbar}\hat{H}'(t)e^{-i\hat{H}_0 t/\hbar}$$

and the wave functions satisfy

$$|\Psi_I(t)\rangle = e^{i\hat{H}_0 t/\hbar}\,|\Psi(t)\rangle\,.$$

If you apply the Dyson series to the Schrödinger equation in the interaction picture, you end up with precisely the perturbation series derived in Section 11.1.2. For more details see Ramamurti Shankar, *Principles of Quantum Mechanics*, 2nd edn, Springer, New York (1994), Section 18.3.

(a) Generalize Equation 11.10 to read

$$\Psi(t)=\sum_n c_n(t)\psi_n e^{-iE_n t/\hbar}, \tag{11.115}$$

and show that

$$\dot{c}_m=-\frac{i}{\hbar}\sum_n c_n H'_{mn} e^{i(E_m-E_n)t/\hbar}, \tag{11.116}$$

where

$$H'_{mn}\equiv\left\langle\psi_m\left|\hat{H}'\right|\psi_n\right\rangle. \tag{11.117}$$

(b) If the system starts out in the state ψ_N, show that (in first-order perturbation theory)

$$c_N(t)\approx 1-\frac{i}{\hbar}\int_0^t H'_{NN}\left(t'\right)dt'. \tag{11.118}$$

and

$$c_m(t)\approx-\frac{i}{\hbar}\int_0^t H'_{mN}\left(t'\right)e^{i(E_m-E_N)t'/\hbar}\,dt',\quad (m\neq N). \tag{11.119}$$

(c) For example, suppose $\hat{H}'$ is *constant* (except that it was turned on at $t=0$, and switched off again at some later time T). Find the probability of transition from state N to state M ($M\neq N$), as a function of T. *Answer:*

$$4\left|H'_{MN}\right|^2\frac{\sin^2\left[(E_N-E_M)T/2\hbar\right]}{(E_N-E_M)^2}. \tag{11.120}$$

(d) Now suppose $\hat{H}'$ is a sinusoidal function of time: $\hat{H}'=V\cos(\omega t)$. Making the usual assumptions, show that transitions occur only to states with energy $E_M=E_N\pm\hbar\omega$, and the transition probability is

$$P_{N\to M}=|V_{MN}|^2\frac{\sin^2\left[(E_N-E_M\pm\hbar\omega)T/2\hbar\right]}{(E_N-E_M\pm\hbar\omega)^2}. \tag{11.121}$$

(e) Suppose a multi-level system is immersed in incoherent electromagnetic radiation. Using Section 11.2.3 as a guide, show that the transition rate for stimulated emission is given by the same formula (Equation 11.54) as for a two-level system.

Problem 11.25 For the examples in Problem 11.24 (c) and (d), calculate $c_m(t)$, to first order. Check the normalization condition:

$$\sum_m |c_m(t)|^2=1, \tag{11.122}$$

and comment on any discrepancy. Suppose you wanted to calculate the probability of *remaining* in the original state ψ_N; would you do better to use $|c_N(t)|^2$, or $1-\sum_{m\neq N}|c_m(t)|^2$?

Problem 11.26 A particle starts out (at time $t = 0$) in the Nth state of the infinite square well. Now the "floor" of the well rises temporarily (maybe water leaks in, and then drains out again), so that the potential inside is uniform but time dependent: $V_0(t)$, with $V_0(0) = V_0(T) = 0$.

(a) Solve for the *exact* $c_m(t)$, using Equation 11.116, and show that the wave function changes *phase*, but no transitions occur. Find the phase change, $\phi(T)$, in terms of the function $V_0(t)$.

(b) Analyze the same problem in first-order perturbation theory, and compare your answers.

Comment: The same result holds *whenever* the perturbation simply adds a constant (constant in x, that is, not in t) to the potential; it has nothing to do with the infinite square well, as such. Compare Problem 1.8.

* **Problem 11.27** A particle of mass m is initially in the ground state of the (one-dimensional) infinite square well. At time $t = 0$ a "brick" is dropped into the well, so that the potential becomes

$$V(x) = \begin{cases} V_0, & 0 \le x \le a/2, \\ 0, & a/2 < x \le a, \\ \infty, & \text{otherwise}, \end{cases}$$

where $V_0 \ll E_1$. After a time T, the brick is removed, and the energy of the particle is measured. Find the probability (in first-order perturbation theory) that the result is now E_2.

Problem 11.28 We have encountered stimulated emission, (stimulated) absorption, and spontaneous emission. How come there is no such thing as spontaneous *absorption*?

*** **Problem 11.29 Magnetic resonance.** A spin-1/2 particle with gyromagnetic ratio γ, at rest in a static magnetic field $B_0\hat{k}$, precesses at the Larmor frequency $\omega_0 = \gamma B_0$ (Example 4.3). Now we turn on a small transverse radiofrequency (rf) field, $B_{\text{rf}}\left[\cos(\omega t)\,\hat{\imath} - \sin(\omega t)\,\hat{\jmath}\right]$, so that the total field is

$$\mathbf{B} = B_{\text{rf}}\cos(\omega t)\,\hat{\imath} - B_{\text{rf}}\sin(\omega t)\,\hat{\jmath} + B_0\,\hat{k}. \tag{11.123}$$

(a) Construct the 2×2 Hamiltonian matrix (Equation 4.158) for this system.

(b) If $\chi(t) = \begin{pmatrix} a(t) \\ b(t) \end{pmatrix}$ is the spin state at time t, show that

$$\dot{a} = \frac{i}{2}\left(\Omega e^{i\omega t} b + \omega_0 a\right); \quad \dot{b} = \frac{i}{2}\left(\Omega e^{-i\omega t} a - \omega_0 b\right), \tag{11.124}$$

where $\Omega \equiv \gamma B_{\text{rf}}$ is related to the strength of the rf field.

(c) Check that the general solution for $a(t)$ and $b(t)$, in terms of their initial values a_0 and b_0, is

$$a(t) = \left\{a_0 \cos\left(\omega' t/2\right) + \frac{i}{\omega'} \left[a_0 \left(\omega_0 - \omega\right) + b_0 \Omega\right] \sin\left(\omega' t/2\right)\right\} e^{i\omega t/2}$$

$$b(t) = \left\{b_0 \cos\left(\omega' t/2\right) + \frac{i}{\omega'} \left[b_0 \left(\omega - \omega_0\right) + a_0 \Omega\right] \sin\left(\omega' t/2\right)\right\} e^{-i\omega t/2}$$

where

$$\omega' \equiv \sqrt{(\omega - \omega_0)^2 + \Omega^2}. \tag{11.125}$$

(d) If the particle starts out with spin up (i.e. $a_0 = 1$, $b_0 = 0$), find the probability of a transition to spin down, as a function of time. *Answer:* $P(t) = \left\{\Omega^2/\left[(\omega - \omega_0)^2 + \Omega^2\right]\right\} \sin^2\left(\omega' t/2\right)$.

(e) Sketch the **resonance curve**,

$$P(\omega) = \frac{\Omega^2}{(\omega - \omega_0)^2 + \Omega^2}, \tag{11.126}$$

as a function of the driving frequency ω (for fixed ω_0 and Ω). Note that the maximum occurs at $\omega = \omega_0$. Find the "full width at half maximum," $\Delta\omega$.

(f) Since $\omega_0 = \gamma B_0$, we can use the experimentally observed resonance to determine the magnetic dipole moment of the particle. In a **nuclear magnetic resonance** (nmr) experiment the g-factor of the proton is to be measured, using a static field of 10,000 gauss and an rf field of amplitude 0.01 gauss. What will the resonant frequency be? (See Section 7.5 for the magnetic moment of the proton.) Find the width of the resonance curve. (Give your answers in Hz.)

** **Problem 11.30** In this problem we will recover the results Section 11.2.1 directly from the Hamiltonian for a charged particle in an electromagnetic field (Equation 4.188). An electromagnetic wave can be described by the potentials

$$\mathbf{A} = \frac{\mathbf{E}_0}{\omega} \sin(\mathbf{k} \cdot \mathbf{r} - \omega t), \qquad \varphi = 0$$

where in order to satisfy Maxwell's equations, the wave must be transverse ($\mathbf{E}_0 \cdot \mathbf{k} = 0$) and of course travel at the speed of light ($\omega = c\ |\mathbf{k}|$).

(a) Find the electric and magnetic fields for this plane wave.

(b) The Hamiltonian may be written as $H^0 + H'$ where H^0 is the Hamiltonian in the absence of the electromagnetic wave and H' is the perturbation. Show that the perturbation is given by

$$\hat{H}'(t) = \frac{e}{2\,i\,m\,\omega} e^{i\,\mathbf{k}\cdot\mathbf{r}}\, \mathbf{E}_0 \cdot \hat{\mathbf{p}}\, e^{-i\,\omega\,t} - \frac{e}{2\,i\,m\,\omega} e^{-i\,\mathbf{k}\cdot\mathbf{r}}\, \mathbf{E}_0 \cdot \hat{\mathbf{p}}\, e^{i\,\omega\,t}, \tag{11.127}$$

plus a term proportional to E_0^2 that we will ignore. *Note*: the first term corresponds to absorption and the second to emission.

(c) In the dipole approximation we set $e^{i\,\mathbf{k}\cdot\mathbf{r}} \approx 1$. With the electromagnetic wave polarized along the z direction, show that the matrix element for absorption is then

$$V_{ba} = -\frac{\omega_0}{\omega}\,\wp\, E_0\,.$$

Compare Equation 11.41. They're not *exactly* the same; would the difference affect our calculations in Section 11.2.3 or 11.3? Why or why not? *Hint*: To turn the matrix element of $\mathbf{p}$ into a matrix element of $\mathbf{r}$, you need to prove the following identity: $i\,m\left[\hat{H}^0, \hat{\mathbf{r}}\right] = \hbar\,\hat{\mathbf{p}}$.

*** **Problem 11.31** In Equation 11.38 I assumed that the atom is so small (in comparison to the wavelength of the light) that spatial variations in the field can be ignored. The *true* electric field would be

$$\mathbf{E}(\mathbf{r}, t) = \mathbf{E}_0 \cos(\mathbf{k}\cdot\mathbf{r} - \omega t). \tag{11.128}$$

If the atom is centered at the origin, then $\mathbf{k}\cdot\mathbf{r} \ll 1$ over the relevant volume ($|\mathbf{k}| = 2\pi/\lambda$, so $\mathbf{k}\cdot\mathbf{r} \sim r/\lambda \ll 1$), and that's why we could afford to drop this term. Suppose we keep the first-order correction:

$$\mathbf{E}(\mathbf{r}, t) = \mathbf{E}_0\left[\cos(\omega t) + (\mathbf{k}\cdot\mathbf{r})\sin(\omega t)\right]. \tag{11.129}$$

The first term gives rise to the **allowed** (**electric dipole**) transitions we considered in the text; the second leads to so-called **forbidden** (**magnetic dipole** and **electric quadrupole**) transitions (higher powers of $\mathbf{k}\cdot\mathbf{r}$ lead to even *more* "forbidden" transitions, associated with higher multipole moments).[39]

(a) Obtain the spontaneous emission rate for forbidden transitions (don't bother to average over polarization and propagation directions, though this should really be done to complete the calculation). *Answer:*

$$R_{b\to a} = \frac{q^2\omega^5}{\pi\epsilon_0\hbar c^5}\left|\left\langle a\left|(\hat{n}\cdot\mathbf{r})\left(\hat{k}\cdot\mathbf{r}\right)\right|b\right\rangle\right|^2. \tag{11.130}$$

(b) Show that for a one-dimensional oscillator the forbidden transitions go from level n to level $n-2$, and the transition rate (suitably averaged over $\hat{n}$ and $\hat{k}$) is

$$R = \frac{\hbar q^2\omega^3 n\,(n-1)}{15\pi\epsilon_0 m^2 c^5}. \tag{11.131}$$

(*Note:* Here ω is the frequency of the *photon*, not the oscillator.) Find the *ratio* of the "forbidden" rate to the "allowed" rate, and comment on the terminology.

(c) Show that the $2S \to 1S$ transition in hydrogen is not possible even by a "forbidden" transition. (As it turns out, this is true for all the higher multipoles as well; the dominant decay is in fact by two-photon emission, and the lifetime is about a tenth of a second.[40])

---

[39] For a systematic treatment (including the role of the magnetic field) see David Park, *Introduction to the Quantum Theory*, 3rd edn (McGraw-Hill, New York, 1992), Chapter 11.

[40] See Masataka Mizushima, *Quantum Mechanics of Atomic Spectra and Atomic Structure*, Benjamin, New York (1970), Section 5.6.

∗∗∗ **Problem 11.32** Show that the spontaneous emission rate (Equation 11.63) for a transition from n, ℓ to n', ℓ' in hydrogen is

$$\frac{e^2\omega^3 I^2}{3\pi\epsilon_0\hbar c^3} \times \begin{cases} \frac{\ell+1}{2\ell+1}, & \ell' = \ell + 1, \\ \frac{\ell}{2\ell+1} & \ell' = \ell - 1, \end{cases} \tag{11.132}$$

where

$$I \equiv \int_0^\infty r^3 R_{n\ell}(r) R_{n'\ell'}(r)\, dr. \tag{11.133}$$

(The atom starts out with a specific value of m, and it goes to *any* of the states m' consistent with the selection rules: $m' = m+1$, m, or $m-1$. Notice that the answer is independent of m.) *Hint:* First calculate all the nonzero matrix elements of x, y, and z between $|n\ell m\rangle$ and $|n'\ell'm'\rangle$ for the case $\ell' = \ell + 1$. From these, determine the quantity

$$\left|\langle n', \ell+1, m+1 \,|\, \mathbf{r} \,|\, n\ell m\rangle\right|^2 + \left|\langle n', \ell+1, m \,|\, \mathbf{r} \,|\, n\ell m\rangle\right|^2 + \left|\langle n', \ell+1, m-1 \,|\, \mathbf{r} \,|\, n\ell m\rangle\right|^2.$$

Then do the same for $\ell' = \ell - 1$. You may find useful the following recursion formulas (which hold for $m \geq 0$):[41]

$$(2\ell+1)\, x P_\ell^m(x) = (\ell+m)\, P_{\ell-1}^m(x) + (\ell-m+1)\, P_{\ell+1}^m(x), \tag{11.134}$$

$$(2\ell+1)\sqrt{1-x^2}\, P_\ell^m(x) = P_{\ell+1}^{m+1}(x) - P_{\ell-1}^{m+1}(x), \tag{11.135}$$

and the orthogonality relation Equation 4.33.

Problem 11.33 The spontaneous emission rate for the 21-cm hyperfine line in hydrogen (Section 7.5) can be obtained from Equation 11.63, except that this is a *magnetic* dipole transition, not an *electric* one:[42]

$$\wp \to \frac{1}{c}\mathrm{M} = \frac{1}{c}\langle 1 \,|(\boldsymbol{\mu}_e + \boldsymbol{\mu}_p)|\, 0\rangle,$$

where

$$\boldsymbol{\mu}_e = -\frac{e}{m_e}\mathbf{S}_e, \quad \boldsymbol{\mu}_p = \frac{5.59\, e}{2m_p}\mathbf{S}_p$$

are the magnetic moments of the electron and proton (Equation 7.89), and $|0\rangle$, $|1\rangle$ are the singlet and triplet configurations (Equations 4.175 and 4.176). Because $m_p \gg m_e$, the proton contribution is negligible, so

$$A = \frac{\omega_0^3 e^2}{3\pi\epsilon_0\hbar c^5 m_e^2} \left|\langle 1|\, \mathbf{S}_e\, |0\rangle\right|^2.$$

Work out $|\langle 1|\, \mathbf{S}_e\, |0\rangle|^2$ (use whichever triplet state you like). Put in the actual numbers, to determine the transition rate and the lifetime of the triplet state. *Answer:* 1.1×10^7 years.

---

[41] George B. Arfken and Hans J. Weber, *Mathematical Methods for Physicists*, 7th edn, Academic Press, San Diego (2013), p. 744.

[42] Electric and magnetic dipole moments have different units—hence the factor of $1/c$ (which you can check by dimensional analysis).

* **Problem 11.34** A particle starts out in the ground state of the infinite square well (on the interval $0 \leq x \leq a$). Now a wall is slowly erected, slightly off-center:[43]

$$V(x) = f(t)\delta\left(x - \frac{a}{2} - \epsilon\right),$$

where $f(t)$ rises gradually from 0 to ∞. According to the adiabatic theorem, the particle will remain in the ground state of the evolving Hamiltonian.

(a) Find (and sketch) the ground state at $t \to \infty$. *Hint:* This should be the ground state of the infinite square well with an impenetrable barrier at $a/2 + \epsilon$. Note that the particle is confined to the (slightly) larger left "half" of the well.

(b) Find the (transcendental) equation for the ground state energy at time t. *Answer:*

$$z \sin z = T\left[\cos z - \cos(z\delta)\right],$$

where $z \equiv ka$, $T \equiv maf(t)/\hbar^2$, $\delta \equiv 2\epsilon/a$, and $k \equiv \sqrt{2mE}/\hbar$.

(c) Setting $\delta = 0$, solve graphically for z, and show that the smallest z goes from π to 2π as T goes from 0 to ∞. Explain this result.

(d) Now set $\delta = 0.01$ and solve numerically for z, using $T = 0, 1, 5, 20, 100$, and 1000.

(e) Find the probability P_r that the particle is in the right "half" of the well, as a function of z and δ. *Answer:* $P_r = 1/\left[1 + (I_+/I_-)\right]$, where $I_\pm \equiv \left[1 \pm \delta - (1/z)\sin(z(1 \pm \delta))\right]\sin^2\left[z\,(1 \mp \delta)\,/2\right]$. Evaluate this expression numerically for the T's and δ in part (d). Comment on your results.

(e) Plot the ground state wave function for those same values of T and δ. Note how it gets squeezed into the left half of the well, as the barrier grows.[44]

*** **Problem 11.35** The case of an infinite square well whose right wall expands at a *constant* velocity (v) can be solved *exactly*.[45] A complete set of solutions is

$$\Phi_n(x,t) \equiv \sqrt{\frac{2}{w}}\sin\left(\frac{n\pi}{w}x\right)e^{i\left(mvx^2 - 2E_n^i at\right)/2\hbar w}, \tag{11.136}$$

where $w(t) \equiv a + vt$ is the width of the well and $E_n^i \equiv n^2\pi^2\hbar^2/2ma^2$ is the nth allowed energy of the *original* well (width a). The general solution is a linear combination of the Φ's:

$$\Psi(x,t) = \sum_{n=1}^{\infty} c_n \Phi_n(x,t), \tag{11.137}$$

whose coefficients c_n are *independent of t*.

(a) Check that Equation 11.136 satisfies the time-dependent Schrödinger equation, with the appropriate boundary conditions.

43 Julio Gea-Banacloche, *Am. J. Phys.* **70**, 307 (2002) uses a rectangular barrier; the delta-function version was suggested by M. Lakner and J. Peternelj, *Am. J. Phys.* **71**, 519 (2003).

44 Gea-Banacloche (footnote 43) discusses the evolution of the wave function *without* using the adiabatic theorem.

45 S. W. Doescher and M. H. Rice, *Am. J. Phys.* **37**, 1246 (1969).

(b) Suppose a particle starts out ($t = 0$) in the ground state of the initial well:

$$\Psi(x, 0) = \sqrt{\frac{2}{a}} \sin\left(\frac{\pi}{a} x\right).$$

Show that the expansion coefficients can be written in the form

$$c_n = \frac{2}{\pi} \int_0^{\pi} e^{-i\alpha z^2} \sin(nz) \sin(z)\, dz, \tag{11.138}$$

where $\alpha \equiv mva/2\pi^2\hbar$ is a dimensionless measure of the speed with which the well expands. (Unfortunately, this integral cannot be evaluated in terms of elementary functions.)

(c) Suppose we allow the well to expand to twice its original width, so the "external" time is given by $w(T_e) = 2a$. The "internal" time is the *period* of the time-dependent exponential factor in the (initial) ground state. Determine T_e and T_i, and show that the adiabatic regime corresponds to $\alpha \ll 1$, so that $\exp\left(-i\alpha z^2\right) \approx 1$ over the domain of integration. Use this to determine the expansion coefficients, c_n. Construct $\Psi(x, t)$, and confirm that it is consistent with the adiabatic theorem.

(d) Show that the phase factor in $\Psi(x, t)$ can be written in the form

$$\theta(t) = -\frac{1}{\hbar} \int_0^t E_1(t')\, dt', \tag{11.139}$$

where $E_n(t) \equiv n^2\pi^2\hbar^2/2mw^2$ is the nth *instantaneous* eigenvalue, at time t. Comment on this result. What is the geometric phase? If the well now contracts back to its original size, what is Berry's phase for the cycle?

*** **Problem 11.36 The driven harmonic oscillator.** Suppose the one-dimensional harmonic oscillator (mass m, frequency ω) is subjected to a driving force of the form $F(t) = m\omega^2 f(t)$, where $f(t)$ is some specified function. (I have factored out $m\omega^2$ for notational convenience; $f(t)$ has the dimensions of *length*.) The Hamiltonian is

$$H(t) = -\frac{\hbar^2}{2m}\frac{\partial^2}{\partial x^2} + \frac{1}{2}m\omega^2 x^2 - m\omega^2 x f(t). \tag{11.140}$$

Assume that the force was first turned on at time $t = 0$: $f(t) = 0$ for $t \le 0$. This system can be solved exactly, both in classical mechanics and in quantum mechanics.[46]

(a) Determine the *classical* position of the oscillator, assuming it started from rest at the origin ($x_c(0) = \dot{x}_c(0) = 0$). *Answer:*

$$x_c(t) = \omega \int_0^t f(t') \sin\left[\omega\left(t - t'\right)\right] dt'. \tag{11.141}$$

(b) Show that the solution to the (time-dependent) Schrödinger equation for this oscillator, assuming it started out in the nth state of the *undriven* oscillator ($\Psi(x, 0) = \psi_n(x)$ where $\psi_n(x)$ is given by Equation 2.62), can be written as

[46] See Y. Nogami, *Am. J. Phys.* **59**, 64 (1991), and references therein.

$$\Psi(x,t) = \psi_n(x - x_c)e^{\frac{i}{\hbar}\left[-\left(n+\frac{1}{2}\right)\hbar\omega t + m\dot{x}_c\left(x-\frac{x_c}{2}\right) + \frac{m\omega^2}{2}\int_0^t f(t')x_c(t')dt'\right]}. \quad (11.142)$$

(c) Show that the eigenfunctions and eigenvalues of $H(t)$ are

$$\psi_n(x,t) = \psi_n(x - f); \quad E_n(t) = \left(n + \frac{1}{2}\right)\hbar\omega - \frac{1}{2}m\omega^2 f^2. \quad (11.143)$$

(d) Show that in the adiabatic approximation the classical position (Equation 11.141) reduces to $x_c(t) \approx f(t)$. State the precise criterion for adiabaticity, in this context, as a constraint on the time derivative of f. *Hint:* Write $\sin\left[\omega\left(t - t'\right)\right]$ as $(1/\omega)\left(d/dt'\right)\cos\left[\omega\left(t - t'\right)\right]$ and use integration by parts.

(e) Confirm the adiabatic theorem for this example, by using the results in (c) and (d) to show that

$$\Psi(x,t) \approx \psi_n(x,t)e^{i\theta_n(t)}e^{i\gamma_n(t)}. \quad (11.144)$$

Check that the dynamic phase has the correct form (Equation 11.92). Is the geometric phase what you would expect?

Problem 11.37 Quantum Zeno Paradox.[47] Suppose a system starts out in an excited state ψ_b, which has a natural lifetime τ for transition to the ground state ψ_a. Ordinarily, for times substantially less than τ, the probability of a transition is proportional to t (Equation 11.49):

$$P_{b\to a} = \frac{t}{\tau}. \quad (11.145)$$

If we make a measurement after a time t, then, the probability that the system is still in the *upper* state is

$$P_b(t) = 1 - \frac{t}{\tau}. \quad (11.146)$$

Suppose we *do* find it to be in the upper state. In that case the wave function collapses back to ψ_b, and the process starts all over again. If we make a *second* measurement, at $2t$, the probability that the system is *still* in the upper state is

$$\left(1 - \frac{t}{\tau}\right)^2 \approx 1 - \frac{2t}{\tau}, \quad (11.147)$$

which is the same as it would have been had we never made the first measurement at t (as one would naively expect).

However, for *extremely* short times, the probability of a transition is *not* proportional to t, but rather to t^2 (Equation 11.46):[48]

$$P_{b\to a} = \alpha t^2. \quad (11.148)$$

[47] This phenomenon doesn't have much to do with Zeno, but it *is* reminiscent of the old adage, "a watched pot never boils," so it is sometimes called the **watched pot effect**.

[48] In the argument leading to linear time dependence, we assumed that the function $\sin^2(\Omega t/2)/\Omega^2$ in Equation 11.46 was a sharp spike. However, the *width* of the "spike" is of order $\Delta\omega = 4\pi/t$, and for *extremely* short t this assumption fails, and the integral becomes $\left(t^2/4\right)\int\rho(\omega)\,d\omega$.

(a) In this case what is the probability that the system is still in the upper state after the two measurements? What *would* it have been (after the same elapsed time) if we had never made the first measurement?

(b) Suppose we examine the system at n regular (extremely short) intervals, from $t = 0$ out to $t = T$ (that is, we make measurements at $T/n, 2T/n, 3T/n, \ldots, T$). What is the probability that the system is still in the upper state at time T? What is its limit as $n \to \infty$? *Moral of the story:* Because of the collapse of the wave function at every measurement, a *continuously* observed system never decays at all![49]

*** **Problem 11.38** The numerical solution to the time-independent Schrödinger equation in Problem 2.61 can be extended to solve the time-*dependent* Schrödinger equation. When we discretize the variable x, we obtain the matrix equation

$$\mathsf{H}\,\boldsymbol{\Psi} = i\hbar \frac{d}{dt}\boldsymbol{\Psi}. \tag{11.149}$$

The solution to this equation can be written

$$\boldsymbol{\Psi}(t + \Delta t) = \mathsf{U}(\Delta t)\,\boldsymbol{\Psi}(t). \tag{11.150}$$

If H is time independent, the exact expression for the time-evolution operator is[50]

$$\mathsf{U}(\Delta t) = e^{-i\,\mathsf{H}\,\Delta t/\hbar} \tag{11.151}$$

and for Δt small enough, the time-evolution operator can be approximated as

$$\mathsf{U}(\Delta t) \approx 1 - i\,\mathsf{H}\frac{\Delta t}{\hbar}. \tag{11.152}$$

While Equation 11.152 is the most obvious way to approximate U, a numerical scheme based on it is unstable, and it is preferable to use **Cayley's form** for the approximation:[51]

$$\mathsf{U}(\Delta t) \approx \frac{1 - \frac{1}{2}i\frac{\Delta t}{\hbar}\mathsf{H}}{1 + \frac{1}{2}i\frac{\Delta t}{\hbar}\mathsf{H}}. \tag{11.153}$$

Combining Equations 11.153 and 11.150 we have

$$\left(1 + \frac{1}{2}i\frac{\Delta t}{\hbar}\mathsf{H}\right)\boldsymbol{\Psi}(t + \Delta t) = \left(1 - \frac{1}{2}i\frac{\Delta t}{\hbar}\mathsf{H}\right)\boldsymbol{\Psi}(t). \tag{11.154}$$

[49] This argument was introduced by B. Misra and E. C. G. Sudarshan, *J. Math. Phys.* **18**, 756 (1977). The essential result has been confirmed in the laboratory: W. M. Itano, D. J. Heinzen, J. J. Bollinger, and D. J. Wineland, *Phys. Rev. A* **41**, 2295 (1990). Unfortunately, the experiment is not as compelling a test of the collapse of the wave function as its designers hoped, for the observed effect can perhaps be accounted for in other ways—see L. E. Ballentine, *Found. Phys.* **20**, 1329 (1990); T. Petrosky, S. Tasaki, and I. Prigogine, *Phys. Lett. A* **151**, 109 (1990).

[50] If you choose Δt small enough, you can actually use this exact form. Routines such as Mathematica's **MatrixExp** can be used to find (numerically) the exponential of a matrix.

[51] See A. Goldberg *et al.*, *Am. J. Phys.* **35**, 177 (1967) for further discussion of these approximations.

This has the form of a matrix equation $\mathsf{M}\,\mathbf{x} = \mathbf{b}$ which can be solved for the unknown $\mathbf{x} = \mathbf{\Psi}(t + \Delta t)$. Because the matrix $\mathsf{M} = 1 + \frac{1}{2}\,i\,\frac{\Delta t}{\hbar}\,\mathsf{H}$ is **tri-diagonal**,[52] efficient algorithms exist for doing so.[53]

(a) Show that the approximation in Equation 11.153 is accurate to second order. That is, show that Equations 11.151 and 11.153, expanded as power series in Δt, agree up through terms of order $(\Delta t)^2$. Verify that the matrix in Equation 11.153 is unitary.

As an example, consider a particle of mass m moving in one dimension in a simple harmonic oscillator potential. For the numerical part set $m = 1$, $\omega = 1$, and $\hbar = 1$ (this just defines the units of mass, time, and length).

(b) Construct the Hamiltonian matrix H for $N + 1 = 100$ spatial grid points. Set the spatial boundaries where the dimensionless length is $\xi = \pm 10$ (far enough out that we can assume that the wave function vanishes there for low-energy states). By computer, find the lowest two eigenvalues of H, and compare the exact values. Plot the corresponding eigenfunctions. Are they normalized? If not, normalize them before doing part (c).

(c) Take $\Psi(0) = (\psi_0 + \psi_1)/\sqrt{2}$ (from part (b)) and use Equation 11.154 to evolve the wave function from time $t = 0$ to $t = 4\pi/\omega$. Create a movie (**Animate**, in Mathematica) showing $\mathrm{Re}(\Psi(t))$, $\mathrm{Im}(\Psi(t))$, and $|\Psi(t)|$, together with the exact result. *Hint:* You need to decide what to use for Δt. In terms of the number of time steps N_t, $N_t\,\Delta t = 4\pi/\omega$. In order for the approximation of the exponential to hold, we need to have $E\,\Delta t/\hbar \ll 1$. The energy of our state is of order $\hbar\omega$, and therefore $N_t \gg 4\pi$. So you will need at least (say) 100 time steps.

∗∗ **Problem 11.39** We can use the technique of Problem 11.38 to investigate time evolution when the Hamiltonian *does* depend on time, as long as we choose Δt small enough. Evaluating H at the midpoint of each time step we simply replace Equation 11.154 with[54]

$$\left[1 + \frac{1}{2}\,i\,\frac{\Delta t}{\hbar}\,\mathsf{H}\left(t + \frac{\Delta t}{2}\right)\right]\mathbf{\Psi}(t + \Delta t) = \left[1 - \frac{1}{2}\,i\,\frac{\Delta t}{\hbar}\,\mathsf{H}\left(t + \frac{\Delta t}{2}\right)\right]\mathbf{\Psi}(t)\,. \tag{11.155}$$

Consider the driven harmonic oscillator of Problem 11.36 with

$$f(t) = A\,\sin(\Omega\,t)\,, \tag{11.156}$$

where A is a constant with the units of length and Ω is the driving frequency. In the following we will set $m = \omega = \hbar = A = 1$ and look at the effect of varying Ω. Use the same parameters for the spatial discretization as in Problem 11.38, but set

---

[52] A tri-diagonal matrix has nonzero entries only along the diagonal and one space to the right or left of the diagonal.

[53] Use your computing enviroment's built-in linear equation solver; in Mathematica that would be **x = LinearSolve[M, b]**. To learn how it actually works, see A. Goldberg *et al.*, footnote 51.

[54] C. Lubich, in *Quantum Simulations of Complex Many-Body Systems: From Theory to Algorithms,* edited by J. Grotendorst, D. Marx, and A. Muramatsu (John von Neumann Institute for Computing, Jülich, 2002), Vol. 10, p. 459. Available for download from the Neumann Institute for Computing (NIC) website.

$N_t = 1000$. For a particle that starts off in the ground state at $t = 0$, create a movie showing the numerical and exact solutions as well as the instantaneous ground state from $t = 0$ to $t = 2\pi/\Omega$ for

(a) $\Omega = \omega/5$. In line with the adiabatic theorem, you should see that the numerical solution is close (up to a phase) to the instantaneous ground state.

(b) $\Omega = 5\,\omega$. In line with what you've learned about sudden perturbations, you should see that the numerical solution is barely affected by the driving force.

(c) $\Omega = 6\,\omega/5$.

导读 / Guidance

第 12 章　跋

这部分介绍量子力学存在的一些争论（一些涉及哲学问题）和一些新的进展，内容有 EPR 佯谬、贝尔定理（贝尔不等式）、混合态和密度矩阵（纯态、混合态、子系统）、量子不可克隆定理、薛定谔猫等。

习题特色

拓展量子力学的一些前沿问题：如对纠缠态的概念和相关问题的讨论，见习题 12.1；对爱因斯坦盒子问题的讨论，见习题 12.2；对隐变量问题的讨论，见习题 12.3。

12 AFTERWORD

Now that you have a sound understanding of what quantum mechanics *says*, I would like to return to the question of what it *means*—continuing the story begun in Section 1.2. The source of the problem is the indeterminacy associated with the statistical interpretation of the wave function. For Ψ (or, more generally, the *quantum state*—it could be a spinor, for example) does not uniquely determine the outcome of a measurement; all it tells us is the statistical distribution of possible results. This raises a profound question: Did the physical system "actually have" the attribute in question *prior* to the measurement (the so-called **realist** viewpoint), or did the act of measurement itself "create" the property, limited only by the statistical constraint imposed by the wave function (the **orthodox** position)—or can we duck the issue entirely, on the grounds that it is "metaphysical" (the **agnostic** response)?

According to the realist, quantum mechanics is an *incomplete* theory, for even if you know *everything quantum mechanics has to tell you* about the system (to wit: its wave function), still you cannot determine all of its features. Evidently there is some *other* information, unknown to quantum mechanics, which (together with Ψ) is required for a complete description of physical reality.

The orthodox position raises even more disturbing problems, for if the act of measurement forces the system to "take a stand," helping to *create* an attribute that was not there previously,[1] then there is something very peculiar about the measurement process. Moreover, in order to account for the fact that an immediately repeated measurement yields the same result, we are forced to assume that the act of measurement **collapses** the wave function, in a manner that is difficult, at best, to reconcile with the normal evolution prescribed by the Schrödinger equation.

In light of this, it is no wonder that generations of physicists retreated to the agnostic position, and advised their students not to waste time worrying about the conceptual foundations of the theory.

[1] This may be *strange*, but it is not *mystical*, as some popularizers would like to suggest. The so-called **wave–particle duality**, which Niels Bohr elevated to the status of a cosmic principle (**complementarity**), makes electrons sound like unpredictable adolescents, who sometimes behave like adults, and sometimes, for no particular reason, like children. I prefer to avoid such language. When I say that a particle does not have a particular attribute before its measurement, I have in mind, for example, an electron in the spin state $\chi = \begin{pmatrix} 1 \\ 0 \end{pmatrix}$; a measurement of the x-component of its angular momentum could return the value $\hbar/2$, or (with equal probability) the value $-\hbar/2$, but until the measurement is made it simply *does not have* a well-defined value of S_x.

12.1 THE EPR PARADOX

In 1935, Einstein, Podolsky, and Rosen[2] published the famous **EPR paradox**, which was designed to prove (on purely theoretical grounds) that the realist position is the only tenable one. I'll describe a simplified version of the EPR paradox, due to David Bohm (call it EPRB). Consider the decay of the neutral pi meson into an electron and a positron:

$$\pi^0 \rightarrow e^- + e^+.$$

Assuming the pion was at rest, the electron and positron fly off in opposite directions (Figure 12.1). Now, the pion has spin zero, so conservation of angular momentum requires that the electron and positron occupy the singlet spin configuration:

$$\frac{1}{\sqrt{2}}\left(|\uparrow\downarrow\rangle - |\downarrow\uparrow\rangle\right). \tag{12.1}$$

If the electron is found to have spin up, the positron must have spin down, and vice versa. Quantum mechanics can't tell you *which* combination you'll get, in any particular pion decay, but it does say that the measurements will be *correlated*, and you'll get each combination half the time (on average). Now suppose we let the electron and positron fly *far* off—10 meters, in a practical experiment, or, in principle, 10 light years—and then you measure the spin of the electron. Say you get spin up. Immediately you know that someone 20 meters (or 20 light years) away will get spin down, if that person examines the positron.

To the realist, there's nothing surprising about this—the electron *really had* spin up (and the positron spin down) from the moment they were created . . . it's just that quantum mechanics didn't know about it. But the "orthodox" view holds that neither particle had either spin up *or* spin down until the act of measurement intervened: Your measurement of the electron collapsed the wave function, and instantaneously "produced" the spin of the positron 20 meters (or 20 light years) away. Einstein, Podolsky, and Rosen considered such "spooky action-at-a-distance" (Einstein's delightful term) preposterous. They concluded that the orthodox position is untenable; the electron and positron must have had well-defined spins all along, whether quantum mechanics knows it or not.

The fundamental assumption on which the EPR argument rests is that no influence can propagate faster than the speed of light. We call this the principle of **locality**. You might be tempted to propose that the collapse of the wave function is *not* instantaneous, but "travels" at some finite velocity. However, this would lead to violations of angular momentum conservation, for if we measured the spin of the positron before the news of the collapse had reached it, there would be a fifty–fifty probability of finding *both* particles with spin up. Whatever you might think of such a theory in the abstract, the experiments are unequivocal: No such violation occurs—the (anti-)correlation of the spins is

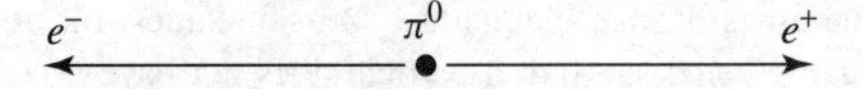

Figure 12.1: Bohm's version of the EPR experiment: A π^0 at rest decays into an electron–positron pair.

---

[2] A. Einstein, B. Podolsky, and N. Rosen, *Phys. Rev.* **47**, 777 (1935).

perfect. Evidently the collapse of the wave function—whatever its ontological status—is instantaneous.[3]

Problem 12.1 Entangled states. The singlet spin configuration (Equation 12.1) is the classic example of an *entangled state*—a two-particle state that cannot be expressed as the product of two one-particle states, and for which, therefore, one cannot really speak of "the state" of either particle separately.[4] You might wonder whether this is somehow an artifact of bad notation—maybe some linear combination of the one-particle states would disentangle the system. Prove the following theorem:

> Consider a two-level system, $|\phi_a\rangle$ and $|\phi_b\rangle$, with $\langle\phi_i|\phi_j\rangle = \delta_{ij}$. (For example, $|\phi_a\rangle$ might represent spin up and $|\phi_b\rangle$ spin down.) The two-particle state
>
> $$\alpha|\phi_a(1)\rangle|\phi_b(2)\rangle + \beta|\phi_b(1)\rangle|\phi_a(2)\rangle$$
>
> (with $\alpha \neq 0$ and $\beta \neq 0$) *cannot* be expressed as a product
>
> $$|\psi_r(1)\rangle|\psi_s(2)\rangle,$$
>
> for *any* one-particle states $|\psi_r\rangle$ and $|\psi_s\rangle$.

Hint: Write $|\psi_r\rangle$ and $|\psi_s\rangle$ as linear combinations of $|\phi_a\rangle$ and $|\phi_b\rangle$.

Problem 12.2 Einstein's Boxes. In an interesting precursor to the EPR paradox, Einstein proposed the following gedanken experiment:[5] Imagine a particle confined to a box (make it a one-dimensional infinite square well, if you like). It's in the ground state, when an impenetrable partition is introduced, dividing the box into separate halves, B_1 and B_2, in such a way that the particle is equally likely to be found in either one.[6] Now the two boxes are moved very far apart, and a measurement is made on B_1 to see if the particle is in that box. Suppose the answer is *yes*. Immediately we know that the particle will *not* be found in the (distant) box B_2.

(a) What would Einstein say about this?

(b) How does the Copenhagen interpretation account for it? What is the wave function in B_2, right after the measurement on B_1?

---

[3] Bohr wrote a famous rebuttal to the EPR paradox (*Phys. Rev.* **48**, 696 (1935)). I doubt many people read it, and certainly very few understood it (Bohr himself later admitted that he had trouble making sense of his own argument), but it was a relief that the great man had solved the problem, and everybody else could go back to business. It was not until the mid-1960s that most physicists began to worry seriously about the EPR paradox.

[4] Although the term "entanglement" is usually applied to systems of two (or more) particles, the same basic notion can be extended to *single* particle states (Problem 12.2 is an example). For an interesting discussion see D. V. Schroeder, *Am. J. Phys.* **85**, 812 (2017).

[5] See T. Norsen, *Am. J. Phys.* **73**, 164 (2005).

[6] The partition is inserted rapidly; if it is done adiabatically the particle may be forced into the (however slightly) larger of the two, as you found in Problem 11.34.

12.2 BELL'S THEOREM

Einstein, Podolsky, and Rosen did not doubt that quantum mechanics is *correct*, as far as it goes; they only claimed that it is an *incomplete* description of physical reality: The wave function is not the whole story—some *other* quantity, λ, is needed, in addition to Ψ, to characterize the state of a system fully. We call λ the "hidden variable" because, at this stage, we have no idea how to calculate or measure it.[7] Over the years, a number of hidden variable theories have been proposed, to supplement quantum mechanics;[8] they tend to be cumbersome and implausible, but never mind—until 1964 the program seemed eminently worth pursuing. But in that year J. S. Bell proved that *any* local hidden variable theory is *incompatible* with quantum mechanics.[9]

Bell suggested a generalization of the EPRB experiment: Instead of orienting the electron and positron detectors along the *same* direction, he allowed them to be rotated independently. The first measures the component of the electron spin in the direction of a unit vector **a**, and the second measures the spin of the positron along the direction **b** (Figure 12.2). For simplicity, let's record the spins in units of $\hbar/2$; then each detector registers the value $+1$ (for spin up) or -1 (spin down), along the direction in question. A table of results, for many π^0 decays, might look like this:

electron	positron	product
$+1$	-1	-1
$+1$	$+1$	$+1$
-1	$+1$	-1
$+1$	-1	-1
-1	-1	$+1$
$\vdots$	$\vdots$	$\vdots$

Bell proposed to calculate the *average* value of the *product* of the spins, for a given set of detector orientations. Call this average $P(\mathbf{a}, \mathbf{b})$. If the detectors are parallel ($\mathbf{b} = \mathbf{a}$), we recover the original EPRB configuration; in this case one is spin up and the other spin down, so the product is always -1, and hence so too is the average:

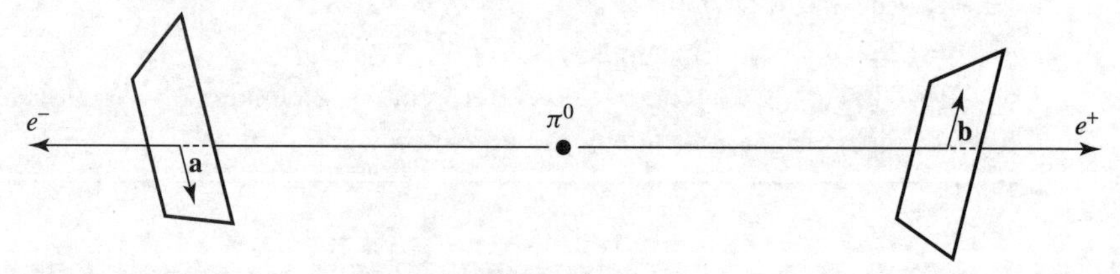

Figure 12.2: Bell's version of the EPRB experiment: detectors independently oriented in directions **a** and **b**.

---

[7] The hidden variable could be a single number, or it could be a whole *collection* of numbers; perhaps λ is to be calculated in some future theory, or maybe it is for some reason of principle incalculable. It hardly matters. All I am asserting is that there must be *something*—if only a *list* of the outcomes of every possible experiment—associated with the system prior to a measurement.

[8] D. Bohm, *Phys. Rev.* **85**, 166, 180 (1952).

[9] Bell's original paper (*Physics* **1**, 195 (1964), reprinted as Chapter 2 in John S. Bell, *Speakable and Unspeakable in Quantum Mechanics*, Cambridge University Press, UK (1987)) is a gem: brief, accessible, and beautifully written.

$$P(\mathbf{a}, \mathbf{a}) = -1. \tag{12.2}$$

By the same token, if they are *anti*-parallel ($\mathbf{b} = -\mathbf{a}$), then every product is $+1$, so

$$P(\mathbf{a}, -\mathbf{a}) = +1. \tag{12.3}$$

For arbitrary orientations, quantum mechanics predicts

$$\boxed{P(\mathbf{a}, \mathbf{b}) = -\mathbf{a} \cdot \mathbf{b}} \tag{12.4}$$

(see Problem 4.59). What Bell discovered is that this result is *incompatible with any local hidden variable theory*.

The argument is stunningly simple. Suppose that the "complete" state of the electron–positron system is characterized by the hidden variable(s) λ (λ varies, in some way that we neither understand nor control, from one pion decay to the next). Suppose further that the outcome of the *electron* measurement is independent of the orientation ($\mathbf{b}$) of the *positron* detector—which may, after all, be chosen by the experimenter at the positron end just before the electron measurement is made, and hence far too late for any subluminal message to get back to the electron detector. (This is the locality assumption.) Then there exists some function $A(\mathbf{a}, \lambda)$ which determines the result of an electron measurement, and some other function $B(\mathbf{b}, \lambda)$ for the positron measurement. These functions can only take on the values ± 1:[10]

$$A(\mathbf{a}, \lambda) = \pm 1; \quad B(\mathbf{b}, \lambda) = \pm 1. \tag{12.5}$$

When the detectors are aligned, the results are perfectly (anti)-correlated:

$$A(\mathbf{a}, \lambda) = -B(\mathbf{a}, \lambda), \tag{12.6}$$

regardless of the value of λ.

Now, the average of the product of the measurements is

$$P(\mathbf{a}, \mathbf{b}) = \int \rho(\lambda) A(\mathbf{a}, \lambda) B(\mathbf{b}, \lambda)\, d\lambda, \tag{12.7}$$

where $\rho(\lambda)$ is the probability density for the hidden variable. (Like any probability density, it is real, nonnegative, and satisfies the normalization condition $\int \rho(\lambda)\, d\lambda = 1$, but beyond this we make no assumptions about $\rho(\lambda)$; different hidden variable theories would presumably deliver quite different expressions for ρ.) In view of Equation 12.6, we can eliminate B:

$$P(\mathbf{a}, \mathbf{b}) = -\int \rho(\lambda) A(\mathbf{a}, \lambda) A(\mathbf{b}, \lambda)\, d\lambda. \tag{12.8}$$

If $\mathbf{c}$ is any *other* unit vector,

$$P(\mathbf{a}, \mathbf{b}) - P(\mathbf{a}, \mathbf{c}) = -\int \rho(\lambda) \left[A(\mathbf{a}, \lambda) A(\mathbf{b}, \lambda) - A(\mathbf{a}, \lambda) A(\mathbf{c}, \lambda) \right] d\lambda. \tag{12.9}$$

[10] This already concedes far more than a *classical* determinist would be prepared to allow, for it abandons any notion that the particles could have well-defined angular momentum vectors with simultaneously determinate components. The point of Bell's argument is to demonstrate that quantum mechanics is incompatible with *any* local deterministic theory—even one that bends over backwards to be accommodating. Of course, if you reject Equation 12.5, then the theory is *manifestly* incompatible with quantum mechanics.

Or, since $[A(\mathbf{b}, \lambda)]^2 = 1$:

$$P(\mathbf{a}, \mathbf{b}) - P(\mathbf{a}, \mathbf{c}) = -\int \rho(\lambda)\left[1 - A(\mathbf{b}, \lambda)A(\mathbf{c}, \lambda)\right]A(\mathbf{a}, \lambda)A(\mathbf{b}, \lambda)\, d\lambda. \tag{12.10}$$

But it follows from Equation 12.5 that $|A(\mathbf{a}, \lambda)A(\mathbf{b}, \lambda)| = 1$; moreover $\rho(\lambda)[1 - A(\mathbf{b}, \lambda)A(\mathbf{c}, \lambda)] \geq 0$, so

$$|P(\mathbf{a}, \mathbf{b}) - P(\mathbf{a}, \mathbf{c})| \leq \int \rho(\lambda)\left[1 - A(\mathbf{b}, \lambda)A(\mathbf{c}, \lambda)\right] d\lambda, \tag{12.11}$$

or, more simply:

$$\boxed{|P(\mathbf{a}, \mathbf{b}) - P(\mathbf{a}, \mathbf{c})| \leq 1 + P(\mathbf{b}, \mathbf{c}).} \tag{12.12}$$

This is the famous **Bell inequality**. It holds for *any* local hidden variable theory (subject only to the minimal requirements of Equations 12.5 and 12.6), for we have made no assumptions whatever as to the nature or number of the hidden variable(s), or their distribution (ρ).

But it is easy to show that the quantum mechanical prediction (Equation 12.4) is incompatible with Bell's inequality. For example, suppose the three vectors lie in a plane, and **c** makes a 45° angle with **a** and **b** (Figure 12.3); in this case quantum mechanics says

$$P(\mathbf{a}, \mathbf{b}) = 0, \quad P(\mathbf{a}, \mathbf{c}) = P(\mathbf{b}, \mathbf{c}) = -0.707,$$

which is patently inconsistent with Bell's inequality:

$$0.707 \not\leq 1 - 0.707 = 0.293.$$

With Bell's modification, then, the EPR paradox proves something far more radical than its authors imagined: If they are right, then not only is quantum mechanics *incomplete*, it is downright *wrong*. On the other hand, if quantum mechanics is right, then *no* hidden variable theory is going to rescue us from the nonlocality Einstein considered so preposterous. Moreover, we are provided with a very simple experiment to settle the issue once and for all.[11]

Many experiments to test Bell's inequality were performed in the 1960s and 1970s, culminating in the work of Aspect, Grangier, and Roger.[12] The details do not concern us here (they

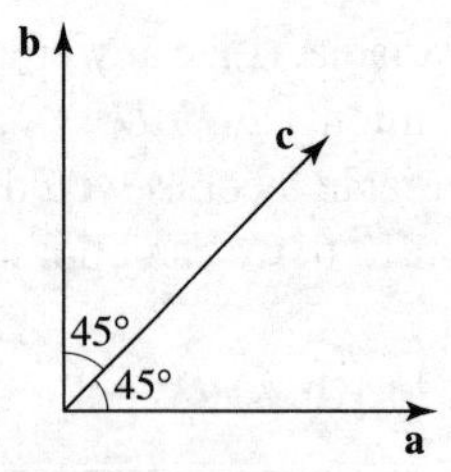

Figure 12.3: An orientation of the detectors that demonstrates quantum violations of Bell's inequality.

[11] It is an embarrassing historical fact that Bell's theorem, which is now universally recognized as one of the most profound discoveries of the twentieth century, was barely noticed at the time, with the exception of an inspired fringe element. For a fascinating account, see David Kaiser, *How the Hippies Saved Physics*, W. W. Norton, New York, 2011.

[12] A. Aspect, P. Grangier, and G. Roger, *Phys. Rev. Lett.* **49**, 91 (1982). There were logically possible (if implausible) loopholes in the Aspect experiment, which were gradually closed over the ensuing years; see J. Handsteiner et al., *Phys. Rev. Lett.* **118**, 060401 (2017). It is now possible to test Bell's inequality in the undergraduate laboratory: D. Dehlinger and M. W. Mitchell, *Am. J. Phys.* **70**, 903 (2002).

actually used two-photon atomic transitions, not pion decays). To exclude the remote possibility that the positron detector might somehow "sense" the orientation of the electron detector, both orientations were set quasi-randomly *after* the photons were already in flight. The results were in excellent agreement with the predictions of quantum mechanics, and inconsitent with Bell's inequality by a wide margin.[13]

Ironically, the experimental confirmation of quantum mechanics came as something of a shock to the scientific community. But not because it spelled the demise of "realism"—most physicists had long since adjusted to this (and for those who could not, there remained the possibility of *nonlocal* hidden variable theories, to which Bell's theorem does not apply).[14] The real shock was the demonstration that *nature itself is fundamentally nonlocal*. Nonlocality, in the form of the instantaneous collapse of the wave function (and for that matter also in the symmetrization requirement for identical particles) had always been a feature of the orthodox interpretation, but before Aspect's experiment it was possible to hope that quantum nonlocality was somehow a nonphysical artifact of the formalism, with no detectable consequences. That hope can no longer be sustained, and we are obliged to reexamine our objection to instantaneous action-at-a-distance.

Why *are* physicists so squeamish about superluminal influences? After all, there are many things that travel faster than light. If a bug flies across the beam of a movie projector, the speed of its shadow is proportional to the distance to the screen; in principle, that distance can be as large as you like, and hence the *shadow* can move at arbitrarily high velocity (Figure 12.4). However, the shadow does not carry any *energy*, nor can it transmit any *information* from one point on the screen to another. A person at point X cannot *cause anything to happen* at point Y by manipulating the passing shadow.

On the other hand, a *causal* influence that propagated faster than light would carry unacceptable implications. For according to special relativity there exist inertial frames in which such a signal propagates *backward in time*—the effect preceding the cause—and this leads to inescapable logical anomalies. (You could, for example, arrange to kill your infant grandfather. Think about it ... not a good idea.) The question is, are the superluminal influences predicted by quantum mechanics and detected by Aspect *causal*, in this sense, or are they somehow ethereal enough (like the bug's shadow) to escape the philosophical objection?

---

[13] Bell's theorem involves *averages* and it is conceivable that an apparatus such as Aspect's contains some secret bias which selects out a nonrepresentative sample, thus distorting the average. In 1989, an improved version of Bell's theorem was proposed, in which the contrast between the quantum prediction and that of any local hidden variable theory is even more dramatic. See D. Greenberger, M. Horne, A. Shimony, and A. Zeilinger, *Am. J. Phys.* **58**, 1131 (1990) and N. D. Mermin, *Am. J. Phys.* **58**, 731, (1990). An experiment of this kind suitable for an undergraduate laboratory has been carried out by Mark Beck and his students: *Am. J. Phys.* **74**, 180 (2006).

[14] It is a curious twist of fate that the EPR paradox, which *assumed* locality in order to *prove* realism, led finally to the demise of locality and left the issue of realism undecided—the outcome (as Bell put it) Einstein would have liked *least*. Most physicists today consider that if they can't have *local* realism, there's not much point in realism at *all*, and for this reason nonlocal hidden variable theories occupy a rather peripheral niche. Still, some authors—notably Bell himself, in *Speakable and Unspeakable in Quantum Mechanics* (footnote 9 in this chapter)—argue that such theories offer the best hope of bridging the conceptual gap between the measured system and the measuring apparatus, and for supplying an intelligible mechanism for the collapse of the wave function.

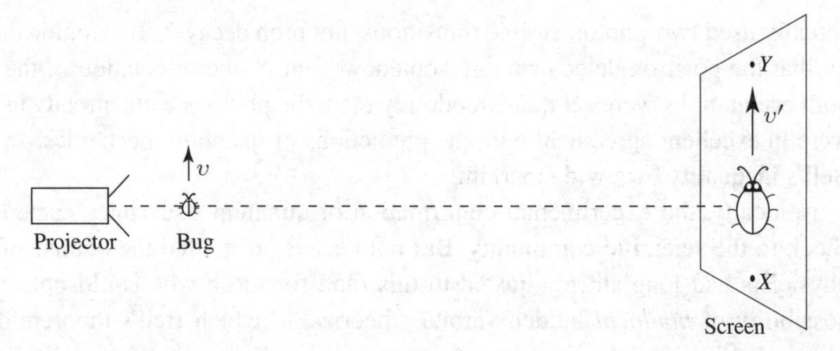

Figure 12.4: The shadow of the bug moves across the screen at a velocity v' greater than c, provided the screen is far enough away.

Well, let's consider Bell's experiment. Does the measurement of the electron *influence* the outcome of the positron measurement? Assuredly it *does*—otherwise we cannot account for the correlations in the data. But does the measurement of the electron *cause* a particular outcome for the positron? Not in any ordinary sense of the word. There is no way the person manning the electron detector could use his measurement to send a signal to the person at the positron detector, since he does not control the outcome of his own measurement (he cannot *make* a given electron come out spin up, any more than the person at X can affect the passing shadow of the bug). It is true that he can decide *whether to make a measurement at all*, but the positron monitor, having immediate access only to data at his end of the line, cannot tell whether the electron has been measured or not. The lists of data compiled at the two ends, considered separately, are completely random. It is only later, when we *compare* the two lists, that we discover the remarkable correlations. In another reference frame the positron measurements occur *before* the electron measurements, and yet this leads to no logical paradox—the observed correlation is entirely symmetrical in its treatment, and it is a matter of indifference whether we say the observation of the electron influenced the measurement of the positron, or the other way around. This is a wonderfully delicate kind of influence, whose only manifestation is a subtle correlation between two lists of otherwise random data.

We are led, then, to distinguish two types of influence: the "causal" variety, which produce actual changes in some physical property of the receiver, detectable by measurements on that subsystem alone, and an "ethereal" kind, which do not transmit energy or information, and for which the only evidence is a correlation in the data taken on the two separate subsystems—a correlation which by its nature cannot be detected by examining either list alone. Causal influences *cannot* propagate faster than light, but there is no compelling reason why ethereal ones should not. The influences associated with the collapse of the wave function are of the latter type, and the fact that they "travel" faster than light may be surprising, but it is not, after all, catastrophic.[15]

---

[15] An enormous amount has been written about Bell's theorem. My favorite is an inspired essay by David Mermin in *Physics Today* (April 1985, page 38). An extensive bibliography will be found in L. E. Ballentine, *Am. J. Phys.* **55**, 785 (1987).

** **Problem 12.3** One example[16] of a (local) deterministic ("hidden variable") theory is ...classical mechanics! Suppose we carried out the Bell experiment with classical objects (baseballs, say) in place of the electron and proton. They are launched (by a kind of double pitching machine) in opposite directions, with equal and opposite spins (angular momenta), $\mathbf{S}_a$ and $\mathbf{S}_b = -\mathbf{S}_a$. Now, these are *classical* objects—their angular momenta can point in any direction, and this direction is set (let's say randomly) at the moment of launch. Detectors placed 10 meters or so on either side of the launch point measure the spin vectors of their respective baseballs. However, in order to match the conditions for Bell's theorem, they only *record* the *sign* of the *component* of $\mathbf{S}$ along the directions $\mathbf{a}$ and $\mathbf{b}$:

$$A \equiv \text{sign}\,(\mathbf{a} \cdot \mathbf{S}_a)\,, \quad B \equiv \text{sign}\,(\mathbf{b} \cdot \mathbf{S}_b)\,.$$

Thus each detector records either $+1$ or -1, in any given trial.

In this example the "hidden variable" is the *actual* orientation of $\mathbf{S}_a$, specified (say) by the polar and azimuthal angles θ and ϕ: $\lambda = (\theta, \phi)$:

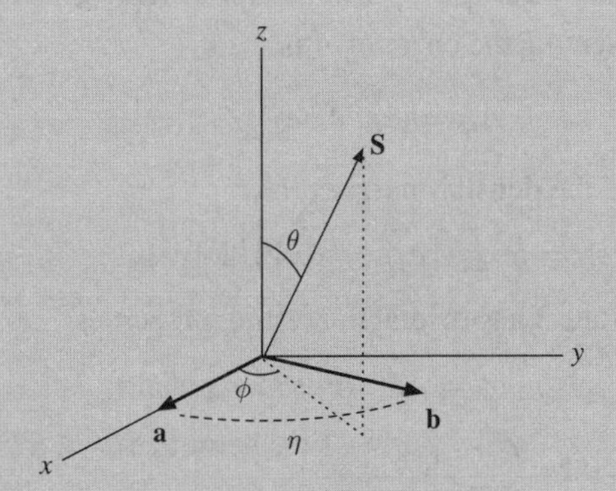

Figure 12.5: Axes for Problem 12.3.

(a) Choosing axes as in the figure, with $\mathbf{a}$ and $\mathbf{b}$ in the x–y plane and $\mathbf{a}$ along the x axis, verify that

$$A(\mathbf{a}, \lambda)\; B(\mathbf{b}, \lambda) = -\text{sign}\,[\cos(\phi)\,\cos(\phi - \eta)]\,,$$

where η is the angle between $\mathbf{a}$ and $\mathbf{b}$ (take it to run from $-\pi$ to $+\pi$).

(b) Assuming the baseballs are launched in such a way that $\mathbf{S}_a$ is equally likely to point in any direction, compute $P(\mathbf{a}, \mathbf{b})$. *Answer:* $(2|\eta|/\pi) - 1$.

(c) Sketch the graph of $P(\mathbf{a}, \mathbf{b})$, from $\eta = -\pi$ to $+\pi$, and (on the same graph) the quantum formula (Equation 12.4, with $\theta \to \eta$). For what values of η does this hidden variable theory agree with the quantum-mechanical result?

(d) Verify that your result satisfies Bell's inequality, Equation 12.12. *Hint*: The vectors $\mathbf{a}$, $\mathbf{b}$, and $\mathbf{c}$ define three points on the surface of a unit sphere; the inequality can be expressed in terms of the distances between those points.

[16] This problem is based on George Greenstein and Arthur G. Zajonc, *The Quantum Challenge*, 2nd edn., Jones and Bartlett, Sudbury, MA (2006), Section 5.3.

12.3 MIXED STATES AND THE DENSITY MATRIX

12.3.1 Pure States

In this book we have dealt with particles in **pure states**, $|\Psi\rangle$—a harmonic oscillator in its nth stationary state, for instance, or in a specific linear combination of stationary states, or a free particle in a gaussian wave packet. The expectation value of some observable A is then

$$\langle A\rangle = \left\langle \Psi \left| \hat{A} \right| \Psi \right\rangle; \tag{12.13}$$

it's the average of measurements on an ensemble of identically-prepared systems, all of them in the same state $|\Psi\rangle$. We developed the whole theory in terms of $|\Psi\rangle$ (a vector in Hilbert space, or, in the position basis, the wave function).

But there are other ways to formulate the theory, and a particularly useful one starts by defining the **density operator**,[17]

$$\hat{\rho} \equiv |\Psi\rangle\langle\Psi|. \tag{12.14}$$

With respect to an orthonormal basis $\{|e_j\rangle\}$ an operator is represented by a *matrix*; the ij element of the matrix A representing the operator $\hat{A}$ is

$$A_{ij} = \left\langle e_i \left| \hat{A} \right| e_j \right\rangle. \tag{12.15}$$

In particular, the ij element of the **density matrix** ρ is

$$\rho_{ij} = \langle e_i \,|\, \hat{\rho} \,|\, e_j\rangle = \langle e_i|\Psi\rangle\langle\Psi \,|\, e_j\rangle. \tag{12.16}$$

The density matrix (for pure states) has several interesting properties:

$$\rho^2 = \rho, \quad \text{(it is idempotent)}, \tag{12.17}$$

$$\rho^\dagger = \rho, \quad \text{(it is hermitian)}, \tag{12.18}$$

$$\operatorname{Tr}(\rho) = \sum_i \rho_{ii} = 1, \quad \text{(its trace is 1)}. \tag{12.19}$$

The expectation value of an observable A is

$$\langle A\rangle = \operatorname{Tr}(\rho \mathsf{A}). \tag{12.20}$$

We could do everything using the density matrix, instead of the wave function, to represent the state of a particle.

Example 12.1

In the standard basis

$$e_1 = \begin{pmatrix}1\\0\end{pmatrix}, \quad e_2 = \begin{pmatrix}0\\1\end{pmatrix}, \tag{12.21}$$

representing spin up and spin down along the z direction (Equation 4.149), construct the density matrix for an electron with spin up along the x direction.

Solution: In this case

$$|\Psi\rangle = \begin{pmatrix}1/\sqrt{2}\\1/\sqrt{2}\end{pmatrix} \tag{12.22}$$

[17] It's actually the "projection operator" onto the state $|\Psi\rangle$—see Equation 3.91.

(Equation 4.151). So

$$\rho_{11} = \left[\begin{pmatrix}1 & 0\end{pmatrix}\begin{pmatrix}1/\sqrt{2}\\1/\sqrt{2}\end{pmatrix}\right]\left[\begin{pmatrix}1/\sqrt{2} & 1/\sqrt{2}\end{pmatrix}\begin{pmatrix}1\\0\end{pmatrix}\right] = \frac{1}{\sqrt{2}}\frac{1}{\sqrt{2}} = \frac{1}{2},$$

$$\rho_{12} = \left[\begin{pmatrix}1 & 0\end{pmatrix}\begin{pmatrix}1/\sqrt{2}\\1/\sqrt{2}\end{pmatrix}\right]\left[\begin{pmatrix}1/\sqrt{2} & 1/\sqrt{2}\end{pmatrix}\begin{pmatrix}0\\1\end{pmatrix}\right] = \frac{1}{\sqrt{2}}\frac{1}{\sqrt{2}} = \frac{1}{2},$$

$$\rho_{21} = \left[\begin{pmatrix}0 & 1\end{pmatrix}\begin{pmatrix}1/\sqrt{2}\\1/\sqrt{2}\end{pmatrix}\right]\left[\begin{pmatrix}1/\sqrt{2} & 1/\sqrt{2}\end{pmatrix}\begin{pmatrix}1\\0\end{pmatrix}\right] = \frac{1}{\sqrt{2}}\frac{1}{\sqrt{2}} = \frac{1}{2},$$

$$\rho_{22} = \left[\begin{pmatrix}0 & 1\end{pmatrix}\begin{pmatrix}1/\sqrt{2}\\1/\sqrt{2}\end{pmatrix}\right]\left[\begin{pmatrix}1/\sqrt{2} & 1/\sqrt{2}\end{pmatrix}\begin{pmatrix}0\\1\end{pmatrix}\right] = \frac{1}{\sqrt{2}}\frac{1}{\sqrt{2}} = \frac{1}{2},$$

and hence

$$\rho = \begin{pmatrix}1/2 & 1/2\\1/2 & 1/2\end{pmatrix}. \tag{12.23}$$

Or, more efficiently,

$$\rho = |\Psi\rangle\langle\Psi| = \begin{pmatrix}1/\sqrt{2}\\1/\sqrt{2}\end{pmatrix}\begin{pmatrix}1/\sqrt{2} & 1/\sqrt{2}\end{pmatrix} = \begin{pmatrix}1/2 & 1/2\\1/2 & 1/2\end{pmatrix}. \tag{12.24}$$

Note that ρ is hermitian, its trace is 1, and

$$\rho^2 = \begin{pmatrix}1/2 & 1/2\\1/2 & 1/2\end{pmatrix}\begin{pmatrix}1/2 & 1/2\\1/2 & 1/2\end{pmatrix} = \begin{pmatrix}1/2 & 1/2\\1/2 & 1/2\end{pmatrix} = \rho. \tag{12.25}$$

Problem 12.4

(a) Prove properties 12.17, 12.18, 12.19, and 12.20.

(b) Show that the time evolution of the density operator is governed by the equation

$$i\hbar\frac{d\hat{\rho}}{dt} = \left[\hat{H}, \hat{\rho}\right]. \tag{12.26}$$

(This is the Schrödinger equation, expressed in terms of $\hat{\rho}$.)

Problem 12.5 Repeat Example 12.1 for an electron with spin down along the y direction.

12.3.2 Mixed States

In practice it is often the case that we simply don't *know* the state of the particle. Suppose, for example, we are interested in an electron emerging from the Stanford Linear Accelerator. It might have spin up (along some chosen direction), or it might have spin down, or it might be in some linear combination of the two—we just don't know.[18] We say that the particle is in a **mixed state**.[19]

---

[18] I'm not talking about any fancy quantum phenomenon (Heisenberg uncertainty or Born indeterminacy, which would apply even if we knew the precise state); I'm talking here about good old-fashioned *ignorance*.

[19] Do not confuse a *linear combination* of two pure states, which itself is still a pure state (the sum of two vectors in Hilbert space is another vector in Hilbert space) with a *mixed* state, which is not represented by *any* (single) vector in the Hilbert space.

How should we describe such a particle? I could simply list the *probability*, p_k, that it's in each possible state $|\Psi_k\rangle$. The expectation value of an observable would now be the average of measurements taken over an ensemble of systems that are *not* identically prepared (they are *not* all in the same state); rather, a fraction p_k of them is in each (pure) state $|\Psi_k\rangle$:

$$\langle A\rangle = \sum_k p_k \left\langle \Psi_k \middle| \hat{A} \middle| \Psi_k \right\rangle. \tag{12.27}$$

There's a slick way to package this information, by generalizing the density operator:

$$\hat{\rho} \equiv \sum_k p_k |\Psi_k\rangle\langle\Psi_k|. \tag{12.28}$$

Again, it becomes a matrix when referred to a particular basis:

$$\rho_{ij} = \sum_k p_k \langle e_i \mid \Psi_k\rangle \left\langle \Psi_k \mid e_j \right\rangle. \tag{12.29}$$

The density matrix encodes all the information available to us about the system.

Like any probabilities,

$$0 \leq p_k \leq 1 \quad \text{and} \quad \sum_k p_k = 1. \tag{12.30}$$

The density matrix for mixed states retains most of the properties we identified for pure states:

$$\rho^\dagger = \rho, \tag{12.31}$$

$$\mathrm{Tr}\,(\rho) = 1, \tag{12.32}$$

$$\langle A\rangle = \mathrm{Tr}(\rho \mathsf{A}). \tag{12.33}$$

$$i\hbar\frac{d\hat{\rho}}{dt} = \left[\hat{H}, \hat{\rho}\right], \; (\text{if } \frac{dp_k}{dt} = 0 \text{ for all } k) \tag{12.34}$$

but ρ is idempotent *only* if it represents a pure state:

$$\rho^2 \neq \rho \tag{12.35}$$

(indeed, this is a quick way to test whether the state is pure).

Example 12.2

Construct the density matrix for an electron that is either in the spin-up state or the spin-down state (along z), with equal probability.

Solution: In this case $p_1 = p_2 = 1/2$, so

$$\begin{aligned}\rho &= \sum_k p_k|\Psi_k\rangle\langle\Psi_k| = \frac{1}{2}\begin{pmatrix}1\\0\end{pmatrix}\begin{pmatrix}1 & 0\end{pmatrix} + \frac{1}{2}\begin{pmatrix}0\\1\end{pmatrix}\begin{pmatrix}0 & 1\end{pmatrix}\\ &= \begin{pmatrix}1/2 & 0\\ 0 & 0\end{pmatrix} + \begin{pmatrix}0 & 0\\ 0 & 1/2\end{pmatrix} = \begin{pmatrix}1/2 & 0\\ 0 & 1/2\end{pmatrix}.\end{aligned} \tag{12.36}$$

Note that ρ is hermitian, and its trace is 1, but

$$\rho^2 = \begin{pmatrix}1/2 & 0\\ 0 & 1/2\end{pmatrix}\begin{pmatrix}1/2 & 0\\ 0 & 1/2\end{pmatrix} = \begin{pmatrix}1/4 & 0\\ 0 & 1/4\end{pmatrix} \neq \rho; \tag{12.37}$$

this is *not* a pure state.

Problem 12.6

(a) Prove properties 12.31, 12.32, 12.33, and 12.34.

(b) Show that $\text{Tr}(\rho^2) \leq 1$, and equal to 1 only if ρ represents a pure state.

(c) Show that $\rho^2 = \rho$ if and only if ρ represents a pure state.

Problem 12.7

(a) Construct the density matrix for an electron that is either in the state spin up along x (with probability 1/3) or in the state spin down along y (with probability 2/3).

(b) Find $\langle S_y \rangle$ for the electron in (a).

Problem 12.8

(a) Show that the most general density matrix for a spin-1/2 particle can be written in terms of three real numbers (a_1, a_2, a_3):

$$\rho = \frac{1}{2}\begin{pmatrix} (1+a_3) & (a_1 - ia_2) \\ (a_1 + ia_2) & (1-a_3) \end{pmatrix} = \frac{1}{2}(1 + \mathbf{a}\cdot\boldsymbol{\sigma}), \tag{12.38}$$

where $\sigma_1, \sigma_2, \sigma_3$ are the three Pauli matrices. *Hint:* It has to be hermitian, and its trace must be 1.

(b) In the literature, **a** is known as the **Bloch vector**. Show that ρ represents a pure state if and only if $|\mathbf{a}| = 1$, and for a mixed state $|\mathbf{a}| < 1$. *Hint:* Use Problem 12.6(c). Thus every density matrix for a spin-1/2 particle corresponds to a point in the **Bloch sphere**, of radius 1. Points on the surface are pure states, and points inside are mixed states.

(c) What is the probability that a measurement of S_z would return the value $+\hbar/2$, if the tip of the Bloch vector is at (i) the north pole ($\mathbf{a} = (0, 0, 1)$), (ii) the center of the sphere ($\mathbf{a} = (0, 0, 0)$), (iii) the south pole ($\mathbf{a} = (0, 0, -1)$)?

(d) Find the spinor χ representing the (pure) state of the system, if the Bloch vector lies on the equator, at azimuthal angle ϕ.

12.3.3 Subsystems

There is another context in which one might invoke the density matrix formalism: an entangled state, such as the singlet spin configuration of an electron/positron pair,

$$\frac{1}{\sqrt{2}}\left(|\uparrow\downarrow\rangle - |\downarrow\uparrow\rangle\right). \tag{12.39}$$

Suppose we are interested only in the positron: what is it's state? I cannot say ... a measurement could return spin up (fifty–fifty probability) or spin down. This has nothing to do with *ignorance*; I know the state of the system precisely. But the *subsystem* (the positron) by itself does *not* occupy a pure state. If I insist on talking about the positron alone, the best I can do is to tell you its density matrix:

$$\rho = \begin{pmatrix} 1/2 & 0 \\ 0 & 1/2 \end{pmatrix}, \tag{12.40}$$

representing the 50/50 mixture.

Of course, this is the *same* as the density matrix representing a positron in a specific (but unknown) spin state (Example 12.2). I'll call it a **subsystem** density matrix, to distinguish it from an **ignorance** density matrix. The EPRB paradox illustrates the difference. Before the electron spin was measured, the positron (alone) was represented by the "subsystem" density matrix (Equation 12.40); when the electron is measured the positron is knocked into a definite state ... but we (at the distant positron detector) don't know which. The positron is now represented by the "ignorance" density matrix (Equation 12.36). But the two density matrices are identical! Our description of the state of the positron has not been altered by the measurement of the electron—all that has changed is our reason for using the density matrix formalism.

12.4 THE NO-CLONE THEOREM

Quantum measurements are typically **destructive**, in the sense that they alter the state of the system measured. This is how the uncertainty principle is enforced in the laboratory. You might wonder why we don't just make a bunch of identical copies (**clones**) of the original state, and measure *them*, leaving the system itself unscathed. It can't be done. Indeed, if you *could* build a cloning device (a "quantum Xerox machine"), quantum mechanics would be out the window.

For example, it would then be possible to send superluminal messages using the EPRB apparatus.[20] Say the message to be transmitted, from the operator of the electron detector (conventionally "Alice") to the operator of the positron detector ("Bob"), is either "yes" ("drop the bomb") or "no." If the message is to be "yes," Alice measures S_z (of the electron). Never mind what result she gets—all that matters is that she makes the measurement, for this means that the positron is now in the pure state $\uparrow$ or $\downarrow$ (never mind which). If she wants to say "no," she measures S_x, and that means the positron is now in the definite state $\leftarrow$ or $\rightarrow$ (never mind which). In any case, Bob makes a million clones of the positron, and measures S_z on half of them, and S_x on the other half. If the first group are all in the same state (all $\uparrow$ or all $\downarrow$), then Alice must have measured S_z, and the message is "yes" (the S_x group should be a 50/50 mixture). If all the S_x measurements yield the same answer (all $\leftarrow$ or all $\rightarrow$), then Alice must have measured S_x, and the message is "no" (in that case the S_z measurements should be a 50/50 mixture).

It doesn't work, because you *can't make* a quantum Xerox machine, as Wootters, Zurek, and Dieks proved in 1982.[21] Schematically, we want the machine to take as input a particle in state $|\psi\rangle$ (the one to be copied), plus a second particle in state $|X\rangle$ (the "blank sheet of paper"), and spit out *two* particles in the state $|\psi\rangle$ (original plus copy):

$$|\psi\rangle|X\rangle \rightarrow |\psi\rangle|\psi\rangle. \tag{12.41}$$

Suppose we have made a device that successfully clones the state $|\psi_1\rangle$:

$$|\psi_1\rangle|X\rangle \rightarrow |\psi_1\rangle|\psi_1\rangle, \tag{12.42}$$

[20] Starting around 1975, members of the so-called "Fundamental Fysiks Group" proposed a series of increasingly ingenious schemes for faster-than-light communication—inspiring in turn a succession of increasingly sophisticated rebuttals, culminating in the no-clone theorem, which finally put a stop to the whole misguided enterprise. For a fascinating account, see Chapter 11 of Kaiser's *How the Hippies Saved Physics* (footnote 11, page 451).

[21] W. K. Wootters and W. H. Zurek, *Nature* **299**, 802 (1982); D. Dieks, *Phys. Lett.* **A 92**, 271 (1982).

and also works for state $|\psi_2\rangle$:

$$|\psi_2\rangle|X\rangle \rightarrow |\psi_2\rangle|\psi_2\rangle \tag{12.43}$$

($|\psi_1\rangle$ and $|\psi_2\rangle$ might be spin up and spin down, for example, if the particle is an electron). So far, so good. But what happens when we feed in a linear combination $|\psi\rangle = \alpha|\psi_1\rangle + \beta|\psi_2\rangle$? Evidently we get[22]

$$|\psi\rangle|X\rangle \rightarrow \alpha|\psi_1\rangle|\psi_1\rangle + \beta|\psi_2\rangle|\psi_2\rangle, \tag{12.44}$$

which is not at all what we wanted—what we *wanted* was

$$\begin{aligned} |\psi\rangle|X\rangle \rightarrow |\psi\rangle|\psi\rangle &= [\alpha|\psi_1\rangle + \beta|\psi_2\rangle][\alpha|\psi_1\rangle + \beta|\psi_2\rangle] \\ &= \alpha^2|\psi_1\rangle|\psi_1\rangle + \beta^2|\psi_2\rangle|\psi_2\rangle + \alpha\beta[|\psi_1\rangle|\psi_2\rangle + |\psi_2\rangle|\psi_1\rangle]. \end{aligned} \tag{12.45}$$

You can make a machine to clone spin-up electrons and spin-down electrons, but it will fail for any nontrivial linear combinations (such as eigenstates of S_x). It's as though you bought a Xerox machine that copies vertical lines perfectly, and also horizontal lines, but completely distorts diagonals.

The no-clone theorem turned out to have an importance well beyond "merely" protecting quantum mechanics from superluminal communication (and hence an inescapable conflict with special relativity).[23] In particular, it opened up the field of **quantum cryptography**, which exploits the theorem to detect eavesdropping.[24] This time Alice and Bob want to agree on a key for decoding messages, without the cumbersome necessity of actually meeting face-to-face. Alice is to send the key (a string of numbers) to Bob via a stream of carefully prepared photons.[25] But they are worried that their nemesis, Eve, might try to intercept this communication, and thereby crack the code, without their knowledge. Alice prepares a string of photons in four different states: linearly polarized (horizontal $|H\rangle$ and vertical $|V\rangle$), and circularly polarized (left $|L\rangle$ and right $|R\rangle$), which she sends to Bob. Eve hopes to capture and clone the photons en route, sending the originals along to Bob, who will be none the wiser. (Later on, she knows, Alice and Bob will compare notes on a sample of the photons, to make sure there has been no tampering—that's why she has to clone them perfectly, to go undetected.) But the no-clone theorem guarantees that Eve's Xerox machine will fail;[26] Alice and Bob will catch the eavesdropping when they compare the samples. (They will then, presumably, discard that key.)

---

22 This assumes that the device acts *linearly* on the state $|\psi\rangle$, as it must, since the time-dependent Schrödinger equation (which presumably governs the process) is linear.

23 The no-clone theorem is one of the foundations for **quantum information theory**, **"teleportation,"** and **quantum computation**. For a brief history and a comprehensive bibliography, see F. W. Strauch, *Am. J. Phys.* **84**, 495 (2016).

24 For a brief summary, see W. K. Wootters and W. H. Zurek, *Physics Today*, February 2009, page 76.

25 Electrons would do, but traditionally the story is told using photons. By the way, there is no entanglement involved, and they're not in a hurry—this has nothing to do with EPR or superluminal signals.

26 If Alice and Bob were foolish enough to use just *two* orthogonal photon states (say, $|H\rangle$ and $|V\rangle$), then Eve might get lucky, and use a quantum Xerox machine that does faithfully clone those two states. But as long as they include nontrivial linear combinations (such as $|R\rangle$ and $|L\rangle$), the cloning is certain to fail, and the eavesdropping will be detected.

12.5 SCHRÖDINGER'S CAT

The measurement process plays a mischievous role in quantum mechanics: It is here that indeterminacy, nonlocality, the collapse of the wave function, and all the attendant conceptual difficulties arise. Absent measurement, the wave function evolves in a leisurely and deterministic way, according to the Schrödinger equation, and quantum mechanics looks like a rather ordinary field theory (much simpler than classical electrodynamics, for example, since there is only *one* field (Ψ), instead of *two* (**E** and **B**), and it's a *scalar*). It is the bizarre role of the measurement process that gives quantum mechanics its extraordinary richness and subtlety. But what, exactly, *is* a measurement? What makes it so different from other physical processes?[27] And how can we tell when a measurement has occurred?

Schrödinger posed the essential question most starkly, in his famous **cat paradox**:[28]

> A cat is placed in a steel chamber, together with the following hellish contraption.... In a Geiger counter there is a tiny amount of radioactive substance, so tiny that maybe within an hour one of the atoms decays, but equally probably none of them decays. If one decays then the counter triggers and via a relay activates a little hammer which breaks a container of cyanide. If one has left this entire system for an hour, then one would say the cat is living if no atom has decayed. The first decay would have poisoned it. The wave function of the entire system would express this by containing equal parts of the living and dead cat.

At the end of the hour, then, the wave function of the cat has the schematic form

$$\psi = \frac{1}{\sqrt{2}}(\psi_{\text{alive}} + \psi_{\text{dead}}). \tag{12.46}$$

The cat is neither alive nor dead, but rather a linear combination of the two, until a measurement occurs—until, say, you peek in the window to check. At that moment your observation forces the cat to "take a stand": dead or alive. And if you find him to be dead, then it's really *you* who killed him, by looking in the window.

Schrödinger regarded this as patent nonsense, and I think most people would agree with him. There is something absurd about the very idea of a *macroscopic* object being in a linear combination of two palpably different states. An electron can be in a linear combination of spin up and spin down, but a cat simply cannot *be* in a linear combination of alive and dead. But how are we to reconcile this with quantum mechanics?

---

[27] There is a school of thought that rejects this distinction, holding that the system and the measuring apparatus should be described by one great big wave function which itself evolves according to the Schrödinger equation. In such theories there is no collapse of the wave function, but one must typically abandon any hope of describing individual events—quantum mechanics (in this view) applies only to *ensembles* of identically prepared systems. See, for example, Philip Pearle, *Am. J. Phys.* **35**, 742 (1967), or Leslie E. Ballentine, *Quantum Mechanics: A Modern Development*, 2nd edn, World Scientific, Singapore (1998).

[28] E. Schrödinger, *Naturwiss.* **48**, 52 (1935); translation by Josef M. Jauch, *Foundations of Quantum Mechanics*, Addison-Wesley, Reading, MA (1968), page 185.

The Schrödinger cat paradox forces us to confront the question "What constitutes a 'measurement,' in quantum mechanics"? Does the "measurement" really occur when we peek in the keyhole? Or did it happen much earlier, when the atom did (or did not) decay? Or was it when the Geiger counter registered (or did not) the decay, or when the hammer did (or did not) hit the vial of cyanide? Historically, there have been many answers to this question. Wigner held that measurement requires the intervention of human consciousness; Bohr thought it meant the interaction between a microscopic system (subject to the laws of quantum mechanics) and a macroscopic measuring apparatus (described by classical laws); Heisenberg maintained that a measurement occurs when a permanent record is left; others have pointed to the *irreversible* nature of a measurement. The embarrassing fact is that none of these characterizations is entirely satisfactory. Most physicists would say that the measurement occurred (and the cat became either alive or dead) well before we looked in the window, but there is no real consensus as to when or why.

And this still leaves the deeper question of why a macroscopic system cannot occupy a linear combination of two clearly distinct states—a baseball, say, in a linear combination of Seattle and Toronto. Suppose you *could* get a baseball into such a state, what would happen to it? In some ultimate sense the macroscopic system must itself be described by the laws of quantum mechanics. But wave functions, in the first instance, represent individual elementary particles; the wave function of a macroscopic object would be a monstrously complicated composite structure, built out of the wave functions of its 10^{23} constituent particles. And it is subject to constant bombardment from the environment[29]—subject, that is, to continuous "measurement" and the attendant collapse. In this process, presumably, "classical" states are statistically favored, and in practice the linear combination devolves almost instantaneously into one of the ordinary configurations we encounter in everyday life. This phenomenon is called **decoherence**, and although it is still not entirely understood it appears to be the fundamental mechanism by which quantum mechanics reduces to classical mechanics in the macroscopic realm.[30]

• • • • • •

In this book I have tried to tell a consistent and coherent story: The wave function (Ψ) represents the state of a particle (or system); particles do not in general possess specific dynamical properties (position, momentum, energy, angular momentum, etc.) until an act of measurement intervenes; the probability of getting a particular value in any given experiment is determined by the statistical interpretation of Ψ; upon measurement the wave function collapses, so that an immediately repeated measurement is certain to yield the same result. There are other possible interpretations—nonlocal hidden variable theories, the "many worlds" picture, "consistent histories," ensemble models, and others—but I believe this one is conceptually the *simplest*,

29 This is true even if you put it in an almost complete vacuum, cool it down practically to absolute zero, and somehow shield out the cosmic background radiation. It is possible to imagine a single electron avoiding all contact for a significant time, but not a macroscopic object.

30 See, for example, M. Schlosshauer, *Decoherence and the Quantum-to-Classical Transition*, Springer, (2007), or W. H. Zurek, *Physics Today*, October, 2014, page 44.

and certainly it is the one shared by most physicists today.[31] It has stood the test of time, and emerged unscathed from every experimental challenge. But I cannot believe this is the end of the story; at the very least, we have much to learn about the nature of measurement and the mechanism of collapse. And it is entirely possible that future generations will look back, from the vantage point of a more sophisticated theory, and wonder how we could have been so gullible.

[31] See Daniel Styer *et al.*, *Am. J. Phys.* **70**, 288 (2002).

APPENDIX

LINEAR ALGEBRA

Linear algebra abstracts and generalizes the arithmetic of ordinary vectors, as we encounter them in first-year physics. The generalization is in two directions: (1) we allow the scalars to be *complex* numbers, and (2) we do not restrict ourselves to three dimensions.

A.1 VECTORS

A **vector space** consists of a set of **vectors** ($|\alpha\rangle$, $|\beta\rangle$, $|\gamma\rangle, \ldots$), together with a set of **scalars** (a, b, $c, \ldots$),[1] which is **closed**[2] under two operations: vector addition and scalar multiplication.

- **Vector Addition**

The "sum" of any two vectors is another vector:

$$|\alpha\rangle + |\beta\rangle = |\gamma\rangle . \tag{A.1}$$

Vector addition is **commutative**:

$$|\alpha\rangle + |\beta\rangle = |\beta\rangle + |\alpha\rangle , \tag{A.2}$$

and **associative**:

$$|\alpha\rangle + (|\beta\rangle + |\gamma\rangle) = (|\alpha\rangle + |\beta\rangle) + |\gamma\rangle . \tag{A.3}$$

There exists a **zero (or null) vector**,[3] $|0\rangle$, with the property that

$$|\alpha\rangle + |0\rangle = |\alpha\rangle , \tag{A.4}$$

for every vector $|\alpha\rangle$. And for every vector $|\alpha\rangle$ there is an associated **inverse vector** ($|-\alpha\rangle$),[4] such that

$$|\alpha\rangle + |-\alpha\rangle = |0\rangle . \tag{A.5}$$

[1] For our purposes, the scalars will be ordinary complex numbers. Mathematicians can tell you about vector spaces over more exotic fields, but such objects play no role in quantum mechanics. Note that $\alpha, \beta, \gamma \ldots$ are *not* (ordinarily) numbers; they are *names* (labels)—"Charlie," for instance, or "F43A-9GL," or whatever you care to use to identify the vector in question.

[2] That is to say, these operations are always well-defined, and will never carry you outside the vector space.

[3] It is customary, where no confusion can arise, to write the null vector without the adorning bracket: $|0\rangle \to 0$.

[4] This is funny notation, since α is not a number. I'm simply adopting the name "−Charlie" for the inverse of the vector whose name is "Charlie." More natural terminology will suggest itself in a moment.

- **Scalar Multiplication**

  The "product" of any scalar with any vector is another vector:

$$a|\alpha\rangle = |\gamma\rangle . \tag{A.6}$$

Scalar multiplication is **distributive** with respect to vector addition:

$$a\left(|\alpha\rangle + |\beta\rangle\right) = a|\alpha\rangle + a|\beta\rangle, \tag{A.7}$$

and with respect to scalar addition:

$$(a+b)\,|\alpha\rangle = a|\alpha\rangle + b|\alpha\rangle. \tag{A.8}$$

It is also **associative** with respect to the ordinary multiplication of scalars:

$$a\left(b|\alpha\rangle\right) = (ab)\,|\alpha\rangle. \tag{A.9}$$

Multiplication by the scalars 0 and 1 has the effect you would expect:

$$0|\alpha\rangle = |0\rangle\,; \quad 1|\alpha\rangle = |\alpha\rangle\,. \tag{A.10}$$

Evidently $|-\alpha\rangle = (-1)\,|\alpha\rangle$ (which we write more simply as $-|\alpha\rangle$).

There's a lot less here than meets the eye—all I have done is to write down in abstract language the familiar rules for manipulating vectors. The virtue of such abstraction is that we will be able to apply our knowledge and intuition about the behavior of ordinary vectors to other systems that happen to share the same formal properties.

A **linear combination** of the vectors $|\alpha\rangle$, $|\beta\rangle$, $|\gamma\rangle$, ..., is an expression of the form

$$a|\alpha\rangle + b|\beta\rangle + c|\gamma\rangle + \cdots . \tag{A.11}$$

A vector $|\lambda\rangle$ is said to be **linearly independent** of the set $|\alpha\rangle$, $|\beta\rangle$, $|\gamma\rangle$, ..., if it cannot be written as a linear combination of them. (For example, in three dimensions the unit vector $\hat{k}$ is linearly independent of $\hat{\imath}$ and $\hat{\jmath}$, but any vector in the xy plane is linearly *dependent* on $\hat{\imath}$ and $\hat{\jmath}$.) By extension, a *set* of vectors is "linearly independent" if each one is linearly independent of all the rest. A collection of vectors is said to **span** the space if *every* vector can be written as a linear combination of the members of this set.[5] A set of *linearly independent* vectors that spans the space is called a **basis**. The number of vectors in any basis is called the **dimension** of the space. For the moment we shall assume that the dimension (n) is *finite*.

With respect to a prescribed basis

$$|e_1\rangle\,,\ |e_2\rangle\,,\ \ldots,\ |e_n\rangle, \tag{A.12}$$

any given vector

$$|\alpha\rangle = a_1|e_1\rangle + a_2|e_2\rangle + \cdots + a_n|e_n\rangle, \tag{A.13}$$

is uniquely represented by the (ordered) n-tuple of its **components**:

$$|\alpha\rangle \leftrightarrow (a_1, a_2, \ldots, a_n)\,. \tag{A.14}$$

[5] A set of vectors that spans the space is also called **complete**, though I personally reserve that word for the infinite-dimensional case, where subtle questions of convergence may arise.

It is often easier to work with the components than with the abstract vectors themselves. To add vectors, you add their corresponding components:

$$|\alpha\rangle + |\beta\rangle \leftrightarrow (a_1 + b_1, a_2 + b_2, \ldots, a_n + b_n); \tag{A.15}$$

to multiply by a scalar you multiply each component:

$$c|\alpha\rangle \leftrightarrow (ca_1, ca_2, \ldots, ca_n); \tag{A.16}$$

the null vector is represented by a string of zeroes:

$$|0\rangle \leftrightarrow (0, 0, \ldots, 0); \tag{A.17}$$

and the components of the inverse vector have their signs reversed:

$$|-\alpha\rangle \leftrightarrow (-a_1, -a_2, \ldots, -a_n). \tag{A.18}$$

The only *dis*advantage of working with components is that you have to commit yourself to a particular basis, and the same manipulations will look very different to someone using a different basis.

Problem A.1 Consider the ordinary vectors in three dimensions $\left(a_x\hat{\imath} + a_y\hat{\jmath} + a_z\hat{k}\right)$, with complex components.

(a) Does the subset of all vectors with $a_z = 0$ constitute a vector space? If so, what is its dimension; if not, why not?

(b) What about the subset of all vectors whose z component is 1? *Hint:* Would the sum of two such vectors be in the subset? How about the null vector?

(c) What about the subset of vectors whose components are all equal?

* **Problem A.2** Consider the collection of all polynomials (with complex coefficients) of degree $< N$ in x.

(a) Does this set constitute a vector space (with the polynomials as "vectors")? If so, suggest a convenient basis, and give the dimension of the space. If not, which of the defining properties does it lack?

(b) What if we require that the polynomials be *even* functions?

(c) What if we require that the leading coefficient (i.e. the number multiplying x^{N-1}) be 1?

(d) What if we require that the polynomials have the value 0 at $x = 1$?

(e) What if we require that the polynomials have the value 1 at $x = 0$?

Problem A.3 Prove that the components of a vector with respect to a given basis are *unique*.

A.2 INNER PRODUCTS

In three dimensions we encounter two kinds of vector products: the dot product and the cross product. The latter does not generalize in any natural way to n-dimensional vector spaces, but the former *does*—in this context it is usually called the **inner product**. The inner product of vectors $|\alpha\rangle$ and $|\beta\rangle$ is a complex number (which we write as $\langle\alpha|\beta\rangle$), with the following properties:

$$\langle\beta\,|\,\alpha\rangle = \langle\alpha\,|\,\beta\rangle^*, \tag{A.19}$$

$$\langle\alpha\,|\,\alpha\rangle \geq 0, \quad \text{and } \langle\alpha\,|\,\alpha\rangle = 0 \Leftrightarrow |\alpha\rangle = |0\rangle, \tag{A.20}$$

$$\langle\alpha|\left(b\,|\beta\rangle + c\,|\gamma\rangle\right) = b\,\langle\alpha\,|\,\beta\rangle + c\,\langle\alpha\,|\,\gamma\rangle. \tag{A.21}$$

Apart from the generalization to complex numbers, these axioms simply codify the familiar behavior of dot products. A vector space with an inner product is called an **inner product space**.

Because the inner product of any vector with itself is a non-negative number (Equation A.20), its square root is *real*—we call this the **norm** of the vector:

$$\|\alpha\| \equiv \sqrt{\langle\alpha|\alpha\rangle}; \tag{A.22}$$

it generalizes the notion of "length." A **unit vector** (one whose norm is 1) is said to be **normalized** (the word should really be "normal," but I guess that sounds too judgmental). Two vectors whose inner product is zero are called **orthogonal** (generalizing the notion of "perpendicular"). A collection of mutually orthogonal normalized vectors,

$$\left\langle\alpha_i\,\middle|\,\alpha_j\right\rangle = \delta_{ij}, \tag{A.23}$$

is called an **orthonormal set**. It is always possible (see Problem A.4), and almost always convenient, to choose an **orthonormal basis;** in that case the inner product of two vectors can be written very neatly in terms of their components:

$$\langle\alpha|\beta\rangle = a_1^* b_1 + a_2^* b_2 + \cdots + a_n^* b_n, \tag{A.24}$$

the norm (squared) becomes

$$\langle\alpha\,|\,\alpha\rangle = |a_1|^2 + |a_2|^2 + \cdots + |a_n|^2, \tag{A.25}$$

and the components themselves are

$$a_i = \langle e_i|\alpha\rangle\,. \tag{A.26}$$

(These results generalize the familiar formulas $\mathbf{a}\cdot\mathbf{b} = a_x b_x + a_y b_y + a_z b_z$, $|\mathbf{a}|^2 = a_x^2 + a_y^2 + a_z^2$, and $a_x = \hat{\imath}\cdot\mathbf{a}$, $a_y = \hat{\jmath}\cdot\mathbf{a}$, $a_z = \hat{k}\cdot\mathbf{a}$, for the three-dimensional orthonormal basis $\hat{\imath}$, $\hat{\jmath}$, $\hat{k}$.) From now on we shall *always* work in orthonormal bases, unless it is explicitly indicated otherwise.

Another geometrical quantity one might wish to generalize is the *angle* between two vectors. In ordinary vector analysis $\cos\theta = (\mathbf{a}\cdot\mathbf{b})\,/\,|\mathbf{a}|\,|\mathbf{b}|$, but because the inner product is in general a complex number, the analogous formula (in an arbitrary inner product space) does not define a (real) angle θ. Nevertheless, it is still true that the *absolute value* of this quantity is a number no greater than 1,

$$|\langle\alpha\,|\,\beta\rangle|^2 \leq \langle\alpha\,|\,\alpha\rangle\,\langle\beta\,|\,\beta\rangle\,. \tag{A.27}$$

This important result is known as the **Schwarz inequality**; the proof is given in Problem A.5. So you can, if you like, define the angle between $|\alpha\rangle$ and $|\beta\rangle$ by the formula

$$\cos\theta = \sqrt{\frac{\langle\alpha\,|\,\beta\rangle\,\langle\beta\,|\,\alpha\rangle}{\langle\alpha\,|\,\alpha\rangle\,\langle\beta\,|\,\beta\rangle}}. \tag{A.28}$$

* **Problem A.4** Suppose you start out with a basis $(|e_1\rangle, |e_2\rangle, \ldots, |e_n\rangle)$ that is *not* orthonormal. The **Gram–Schmidt procedure** is a systematic ritual for generating from it an orthonormal basis $(|e_1'\rangle, |e_2'\rangle, \ldots, |e_n'\rangle)$. It goes like this:

(i) Normalize the first basis vector (divide by its norm):

$$|e_1'\rangle = \frac{|e_1\rangle}{\|e_1\|}.$$

(ii) Find the projection of the second vector along the first, and subtract it off:

$$|e_2\rangle - \langle e_1' \mid e_2\rangle |e_1'\rangle.$$

This vector is orthogonal to $|e_1'\rangle$; normalize it to get $|e_2'\rangle$.

(iii) Subtract from $|e_3\rangle$ its projections along $|e_1'\rangle$ and $|e_2'\rangle$:

$$|e_3\rangle - \langle e_1' \mid e_3\rangle |e_1'\rangle - \langle e_2' \mid e_3\rangle |e_2'\rangle.$$

This is orthogonal to $|e_1'\rangle$ and $|e_2'\rangle$; normalize it to get $|e_3'\rangle$. And so on.

Use the Gram–Schmidt procedure to orthonormalize the 3-space basis $|e_1\rangle = (1+i)\,\hat{\imath} + (1)\,\hat{\jmath} + (i)\,\hat{k}$, $|e_2\rangle = (i)\,\hat{\imath} + (3)\,\hat{\jmath} + (1)\,\hat{k}$, $|e_3\rangle = (0)\,\hat{\imath} + (28)\,\hat{\jmath} + (0)\,\hat{k}$.

Problem A.5 Prove the Schwarz inequality (Equation A.27). *Hint:* Let $|\gamma\rangle = |\beta\rangle - (\langle\alpha|\beta\rangle / \langle\alpha|\alpha\rangle)\,|\alpha\rangle$, and use $\langle\gamma|\gamma\rangle \geq 0$.

Problem A.6 Find the angle (in the sense of Equation A.28) between the vectors $|\alpha\rangle = (1+i)\,\hat{\imath} + (1)\,\hat{\jmath} + (i)\,\hat{k}$ and $|\beta\rangle = (4-i)\,\hat{\imath} + (0)\,\hat{\jmath} + (2-2i)\,\hat{k}$.

Problem A.7 Prove the **triangle inequality**: $\|\,(|\alpha\rangle + |\beta\rangle)\,\| \leq \|\alpha\| + \|\beta\|$.

A.3 MATRICES

Suppose you take every vector (in 3-space) and multiply it by 17, or you rotate every vector by 39° about the z axis, or you reflect every vector in the xy plane—these are all examples of **linear transformations**. A linear transformation[6] $(\hat{T})$ takes each vector in a vector space and "transforms" it into some other vector $(|\alpha\rangle \to |\alpha'\rangle = \hat{T}|\alpha\rangle)$, subject to the condition that the operation be *linear:*

$$\hat{T}\left(a|\alpha\rangle + b|\beta\rangle\right) = a\left(\hat{T}|\alpha\rangle\right) + b\left(\hat{T}|\beta\rangle\right), \tag{A.29}$$

for any vectors $|\alpha\rangle$, $|\beta\rangle$ and any scalars a, b.

If you know what a particular linear transformation does to a set of *basis* vectors, you can easily figure out what it does to *any* vector. For suppose that

---

[6] In this chapter I'll use a hat (ˆ) to denote linear transformations; this is not inconsistent with my convention in the text (putting hats on operators), for (as we shall see) quantum operators *are* linear transformations.

$$\hat{T}|e_1\rangle = T_{11}|e_1\rangle + T_{21}|e_2\rangle + \cdots + T_{n1}|e_n\rangle,$$
$$\hat{T}|e_2\rangle = T_{12}|e_1\rangle + T_{22}|e_2\rangle + \cdots + T_{n2}|e_n\rangle,$$
$$\cdots$$
$$\hat{T}|e_n\rangle = T_{1n}|e_1\rangle + T_{2n}|e_2\rangle + \cdots + T_{nn}|e_n\rangle,$$

or, more compactly,

$$\hat{T}\left|e_j\right\rangle = \sum_{i=1}^{n} T_{ij}|e_i\rangle, \quad (j = 1, 2, \ldots, n). \tag{A.30}$$

If $|\alpha\rangle$ is an arbitrary vector,

$$|\alpha\rangle = a_1|e_1\rangle + a_2|e_2\rangle + \cdots + a_n|e_n\rangle = \sum_{j=1}^{n} a_j|e_j\rangle, \tag{A.31}$$

then

$$\hat{T}|\alpha\rangle = \sum_{j=1}^{n} a_j\left(\hat{T}|e_j\rangle\right) = \sum_{j=1}^{n}\sum_{i=1}^{n} a_j T_{ij}|e_i\rangle = \sum_{i=1}^{n}\left(\sum_{j=1}^{n} T_{ij}a_j\right)|e_i\rangle. \tag{A.32}$$

Evidently $\hat{T}$ takes a vector with components $a_1, a_2, \ldots, a_n$ into a vector with components[7]

$$a_i' = \sum_{j=1}^{n} T_{ij}a_j. \tag{A.33}$$

Thus the n^2 **elements** T_{ij} uniquely characterize the linear transformation $\hat{T}$ (with respect to a given basis), just as the n components a_i uniquely characterize the vector $|\alpha\rangle$ (with respect to that basis):

$$\hat{T} \leftrightarrow (T_{11}, T_{12}, \ldots, T_{nn}). \tag{A.34}$$

If the basis is orthonormal, it follows from Equation A.30 that

$$T_{ij} = \left\langle e_i \middle| \hat{T} \middle| e_j \right\rangle. \tag{A.35}$$

It is convenient to display these complex numbers in the form of a **matrix**:[8]

$$\mathsf{T} = \begin{pmatrix} T_{11} & T_{12} & \ldots & T_{1n} \\ T_{21} & T_{22} & \ldots & T_{2n} \\ \vdots & \vdots & & \vdots \\ T_{n1} & T_{n2} & \ldots & T_{nn} \end{pmatrix}. \tag{A.36}$$

The study of linear transformations reduces, then, to the theory of matrices. The *sum* of two linear transformations $\left(\hat{S} + \hat{T}\right)$ is defined in the natural way:

$$\left(\hat{S} + \hat{T}\right)|\alpha\rangle = \hat{S}|\alpha\rangle + \hat{T}|\alpha\rangle; \tag{A.37}$$

---

7 Notice the reversal of indices between Equations A.30 and A.33. This is not a typographical error. Another way of putting it (switching $i \leftrightarrow j$ in Equation A.30) is that if the *components* transform with T_{ij}, the *basis* vectors transform with T_{ji}.

8 I'll use boldface capital letters, sans serif, to denote square matrices.

this matches the usual rule for adding matrices (you add the corresponding elements):

$$\mathsf{U} = \mathsf{S} + \mathsf{T} \Leftrightarrow U_{ij} = S_{ij} + T_{ij}. \tag{A.38}$$

The *product* of two linear transformations $\left(\hat{S}\hat{T}\right)$ is the net effect of performing them in succession—first $\hat{T}$, then $\hat{S}$:

$$|\alpha'\rangle = \hat{T}|\alpha\rangle; \quad |\alpha''\rangle = \hat{S}|\alpha'\rangle = \hat{S}\left(\hat{T}|\alpha\rangle\right) = \hat{S}\hat{T}|\alpha\rangle. \tag{A.39}$$

What matrix U represents the combined transformation $\hat{U} = \hat{S}\hat{T}$? It's not hard to work it out:

$$a_i'' = \sum_{j=1}^{n} S_{ij} a_j' = \sum_{j=1}^{n} S_{ij} \left(\sum_{k=1}^{n} T_{jk} a_k \right) = \sum_{k=1}^{n} \left(\sum_{j=1}^{n} S_{ij} T_{jk} \right) a_k = \sum_{k=1}^{n} U_{ik} a_k.$$

Evidently

$$\mathsf{U} = \mathsf{ST} \Leftrightarrow U_{ik} = \sum_{j=1}^{n} S_{ij} T_{jk} \tag{A.40}$$

—this is the standard rule for matrix multiplication: to find the ikth element of the product, you look at the ith row of S, and the kth column of T, multiply corresponding entries, and add. The same prescription allows you to multiply *rectangular* matrices, as long as the number of columns in the first matches the number of rows in the second. In particular, if we write the n-tuple of components of $|\alpha\rangle$ as an $n \times 1$ **column matrix** (or "column vector"):[9]

$$\mathsf{a} \equiv \begin{pmatrix} a_1 \\ a_2 \\ \vdots \\ a_n \end{pmatrix}, \tag{A.41}$$

the transformation rule (Equation A.33) can be expressed as a matrix product:

$$\mathsf{a}' = \mathsf{Ta}. \tag{A.42}$$

Now some matrix terminology:

- The **transpose** of a matrix (which we shall write with a tilde: $\tilde{\mathsf{T}}$) is the same set of elements, but with rows and columns interchanged. In particular, the transpose of a *column* matrix is a **row matrix**:

$$\tilde{\mathsf{a}} = \begin{pmatrix} a_1 & a_2 & \ldots & a_n \end{pmatrix}. \tag{A.43}$$

For a *square* matrix taking the transpose amounts to reflecting in the **main diagonal** (upper left to lower right):

$$\tilde{\mathsf{T}} = \begin{pmatrix} T_{11} & T_{21} & \ldots & T_{n1} \\ T_{12} & T_{22} & \ldots & T_{n2} \\ \vdots & \vdots & & \vdots \\ T_{1n} & T_{2n} & \ldots & T_{nn} \end{pmatrix}. \tag{A.44}$$

[9] I'll use boldface lower-case letters, sans serif, for row and column matrices.

A (square) matrix is **symmetric** if it is equal to its transpose; it is **antisymmetric** if this operation reverses the sign:

$$\text{symmetric} : \tilde{\mathsf{T}} = \mathsf{T}; \quad \text{antisymmetric} : \tilde{\mathsf{T}} = -\mathsf{T}. \tag{A.45}$$

- The (complex) **conjugate** of a matrix (which we denote, as usual, with an asterisk, T^*), consists of the complex conjugate of every element:

$$\mathsf{T}^* = \begin{pmatrix} T_{11}^* & T_{12}^* & \dots & T_{1n}^* \\ T_{21}^* & T_{22}^* & \dots & T_{2n}^* \\ \vdots & \vdots & & \vdots \\ T_{n1}^* & T_{n2}^* & \dots & T_{nn}^* \end{pmatrix}; \quad \mathsf{a}^* = \begin{pmatrix} a_1^* \\ a_2^* \\ \vdots \\ a_n^* \end{pmatrix}. \tag{A.46}$$

A matrix is **real** if all its elements are real, and **imaginary** if they are all imaginary:

$$\text{real} : \mathsf{T}^* = \mathsf{T}; \quad \text{imaginary} : \mathsf{T}^* = -\mathsf{T}. \tag{A.47}$$

- The **hermitian conjugate** (or **adjoint**) of a matrix (indicated by a dagger, $\mathsf{T}^\dagger$) is the transpose conjugate:

$$\mathsf{T}^\dagger \equiv \tilde{\mathsf{T}}^* = \begin{pmatrix} T_{11}^* & T_{21}^* & \dots & T_{n1}^* \\ T_{12}^* & T_{22}^* & \dots & T_{n2}^* \\ \vdots & \vdots & & \vdots \\ T_{1n}^* & T_{2n}^* & \dots & T_{nn}^* \end{pmatrix}; \quad \mathsf{a}^\dagger \equiv \tilde{\mathsf{a}}^* = \begin{pmatrix} a_1^* & a_2^* & \dots & a_n^* \end{pmatrix}. \tag{A.48}$$

A square matrix is **hermitian** (or **self-adjoint**) if it is equal to its hermitian conjugate; if hermitian conjugation introduces a minus sign, the matrix is **skew hermitian** (or **anti-hermitian**):

$$\text{hermitian} : \mathsf{T}^\dagger = \mathsf{T}; \quad \text{skew hermitian} : \mathsf{T}^\dagger = -\mathsf{T}. \tag{A.49}$$

In this notation the inner product of two vectors (with respect to an orthonormal basis—Equation A.24), can be written very neatly as a matrix product:

$$\langle \alpha | \beta \rangle = \mathsf{a}^\dagger \mathsf{b}. \tag{A.50}$$

Notice that each of the three operations defined in this paragraph, if applied twice, returns you to the original matrix.

Matrix multiplication is not, in general, commutative ($\mathsf{ST} \neq \mathsf{TS}$); the *difference* between the two orderings is called the **commutator**:[10]

$$[\mathsf{S}, \mathsf{T}] \equiv \mathsf{ST} - \mathsf{TS}. \tag{A.51}$$

The transpose of a product is the product of the transposes *in reverse order*:

$$\left(\widetilde{\mathsf{ST}}\right) = \tilde{\mathsf{T}}\tilde{\mathsf{S}}, \tag{A.52}$$

(see Problem A.11), and the same goes for hermitian conjugates:

$$(\mathsf{ST})^\dagger = \mathsf{T}^\dagger \mathsf{S}^\dagger. \tag{A.53}$$

[10] The commutator only makes sense for *square* matrices, of course; for rectangular matrices the two orderings wouldn't even be the same size.

The **identity matrix** (representing a linear transformation that carries every vector into itself) consists of ones on the main diagonal, and zeroes everywhere else:

$$\mathsf{I} \equiv \begin{pmatrix} 1 & 0 & \dots & 0 \\ 0 & 1 & \dots & 0 \\ \vdots & \vdots & & \vdots \\ 0 & 0 & \dots & 1 \end{pmatrix}. \tag{A.54}$$

In other words,

$$\mathsf{I}_{ij} = \delta_{ij}. \tag{A.55}$$

The **inverse** of a (square) matrix (written T^{-1}) is defined in the obvious way:[11]

$$\mathsf{T}^{-1}\mathsf{T} = \mathsf{T}\mathsf{T}^{-1} = \mathsf{I}. \tag{A.56}$$

A matrix has an inverse if and only if its **determinant**[12] is nonzero; in fact,

$$\mathsf{T}^{-1} = \frac{1}{\det \mathsf{T}}\tilde{\mathsf{C}}, \tag{A.57}$$

where C is the matrix of **cofactors** (the cofactor of element T_{ij} is $(-1)^{i+j}$ times the determinant of the submatrix obtained from T by erasing the ith row and the jth column). A matrix that has no inverse is said to be **singular**. The inverse of a product (assuming it exists) is the product of the inverses *in reverse order*:

$$(\mathsf{ST})^{-1} = \mathsf{T}^{-1}\mathsf{S}^{-1}. \tag{A.58}$$

A matrix is **unitary** if its inverse is equal to its hermitian conjugate:[13]

$$\text{unitary}: \mathsf{U}^\dagger = \mathsf{U}^{-1}. \tag{A.59}$$

Assuming the basis is orthonormal, the columns of a unitary matrix constitute an orthonormal set, and so too do its rows (see Problem A.12). Linear transformations represented by unitary matrices preserve inner products, since (Equation A.50)

$$\langle \alpha' \mid \beta' \rangle = \mathsf{a}'^\dagger \mathsf{b}' = (\mathsf{Ua})^\dagger (\mathsf{Ub}) = \mathsf{a}^\dagger \mathsf{U}^\dagger \mathsf{U} \mathsf{b} = \mathsf{a}^\dagger \mathsf{b} = \langle \alpha | \beta \rangle. \tag{A.60}$$

* **Problem A.8** Given the following two matrices:

$$\mathsf{A} = \begin{pmatrix} -1 & 1 & i \\ 2 & 0 & 3 \\ 2i & -2i & 2 \end{pmatrix}, \quad \mathsf{B} = \begin{pmatrix} 2 & 0 & -i \\ 0 & 1 & 0 \\ i & 3 & 2 \end{pmatrix},$$

compute: (a) $\mathsf{A}+\mathsf{B}$, (b) $\mathsf{A}\,\mathsf{B}$, (c) $[\mathsf{A},\mathsf{B}]$, (d) $\tilde{\mathsf{A}}$, (e) A^*, (f) $\mathsf{A}^\dagger$, (g) $\det(\mathsf{B})$, and (h) B^{-1}. Check that $\mathsf{B}\,\mathsf{B}^{-1} = \mathbf{1}$. Does A have an inverse?

---

[11] Note that the left inverse is equal to the right inverse, for if $\mathsf{AT} = \mathsf{I}$ and $\mathsf{TB} = \mathsf{I}$, then (multiplying the second on the left by A and invoking the first) we get $\mathsf{B} = \mathsf{A}$.

[12] I assume you know how to evaluate determinants. If not, see Mary L. Boas, *Mathematical Methods in the Physical Sciences*, 3rd edn (John Wiley, New York, 2006), Section 3.3.

[13] In a *real* vector space (that is, one in which the scalars are real) the hermitian conjugate is the same as the transpose, and a unitary matrix is **orthogonal**: $\tilde{\mathsf{O}} = \mathsf{O}^{-1}$. For example, rotations in ordinary 3-space are represented by orthogonal matrices.

* **Problem A.9** Using the square matrices in Problem A.8, and the column matrices

$$\mathsf{a} = \begin{pmatrix} i \\ 2i \\ 2 \end{pmatrix}, \quad \mathsf{b} = \begin{pmatrix} 2 \\ (1-i) \\ 0 \end{pmatrix},$$

find: (a) Aa, (b) $\mathsf{a}^\dagger\mathsf{b}$, (c) $\tilde{\mathsf{a}}\mathsf{Bb}$, (d) $\mathsf{a}\,\mathsf{b}^\dagger$.

Problem A.10 By explicit construction of the matrices in question, show that any matrix T can be written

(a) as the sum of a symmetric matrix S and an antisymmetric matrix A;
(b) as the sum of a real matrix R and an imaginary matrix M;
(c) as the sum of a hermitian matrix H and a skew-hermitian matrix K.

* **Problem A.11** Prove Equations A.52, A.53, and A.58. Show that the product of two unitary matrices is unitary. Under what conditions is the product of two hermitian matrices hermitian? Is the sum of two unitary matrices necessarily unitary? Is the sum of two hermitian matrices always hermitian?

Problem A.12 Show that the rows and columns of a unitary matrix constitute orthonormal sets.

Problem A.13 Noting that $\det(\tilde{\mathsf{T}}) = \det(\mathsf{T})$, show that the determinant of a hermitian matrix is real, the determinant of a unitary matrix has modulus 1 (hence the name), and the determinant of an orthogonal matrix (footnote 13) is either $+1$ or -1.

A.4 CHANGING BASES

The components of a vector depend, of course, on your (arbitrary) choice of basis, and so do the elements of the matrix representing a linear transformation. We might inquire how these numbers change when we switch to a different basis. The old basis vectors, $|e_i\rangle$ are—like *all* vectors—linear combinations of the new ones, $|f_i\rangle$:

$$\begin{aligned}
|e_1\rangle &= S_{11}|f_1\rangle + S_{21}|f_2\rangle + \cdots + S_{n1}|f_n\rangle, \\
|e_2\rangle &= S_{12}|f_1\rangle + S_{22}|f_2\rangle + \cdots + S_{n2}|f_n\rangle, \\
&\cdots \\
|e_n\rangle &= S_{1n}|f_1\rangle + S_{2n}|f_2\rangle + \cdots + S_{nn}|f_n\rangle,
\end{aligned}$$

(for some set of complex numbers S_{ij}), or, more compactly,

$$|e_j\rangle = \sum_{i=1}^{n} S_{ij}|f_i\rangle, \quad (j = 1, 2, \ldots, n). \tag{A.61}$$

This is *itself* a linear transformation (compare Equation A.30),[14] and we know immediately how the components transform:

$$a_i^f = \sum_{j=1}^{n} S_{ij} a_j^e, \tag{A.62}$$

(where the superscript indicates the basis). In matrix form

$$\mathsf{a}^f = \mathsf{S}\mathsf{a}^e. \tag{A.63}$$

What about the matrix representing a linear transformation $\hat{T}$—how is *it* modified by a change of basis? Well, in the old basis we had (Equation A.42)

$$\mathsf{a}'^e = \mathsf{T}^e \mathsf{a}^e,$$

and Equation A.63—multiplying both sides by S^{-1}—entails[15] $\mathsf{a}^e = \mathsf{S}^{-1}\mathsf{a}^f$, so

$$\mathsf{a}'^f = \mathsf{S}\mathsf{a}'^e = \mathsf{S}\left(\mathsf{T}^e\mathsf{a}^e\right) = \mathsf{S}\mathsf{T}^e\mathsf{S}^{-1}\mathsf{a}^f.$$

Evidently

$$\mathsf{T}^f = \mathsf{S}\mathsf{T}^e\mathsf{S}^{-1}. \tag{A.64}$$

In general, two matrices (T_1 and T_2) are said to be **similar** if $\mathsf{T}_2 = \mathsf{S}\mathsf{T}_1\mathsf{S}^{-1}$ for some (non-singular) matrix S. What we have just found is that *matrices representing the same linear transformation, with respect to different bases, are similar*. Incidentally, if the first basis is orthonormal, the second will also be orthonormal if and only if the matrix S is *unitary* (see Problem A.16). Since we always work in orthonormal bases, we are interested mainly in *unitary* similarity transformations.

While the *elements* of the matrix representing a given linear transformation may look very different in the new basis, two special numbers associated with the matrix are unchanged: the determinant and the **trace**. For the determinant of a product is the product of the determinants, and hence

$$\det\left(\mathsf{T}^f\right) = \det\left(\mathsf{S}\mathsf{T}^e\mathsf{S}^{-1}\right) = \det(\mathsf{S})\det(\mathsf{T}^e)\det\left(\mathsf{S}^{-1}\right) = \det\mathsf{T}^e. \tag{A.65}$$

And the trace, which is the *sum of the diagonal elements*,

$$\mathrm{Tr}(\mathsf{T}) \equiv \sum_{i=1}^{m} T_{ii}, \tag{A.66}$$

has the property (see Problem A.17) that

$$\mathrm{Tr}(\mathsf{T}_1\mathsf{T}_2) = \mathrm{Tr}(\mathsf{T}_2\mathsf{T}_1), \tag{A.67}$$

---

[14] Notice, however, the radically different perspective: In this case we're talking about one and the same *vector*, referred to two completely different *bases*, whereas before we were thinking of a completely *different* vector, referred to the *same* basis.

[15] Note that S^{-1} certainly exists—if S were singular, the $|f_i\rangle$s would not span the space, so they wouldn't constitute a basis.

(for any two matrices T_1 and T_2), so

$$\mathrm{Tr}\left(\mathsf{T}^f\right) = \mathrm{Tr}\left(\mathsf{S}\mathsf{T}^e\mathsf{S}^{-1}\right) = \mathrm{Tr}\left(\mathsf{T}^e\mathsf{S}^{-1}\mathsf{S}\right) = \mathrm{Tr}(\mathsf{T}^e). \tag{A.68}$$

Problem A.14 Using the standard basis $\left(\hat{\imath}, \hat{\jmath}, \hat{k}\right)$ for vectors in three dimensions:

(a) Construct the matrix representing a rotation through angle θ (counterclockwise, looking down the axis toward the origin) about the z axis.

(b) Construct the matrix representing a rotation by 120° (counterclockwise, looking down the axis) about an axis through the point (1,1,1).

(c) Construct the matrix representing reflection through the xy plane.

(d) Check that all these matrices are orthogonal, and calculate their determinants.

Problem A.15 In the usual basis $\left(\hat{\imath}, \hat{\jmath}, \hat{k}\right)$, construct the matrix T_x representing a rotation through angle θ about the x axis, and the matrix T_y representing a rotation through angle θ about the y axis. Suppose now we change bases, to $\hat{\imath}' = \hat{\jmath},\ \hat{\jmath}' = -\hat{\imath},\ \hat{k}' = \hat{k}$. Construct the matrix S that effects this change of basis, and check that $\mathsf{S}\mathsf{T}_x\mathsf{S}^{-1}$ and $\mathsf{S}\mathsf{T}_y\mathsf{S}^{-1}$ are what you would expect.

Problem A.16 Show that similarity preserves matrix multiplication (that is, if $\mathsf{A}^e\mathsf{B}^e = \mathsf{C}^e$, then $\mathsf{A}^f\mathsf{B}^f = \mathsf{C}^f$). Similarity does *not*, in general, preserve symmetry, reality, or hermiticity; show, however, that if S is *unitary*, and H^e is hermitian, then H^f is hermitian. Show that S carries an orthonormal basis into another orthonormal basis if and only if it is unitary.

* **Problem A.17** Prove that $\mathrm{Tr}(\mathsf{T}_1\mathsf{T}_2) = \mathrm{Tr}(\mathsf{T}_2\mathsf{T}_1)$. It follows immediately that $\mathrm{Tr}(\mathsf{T}_1\mathsf{T}_2\mathsf{T}_3) = \mathrm{Tr}(\mathsf{T}_2\mathsf{T}_3\mathsf{T}_1)$, but is it the case that $\mathrm{Tr}(\mathsf{T}_1\mathsf{T}_2\mathsf{T}_3) = \mathrm{Tr}(\mathsf{T}_2\mathsf{T}_1\mathsf{T}_3)$, in general? Prove it, or disprove it. *Hint:* The best disproof is always a counterexample—the simpler the better!

A.5 EIGENVECTORS AND EIGENVALUES

Consider the linear transformation in 3-space consisting of a rotation, about some specified axis, by an angle θ. Most vectors (with tails at the origin) will change in a rather complicated way (they ride around on a cone about the axis), but vectors that happen to lie *along* the axis have very simple behavior: They don't change at all ($\hat{T}\,|\alpha\rangle = |\alpha\rangle$). If θ is 180°, then vectors which lie in the the "equatorial" plane reverse signs ($\hat{T}\,|\alpha\rangle = -\,|\alpha\rangle$). In a complex vector space[16] *every* linear transformation has "special" vectors like these, which are transformed into scalar multiples of themselves:

[16] This is *not* always true in a *real* vector space (where the scalars are restricted to real values). See Problem A.18.

$$\hat{T}|\alpha\rangle = \lambda|\alpha\rangle; \tag{A.69}$$

they are called **eigenvectors** of the transformation, and the (complex) number λ is their **eigenvalue**. (The *null* vector doesn't count, even though in a trivial sense it obeys Equation A.69 for *any* $\hat{T}$ and *any* λ; technically, an eigenvector is any *nonzero* vector satisfying Equation A.69.) Notice that any (nonzero) *multiple* of an eigenvector is still an eigenvector, with the same eigenvalue.

With respect to a particular basis, the eigenvector equation assumes the matrix form

$$\mathsf{T}\mathsf{a} = \lambda\mathsf{a}, \qquad (\mathsf{a} \neq \mathsf{0}), \tag{A.70}$$

or

$$(\mathsf{T} - \lambda\mathsf{I})\,\mathsf{a} = \mathsf{0}. \tag{A.71}$$

(Here $\mathsf{0}$ is the (column) matrix whose elements are all zero.) Now, if the matrix $(\mathsf{T} - \lambda\mathsf{I})$ had an *inverse*, we could multiply both sides of Equation A.71 by $(\mathsf{T} - \lambda\mathsf{I})^{-1}$, and conclude that $\mathsf{a} = \mathsf{0}$. But by assumption a is *not* zero, so the matrix $(\mathsf{T} - \lambda\mathsf{I})$ must in fact be singular, which means that its determinant is zero:

$$\det(\mathsf{T} - \lambda\mathsf{I}) = \begin{vmatrix} (T_{11} - \lambda) & T_{12} & \dots & T_{1n} \\ T_{21} & (T_{22} - \lambda) & \dots & T_{2n} \\ \vdots & \vdots & & \vdots \\ T_{n1} & T_{n2} & \dots & (T_{nn} - \lambda) \end{vmatrix} = 0. \tag{A.72}$$

Expansion of the determinant yields an algebraic equation for λ:

$$C_n\lambda^n + C_{n-1}\lambda^{n-1} + \cdots + C_1\lambda + C_0 = 0, \tag{A.73}$$

where the coefficients C_i depend on the elements of T (see Problem A.20). This is called the **characteristic equation** for the matrix; its solutions determine the eigenvalues. Notice that it's an nth-order equation, so (by the **fundamental theorem of algebra**) it has n (complex) roots.[17] However, some of these may be multiple roots, so all we can say for certain is that an $n \times n$ matrix has *at least one* and *at most n* distinct eigenvalues. The collection of all the eigenvalues of a matrix is called its **spectrum**; if two (or more) linearly independent eigenvectors share the same eigenvalue, the spectrum is said to be **degenerate**.

To construct the eigen*vectors* it is generally easiest simply to plug each λ back into Equation A.70 and solve "by hand" for the components of a. I'll show you how it goes by working out an example.

Example A.1

Find the eigenvalues and eigenvectors of the following matrix:

$$\mathsf{M} = \begin{pmatrix} 2 & 0 & -2 \\ -2i & i & 2i \\ 1 & 0 & -1 \end{pmatrix}. \tag{A.74}$$

[17] It is here that the case of *real* vector spaces becomes more awkward, because the characteristic equation need not have any (real) solutions at all. See Problem A.18.

Solution: The characteristic equation is

$$\begin{vmatrix} (2-\lambda) & 0 & -2 \\ -2i & (i-\lambda) & 2i \\ 1 & 0 & (-1-\lambda) \end{vmatrix} = -\lambda^3 + (1+i)\,\lambda^2 - i\lambda = 0, \tag{A.75}$$

and its roots are 0, 1, and i. Call the components of the first eigenvector (a_1, a_2, a_3); then

$$\begin{pmatrix} 2 & 0 & -2 \\ -2i & i & 2i \\ 1 & 0 & -1 \end{pmatrix} \begin{pmatrix} a_1 \\ a_2 \\ a_3 \end{pmatrix} = 0 \begin{pmatrix} a_1 \\ a_2 \\ a_3 \end{pmatrix} = \begin{pmatrix} 0 \\ 0 \\ 0 \end{pmatrix},$$

which yields three equations:

$$\begin{aligned} 2a_1 - 2a_3 &= 0, \\ -2ia_1 + ia_2 + 2ia_3 &= 0, \\ a_1 - a_3 &= 0. \end{aligned}$$

The first determines a_3 (in terms of a_1): $a_3 = a_1$; the second determines a_2: $a_2 = 0$; and the third is redundant. We may as well pick $a_1 = 1$ (since any multiple of an eigenvector is still an eigenvector):

$$\mathsf{a}^{(1)} = \begin{pmatrix} 1 \\ 0 \\ 1 \end{pmatrix}, \text{ for } \lambda_1 = 0. \tag{A.76}$$

For the second eigenvector (recycling the same notation for the components) we have

$$\begin{pmatrix} 2 & 0 & -2 \\ -2i & i & 2i \\ 1 & 0 & -1 \end{pmatrix} \begin{pmatrix} a_1 \\ a_2 \\ a_3 \end{pmatrix} = 1 \begin{pmatrix} a_1 \\ a_2 \\ a_3 \end{pmatrix} = \begin{pmatrix} a_1 \\ a_2 \\ a_3 \end{pmatrix},$$

which leads to the equations

$$\begin{aligned} 2a_1 - 2a_3 &= a_1, \\ -2ia_1 + ia_2 + 2ia_3 &= a_2, \\ a_1 - a_3 &= a_3, \end{aligned}$$

with the solution $a_3 = (1/2)\,a_1$, $a_2 = \left[(1-i)/2\right] a_1$; this time I'll pick $a_1 = 2$, so

$$\mathsf{a}^{(2)} = \begin{pmatrix} 2 \\ 1-i \\ 1 \end{pmatrix}, \text{ for } \lambda_2 = 1. \tag{A.77}$$

Finally, for the third eigenvector,

$$\begin{pmatrix} 2 & 0 & -2 \\ -2i & i & 2i \\ 1 & 0 & -1 \end{pmatrix} \begin{pmatrix} a_1 \\ a_2 \\ a_3 \end{pmatrix} = i \begin{pmatrix} a_1 \\ a_2 \\ a_3 \end{pmatrix} = \begin{pmatrix} ia_1 \\ ia_2 \\ ia_3 \end{pmatrix},$$

which gives the equations

$$\begin{aligned} 2a_1 - 2a_3 &= ia_1, \\ -2ia_1 + ia_2 + 2ia_3 &= ia_2, \\ a_1 - a_3 &= ia_3, \end{aligned}$$

whose solution is $a_3 = a_1 = 0$, with a_2 undetermined. Choosing $a_2 = 1$, we conclude

$$\mathsf{a}^{(3)} = \begin{pmatrix} 0 \\ 1 \\ 0 \end{pmatrix}, \text{ for } \lambda_3 = i. \tag{A.78}$$

If the eigenvectors span the space (as they do in the preceding example), we are free to use *them* as a basis:

$$\begin{aligned} \hat{T}|f_1\rangle &= \lambda_1|f_1\rangle, \\ \hat{T}|f_2\rangle &= \lambda_2|f_2\rangle, \\ &\dots \\ \hat{T}|f_n\rangle &= \lambda_n|f_n\rangle. \end{aligned}$$

In this basis the matrix representing $\hat{T}$ takes on a very simple form, with the eigenvalues strung out along the main diagonal, and all other elements zero:

$$\mathsf{T} = \begin{pmatrix} \lambda_1 & 0 & \dots & 0 \\ 0 & \lambda_2 & \dots & 0 \\ \vdots & \vdots & & \vdots \\ 0 & 0 & \dots & \lambda_n \end{pmatrix}, \tag{A.79}$$

and the (normalized) eigenvectors are

$$\begin{pmatrix} 1 \\ 0 \\ 0 \\ \vdots \\ 0 \end{pmatrix}, \begin{pmatrix} 0 \\ 1 \\ 0 \\ \vdots \\ 0 \end{pmatrix}, \quad \dots \quad , \begin{pmatrix} 0 \\ 0 \\ 0 \\ \vdots \\ 1 \end{pmatrix}. \tag{A.80}$$

A matrix that can be brought to **diagonal form** (Equation A.79) by a change of basis is said to be **diagonalizable** (evidently a matrix is diagonalizable if and only if its eigenvectors span the space). The similarity matrix that effects the diagonalization can be constructed by using the eigenvectors (in the old basis) as the columns of S^{-1}:

$$\left(\mathsf{S}^{-1}\right)_{ij} = \left(\mathsf{a}^{(j)}\right)_i. \tag{A.81}$$

Example A.2

In Example A.1,

$$\mathsf{S}^{-1} = \begin{pmatrix} 1 & 2 & 0 \\ 0 & (1-i) & 1 \\ 1 & 1 & 0 \end{pmatrix},$$

so (using Equation A.57)

$$\mathsf{S} = \begin{pmatrix} -1 & 0 & 2 \\ 1 & 0 & -1 \\ (i-1) & 1 & (1-i) \end{pmatrix};$$

you can check for yourself that

$$\mathsf{S}\mathsf{a}^{(1)} = \begin{pmatrix} 1 \\ 0 \\ 0 \end{pmatrix}, \quad \mathsf{S}\mathsf{a}^{(2)} = \begin{pmatrix} 0 \\ 1 \\ 0 \end{pmatrix}, \quad \mathsf{S}\mathsf{a}^{(3)} = \begin{pmatrix} 0 \\ 0 \\ 1 \end{pmatrix},$$

and

$$\mathsf{S}\mathsf{M}\mathsf{S}^{-1} = \begin{pmatrix} 0 & 0 & 0 \\ 0 & 1 & 0 \\ 0 & 0 & i \end{pmatrix}.$$

There's an obvious advantage in bringing a matrix to diagonal form: it's much easier to work with. Unfortunately, not every matrix *can* be diagonalized—the eigenvectors have to span the space. If the characteristic equation has n distinct roots, then the matrix is certainly diagonalizable, but it *may* be diagonalizble even if there are multiple roots. (For an example of a matrix that *cannot* be diagonalized, see Problem A.19.) It would be handy to know in advance (before working out all the eigenvectors) whether a given matrix is diagonalizable. A useful sufficient (though not necessary) condition is the following: A matrix is said to be **normal** if it commutes with its hermitian conjugate:

$$\text{normal}: \left[\mathsf{N}^{\dagger}, \mathsf{N}\right] = 0. \tag{A.82}$$

Every normal matrix is diagonalizable (its eigenvectors span the space). In particular, every *hermitian* matrix is diagonalizable, and so is every *unitary* matrix.

Suppose we have *two* diagonalizable matrices; in quantum applications the question often arises: Can they be **simultaneously diagonalized** (by the *same* similarity matrix S)? That is to say, does there exist a basis all of whose members are eigenvectors of *both* matrices? In this basis, both matrices would be diagonal. The answer is yes *if and only if the two matrices commute*, as we shall now prove. (By the way, if two matrices commute with respect to *one* basis, they commute with respect to *any* basis—see Problem A.23.)

We first show that if a basis of simultaneous eigenvectors exists then the matrices commute. Actually, it's trivial in the (simultaneously) diagonal form:

$$\mathsf{T}\mathsf{V} = \begin{pmatrix} \lambda_1 & 0 & \cdots & 0 \\ 0 & \lambda_2 & \cdots & 0 \\ \vdots & \vdots & \ddots & \vdots \\ 0 & 0 & \cdots & \lambda_n \end{pmatrix} \begin{pmatrix} \nu_1 & 0 & \cdots & 0 \\ 0 & \nu_2 & \cdots & 0 \\ \vdots & \vdots & \ddots & \vdots \\ 0 & 0 & \cdots & \nu_n \end{pmatrix}$$

$$= \begin{pmatrix} \lambda_1\nu_1 & 0 & \cdots & 0 \\ 0 & \lambda_2\nu_2 & \cdots & 0 \\ \vdots & \vdots & \ddots & \vdots \\ 0 & 0 & \cdots & \lambda_n\nu_n \end{pmatrix} = \mathsf{V}\mathsf{T}. \tag{A.83}$$

The converse is trickier. We start with the special case where the spectrum of T is nondegenerate. Let the basis of eigenvectors of T be labeled $\mathsf{a}^{(i)}$

$$\mathsf{T}\,\mathsf{a}^{(i)} = \lambda_i\,\mathsf{a}^{(i)}. \tag{A.84}$$

We assume $\left[\mathsf{T}, \mathsf{V}\right] = 0$ and we want to prove that $\mathsf{a}^{(i)}$ is also an eigenvector of V.

$$\left[\mathsf{T}, \mathsf{V}\right]\mathsf{a}^{(i)} = 0, \qquad \text{or} \qquad \mathsf{T}\,\mathsf{V}\,\mathsf{a}^{(i)} - \mathsf{V}\,\mathsf{T}\,\mathsf{a}^{(i)} = 0 \tag{A.85}$$

and from Equation A.84

$$\mathsf{T}\left(\mathsf{V}\,\mathsf{a}^{(i)}\right) = \lambda_i \left(\mathsf{V}\,\mathsf{a}^{(i)}\right). \tag{A.86}$$

Equation A.86 says that the vector $\mathsf{b}^{(i)} \equiv \mathsf{V}\,\mathsf{a}^{(i)}$ is an eigenvector of T with eigenvalue λ_i. But by assumption, the spectrum of T is nondegenerate and that means that $\mathsf{b}^{(i)}$ must be (up to a constant) $\mathsf{a}^{(i)}$ itself. If we call the constant ν_i,

$$\mathsf{b}^{(i)} = \mathsf{V}\,\mathsf{a}^{(i)} = \nu_i\,\mathsf{a}^{(i)}, \tag{A.87}$$

so $\mathsf{a}^{(i)}$ is an eigenvector of V.

All that remains is to relax the assumption of nondegeneracy. Assume now that T has at least one degenerate eigenvalue such that both $\mathsf{a}^{(1)}$ and $\mathsf{a}^{(2)}$ are eigenvectors of T with the same eigenvalue λ_0:

$$\begin{aligned} \mathsf{T}\,\mathsf{a}^{(1)} &= \lambda_0\,\mathsf{a}^{(1)} \\ \mathsf{T}\,\mathsf{a}^{(2)} &= \lambda_0\,\mathsf{a}^{(2)}. \end{aligned}$$

We again assume that the matrices T and V commute, so

$$\begin{aligned} \left[\mathsf{T},\mathsf{V}\right]\,\mathsf{a}^{(1)} &= 0 \\ \left[\mathsf{T},\mathsf{V}\right]\,\mathsf{a}^{(2)} &= 0, \end{aligned}$$

which leads to the conclusion (as in the nondegenerate case) that both $\mathsf{b}^{(1)} \equiv \mathsf{V}\,\mathsf{a}^{(1)}$ and $\mathsf{b}^{(2)} \equiv \mathsf{V}\,\mathsf{a}^{(2)}$ are eigenvectors of T with eigenvalue λ_0. But this time we can't say that $\mathsf{b}^{(1)}$ is a constant times $\mathsf{a}^{(1)}$ since any linear combination of $\mathsf{a}^{(1)}$ and $\mathsf{a}^{(2)}$ is an eigenvector of T with eigenvalue λ_0. All we know is that

$$\begin{aligned} \mathsf{b}^{(1)} &= \mathsf{V}\,\mathsf{a}^{(1)} = c_{11}\,\mathsf{a}^{(1)} + c_{21}\,\mathsf{a}^{(2)} \\ \mathsf{b}^{(2)} &= \mathsf{V}\,\mathsf{a}^{(2)} = c_{12}\,\mathsf{a}^{(1)} + c_{22}\,\mathsf{a}^{(2)} \end{aligned}$$

for some constants c_{ij}. So $\mathsf{a}^{(1)}$ and $\mathsf{a}^{(2)}$ are not eigenvectors of V (unless the constants c_{12} and c_{21} just happen to vanish). But suppose we choose a different basis of eigenvectors $\tilde{\mathsf{a}}$,

$$\begin{aligned} \tilde{\mathsf{a}}^{(1)} &= d_{11}\,\mathsf{a}^{(1)} + d_{21}\,\mathsf{a}^{(2)} \\ \tilde{\mathsf{a}}^{(2)} &= d_{12}\,\mathsf{a}^{(1)} + d_{22}\,\mathsf{a}^{(2)}, \end{aligned} \tag{A.88}$$

for some constants d_{ij}, such that $\tilde{\mathsf{a}}^{(1)}$ and $\tilde{\mathsf{a}}^{(2)}$ *are* eigenvectors of V:

$$\mathsf{V}\,\tilde{\mathsf{a}}^{(i)} = \nu_i\,\tilde{\mathsf{a}}^{(i)}. \tag{A.89}$$

The $\tilde{\mathsf{a}}$s are still eigenvectors of T, with the same eigenvalue λ_0, since any linear combinations of $\mathsf{a}^{(1)}$ and $\mathsf{a}^{(2)}$ are. But *can* we construct linear combinations (A.88) that are eigenvectors of V—how do we get the appropriate coefficients d_{ij}? *Answer:* We solve the eigenvalue problem[18]

$$\underbrace{\begin{pmatrix} c_{11} & c_{12} \\ c_{21} & c_{22} \end{pmatrix}}_{\mathsf{C}} \begin{pmatrix} d_{1i} \\ d_{2i} \end{pmatrix} = \nu_i \begin{pmatrix} d_{1i} \\ d_{2i} \end{pmatrix}. \tag{A.90}$$

[18] You might worry that the matrix C is not diagonalizable, but you need not. The matrix C is a 2×2 block of the transformation $\hat{V}$ written in the basis $\mathsf{a}^{(i)}$; it is diagonalizable by virtue of the fact that V itself is diagonalizable.

I'll let you show (Problem A.24) that the eigenvectors $\vec{a}^{(i)}$ constructed in this way satisfy Equation A.88, completing the proof.[19] What we have seen is that, when the spectrum contains degeneracy, the eigenvectors of one matrix aren't *automatically* eigenvectors of a second commuting matrix, but we can always *choose* a linear combination of them to form a simultaneous basis of eigenvectors.

* **Problem A.18** The 2 × 2 matrix representing a rotation of the xy plane is

$$\mathsf{T} = \begin{pmatrix} \cos\theta & -\sin\theta \\ \sin\theta & \cos\theta \end{pmatrix}. \tag{A.91}$$

Show that (except for certain special angles—what are they?) this matrix has no real eigenvalues. (This reflects the geometrical fact that no vector in the plane is carried into itself under such a rotation; contrast rotations in *three* dimensions.) This matrix *does*, however, have *complex* eigenvalues and eigenvectors. Find them. Construct a matrix S that diagonalizes T. Perform the similarity transformation $\left(\mathsf{STS}^{-1}\right)$ explicitly, and show that it reduces T to diagonal form.

Problem A.19 Find the eigenvalues and eigenvectors of the following matrix:

$$\mathsf{M} = \begin{pmatrix} 1 & 1 \\ 0 & 1 \end{pmatrix}.$$

Can this matrix be diagonalized?

Problem A.20 Show that the first, second, and last coefficients in the characteristic equation (Equation A.73) are:

$$C_n = (-1)^n, \; C_{n-1} = (-1)^{n-1}\,\mathrm{Tr}(\mathsf{T}), \text{ and } C_0 = \det(\mathsf{T}). \tag{A.92}$$

For a 3 × 3 matrix with elements T_{ij}, what is C_1?

Problem A.21 It's obvious that the trace of a *diagonal* matrix is the sum of its eigenvalues, and its determinant is their product (just look at Equation A.79). It follows (from Equations A.65 and A.68) that the same holds for any *diagonalizable* matrix. Prove that in fact

$$\det(\mathsf{T}) = \lambda_1\lambda_2\cdots\lambda_n, \quad \mathrm{Tr}(\mathsf{T}) = \lambda_1 + \lambda_2 + \cdots + \lambda_n, \tag{A.93}$$

for *any* matrix. (The λ's are the n solutions to the characteristic equation—in the case of multiple roots, there may be fewer linearly-independent eigen*vectors* than there are solutions, but we still count each λ as many times as it occurs.) *Hint:* Write the characteristic equation in the form

$$(\lambda_1 - \lambda)(\lambda_2 - \lambda)\cdots(\lambda_n - \lambda) = 0,$$

and use the result of Problem A.20.

[19] I've only proved it for a two-fold degeneracy, but the argument extends in the obvious way to a higher-order degeneracy; you simply need to diagonalize a bigger matrix C.

Problem A.22 Consider the matrix

$$\mathsf{M} = \begin{pmatrix} 1 & 1 \\ 1 & i \end{pmatrix}.$$

(a) Is it normal?
(b) Is it diagonalizable?

Problem A.23 Show that if two matrices commute in *one* basis, then they commute in *any* basis. That is:

$$\left[\mathsf{T}_1^e, \mathsf{T}_2^e\right] = 0 \Rightarrow \left[\mathsf{T}_1^f, \mathsf{T}_2^f\right] = 0. \tag{A.94}$$

Hint: Use Equation A.64.

Problem A.24 Show that the $\tilde{\mathsf{a}}$ computed from Equations A.88 and A.90 are eigenvectors of V.

* **Problem A.25** Consider the matrices

$$\mathsf{A} = \begin{pmatrix} 1 & 4 & 1 \\ 4 & -2 & 4 \\ 1 & 4 & 1 \end{pmatrix}, \quad \mathsf{B} = \begin{pmatrix} 1 & -2 & -1 \\ -2 & 2 & -2 \\ -1 & -2 & 1 \end{pmatrix}.$$

(a) Verify that they are diagonalizable and that they commute.
(b) Find the eigenvalues and eigenvectors of A and verify that its spectrum is nondegenerate.
(c) Show that the eigenvectors of A are eigenvectors of B as well.

Problem A.26 Consider the matrices

$$\mathsf{A} = \begin{pmatrix} 2 & 2 & -1 \\ 2 & -1 & 2 \\ -1 & 2 & 2 \end{pmatrix}, \quad \mathsf{B} = \begin{pmatrix} 2 & -1 & 2 \\ -1 & 5 & -1 \\ 2 & -1 & 2 \end{pmatrix}.$$

(a) Verify that they are diagonalizable and that they commute.
(b) Find the eigenvalues and eigenvectors of A and verify that its spectrum is degenerate.
(c) Are the eigenvectors that you found in part (b) also eigenvectors of B? If not, find the vectors that are simultaneous eigenvectors of both matrices.

A.6 HERMITIAN TRANSFORMATIONS

In Equation A.48 I defined the hermitian conjugate (or "adjoint") of a *matrix* as its transpose-conjugate: $\mathsf{T}^\dagger = \tilde{\mathsf{T}}^*$. Now I want to give you a more fundamental definition for the hermitian conjugate of a *linear transformation*: It is that transformation $\hat{T}^\dagger$ which, when applied to the *first* member of an inner product, gives the same result as if $\hat{T}$ itself had been applied to the *second* vector:

$$\left\langle \hat{T}^{\dagger}\alpha \middle| \beta \right\rangle = \left\langle \alpha \middle| \hat{T}\beta \right\rangle, \tag{A.95}$$

(for all vectors $|\alpha\rangle$ and $|\beta\rangle$).[20] I have to warn you that although everybody uses it, this is lousy notation. For α and β are not *vectors* (the *vectors* are $|\alpha\rangle$ and $|\beta\rangle$), they are *names*. In particular, they are endowed with no mathematical properties at all, and the expression "$\hat{T}\beta$" is literally *nonsense*: Linear transformations act on *vectors*, not *labels*. But it's pretty clear what the notation *means*: $\hat{T}\beta$ is the name of the vector $\hat{T}|\beta\rangle$, and $\langle \hat{T}^{\dagger}\alpha|\beta\rangle$ is the inner product of the vector $\hat{T}^{\dagger}|\alpha\rangle$ with the vector $|\beta\rangle$. Notice in particular that

$$\langle \alpha \mid c\beta \rangle = c \langle \alpha \mid \beta \rangle, \tag{A.96}$$

whereas

$$\langle c\alpha \mid \beta \rangle = c^{*} \langle \alpha \mid \beta \rangle, \tag{A.97}$$

for any scalar c.

If you're working in an orthonormal basis (as we always do), the hermitian conjugate of a linear transformation is represented by the hermitian conjugate of the corresponding matrix; for (using Equations A.50 and A.53),

$$\left\langle \alpha \middle| \hat{T}\beta \right\rangle = \mathsf{a}^{\dagger}\mathsf{T}\,\mathsf{b} = \left(\mathsf{T}^{\dagger}\mathsf{a}\right)^{\dagger}\mathsf{b} = \left\langle \hat{T}^{\dagger}\alpha \middle| \beta \right\rangle. \tag{A.98}$$

So the terminology is consistent, and we can speak interchangeably in the language of transformations or of matrices.

In quantum mechanics, a fundamental role is played by **hermitian transformations** $\left(\hat{T}^{\dagger} = \hat{T}\right)$. The eigenvectors and eigenvalues of a hermitian transformation have three crucial properties:

1. **The eigenvalues of a hermitian transformation are real.**

   Proof: Let λ be an eigenvalue of $\hat{T}$: $\hat{T}|\alpha\rangle = \lambda|\alpha\rangle$, with $|\alpha\rangle \neq |0\rangle$. Then

   $$\left\langle \alpha \middle| \hat{T}\alpha \right\rangle = \langle \alpha \mid \lambda\alpha \rangle = \lambda \langle \alpha \mid \alpha \rangle .$$

   Meanwhile, if $\hat{T}$ is hermitian, then

   $$\left\langle \alpha \middle| \hat{T}\alpha \right\rangle = \left\langle \hat{T}\alpha \middle| \alpha \right\rangle = \langle \lambda\alpha \mid \alpha \rangle = \lambda^{*} \langle \alpha \mid \alpha \rangle .$$

   But $\langle \alpha|\alpha\rangle \neq 0$ (Equation A.20), so $\lambda = \lambda^{*}$, and hence λ is real. QED

2. **The eigenvectors of a hermitian transformation belonging to distinct eigenvalues are orthogonal.**

   Proof: Suppose $\hat{T}|\alpha\rangle = \lambda|\alpha\rangle$ and $\hat{T}|\beta\rangle = \mu|\beta\rangle$, with $\lambda \neq \mu$. Then

   $$\left\langle \alpha \middle| \hat{T}\beta \right\rangle = \langle \alpha \mid \mu\beta \rangle = \mu \langle \alpha \mid \beta \rangle ,$$

[20] If you're wondering whether such a transformation necessarily *exists*, that's a good question, and the answer is "yes." See, for instance, Paul R. Halmos, *Finite Dimensional Vector Spaces*, 2nd edn, van Nostrand, Princeton (1958), Section 44.

and if $\hat{T}$ is hermitian,

$$\left\langle\alpha \middle| \hat{T}\beta\right\rangle = \left\langle\hat{T}\alpha \middle| \beta\right\rangle = \langle\lambda\alpha \mid \beta\rangle = \lambda^* \langle\alpha \mid \beta\rangle .$$

But $\lambda = \lambda^*$ (from (1)), and $\lambda \neq \mu$, by assumption, so $\langle\alpha|\beta\rangle = 0$. QED

3. **The eigenvectors of a hermitian transformation span the space.**
As we have seen, this is equivalent to the statement that any hermitian matrix can be diagonalized. This rather technical fact is, in a sense, the mathematical support on which much of quantum mechanics leans. It turns out to be a thinner reed than one might have hoped, because the proof does not carry over to infinite-dimensional vector spaces.

Problem A.27 A hermitian linear transformation must satisfy $\langle\alpha|\hat{T}\beta\rangle = \langle\hat{T}\alpha|\beta\rangle$ for all vectors $|\alpha\rangle$ and $|\beta\rangle$. Prove that it is (surprisingly) sufficient that $\langle\gamma|\hat{T}\gamma\rangle = \langle\hat{T}\gamma|\gamma\rangle$ for all vectors $|\gamma\rangle$. *Hint:* First let $|\gamma\rangle = |\alpha\rangle + |\beta\rangle$, and then let $|\gamma\rangle = |\alpha\rangle + i|\beta\rangle$.

* **Problem A.28** Let

$$\mathsf{T} = \begin{pmatrix} 1 & 1-i \\ 1+i & 0 \end{pmatrix}.$$

(a) Verify that T is hermitian.
(b) Find its eigenvalues (note that they are real).
(c) Find and normalize the eigenvectors (note that they are orthogonal).
(d) Construct the unitary diagonalizing matrix S, and check explicitly that it diagonalizes T.
(e) Check that det(T) and Tr(T) are the same for T as they are for its diagonalized form.

** **Problem A.29** Consider the following hermitian matrix:

$$\mathsf{T} = \begin{pmatrix} 2 & i & 1 \\ -i & 2 & i \\ 1 & -i & 2 \end{pmatrix}.$$

(a) Calculate det (T) and Tr(T).
(b) Find the eigenvalues of T. Check that their sum and product are consistent with (a), in the sense of Equation A.93. Write down the diagonalized version of T.
(c) Find the eigenvectors of T. Within the degenerate sector, construct two linearly independent eigenvectors (it is this step that is always possible for a *hermitian* matrix, but not for an *arbitrary* matrix—contrast Problem A.19). Orthogonalize them, and check that both are orthogonal to the third. Normalize all three eigenvectors.
(d) Construct the unitary matrix S that diagonalizes T, and show explicitly that the similarity transformation using S reduces T to the appropriate diagonal form.

Problem A.30 A **unitary transformation** is one for which $\hat{U}^\dagger\hat{U} = 1$.
(a) Show that unitary transformations preserve inner products, in the sense that $\langle\hat{U}\alpha|\hat{U}\beta\rangle = \langle\alpha \mid \beta\rangle$, for all vectors $|\alpha\rangle$, $|\beta\rangle$.

(b) Show that the eigenvalues of a unitary transformation have modulus 1.
(c) Show that the eigenvectors of a unitary transformation belonging to distinct eigenvalues are orthogonal.

*** **Problem A.31** *Functions* of matrices are typically defined by their Taylor series expansions. For example,

$$e^{\mathsf{M}} \equiv \mathsf{I} + \mathsf{M} + \frac{1}{2}\mathsf{M}^2 + \frac{1}{3!}\mathsf{M}^3 + \cdots. \tag{A.99}$$

(a) Find exp (M), if

$$\text{(i) } \mathsf{M} = \begin{pmatrix} 0 & 1 & 3 \\ 0 & 0 & 4 \\ 0 & 0 & 0 \end{pmatrix}; \quad \text{(ii) } \mathsf{M} = \begin{pmatrix} 0 & \theta \\ -\theta & 0 \end{pmatrix}.$$

(b) Show that if M is diagonalizable, then

$$\det\left(e^{\mathsf{M}}\right) = e^{\mathrm{Tr}(\mathsf{M})}. \tag{A.100}$$

Comment: This is actually true even if M is *not* diagonalizable, but it's harder to prove in the general case.

(c) Show that if the matrices M and N *commute*, then

$$e^{\mathsf{M}+\mathsf{N}} = e^{\mathsf{M}}e^{\mathsf{N}}. \tag{A.101}$$

Prove (with the simplest counterexample you can think up) that Equation A.101 is *not* true, in general, for *non*-commuting matrices.[21]

(d) If H is hermitian, show that $e^{i\mathsf{H}}$ is unitary.

[21] See Problem 3.29 for the more general "Baker–Campbell–Hausdorff" formula.